Lecture Notes in Computer Science 4076

Commenced Publication in 1973
Founding and Former Series Editors:
Gerhard Goos, Juris Hartmanis, and Jan van Leeuwen

T0254180

Florian Hess   Sebastian Pauli
Michael Pohst (Eds.)

Algorithmic Number Theory

7th International Symposium, ANTS-VII
Berlin, Germany, July 23-28, 2006
Proceedings

Springer

Volume Editors

Florian Hess
Sebastian Pauli
Michael Pohst
Technische Universität Berlin
Fakultät II, Institut für Mathematik MA 8-1
Strasse des 17. Juni 136, 10623 Berlin, Germany
E-mail: {hess, pauli, pohst}@math.tu-berlin.de

Library of Congress Control Number: Applied for

CR Subject Classification (1998): F.2, G.2, E.3, I.1

LNCS Sublibrary: SL 1 – Theoretical Computer Science and General Issues

ISSN   0302-9743
ISBN-10   3-540-36075-1 Springer Berlin Heidelberg New York
ISBN-13   978-3-540-36075-9 Springer Berlin Heidelberg New York

Springer is a part of Springer Science+Business Media

springer.com

© Springer-Verlag Berlin Heidelberg 2006
Printed in Germany

Typesetting: Camera-ready by author, data conversion by Scientific Publishing Services, Chennai, India
Printed on acid-free paper   SPIN: 11792086   06/3142   5 4 3 2 1 0

Preface

The first Algorithmic Number Theory Symposium (ANTS) conference was hosted by Cornell University, Ithaca, New York, USA in 1994. The goal of the conference was to bring together number theorists from around the world, and to advance theoretical and practical research in the field. ANTS I was soon followed by conferences in Bordeaux, France in 1996, Portland, Oregon, USA in 1998, Leiden, in the Netherlands in 2000, Sydney, Australia in 2002, and Burlington, Vermont, USA in 2004. Technische Universität Berlin in Germany hosted ANTS VII during July, 23-28 2006.

Five invited speakers attended ANTS VII. Thirty seven contributed papers were presented and a poster session was held. The invited speakers were Nigel Boston of the University of Wisconsin at Madison, John Cremona of the University of Nottingham, Bas Edixhoven of Universteit Leiden, Jürgen Klüners of Universität Kassel, and Don Zagier from the Max-Planck-Institut für Mathematik, Bonn.

Each submitted paper was reviewed by at least two experts external to the Program Committee which decided about acceptance or rejection on the basis of their recommendations. The Selfridge prize in computational number theory was awarded to the authors of the best contributed paper presented at the conference.

The organizers of ANTS VII express their gratitude and thanks to former organizers Duncan Buell, John Cannon, Henri Cohen, David Joyner, Blair Kelly III and Peter Stevenhagen for their important and valuable advice. We also appreciate the sponsorships by the Deutsche Forschungsgemeinschaft and the Mathematical Institute of Technische Unversität Berlin.

July 2006

Florian Hess
Sebastian Pauli
Michael Pohst

The ANTS VII Organizers

Organization

Program Committee

Karim Belabas, Université Bordeaux 1
Johannes Buchmann, Universität Darmstadt
John Cannon, University of Sydney
Gerhard Frey, Universität Duisburg-Essen
István Gaál, Debreceni Egyetem
François Morain, École Polytechnique Paris
Ken Nakamula, Tokyo Metropolitan University
Enric Nart, Universitat Autònoma de Barcelona
Takakazu Satoh, Tokyo Institute of Technology
Peter Stevenhagen, Universiteit Leiden
Fernando Villegas, University of Texas
Hugh Williams, University of Calgary

Conference Website

The names of the winners of the Selfridge prize, material supplementing the
contributed papers and errata for the proceedings, as well as the abstracts of
the posters and the posters presented at ANTS VII, can be found under

http://www.math.tu-berlin.de/~kant/ants.

Table of Contents

Invited Talks

Algebraic Number Theory

Analytic and Elementary Number Theory

Lattices

Curves and Varieties over Fields of Characteristic Zero

Computing Pro-P Galois Groups[*]

Nigel Boston and Harris Nover

Department of Mathematics, University of Wisconsin, Madison, WI 53706
{boston, nover}@math.wisc.edu

Abstract. We describe methods for explicit computation of Galois
groups of certain tamely ramified p-extensions. In the finite case this
yields a short list of candidates for the Galois group. In the infinite case it
produces a family or few families of likely candidates.

1 Introduction

Throughout this paper K will denote a number field, p a rational prime, and
S a finite set of primes, none of which lies above p. Furthermore, $K^{un,p}$ will
denote the maximal everywhere unramified p-extension of K. Our aim is to
compute the Galois group G of the maximal p-extension of K unramified outside
S. Whereas much is known about p-extensions ramified at p [16],[23], those
unramified at p are poorly understood. Wingberg [26] calls them among the
most mysterious objects in number theory. They are, however, important test
cases for the Fontaine-Mazur conjecture [13], which here implies that G has no
infinite p-adic analytic quotient. The first author's work on this conjecture [5],
[6] suggests that in fact G should have nontrivial actions on locally finite, rooted
trees, providing glimpses of a theory of arboreal Galois representations in parallel
to the well-developed theory of p-adic Galois representations.

We will focus on 2-extensions. The sort of information available is the abelianization of low index (usually $1, 2, 4$, and 8) subgroups, computed as quotients of
ray class groups thanks to class field theory, and exact values of, or at least
bounds on, generator and relation ranks of G. In addition, in the cases that
G is infinite, class field theory gives the further information that every subgroup
of finite index has finite abelianization (such a group is called FAb). In certain
cases, such as $K = \mathbb{Q}$, something is known about the form of the relations.

If a finite index subgroup H has cyclic abelianization, then Burnside's basis
theorem forces $H' = \{1\}$ and so G is finite. Moreover, a finite index subgroup H
with abelianization the Klein 4-group forces G to be finite since by an old result
of Taussky H has a cyclic subgroup of index 2. This allowed Boston and Perry
[8] to find the Galois groups of several 2-extensions of $\mathbb{Q}$. On the other hand, the
method is limited since in most cases these conditions do not hold at low index.

Boston and Leedham-Green [7] next introduced a new method for computing
G in more general circumstances. The idea is to search for G in O'Brien's tree [20],

---

[*] The authors would like to thank Rafe Jones, Jeremy Rouse, Rob Rhoades and Jayce
Getz for useful discussions. Nigel Boston was partially supported by the NSF. Harris
Nover was supported by the Office of Naval Research through an NDSEG fellowship.

which contains every 2-group H with d generators once (up to isomorphism), organized in such a way that edges go from $H/P_k(H)$ to its immediate descendant $H/P_{k+1}(H)$, where $P_k(H)$ is the kth group in the p-central series of H. The computer software, Magma [4], provides an implementation that computes all immediate descendants of a given 2-group.

By imposing filters that passed only those immediate descendants satisfying certain number-theoretical constraints, this tree was drastically pruned and in many cases the process terminated in a small collection of groups that could be G. The number-theoretical constraints included consequences of G having as many generators as relations (this occurs in the case $K = \mathbb{Q}$), the form of the relations (for $K = \mathbb{Q}$ they all come from local tame relations), and the structure of the abelianizations of index 1 and 2 subgroups. For many finite sets S of rational primes, the Galois group of the maximal 2-extension of $\mathbb{Q}$ unramified outside S was computed, up to a small ambiguity. Often more than one group (usually two) resulted but these groups were so similar that the answer to most questions regarding them would be the same regardless.

Eick and Koch [12] then proved results on the abelianizations of index 2 subgroups for entire families of S, partially verifying conjectures of Boston and Leedham-Green on the structure of G. These conjectures suggested that the order and other properties of G should depend quite simply on the form of S. It is notable that whereas G cannot be determined uniquely, there are still apparently few 2-groups that can arise as G for some choice of S (with $K = \mathbb{Q}$).

The next development was the work of Bush [10] on extending the method of Boston and Leedham-Green to apply to 2-class towers of imaginary quadratic fields. In particular, Stark had asked whether the 2-class tower (i.e. $S = \emptyset$) of $K = \mathbb{Q}(\sqrt{-2379})$ is infinite. If so, then the upper bound on root discriminants in infinite families of totally complex number fields [15] would be drastically reduced. Bush showed, however, that this 2-class tower has finite Galois group. He also explored the Galois groups of similar 2-class towers, finding none that were infinite, but obtaining ones of derived length 3, answering a question of Benjamin, Lemmermeyer, and Snyder [1],[2]. It is still an open problem whether any such Galois group can have finite derived length greater than 3. Steurer [25] has refined the results of Boston and Leedham-Green and of Bush by closer analysis of the candidates' lattice of subgroups. For instance, in the case of $\mathbb{Q}(\sqrt{-2379})$ and $S = \emptyset$, she eliminates half the candidates for G.

In this paper we describe two recent developments. First, the work of the second author systematizing computation of Galois groups of 2-class towers of imaginary quadratic fields is presented. After Bush's work the question arose as to whether other such fields might have infinite 2-class tower but small enough root discriminant to lower the known upper bound. The next likely candidate was $\mathbb{Q}(\sqrt{-3135})$, but existing methods led to a combinatorial explosion even in the pruned O'Brien tree. Nover's improved techniques, however, showed that this and several other fields have finite 2-class tower. Work by Benjamin, Lemmermeyer, and Snyder [3] on imaginary quadratic fields whose class group has order 8 and exponent 2 had found G explicitly in all but one situation. Nover's work tackles many examples of that remaining case.

The second new development is the work of the first author [6] on computing G in cases where it is known to be infinite. For instance, if $K = \mathbb{Q}$ and $S = \{q, r\}$ where q and r are odd primes 5 (mod 8) such that q is a 4th power modulo r but not vice versa, then G is known to be infinite by a refinement of the Golod-Shafarevich criterion due to Kuhnt [17]. In this case, G has two generators and two relations, whose general form is known. Relations of this form are chosen at random and filters employed that save those groups for which, within the computing range of Magma, the 2-central quotients do not stabilize and the abelianizations of low index subgroups are finite.

Amazingly, one special family of candidates for G emerges. This echoes the finite case in that we expect to obtain a short list of very similar groups. The groups that arise have interesting properties, besides being an apparently new class of FAb pro-2 groups. In particular their Lie algebras all appear isomorphic to a Lie algebra arising from quantum field theory [9],[22]. Also, they have torsion but all have a normal subgroup H of index 4 with Lie algebra isomorphic to that of $F \times F$, where F is the free pro-2 group on 2 generators. This group H is conjecturally a mild group. This comes out of the recent surprising work of Labute [18], who proved that for various S, p the Galois group G of the maximal p-extension (p odd) of $\mathbb{Q}$ unramified outside S is mild. This implies that such a G is of cohomological dimension 2 and thus torsion-free. It also has implications for the Fontaine-Mazur conjecture [19]. Schmidt [21] and Bush-Labute [11] provided refinements of Labute's work.

We begin with background on the tools that will be employed in our search for G. There follows a section on Nover's work establishing new results on finite Galois groups of 2-class towers. Finally, Boston's work on the infinite case is expounded upon, leading to various speculations as to where the future development of the subject may lead.

2 Mathematical Background

We make use primarily of two theoretical tools: class field theory and the O'Brien tree. We review each here. Let K be a field and $\mathrm{Cl}_p(K)$ the p-primary part of its class field. Then class field theory tells us that there is a correspondence between unramified abelian p-extensions of K and subgroups of $\mathrm{Cl}_p(K)$, and further that the maximal unramified abelian p-extension has $\mathrm{Cl}_p(K)$ as its Galois group over K. Put another way, if G is $\mathrm{Gal}(K^{un,p}/K)$ (that is, $S = \emptyset$) and G' the closure of its commutator subgroup, then the abelianization of G, $G/G' \cong \mathrm{Cl}_p(K)$. As the Galois correspondence associates closed subgroups of G with unramified p-extensions of K, we further have that the lattice of subgroups of G with abelianizations attached corresponds to the lattice of unramified p-extensions of K with Cl_p's attached. This simple observation will be quite powerful.

The second theoretical tool we use is the O'Brien tree. We start with a definition. The lower p-central series of a p-group G is defined recursively as $P_0(G) = G$ and $P_{k+1}(G) = P_k(G)^p[G, P_k(G)]$.[1] The smallest c such that $P_c(G) = \{1\}$ is called the p-class of G. It follows easily from our definitions that the p-class of $G/P_k(G)$ is the smaller of k and the p-class of G. Note that if G is a pro-p group, then $P_k(G)$ is the closed subgroup defined as above and in particular $P_1(G)$ is the Frattini subgroup of G. For reasons that will shortly become apparent, we say that a p-group G is a descendant of H, and that H is an ancestor of G, if $G/P_k(G) = H$ for some k. Further, G is an immediate descendant of H if additionally the p-class of G is $k + 1$.

Any group has a finite number of immediate descendants, and so given some starting group we can compute all of its descendants by computing its immediate descendants, their immediate descendants, and so forth. We may organize this information as a tree where the root is our starting group, the next level is its immediate descendants, the level after that their immediate descendants, and so on, where we put an edge between groups exactly when one is an immediate descendant of the other. We call this the O'Brien tree for the initial group. An algorithm described by O'Brien [19] allows us to compute the immediate descendants of a p-group and correspondingly compute the tree to finite depth. By Burnside's basis theorem, given a p-group G with exactly d generators, we know that $G/P_1(G)$ will be elementary abelian of rank d. Thus G will appear in the O'Brien tree with root the rank d elementary abelian p-group. Further, infinite pro-p groups with exactly d topological generators correspond to infinite ends of that tree.

3   Finite Galois Groups

Let $G = \mathrm{Gal}(K^{un,2}/K)$. We want to determine G. Our algorithm can be broken down into 3 steps.

3.1   Step 1: Generating Lattice Data

Galois theory tells us that the low index subgroups of G correspond to unramified low degree extensions of K. Explicit class field theory gives us a method of producing these extensions. First we compute the 2-class group of K, and use explicit class field theory to construct the corresponding unramified degree 2 extensions. Then for each of these extensions we compute its 2-class group and its degree 2 extensions, checking for duplicates as some degree 4 extensions will have more than 1 intermediate subfield. Then we compute the 2-class groups of the degree 4 extensions. As described in Sect. 2, by class field theory this tells us quite a bit of information about G. We now know the lattice of subgroups with abelianizations for index at most 4.

---

[1] Be aware that some authors define $P_1(G) = G$, and arrive correspondingly at different definitions of p-class, etc.

3.2 Step 2: Pruning the Tree

Now we use the O'Brien tree to get a (hopefully finite) list of candidates for G. If $\mathrm{Cl}_2(K)$ has 2-rank d, then we already know that $G/P_1(G)$ is an elementary abelian 2-group of rank d, which is to say that we know the root of an O'Brien tree which must contain G. We proceed with a recursive algorithm to pick out a subtree which will contain G. Assume we are working on a group Q of p-class $k-1$ which might have G as a descendant. We first use O'Brien's algorithm to generate each immediate descendant P, which will be of p-class k.

Now for each P, we are interested in two questions. The first question is whether any descendant of P, immediate or not, could be G, that is, could $P \cong G/P_k(G)$. We have two necessary conditions on P for this to be possible. The first condition is to see if P is compatible with the lattice data we collected earlier. Essentially, we compute the subgroup lattice of P up to index 4 with abelianizations, and then try to match up the subgroups of P with the extension fields of the lattice data. For each subgroup H of G, if $P_k(G) \leq H$, then the isomorphism theorems tell us that the lattice of P must contain an entry for H whose abelianization is a quotient of the abelianization of H. Further, for large enough k, $P_k(G) \leq H'$ and so the abelianizations must in fact be equal. Given the index of H and the known size of its abelianization, it is straightforward to compute the k at which $P_k(G) \leq H, H'$. For example, if H is an index 4 subgroup of G, then $P_k(G) \leq H$ if $k \geq 3$. We must have equality of abelianizations if $k \geq 2v([H:H']) - v([H_0:H_0']) + 3$, where H_0 is an index 2 subgroup containing H and v is the 2-adic valuation. We can improve this to 2 and $v([H:H']) + 2$ if H is contained in 3 subgroups (that is, if there are three intermediate fields between the corresponding field and K), as this implies H is normal.

The second condition is essentially cohomological in nature. Let $r(G)$ and $d(G)$ be, respectively, the relation and generator ranks of G. We require that the difference in ranks between the p-multiplicator and the nucleus of P is less than or equal to $d(G) + 1$. These terms are defined in [20]. This follows from the following two propositions.

Proposition 1. *If $G = \mathrm{Gal}(K^{un,2}/K)$ for K a totally imaginary number field, then $r(G) - d(G) \leq [K:\mathbb{Q}]/2$.*

Proof. This follows easily from [23]. □

Proposition 2. *Let G be a pro-p group with finite abelianization and let $G_i = G/P_i(G)$. Then for $i \geq 1$ the difference between the ranks of the p-multiplicator and the nucleus of G_i is at most $r(G)$.*

Proof. This is a generalization of a proposition in [12] inspired by a lemma in [7]. We follow the proof in the former.

Let $M(G_i)$ and $N(G_i)$ be, respectively, the p-multiplicator and nucleus of G_i. $M(G_i)$ is elementary p-abelian, so it suffices to bound the rank of $M(G_i)/N(G_i)$. Let $G = F/R$ be a free presentation where R is a normal subgroup of F. For any subgroup $U \leq F$, let $U^* = [U, F]U^p$. Let $K = P_i(F)$. We have that $G_i = F/KR$, which implies that $M(G_i) = KR/(KR)^*$ and $N(G_i) = K(KR)^*/(KR)^*$. So

$$M(G_i)/N(G_i) = (KR/(KR)^*) \big/ (K(KR)^*/(KR)^*) \cong KR/K(KR)^* .$$

As K and R are normal in F, the latter group is $KR/KR^* \cong R/(R \cap KR^*)$, which is a quotient of R/R^*, the p-multiplicator of G. This has rank at most $r(G)$ (by, for example, a group homological argument using the Universal Coefficient Theorem), as desired.   □

If P meets these two conditions it might be an ancestor of G. In that case, we ask if P might itself be G. There are again two conditions, corresponding once again to lattice and cohomological data. First, if k is small enough that we have not already done so, we check that the lattice for P matches our lattice data exactly, requiring equality of abelianizations. Second, for each subgroup in P up to some chosen index, we compute the generator and relation ranks by computing the abelianization and the second homology group. We then check to see if this is compatible with Proposition 1 above. We say that groups which pass these two conditions are candidates to be G. Given how quickly the bound of the proposition grows, we did not find it profitable to check subgroups of particularly large index. In fact, we did not find any otherwise candidates which had failing subgroups of index larger than 2, although we check up to index 16 for completeness.

We have now evaluated for each immediate descendant P of Q if it is a potential ancestor or candidate for G. If it is a candidate, we save it to a list of candidates. If it might be an ancestor, then we use it for Q, computing its immediate descendants, checking them, and recursing accordingly. Each of these calls on potential ancestor P's, if they terminate, will return a list of candidates, which we append to our current list of candidates. When we have tried all the P's, we return the list of candidates we have compiled, which are all the descendants of Q which might be G. By calling this algorithm on the known $G/P_1(G)$, we are guaranteed that if it terminates, it will return a finite list of candidates for G. In the event that it does not terminate, we still gain some information about G as we at least know its possible quotients by the lower p-central series, particularly if we modify the algorithm to only produce the subtree to some specified depth.

3.3   Step 3: Narrowing Candidates

If step 2 terminates, we know that G is on the list of candidates and in particular that it, and so the 2-tower, is finite. The number of candidates for G may still be somewhat large. We narrow it as follows. All candidates at this point have the same index 4 lattice, so we compute the index 8 lattices (with abelianizations) of each candidate group. We then find an index 4 subgroup whose subgroups in the index 8 lattice differ among the different candidates. We then map this back to our lattice data and determine which degree 4 extension of K corresponds to this subgroup. As all we have are the lattice data it is possible several degree 4 extensions might conceivably correspond. For each of them, we use their class group to compute their degree 2 extensions, and then compute the 2-class groups of those extensions. This lets us expand our lattice data and eliminate some of the candidates. After culling the candidates, we repeat the process until all our candidates have the same index 8 data.

3.4   An Example

The algorithm has been entirely implemented in a collection of Magma programs controlled by BASH scripts. As an example, we report the results of the algorithm on $K = \mathbb{Q}(\sqrt{-3135})$. We chose this group since, by [3], among imaginary quadratic fields with 2-class group $(2,2,2)$ (by which we mean $C_2 \times C_2 \times C_2$) and derived length at least 3 it has the smallest discriminant. If this group had infinite 2-tower, it would provide a much reduced upper bound of 56.0 on the root discriminant problem, as the current best bound, found in [15], is 82.2. Unfortunately, this is not the case. The algorithm narrows G down to one of 84 possible groups, each of derived length 3. In particular, each of these groups is finite, which shows that $\mathbb{Q}(\sqrt{-3135})$ has finite 2-class tower, a previously unknown result. Assuming GRH, we are further able to narrow G to one of 4 groups. We have also used the algorithm to show the finiteness of the 2-class tower for over a dozen other fields with 2-class group $(2,2,2)$ where this was unknown, such as $\mathbb{Q}(\sqrt{-966})$, $\mathbb{Q}(\sqrt{-969})$, $\mathbb{Q}(\sqrt{-1554})$, and $\mathbb{Q}(\sqrt{-16296})$.

First we compute the lattice data. The 2-class group of K is $(2,2,2)$ and so we know that, for $G = \text{Gal}(K^{un,2}/K)$, $G/P_1(G)$ is $(2,2,2)$. Continuing, we find that there are 7 degree 2 extensions. Three have 2-class field $(2,2,2)$, two have $(2,8)$, and there is one each of $(2,2,2,2)$ and $(2,16)$. Taking further degree 2 extensions and eliminating duplicates, we find 31 degree 4 fields and compute their class groups.

Next we run our modified O'Brien algorithm to produce the subtree of the O'Brien tree with root $(2,2,2)$. Although $(2,2,2)$ has 67 descendants, 4 of them fail the cohomological condition, another 44 have too large an abelianization, and another 18 have subgroups which will not fit with our lattice up to index 2. This leaves 1 descendant group, which must be an ancestor of G. This group, of 2-class 2 and order 32, has 186 descendants. 82 of these fail the cohomological condition and another 88 have lattices which cannot be quotient lattices in the appropriate sense at index 4. This leaves 16 descendant groups, each of order 256, to explore. At this point the width of our subtree broadens considerably and so we cease our description of this stage of the algorithm. After examining several thousand groups we are left with a finite subtree in which 240 groups pass the lattice conditions for candidacy.

For implementation reasons we leave the test on relation ranks of subgroups until after the tree portion of the algorithm is completed. Applying that test leaves us with 84 candidates with 24 unique index 8 lattices. At this point, without GRH, we are stuck, as we cannot unconditionally compute the class groups of the extensions of degree 4 extensions of K due to memory limitations. However, Magma allows a more efficient computation of the class group whose correctness is conditional on GRH, and so we use that from here on. Computing the 7 extensions with 2-class groups of a particular degree 4 extension of K leaves only 30 compatible candidates with 8 different index 8 lattices. Computing another set of extensions for a particular field yields 8 candidates with 2 different lattices. We then compute extensions of another 4 fields which leaves us with 4 groups, all sharing the same index 8 lattice. These 4 groups are similar in many other ways as well. They are all of order 8192, derived length 3, and 2-class 7. They also all have G/G_3 isomorphic to the group 32.033 (Hall-Senior notation), consistent with the results in [3], where $G_3 = [G, G']$. In fact, they are all immediate descendants of the same group, and so their quotients by the non-trivial terms in the lower p-central series are all isomorphic, and this is also true for the lower central series.

4 Infinite Galois Groups

If L/K is an infinite p-extension unramified at finitely many primes, none above p, then the structure of $\mathrm{Gal}(L/K)$ is mysterious, although the conjecture of Fontaine-Mazur [13] says that it should not be p-adic analytic. In fact, no explicit presentation of such a group is known.

One might try to shed some light on the structure of such a Galois group (and simultaneously put the Fontaine-Mazur conjecture to test) by the methods of the previous section, since finitely generated pro-p groups arise as the ends of O'Brien trees. This, however, is fruitless, since typically there will be combinatorial explosion in the pruned tree. As we shall see, this illustrates the fact that there are too many candidate groups to organize this way. On the other hand, the results in the finite case suggest that we should expect the possible groups to have some restricted form and to occur in families.

It follows that the computation should be organized differently. In [6], Boston introduced the method of picking relations at random (of the permitted form) and then examining the pro-2 groups so presented. Magma allows us to filter out groups H such that $H/P_n(H)$ is no larger than $H/P_{n-1}(G)$ for n up to about 63. This first filter eliminates many pro-2 groups that are actually finite. Secondly we filter out groups H having a subgroup of index 1, 2, 4, 8, or 16 with infinite abelianization, since such a group cannot satisfy FAb. "Drilling" into the group by computing successive index 2 subgroups and checking their abelianizations allows random tests of subgroups of index up to 2048 to see if FAb is violated.

In particular, let $K = \mathbb{Q}$ and $S = \{q, r\}$ where q and r are distinct odd primes. The Galois group of the maximal 2-extension of $\mathbb{Q}$ unramified outside S, G, is the finitely presented pro-2 group with presentation $< x, y | x^a = x^q, y^b = y^r >$ for some a, b in the free pro-2 group on x, y [16]. The simplest situation in which we can ensure that G is infinite arises by requiring that $q, r \equiv 5 \pmod 8$, so that the abelianization is $(4, 4)$, and additionally requiring that q is a 4th power modulo r but not vice versa (for example, $\{61, 5\}$). In this case a refinement of Golod-Shafarevich due to Kuhnt [17] shows that G is infinite. The filtering process described above works well, leaving one family of groups.

Conjecture 1. There exists a subset $\mathcal{F}$ of the free pro-2 group on x, y such that the Galois groups G above have presentations $< x, y | x^c = x^5, y^4 = 1 >$ with $c \in \mathcal{F}$.

The shortest elements in $\mathcal{F}$ have length 6. There are 48 of them of this length, for instance $y^2 xyxy, y^2 xyx^{-1}y^{-1}, \ldots$. Moreover, the sequence $\log_2(|G/P_n(G)|)$ is always the same, namely

$$2, 5, 8, 11, 14, 16, 20, 24, 30, 36, 44, 52, 64, 76, 93, 110, 135, 160, 196, 232, 286,$$
$$340, 419, 498, 617, 736, 913, 1090, 1357, 1634, \ldots$$

$\Delta_n := \log_2 |P_n(G)/P_{n+1}(G)|$ is identified by inputting it in Sloane's On-Line Encyclopedia of Integer Sequences [24]. It arises in [9], namely as, for $n \geq 4$, $\Delta_{2n-2} = \Delta_{2n-1} = \sum_{m=1}^n (1/m) \sum_{d|m} \mu(m/d)(F_{d-1} + F_{d+1})$, where μ is the usual Möbius function and F_n the nth Fibonacci number (so that in fact $F_{d-1} + F_{d+1}$ is the dth Lucas number). Letting $L(G) = \oplus P_n(G)/P_{n+1}(G)$, the $\mathbb{F}_p$-Lie algebra of G, its graded pieces have the same dimensions as (i) the free Lie algebra generated by one generator in degree 1 and one in degree 2 (arising in work on multi-zeta values and quantum field theory [9]) and (ii) Cameron's permutation group algebra of $C_2 \wr A$, where A is the group of all order-preserving permutations of the rationals. This suggests the following amazing possibility.

Conjecture 2. If G is in the family of groups above, then $L(G)$ is the $\mathbb{F}_p$-Lie algebra in (i) or (ii) above.

The groups G just found appear always to have a normal subgroup H satisfying $G/H \cong C_4$ and having generator and relation ranks both equal to 4. We call this the critical subgroup of G. It confirms the suggestion of Boston that these Galois groups should always have a subgroup of finite index that violates the Golod-Shafarevich criterion. Moreover, the growth of $H/P_n(H)$ is consistent with the conjecture that H is a mild pro-2 group in the sense of Labute [18], which would then imply that G has virtual cohomological dimension 2.

This method has been tried in other, similar circumstances. For example, it appears that the only infinite pro-3 group satisfying FAb and having presentation of the form $< x, y | x^a = x^4, y^b = y^4 >$ is the Sylow pro-3 subgroup S of $PSL_2(\mathbb{Z}_3)$. To improve this process, we are now investigating the degree to which a candidate's presentation can be "deformed" to yield another candidate - S would be rigid to this sort of deformation.

References

1. E. Benjamin, F. Lemmermeyer, and C. Snyder, Imaginary quadratic number fields with cyclic $Cl_2(k^1)$, J. Number Theory **67** (1997), 229–245.
2. E. Benjamin, F. Lemmermeyer, and C. Snyder, Imaginary quadratic fields k with $Cl_2(k) = (2, 2^m)$ and $d(Cl_2(k^1)) = 2$, Pac. J. Math. **198** (2001), 15–31.
3. E. Benjamin, F. Lemmermeyer, and C. Snyder, Imaginary quadratic fields with $Cl_2(k) = (2, 2, 2)$, J. Number Theory **103** (2003), 38–70.

4. W. Bosma, J. J. Cannon, *Handbook of Magma functions*, School of Mathematics and Statistics, University of Sydney (1996).
5. N.Boston, Some Cases of the Fontaine-Mazur Conjecture II, J. Number Theory **75** (1999), 161–169.
6. N.Boston, Reducing the Fontaine-Mazur conjecture to group theory, in "Progress in Galois Theory," Proceedings of Thompson's 70th Birthday Conference (2005), 39–50.
7. N.Boston and C.R.Leedham-Green, Explicit computation of Galois p-groups unramified at p, J. Algebra **256** (2002), 402–413.
8. N.Boston and D.Perry, Maximal 2-extensions with restricted ramification, J. Algebra **232** (2000), 664–672.
9. D.J.Broadhurst, On the enumeration of irreducible k-fold Euler sums and their roles in knot theory and field theory, preprint.
10. M.R.Bush, Computation of Galois groups associated to the 2-class towers of some quadratic fields, J. Number Theory **100** (2003), 313–325.
11. M. R. Bush and J. Labute, Mild pro-p groups with 4-generators, preprint.
12. B.Eick and H.Koch, On maximal 2-extensions of $\mathbb{Q}$ with given ramification, to appear in the Proc. St.Petersburg Math. Soc. (Russian), AMS Translations (English version).
13. J.-M.Fontaine and B.Mazur, Geometric Galois representations, *in* "Elliptic curves and modular forms, Proceedings of a conference held in Hong Kong, December 18-21, 1993," International Press, Cambridge, MA and Hong Kong.
14. F.Gerth, A density result for some imaginary quadratic fields with infinite Hilbert 2-class field towers, Arch. Math. (Basel) **82** (2004), 23–27.
15. F.Hajir and C.Maire, Tamely ramified towers and discriminant bounds for number fields. II, J. Symbolic Comput. **33** (2002), 415–423.
16. H.Koch, Galois theory of p-extensions. With a foreword by I. R. Shafarevich. Translated from the 1970 German original by Franz Lemmermeyer. With a postscript by the author and Lemmermeyer. Springer Monographs in Mathematics. Springer-Verlag, Berlin, 2002.
17. T.Kuhnt, Generalizations of the theorem of Golod-Shafarevich and applications, UIUC Ph.D. thesis (2002).
18. J.Labute, Mild pro-p-groups and Galois groups of p-Extensions of $\mathbb{Q}$ (to appear in J. Reine Angew. Math.)
19. C.Maire, Cohomology of pro-p extensions of number fields and the Fontaine-Mazur conjecture, preprint.
20. E.A.O'Brien, The p-group generation algorithm, J. Symbolic Comput. **9** (1990), 677–698.
21. A.Schmidt, Circular sets of prime numbers and p-extensions of the rationals (to appear in J. Reine Angew. Math.)
22. L.Schneps, Five Lie algebras, Lecture at AIM (available from www.aimath.org), April 2004.
23. I.R.Shafarevich, Extensions with prescribed ramification points (Russian), IHES Publ. Math. 18, 71–95 (1964).
24. N.J.A.Sloane, On-Line Encyclopedia of Integer Sequences, http://www.research.att.com/~njas/sequences/Seis.html
25. A.Steurer, On the Galois group of the 2-class tower of a quadratic field, UMD Ph.D. thesis (2006).
26. K.Wingberg, On the maximal unramified p-extension of an algebraic number field, J.Reine Angew.Math. **440** (1993), 129–156.

The Elliptic Curve Database for Conductors to 130000

John Cremona

School of Mathematical Sciences, University of Nottingham
University Park, Nottingham NG7 2RD, UK
John.Cremona@nottingham.ac.uk

Abstract. Tabulating elliptic curves has been carried out since the earliest days of machine computation in number theory. After some historical remarks, we report on significant recent progress in enlarging the database of elliptic curves defined over $\mathbb{Q}$ to include all those of conductor $N \leq 130000$. We also give various statistics, summarize the data, describe how it may be obtained and used, and mention some recent work regarding the verification of Manin's "$c = 1$" conjecture.

1 Background and History

Tabulating elliptic curves has been carried out since the earliest days of machine computation in number theory. In this article we concentrate on tables which claim to contain complete lists of elliptic curves with conductors in certain ranges. Other tables exist, notably tables of curves with prime conductor by Brumer and McGuinness [4] and, more recently, by Stein and Watkins [21].

We first review the tables existing before 1990, and then describe the tables we have compiled since then, concentrating on the large increase in the data available since mid-2005. We will describe the origins of the tables and give some information on the methods used to compile them. We give a summary of the data obtained to date, describe how to obtain and use the data, and mention some recent work regarding the verification of Manin's "$c = 1$" conjecture.

1.1 The Antwerp Tables

For many years the only published tables giving data on elliptic curves of small conductors were those in the volume [2], popularly known as "Antwerp IV", which forms part of the Proceedings of an International Summer School in Antwerp, July/August 1972, with the title Modular Functions of One Variable IV (edited by Birch and Kuyk).

The Antwerp tables consist of the following:

Table 1: All elliptic curves of conductor $N \leq 200$, arranged into isogeny classes, with the structure of the Mordell-Weil group (in most cases) and local data for primes of bad reduction. The origin of this table is discussed below.

Table 2: Generators for the curves of positive rank (one in each isogeny class) in Table 1, which all have rank 1. These were determined independently by Nelson Stephens and James Davenport; there are two omissions (143A and 154C) and two errors (155D and 170A).

Table 3: Hecke eigenvalues for primes $p < 100$ for the newforms associated to the elliptic curves in Table 1; due to Vélu, Stephens and Tingley.

Table 4: All elliptic curves whose conductor has the form $N = 2^a 3^b$, arranged in isogeny classes (with no information on the Mordell-Weil groups); due to Coghlan.

Table 5: Dimensions of the space of newforms for $\Gamma_0(N)$ for $N \leq 300$, including the dimensions of eigenspaces for the Atkin-Lehner involutions W_q and the splitting of the space of newforms over $\mathbb{Q}$; due to Atkin and Tingley.

Table 6: Factorized polynomials in $\mathbb{F}_p[j]$, for primes $p \leq 307$, whose roots are the supersingular values of j in characteristic p; due to Atkin.

To quote [2], "The origins of Table 1 are more complicated"; two pages of [2] are devoted to explaining this further. Briefly, the list 749 curves in this table evolved as follows.

- Swinnerton-Dyer searched for curves with small coefficients and kept those with conductor $N \leq 200$; he added curves obtained via a succession of 2- and 3-isogenies. Only the coefficients, discriminant and conductor were tabulated at first.
- Higher degree isogenies were checked using Vélu's method [24], adding some curves.
- Tingley used modular symbols to compute the space of newforms for $N \leq 300$, together with the action of the Hecke algebra and hence its splitting into eigenspaces. This revealed 30 "gaps", isogeny classes which had previously been missed. These were then filled, either by twisting known curves or by extending the original search region. For example, in isogeny class 78A the curve with smallest coefficients is[1] $78a1 = [1, 1, 0, -19, 685]$ which is unlikely to have been found by a search. Subsequently, Tingley went on to find equations of the associated elliptic curves directly from the newforms, using a method very similar to the one which we later developed, as described in [7]. Much of Tingley's work was never published except in the contribution to the Antwerp tables, and can only be found in one of three existing typescript copies of his thesis [23] (Oxford 1975). For the higher levels in the range $N \leq 300$, Tingley's 1975 program was slow and he only computed the elliptic curves for newforms where there was no corresponding curve yet known. By contrast (and to show how both the algorithms and hardware have improved in 30 years), in 2006 our program can find these curves (for $N \leq 300$) in around 20 seconds.

---

[1] We always specify curves by giving the coefficients $[a_1, a_2, a_3, a_4, a_6]$ of a minimal Weierstrass model. See section 3.2 below for more on labelling conventions.

- The Mordell-Weil ranks were computed by James Davenport, using the method of 2-descent as described in [3]. In eight cases these were not certain; in seven cases the rank is given as "0?" and is in fact 0; in one case it is given as "1?" but is 0.
- The list was known to be complete for certain conductors N, such as $N = 2^a 3^b$ and several prime N.
- Tingley's 1975 thesis [23] contains further curves with $200 < N \leq 320$ found via modular symbols, newforms and periods.

In the review by Vélu for *Mathematical Reviews* (MR0389726 (52 #10557)) a number of other minor errors in these tables are corrected.

No more systematic enumeration of elliptic curves by conductor occurred (as far as we are aware) between 1972 and the mid 1980s.

1.2 The 1992 Tables

During the 1980s my research and computations mainly concerned modular symbols and elliptic curves over imaginary quadratic fields. For this, methods were developed and implemented for handling modular symbols over such fields (initially, only those of class number one), including the computation of Hecke eigenvalues and periods, and also for dealing with the easier aspects of the arithmetic of elliptic curves (conductors and point searching, but not ranks). This work included a need to have information concerning elliptic curves defined over $\mathbb{Q}$ whose conductor lay beyond the range of the Antwerp tables, which led to the development of a new implementation of the modular symbol method over $\mathbb{Q}$. At around this time, conversations with Richard Pinch led me to implement modular symbols over $\mathbb{Q}$ with quadratic character (as described in [5]).

One obstacle to the writing up of much of this work was the lack of any suitable reference in the literature to the modular symbol method over $\mathbb{Q}$ for $\Gamma_0(N)$. The new implementation was now not only recomputing from scratch all the curves listed in Antwerp IV, but also extending the list to larger conductors. Although these tables did not at this point include isogenous curves or ranks or generators, they did contain some data not in the Antwerp tables pertaining to the Birch–Swinnerton-Dyer conjectures: specifically, they contained for each curve E the rational number $L(E, 1)/\Omega_E$ (where Ω_E is the least real period of E), whose value is conjectured to be 0 if and only if $E(\mathbb{Q})$ has positive rank, and is given by a conjectural formula involving the order of the Tate-Shafarevich group when $E(\mathbb{Q})$ has rank 0.

As a result, although the use of modular symbols to compute elliptic curves over $\mathbb{Q}$ was not in itself original, I decided that there was enough new material here to be worthy of publication, and in 1988 submitted a paper to *Mathematics of Computation* containing a table of elliptic curves of conductor up to 600. At this point only one curve was listed for each newform: no isogenies, ranks or generators. This paper was rejected in 1989, on two grounds: there were too many implementation details, and the referee wanted fuller information to be given for each curve – including the isogenous curves, and their ranks and generators. I was invited to resubmit the paper with this extra data included. Carrying this out required considerable effort, most significantly in re-implementing the 2-descent method of Birch and Swinnerton-Dyer to compute the ranks. James Davenport had been asked if his program for this still existed, but had replied that the only copy in existence was on a magnetic tape containing machine code for a computer which no longer existed; so this had to be done from scratch. Programs to compute isogenies and find Mordell-Weil group generators also had to be developed and written.

In 1990 I resubmitted the paper to *Mathematics of Computation*. The tables now covered all conductors to 1000, as well as containing all the requested information on ranks and generators. The text of the paper was still only 27 pages long, but it was accompanied by more than 200 pages of tables. The journal did offer to publish the paper, but with the tables as a microfiche supplement. However, while the refereeing was taking place, I was approached by several publishers who had seen the spiral-bound preprint and were interested in publishing it as a book. As nobody wanted the tables to be available only in microfiche format (which was rather old-fashioned even in 1990) I therefore withdrew it from *Mathematics of Computation* and signed up with Cambridge University Press.

Now, of course, 27 pages of text were insufficient for a book. In the first edition [6] of "Algorithms for Modular Elliptic Curves" the text was expanded to around 90 pages, with tables for curves to conductor 1000. It was published on 8 October 1992 and contained 5089 curves (those for $N = 702$ were missing through a stupid error: the number should have been 5113).

1.3   The 1997 Tables

By around 1995 the book [6] was out of print and CUP asked me to prepare a revised version. This duly appeared as [7] in 1997. As well as containing corrections and the missing curves of conductor 702, some sections were rewritten and a new section and table on the degree of the modular parametrization were added. However, the range of the printed tables was not extended, though links were given to online data which extended the range to $N \leq 5077$. In addition, the period between 1992 and 1997 also saw the proof of the Shimura-Taniyama-Weil conjecture, which changed the status of some of the statements in the text as well (obviously) as the status of the tables themselves, which could now be described as listing *all* elliptic curves of conductor $N \leq 1000$ rather than just those which were modular.

The full text of [7] has been available online since around 2002.

2   Algorithms and Implementation

The method we use to find all (modular) elliptic curves of a given conductor N uses modular symbols for $\Gamma_0(N)$, as is explained in detail in [7]. The original method was similar to that used by Tingley, though with certain improvements. Moreover, there have been many improvements in the details of the algorithm since the publication of [7], some of which have been developed in collaboration with William Stein. As these are rather technical we do not go into details, but give a brief summary. For some more details (but not the more technical and recent improvements), see Chapter 2 of [7].

2.1 Finding the Newforms

For each level N, one first computes the space of $\Gamma_0(N)$-modular symbols, and the action of the Hecke algebra on this space, to find one-dimensional eigenspaces with rational integer eigenvalues. Each of these corresponds to a rational newform f, where "rational" means that the Hecke eigenvalues, and hence the Fourier coefficients, are rational integers. Actually constructing the space of modular symbols is fast, though (for large levels) requires sparse matrix methods in order to fit in available machine memory. Sparse methods are also crucial when finding Hecke eigenspaces; this step is the most expensive in terms of memory requirements, and is also time-consuming, when the dimension of the space of modular symbols is large.

2.2 Finding the Curves

Given the newform f, we then integrate $2\pi i f(z)dz$ along certain paths in the upper half-plane, which are also given in terms of modular symbols, to obtain first the periods and then the equation of the associated elliptic curve of conductor N and L-series $L(E, s) = L(f, s)$. Finding E in practice involves computing the period lattice of f to sufficiently high precision; which in turn requires knowing many terms of the Fourier expansion of f, i.e. many Hecke eigenvalues. From the (approximate) period lattice of E, we obtain the invariants c_4, c_6 of E, at least approximately; but they are known to be integers. [This was first made explicit by Edixhoven in [13], following Katz-Mazur; (see also [1]).] Hence c_4 and c_6 can be determined exactly if we have sufficient precision. The precision requirement means that many Hecke eigenvalues are needed (up to 3500 for levels around 130000), so for this step it is also important for the implementation to be very efficient. The memory requirements for this step, and the time to compute the periods themselves, are negligible.

2.3 Reliability of the Data

Clearly no large-scale computation such as this can every guarantee 100% accuracy, and the software undoubtedly will always have bugs. Most errors to date have arisen through data processing mistakes: much of the handling of the large data files produced by our programs was done manually. More recently we have automated most of this and incorporated checks into our scripts wherever possible. Occasionally, at certain levels we missed newforms and hence elliptic curves; this has happened most often just after major rewriting of the code. When curves are missed at level N, we usually discover the fact when processing level $2N$, since then certain oldforms are not recognised as such. The online data is updated regularly and such corrections are logged; the data imported into packages (see below) may not be quite so up-to-date.

2.4   Obtaining Information About the Curves

For each elliptic curve found, we determine the analytic rank from the newform; when this is greater than 1 we check that it equals the Mordell-Weil rank using 2-descent.

Generators are found using a combination of the traditional methods: (1) search; (2) 2-descent, using our program mwrank [10]; (3) Heegner points (we now use MAGMA [16] for these, as the current implementation by Watkins, based on earlier versions by Cremona, Womack, Watkins and Delaunay, is extremely efficient); plus saturation methods.

We also compute isogenies, and all data on the isogenous curves. Since the computation of isogenies is rather delicate (it is easy to miss some if the precision is insufficient) this is done independently, as a check, using a program of Mark Watkins; as a benefit, Watkins's program also computes the degree of the modular parametrization and determines the curve in each isogeny class of minimal Faltings height. This method of computing the modular degree (described in [25]) is very much more efficient than the original one described in [8], which we stopped using at around $N = 14000$. The Faltings height information also allows verification of Stevens's Conjecture [22], that the curve with minimal Faltings height in each isogeny class is the one associated with $\Gamma_1(N)$ (which is usually, though not always (especially for smaller N), the same as the curve associated with $\Gamma_0(N)$).

2.5   Software

The original program was written in the 1980s in Algol68, and converted to C++ in the early 1990s. We use either Victor Shoup's NTL library (see [15]) or the LiDIA library (see [14]) for high-precision arithmetic, as well as STL (the Standard Template Library for C++). The sparse matrix code has been completely redeveloped, based on an earlier version by Luiz Figueiredo. This is probably the most important single programming improvement, and is essential both to physically allow levels as high as 100000 to be run on a machine with 2GB of RAM, and also for greatly increased speed of execution. Even so, some levels around 130000 require more than 2GB or RAM in which to run.

Without many low-level efficiency and algorithmic improvements it would not have been possible to have progressed so far. Some of these have been developed in collaboration with William Stein, who has written more general programs for computing with higher weights and characters: implemented originally in C++, then in MAGMA, and most recently in his package SAGE (see [18] and [19]).

One example: in [21] an example is given of a curve of rank 2 and rational 5-torsion of conductor 13881, which was then (2002) "beyond the range of

Cremona's tables"; computing the four curves (up to isogeny) of this conductor now takes less than 2 minutes to run, requiring about 60MB of RAM. Most of the computation time is taken up finding the eigenspaces for the first Hecke operator T_2 on the modular symbol space of dimension 1768.

2.6   Hardware

The other factor which has had an enormous impact on the expansion of the tables since spring 2005 is the availability at the University in Nottingham of a 1024-processor High Performance Computing "GRID" cluster, on which each user may (normally) use up to 256 processors simultaneously. This has enabled the processing of a hundred or more levels at a time. The GRID processors are arranged in pairs in 512 nodes, with each node (a "V20z dual Opteron") having access to its own 2GB of RAM. No parallel code is used (yet), so the advantage of the cluster is simply that of having a large number of machines controlled via a scheduling system to keep them all busy with the minimum amount of human intervention.

The nodes in the cluster have "only" 2GB or RAM each; hence for some larger levels it is necessary to perform separate runs on a different machine, with more RAM (8GB). So far this has sufficed, but further developments in the code are under way to enable the current upper bound of 130000 to be passed.

2.7   Milestones

Before using the HPC GRID we used between 0 and 3 machines, all shared with other users and jobs.

Date	Conductor reached
Mar 2001	10000
Nov 2001	12000
Aug 2002	13000
Oct 2002	15000
Jan 2003	16000
Feb 2003	18000
Mar 2003	19000
Apr 2003	20000
Mar 2004	21000
Apr 2004	23000
May 2004	24000
Jun 2004	25000
Oct 2004	26000
Nov 2004	27000
Jan 2005	29000
Feb 2005	30000

After starting to use the HPC GRID, the pace increased considerably:

Date	Conductor reached
22 Apr 2005	40000
27 May 2005	50000
9 Jun 2005	60000
20 Jun 2005	70000
14 Jul 2005	80000
26 Aug 2005	90000
31 Aug 2005	100000
18 Sep 2005	120000
3 Nov 2005	130000

Currently the program is undergoing further refinements in the expectation that it will be able to make further progress without moving wholesale to machines with more RAM. It would be interesting to cover all levels to $N = 234446$, which is the smallest known conductor of a curve with rank 4, namely $234446b1 = [1, -1, 0, -79, 289]$. Level $N = 234446$ itself has been run successfully; as well as the rank 4 curve there are two others with this conductor, both of which have rank 3: $234446a1 = [1, 1, 0, -696, 6784]$ and $234446c1 = [1, 1, 1, -949, -7845]$.

2.8   Using the GRID

To use the HPC GRID we use a fairly simple shell script, which loops over a range of values of N. This script runs simultaneously on however many nodes are available. At each pass through the loop, shell commands are used to detect the existence of a log file associated with the value of N in question, which would indicate that another node was already working on this level. If so, N is incremented; otherwise a series of C++ programs is run with N (and other parameters) as input, which result in all the necessary computations being carried out for that level with the output suitably recorded. One minor technical issue here is that the system has to be able to handle several hundreds of thousands of data files, something of which system administrators may disapprove. [We keep the data for each level accessible for later runs, since our method of eliminating oldforms currently involves accessing the data at levels M dividing N, rather than using degeneracy maps.]

A typical extract from the log file of one node follows:

```
running nfhpcurve on level 120026 at Fri Sep 23 18:26:48 BST 2005
running nfhpcurve on level 120197 at Fri Sep 23 20:12:31 BST 2005
running nfhpcurve on level 120224 at Fri Sep 23 20:58:18 BST 2005
running nfhpcurve on level 120312 at Fri Sep 23 23:35:19 BST 2005
running nfhpcurve on level 120431 at Sat Sep 24 04:19:54 BST 2005
running nfhpcurve on level 120568 at Sat Sep 24 10:42:18 BST 2005
running nfhpcurve on level 120631 at Sat Sep 24 13:56:49 BST 2005
running nfhpcurve on level 120646 at Sat Sep 24 14:48:21 BST 2005
```

```
running nfhpcurve on level 120679 at Sat Sep 24 15:59:54 BST 2005
running nfhpcurve on level 120717 at Sat Sep 24 18:11:20 BST 2005
running nfhpcurve on level 120738 at Sat Sep 24 19:13:11 BST 2005
running nfhpcurve on level 120875 at Sun Sep 25 02:20:27 BST 2005
running nfhpcurve on level 120876 at Sun Sep 25 02:20:28 BST 2005
running nfhpcurve on level 120918 at Sun Sep 25 04:58:32 BST 2005
running nfhpcurve on level 120978 at Sun Sep 25 08:08:00 BST 2005
```

The program being run here is called "nfhpcurve", where "nf" stands for newform, "hp" for "H_1^+" indicates that we use the plus part of the modular symbol space, and "curve" that we compute the equations for the curve from each newform. Separate programs are run to find isogenous curves and Mordell-Weil generators and other data.

The levels here are in the range 120000–121000; those not listed are being run on other nodes. Approximately 10 levels per processor per day are completed, though the time for each individual level varies greatly, depending on several factors: highly composite N have modular symbol spaces of higher dimension, which has a major effect on the time required for linear algebra; levels with no newforms obviously save on the time required to compute many Hecke eigenvalues a_p; and curves with very large c_4, c_6 invariants require working to higher precision with more a_p needing to be computed.

Certain values of N are known not to be possible conductors (specifically, N which are divisible by 2^9 or by 3^6 or by p^3 with $p \geq 5$) and these are skipped.

3   Summary of Data and Highlights of Results

3.1   Availability of the Data

Full data is available from [9]. The data is mostly in plain ascii files for ease of use by other programs, rather than in typeset tables as in the book. A mirror is maintained by William Stein at http://modular.math.washington.edu/cremona/. (Stein's Modular Forms Database at http://modular.math.washington.edu/Tables/ also has links to many other tables.) Currently there is approximately 106MB of data (as a gzipped tar file) which unpacks to 260MB. This only includes a_p for $p < 100$, as further values can obviously be recomputed from the curve itself.

Recently, in collaboration with various other people, other more convenient ways of accessing and processing the data have been developed.

- A web-based interface by Gonzalo Tornaria is at http://www.math.utexas.edu/users/tornaria/cnt/cremona.html, covering $N < 100000$. This provides an attractive interactive interface to the data; as a bonus, information on quadratic twists is included.
- The free open-source number theory package pari/gp (see [17]) makes the full elliptic curve database available (though not installed by default). For example

```
(12:05) gp > ellsearch(5077)
%1=[["5077a1", [0, 0, 1, -7, 6], [[-2, 3], [-1, 3], [0, 2]]]]
(12:05) gp > ellinit("5077a1")
%2 = [0, 0, 1, -7, 6, 0, -14, 25, -49, 336, -5400, 5077, ...
(12:05) gp > ellidentify(ellinit([1,2,3,4,5]))
%3 = [["10351a1", [1, -1, 0, 4, 3], [[2, 3]]], [1, -1, 0, -1]]
```

The output of `ellsearch` contains all matching curves with their generators.
The output of `ellidentify`, whose input need not be given in minimal or
standardised form, includes the standard transformation $[u, r, s, t]$ mapping
the input curve to standard minimal form. Full integration of this capability
with standard `pari/gp` elliptic curve functions is ongoing (thanks to Bill
Allombert).

- William Stein's free open-source package SAGE (Software for Algebra and
  Geometry Experimentation, see [18] and [19]) also has all our data available
  and many ways of working with it, including a transparent interface to many
  other pieces of elliptic curve software. For example:

```
sage: E = EllipticCurve("389a")
sage: E
Elliptic Curve defined by y^2 + y = x^3 + x^2 - 2*x  over Rational Field
sage: E.rank()
2
sage: E.gens()                          # Cremona's mwrank
[(-1 : 1 : 1), (0 : 0 : 1)]
sage: L = E.Lseries_dokchitser(); L(1+I)   # Tim Dokchitser's program
-0.63840993858803874 + 0.71549523920466740*I
sage: E.Lseries_zeros(4)                 # Mike Rubinstein's program
[0.00000000000, 0.00000000000, 2.8760990715, 4.4168960843]
```

- MAGMA has the database for conductors up to 70000 (as of version 2.12-16):

```
> ECDB:=CremonaDatabase();
> NumberOfCurves(ECDB);
462968
> LargestConductor(ECDB);
70000
> E:=EllipticCurve(ECDB,"389A1");
> E;
Elliptic Curve defined by y^2 + y = x^3 + x^2 - 2*x over Rational Field
> Rank(E);
2
```

3.2 The Naming of Curves

Since many authors refer to individual elliptic curves by means of their label in
the database, it is desirable to use a sensible naming convention which is concise,
informative and only changes when absolutely necessary.

The Antwerp tables use a labelling system for the elliptic curves which consists
of the conductor followed by a single upper case letter. The order of these is
not easy to define; the curves are grouped into isogeny classes, but one cannot

determine this from the label alone. For example, the curves of conductor 37 are in two classes, $\{37A\}$ and $\{37B, 37C, 37D\}$. Clearly this system cannot be used once we have more than 26 curves per conductor.

For the tables of [7] we introduced an additional layer into the notation. The isogeny classes have labels similar to those of individual curves in the Antwerp system, consisting of a single uppercase letter following the conductor. The curves in the class are indicated by suffixing (or occasionally subscripting) the class code with an integer. For example at conductor 37 the classes are $\{37A1\}$ and $\{37B1, 37B2, 37B3\}$.

The ordering of the isogeny classes is determined by the order in which the newforms are found with our modular symbols program; this has changed over the years and so is now, unfortunately, almost impossible to define precisely. However for all levels between 451 and 130000 the order is lexicographical order of the Hecke eigenvalues of the newforms, with the eigenvalues of the Atkin-Lehner involutions W_q (for bad primes q) listed first, and the eigenvalues for W_q ordered $+1, -1$ those for T_p as $0, +1, -1, +2, -2, \ldots$. It is planned to change this system for $N > 130000$ to one based on simple lexicographical order of the complete eigenvalue sequence, with all primes in their natural order; but there will be no further change in the labels for $N \leq 130000$!

The order of the curves within each isogeny class is likewise difficult to define precisely. The first curve in each class is the curve variously called the "strong Weil curve" or the $\Gamma_0(N)$-optimal curve; that is, the curve whose period lattice (of a minimal model) is exactly that of the normalised newform. After that, the order is determined by our algorithm for finding isogenies.

In the tables in [7], for $N \leq 200$ the Antwerp codes were given alongside the new ones.

When the tabulation reached $N = 1728$, where there are for the first time more than 26 isogeny classes (there are 28), something new was required. Without sufficient thought for "future-proofing" we simply followed the sequence $A, B, \ldots, Z$ by $AA, BB, \ldots, ZZ$ and then (at level $N = 4800$ which has 72 rational newforms) $AAA, BBB, \ldots$ and so on. In 2005 this system was becoming unworkable. At level 100800 there are 418 rational newforms with codes from $100800A$ to $100800BBBBBBBBBBBBBBBBBBBBBBBBBBBBBBB$.

It was therefore decided to use a new coding system for the isogeny class labels, and after widespread consultation the following scheme was decided upon (thanks to David Kohel in particular). We now use a base 26 number system, with the letters $a, \ldots, z$ for the "digits" $0, \ldots, 25$ and leading as omitted. So after z comes ba, and the last class at level 100800 has label $100800qb$. For conductor 37, the classes are now $\{37a1\}$ and $\{37b1, 37b2, 37b3\}$. When we reach a conductor where the number of classes is more than $26^2 = 676$, all we need do is follow zz with baa. [In the Stein-Watkins database of elliptic curves there are conductors with many thousands of isogeny classes.]

Lower case letters were used to avoid confusion between old and new coding systems; so (happily) the only difference for curves of conductor less than 1728 is the change of case.

The online tables have been altered to reflect this change of coding, as have the databases available in SAGE and pari/gp, but MAGMA V2.12 still uses the old system.

3.3 Numbers of Curves

In Table 1 we give the numbers of isogeny classes of curves for ranges of conductors of the form $10000k \leq N < 10000(k+1)$, together with the numbers for each value of the rank. One very remarkable feature is that the number in each range is close to constant. This feature is maintained in smaller ranges: in each range of 1000 consecutive conductors there are very close to 4400 isogeny classes of curves.

Table 1. Numbers of isogeny classes of curves, by rank

range of N	#	$r = 0$	$r = 1$	$r = 2$	$r = 3$
0-9999	38042	16450	19622	1969	1
10000-19999	43175	17101	22576	3490	8
20000-29999	44141	17329	22601	4183	28
30000-39999	44324	16980	22789	4517	38
40000-49999	44519	16912	22826	4727	54
50000-59999	44301	16728	22400	5126	47
60000-69999	44361	16568	22558	5147	88
70000-79999	44449	16717	22247	5400	85
80000-89999	44861	17052	22341	5369	99
90000-99999	43651	16370	21756	5442	83
100000-109999	44274	16599	22165	5369	141
110000-119999	44071	16307	22173	5453	138
120000-129999	44655	16288	22621	5648	98
0-129999	568824	217401	288675	61840	908

The chart in Figure 1 shows the overall distribution of ranks.

In Table 2 we give the total number of curves up to isomorphism. This reveals that the average size of the isogeny classes found is currently about 1.487. This average seems to be steadily but gradually decreasing (the value for $N \leq 1000$ was just over 2.0). Mark Watkins has pointed out that if one considers curves in a large box with $|c_4| < X^2$ and $|c_6| < X^3$, then the average size of the isogeny class tends to 1 as $X \to \infty$. Also, Duke has shown in [12] that almost all curves (ordered in this way) have no exceptional primes, and in particular no rational isogenies.

The sizes of individual isogeny classes are given in Table 3. Here we classify isogeny classes by the maximal degree D of an isogeny (with cyclic kernel)

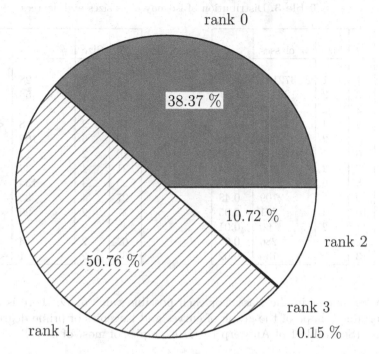

Fig. 1. Overall distribution of ranks

Table 2. Numbers of isogeny and isomorphism classes of curves

range of N	# isogeny classes	# isomorphism classes
0-9999	38042	64687
10000-19999	43175	67848
20000-29999	44141	66995
30000-39999	44324	66561
40000-49999	44519	66275
50000-59999	44301	65393
60000-69999	44361	65209
70000-79999	44449	64687
80000-89999	44861	64864
90000-99999	43651	63287
100000-109999	44274	63410
110000-119999	44071	63277
120000-129999	44655	63467
0-129999	568824	845960

Table 3. Distribution of isogeny class sizes and degrees

D	Size	# classes	%		D	Size	# classes	%
1	1	372191	65.43		14	4	28	< 0.01
2	2	123275	21.67		15	4	58	0.01
3	2	31372	5.52		16	8	270	0.05
4	4	27767	4.88		17	2	8	< 0.01
5	2	2925	0.51		18	6	162	0.03
6	4	3875	0.68		19	2	12	< 0.01
7	2	808	0.14		21	4	30	0.01
8	6	2388	0.42		25	3	134	0.02
9	3	2709	0.48		27	4	33	0.01
10	4	271	0.05		37	2	20	< 0.01
11	2	60	0.01		43	2	7	< 0.01
12	8	286	0.05		67	2	4	< 0.01
13	2	130	0.02		163	2	1	< 0.01

between curves in the class. For each possible value of D, there is a uniquely determined shape of the graph of curves and isogenies of prime degree between them. (See Table 1 of Antwerp IV for examples of most of these.)

3.4 Mordell-Weil Groups

For almost all the elliptic curves found we have determined the full Mordell-Weil group. In a very small number of cases we can only (at present) guarantee that the generators listed in the tables generate a subgroup of finite index. In all cases where the analytic rank is 2 or 3 we have verified by 2-descent that the rank is equal to the analytic rank. When the analytic rank is 0 or 1 this is known to be true by results of Rubin, Kolyvagin and Gross and Zagier.

In most cases of positive rank, searching for points suffices to find the expected number of independent generators, following which we apply a saturation procedure to obtain the full Mordell-Weil group. The exceptional cases are those for which we were not able to determine a bound on the index, on account of the bound on the difference between the logarithmic and canonical heights being rather large. This situation should improve after full implementation of the improved height bound algorithm described elsewhere in this volume: see [11].

For curves where searching for points was insufficient, most had rank 1 and generators could be found using Heegner points. Since 2004, techniques for computing Heegner points of large height have improved very significantly, thanks to work of Delaunay and Watkins. The MAGMA implementation is now extremely fast and we have used it extensively for all the larger generators in the tables.

The current record is curve $108174c2$, whose generator P has canonical height $\hat{h}(P) = 1193.35$. Here $P = (a/c^2, b/c^3)$ where

$a =$ −1363283370314068103350302367912867052955821842006343239797143928187616893692560809927868610376827116575 1

437633556213041024136275990157472508801182302454436678900455860307034813576105868447511602833327656978462

242557413116494486538310447476190358439933060717111176029723557330999410077664104893597013481236052075987

425547135210992941868374222370098962971095497629371786841015352894106057367293353077806131982247703253651 11

29607075613734924952158278253743039282375024853516001988744749085116423499171358836518920399114139315005,

$b =$ 77684538615967858963507761534649218160103504276800201439664696233377268844630389216260652695579081249211

185106671917236143678971202347339963247386055808925185619325909681380265508543158979491984235466881248491

978341526711100575326744746030922470291782156359389005809065313914236892470866399096616908015986267206085

816145609347461468770147859622405813347969542380216159923828490925517451952455079424426512616714569247069

065790676549942365146817589522964032348349807255751358289869629122053879780510640219504941970766697032823

589255263953926885142009701275092664710953135501372398976396568319085695054751879368605289437600720585853

465424006259176930980665902501637183477157293942231705607887213321716750749368884791336280387610317598902

03302543264770366827148378274013771150847966 91,

$c =$ 1139668556693332928963288336905529439332124222622872858583364718432796440766474865924602420890490333 70292

485250756121056680073078113806049657487759641390843477809887412203584409641844116068236428572188929747

769498615000931961765366269300665024812605970444 1347.

In MAGMA, finding this generator is as easy (and quick) as this:

```
> E:=EllipticCurve([1,1,0,-330505909530535,-231268766069798 6706251]);
> time HeegnerPoint(E);
true (-13632833.../12988444...  :  77684538.../14802521...  :  1)
Time: 36.340
```

At the other extreme, the minimal height of a generator is for curve $61050cs1$ $= [1, 0, 0, -23611588, 39078347792]$, whose generator $(-3718, 276584)$ has canonical height 0.0148.

In the small number of cases where curves of rank greater than 1 have generators too large to be found easily by searching, we found the generator using 2-descent and in some cases 4-descent. The latter, for which algorithms were developed by Siksek and Womack, is now efficiently implemented in MAGMA.

3.5 Torsion Structures

The distribution of the 15 possible structures for the torsion subgroups of the curves is given in Table 4. Here C_n denotes a cyclic group of order n.

3.6 Degrees of Modular Parametrizations

As already mentioned, for most of the range the modular parametrization degrees were computed using a program of Mark Watkins (see [25] for the method).

Table 4. Torsion structures

Structure	# curves	%
C_1	432622	51.14
C_2	344010	40.67
C_3	18512	2.19
C_4	12832	1.52
$C_2 \times C_2$	33070	3.91
C_5	698	0.08
C_6	3155	0.37
C_7	50	< 0.01
C_8	101	0.01
$C_2 \times C_4$	793	0.09
C_9	16	< 0.01
C_{10}	28	< 0.01
C_{12}	11	< 0.01
$C_2 \times C_6$	58	< 0.01
C_{16}	4	< 0.01
Odd	451898	53.42
Even	394062	46.58
All	845960	100.00

Here we only mention that the largest degree so far is for 96054k1, for which $\deg(\varphi) = 32035843840 = 2^8 \cdot 5 \cdot 7 \cdot 11^2 \cdot 13 \cdot 2273$.

3.7   (Analytic) Orders of Ш

For each curve E in the tables, we have computed all the quantities appearing in the Birch–Swinnerton-Dyer conjecture for E, with the exception of the order of the Tate-Shafarevich group Ш. It is customary to define the "analytic order of Ш" to be the order predicted by the Birch–Swinnerton-Dyer conjecture, which we can determine from this data. In the case of curves of rank 0 this is computed as an exact rational number, which turns out in every case to be an integral perfect square. For curves of positive rank it is computed as a floating-point approximation (using approximations for the regulator, the real period and $L^{(r)}(E, 1)$); we again always find a value close to an integer which is a perfect square. In Table 5 below, we do not distinguish between the different status of these values. The current record is $676 = 26^2$ for curve 95438a1 (which has rank 0 so this is an exact value).

We should also mention here recent work of Stein and others (see [20]) towards verifying precisely the Birch–Swinnerton-Dyer conjecture for non-CM curves of rank at most 1 and conductor up to 1000.

Table 5. Analytic orders of Ⅲ

$|Ⅲ|$	#
2^2	37074
3^2	11512
4^2	4013
5^2	1954
6^2	426
7^2	468
8^2	250
9^2	85
10^2	52
11^2	73
12^2	20
13^2	19

$|Ⅲ|$	#
14^2	9
15^2	2
16^2	6
17^2	4
19^2	2
20^2	3
21^2	2
23^2	4
26^2	1
all > 1	55979

3.8   The Manin Constant

Recall that the Manin constant for an elliptic curve E of conductor N is the rational number c such that

$$\varphi^*(\omega_E) = c(2\pi i f(z)dz),$$

where ω_E is a Néron differential on E, f is the normalized newform for $\Gamma_0(N)$ associated to E, and $\varphi : X_0(N) \to E$ is the modular parametrization. A long-standing conjecture is that $c = 1$ for all elliptic curves over $\mathbb{Q}$ which are optimal quotients of $J_0(N)$ (or "strong Weil curves" in the older terminology). A result already cited [13] is that $c \in \mathbb{Z}$, and there are many results restricting the primes which may divide c.

Recent developments, described in [1], have strengthened these conditions considerably. Also in [1] there is an account of numerical verifications we have carried out which establish the conjecture for most of the curves in our tables. The following result is taken from [1].

Theorem 1. *(a) For all $N \le 60000$, every optimal elliptic quotient of $J_0(N)$ has Manin constant equal to 1.*

(b) For all N in the range $60000 < N \le 130000$, every optimal elliptic quotient of $J_0(N)$ has Manin constant equal to 1, except for the following cases where the Manin constant is either 1 or 2:

$$67664a, 71888e, 72916a, 75092a, 85328d, 86452a, 96116a,$$

$$106292b, 111572a, 115664a, 121168e, 125332a.$$

In each of the 12 undecided cases listed, the isogeny class consists of two curves linked by 2-isogenies, and we have not yet verified which of the two curves is the optimal quotient of $J_0(N)$. See [1] for details.

Acknowledgements

Thanks to William Stein and Mark Watkins for comments on an earlier draft of this paper.

References

1. A. Agashe, K. Ribet and W. A. Stein (with an appendix by J. E. Cremona), *The Manin Constant, Congruence Primes and the Modular Degree*, preprint 2006.
2. B. J. Birch and W. Kuyk (eds.), *Modular Functions of One Variable IV*, Lecture Notes in Mathematics **476**, Springer-Verlag 1975. See also http://modular.math.washington.edu/scans/antwerp/.
3. B. J. Birch and H. P. F. Swinnerton-Dyer, *Notes on Elliptic Curves I*, J. Reine Angew. Math. **212** (1963), 7–25.
4. A. Brumer and O. McGuinness, *The behaviour of the Mordell-Weil group of elliptic curves*, Bull. Amer. Math. Soc. (N.S.) **23** (1990), no. 2, 375–382.
5. J. E. Cremona, *Modular symbols for $\Gamma_1(N)$ and elliptic curves with everywhere good reduction*, Math. Proc. Cambridge Philos. Soc., **111** (1992) no. 2, 199–218.
6. J. E. Cremona, *Algorithms for modular elliptic curves*, Cambridge University Press, 1992.
7. J. E. Cremona, *Algorithms for modular elliptic curves (2nd edition)*, Cambridge University Press, 1997. See also http://www.maths.nott.ac.uk/personal/jec/book/fulltext/.
8. J. E. Cremona, *Computing the degree of the modular parametrization of a modular elliptic curve*, Math. Comp. **64** no. 211 (1995), 1235–1250.
9. J. E. Cremona, Tables of Elliptic Curves, http://www.maths.nott.ac.uk/personal/jec/ftp/data/.
10. J. E. Cremona, mwrank, a program for 2-descent on elliptic curves over $\mathbb{Q}$, http://www.maths.nott.ac.uk/personal/jec/mwrank/.
11. J. E. Cremona and S. Siksek, *Computing a Lower Bound for the Canonical Height on Elliptic Curves over $\mathbb{Q}$*, ANTS VII Proceedings 2006, Springer 2006.
12. W. Duke, *Elliptic curves with no exceptional primes*, C. R. Acad. Sci. Paris Sér. I Math. **325** (1997), no. 8, 813–818.
13. B. Edixhoven , *On the Manin constants of modular elliptic curves*, in *Arithmetic algebraic geometry (Texel, 1989)*, Progr. Math. **89**, Birkhäuser, Boston (1991), 25–39.
14. LiDIA: A C++ Library For Computational Number Theory, http://www.informatik.tu-darmstadt.de/TI/LiDIA/.
15. NTL: A Library for doing Number Theory, http://www.shoup.net/ntl/.
16. The MAGMA Computational Algebra System, version 2.12-16, http://magma.maths.usyd.edu.au/magma/.
17. pari/gp, version 2.2.13, Bordeaux (2006), http://pari.math.u-bordeaux.fr/.
18. W. A. Stein and D. Joyner, SAGE: *System for Algebra and Geometry Experimentation*, Comm. Computer Algebra **39** (2005), 61–64.
19. W. A. Stein, SAGE, Software for Algebra and Geometry Experimentation, http://sage.scipy.org/sage.
20. W. A. Stein, G. Grigorov, A. Jorza, S. Patrikis and C. Patrascu, *Verification of the Birch and Swinnerton-Dyer Conjecture for Specific Elliptic Curves*, preprint 2006.

21. W. A. Stein and M. Watkins, *A database of elliptic curves—first report*, in *Algorithmic number theory (Sydney, 2002)*, Lecture Notes in Comput. Sci., **2369**, Springer, Berlin, 2002, 267–275.
22. G. Stevens, *Stickelberger elements and modular parametrizations of elliptic curves*, Invent. Math. **98** (1989), no. 1, 75–106.
23. D. J. Tingley, *Elliptic curves uniformized by modular functions*, University of Oxford D. Phil. thesis (1975).
24. J. Vélu, *Isogénies entre courbes elliptiques*, C. R. Acad. Sci. Paris Sér. A-B, **273** (1971), A238–A241.
25. M. Watkins, *Computing the modular degree of an elliptic curve*, Experiment. Math. **11** no. 4 (2002), 487–502.

On the Computation of the Coefficients of a Modular Form

Bas Edixhoven

Mathematisch Instituut, Universiteit Leiden
Postbus 9512, 2300 RA Leiden, The Netherlands
edix@math.leidenuniv.nl
http://www.math.leidenuniv.nl/~edix

Abstract. We give an overview of the recent result by Jean-Marc Couveignes, Bas Edixhoven and Robin de Jong that says that for l prime the mod l Galois representation associated to the discriminant modular form Δ can be computed in time polynomial in l. As a consequence, Ramanujan's $\tau(p)$ for prime numbers p can be computed in time polynomial in $\log p$.

The mod l Galois representation occurs in the Jacobian of the modular curve $X_1(l)$, whose genus grows quadratically with l. The challenge therefore is to do the necessary computations in time polynomial in the dimension of this Jacobian. The field of definition of the l^2 torsion points of which the representation consists is found via a height estimate, obtained from Arakelov theory, combined with numerical approximation. The height estimate implies that the required precision for the approximation grows at most polynomially in l.

The results in this note have been obtained in collaboration with Jean-Marc Couveignes and Robin de Jong. A rather lengthy report on our work, written in the context of a contract between the University of Leiden and the French CELAR (Centre Électronique de l'Armement) will be available on arxiv very soon from today (April 14, 2006), see [4]. A concise research article by Couveignes, Edixhoven and de Jong is in preparation. The length of this note is kept rather short because of the existence of the lengthy report.

Edixhoven is partially supported by the European Marie Curie Research Training Network "arithmetic algebraic geometry", contract MRTN-CT2003-504917, and, since January 2005, supported by NWO.

1 Statement of the Main Results

We recall that Ramanujan's τ-function is defined by the following identity of formal power series with integer coefficients:

$$x \prod_{n \geq 1} (1 - x^n)^{24} = \sum_{n \geq 1} \tau(n) x^n. \tag{1}$$

F. Hess, S. Pauli, and M. Pohst (Eds.): ANTS 2006, LNCS 4076, pp. 30–39, 2006.
© Springer-Verlag Berlin Heidelberg 2006

Theorem 1. *There exists a probabilistic algorithm that on input a prime number p gives $\tau(p)$, in expected running time polynomial in $\log p$.*

We note that from the definition of $\tau(n)$ above it is not clear how to compute $\tau(p)$ as fast as in the theorem. Behind the theorem is the existence of certain Galois representations. The function Δ on the complex upper half plane $\mathbb{H}$ give by:

$$\Delta\colon \mathbb{H} \to \mathbb{C}, \quad z \mapsto \sum_{n \geq 1} \tau(n) e^{2\pi i n z} \qquad (2)$$

is a modular form, the so-called discriminant modular form. It is a new-form of level 1 and weight 12. Deligne showed in [3] that, as conjectured by Serre, for each prime number l there is a (necessarily unique) semi-simple continuous representation:

$$\rho_l\colon \mathrm{Gal}(\overline{\mathbb{Q}}/\mathbb{Q}) \to \mathrm{Aut}(V_l), \qquad (3)$$

with V_l a two-dimensional $\mathbb{F}_l$-vector space, such that ρ_l is unramified at all primes $p \neq l$, and such that for all $p \neq l$ the characteristic polynomial of $\rho_l(\mathrm{Frob}_p)$ is given by:

$$\det(1 - x\mathrm{Frob}_p, V_l) = 1 - \tau(p)x + p^{11}x^2. \qquad (4)$$

In particular, we have $\mathrm{trace}(\rho_l\mathrm{Frob}_p) = \tau(p) \bmod l$ for all primes $p \neq l$. Theorem 1 above is then a rather easy consequence of the next theorem.

Theorem 2. *There exists a probabilistic algorithm that computes the mod l Galois representation associated to Δ in time polynomial in l. More precisely, on input a prime number l it gives:*

1. *a Galois extension K_l of $\mathbb{Q}$, given as a $\mathbb{Q}$-basis e and the products $e_i e_j$ (i.e., the $a_{i,j,k}$ in $\mathbb{Q}$ such that $e_i e_j = \sum_k a_{i,j,k} e_k$ are given);*
2. *a list of the elements σ of $\mathrm{Gal}(K_l/\mathbb{Q})$, where each σ is given as its matrix with respect to e;*
3. *an injective morphism ρ_l from $\mathrm{Gal}(K_l/\mathbb{Q})$ into $\mathrm{GL}_2(\mathbb{F}_l)$,*

such that K_l is unramified outside $\{l\}$, and for all prime numbers p different from l we have $\mathrm{trace}(\rho_l\mathrm{Frob}_p) = \tau(p) \bmod l$ and $\det(\rho_l\mathrm{Frob}_p) = p^{11} \bmod l$. The expected running time of the algorithm is polynomial in l.

The reader notices that the algorithms in Theorems 1 and 2 are probabilistic. It is almost certain that these two results can be strengthened from probabilistic polynomial time to deterministic polynomial time (for this, it suffices that [1] be generalised from the modular curves $X_0(l)$ to for example $X_1(5l)$).

2   Some Context

Before we discuss the proofs of the main results, we provide some context for them.

First of all, there is a relation with Schoof's algorithm for counting points on elliptic curves over finite fields. Let E be an elliptic curve over $\mathbb{Q}$. For p a prime of good reduction for E, Schoof's algorithm computes $\#E(\mathbb{F}_p)$ in time polynomial in $\log p$. This is done by writing $\#E(\mathbb{F}_p)$ as $p + 1 - a_p(E)$, interpreting $a_p(E)$ as the trace of the Frobenius endomorphism of the reduction $E_{\mathbb{F}_p}$ of E mod p, and computing $a_p(E)$ modulo small prime numbers l via the action of the Frobenius endomorphism on the l-torsion points of $E_{\mathbb{F}_p}$. The Hasse bound $|a_p(E)| \leq 2p^{1/2}$ implies that only primes l of size $O(\log p)$ are needed. Nowadays, since Wiles, modularity of elliptic curves over $\mathbb{Q}$ is known, and the $a_p(E)$ are indeed the coefficients at q^p of the new-form f_E associated to E. Hence Schoof's algorithm shows that the coefficients $a_p(E)$ of f_E can be computed in time polynomial in $\log p$. The new-form f_E has weight 2 and its level is the conductor of E. Pila's generalisation of Schoof's algorithm in [7] implies that for f a new-form of weight 2 the coefficients $a_p(f)$ can be computed in time polynomial in $\log p$ (we stress that f is fixed here; the dependence of the running time on the degree over $\mathbb{Q}$ of the field generated by the coefficients of f is exponential).

A general method to count rational points on varieties over finite fields is to apply the Lefschetz fix-point theorem for the Frobenius endomorphism, acting on the cohomology of some suitable type. Schoof's algorithm is no exception to this, as the two-dimensional $\mathbb{F}_l$-vector space $E_{\mathbb{F}_p}(\overline{\mathbb{F}}_p)[l]$ is the dual of $\mathrm{H}^1(E_{\overline{\mathbb{F}}_p,\mathrm{et}}, \mathbb{F}_l)$, the first étale cohomology group with $\mathbb{F}_l$-coefficients of $E_{\overline{\mathbb{F}}_p}$. A natural question that arises is then the following:

> are there other interesting cases where cohomology groups can be used to construct polynomial time algorithms for counting rational points of varieties over finite fields?

The answer to this question is a clear *yes*. Since about 6 years ago, methods based on p-adic cohomology theories have been developed mainly by Satoh, Kedlaya, Lauder and Wan. An appropriate reference for this is certainly [6]. In 2001, Kedlaya gave an algorithm for counting the rational points on hyper-elliptic curves over finite fields $\mathbb{F}_{p^m}$ with $p > 2$, with running time polynomial in m and the genus g. Higher degree cohomology groups, and hence higher dimensional varieties can also be treated. Summarising this recent progress, one can say that, at least from a theoretical point of view, the problem of counting the solutions of systems of polynomial equations over finite fields of a fixed characteristic p and in a fixed number of variables has been solved. On the other hand, the algorithms obtained from p-adic methods have in common that their running time is at least linear in the characteristic p. If p is not bounded, then almost nothing is known about the existence of polynomial time algorithms.

The main results of the previous section prove that there are interesting higher degree cohomology groups that are accessible to computation for point counting, i.e., where the characteristic polynomials of Frobenius elements can be computed efficiently. Indeed, by the so-called Eichler-Shimura isomorphism, the modular forms of a fixed weight $k > 1$ and of a fixed level n form a piece of the de Rham cohomology in degree $k - 1$ of a $k - 1$-dimensional variety (the $k - 2$ self-product of the universal elliptic curve with level n structure). Each new-form gives rise to a 2-dimensional subspace of this cohomology group. Therefore, the direction in which our main results go is that of the computation of the mod l étale realisations of motives over $\mathbb{Q}$ of fixed rank (rank 2 in the present case), but of arbitrary weight (more precisely, the length of the Hodge filtration is arbitrary).

An important difference between our main results, using étale cohomology with coefficients in $\mathbb{F}_l$, and the p-adic methods, is that the Galois representations on $\mathbb{F}_l$-vector spaces that we obtain are *global* in the sense that they are representations of the absolute Galois group of the global field $\mathbb{Q}$. The field extensions such as the $K_l = \mathbb{Q}[x]/(f_l)$ arising from Δ discussed in the previous section have the advantage that one can choose to do the required computations over the complex numbers, approximating f_l, or p-adically at some suitable prime p, or in $\mathbb{F}_p$ for sufficiently many small p. Also, as we have already said, being able to compute such field extensions K_l, that give mod l information on the Frobenius elements at all primes $p \neq l$, is very interesting. On the other hand, the p-adic methods force one to compute with p-adic numbers, or, actually, modulo some sufficiently high power of p, and it gives information only on the Frobenius at p. The main drawback of the étale cohomology with $\mathbb{F}_l$-coefficients seems to be that the degree of the field extensions as K_l to be dealt with grows exponentially in the dimension of the cohomology groups; for that reason, we do not know how to use étale cohomology to compute $\#X(\mathbb{F}_q)$ for X a curve of arbitrary genus in a time polynomial in $\log q$ and the genus of X.

Another piece of context for our main results are the well known congruences (see [8]) for Ramanujan's τ-function, of which the most famous one is:

$$\tau(p) \equiv 1 + p^{11} \bmod 691. \tag{5}$$

The primes l modulo which one has these congruences are precisely those for which V_l is exceptional, in the sense that the image of $\mathrm{Gal}(\overline{\mathbb{Q}}/\mathbb{Q})$ in $\mathrm{Aut}(V_l)$ does not contain $\mathrm{SL}(V_l)$; these are 2, 3, 5, 7, 23 and 691. In what follows we will assume that l is not exceptional.

3   The Strategy for the Computation

Deligne's construction of V_l takes place in the degree 11 étale cohomology group with coefficients in $\mathbb{F}_l$ of the 11-dimensional variety obtained by taking the 10-fold self-product of the universal elliptic curve. It seems to be unknown how to treat such cohomology groups directly in terms of their definition.

Via the fibration of the 11-dimensional variety over the j-line one can see the dual of V_l in the first degree cohomology group of a locally constant sheaf of 11-dimensional $\mathbb{F}_l$-vector spaces on the j-line. This cohomology group is made explicit more easily: it is the set of isomorphism classes of torsors under the sheaf in question. However, it seems to be unknown how to deal with this computationally (it is possible that *using* the algorithm of Theorem 2 one can deal better with such a group).

A standard technique in étale cohomology is to make locally constant sheaves constant by passing to a cover. In our case, this boils down to the fact that V_l occurs in the l-torsion of the Jacobian $J_1(l)$ of the modular curve $X_1(l)$. Using this, one is back in the familiar situation of torsion points on Abelian varieties. The main problem, however, is that now the dimension of $J_1(l)$ is not bounded, so that one cannot apply the results of Jonathan Pila in [7]. In fact, the dimension of $J_1(l)$ grows quadratically in l. This means that our 2-dimensional $\mathbb{F}_l$-vector space V_l is embedded in the very large space consisting of the l-torsion of $J_1(l)$. It is now easy to write down equations for V_l, but standard methods from computer algebra for solving such equations usually take an amount of time that is exponential in the dimension, hence in l^2.

In February of 1999, Edixhoven discussed this problem with Couveignes when Couveignes visited Rennes as a speaker in the algebraic geometry seminar. He suggested another method. He said that as the goal was to compute the number field K_l, one should try to construct a generator of K_l, and approximate it, numerically, with a precision so high that the exact minimal polynomial P_l of the generator is determined by the approximation. The precision that is required is determined by the height of the generator. A way to construct a generator would be to take a function on $J_1(l)$, defined over $\mathbb{Q}$, and evaluate it at the points of V_l. However, it was not clear at that time what function one could take that would give values of small enough height, and that could be approximated easily enough.

Finally, in October 2000 Edixhoven had an idea that allows to carry out Couveignes approach. The idea was simply to consider a suitable function on the curve $X_1(l)$ instead of on its Jacobian $J_1(l)$, and use the description of points on $J_1(l)$ in terms of divisors on $X_1(l)$; the function can then be used to push divisors to the projective line, and get equations for their image. There are then two problems to be dealt with. The first problem is to give a construction of a generator of the field to be computed and show that the height of that generator is polynomial in l when l varies. It was decided that Edixhoven would study this problem, using Arakelov theory. The second problem is to show that the necessary approximations can be done in time polynomial in l. This seems reasonable, using the complex uniformisation of $X_1(l)$ and of $J_1(l)$, however, to really *prove* that it works is a very different matter. Couveignes would try to solve the approximation problem.

Let us now discuss in some more detail how generators for K_l are produced. In order to describe such generators it is good to use the modern version of Galois theory that says that the functor $A \mapsto \mathrm{Hom}_{\mathbb{Q}}(A, \overline{\mathbb{Q}})$ is an anti-equivalence from the category of finite separable $\mathbb{Q}$-algebras to that of finite discrete (continuous) $\mathrm{Gal}(\overline{\mathbb{Q}}/\mathbb{Q})$-sets. An inverse is given by the map that sends X to $\mathrm{Hom}_{\mathrm{Gal}(\overline{\mathbb{Q}}/\mathbb{Q})}(X, \overline{\mathbb{Q}})$, the $\mathbb{Q}$-algebra of functions f from X to $\overline{\mathbb{Q}}$ such that $f(gx) = g(f(x))$ for all g in $\mathrm{Gal}(\overline{\mathbb{Q}}/\mathbb{Q})$ and all x in X. Under this correspondence, fields correspond to transitive $\mathrm{Gal}(\overline{\mathbb{Q}}/\mathbb{Q})$-sets, and the field K_l corresponds to an orbit in the $\mathrm{Gal}(\overline{\mathbb{Q}}/\mathbb{Q})$-set $\mathrm{Isom}(\mathbb{F}_l^2, V_l)$ of $\mathbb{F}_l$-bases of V_l.

Instead of looking at K_l, we look at $V_l - \{0\}$ and let K_l' denote the $\mathbb{Q}$-algebra that corresponds to it; it is a field extension of degree $l^2 - 1$ of $\mathbb{Q}$, and K_l is its splitting field. If the minimal polynomial of a generator of K_l' is known, then we get the field K_l using factoring algorithms, or by more direct methods.

For technical reasons, we consider V_l embedded into $J_1(5l)$. We show that we can take an effective cuspidal divisor D of degree g_l (the genus of $X_1(5l)$) on $X_1(5l)$, defined over $\mathbb{Q}(\zeta_l)$, such that for every $x \neq 0$ in V_l there is a unique effective divisor $D'_x = Q_{x,1} + \cdots + Q_{x,g_l}$ on $X_1(5l)_{\overline{\mathbb{Q}}}$ such that x is the divisor class of $D'_x - D$. We choose a function $f \colon X_1(5l)_{\mathbb{Q}} \to \mathbb{P}^1_{\mathbb{Q}}$ such that all $f_* D'_x$ are distinct (we can find such a function of small degree and with small coefficients by first choosing such an f over a suitable finite field and then lifting it). The f that we choose has its poles at the cusps. As the D'_x are permuted transitively under $\mathrm{Gal}(\overline{\mathbb{Q}}/\mathbb{Q}(\zeta_l))$ they all have the same cuspidal part, and we let $D''_x = \sum_{1 \leq i \leq d_l} Q_{x,i}$ denote their non-cuspidal parts.

With these two choices, D and f, and a choice of $d \in \{1, \ldots, d_l\}$, we get an element $k_{D,f,d}$ of the $\mathbb{Q}(\zeta_l)$-algebra $K'_{l,\mathbb{Q}(\zeta_l)}$ corresponding to the $\mathrm{Gal}(\overline{\mathbb{Q}}/\mathbb{Q}(\zeta_l))$-set $V_l - \{0\}$ as follows. We put:

$$k_{D,f,d} \colon V_l - \{0\} \to \overline{\mathbb{Q}}, \quad x \mapsto \Sigma_d(f(Q_{x,1}), \ldots, f(Q_{x,d_l})), \tag{6}$$

where Σ_d is the elementary symmetric polynomial of degree d in d_l variables. By construction, these $k_{D,f,d}$ generate the field $K'_{l,\mathbb{Q}(\zeta_l)}$ as $\mathbb{Q}(\zeta_l)$-algebra. Then a suitable $\mathbb{Z}[\zeta_l]$-linear combination:

$$k := \sum_{1 \leq d \leq d_l} a_d k_{D,f,d} \tag{7}$$

will be a generator of $K'_{l,\mathbb{Q}(\zeta_l)}$ over $\mathbb{Q}(\zeta_l)$. Finding such a_d can be done using reduction modulo the same finite place as the one that was used to find f. We let:

$$P_l := \prod_{x \in V_l - \{0\}} (T - k(x)) \tag{8}$$

in $\mathbb{Q}(\zeta_l)[T]$ be the minimal polynomial over $\mathbb{Q}(\zeta_l)$ of k.

4   The Height Bound

The work on this part was done in collaboration with Robin de Jong. The main result here is the following.

Theorem 3. *There is an integer c such that for all l we can take D, f and the a_d in such a way that the (logarithmic) heights of the coefficients of P_l in $\mathbb{Q}(\zeta_l)[T]$ as above are bounded above by l^c.*

In order to prove this result, heights are related to Arakelov intersection numbers on the arithmetic surface $\mathcal{X}$ with generic fibre $X_1(5l)$, over the spectrum B of the ring of integers of a sufficiently large number field K. The next step, the most interesting one, is an application of Faltings's arithmetic Riemann-Roch theorem on $\mathcal{X}$, leading to the following inequality (we state it here just in order to show what such things look like, there is no place to explain what all the terms are):

$$(D'_x, P) + \log \#\mathrm{R}^1 p_* O_\mathcal{X}(D'_x) \leq -\frac{1}{2}(D, D - \omega_{\mathcal{X}/B}) + 2g_l^2 \sum_{s \in B} \delta_s \log \#k(s)$$

$$+ \sum_\sigma \log \|\vartheta\|_{\sigma,\mathrm{sup}} + \frac{g_l}{2}[K : \mathbb{Q}] \log(2\pi) \qquad (9)$$

$$+ \frac{1}{2} \deg \det p_* \omega_{\mathcal{X}/B} + (D, P),$$

where P is in $\mathcal{X}(B)$, where s runs through the closed points of B, and where the sup-norm $\|\vartheta\|_{\sigma,\mathrm{sup}}$ is taken over $\mathrm{Pic}^{g_l-1}(X_\sigma)$.

Once this inequality is known, it is clear which quantities need to be bounded polynomially in l in order to prove Theorem 3: the Faltings height of $X_1(5l)$, the log of the sup-norm of a certain theta function, and, finally, the sup of the so-called Arakelov-Green functions $g_{a,\mu}$ associated to points a in $X_1(5l)(\mathbb{C})$.

Sinnou David, who visited Leiden for a workshop during the Summer of 2003, provided a suitable upper bound for the sup-norm of the theta function. His arguments, based on results by him and Philippon are not the ones that are now in [4], as Edixhoven and de Jong found more direct arguments for it, using ingredients that are also used for bounding the other two quantities. The result is:

$$\log \|\vartheta\|_{\mathrm{sup}} = O(l^6). \qquad (10)$$

Getting a bound for the height of $X_1(5l)$ was not so much of a problem, as it would suffice to generalise the method used in a previous article by Coleman and Edixhoven for estimating the height of $X_0(l)$. The result is:

$$h_{\mathrm{abs}}(X_1(5l)) = O(l^2 \log(l)). \qquad (11)$$

The last main hurdle to overcome was to get suitable upper bounds for the Arakelov-Green functions. Suitable upper bounds for these were supplied by Franz Merkl (see Section 18.2 in [4]), and by Jorgenson and Kramer in [5]. Applying Merkl's general results to the case of $X_1(5l)$ we have:

$$\sup_{a \neq b} g_{a,\mu}(b) = O(l^6). \qquad (12)$$

A nice byproduct of our estimates from Arakelov theory is an upper bound for the number of prime numbers where the geometry of our methods (uniqueness of divisors on $X_1(5l)$, poles of certain functions on $X_1(5l)$) does not specialise well. Here is the technical statement.

Theorem 4. *There is an integer c with the following property. Let $l > 5$ be a prime number, and let D and the D'_x be as before. We recall that $X_1(5l)$ has good reduction outside $5l$. A prime number $p \nmid 5l$ is said to be l-good if for all x in $V_l - \{0\}$ the following two conditions are satisfied:*

1. *at all places v of $\overline{\mathbb{Q}}$ over p the specialisation $(D'_x)_{\overline{\mathbb{F}}_p}$ at v is the unique effective divisor on the reduction $X_1(5l)_{\overline{\mathbb{F}}_p}$ such that the difference with $D_{\overline{\mathbb{F}}_p}$ represents the specialisation of x;*
2. *the specialisations of the non-cuspidal part D''_x of D'_x at all v above p are disjoint from the cusps.*

Then we have:

$$\sum_{p \text{ not } l\text{-good}} \log p = O(l^{14}). \tag{13}$$

As a consequence, it is then possible to do all necessary computations over finite fields. Couveignes results on the feasibility of such computations are described in the next section.

5 The Approximation Methods

Theorem 3 says that the heights of the coefficients of the polynomials P_l that we want to compute grow at most polynomially with l. Hence, in order to find the coefficients of P_l, it is enough to have complex or p-adic approximations of them with a precision that is polynomial in l (with precision we mean the number of significant digits). Likewise, it is enough to know the specialisations of those coefficients to sufficiently many finite fields; the required number of such specialisations is at most polynomial in l.

Couveignes has developed methods for dealing with such approximation problems in time polynomial in l and the required precision for all of the three approaches just mentioned: complex, p-adic or finite fields. Moreover, it is he who suggested the last two alternatives to the complex approach in August of 2003.

Let us first discuss the complex approach. Complex points of a complex modular curve X can be represented by elements of $\mathbb{H} \cup \mathbb{P}^1(\mathbb{Q})$. Each point lies in a disk around some cusp, and in such a disk we have an appropriate notion of precision. A basis $\omega = (\omega_1, \ldots, \omega_g)$ of the space of holomorphic differentials is given by the theory of new-forms. With such a basis, one identifies the Jacobian J of X with $\mathbb{C}^g$ modulo the period lattice Λ, hence there too one has an appropriate notion of precision. Given g points $P_1, \ldots, P_g$ on X the map ϕ that sends a g-tuple $Q = (Q_1, \ldots, Q_g)$ of points of X to the divisor class $\sum_i Q_i - \sum_i P_i$ in J is given by integration:

$$\phi \colon X^g \longrightarrow J = \mathbb{C}^g/\Lambda \tag{14}$$

$$Q \longmapsto [Q_1 + \cdots + Q_g - P_1 - \cdots - P_g] = \sum_{i=1}^{g} \int_{P_i}^{Q_i} (\omega_1, \ldots, \omega_g).$$

In the case of the modular curves $X_0(l)$ with l prime he has proved the following results (Theorems 1 and 2 of [1]).

Theorem 5 (Couveignes). *The operations of addition and subtraction in the complex Jacobian $J_0(l)(\mathbb{C})$ of $X_0(l)$ can be done in deterministic polynomial time in l and the required precision. More precisely, given elements P, Q and R of $X_0(l)^g$, elements S and D of $X_0(l)^g$ can be computed in time polynomial in l and the required precision, such that $\phi(S) = \phi(Q) + \phi(R)$ and $\phi(D) = \phi(Q) - \phi(R)$ hold within the required precision. Moreover, for x in $\mathbb{C}^g/\Lambda$, one can compute Q in $X_0(l)^g$ in time polynomial in l and the required precision, such that $\phi(Q) = x$ holds within the required precision.*

Of course, for our purposes, we need to show that we can approximate the divisors $D'_x = Q_{x,1} + \cdots + Q_{x,g_l}$ with the required precision in polynomial time. For that, one must control the difference between the notions of precision in X^g and J, i.e., one must control the norm of the inverse of the tangent map of ϕ at Q. This question is also addressed in [1]. Using the height bounds on D'_x from the previous section, one finds that the difference between the notions of precision in $X_1(5l)^{g_l}$ and $J_1(5l)$ at Q_x and $\phi(Q_x)$ is at most polynomial in l.

The results of [1] will almost certainly be generalised to all curves $X_1(n)$, so that the deterministic variant of Theorem 2 is then a consequence of that generalisation and Theorem 3.

For the moment, we do not have such a generalisation, but Couveignes has the following result (Theorem 2 of [2]) on the finite field approach for the modular curves $X_1(5l)$.

Theorem 6 (Couveignes). *There is a probabilistic algorithm that on input l computes for p a prime that is l-good as in Theorem 4, the reductions $(D'_x)_{\mathbb{F}}$ of the divisors D'_x on $X_1(5l)_{\mathbb{F}}$, where $\mathbb{F}$ is a suitable extension of any residue field of $\mathbb{Z}[\zeta_l]$ at p, with an expected running time that is polynomial in l and p.*

This theorem, together with Theorems 4 and 3 easily imply Theorem 2.

For proving Theorem 6, Couveignes starts by showing that the standard operations on divisors on $X_1(5l)_{\mathbb{F}}$, such as computing $H^0(X_1(5l)_{\mathbb{F}}, O(D_1 - D_2))$, as well as the standard operations on $J_1(5l)_{\mathbb{F}}(\mathbb{F})$ such as addition and applying Hecke operators T_n, can be computed in probabilistic polynomial time in l, $\log \#\mathbb{F}$, $\deg D_i$, and n.

The reason for which the expected running time of Couveignes algorithm is not polynomial in l and $\log p$ is simply that he needs to compute the numerator of the zeta function of $X_1(5l)_{\mathbb{F}}$. Modular symbols methods are used to compute the characteristic polynomial of the Hecke operator T_p on the Jacobian $J_1(5l)$; the Eichler-Shimura relation then gives the characteristic polynomial of the Frobenius endomorphism of $J_1(5l)_{\mathbb{F}_p}$. It is only the modular symbols part of Couveignes algorithm of which the running time is not polynomial in $\log p$.

6   Examples

Johan Bosman, who started working as a PhD-student in June 2004, has done computations using modular symbols packages and the complex uniformisation of $X_1(l)$ and $J_1(l)$, without using special properties of these Jacobians. For the primes $l = 13$, 17 and 19, he has been able to find good approximations of the polynomials of degree $l^2 - 1$ and $l + 1$ describing $V_l - \{0\}$ and $\mathbb{P}(V_l)$. The genus of $X_1(l)$ being $(l - 5)(l - 7)/24$, the curves with which he has been computing have genus 2, 5 and 7, respectively.

These approximations lead to polynomials with rational coefficients, of which we have no proof that they are correct, but which pass the following tests: the ring of integers of the corresponding number field ramifies only at l, the reductions modulo small primes p correspond to the orbit structures of $\rho_l(\mathrm{Frob}_p)$ on $V_l - \{0\}$ and $\mathbb{P}(V_l)$.

In the three cases, the required precision as suggested by Bosman's computations is about 80 digits for $l = 13$, 400 digits for $l = 17$ and 830 digits for $l = 19$. For $l = 19$ the computations were distributed over several machines and still took a couple of months. Hence it seems that it is hard to get much further.

Acknowledgements

Thanks are due to my collaborators Jean-Marc Couveignes and Robin de Jong. I thank Johan Bosman for his examples. I also thank those who have helped me during this project: Franz Merkl, Jay Jorgenson and Jürg Kramer, Sinnou David, Johan Bosman, and Hendrik Lenstra.

References

1. Couveignes, J-M.: Jacobiens, jacobiennes et stabilité numérique. Preprint, available on the author's home page: http://www.univ-tlse2.fr/grimm/couveignes/
2. Couveignes,J-M.: Linearizing torsion classes in the Picard group of algebraic curves over finite fields. In preparation.
3. Deligne, P.: Formes modulaires et représentations l-adiques. Séminaire Bourbaki, 355, Février 1969.
4. Edixhoven, S.J.: On the computation of coefficients of a modular form (in collaboration with J-M. Couveignes, R.S. de Jong, F. Merkl and J.G. Bosman). Will appear on arxiv soon.
5. Jorgenson, J. and Kramer, J.: Bounds on canonical Green's functions. To appear in Compositio Mathematica.
6. Kedlaya, K.S.: Computing zeta functions via p-adic cohomology. Algorithmic number theory, 1–17, Lecture Notes in Comput. Sci., 3076, Springer, Berlin, 2004 (also available on arxiv).
7. Pila, J.: Frobenius maps of abelian varieties and finding roots of unity in finite fields. Math. Comp. **55** (1990), no. 192, 745–763.
8. Swinnerton-Dyer, H.P.F.: On l-adic representations and congruences for coefficients of modular forms. Modular functions of one variable, III (Proc. Internat. Summer School, Univ. Antwerp, 1972), pp. 1–55. Lecture Notes in Math., Vol. 350, Springer, Berlin, 1973.

Cohen–Lenstra Heuristics of Quadratic Number Fields

Étienne Fouvry [1] and Jürgen Klüners [2]

[1] Mathématique, Bât. 425, Univ. Paris–Sud, Campus d'Orsay
F–91405 ORSAY Cedex, France
Etienne.Fouvry@math.u-psud.fr
[2] Universität Kassel, Fachbereich Mathematik/Informatik
Heinrich-Plett-Str. 40, 34132 Kassel, Germany
klueners@mathematik.uni-kassel.de

Abstract. We establish a link between some heuristic asymptotic formulas (due to Cohen and Lenstra) concerning the moments of the p–part of the class groups of quadratic fields and formulas giving the frequency of the values of the p–rank of these class groups.

Furthermore we report on new results for 4–ranks of class groups of quadratic number fields.

1 Introduction and Notations

In [1], Cohen and Lenstra have built a probabilistic model to guess the frequency of some algebraic properties of the narrow class group C_D of the ring of integers of the quadratic fields $\mathbb{Q}(\sqrt{D})$, where the letter D is reserved to fundamental discriminants, throughout this paper. Their idea was, roughly speaking, to attach to each abelian group a weight which is the inverse of the number of its automorphisms. These *heuristics*, the proof of which must lie very deep, are strongly supported by numerical evidence and explain why, for instance, the odd part of C_D is a cyclic group with a higher frequency than one could think at first approach. From these heuristics, they deduce several facts and the aim of our work is to show that some of these deductions imply another ones.

To present the results, we shall use the following notations. The letter p is reserved to prime numbers. For A a finitely generated abelian group, the p–rank of A is defined as $\mathrm{rk}_p(A) = \dim_{\mathbb{F}_p}(A/A^p)$. For an integer $k \geq 0$ and $t > 1$, we introduce the functions η_k and η_∞ defined by

$$\eta_k(t) = \prod_{1 \leq j \leq k} \left(1 - t^{-j}\right)$$

and

$$\eta_\infty(t) = \prod_{j \geq 1} \left(1 - t^{-j}\right).$$

If $f(D)$ is a real valued function defined on the set of positive or negative discriminants, we say that $f(D)$ has the *average value* $c_0 (\in \mathbb{R})$, if, as $X \to +\infty$, we have

F. Hess, S. Pauli, and M. Pohst (Eds.): ANTS 2006, LNCS 4076, pp. 40–55, 2006.

$$\sum_{0<\pm D<X} f(D) = (c_0 + o(1)) \sum_{0<\pm D<X} 1. \tag{1}$$

In the particular case, where $f(D)$ is the characteristic function of the set of discriminants satisfying some indicated property, we say that c_0 is the *density* of this set.

One of the consequences of the Cohen–Lenstra heuristics is to describe the distribution of the values of $\mathrm{rk}_p(C_D)$ as D ranges over the set of positive or negative discriminants, and p a fixed odd prime. These heuristics do not concern the special prime $p = 2$. To circumvent this defect, Gerth [4], [5] had the idea to generalize these heuristics to the group C_D^2. He was led to this generalization by meeting with the densities quoted in Conjectures 2 and 4 (for $p = 2$) below again, when studying the sets of D with a fixed number of prime factors and a fixed value of the 2–rank for C_D^2 (see [4, (1.5) & (1.6)]). From [1], generalized by Gerth, we extract the four conjectures, anyone of which is a consequence of these heuristics.

Conjecture 1. *For every positive integer α, for every prime p, the average value of*

$$\prod_{0\leq i<\alpha} \left(p^{\mathrm{rk}_p(C_D^2)} - p^i\right)$$

is equal to 1, when D ranges over the set of negative fundamental discriminants.

Conjecture 2. *For every non-negative integer r, for every prime p, the density of negative fundamental discriminants D such that $\mathrm{rk}_p(C_D^2) = r$ is equal to*

$$p^{-r^2}\eta_\infty(p)\,\eta_r(p)^{-2}.$$

Conjecture 3. *For every positive integer α, for every prime p, the average value of*

$$\prod_{0\leq i<\alpha} \left(p^{\mathrm{rk}_p(C_D^2)} - p^i\right)$$

is equal to $p^{-\alpha}$, when D ranges over the set of positive fundamental discriminants.

Conjecture 4. *For every non-negative integer r, for every prime p, the density of positive fundamental discriminant D such that $\mathrm{rk}_p(C_D^2) = r$ is equal to*

$$p^{-r(r+1)}\eta_\infty(p)\eta_r(p)^{-1}\eta_{r+1}(p)^{-1}.$$

For $p \geq 3$, Conjectures 1, 2, 3 and 4 are the conjectures (C.6), (C.5), (C.10) and (C.9) of [1, p. 56 & 57], respectively. Note that for $p \geq 3$, we have the equality $\mathrm{rk}_p(C_D^2) = \mathrm{rk}_p(C_D)$, and, by definition we have $\mathrm{rk}_2(C_D^2) = \mathrm{rk}_4(C_D)$, the 4–rank of C_D.

Very little is known about these conjectures : Conjectures 1 and 3 are trivially true for any p and $\alpha = 0$. These conjectures are also proved for $p = 3$ and $\alpha = 1$, this is the famous work of Davenport and Heilbronn [2]. Both authors of this paper recently proved that Conjectures 1 and 3 are true for $p = 2$ and any $\alpha \geq 0$ (see [3, Theorem 1]) and that they remain true if the narrow class group C_D is replaced by the ordinary class group Cl_D.

The aim of this paper, roughly speaking, is to prove that if for some p, Conjecture 1 is true for every α, then Conjecture 3 is also true for this p and every r. The same implication holds between Conjecture 2 and Conjecture 4. More precisely, we shall prove

Theorem 1. *Let p a prime and assume that for every integer $\alpha \geq 0$ the average value of*

$$\prod_{0 \leq i < \alpha} \left(p^{\mathrm{rk}_p(C_D^2)} - p^i \right)$$

is equal to 1, when D ranges over the set of negative fundamental discriminants. Then for every integer $r \geq 0$ the density of the set of negative fundamental discriminants D such that $\mathrm{rk}_p(C_D^2) = r$ is equal to

$$p^{-r^2} \eta_\infty(p) \eta_r(p)^{-2}.$$

and

Theorem 2. *Let p a prime and assume that for every integer $\alpha \geq 0$ the average value of*

$$\prod_{0 \leq i < \alpha} \left(p^{\mathrm{rk}_p(C_D^2)} - p^i \right)$$

is equal to $p^{-\alpha}$, when D ranges over the set of positive fundamental discriminants. Then, for every integer $r \geq 0$, the density of the set of positive fundamental discriminants D such that $\mathrm{rk}_p(C_D^2) = r$ is equal to

$$p^{-r(r+1)} \eta_\infty(p) \eta_r(p)^{-1} \eta_{r+1}(p)^{-1}.$$

Since Conjectures 1 and 3 are proved in the particular case $p = 2$ and for every α, see Theorem 3 and [3, Theorem 1], we now state the following corollary.

Corollary 1. *For ever integer $r \geq 0$ the density of the set of negative fundamental discriminants such that $\mathrm{rk}_4(C_D) = r$ is equal to*

$$2^{-r^2} \eta_\infty(2) \eta_r(2)^{-2}.$$

and for positive discriminants, this density is equal to

$$2^{-r(r+1)} \eta_\infty(2) \eta_r(2)^{-1} \eta_{r+1}(2)^{-1}.$$

We remark that this corollary implies that the probability for a discriminant D to satisfy $\mathrm{rk}_4(C_D) = 0$ is twice larger when it is positive than when it is negative. It would be interesting to have a direct proof of that phenomenon.

Reciprocally, it seems difficult to deduce Conjecture 1 from Conjecture 2 or Conjecture 3 from Conjecture 4 in the form they are written above. Such implications may require a more precise statement for Conjectures 2 and 4 (for instance, with a control of the term $o(1)$ in the formulas (1), corresponding to the densities in question).

In Section 3 we report on results obtained in [3]. We show that Conjectures 1 and 3 are true for all $\alpha \geq 0$ in the case $p = 2$.

2   A Transition to Moments

In [3] an equivalent form of Conjectures 1 and 3 is proved in terms of the function $\mathcal{N}(\alpha, p)$ which denotes the total number of vector subspaces of $\mathbb{F}_p^\alpha$. This equivalent form was an important step in our proof of Conjectures 1 and 3, for $p = 2$ and appears to be more natural in terms of analytic methods : to study the values of an arithmetic function f. These methods are more adapted to deal with the moments f^α of this function f rather than with expressions of the form $\prod_{0 \leq i < \alpha} (f - p^i)$ even if these expressions have been introduced to seize the algebraic properties of an abelian group (see [1, p.50], for the particular case $f(D) = p^{\mathrm{rk}_p(C_D)}$).

We have

Proposition 1. *[3, Prop.1] Let p be a fixed prime and α_0 be a fixed positive integer. Then the average value of*

$$\prod_{0 \leq i < \alpha} \left(p^{\mathrm{rk}_p(C_D^2)} - p^i \right)$$

is equal to 1, for every $0 \leq \alpha \leq \alpha_0$, when D ranges over the set of negative fundamental discriminants, if and only if, under the same conditions, the average value of

$$p^{\alpha \, \mathrm{rk}_p(C_D^2)}$$

is equal to $\mathcal{N}(\alpha, p)$, for every $0 \leq \alpha \leq \alpha_0$.

The same holds for positive discriminants if the above average values 1 and $\mathcal{N}(\alpha, p)$ are replaced by $p^{-\alpha}$ and $p^{-\alpha}(\mathcal{N}(\alpha + 1, p) - \mathcal{N}(\alpha, p))$, respectively.

We now give expressions of the function $\mathcal{N}(\alpha, p)$ in terms of the function η. Since the number of vector subspaces of dimension ℓ of $\mathbb{F}_p^\alpha$ is equal to

$$\prod_{i=0}^{\ell-1} \frac{p^\alpha - p^i}{p^\ell - p^i} = \prod_{i=1}^{\ell} \frac{p^{\alpha-i+1} - 1}{p^i - 1} = p^{\ell(\alpha-\ell)} \frac{\eta_\alpha(p)}{\eta_\ell(p) \cdot \eta_{\alpha-\ell}(p)},$$

and since, uniformly in $k \geq 0$, we have

$$1 \ll_p \eta_k(p) \leq 1,$$

we get

Lemma 1. *For every integer $\alpha \geq 0$ and every $p \geq 2$, we have the equalities*

$$\mathcal{N}(\alpha, p) = \eta_\alpha(p) \sum_{\ell=0}^{\alpha} \frac{p^{\ell(\alpha-\ell)}}{\eta_\ell(p) \cdot \eta_{\alpha-\ell}(p)},$$

In particular, the function $\mathcal{N}(\alpha, p)$ *satisfies*

$$\mathcal{N}(\alpha, p) = O_p\left(p^{\frac{\alpha^2}{4}}\right).$$

3   4–Ranks of Class Groups

The aim of this section is to report on results on 4–ranks of the class group obtained in [3]. We do not give proofs in this section and refer the reader to [3].

We remark that the 4–rank of an abelian group A is the same as the 2–rank of A^2. Therefore we like to study $\mathrm{rk}_2(C_D^2)$ and $\mathrm{rk}_2(\mathrm{Cl}_D^2)$, respectively. We remark that the ordinary class group Cl_D and the narrow class group C_D are the same when $D < 0$. For positive discriminants they are the same if and only if the fundamental unit of $\mathbb{Q}(\sqrt{D})$ has norm -1. In order to simplify we will consider 4–ranks of the narrow class group C_D.

In order to present our results we need the following definition.

Definition 1. *Let* $(a|b) : \mathbb{Q}^* \times \mathbb{Q}^* \to \{0,1\}$, *where* $(a|b) = 1$ *if and only if the equation* $x^2 - ay^2 - bz^2 = 0$ *has a solution* $(x, y, z) \in \mathbb{Q}^3 \setminus \{(0,0,0)\}$.

The 4–rank of the narrow class group can be described by the following theorem which is already implicitly contained in [8, p. 56].

Theorem 3

$$2^{\mathrm{rk}_4(C_D)} = \frac{1}{2}\#\{a \mid a > 0 \text{ squarefree}, a \mid D, (a| - b) = 1\},$$

where $b \in \mathbb{Z}$ is squarefree such that $aD = bc^2$ for a suitable $c \in \mathbb{Z}$. Let us further simplify and concentrate on the case of negative discriminants which are congruent to 1 modulo 4. Then $|D|$ is squarefree as well as the numbers a, b occurring in Theorem 3. Furthermore $b < 0$ in this case and therefore $-b > 0$. It is an easy exercise to see that for coprime integers a and b the symbol $(a|b) = 1$ if and only if a is a square mod b and b is a square mod a. Therefore we get.

Lemma 2. *Let* $D < 0$ *be a fundamental discriminant with* $D \equiv 1 \mod 4$. *Then we have the equality*

$$2^{\mathrm{rk}_4(D)} = \frac{1}{2} \#\{(a,b) \mid a, b \geq 1, -D = ab, a \text{ is a square } \mod b$$

and b *is a square* $\mod a\}$.

Now we use the Jacobi symbol $\left(\frac{a}{b}\right)$ (for odd $b \geq 1$) to detect if a is a square mod b with the formula

$$\frac{1}{2^{\omega(b)}} \prod_{p|b} \left(1 + \left(\frac{a}{p}\right)\right) = \frac{1}{2^{\omega(b)}} \sum_{c|b} \left(\frac{a}{c}\right).$$

By Lemma 2, we get

$$2^{\mathrm{rk}_4(C_D)} = \frac{1}{2 \cdot 2^{\omega(-D)}} \sum_{-D=ab} \left(\sum_{c|b} \left(\frac{a}{c}\right) \right) \left(\sum_{d|a} \left(\frac{b}{d}\right) \right)$$

which gives us with the change of variables $a = D_2 D_3$, $b = D_0 D_1$, $c = D_0$, and $d = D_3$ the following:

$$2^{\mathrm{rk}_4(D)} = \frac{1}{2 \cdot 2^{\omega(-D)}} \sum_{-D=D_0 D_1 D_2 D_3} \left(\frac{D_2}{D_0}\right) \left(\frac{D_1}{D_3}\right) \left(\frac{D_3}{D_0}\right) \left(\frac{D_0}{D_3}\right),$$

always under the assumption that $D < 0$ is congruent to 1 modulo 4.

In [3] we show how to do the summation over all D_0, D_1, D_2, D_3 such that $-D_0 D_1 D_2 D_3$ is a fundamental discriminant. We show that this sum has linear asymptotics, where the main term can be obtained by choosing ($D_0 = 1$ or $D_2 = 1$) and ($D_1 = 1$ or $D_3 = 1$). This choice implies that all the four symbols are 1 and the summation can be easily done. In all the other cases we get an oscillating sum. By using large sieve techniques and Siegel-Walfisz theorem, respectively, we are able to show that those oscillating sums are bounded by $O_\epsilon(X \log(X)^{-\frac{1}{2}+\epsilon})$ for all $\epsilon > 0$.

For the higher moments, i.e. the average of $2^{\mathrm{rk}_4(C_D)}$ we use many tricks described in [7], where geometry over $\mathbb{F}_2$ plays a crucial role.

Let us state the main results of [3]. For this we introduce the sums:

$$S^-(X, k, a, b) := \sum_{\substack{0 < -D < X \\ D \equiv a \bmod b}} 2^{k \, \mathrm{rk}_4(C_D)}$$

and

$$S^+(X, k, a, b) := \sum_{\substack{0 < D < X \\ D \equiv a \bmod b}} 2^{k \, \mathrm{rk}_4(C_D)}.$$

Theorem 4. For every positive integer k and every positive ϵ the following equalities are true, where $R(X, \epsilon, k) := X (\log X)^{-2^{-k}+\epsilon}$:

$$S^-(X, k, 1, 4) = \mathcal{N}(k, 2) \left(\sum_{\substack{0 < -D < X \\ D \equiv 1 \bmod 4}} 1 \right) + O_{\epsilon,k}(R(X, \epsilon, k))$$

$$S^+(X, k, 1, 4) = \frac{1}{2^k} (\mathcal{N}(k+1, 2) - \mathcal{N}(k, 2)) \left(\sum_{\substack{0 < D < X \\ D \equiv 1 \bmod 4}} 1 \right) + O_{\epsilon,k}(R(X, \epsilon, k))$$

$$S^-(X, k, 0, 8) = \mathcal{N}(k, 2) \left(\sum_{\substack{0 < -D < X \\ D \equiv 0 \bmod 8}} 1 \right) + O_{\epsilon,k}(R(X, \epsilon, k))$$

$$S^+(X, k, 0, 8) = \frac{1}{2^k} (\mathcal{N}(k+1, 2) - \mathcal{N}(k, 2)) \left(\sum_{\substack{0 < D < X \\ D \equiv 0 \bmod 8}} 1 \right) + O_{\epsilon,k}(R(X, \epsilon, k))$$

$$S^-(X,k,4,8) = \mathcal{N}(k,2)\Big(\sum_{\substack{0<-D<X \\ D\equiv 4 \bmod 8}} 1\Big) + O_{\epsilon,k}(R(X,\epsilon,k))$$

$$S^+(X,k,4,8) = \frac{1}{2^k}\big(\mathcal{N}(k+1,2) - \mathcal{N}(k,2)\big)\Big(\sum_{\substack{0<D<X \\ D\equiv 4 \bmod 8}} 1\Big) + O_{\epsilon,k}(R(X,\epsilon,k)).$$

Now can we apply Proposition 1 and get

Theorem 5. Conjectures 1 and 3 are true for $p = 2$ and all $\alpha \geq 0$.

The results remain true when we replace the narrow class group by the ordinary class group in the definition of $S^+(X,k,a,b)$.

4 Proof of Theorem 1

We consider first the case of negative discriminants and prove Theorem 1. We postpone the proof of Theorem 2 to §5, where we shall omit details. We follow some ideas contained in [7, p. 359–362]. Since $p \geq 2$ is considered as fixed, we shall forget the dependence on this number in several quantities. Under the hypothesis of Theorem 1 and by Proposition 1, we deduce that for each $k \geq 0$, the average value of $p^{k\,\mathrm{rk}_p(C_D^2)}$ is equal to $\mathcal{N}(k,p)$.

For $X \geq 1$, let

$$N(x) := \sharp\{D\,;\, 0 < -D < X\} \text{ and } N(X,r) := \sharp\{D \in N(X)\,;\, \mathrm{rk}_p(C_D^2) = r\}.$$

For every $X \geq 1$ and every $k \geq 0$, the definition of $N(X,r)$ and the assertion of Theorem 1 implies

$$\sum_{r=0}^{\infty} \frac{N(X,r)}{N(X)} p^{k\,r} = \frac{1}{N(X)} \sum_{0<-D<X} p^{k\,\mathrm{rk}_p(C_D^2)} = \mathcal{N}(k,p) + o_k(1). \qquad (2)$$

We apply (2) with k replaced by $2k + 1$ and appeal to Lemma 1 to write

$$\frac{N(X,r)}{N(X)} p^{(2k+1)r} \leq \sum_{\ell=0}^{\infty} \frac{N(X,\ell)}{N(X)} p^{(2k+1)\ell} = O_k(1),$$

from which we deduce that $N(X,r)/N(X)$ goes quickly to 0 as $r \to +\infty$ under the form

$$\frac{N(X,r)}{N(X)} \ll_k p^{-(2k+1)r}, \qquad (3)$$

uniformly in $X \geq 1$ and $r \geq 0$.

For each r, the sequence $n \mapsto N(n,r)/N(n)$ is a real sequence in the compact set $[0,1]$. By a diagonal process, there exists real numbers $d_r \in [0,1]$ ($r \geq 0$) and an infinite subset $\mathcal{M}$ of positive integers such that

$$N(m,r)/N(m) \to d_r \ (m \in \mathcal{M},\ m \to \infty),$$

for each $r \geq 0$. Write (2) in the particular form

$$\sum_{r=0}^{\infty} \frac{N(m,r)}{N(m)} p^{k\,r} = \mathcal{N}(k,p) + o_k(1), \tag{4}$$

for $m \in \mathcal{M}$, note that (3) implies

$$\sum_{r=0}^{\infty} \frac{N(m,r)}{N(m)} p^{k\,r} = O_k(1)$$

uniformly in $m \in \mathcal{M}$, then apply the Lebesgue Dominated Convergence Theorem (see for instance [9, p. 27]) to (4) to finally write, by definition of the d_r, the equality

$$\sum_{r=0}^{\infty} d_r p^{k\,r} = \mathcal{N}(k,p),$$

which is true for every integer k.

Let (S^-) be the infinite linear system

$$(S^-) \begin{cases} x_0 & +x_1 & +x_2 & +x_3 & +\cdots & +\cdots & =\mathcal{N}(0,p) \\ x_0 & +x_1 p & +x_2 p^2 & +x_3 p^3 & +\cdots & +\cdots & =\mathcal{N}(1,p) \\ x_0 & +x_1 p^2 & +x_2 p^4 & +x_3 p^6 & +\cdots & +\cdots & =\mathcal{N}(2,p) \\ x_0 & +x_1 p^3 & +x_2 p^6 & +x_3 p^9 & +\cdots & +\cdots & =\mathcal{N}(3,p) \\ \cdots & & & & & & \end{cases}$$

in positive unknowns $(x_i)_{i \geq 0}$. Note that each $(d_r)_{r \geq 0}$ obtained by the above diagonal procedure is a solution to (S^-). Hence this system has at least one solution. We shall first give an explicit solution to (S^-): the numbers appearing in Theorem 1 (see Proposition 2), and prove that (S^-) has at most one system of solutions (see Proposition 3). This will imply that for each $r \geq 0$, the sequence $N(X,r)/N(X)$ has only one limit point as X tends to infinty and that this limit point is $p^{-r^2}\eta_\infty(p)\eta_r^{-2}(p)$.

4.1   A Special Solution of (S^-)

We shall prove

Proposition 2. *The sequence of numbers $(x_r)_{r \geq 0}$ with $x_r = p^{-r^2}\eta_\infty(p)\eta_r^{-2}(p)$ is a solution to (S^-).*

The proof of this is based on formulas around the theory of partitions. Let $p(n)$ be the partition function, then classically for any x with $|x| < 1$ we have the equality

$$\sum_{n \geq 0} p(n)x^n = \frac{1}{(1-x)(1-x^2)(1-x^3)\cdots} = \eta_\infty(1/x)^{-1}.$$

This formula has been extended into

Lemma 3. *[6, Thm 351] For any $|x| < 1$, we have*

$$\frac{1}{(1-x)(1-x^2)(1-x^3)\cdots}$$
$$= 1 + \frac{x}{(1-x)^2} + \frac{x^4}{(1-x)^2(1-x^2)^2} + \frac{x^9}{(1-x)^2(1-x^2)^2(1-x^3)^2} + \cdots$$

In other words, we have the formula $\eta_\infty(1/x)^{-1} = \sum_{k=0}^{\infty} \frac{x^{k^2}}{\eta_k^2(1/x)}$. By choosing $x = 1/p$, we proved that the sequence (x_r) satisfies the first equation of (S^-). We must continue this checking to the other equations of (S^-).

We shall first generalize Lemma 3 in

Lemma 4. *Let $t \geq 0$ be an integer. Then for any $|x| < 1$, we have*

$$\frac{1}{(1-x)(1-x^2)(1-x^3)\cdots}$$
$$= \sum_{r=t}^{\infty} \frac{x^{r(r-t)}}{(1-x)^2 \cdots (1-x^{r-t})^2(1-x^{r-t+1}) \cdots (1-x^r)}.$$

In other words, we have the formula

$$\eta_\infty(1/x)^{-1} = \sum_{r=t}^{\infty} \frac{x^{r(r-t)}}{\eta_{r-t}(1/x)\eta_r(1/x)}.$$

Proof. This formula for instance is in [1, Cor. 6.7,p.51], where the authors say that a proof can be given directly or as a consequence of combination of theorems of their work. For sake of completeness, we give a proof which follows the proof of [6, Thm 351]. The integer $t \geq 0$ is now fixed, and we define the Durfee rectangle with defect $-t$ of a partition of an integer n as the largest rectangle of size $(r, r - t)$ that can be inserted in the northwest corner of this partition. For instance, choose $t = 1$ and $n = 29$, and consider

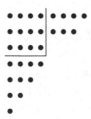

The above drawing explains for the partition

$$29 = 8 + 7 + 4 + 4 + 3 + 2 + 1,$$

what is the Durfee rectangle of defect -1. It has size $(4,3)$. Note that there are $\ell = 10$ points out of this Durfee rectangle, southwards, and this ℓ appears as decomposed in partition with summands $\leq r = 4$ ($10 = 4 + 3 + 2 + 1$). Similarly, eastwards, it remains $m = 7$ points, written in partition with summands $\leq r - 1 = 3$ ($7 = 2 + 2 + 2 + 1$).

More generally, given a partition of n, with a Durfee rectangle of defect $-t$, with dimension $(r, r - t)$, we write $n = r(r - t) + \ell + m$ and the number of partitions of ℓ in parts $\leq r$ is the coefficient of x^ℓ in

$$\frac{1}{(1 - x)(1 - x^2) \cdots (1 - x^r)}.$$

Similarly, the number of partitions of m in parts $\leq r - t$ is the coefficient of x^m in

$$\frac{1}{(1 - x)(1 - x^2) \cdots (1 - x^{r-t})}.$$

Hence the number of partitions of n, with Durfee rectangle of size $(r, r - t)$ is the coefficient of $x^{n - r(r-t)}$ in the fraction

$$\frac{1}{(1 - x)^2(1 - x^2)^2 \cdots (1 - x^{r-t})^2(1 - x^{r-t+1}) \cdots (1 - x^r)}.$$

Summing over all the possible of $r \geq t$, we obtain the expected expression of the function $\sum p(n)x^n$. $\qquad\qquad \Box$

We shall also prove

Lemma 5. *Let $r \geq k \geq 0$ be integers. Then for every $|x| < 1$ we have the equality*

$$\frac{x^{r(r-k)}}{(1 - x)^2 \cdots (1 - x^r)^2} = \sum_{\ell=0}^{k} \frac{\mathfrak{n}(k, \ell, 1/x)x^{r(r-\ell)}}{(1 - x)^2 \cdots (1 - x^{r-\ell})^2(1 - x^{r-(\ell-1)}) \cdots (1 - x^r)},$$

where

$$\mathfrak{n}(k, \ell, 1/x) = \begin{cases} \displaystyle\prod_{i=1}^{\ell} \frac{(1/x)^{k-i+1} - 1}{(1/x)^i - 1}, & for\ 0 \leq \ell \leq k \\[4mm] 0, & for\ \ell > k. \end{cases}$$

Proof. Remark first that $\mathfrak{n}(k, \ell, p)$ is equal to the number of vector subspaces of $\mathbb{F}_p^k$ with dimension ℓ, and that this function satisfies the recursive formula (see [3, Lemma 1]):

$$\mathfrak{n}(k + 1, \ell, 1/x) = \mathfrak{n}(k, \ell - 1, 1/x) + \frac{1}{x^\ell}\mathfrak{n}(k, \ell, 1/x). \qquad (5)$$

By multiplication, we see that Lemma 5 is proved if and only if we proved

$$x^{r(r-k)} = \sum_{\ell=0}^{k} \mathfrak{n}(k, \ell, 1/x)x^{r(r-\ell)}(1 - x^{r-(\ell-1)}) \cdots (1 - x^r),$$

or equivalently

$$\sum_{\ell=0}^{k} \mathfrak{n}(k,\ell,1/x)x^{r(k-\ell)}(1-x^{r-(\ell-1)})\cdots(1-x^r) = 1. \tag{6}$$

Actually, the fact that r is an integer is useless in the proof of (6), and defining $y = x^r$, we shall prove

$$\sum_{\ell=0}^{k} \mathfrak{n}(k,\ell,1/x)y^{k-\ell}\left(1 - \frac{y}{x^{\ell-1}}\right)\cdots\left(1 - \frac{y}{x}\right)(1-y) = 1, \tag{7}$$

for every real positive numbers x and y and any positive integer $k \geq 0$. The proof of (7) works by induction on k.

This formula is true for $k = 0$ and $k = 1$, since $\mathfrak{n}(0,0,1/x) = \mathfrak{n}(1,0,1/x) = \mathfrak{n}(1,1,1/x) = 1$. It is also true for $k = 2$, since $\mathfrak{n}(2,0,1/x) = \mathfrak{n}(2,2,1/x) = 1$ and $\mathfrak{n}(2,1,1/x) = 1+1/x$. Suppose now that (7) is true for some value $k \geq 3$. So we now study

$$\sum_{\ell=0}^{k+1} \mathfrak{n}(k+1,\ell,1/x)y^{k+1-\ell}\left(1 - \frac{y}{x^{\ell-1}}\right)\cdots\left(1 - \frac{y}{x}\right)(1-y), \tag{8}$$

and replace the term $\mathfrak{n}(k+1,\ell,1/x)$ by the recursive formula (5). The contribution of the second term on the right–hand side of (5) is equal to

$$\sum_{\ell=0}^{k} \mathfrak{n}(k,\ell,1/x)\frac{y^{k+1-\ell}}{x^\ell}\left(1 - \frac{y}{x^{\ell-1}}\right)\cdots\left(1 - \frac{y}{x}\right)(1-y)$$

$$= -\sum_{\ell=0}^{k} \mathfrak{n}(k,\ell,1/x)y^{k-\ell}\left(1 - \frac{y}{x^\ell}\right)\cdots\left(1 - \frac{y}{x}\right)(1-y)$$

$$+ \sum_{\ell=0}^{k} \mathfrak{n}(k,\ell,1/x)y^{k-\ell}\left(1 - \frac{y}{x^{\ell-1}}\right)\cdots\left(1 - \frac{y}{x}\right)(1-y).$$

By hypothesis, the last sum of the above equation is equal to 1, and the first sum annihilates with the contribution to (8) of the first term $\mathfrak{n}(k,\ell-1,1/x)$ coming from the right–hand side of (5) (make the change of variable $\ell \mapsto \ell - 1$). Hence (7) is proved, and subsequently (6). The proof of Lemma 5 is complete.   □

We now turn to the proof of Proposition 2. To check that the equation of order $k + 1$ of (S^-) is satisfied by the values of (x_r) given in Proposition 2, we have to compute, for $x = 1/p$ the quantity

$$S_k := \sum_{r=0}^{\infty} x_r p^{kr} = \eta_\infty(p)\sum_{r=0}^{\infty} \frac{x^{r(r-k)}}{(1-x)^2\cdots(1-x^r)^2}. \tag{9}$$

By Lemma 5, this is equal to $S_k =$

$$\eta_\infty(p) \sum_{\ell=0}^{k} \mathfrak{n}(k,\ell,1/x) \sum_{r=0}^{\infty} \frac{x^{r(r-\ell)}}{(1-x)^2 \cdots (1-x^{r-\ell})^2(1-x^{r-(\ell-1)}) \cdots (1-x^r)},$$

and finally, by Lemma 4, we obtain the equality (still having $x = 1/p$)

$$S_k = \eta_\infty(p) \sum_{\ell=0}^{k} \frac{\mathfrak{n}(k,\ell,1/x)}{\eta_\infty(1/x)} = \mathcal{N}(k,p). \tag{10}$$

This completes the proof of Proposition 2.

4.2 Unicity of Solutions of an Infinite Linear System

Let a be a real integer > 1, and $(C_k)_{k \geq 0}$ an infinite sequence of positive real numbers. We are searching for growth conditions on (C_k) to ensure that the linear system with infinitely many equations

$$\sum_{s=0}^{\infty} x_s a^{sk} = C_k \quad (k = 0, 1, \dots) \tag{11}$$

has at most one solution $(x_i)_{i \geq 0}$ with $x_i \geq 0$. Such a system was considered by Heath–Brown [7, Lemmas 17&18] in the particular case $a = 4$, with an appeal to the properties of Vandermonde determinants. We shall rather use Jensen's formula (see Lemma 6 below).

A condition on the growth of C_k is obligatory to ensure the unicity of solutions of (11) in non-negative x_s as can be seen in the following example.

Example 1. Let a be a positive integer and $C_k = \sinh(\pi a^k)$. Then define the coefficients x_s and x_s' by the Taylor expansions $\sinh(\pi t) = \sum x_s t^s$, and $\sin(\pi t) + \sinh(\pi t) = \sum x_s' t^s$. Both sequences (x_s) and (x_s') consist of non-negative numbers and are solutions of (11).

However the particular coefficients C_k chosen before do not satisfy (13) below.

So let $(x_i)_{i \geq 0}$ be a positive solution to (11). By positivity we deduce the inequality

$$x_s \leq a^{-sk} C_k, \tag{12}$$

for any $s \geq 0$ and $k \geq 0$. To push further the computations, we suppose that there exists an absolute c_0 such that

$$C_k \leq c_0 a^{\frac{k^2}{2}} \quad (k = 0, 1, \dots). \tag{13}$$

By choosing $k = s$ in (12), we get

$$0 \leq x_s \leq c_0 a^{-\frac{s^2}{2}}. \tag{14}$$

Now let (x_s) and (x'_s) be two solutions of (11) and consider

$$f(z) = \sum_{s=0}^{\infty}(x_s - x'_s)z^s, \tag{15}$$

considered as a function of the complex variable z. The radius of convergence of this power series is $+\infty$, by (14). It is an entire function, which is zero at each a^k ($k = 0, 1, \dots$). It also satisfies the inequality

$$|f(z)| \leq 2c_0 \sum_{s=0}^{\infty} a^{-\frac{s^2}{2}}|z|^s.$$

In particular, if $|z| = a^k$, for some absolute c'_0, we get

$$|f(z)| \leq 2c_0 \sum_{s=0}^{\infty} a^{-\frac{s^2}{2}}a^{ks} \leq c'_0 a^{\frac{k^2}{2}}. \tag{16}$$

We shall first prove

Lemma 6. *Let $\ell \geq 0$ be an integer and $a \in \mathbb{C}$ such that $|a| > 1$. Furthermore let $g(z)$ be an entire function which has a zero of order ℓ at $z = 0$ and satisfying $g(a^k) = 0$ for any $k \geq 0$. Then for every $k \geq 0$ the function g satisfies the inequality*

$$\sup_{|z|=|a|^k} |g(z)| \geq \frac{|g^{(\ell)}(0)|}{\ell!}|a|^{\frac{k(k+1)}{2}+k\ell}.$$

Proof. This is an application of Jensen's formula (see for instance [9, Thm 15.18]), applied to the function $h(z) = z^{-\ell}g(z)$. With ρ denoting any zero of h, we have the relations

$$|h(0)||a|^{\frac{k(k+1)}{2}} = |h(0)| \prod_{0 \leq \ell \leq k} \frac{|a|^k}{|a^\ell|} \leq |h(0)| \prod_{\rho,\, |\rho| \leq |a|^k} \frac{|a|^k}{|\rho|}$$

$$= \exp\{\frac{1}{2\pi} \int_{-\pi}^{\pi} \log|h(|a|^k e^{i\theta})| \, d\theta\} \leq \sup_{|z|=|a|^k} |h(z)| = |a|^{-k\ell} \sup_{|z|=|a|^k} |g(z)|,$$

which gives the result.   □

Now suppose that we have two distinct solutions (x_s) and (x'_s) to (11). Let ℓ be the least index s such that $x_s \neq x'_s$. Hence the function defined in (15) is not identically equal to 0. Then we apply Lemma 6 to $f(z)$, and by comparing with (16), we are led to the lower bound $a^{\frac{k^2}{2}} \gg a^{\frac{k(k+1+2\ell)}{2}}$. This is impossible for k sufficiently large. Hence $f \equiv 0$. In conclusion we proved

Proposition 3. *If the coefficients (C_k) satisfy the conditions (13), then the infinite linear system (11) has at most one solution in positive $(x_s)_{s \geq 0}$.*

To prove Theorem 1, it remains to combine Proposition 1, 2, 3, and Lemma 1 in order to deduce that, under the hypothesis of this theorem, for each $r \geq 0$, for $X \to +\infty$, the function $N(X, r)/N(X)$ has only one limit point which is equal to $p^{-r^2}\eta_\infty(p)\eta_r(p)^{-2}$.

5   The Case of Positive Discriminants

The strategy is the same. We study the limit points of the sequence $N(X, r)/N(X)$, where

$$N(x) := \sharp\{D\,;\, 0 < D < X\} \text{ and } N(X, r) := \sharp\{D \in N(X)\,;\, \mathrm{rk}_p(C_D^2) = r\}.$$

These limit points are solutions to (S^+) where S^+ the infinite linear system

$$(S^+) \quad \begin{cases} x_0 & +x_1 & +x_2 & +x_3 & +\cdots & +\cdots & =\mathcal{M}_0(p) \\ x_0 & +x_1 p & +x_2 p^2 & +x_3 p^3 & +\cdots & +\cdots & =\mathcal{M}_1(p) \\ x_0 & +x_1 p^2 & +x_2 p^4 & +x_3 p^6 & +\cdots & +\cdots & =\mathcal{M}_2(p) \\ x_0 & +x_1 p^3 & +x_2 p^6 & +x_3 p^9 & +\cdots & +\cdots & =\mathcal{M}_3(p) \\ \cdots \end{cases}$$

where we defined

$$\mathcal{M}_k(p) = \frac{1}{p^k}\left(\mathcal{N}(k+1, p) - \mathcal{N}(k, p)\right).$$

We first notice

Lemma 7. *For every $k \geq 1$, we have*

$$\mathcal{M}_k(p) = \frac{\mathcal{M}_{k-1}(p)}{p} + \mathcal{N}(k-1, p).$$

Proof. This is an easy consequence of the equality

$$\mathcal{N}(k+1, p) = 2\mathcal{N}(k, p) + (p^k - 1)\mathcal{N}(k-1, p) \ (k \geq 1).$$

which is proved in [3, Lemma 3].   □

We are now in position to prove

Proposition 4. *The sequence of numbers $(x_r)_{r \geq 0}$ with*

$$x_r = p^{-r(r+1)} \eta_\infty(p) \eta_r(p)^{-1} \eta_{r+1}(p)^{-1}$$

is a solution of (S^+).

Proof. By linear combination and by Lemma 7, we see that (S^+) is equivalent to the system (Σ^+) defined by

$$(\Sigma^+) \quad \begin{cases} x_0 & +x_1 & +x_2 & +\cdots = \mathcal{M}_0(p) \\ x_0(1 - p^{-1}) & +x_1(p - p^{-1}) & +x_2(p^2 - p^{-1}) & +\cdots = \mathcal{N}(0, p) \\ x_0(1 - p^{-1}) & +x_1(p^2 - 1) & +x_2(p^4 - p) & +\cdots = \mathcal{N}(1, p) \\ x_0(1 - p^{-1}) & +x_1(p^3 - p) & +x_2(p^6 - p^3) & +\cdots = \mathcal{N}(2, p) \\ \cdots \end{cases}$$

where the line of order $k + 1$ $(k \geq 1)$ is given by

$$\sum_{r=0}^{\infty} x_r \left(p^{kr} - p^{(k-1)r-1} \right) = \mathcal{N}(k-1, p). \qquad (17)$$

The first line is satisfied with the above choice of the (x_r), since we have the equality

$$\sum_{r=0}^{\infty} x_r = \eta_{\infty}(p) \sum_{r=0}^{\infty} \frac{(1/p)^{r(r+1)}}{\eta_r(p)\eta_{r+1}(p)} = \eta_{\infty}(p) \sum_{r=1}^{\infty} \frac{(1/p)^{r(r-1)}}{\eta_{r-1}(p)\eta_r(p)}$$

$$= \eta_{\infty}(p)\eta_{\infty}(p)^{-1} = 1 = \mathcal{M}_0(p)$$

by Lemma 4.

To study the line of order $k + 1$ $(k \geq 1)$ of (Σ^+), we write the equalities

$$\sum_{r=0}^{\infty} x_r \left(p^{kr} - p^{(k-1)r-1} \right)$$

$$= \eta_{\infty}(p) \sum_{r=0}^{\infty} \frac{(1/p)^{r(r+1)} \cdot (1/p)^{-kr}(1 - (1/p)^{r+1})}{(1 - (1/p))^2 \cdots (1 - (1/p^r))^2(1 - (1/p^{r+1}))}$$

$$= S_{k-1},$$

where S_{k-1} is the expression introduced in (9) to study the linear system (S^-). By (10), we know that this is equal to $\mathcal{N}(k-1, p)$. This ends the proof of (17) for all the values of k, hence (Σ^+) and (S^+) are satisfied with the chosen values of x_r. The proof of Proposition 4 is now complete. $\qquad \square$

5.1 Unicity of Solutions

By definition of $\mathcal{M}_k(p)$ and by Lemma 1, we get the relation

$$\mathcal{M}_k(p) = O(p^{\frac{k^2}{4} - \frac{k}{2}}).$$

Hence $\mathcal{M}_k(p)$ satisfy the conditions (13). By Proposition 3, the infinite linear (S^+) system has at most one solution. As for the case of negative discriminants, we deduce that, for any $r \geq 0$, the function $N(X, r)/N(X)$ has only one limit point as $X \to +\infty$. By Proposition 4, these limit points have the values announced in Theorem 2. The proof of this theorem is complete.

Acknowledgments

The first author thanks P. Gérard for interesting conversations about §4.2.

References

1. Cohen, H., Lenstra, H.W.: Heuristics on class groups of number fields. In: Number theory, Noordwijkerhout 1983, volume 1068 of Lecture Notes in Math., pages 33–62. Springer, Berlin (1984)
2. Davenport, H.,Heilbronn, H: On the density of discriminants of cubic fields II. Proc. Roy. Soc. London Ser. A **322**(1551),405–420 (1971)
3. Fouvry, E., Klüners, J.: On the 4–rank of class groups of quadratic number fields. Preprint, (2006)
4. Gerth III, F.: The 4-class ranks of quadratic fields. Invent. Math. **77**(3),489–515 (1984)
5. Gerth III, F.: Extension of conjectures of Cohen and Lenstra. Exposition. Math. **5**(2),181–184 (1987)
6. Hardy, G.H., Wright, E.M.: An Introduction to the Theory of Numbers. Oxford University Press (1975)
7. Heath–Brown, D.R.: The size of Selmer groups for the congruent number problem II. Inv. Math., **118**, 331–370, (1994)
8. Redei, L.: Arithmetischer Beweis des Satzes über die Anzahl der durch vier teilbaren Invarianten der absoluten Klassengruppe im quadratischen Zahlkörper. J. Reine Angew. Math. **171**, 55–60 (1934)
9. Rudin, W.: Real and Complex Analysis, second edition. McGraw–Hill Book Company, (1974)

An Algorithm for Computing p-Class Groups of Abelian Number Fields

Miho Aoki[1,*] and Takashi Fukuda[2]

[1] Department of Mathematics, Tokyo Institute of Technology
2-12-1 Oh-okayama, Meguro-ku, Tokyo, Japan
m-aoki@math.titech.ac.jp
[2] Department of Mathematics, College of Industrial Technology
Nihon University, 2-11-1 Shin-ei, Narashino, Chiba, Japan
fukuda@math.cit.nihon-u.ac.jp

Abstract. For an abelian number field F and an odd prime number p which does not divide the degree $[F : \mathbb{Q}]$, we propose a new algorithm for computing the p-primary part of the ideal class group of F using Gauss sums and cyclotomic units.

1 Introduction

It is a fundamental task in algebraic number theory to compute explicitly the structure of the ideal class group Cl_F for a number field F of finite degree and especially important in Iwasawa theory to compute the p-primary component $Cl_F\{p\}$ of Cl_F for a prime number p.

There is a well-known Buchmann's algorithm which computes Cl_F for an arbitrary number field F and is implemented in several software packages. A feature of Buchmann's algorithm resides in a parallel computation of the ideal class group and the fundamental units. It sometimes needs GRH (Generalized Riemann Hypothesis) to finish the computation in a reasonable time for a field of large discriminant.

For a cyclotomic field F of prime conductor, there are some preceding works by Schoof [9], [10] and Cornacchia [2]. In [9] and [2], the cyclicty of Cl_F as a Galois module for such a field was studied. As they wrote in [9] and [2, p.2], if Cl_F is cyclic as a Galois module, the Galois module structure of Cl_F is described by using the Stickelberger ideal for the minus part and the certain annihilator ideal for the plus part which is constructed by developing the Kolyvagin-Rubin-Thaine's method. In [2], Cornacchia treated only the 2-part, but his method seems to be adaptable for the odd p-part. Further in [10], Schoof investigated the order of a certain subgroup of the plus part of Cl_F by using cyclotomic units.

In this paper, we will give a new algorithm for computing the structure of $Cl_F\{p\}$ for an abelian number field F and a prime number p which satisfies the following assumptions.

---

* This paper is supported by the 21 COE program "Constitution of wide-angle mathematical basis focused on knots".

F. Hess, S. Pauli, and M. Pohst (Eds.): ANTS 2006, LNCS 4076, pp. 56–71, 2006.
© Springer-Verlag Berlin Heidelberg 2006

Assumptions.
 I. $p \neq 2$.
 II. p does not divide the degree $[F : \mathbb{Q}]$.

Fix a prime number p satisfying the above assumptions and set $A_F = Cl_F\{p\}$. By the assumption I, A_F is decomposed to $A_F = A_F^+ \oplus A_F^-$, where $A_F^\pm = \{x \in A_F \mid Jx = \pm x\}$ and J is the complex conjugation. We study the minus part A_F^- and the plus part A_F^+ separately. It is known that Stickelberger element is a good annihilator for the minus part A_F^- and Gauss sum gives a generator of the principal ideal obtained by acting Stickelberger element to a representative ideal of the class. For the plus part A_F^+, there are Kolyvagin-Rubin-Thaine's annihilators and cyclotomic units, which are analogue to Stickelberger elements and Gauss sums. We compute Gauss sums for A_F^- and cyclotomic units for A_F^+ to give explicitly the generators of A_F using auxiliary prime numbers ℓ and ℓ^*, and finally determine the structure of A_F.

In Iwasawa theory, it is important to decompose A_F in a more precise manner into the χ-parts $A_{F,\chi}$ (definition is given in §2). In the case of Buchmann's algorithm, it is necessary to compute the whole ideal class group Cl_F to determine $A_{F,\chi}$. On the other hand, our algorithm determines $A_{F,\chi}$ directly.

Our algorithm is carried out in $\mathbb{Z}$ and determines rigorously the structure of A_F without assuming GRH and without computing the fundamental units. This work was motivated by a recent result of Sumida [11] which enables us to compute the order of A_F^+. The authors express our gratitude for his generosity providing us with his manuscript in an early stage. The authors are also grateful to referees for their careful reading and many apt suggestions. It is entirely thanks to referees that we were able to make the manuscript much clearer.

2   Notations and Main Results

Let F be an abelian number field and set $\Delta = \mathrm{Gal}(F/\mathbb{Q})$. Fix a prime number p satisfying the assumptions I and II in §1. We denote by $\mathbb{Z}_p$, $\mathbb{Q}_p$ and $\overline{\mathbb{Q}}_p$ the ring of p-adic integers, the field of p-adic numbers and the algebraic closure of $\mathbb{Q}_p$ respectively.

For a character $\chi : \Delta \to \overline{\mathbb{Q}}_p^\times$, we define the idempotent e_χ by

$$e_\chi = \frac{1}{|\Delta|} \sum_{\sigma \in \Delta} \mathrm{Tr}(\chi^{-1}(\sigma))\sigma \in \mathbb{Z}_p[\Delta],$$

where $\mathrm{Tr} : \mathbb{Q}_p(\chi(\Delta)) \to \mathbb{Q}_p$ is the trace map. For any $\mathbb{Z}_p[\Delta]$-module M, we define the χ-part M_χ of M by $M_\chi = e_\chi M$. Let $\mathcal{O}_\chi$ be the extension ring of $\mathbb{Z}_p$ generated by the values of χ. Then M_χ is $\mathcal{O}_\chi$-module in the following way:

$$\chi(\sigma)a = \sigma a \quad \text{for any } a \in M_\chi \text{ and } \sigma \in \Delta.$$

By the assumption II, we have $A_F = \oplus_\chi A_{F,\chi}$, where χ runs over all representatives of $\mathbb{Q}_p$-conjugacy classes of characters of Δ. Hence it is enough to study the $\mathcal{O}_\chi$-module $A_{F,\chi}$ for each χ. For a character χ of Δ, we denote by F^χ the field fixed by the kernel of χ. We can regard χ as a character of $\mathrm{Gal}(F^\chi/\mathbb{Q})$. By an easy argument, we get an isomorphism $A_{F,\chi} \simeq A_{F^\chi,\chi}$ as $\mathcal{O}_\chi$-module. Hence it is enough to study $A_{F,\chi}$ in the case $F = F^\chi$.

From now on, for a fixed character χ of Δ, we put $K = F^\chi$, $\Delta = \mathrm{Gal}(K/\mathbb{Q})$ and $A = A_K$. Note that K is totally imaginary or totally real according as χ is odd or even. For the Teichmüller character ω and trivial character 1, we know $A_\omega = A_1 = 0$. Hence we assume $\chi \neq \omega, 1$. We denote the conductor of χ (it is also the conductor of K) by $f = p^e f_0$ with $p \nmid f_0$ ($e \leq 1$ by the assumption II), and define Gauss sums and cyclotomic units as follows. For a positive rational integer n, we denote by ζ_n a primitive n-th root of unity and by μ_n the group generated by ζ_n.

Definition 1 (Gauss sum). *Let ℓ be a rational prime satisfying $\ell \equiv 1 \pmod{f}$ and $\widetilde{\mathcal{L}}$ (resp. $\mathcal{L}$) a prime ideal of $\mathbb{Q}(\mu_f)$ (resp. K) satisfying the inclusion of prime ideals $\widetilde{\mathcal{L}} \supset \mathcal{L} \supset (\ell)$. We define the Gauss sums $\tau_{\widetilde{\mathcal{L}}} \in \mathbb{Q}(\mu_{f\ell})^\times$ and $\tau_{\mathcal{L}} \in K(\mu_\ell)^\times$ by*

$$\tau_{\widetilde{\mathcal{L}}} = \sum_{a=1}^{\ell-1} \chi_{\widetilde{\mathcal{L}}}(a)\zeta_\ell^a, \quad \tau_{\mathcal{L}} = N_{\mathbb{Q}(\mu_{f\ell})/K(\mu_\ell)}\left(\tau_{\widetilde{\mathcal{L}}}\right),$$

where $\chi_{\widetilde{\mathcal{L}}} : (\mathbb{Z}/\ell\mathbb{Z})^\times \to \mu_f$ is the character given by $\chi_{\widetilde{\mathcal{L}}}(a) \equiv a^{-(\ell-1)/f} \pmod{\widetilde{\mathcal{L}}}$ and $N_{\mathbb{Q}(\mu_{f\ell})/K(\mu_\ell)} : \mathbb{Q}(\mu_{f\ell})^\times \to K(\mu_\ell)^\times$ is the norm map. Note that $\tau_{\mathcal{L}}$ is defined independently of a choice of $\widetilde{\mathcal{L}}$.

Definition 2 (Cyclotomic unit). *Let n be a positive rational integer. We define the cyclotomic unit $\xi_n \in K(\mu_n)^\times$ by*

$$\xi_n = N_{\mathbb{Q}(\mu_{fn})/K(\mu_n)}(\zeta_{fn} - 1).$$

For $n = 1$, we abbreviate ξ_1 as ξ.

Let $d_0^{[\mathcal{O}_\chi : \mathbb{Z}_p]} = |A_{K,\chi}|$. For our algorithm, we choose d which is a power of p satisfying the following condition.

$$\boxed{\text{The condition on } d} : \quad d = d_0 \text{ if } \chi \text{ is odd and } d \geq d_0 \text{ if } \chi \text{ is even.} \tag{1}$$

Remark 1. It is easy to obtain such d. In fact, we quickly get d_0 from the generalized Bernoulli number $B_{1,\chi^{-1}}$ in the odd case and get an upper bound of d_0 using Lemma 1 below in the even case. Note the following remarks on d_0.

1) If χ is odd, it is well known that the equality

$$|A_\chi| = |B_{1,\chi^{-1}}|_p^{-[\mathcal{O}_\chi : \mathbb{Z}_p]} \tag{2}$$

holds, where $|\ \ |_p$ is the p-adic valuation normalized by $|p|_p = p^{-1}$. This is a direct consequence of Iwasawa main conjecture proved by Mazur-Wiles (cf. [6, p.216 Theorem 2]). Hence d_0 is the maximal power of p dividing $B_{1,\chi^{-1}}$.

2) If χ is even, the computation of d_0 is difficult. Let E denote the group of units in K and set $E_\chi = (E \otimes_{\mathbb{Z}} \mathbb{Z}_p)_\chi$. By the Dirichlet unit theorem and the assumption $\chi \neq 1$, we have $E_\chi \simeq \mathcal{O}_\chi$. We define the χ-part of the group of cyclotomic units C_χ to be the $\mathcal{O}_\chi$-module generated by $\xi^{e_\chi} \in E_\chi$. Then it is well known that the equality

$$|A_\chi| = |E_\chi/C_\chi|,   \tag{3}$$

which is a part of Gras conjecture, is proved also as a consequence of Iwasawa main conjecture (cf. [3, Proposition 9]). Hence d_0 is the maximal power of p satisfying $\xi^{e_\chi} \in E_\chi^{d_0}$. It is easy to see that $\xi^{e_\chi} \notin E_\chi^d$ but difficult to see that $\xi^{e_\chi} \in E_\chi^d$ because it needs computations in a global field.

3) There are another proofs of (2) and (3) for cyclotomic cases using Euler system (cf. [7] and [8]). It is not difficult to modify these proofs applicable to a general abelian number field F and an odd prime number p satisfying $p \nmid [F : \mathbb{Q}]$.

4) Recently, Sumida [11] gave a nice algorithm for computing d_0 when χ is even and satisfies $\chi^{-1}\omega(p) \neq 1$. On the other hand, Theorem 2 below enables us to compute the structure of A_χ even when $\chi^{-1}\omega(p) = 1$.

For any $x \in \mathbb{Z}_p[\Delta]$, we define a truncation x_{p^n} of x to be an element of $\mathbb{Z}[\Delta]$ such that $x_{p^n} \equiv x \pmod{p^n}$. Of course x_{p^n} is not unique. The choice of x_{p^n} has no influences on the following arguments unless a particular one is not specified. When the truncation for x is clear, we will substitute x for x_{p^n} to avoid complicated expressions.

Lemma 1. *Let $\chi(\neq 1)$ be an even character. If there exists a prime number ℓ which is congruent to 1 modulo p^{n+1} and totally decomposed in K and satisfies*

$$(\xi^{e_{\chi,p^{n+1}}})^{\frac{\ell-1}{p^{n+1}}} \not\equiv 1 \pmod{\mathcal{L}}   \tag{4}$$

for some prime ideal $\mathcal{L}$ of K lying above ℓ, then we have $|A_\chi| \leq p^{n[\mathcal{O}_\chi : \mathbb{Z}_p]}$.

Proof. Since ℓ is totally decomposed in K, the residue field $\mathcal{O}_K/\mathcal{L}$ ($\mathcal{O}_K$ is the ring of integers of K) is isomorphic to $\mathbb{Z}/\ell\mathbb{Z}$. Hence its multiplicative group has order $\ell - 1$ and the condition (4) implies $\xi^{e_\chi} \notin E_\chi^{p^{n+1}}$. Hence $|A_\chi| = |E_\chi/C_\chi| \leq p^{n[\mathcal{O}_\chi : \mathbb{Z}_p]}$.   □

Next, we choose two finite sets L, L^* of rational primes which satisfy the following conditions.

In the case that χ is odd

- L is a finite set of primes ℓ satisfying $\ell \equiv 1 \pmod{f}$.
- L^* is a finite set of primes ℓ^* satisfying $\ell^* \equiv 1 \pmod{df_0\ell}$ for every ℓ in L.

In the case that χ is even

- L is a finite set of primes ℓ which are congruent to 1 modulo d^2 satisfying the following conditions:

$$\chi(\ell) = 1,   \tag{5}$$

$$(\xi^{e_\chi})^{\frac{\ell-1}{d_p}} \not\equiv 1 \pmod{\mathcal{L}}   \tag{6}$$

for some prime ideal $\mathcal{L}$ of K lying above ℓ and

$$(\xi^{e_\chi})^{\frac{\ell-1}{d}} \equiv 1 \pmod{\mathcal{L}} \tag{7}$$

for any prime ideal $\mathcal{L}$ of K lying above ℓ.

- L^* is a finite set of primes ℓ^* satisfying $\ell^* \equiv 1 \pmod{d^2 f_0 \ell}$ for every ℓ in L.

Note that e_χ in (6) or (7) means $e_{\chi,dp}$ or $e_{\chi,d}$ respectively. A prime $\ell \in L$ is totally decomposed in $\mathbb{Q}(\mu_f)$ in the odd case and in K in the even case because of (5). On the other hand, a prime $\ell^* \in L^*$ is totally decomposed in $\mathbb{Q}(\mu_{ft})$ in both cases, where $t = \prod_{\ell \in L} \ell$. According to these properties on ℓ and ℓ^*, we can pursue almost all calculations in $\mathbb{Z}$.

As we show later, we try to choose L so that its elements generate A_χ and use L^* to guarantee that primes in L actually generate A_χ. Put

$$r = \begin{cases} 1 & \text{if } \chi \text{ is odd,} \\ \prod_{\ell \in L} \ell & \text{if } \chi \text{ is even,} \end{cases} \quad \text{and} \quad J_{L^*} = \prod_{\ell^* \in L^*} (\mathcal{O}_{K(\mu_r)} / \mathcal{L}^*)^\times,$$

where ℓ^* runs over all primes of L^* and $\mathcal{L}^*$ is a prime ideal of $K(\mu_r)$ lying above ℓ^* (we choose one $\mathcal{L}^*$ for each ℓ^*). We denote by $W_{L^*,\chi}$ the $\mathcal{O}_\chi$-submodule of $(K(\mu_r)^\times / K(\mu_r)^{\times d} E)_\chi$ generated by all elements whose representatives are prime to ℓ^* for all $\ell^* \in L^*$. If χ is odd, then we have $(K(\mu_r)^\times / K(\mu_r)^{\times d} E)_\chi = (K^\times / K^{\times d})_\chi$ because $r = 1$ and $\chi \neq \omega$. If $\chi (\neq 1)$ is even, then the $\mathcal{O}_\chi$-module $(E/E^d)_\chi$ is isomorphic to $\mathcal{O}_\chi / d\mathcal{O}_\chi$. Let $\varepsilon \in E$ be a generator of $(E/E^d)_\chi$ as $\mathcal{O}_\chi$-module. We define the subgroup $\overline{E}$ of J_{L^*} by

$$\overline{E} = \begin{cases} 1 & \text{if } \chi \text{ is odd,} \\ \langle (\varepsilon^\sigma \bmod \mathcal{L}^*)_{\mathcal{L}^*} \in J_{L^*} \mid \sigma \in \Delta \rangle & \text{if } \chi \text{ is even.} \end{cases}$$

Note that $\overline{E}$ is dependent on a choice of ε but $J_{L^*}/J_{L^*}^d \overline{E}$ is determined independently of ε.

Hence we can define the diagonal map $D^* : W_{L^*,\chi} \to J_{L^*}/J_{L^*}^d \overline{E}$ by $a \mapsto (a, \cdots, a) \pmod{J_{L^*}^d \overline{E}}$. We fix a generator σ_ℓ of the cyclic group $\mathrm{Gal}(K(\mu_\ell)/K)$ and set $D_\ell = \sum_{i=0}^{\ell-2} i \sigma_\ell{}^i$.

Before giving the definition of $\mathcal{M}_{L,L^*}$, which is the most important object in this paper, we recall the well-known fact that $\tau_{\mathcal{L}}^{e_{\chi,p^n}}$ $(n > 0)$ is contained in $K^\times$ if a truncation e_{χ,p^n} of e_χ is suitably chosen. This is an immediate consequence of [13, Lemma 6.4] and $p^2 \nmid f$.

In the following definition, e_χ means a well-chosen $e_{\chi,d}$ if χ is odd and any $e_{\chi,d}$ if χ is even.

Definition 3. *We define the $\mathbb{Z}_p$-submodule $\mathcal{M}_{L,L^*}$ of $J_{L^*}/J_{L^*}^d \overline{E}$ by*

$$\mathcal{M}_{L,L^*} = \begin{cases} D^*(\langle \tau_{\mathcal{L}}^{e_\chi} \bmod K^{\times d} \mid \mathcal{L} \mid \ell, \ell \in L \rangle_{\mathcal{O}_\chi}) & \text{if } \chi \text{ is odd,} \\ D^*(\langle \xi_\ell^{D_\ell e_\chi} \bmod K(\mu_r)^{\times d} E \mid \ell \in L \rangle_{\mathcal{O}_\chi}) & \text{if } \chi \text{ is even,} \end{cases}$$

where $\langle * \rangle_{\mathcal{O}_\chi}$ *denotes the* $\mathcal{O}_\chi$*-submodule of* $W_{L^*,\chi}$ *generated by* $*$*, and* $\mathcal{L} \mid \ell$ *means that we choose one prime ideal* $\mathcal{L}$ *of* K *lying above* ℓ.

Remark 2. Note that the Gauss sum $\tau_{\mathcal{L}}^{e_\chi}$ and cyclotomic unit $\xi_\ell^{e_\chi}$ are in $W_{L^*,\chi}$, because ξ_ℓ is a unit and the principal ideal $(\tau_{\mathcal{L}})$ is a product of primes lying above ℓ.

Our results are summarized to the following theorems. Theorem 2 enables us to determine the structure of A_χ via $\mathcal{M}_{L,L^*}$ if we choose d, L and L^* appropriately and Theorem 1 guarantees that there always exist L and L^* for which Theorem 2 holds if we take $d = d_0$, where d_0 is the power of p such that $|A_\chi| = d_0^{[\mathcal{O}_\chi:\mathbb{Z}_p]}$ as before.

Theorem 1. *For* $d = d_0$*, there exist finite sets* L *and* L^* *of rational primes satisfying* $|\mathcal{M}_{L,L^*}| = d_0^{[\mathcal{O}_\chi:\mathbb{Z}_p]}$.

Theorem 2. *For* d*,* L *and* L^* *which satisfy* $|\mathcal{M}_{L,L^*}| = d^{[\mathcal{O}_\chi:\mathbb{Z}_p]}$*, we have* $A_\chi \simeq \mathcal{M}_{L,L^*}$ *as* $\mathbb{Z}_p$*-module.*

As will be seen in §5, the order $|\mathcal{M}_{L,L^*}|$ is computed numerically. It is difficult to determine d_0 beforehand in the even case and we start with an upper bound d of d_0. But, once the equality $|\mathcal{M}_{L,L^*}| = d^{[\mathcal{O}_\chi:\mathbb{Z}_p]}$ holds, we can conclude $d = d_0$. We start with a rough bound d and sharpen it until we reach the equality $|\mathcal{M}_{L,L^*}| = d^{[\mathcal{O}_\chi:\mathbb{Z}_p]}$. In both odd and even cases, there are no known nice methods to seek L and L^*. We first choose the smallest ℓ, ℓ^* and continue to change them until we reach the equality. It may be meaningful to note that we have reached the equality after a few trials (sometimes the first trial) in computations in §5. Note that $\mathbb{Z}_p$-module structure of A_χ describes its $\mathcal{O}_\chi$-module structure because p is a prime element of $\mathcal{O}_\chi$ and $\mathcal{O}_\chi/p^a\mathcal{O}_\chi \simeq (\mathbb{Z}/p^a\mathbb{Z})^{[\mathcal{O}_\chi:\mathbb{Z}_p]}$.

3   Proofs of Theorems 1 and 2

Let d be a power of p satisfying the condition (1), which means $A_\chi^d = 0$. We define the $\mathcal{O}_\chi$-homomorphism ψ as follows.

$$\psi : A_\chi \to (K^\times/K^{\times^d}E)_\chi$$

$$\mathrm{cl}(\mathfrak{a}) \mapsto \alpha \bmod K^{\times^d}E$$

where $\mathrm{cl}(\mathfrak{a})$ means the ideal class containing $\mathfrak{a}$ and $\alpha \in K^\times$ is given by $\mathfrak{a}^d = (\alpha)$. Clearly ψ is well defined and injective. Note that if χ is odd and $\chi \neq \omega$, then we have $(K^\times/K^{\times^d}E)_\chi = (K^\times/K^{\times^d})_\chi$. Set $\nu = |Cl_K/A|$ and define the $\mathcal{O}_\chi$-submodule $B(L)_\chi$ of A_χ by $B(L)_\chi = \langle \mathrm{cl}(\mathcal{L})^{\nu e_\chi} \mid \mathcal{L}|\ell, \ell \in L \rangle_{\mathcal{O}_\chi}$. For a finite set L of primes, we think about the image of $B(L)_\chi$ by the map ψ. We prepare two lemmas. The first is well known result of Stickelberger.

Lemma 2 (Stickelberger). *Let* $\chi(\neq \omega)$ *be an odd character and* $\mathcal{L}$ *a prime ideal of* K *lying above a rational prime* $\ell \in L$*. Then we have* $\mathcal{L}^{u_{d^2} d e_{\chi,d^2}} = (\tau_{\mathcal{L}}^{e_{\chi,d^2}})(b^d)$ *in* K *for some* $b \in K^\times$ *and some* $u \in \mathbb{Z}_p[\Delta]^\times$*, where* u_{d^2} *and* e_{χ,d^2} *are suitably chosen truncations of* u *and* e_χ *respectively.*

As a reference to Lemma 2, we cite Theorem 2.2 of Chapter 1 in [5] for a cyclotomic case and the proof of Theorem 6.10 in [13] for a general abelian case.

For the plus part, we refer to the following result which was proved by Kolyvagin-Rubin-Thaine for cyclotomic case (cf. [4, Theorem 5], [7, Lemma 2.2, Proposition 2.4] and [12]). It is proved quite similarly for an abelian number field K and a prime number p satisfying the assumptions I and II in §1. We sketch out the arguments for convenience to readers.

They constructed an annihilator of $\mathcal{L}$ by using the cyclotomic unit ξ. They showed that for a certain Galois module isomorphism $\varphi_{\mathcal{L}} : (\mathcal{O}_K/\ell\mathcal{O}_K)^\times \simeq (\mathbb{Z}/(\ell-1)\mathbb{Z})[\Delta]$, the element $\varphi_{\mathcal{L}}(\xi)$ of $(\mathbb{Z}/(\ell-1)\mathbb{Z})[\Delta]$ annihilates $\mathcal{L}$ in $I_K/I_K^{\ell-1}$ (I_K is the ideal group of K). Namely they showed $(\kappa_\ell) \equiv \mathcal{L}^{\varphi_{\mathcal{L}}(\xi)} \pmod{I_K^{\ell-1}}$ for some $\kappa_\ell \in K^\times$, and hence $(\kappa_\ell) \equiv \mathcal{L}^{\varphi_{\mathcal{L}}(\xi)} \pmod{I_K^{d^2}}$. By the assumptions (6) and (7), we can show $\varphi_{\mathcal{L}}(\xi^{e_\chi}) = du \bmod d^2\mathcal{O}_\chi$ in $(\mathbb{Z}/d^2\mathbb{Z})[\Delta]e_\chi \simeq \mathcal{O}_\chi/d^2\mathcal{O}_\chi$ for some $u \in \mathcal{O}_\chi^\times$. Further they showed the congruence $\kappa_\ell \equiv \xi_\ell^{D_\ell} \pmod{K(\mu_\ell)^{\times d^2}}$. Putting these results together, we get the following lemma.

Lemma 3 (Kolyvagin-Rubin-Thaine). *Let $\chi(\neq 1)$ be an even character and $\mathcal{L}$ a prime ideal of K lying above a rational prime $\ell \in L$. Then there exists an element $\kappa_\ell \in K^\times$ which is determined modulo $K^{\times d^2}$ and satisfies $\kappa_\ell \equiv \xi_\ell^{D_\ell}$ (mod $K(\mu_\ell)^{\times d^2}$)). Furthermore, we have $\mathcal{L}^{u_{d^2} de_{\chi,d^2}} = (\kappa_\ell^{e_{\chi,d^2}})(b^d)$ in K for some $b \in K^\times$ and some $u \in \mathbb{Z}_p[\Delta]^\times$, where u_{d^2} and e_{χ,d^2} are arbitrary truncations of u and e_χ respectively.*

Lemmas 2 and 3 combining the definition of $B(L)_\chi$ lead to the following.

Lemma 4. *There holds*

$$\psi(B(L)_\chi) = \begin{cases} \langle \tau_{\mathcal{L}}^{e_\chi} \bmod K^{\times d} \mid \mathcal{L}|\ell,\ \ell \in L \rangle_{\mathcal{O}_\chi} & \text{if } \chi \text{ is odd,} \\ \langle \kappa_\ell^{e_\chi} \bmod K^{\times d}E \mid \ell \in L \rangle_{\mathcal{O}_\chi} & \text{if } \chi \text{ is even.} \end{cases}$$

Let r be the integer defined in §2. Since $\chi \neq \omega$ and the natural map $(K^\times/K^{\times d})_\chi \to (K(\mu_r)^\times/K(\mu_r)^{\times d})_\chi$ is injective, the map $i_r : (K^\times/K^{\times d}E)_\chi \to (K(\mu_r)^\times/K(\mu_r)^{\times d}E)_\chi$ is also injective. The composition map $i_r \circ \psi : A_\chi \to (K(\mu_r)^\times/K(\mu_r)^{\times d}E)_\chi$ is an injective $\mathcal{O}_\chi$-homomorphism and we have

$$i_r \circ \psi(B(L)_\chi) = \begin{cases} \langle \tau_{\mathcal{L}}^{e_\chi} \bmod K^{\times d} \mid \mathcal{L}|\ell,\ \ell \in L \rangle_{\mathcal{O}_\chi} & \text{if } \chi \text{ is odd,} \\ \langle \xi_\ell^{D_\ell e_\chi} \bmod K(\mu_r)^{\times d}E \mid \ell \in L \rangle_{\mathcal{O}_\chi} & \text{if } \chi \text{ is even,} \end{cases}$$

because of $\kappa_\ell \equiv \xi_\ell^{D_\ell} \bmod K(\mu_\ell)^{\times d}$ by Lemma 3.

Let $D^* : W_{L^*,\chi} \to J_{L^*}/J_{L^*}^d . \overline{E}$ be the diagonal map (this is not an $\mathcal{O}_\chi$-homomorphism but a $\mathbb{Z}_p$-homomorphism) and $\mathcal{M}_{L,L^*}$ the $\mathbb{Z}_p$-module defined in Definition 3. The following lemma is an immediate consequence of Lemma 4 and plays an essential role in our algorithm.

Lemma 5. *We have* $\mathcal{M}_{L,L^*} = D^* \circ i_r \circ \psi(B(L)_\chi)$.

Now we give the proof of Theorem 1 which depends essentially on the density theorem of Chebotarev.

Proof of Theorem 1. We will choose finite sets L and L^* of rational primes which satisfy the conditions on L and L^* in §2 and $|\mathcal{M}_{L,L^*}| = d_0^{[\mathcal{O}_\chi : \mathbb{Z}_p]}$.

$\boxed{\text{In the case } \chi \text{ is odd}}$

Applying the Chebotarev density theorem to the maximal unramified abelian p-extension over $\mathbb{Q}(\zeta_f)$, we choose prime ideals $\mathcal{L}$ of K which lie above rational primes ℓ satisfying $\ell \equiv 1 \pmod{f}$ and whose classes generate A_χ over $\mathcal{O}_\chi$. Let L be a finite set of such primes ℓ. We have $A_\chi = B(L)_\chi \simeq i_r \circ \psi(B(L)_\chi) = \langle \tau_{\mathcal{L}}^{e_\chi} \bmod K^{\times d_0} \mid \mathcal{L} | \ell, \ \ell \in L \rangle_{\mathcal{O}_\chi}$. Hence it is enough to choose L^* such that the restriction of the diagonal map D^* to $i_r \circ \psi(B(L)_\chi)$ is injective. Set $K' = K(\mu_{d_0 f_0 t})$ with $t = \prod_{\ell \in L} \ell$. We can regard the $\mathcal{O}_\chi$-submodule $i_r \circ \psi(B(L)_\chi)$ of $(K^\times / K^{\times d_0})_\chi$ as the $\mathcal{O}_\chi$-submodule of $([K'^\times / K'^{\times d_0}]^{\mathrm{Gal}(K'/K)})_\chi$ by the natural isomorphism

$$(K^\times / K^{\times d_0})_\chi \simeq ([K'^\times / K'^{\times d_0}]^{\mathrm{Gal}(K'/K)})_\chi. \tag{8}$$

Consider the Kummer extension

$$K'_L = K'(\{ \sqrt[d_0]{a} \mid a \bmod K^{\times d_0} \in i_r \circ \psi(B(L)_\chi) \}).$$

The extension K'_L/K' is a finite abelian extension. By the Chebotarev density theorem, we can choose generators of the Galois group $\mathrm{Gal}(K'_L/K')$ to be the Frobenius substitutions $(\widetilde{\mathcal{L}^*}, K'_L/K')$ with prime ideals $\widetilde{\mathcal{L}^*}$ of K' which are totally decomposed in $K'/\mathbb{Q}$. Let ℓ^* be the rational prime divisible by $\widetilde{\mathcal{L}^*}$, and L^* be a finite set of such primes ℓ^*. Note that L and L^* satisfy the conditions in §2. We will show that the restriction map of D^* to $i_r \circ \psi(B(L)_\chi)$ is injective. Let $\mathcal{L}^*$ be the prime ideal of K satisfying $\widetilde{\mathcal{L}^*} \supset \mathcal{L}^* \supset (\ell^*)$. Let $a \in i_r \circ \psi(B(L)_\chi) = \langle \tau_{\mathcal{L}}^{e_\chi} \mid \mathcal{L} | \ell, \ \ell \in L \rangle_{\mathcal{O}_\chi}$, and suppose that $D^*(a) = 0$, that is, $a \bmod \mathcal{L}^* \in (\mathcal{O}_K/\mathcal{L}^*)^{\times d_0}$ for every $\mathcal{L}^*$. By the natural isomorphism $\mathcal{O}_K/\mathcal{L}^* \simeq \mathcal{O}_{K'}/\widetilde{\mathcal{L}^*}$, we get $a \bmod \widetilde{\mathcal{L}^*} \in (\mathcal{O}_{K'}/\widetilde{\mathcal{L}^*})^{\times d_0}$ for every $\widetilde{\mathcal{L}^*}$. This implies the restriction of $(\widetilde{\mathcal{L}^*}, K'_L/K')$ to $K'(\sqrt[d_0]{a})$ is trivial for every $\widetilde{\mathcal{L}^*}$. Since $K'(\sqrt[d_0]{a})$ is a subfield of K'_L/K' and $\mathrm{Gal}(K'_L/K')$ is generated by the Frobenius substitutions $(\widetilde{\mathcal{L}^*}, K'_L/K')$ with $\widetilde{\mathcal{L}^*}$, we obtain $K' = K'(\sqrt[d_0]{a})$. Hence we conclude $a \in K'^{\times d_0}$, which means $a \in K^{\times d_0}$ by (8), and get the conclusion.

$\boxed{\text{In the case } \chi \text{ is even}}$

For every generator c of A_χ as $\mathcal{O}_\chi$-module, we can choose its representative $\mathcal{L}$ lying above a rational prime ℓ such that $\ell \equiv 1 \pmod{d_0^2 f}$ and satisfying $\xi^{e_\chi} \bmod \mathcal{L} \notin (\mathcal{O}_K/\mathcal{L})^{\times d_0 p}$ because we have $\xi^{e_\chi} \notin K^{\times d_0 p}$ by the definition of

d_0 (cf. [13, Proposition 15.4]). Let L be a finite set of such primes ℓ. Since $(\mathcal{O}_K/\mathcal{L})^\times$ is a cyclic group of order $\ell - 1$, we have $(\xi^{e_\chi})^{(\ell-1)/d_0 p} \not\equiv 1 \pmod{\mathcal{L}}$. Further, since $\xi^{e_\chi} \in E_\chi$ is a generator of the group of cyclotomic units C_χ and $E_\chi/C_\chi \simeq \mathcal{O}_\chi/d_0\mathcal{O}_\chi$ as $\mathcal{O}_\chi$-modules, we have $\xi^{e_\chi} \bmod \mathcal{L} \in (\mathcal{O}_K/\mathcal{L})^{\times d_0}$. Hence we have $(\xi^{e_\chi})^{(\ell-1)/d_0} \equiv 1 \pmod{\mathcal{L}}$. Next we choose a finite set $\mathcal{L}^*$ of primes. Set $K_t = K(\mu_t)$ and $K' = K(\mu_{d_0^2 f_0 t})$ with $t = \prod_{\ell \in L} \ell$. We regard the $\mathcal{O}_\chi$-submodule

$$X = \langle \xi_\ell^{D_\ell e_\chi} \bmod K_t^{\times d_0} \mid \ell \in L \rangle_{\mathcal{O}_\chi} (EK_t^{\times d_0}/K_t^{\times d_0})_\chi$$

of $(K_t^\times/K_t^{\times d_0})_\chi$ as the $\mathcal{O}_\chi$-submodule of $\left((K'^\times/K'^{\times d_0})^{\mathrm{Gal}(K'/K_t)}\right)_\chi$ in the same way as odd case. Set $K'_L = K'(\{\sqrt[d_0]{a} \mid a \bmod K_t^{\times d_0} \in X\})$. We choose generators of the Galois group $\mathrm{Gal}(K'_L/K')$ to be the Frobenius substitutions $(\widetilde{\mathcal{L}^*}, K'_L/K')$ with prime ideals $\widetilde{\mathcal{L}^*}$ of K' which are totally decomposed in $K'/\mathbb{Q}$. Let ℓ^* be the rational prime divisible by $\widetilde{\mathcal{L}^*}$, and L^* be a finite set of such primes ℓ^*. Let $\mathcal{L}^*$ be the prime ideal of K_t satisfying $\widetilde{\mathcal{L}^*} \supset \mathcal{L}^* \supset (\ell^*)$. To show that the restriction map of the diagonal map D^* to $i_r \circ \psi(B(L)_\chi)$ is injective, it is enough to show that $(a \bmod \mathcal{L}^*)_{\mathcal{L}^*} \in J_{L^*}^{d_0}\overline{E}$ implies $a \in K_t^{\times d_0}E$ for every $a \bmod K_t^{\times d_0}E \in \langle \xi_\ell^{D_\ell e_\chi} \bmod K_t^{\times d_0}E \mid \ell \in L \rangle_{\mathcal{O}_\chi}$. Let $a \bmod K_t^{\times d_0}E \in \langle \xi_\ell^{D_\ell e_\chi} \bmod K_t^{\times d_0}E \mid \ell \in L \rangle_{\mathcal{O}_\chi}$ and suppose that for every $\mathcal{L}^*$, $a \equiv b_i^{d_0}u \bmod \mathcal{L}^*$ with some $b_i \in \mathcal{O}_{K_t}$ and some $u \in E$. We get $au^{-1} \bmod \mathcal{L}^* \in (\mathcal{O}_{K_t}/\mathcal{L}^*)^{\times d_0}$ for every $\mathcal{L}^*$. By the same arguments as in the odd case, we obtain $au^{-1} \in K_t^{\times d_0}$. Hence we conclude $a \in K_t^{\times d_0}E$. $\qquad\square$

Next, we prove Theorem 2.

Proof of Theorem 2. We have $|A_\chi| \geq |B(L)_\chi| = |i_r \circ \psi(B(L)_\chi)| \geq |D^* \circ i_r \circ \psi(B(L)_\chi)| = |\mathcal{M}_{L,L^*}|$. By the assumption, we have $|\mathcal{M}_{L,L^*}| = d^{[\mathcal{O}_\chi:\mathbb{Z}_p]}$. On the other hand, we have $d^{[\mathcal{O}_\chi:\mathbb{Z}_p]} \geq |A_\chi|$ by the definition of d. Hence we conclude that $|A_\chi| = |B(L)_\chi| = |D^* \circ i_r \circ \psi(B(L)_\chi)| = |\mathcal{M}_{L,L^*}| = d^{[\mathcal{O}_\chi:\mathbb{Z}_p]}$. Since $B(L)_\chi$ is the submodule of A_χ, we have $A_\chi = B(L)_\chi$. Further, we obtain the $\mathbb{Z}_p$-isomorphism $A_\chi \simeq \mathcal{M}_{L,L^*}$ because the composition map $D^* \circ i_r \circ \psi$ is $\mathbb{Z}_p$-homomorphism. $\qquad\square$

4 Algorithms for $\mathcal{M}_{L,L^*}$

In order to determine the structure of A_χ using Theorem 2, we continue to seek d, L and L^* until we reach the equality $|\mathcal{M}_{L,L^*}| = d^{[\mathcal{O}_\chi:\mathbb{Z}_p]}$. In this section, we explain how to determine the structure of $\mathcal{M}_{L,L^*}$ when we fix d, L and L^*. Several techniques are used to decrease the calculating time and almost all calculations are reduced to those in $\mathbb{Z}$.

4.1 The Odd Case

First we note that we can use any truncation of e_χ in Definition 1 because we work in the large field $\mathbb{Q}(\mu_{f\ell})$ and ℓ^* splits completely in $\mathbb{Q}(\mu_{f\ell})/\mathbb{Q}$.

Let ℓ be a prime in L and ρ a primitive root of ℓ. Then the character $\chi_{\widetilde{\mathcal{L}}}$: $(\mathbb{Z}/\ell\mathbb{Z})^{\times} \longrightarrow \mu_f$ which appears in Definition 1 is determined by the image of ρ. Note that if ζ_f is any primitive f-th root of unity, then there exists some prime ideal $\widetilde{\mathcal{L}}$ of $\mathbb{Q}(\mu_f)$ lying above ℓ such that $\chi_{\widetilde{\mathcal{L}}}(\rho) = \zeta_f$. Since we do not need specify $\widetilde{\mathcal{L}}$, we can set $\chi_{\widetilde{\mathcal{L}}}(\rho) = \zeta_f$ for an arbitrarily given ζ_f.

Let ℓ^* be a prime in L^* and g a primitive root of ℓ^*. Take $s, t \in \mathbb{Z}$ such that $s \equiv g^{(\ell^*-1)/f} \pmod{\ell^*}$ and $t \equiv g^{(\ell^*-1)/\ell} \pmod{\ell^*}$. Then $\zeta_f \equiv s \pmod{\widetilde{\mathcal{L}^*}}$ and $\zeta_\ell \equiv t \pmod{\widehat{\mathcal{L}^*}}$ for some prime ideals $\widetilde{\mathcal{L}^*}$ of $\mathbb{Q}(\mu_f)$ and $\widehat{\mathcal{L}^*}$ of $\mathbb{Q}(\mu_\ell)$ lying above ℓ^*. Let $\overline{\mathcal{L}^*}$ be the prime ideal of $\mathbb{Q}(\mu_{f\ell})$ lying above both $\widetilde{\mathcal{L}^*}$ and $\widehat{\mathcal{L}^*}$. Hence we have

$$\tau_{\widetilde{\mathcal{L}}} = \sum_{i=0}^{\ell-2} \zeta_f^i \zeta_\ell^{\rho^i} \equiv \sum_{i=0}^{\ell-2} s^i t^{\rho^i} \pmod{\overline{\mathcal{L}^*}}$$

in $\mathbb{Q}(\mu_{f\ell})$. In order to get $\tau_{\mathcal{L}}^{ex}$, we need to calculate

$$\tau_{\widetilde{\mathcal{L}}}^{\sigma_r} \equiv \sum_{i=0}^{\ell-2} s^{r i} t^{\rho^i} \pmod{\overline{\mathcal{L}^*}}$$

for each σ_r, where σ_r is defined by $\zeta_f^{\sigma_r} = \zeta_f^r$ and runs over $\mathrm{Gal}(\mathbb{Q}(\mu_f)/\mathbb{Q})$. According to [11], we calculate $\tau_{\widetilde{\mathcal{L}}}^{\sigma_r} \bmod \overline{\mathcal{L}^*}$ $(r \in (\mathbb{Z}/f\mathbb{Z})^{\times})$ in $O(\ell \log \ell)$ times using FFT and get $x \in \mathbb{Z}$ such that $\tau_{\mathcal{L}}^{ex} \equiv x \pmod{\mathcal{L}^*}$ in K, where $\mathcal{L}^*$ is the prime ideal of K divisible by $\widetilde{\mathcal{L}^*}$.

Next we determine the structure of $\mathcal{M}_{L,L^*}$. If $|L| = |L^*| = 1$, it is quite easy because $J_{L^*}/J_{L^*}^d$ is a cyclic group. Let $d = p^a$ and $\tau_{\mathcal{L}}^{ex} \equiv x \pmod{\mathcal{L}^*}$ with $x \in \mathbb{Z}$. If $x^{(\ell^*-1)/p^i} \equiv 1 \pmod{\ell^*}$ and $x^{(\ell^*-1)/p^{i+1}} \not\equiv 1 \pmod{\ell^*}$, then $|\mathcal{M}_{L,L^*}| = p^{a-i}$. This requires $O(a \log \ell^*)$ times.

When $|L^*| > 1$, it becomes slightly difficult. We shall explain the case $|L| = |L^*| = m$ ($|L| \le |L^*|$ in general). Let $L = \{\ell_1, \cdots, \ell_m\}$, $L^* = \{\ell_1^*, \cdots, \ell_m^*\}$ and g_i be a primitive root of ℓ_i^*. We get $x_{i,j} \in \mathbb{Z}$ such that

$$\tau_{\mathcal{L}_i}^{ex} \equiv x_{i,j} \pmod{\mathcal{L}_j^*}, \tag{9}$$

where $\mathcal{L}_i$ (resp. $\mathcal{L}_j^*$) is a prime ideal lying above ℓ_i (resp. ℓ_j^*). An integer $y_{i,j}$ such that

$$(x_{i,j} g_j^{y_{i,j}})^{\frac{\ell_j^*-1}{d}} \equiv 1 \pmod{\ell_j^*}$$

is obtained in $O(d \log \ell_j^*)$ times. If $(d_1, \cdots, d_m)$ is the elementary divisors of the $m * m$ matrix $N = (y_{i,j})$ in $\mathbb{Z}_p$, then we have

$$\mathcal{M}_{L,L^*} \simeq \prod_{i=1}^{m} \mathbb{Z}_p/(d/d_i)\mathbb{Z}_p$$

as abelian group. We can determine $\mathcal{M}_{L,L^*}$ in reasonable time (i.e. $O(p(\log \ell_1^* + \cdots + \log(\ell_m^*)))$) if $d = p$. Since it scarcely occurs that $|L^*| > 1$ and $d = p^a$ ($a > 1$) with large p, we can handle almost all cases in this manner.

4.2   The Even Case

First we seek several d and ℓ which satisfy (5), (6) and (7). Lemma 1 implies that these d give upper bounds of $|A_\chi|$. We choose minimum d in our trial, which probably realizes the equality $d^{[\mathcal{O}_\chi : \mathbb{Z}_p]} = |A_\chi|$.

A slight difficulty lies in this process. We require that ℓ splits completely not in $\mathbb{Q}(\mu_f)$ but in K. This is necessary to pick up a small ℓ and calculate the cyclotomic unit ξ_ℓ in reasonable time. We checked the conditions (6) and (7) as follows. Let $[K : \mathbb{Q}] = n$ and $\{v_1, \cdots, v_n\}$ an integral basis of K. We calculate approximate values of ξ^ρ ($\rho \in \mathrm{Gal}(K/\mathbb{Q})$) and get coefficients $x_i \in \mathbb{Z}$ of ξ with respect to $\{v_i\}$ by solving

$$\sum_{i=0}^{n} x_i v_i^\rho = \xi^\rho \quad (\rho \in \mathrm{Gal}(K/\mathbb{Q}))$$

approximately. It is easy to get $y_{i,\rho} \in \mathbb{Z}$ such that $v_i^\rho \equiv y_{i,\rho} \pmod{\mathcal{L}}$. We have

$$\xi^\rho \equiv \sum_{i=1}^{n} x_i y_{i,\rho} \pmod{\mathcal{L}} \tag{10}$$

and finally get $z \in \mathbb{Z}$ such that $\xi^{e_x} \equiv z \pmod{\mathcal{L}}$. Then the conditions (6) and (7) are equivalent to $z^{(\ell-1)/dp} \not\equiv 1 \pmod{\mathcal{L}}$ and $z^{(\ell-1)/d} \equiv 1 \pmod{\mathcal{L}}$ respectively.

It seems difficult to determine the structure of $\mathcal{M}_{L,L^*}$ in the even case because $\mathcal{M}_{L,L^*}$ is a subgroup of $J_{L^*}/J_{L^*}^d\cdot\overline{E}$ and the calculation of fundamental units is considered as difficult as that of ideal class group. However we can avoid the effects of units as in the following way.

Consider the typical case that $[K : \mathbb{Q}]$ divides $p - 1$ and assume $|A_\chi| = d$. In order to calculate $\mathcal{M}_{L,L^*}$, it is enough to be able to decide whether $\bar{x} = (x, \cdots, x)$ is contained in $J_{L^*}^d\cdot\overline{E}$ for each $x \in \mathcal{O}_{K(\mu_r)}$. Since $|A_\chi| = d$, $\eta = \sqrt[d]{\xi^{e_x}}$ generates $(E/E^d)_\chi$ and

$$\bar{x} \in J_{L^*}^d \cdot \overline{E} \iff \exists\, \varepsilon \in (E/E^d)_\chi, \ \forall\, \mathcal{L}^*, \quad x\varepsilon \in (\mathcal{O}_{K(\mu_r)}/\mathcal{L}^*)^{\times d}$$

$$\iff \exists\, \varepsilon \in (E/E^d)_\chi, \ \forall\, \mathcal{L}^*, \quad (x\varepsilon)^{\frac{\ell^*-1}{d}} \equiv 1 \pmod{\mathcal{L}^*}$$

$$\iff \exists\, i\,(0 \le i < d), \ \forall\, \mathcal{L}^*, \quad (x\eta^i)^{\frac{\ell^*-1}{d}} \equiv 1 \pmod{\mathcal{L}^*}$$

$$\iff \exists\, i\,(0 \le i < d), \ \forall\, \mathcal{L}^*, \quad (x^d\xi^{ie_x})^{\frac{\ell^*-1}{d^2}} \equiv 1 \pmod{\mathcal{L}^*}.$$

Analyzing this procedure when $|L^*| = 1$, one can easily find that the inequality $|\mathcal{M}_{L,L^*}| < d$ always holds if $(\xi^{e_x})^{(\ell^*-1)/d^2} \not\equiv 1 \pmod{\mathcal{L}^*}$. Therefore we request further condition on L^* including the case $|L^*| > 1$:

$$(\xi^{e_x})^{\frac{\ell^*-1}{d^2}} \equiv 1 \pmod{\mathcal{L}^*} \tag{11}$$

for any prime ideal $\mathcal{L}^*$ of K lying above ℓ^* and any ℓ^* in L^*. Once one assumes (11), the structure of $\mathcal{M}_{L,L^*}$ is easily determined because $J_{L^*}/J_{L^*}^d\cdot\overline{E} = J_{L^*}/J_{L^*}^d$.

Note that the inequality $d^{[\mathcal{O}_\chi : \mathbb{Z}_p]} \geq |A_\chi|$ is enough to conclude $\overline{E} \subset J_{\overline{L}^*}^d$ under the assumption (11). We note that the above argument and Theorem 1 guarantees we can always find ℓ^* satisfying (11) when $|L^*| = 1$. There is no theoretical assurance that we are able to find such ℓ^* when $|L^*| > 1$. But we succeeded in finding such ℓ^* in all our practical calculations. So (11) seems to be a reasonable condition in spite of its highly technical looks.

Furthermore, it is not difficult to find ℓ^* which satisfies (11). Indeed, we already have $x_i \in \mathbb{Z}$ such that $\xi = \sum_{i=1}^{n} x_i v_i$ and verify (11) by congruence calculation similar to (10). The seeking time is ignored in comparison with the calculation of ξ_ℓ.

The calculation of ξ_ℓ is straightforward. Let ρ (resp. g) be a primitive root of ℓ (resp. ℓ^*) and ζ_f (resp. ζ_ℓ) a primitive f-th (resp. ℓ-th) root of unity. Then

$$\zeta_f \equiv g^{\frac{\ell^*-1}{f}} \quad (\mathrm{mod}\ \widetilde{\mathcal{L}^*})$$

for some prime ideal $\widetilde{\mathcal{L}^*}$ of $\mathbb{Q}(\mu_f)$ lying over ℓ^* and

$$\zeta_\ell \equiv g^{\frac{\ell^*-1}{\ell}} \quad (\mathrm{mod}\ \widehat{\mathcal{L}^*})$$

for some prime ideal $\widehat{\mathcal{L}^*}$ of $\mathbb{Q}(\mu_\ell)$ lying over ℓ^*. Let $\overline{\mathcal{L}^*}$ be the prime ideal of K such that $\overline{\mathcal{L}^*} \subset \widetilde{\mathcal{L}^*}$ and $\mathcal{L}^*$ a prime ideal of $K(\mu_r)$ lying above both $\overline{\mathcal{L}^*}$ and $\widehat{\mathcal{L}^*}$. Then we have

$$\xi_\ell^{D_\ell} \equiv \prod_{j \in H} \prod_{i=0}^{\ell-2} \left(g^{\frac{\ell^*-1}{f}j + \frac{\ell^*-1}{\ell}\rho^i} - 1 \right)^i \quad (\mathrm{mod}\ \mathcal{L}^*),$$

where H is the subgroup of $(\mathbb{Z}/f\mathbb{Z})^\times$ corresponding to K. This is an $O(f\ell)$ algorithm.

Finally we remark a possibility of fast computation of $\xi_\ell^{D_\ell}$. Let $G(X)$ be the minimal polynomial of ζ_f over K. Then we have

$$\xi_\ell^{D_\ell} = \prod_{j \in H} \prod_{i=0}^{\ell-2} (\zeta_f^j \zeta_\ell^{\rho^i} - 1)^i$$

$$\equiv \prod_{i=1}^{\ell-1} \prod_{j \in H} (\zeta_\ell^{-i} - \zeta_f^j)^{\nu_\ell(i)} \quad (\mathrm{mod}\ K(\mu_r)^{\times d^2})$$

$$= \prod_{i=1}^{\ell-1} G(\zeta_\ell^{-i})^{\nu_\ell(i)},$$

where $\nu_\ell(i)$ is an integer satisfying $i \equiv \rho^{\nu_\ell(i)} \pmod{\ell}$. It is well known that $G(\zeta_\ell^{-i})$ $(1 \leq i \leq \ell-1)$ are computable in $O(\ell \log \ell)$ times using FFT. A discrete logarithm function ν_ℓ is easily calculated because $\nu_\ell(i) \bmod d^2$ is enough for our purpose. Hence $\xi_\ell^{D_\ell}$ is computable in $O(\ell \log \ell)$ times if we can get $G(X)$ quickly.

5   Examples

We show several numerical examples. Let $K = \mathbb{Q}(\sqrt{m}, \mu_5)$ with square-free integer m different from 1 and 5. Then $K/\mathbb{Q}$ is an abelian extension of degree 8. Let χ be the character of $\mathbb{Q}(\sqrt{m})$ and ω the Teichmüller character of $\mathbb{Q}(\mu_5)$. Then the set of characters of Δ is $\{1, \omega, \omega^2, \omega^3, \chi, \chi\omega, \chi\omega^2, \chi\omega^3\}$. Since the class number of $\mathbb{Q}(\mu_5)$ is one, we see $A_\omega = A_{\omega^2} = A_{\omega^3} = 0$. The fixed fields of χ and $\chi\omega^2$ are $\mathbb{Q}(\sqrt{m})$ and $\mathbb{Q}(\sqrt{5m})$ respectively. So we are interested in $\chi\omega$ and $\chi\omega^3$ which correspond to the cyclic subfield of K with degree 4.

In the following typical examples, we try to illustrate how we use Theorem 2 to determine $A = A_K$.

5.1   The Case $p = 5$

When $p = 5$, we have $A = A_\chi \oplus A_{\chi\omega} \oplus A_{\chi\omega^2} \oplus A_{K,\chi\omega^3}$. For each character ψ of Δ, we see $\mathcal{O}_\psi = \mathbb{Z}_5$.

Example 1. Let $m = 36227$ and $\psi = \chi\omega^3$. Then $f = 724540$ and ψ is an odd character. We see $|A_\psi| = 5^3$ by calculating $B_{1,\psi^{-1}}$ and set $d = 5^3$. The prime numbers ℓ which satisfy $\ell \equiv 1 \pmod{f}$ are $8694481, 15939881, 26807981, \ldots$. We choose $\ell = 8694481$ and $L = \{\ell\}$. Though a theoretical condition for ℓ^* is $\ell^* \equiv 1 \pmod{df_0\ell}$, we require a technical condition $\ell^* \equiv 1 \pmod{2df_0\ell}$ to write a FFT program easily. Then ℓ^* are $6614474226927001, 12284023564293001, \ldots$. We choose $L^* = \{6614474226927001\}$. Since $\mathcal{M}_{L,L^*}$ is cyclic in this case, it is quite easy to see $|\mathcal{M}_{L,L^*}| = 5^3$. Hence we have $A_\psi \simeq \mathcal{M}_{L,L^*} \simeq \mathbb{Z}/5^3\mathbb{Z}$ from Theorem 2.

Example 2. Let $m = 1111$ and $\psi = \chi\omega$. Then $f = 22220$ and ψ is an odd character. We see $|A_\psi| = 5^2$ by calculating $B_{1,\psi^{-1}}$ and set $d = 5^2$. The prime numbers ℓ which satisfy $\ell \equiv 1 \pmod{f}$ are $133321, 177761, 266641, \ldots$. First we choose $L = \{133321\}$. Then we see $|\mathcal{M}_{L,L^*}| \leq 5$ for several L^*. We also see $|\mathcal{M}_{L,L^*}| \leq 5$ for $L = \{177761\}$ and several L^*. This suggests a possibility that A_ψ is not cyclic. So we choose $\ell_1 = 133321, \ell_2 = 177761$ and $L = \{\ell_1, \ell_2\}$. Then prime numbers ℓ^* which satisfy $\ell^* \equiv 1 \pmod{2df_0\ell_1\ell_2}$ are $126383489885716801, 221171107300004401, 289628830988101001, 384416448402388601 \ldots$. We choose $L^* = \{126383489885716801, 221171107300004401\}$. Then we have

$$N = \begin{pmatrix} 10 & 5 \\ 5 & 10 \end{pmatrix} \sim \begin{pmatrix} 5 & 0 \\ 0 & 5 \end{pmatrix}$$

and $|\mathcal{M}_{L,L^*}| = 5^2$, where N is the matrix defined in 4.1. Hence Theorem 2 implies $A_\psi \simeq \mathcal{M}_{L,L^*} \simeq \mathbb{Z}/5\mathbb{Z} \oplus \mathbb{Z}/5\mathbb{Z}$.

Example 3. Let $m = -14606$ and $\psi = \chi\omega$. Then $f = 292120$ and ψ is an even character. Using Lemma 1 with $\ell = 43818001$, we see $|A_{K,\psi}| \leq 5^2$ and set $d = 5^2$. As a finite set of ℓ which satisfy $\ell \equiv 1 \pmod{d^2}$ and (5), (6), (7), we choose $L = \{\ell_1 = 11251, \ell_2 = 22501\}$. We pick up the smallest two prime numbers ℓ^* which satisfy $\ell^* \equiv 1 \pmod{d^2 f_0 \ell_1 \ell_2}$ and (11). Namely we set $L^* = \{6868360202024395001, 13662767669706670001\}$. We mention that the checking time for (11) is $O(f + \log \ell^*)$. Since $J_{L^*}/J_{L^*}^d \cdot \overline{E} = J_{L^*}/J_{L^*}^d$ by (11), the calculation of $\mathcal{M}_{L,L^*}$ is same as in the odd case (replace $\tau_{\mathcal{L}_i}$ in (9) by ξ_{ℓ_i}). We have

$$N = \begin{pmatrix} 15 & 15 \\ 0 & 15 \end{pmatrix} \sim \begin{pmatrix} 5 & 0 \\ 0 & 5 \end{pmatrix}$$

and $|\mathcal{M}_{L,L^*}| = 5^2$. Hence Theorem 2 implies $A_\psi \simeq \mathcal{M}_{L,L^*} \simeq \mathbb{Z}/5\mathbb{Z} \oplus \mathbb{Z}/5\mathbb{Z}$.

We show some other examples in Table 1 with calculating time using Alpha 21264 667MHz, where $(5,5)$ means $\mathbb{Z}/5\mathbb{Z} \oplus \mathbb{Z}/5\mathbb{Z}$ for example. We mainly picked up K with non-cyclic 5-ideal class groups to show that Theorem 2 works well.

Table 1. $p = 5$

m	$A_{K,\chi}$	$A_{K,\chi\omega}$	$A_{K,\chi\omega^2}$	$A_{K,\chi\omega^3}$	time
1111	(5)	(5,5)	0	0	24s
7523	0	0	(5)	$(5^2,5)$	7m22s
36227	0	0	0	(5^3)	9m36s
36293	0	0	0	(5^2)	2m54s
36322	(5)	$(5^3,5)$	0	0	29m28s
42853	(5)	$(5^3,5)$	0	(5)	8m47s
−5657	0	0	(5,5)	(5^2)	24m33s
−14606	(5,5)	(5,5)	0	0	6h42m

5.2   The Case $p = 3$

When $p = 3$, we have $A = A_\chi \oplus A_{\chi\omega} \oplus A_{\chi\omega^2}$ because the character $\chi\omega^3$ is conjugate to $\chi\omega$. In this case, $\mathcal{O}_\chi = \mathcal{O}_{\chi\omega^2} = \mathbb{Z}_3$ and $\mathcal{O}_{\chi\omega} = \mathbb{Z}_3[\mu_4]$.

Example 4. Let $m = 15338$ and $\psi = \chi\omega$. Then $f = 306760$ and ψ is an odd character. We see $|A_\psi| = 3^4$ by calculating $B_{1,\psi^{-1}}$ and set $d = 3^2$. We choose $L = \{920281, 1840561\}$ and $L^* = \{5393920089751369845571 5841, 6129454647444738 4608768001, 9071592878218212922 0976641, 1299444385258284553 70588161\}$. Then we have

$$N = \begin{pmatrix} 0 & 6 & 3 & 0 \\ 3 & 6 & 3 & 6 \\ 3 & 0 & 6 & 6 \\ 0 & 3 & 3 & 6 \end{pmatrix} \sim \begin{pmatrix} 3 & 0 & 0 & 0 \\ 0 & 3 & 0 & 0 \\ 0 & 0 & 3 & 0 \\ 0 & 0 & 0 & 3 \end{pmatrix}$$

and $|\mathcal{M}_{L,L^*}| = 3^4 = d^2$. Hence Theorem 2 implies $A_\psi \simeq \mathcal{M}_{L,L^*} \simeq \mathbb{Z}/3\mathbb{Z} \oplus \mathbb{Z}/3\mathbb{Z} \oplus \mathbb{Z}/3\mathbb{Z} \oplus \mathbb{Z}/3\mathbb{Z}$ as abelian group. We see $A_\psi \simeq \mathcal{O}_\psi/3\mathcal{O}_\psi \oplus \mathcal{O}_\psi/3\mathcal{O}_\psi$ as $\mathcal{O}_\psi$-module.

We show other examples in Table 2.

Table 2. $p = 3$

m	$A_{K,\chi}$	$A_{K,\chi\omega}$	$A_{K,\chi\omega^2}$	$A_{K,\chi\omega}$ as $\mathcal{O} = \mathcal{O}_{\chi\omega}$-module
853	0	$(3^2, 3^2)$	0	$\mathcal{O}/3^2\mathcal{O}$
9546	0	$(3^3, 3^3)$	0	$\mathcal{O}/3^3\mathcal{O}$
11703	0	$(3^3, 3^3)$	0	$\mathcal{O}/3^3\mathcal{O}$
13767	(3)	$(3^3, 3^3)$	(3)	$\mathcal{O}/3^3\mathcal{O}$
13894	0	$(3^3, 3^3)$	(3^2)	$\mathcal{O}/3^3\mathcal{O}$
15338	0	$(3, 3, 3, 3)$	(3)	$\mathcal{O}/3\mathcal{O} \oplus \mathcal{O}/3\mathcal{O}$

5.3 Cyclotomic Cases

When the conductor f of K is small, we can use the condition $\ell \equiv 1 \pmod{f}$ in place of (5). In exchange for a growth of the size of ℓ, we get an advantage of verifying (6), (7) and (11) without using an integral basis of K. This situation often happens in cyclotomic cases.

There is an extensive work of Schoof [10] dealing with the order h_ℓ^+ of the ideal class group $\mathbb{Q}(\zeta_\ell + \zeta_\ell^{-1})$ for a prime number ℓ. He introduced the number $\tilde{h}_\ell^+$, which is the order of a certain group and a divisor of h_ℓ^+, and calculated $\tilde{h}_\ell^+$ for $\ell < 10000$. His heuristic arguments lead to the assertion that $\tilde{h}_\ell^+$ coincides with h_ℓ^+ in this range with 98% probability. The whole class number h_ℓ^+ is very hard to compute and $\tilde{h}_\ell^+$ is also. On the other hand, the p-part of h_ℓ^+ is easy. For example, one can verify immediately that h_{167}^+ is not divisible by 3 using Lemma 1 with $n = 0$ and $\ell = 14029$ (cf. [10, p. 914]). It is also easy to verify that the non-trivial p-part of $\tilde{h}_\ell^+$ is the p-part of h_ℓ^+ for any prime number $\ell < 10000$ and any odd prime number p such that $\ell \not\equiv 1 \pmod{p}$ combining a lower bound obtained by Schoof and an upper bound obtained by Lemma 1. The following shows an another approach using Theorem 2.

Example 5. Let $p = 3$ and $F = \mathbb{Q}(\zeta_{521} + \zeta_{521}^{-1})$. We are interested in the 3-part A_F of Cl_F. Let K be the subfield of F with $[K : \mathbb{Q}] = 26$. Fix a primitive 26-th root of unity ζ_{26} in $\overline{\mathbb{Q}}_3$ such that $\zeta_{26} + \zeta_{26}^3 + \zeta_{26}^9 \equiv 4 \pmod{9}$ and define the character $\psi : \mathrm{Gal}(K/\mathbb{Q}) \to \mathbb{Q}_3(\zeta_{26})^\times$ by $\psi(\sigma) = \zeta_{26}$, where σ is the generator of $\mathrm{Gal}(K/\mathbb{Q})$ induced by $\zeta_{521} \mapsto \zeta_{521}^3$. Then, applying Theorem 2 with $L = \{18757\}$ and $L^* = \{\, 9322866739,\ 51363718633,\ 98153955469 \,\}$ (this is the first pair satisfying the conditions for L and L^*), we see $A_{K,\psi} \simeq \mathbb{Z}/3\mathbb{Z} \oplus \mathbb{Z}/3\mathbb{Z} \oplus \mathbb{Z}/3\mathbb{Z}$. For other subfields K' and injective characters ψ' which are not conjugate to ψ over $\mathbb{Q}_3$, Lemma 1 implies $A_{K',\psi'} = 0$. Hence we have $A_F \simeq \mathbb{Z}/3\mathbb{Z} \oplus \mathbb{Z}/3\mathbb{Z} \oplus \mathbb{Z}/3\mathbb{Z}$.

6 Running Time of the Algorithm

We have not yet succeeded in giving explicit upper bounds for L and L^* in the proof of Theorem 1 because we have no idea to evaluate the discriminant and the degree of the maximal unramified abelian p-extension of $\mathbb{Q}(\zeta_f)$, which are needed to apply an effective version of Chebotarev density theorem (cf. [1]). So we can not estimate the running time of the algorithm. The time of computing $\mathcal{M}_{L,L^*}$ essentially depends on the size of ℓ in L and little depends on the size of ℓ^* in L^*. So the estimate for L is important.

Finally we argue how big fields K can be handled by Theorem 2. There are three important objects to be calculated, namely e_χ, $\tau_{\mathcal{L}}$ and $\xi_\ell^{D_\ell}$. The complexity depends on the conductor f and the degree n of K. The calculation of e_χ becomes hard when n is large because we have to calculate $\mathrm{Tr}_{\mathbb{Q}_p(\zeta_n)/\mathbb{Q}_p}(\zeta_n)$. The limit of n may be 10^3 as long as we rely on a brute force method of factoring the n-th cyclotomic polynomial over $\mathbb{Z}_p$. The calculating time of $\tau_{\mathcal{L}}$ and $\xi_\ell^{D_\ell}$ depends only on f, not on n.

For the minus part, the experiments suggest that we will compute $\tau_{\mathcal{L}}$ within a hour if $f < 10^6$.

For the plus part, we need an integral basis of K to verify (6), (7) and (11) as explained in 4.2. The calculating time of an integral basis increases as n increases. An actual limit of n may be 10^2 except for cyclotomic fields of small conductors as in 5.3. The experiments also suggest the limit of f for $\xi_\ell^{D_\ell}$ is about 10^6.

References

1. E. Bach and J. Sorenson, Explicit bounds for primes in residue classes, Math. Comp. **65** (1996), 1717–1735.
2. P. Cornacchia, The 2-ideal class groups of $\mathbb{Q}(\zeta_\ell)$, Nagoya Math. J. **162** (2001), 1–18.
3. R. Greenberg, On p-adic L-functions and cyclotomic fields II, Nagoya Math. J. **67** (1977), 139–158.
4. V. Kolyvagin, Euler systems, in The Grothendieck Festschrift II, Prog. Math. vol. 87 (1990), 435–483.
5. S. Lang, Cyclotomic Fields I and II, Graduate Texts in Mathematics, vol. 121, Springer-Verlag, Berlin/New York, 1990
6. B. Mazur and A. Wiles, Class fields of abelian extensions of $\mathbb{Q}$, Invent. math. **76** (1984), 179–330.
7. K. Rubin, The main conjecture, Appendix to Lang [5].
8. K. Rubin, Kolyvagin's system of Gauss sums, in Arithmetic Algebraic Geometry, Prog. Math. vol. 89, Birkhauser Boston, 1991, 309–324.
9. R. Schoof, Minus class groups of the fields of the ℓ-th roots of unity, Math. Comp. **67** (1998), 1225–1245.
10. R. Schoof, Class numbers of real cyclotomic fields of prime conductor, Math. Comp. **72** (2003), 913–937.
11. H. Sumida, Computation of Iwasawa invariants of certain real abelian fields, J. Number Theory **105** (2004), 235–250.
12. F. Thaine, On the ideal class groups of real abelian number fields, Ann. Math. **128** (1988), 1–18.
13. L. Washington, Introduction to Cyclotomic Fields, 2nd ed., Graduate Texts in Mathematics, vol. 83, Springer-Verlag, Berlin/New York, 1997

Computation of Locally Free Class Groups

Werner Bley[1] and Robert Boltje[2,*]

[1] Universität Kassel, 34132 Kassel, Germany
bley@mathematik.uni-kassel.de
http://www.mathematik.uni-kassel.de/~bley
[2] University of California, Santa Cruz, CA 95064, USA
boltje@ucsc.edu
http://math.ucsc.edu/~boltje

Abstract. We show that the locally free class group of an order in a semisimple algebra over a number field is isomorphic to a certain ray class group. This description is then used to present an algorithm that computes the locally free class group. The algorithm is implemented in MAGMA for the case where the algebra is a group ring over the rational numbers.

Introduction

Throughout this paper we fix a number field K with ring of integers $\mathcal{O}_K$, a finite-dimensional semisimple K-algebra A and an $\mathcal{O}_K$-order $\mathcal{A}$ in A. The purpose of this paper is to give an algorithm that computes the locally free class group $\mathrm{cl}(\mathcal{A})$, cf. [4, (39.12)] for a definition. This was done in [1] in the case that A is commutative, where $\mathrm{cl}(\mathcal{A})$ is isomorphic to the Picard group $\mathrm{Pic}(\mathcal{A})$, cf. [4, (55.26)]. Here we treat the general case. As in [1] we can show that $\mathrm{cl}(\mathcal{A})$ is isomorphic to a quotient of a certain ray class group in the center of A, cf. Corollary 1.9. This is achieved in several steps. We choose a maximal order $\mathcal{M}$ in A containing $\mathcal{A}$, and a full ideal $\mathfrak{f}$ of $\mathcal{M}$ which is contained in $\mathcal{A}$. A canonical pull-back diagram involving $\mathcal{A}$ and $\mathcal{A}/\mathfrak{f}$ gives rise to a Mayer-Vietoris sequence and the induced exact sequence

$$ K_1(\mathcal{A}/\mathfrak{f}) \xrightarrow{\ \partial\ } \mathrm{cl}(\mathcal{A},\mathfrak{f}) \longrightarrow \mathrm{cl}(\mathcal{A}) \longrightarrow 0 \,, $$

cf. (5), with a term $\mathrm{cl}(\mathcal{A},\mathfrak{f})$ coming directly from the Mayer-Vietoris sequence. In Theorem 1.5 we use Wilson's idèle theoretic description of locally free class groups, cf. [14], in order to show that the term $\mathrm{cl}(\mathcal{A},\mathfrak{f})$ is isomorphic to a ray class group.

In the proof of this theorem we make repeatedly use of Theorem 2.2 which is of independent interest. It determines the image under the reduced norm map of higher principal unit groups in the maximal order of a division algebra over a p-adic field.

---

* Research supported by the NSF, DMS-0200592 and 0128969.

F. Hess, S. Pauli, and M. Pohst (Eds.): ANTS 2006, LNCS 4076, pp. 72–86, 2006.

In the last section we present an algorithm that computes $\mathrm{cl}(\mathcal{A})$ using the ray class group description of $\mathrm{cl}(\mathcal{A}, \mathfrak{f})$ in the case of group algebras $A = KG$, where G denotes a finite group. A maximal order $\mathcal{M}$ containing $\mathcal{A}$ can be computed using an algorithm of Friedrichs, cf. [7]. We then describe how one can compute an ideal $\mathfrak{f}$, the relevant ray class group, generators of $K_1(\mathcal{A}/\mathfrak{f})$, and the map ∂ which turns out to be a reduced norm map. We also show how our approach can be used to compute the kernel group $D(\mathcal{A}) := \ker(\mathrm{cl}(\mathcal{A}) \to \mathrm{cl}(\mathcal{M}))$. It is well known that $D(\mathcal{A})$ does not depend on the choice of the maximal order $\mathcal{M}$.

This algorithm is implemented in MAGMA for group algebras $A = \mathbb{Q}G$ over the rational numbers. The program and tables of locally free class groups of integral grouprings $\mathbb{Z}[G]$ for many small groups G are available at http://www.mathematik.uni-kassel.de/~bley/pub.html.

1 Locally Free Class Groups in Terms of Ray Class Groups

Throughout this paper we fix the following notation:

1.1 Notation. Let $\mathcal{O}_K \subset K$ and $\mathfrak{f} \subseteq \mathcal{A} \subseteq \mathcal{M} \subset A$ be as in the introduction. We set $\overline{\mathcal{M}} := \mathcal{M}/\mathfrak{f}$ and $\overline{\mathcal{A}} := \mathcal{A}/\mathfrak{f}$ so that $\overline{\mathcal{A}} \subseteq \overline{\mathcal{M}}$ are finite rings. The canonical map $\mathcal{M} \to \overline{\mathcal{M}}$ will be denoted by $m \mapsto \overline{m}$. We will denote the center of a ring R by $Z(R)$. We set $C := Z(A)$ and denote by $\mathcal{O}_C$ the integral closure of $\mathcal{O}_K$ in C. The primitive idempotents of C will be denoted by $e_1, \ldots, e_r$. For $i = 1, \ldots, r$, we set $A_i := Ae_i$. Then

$$A = A_1 \oplus \cdots \oplus A_r \tag{1}$$

is a decomposition into the indecomposable ideals A_i of A. Each A_i is a K-algebra with identity element e_i. By Wedderburn's Theorem, the centers $K_i := Z(A_i)$ are finite field extensions of K via $K \to K_i$, $\alpha \mapsto \alpha e_i$, and we have K-algebra isomorphisms $A_i \cong \mathrm{Mat}_{n_i}(D_i)$ for each $i = 1, \ldots, r$, where D_i is a division ring with $Z(D_i) \cong K_i$. The Wedderburn decomposition (1) induces decompositions

$$C = K_1 \oplus \cdots \oplus K_r \tag{2}$$

and

$$\mathcal{O}_C = \mathcal{O}_{K_1} \oplus \cdots \oplus \mathcal{O}_{K_r}, \tag{3}$$

where $\mathcal{O}_{K_i}$ denotes the ring of algebraic integers of K_i for $i = 1, \ldots, r$. Since $\mathcal{M}$ is a maximal $\mathcal{O}_K$-order of A, it contains the central idempotents e_i and decomposes into $\mathcal{M} = \mathcal{M}_1 \oplus \cdots \oplus \mathcal{M}_r$ with $\mathcal{M}_i := \mathcal{M}e_i$. As a consequence, the ideal $\mathfrak{f}$ of $\mathcal{M}$ also decomposes into $\mathfrak{f} = \mathfrak{f}_1 \oplus \cdots \oplus \mathfrak{f}_r$ with ideals $\mathfrak{f}_i = \mathfrak{f}e_i$ of $\mathcal{M}_i$.

1.2. We consider the $\mathcal{O}_K$-order

$$\mathcal{D} := \mathcal{D}(\mathcal{A}, \mathfrak{f}) := \{(a_1, a_2) \in \mathcal{A} \times \mathcal{A} \mid a_1 \equiv a_2 \mod \mathfrak{f}\}$$

in $A \times A$ which fits into the pull-back diagram

$$\begin{array}{ccc} \mathcal{D} & \xrightarrow{q_1} & \mathcal{A} \\ {\scriptstyle q_2}\downarrow & & \downarrow{\scriptstyle p_1} \\ \mathcal{A} & \xrightarrow{p_2} & \overline{\mathcal{A}} \end{array}$$

of rings. By [4, Theorem 49.27], this leads to an exact sequence

$$K_1(\mathcal{A}) \times K_1(\mathcal{A}) \xrightarrow{p_1/p_2} K_1(\overline{\mathcal{A}}) \xrightarrow{\partial} \mathrm{cl}(\mathcal{D}) \xrightarrow{(q_1,\,q_2)} \mathrm{cl}(\mathcal{A}) \times \mathrm{cl}(\mathcal{A}) \longrightarrow 0 \,.$$

Here, the first map is given by $(x,y) \mapsto (K_1(p_1))(x) \cdot (K_1(p_2))(y)^{-1}$ and the second map is defined as follows. Every element of $K_1(\overline{\mathcal{A}})$ is represented by an element $u \in \overline{\mathcal{A}}^\times$ and such an element is mapped to the class of the locally free $\mathcal{D}$-module

$$M(u) := \{(a_1, a_2) \in \mathcal{A} \times \mathcal{A} \mid \overline{a}_1 \cdot u = \overline{a}_2\} \,, \tag{4}$$

cf. the proof of [4, Theorem 49.27]. We set

$$\mathrm{cl}(\mathcal{A}, \mathfrak{f}) := \ker\big(q_2 \colon \mathrm{cl}(\mathcal{D}) \to \mathrm{cl}(\mathcal{A})\big)$$

and obtain a short exact sequence

$$K_1(\mathcal{A}) \xrightarrow{p_1} K_1(\overline{\mathcal{A}}) \xrightarrow{\partial} \mathrm{cl}(\mathcal{A}, \mathfrak{f}) \xrightarrow{q_1} \mathrm{cl}(\mathcal{A}) \longrightarrow 0 \,, \tag{5}$$

as can be easily verified.

1.3. In the following $\mathfrak{p}$ will usually stand for a maximal ideal of $\mathcal{O}_K$. For an $\mathcal{O}_K$-module M we write $M_\mathfrak{p}$ for the completion at $\mathfrak{p}$. We let

$$J(A) := \{(a_\mathfrak{p})_\mathfrak{p} \in \prod_\mathfrak{p} A_\mathfrak{p}^\times \mid a_\mathfrak{p} \in \mathcal{A}_\mathfrak{p}^\times \text{ for almost all } \mathfrak{p}\}$$

denote the idèles of A and write $U(\mathcal{A}) = \prod_\mathfrak{p} \mathcal{A}_\mathfrak{p}^\times$ for the subgroup of unit idèles. Here $\mathfrak{p}$ runs through all maximal ideals of $\mathcal{O}_K$. One has canonical isomorphisms

$$A_\mathfrak{p} \cong K_\mathfrak{p} \otimes_K A \cong \bigoplus_{i=1}^{r} K_\mathfrak{p} \otimes_K A_i \cong \bigoplus_{i=1}^{r} K_\mathfrak{p} \otimes_K K_i \otimes_{K_i} A_i$$

$$\cong \bigoplus_{i=1}^{r} \bigoplus_{\mathfrak{P}} (K_i)_\mathfrak{P} \otimes_{K_i} A_i \cong \bigoplus_{i,\mathfrak{P}} A_{i,\mathfrak{P}} \tag{6}$$

involving various completions, where, for given $i \in \{1,\ldots,r\}$, $\mathfrak{P}$ runs through all maximal ideals of $\mathcal{O}_{K_i}$ dividing $\mathfrak{p}$ and $A_{i,\mathfrak{P}}$ is defined as $(A_i)_\mathfrak{P}$. Using the above isomorphism, we will often interpret elements of $J(A)$, resp. $A_\mathfrak{p}$, as tuples $(a_{i,\mathfrak{P}})_{i,\mathfrak{P}}$, where $\mathfrak{P}$ ranges over all maximal ideals of $\mathcal{O}_{K_i}$, resp. over those that contain $\mathfrak{p}$. Note that $J(A)$ does not depend on the order $\mathcal{A}$. Similarly we denote by $J(C)$ the group of idèles of C. Again one has a canonical isomorphism

$$C_{\mathfrak{p}} \cong \bigoplus_{i,\mathfrak{P}} K_{i,\mathfrak{P}}$$

and we will interpret elements of $J(C)$, resp. $C_{\mathfrak{p}}$, often as tuples $(\alpha_{i,\mathfrak{P}})_{i,\mathfrak{P}}$.

By nr: $J(A) \rightarrow J(C)$ we denote the reduced norm map (which translates into the component-wise reduced norm maps nr: $A_{i,\mathfrak{P}}^{\times} \rightarrow K_{i,\mathfrak{P}}^{\times}$ after the above identifications). We recall from [4, Proposition 45.8 and Theorem 7.48] that

$$\mathrm{nr}(J(A)) = J(C) \quad \text{and} \quad \mathrm{nr}(A^{\times}) = C^{\times+},$$

where

$$C^{\times+} := \{c \in C^{\times} \mid c \text{ is positive at quaternionic components}\}.$$

The last condition means that if $c = (c_i)$ with $c_i \in K_i^{\times}$ then $\tau(c_i) > 0$ whenever $i \in \{1, \ldots, r\}$ and $\tau \colon K_i \rightarrow \mathbb{R}$ is a real embedding such that the corresponding scalar extension $A_i \otimes_{K_i, \tau} \mathbb{R}$ is a full matrix ring over the quaternions. Furthermore, we define the subgroup

$$U_{\mathfrak{f}}(A) := \{(a_{\mathfrak{p}})_{\mathfrak{p}} \in U(A) \mid a_{\mathfrak{p}} \equiv 1 \mod \mathfrak{f}_{\mathfrak{p}}\}$$

of $U(A)$.

1.4. Next we define commutative invariants in C of the non-commutative data A, $\mathcal{A}$ and $\mathfrak{f}$. With $\mathfrak{f}$ also the ideal $\mathfrak{g} := \mathfrak{f} \cap C$ of $\mathcal{O}_C$ decomposes into $\mathfrak{g} = \mathfrak{g}_1 \oplus \cdots \oplus \mathfrak{g}_r$ with ideals $\mathfrak{g}_i = \mathfrak{g}e_i$ of $\mathcal{O}_{K_i}$. We denote by $I_{\mathfrak{g}} = I_{\mathfrak{g}}(C)$ the group of fractional $\mathcal{O}_C$-ideals of C that are coprime to $\mathfrak{g}$ and have

$$I_{\mathfrak{g}}(C) = I_{\mathfrak{g}_1}(K_1) \times \cdots \times I_{\mathfrak{g}_r}(K_r).$$

For each $i \in \{1, \ldots, r\}$ we write ∞_i for the formal product over real archimedian places $\tau \colon K_i \rightarrow \mathbb{R}$ such that $A \otimes_{K_i, \tau} \mathbb{R}$ is a full matrix ring over the quaternions, and we define the 'ray modulo $\mathfrak{g}\infty$' by

$$P_{\mathfrak{g}}^{+} := \{(\alpha_i \mathcal{O}_{K_i})_i \in I_{\mathfrak{g}} \mid \alpha_i \equiv 1 \mod^{\times} \mathfrak{g}_i \infty_i, \text{ for all } i = 1, \ldots, r\}.$$

Note that $P_{\mathfrak{g}}^{+}$ is a subgroup of $I_{\mathfrak{g}}$.

The next theorem gives both an idèle and ideal theoretic description of $\mathrm{cl}(\mathcal{A}, \mathfrak{f})$. Note that the ideal theoretic part only involves 'commutative data' located in the center C of A.

1.5 Theorem. *There are canonical isomorphisms*

$$\mathrm{cl}(\mathcal{A}, \mathfrak{f}) \cong J(C)/(C^{\times+}\mathrm{nr}(U_{\mathfrak{f}}(\mathcal{A}))) \cong I_{\mathfrak{g}}/P_{\mathfrak{g}}^{+}.$$

1.6 Remark. It is immediate from the above theorem that $\mathrm{cl}(\mathcal{B}, \mathfrak{f}) \cong \mathrm{cl}(\mathcal{M}, \mathfrak{f})$ for any $\mathcal{O}_K$-order $\mathcal{B}$ such that $\mathfrak{f} \subseteq \mathcal{B} \subseteq \mathcal{M}$. We also note that $U_{\mathfrak{f}}(\mathcal{B}) = U_{\mathfrak{f}}(\mathcal{M})$. In fact, if $b_{\mathfrak{p}} \in \mathcal{M}_{\mathfrak{p}}^{\times}$ and $b_{\mathfrak{p}} \equiv 1 \mod \mathfrak{f}_{\mathfrak{p}}$, then $b_{\mathfrak{p}}^{-1} \equiv 1 \mod \mathfrak{f}_{\mathfrak{p}}$, and since $\mathfrak{f}_{\mathfrak{p}} \subseteq \mathcal{B}_{\mathfrak{p}}$, both $b_{\mathfrak{p}}$ and $b_{\mathfrak{p}}^{-1}$ are elements of $\mathcal{B}_{\mathfrak{p}}$.

Proof. (of Theorem 1.5.)   From [14, Theorem 1] we obtain natural isomorphisms

$$w : \mathrm{cl}(\mathcal{D}) \cong \frac{J(C) \times J(C)}{(C^{\times +} \times C^{\times +})\mathrm{nr}(U(\mathcal{D}))} \tag{7}$$

and

$$\mathrm{cl}(\mathcal{A}) \cong \frac{J(C)}{(C^{\times +})\mathrm{nr}(U(\mathcal{A}))} .$$

Inspecting the definition of these isomorphisms one verifies easily that the map induced by q_2 is translated into a map between the right hand sides of the above equations that is induced by the projection map $J(C) \times J(C) \to J(C)$ onto the second component. Therefore, we obtain an isomorphism

$$\mathrm{cl}(\mathcal{A}, \mathfrak{f}) \cong \frac{J(C) \times C^{\times +}\mathrm{nr}(U(\mathcal{A}))}{(C^{\times +} \times C^{\times +})\mathrm{nr}(U(\mathcal{D}))} .$$

We will show that the map

$$\frac{J(C) \times C^{\times +}\mathrm{nr}(U(\mathcal{A}))}{(C^{\times +} \times C^{\times +})\mathrm{nr}(U(\mathcal{D}))} \xrightarrow{\ \sigma\ } \frac{J(C)}{C^{\times +}\mathrm{nr}(U_{\mathfrak{f}}(\mathcal{A}))} \tag{8}$$

induced by

$$((a_{\mathfrak{p}})_{\mathfrak{p}}, (b_{\mathfrak{p}})_{\mathfrak{p}}) \mapsto (a_{\mathfrak{p}}/b_{\mathfrak{p}})_{\mathfrak{p}} \in J(C)$$

for $((a_{\mathfrak{p}})_{\mathfrak{p}}, (b_{\mathfrak{p}})_{\mathfrak{p}}) \in J(C) \times C^{\times +}\mathrm{nr}(U(\mathcal{A}))$ is an isomorphism. The map is obviously well-defined. Let τ denote the map in the inverse direction induced by

$$J(C) \ni (x_{\mathfrak{p}})_{\mathfrak{p}} \mapsto ((x_{\mathfrak{p}})_{\mathfrak{p}}, (1)_{\mathfrak{p}}) \in J(C) \times C^{\times +}\mathrm{nr}(U(\mathcal{A})) .$$

Again it is straightforward to verify that τ is well-defined. Obviously $\sigma \circ \tau = id$. In order to show that $\tau \circ \sigma = id$ we have to prove that

$$((a_{\mathfrak{p}})_{\mathfrak{p}}, (b_{\mathfrak{p}})_{\mathfrak{p}}) \equiv ((a_{\mathfrak{p}}/b_{\mathfrak{p}})_{\mathfrak{p}}, (1)_{\mathfrak{p}}) \mod (C^{\times +} \times C^{\times +})\mathrm{nr}(U(\mathcal{D})) .$$

But this is equivalent to the statement

$$((b_{\mathfrak{p}})_{\mathfrak{p}}, (b_{\mathfrak{p}})_{\mathfrak{p}}) \in (C^{\times +} \times C^{\times +})\mathrm{nr}(U(\mathcal{D})),$$

which is immediate from $(b_{\mathfrak{p}})_{\mathfrak{p}} \in C^{\times +}\mathrm{nr}(U(\mathcal{A}))$. This concludes the proof of $\mathrm{cl}(\mathcal{A}, \mathfrak{f}) \cong J(C)/(C^{\times +}\mathrm{nr}(U_{\mathfrak{f}}(\mathcal{A})))$.

Next we will show that $J(C)/(C^{\times +}\mathrm{nr}(U_{\mathfrak{f}}(\mathcal{A}))) \cong I_{\mathfrak{g}}/P_{\mathfrak{g}}^{+}$. First note that $U_{\mathfrak{f}}(\mathcal{A}) = U_{\mathfrak{f}}(\mathcal{M})$, by Remark 1.6. Therefore, $J(C)/(C^{\times +}\mathrm{nr}(U_{\mathfrak{f}}(\mathcal{A})))$ breaks up into a direct product of components, one for each $i = 1, \ldots, r$. So obviously does $I_{\mathfrak{g}}/P_{\mathfrak{g}}^{+}$, and we may work component-wise. So we fix $i \in \{1, \ldots, r\}$, set $L := K_i$

and rename $\mathfrak{g}_i$ by $\mathfrak{g}$, an ideal of $\mathcal{O}_L$, $\mathfrak{f}_i$ by $\mathfrak{f}$ and $\mathfrak{M}_i$ by $\mathfrak{M}$ for the remainder of the proof. Then it suffices to show that

$$J(L)/L^{\times+}\mathrm{nr}(U_{\mathfrak{f}}(\mathfrak{M})) \cong I_{\mathfrak{g}}/P_{\mathfrak{g}}^+ .$$

We define a map

$$\psi_0 \colon J(L) \to I_{\mathfrak{g}}/P_{\mathfrak{g}}^+ .$$

as follows. For $\alpha = (\alpha_{\mathfrak{P}}) \in J(L)$, we apply the approximation theorem to choose an element $\beta \in L^{\times+}$ with

$$v_{\mathfrak{P}}(\alpha_{\mathfrak{P}}\beta - 1) \geqslant v_{\mathfrak{P}}(\mathfrak{g}) \quad \text{for all } \mathfrak{P} \text{ with } \mathfrak{P} \mid \mathfrak{g}$$

and set

$$\psi_0(\alpha) := \left(\prod_{\mathfrak{P}} \mathfrak{P}^{v_{\mathfrak{P}}(\alpha_{\mathfrak{P}}\beta)} \right) \cdot P_{\mathfrak{g}}^+ .$$

Then, by [10, Proposition IV.8.1] and its proof, ψ_0 induces an isomorphism

$$\psi \colon J(L)/L^{\times+} \cdot U_{\mathfrak{g}}(\mathcal{O}_L) \to I_{\mathfrak{g}}/P_{\mathfrak{g}}^+ , \tag{9}$$

where

$$U_{\mathfrak{g}}(\mathcal{O}_L) := \{(\alpha_{\mathfrak{P}}) \in \prod_{\mathfrak{P}}(\mathcal{O}_L)_{\mathfrak{P}}^{\times} \mid \alpha_{\mathfrak{P}} \equiv 1 \quad \mathrm{mod}^{\times} \mathfrak{g}_{\mathfrak{P}} \text{ for all } \mathfrak{P} \mid \mathfrak{g}\} .$$

But, by Corollary 2.4, we have $U_{\mathfrak{g}}(\mathcal{O}_L) = \mathrm{nr}(U_{\mathfrak{f}}(\mathfrak{M}))$ and the proof is complete.

$\square$

1.7. Using the isomorphisms in Theorem 1.5, the short exact sequence (5) yields a short exact sequence

$$K_1(\overline{A}) \xrightarrow{\hat{\partial}} I_{\mathfrak{g}}/P_{\mathfrak{g}}^+ \xrightarrow{q_2} \mathrm{cl}(A) \longrightarrow 0 . \tag{10}$$

Since $\overline{A}$ is a semilocal ring, the canonical map $\pi \colon \overline{A}^{\times} \to K_1(\overline{A})$ is surjective (cf. [4, Theorem 40.31]) and the image of $\hat{\partial}$ in (10) is equal to the image of the composition

$$\nu \colon \overline{A}^{\times} \xrightarrow{\pi} K_1(\overline{A}) \xrightarrow{\hat{\partial}} I_{\mathfrak{g}}/P_{\mathfrak{g}}^+ . \tag{11}$$

The map ν is given explicitly by the next proposition.

1.8 Proposition. *Let $x \in \overline{A}^{\times}$ and let $a \in A$ such that $\overline{a} = x$. Then $\nu(x)$ is equal to the class of the ideal $\mathrm{nr}(a)\mathcal{O}_C \in I_{\mathfrak{g}}$ in $I_{\mathfrak{g}}/P_{\mathfrak{g}}^+$.*

Proof. The map ν is the composite

$$\overline{A}^{\times} \xrightarrow{\partial \circ \pi} \mathrm{cl}(\mathcal{D}) \xrightarrow{\omega} \frac{J(C) \times J(C)}{(C^{\times+} \times C^{\times+})\mathrm{nr}(U(\mathcal{D}))} \xrightarrow{\psi \circ \sigma} I_{\mathfrak{g}}/P_{\mathfrak{g}}^+ ,$$

where ∂ originates from the Mayer-Vietoris sequence and ω, σ, ψ are defined in (7), (8), (9), respectively.

By the theory of Mayer-Vietoris sequences, cf. proof of [4, Theorem (49.27)], we have $(\partial \circ \pi)(\bar{a}) = M(\bar{a})$, cf. (4). In order to describe the image of the class of $M(\bar{a})$ under ω we have to find a $\mathcal{D}_{\mathfrak{p}}$-basis $\lambda_{\mathfrak{p}}$ of $M(\bar{a})_{\mathfrak{p}}$ for each $\mathfrak{p}$. For each $\mathfrak{p}$, the ring $\mathcal{A}_{\mathfrak{p}}$ is semilocal, so that we may choose $a_{\mathfrak{p}} \in \mathcal{A}_{\mathfrak{p}}^{\times}$ satisfying $a_{\mathfrak{p}} \equiv a$ mod $\mathfrak{f}_{\mathfrak{p}}$, cf. [4, Lemma 50.7]. One easily shows that one can choose

$$\lambda_{\mathfrak{p}} = \begin{cases} (1,1), & \text{if } \mathfrak{f}_{\mathfrak{p}} = \mathcal{M}_{\mathfrak{p}}, \\ (1, a_{\mathfrak{p}}), & \text{if } \mathfrak{f}_{\mathfrak{p}} \neq \mathcal{M}_{\mathfrak{p}}. \end{cases}$$

By the definition of ω, the image of the class of $M(\bar{a})$ is therefore represented by $(y_{\mathfrak{p}}, z_{\mathfrak{p}}) \in J(C) \times J(C)$ with

$$(y_{\mathfrak{p}}, z_{\mathfrak{p}}) = \begin{cases} (1,1), & \text{if } \mathfrak{f}_{\mathfrak{p}} = \mathcal{M}_{\mathfrak{p}}, \\ (1, \mathrm{nr}(a_{\mathfrak{p}})), & \text{if } \mathfrak{f}_{\mathfrak{p}} \neq \mathcal{M}_{\mathfrak{p}}. \end{cases}$$

It follows from the definition of σ together with Corollary 2.4, that we may choose $\beta = \mathrm{nr}(a)$ in the definition of ψ. The result now follows easily. □

The following corollary is now immediate.

1.9 Corollary. *If $a_1, \ldots, a_s$ are elements in $\mathcal{A}$ such that $\pi(\bar{a}_1), \ldots, \pi(\bar{a}_s)$ are generators of $K_1(\bar{\mathcal{A}})$, and if U is the subgroup of $I_{\mathfrak{g}}/P_{\mathfrak{g}}^+$ generated by the classes of the ideals $\mathrm{nr}(a_j)\mathcal{O}_C$, $j = 1, \ldots, s$, then there exists an isomorphism*

$$\mathrm{cl}(\mathcal{A}) \cong \left(I_{\mathfrak{g}}/P_{\mathfrak{g}}^+ \right)/U.$$

2   A Local Result

The aim of this section is to provide Corollary 2.4 which was needed in the proof of Theorem 1.5.

2.1 Notation. Throughout this section we assume the following notation. We deviate from our general assumption in the introduction and assume (for this section only) that K is a finite extension field of the field $\mathbb{Q}_p$ of p-adic numbers. We write $\mathcal{O}_K$ or just $\mathcal{O}$ for its valuation ring, $\mathfrak{p}$ for its maximal ideal, and choose a prime element π (so that $\mathfrak{p} = \pi\mathcal{O}$).

Furthermore, we denote by D a division ring with $Z(D) = K$. We refer the reader to [11, Section 14] for standard results in this situation. One has $[D : K] = n^2$ for some $n \in \mathbb{N}$. We denote by Δ the maximal order of D, and by $\mathfrak{P}_D$ the unique maximal (two-sided) ideal of Δ. Every non-zero (two-sided) ideal of Δ is of the form $\mathfrak{P}_D^k$ for some $k \in \mathbb{N}_0$.

If $q := |\mathcal{O}/\mathfrak{p}|$, then we can choose a root of unity ω of order $q^n - 1$ in Δ. For given $\pi \in \mathcal{O}$ and ω there exist an element $\pi_D \in \mathfrak{P}_D$ and a natural number $r \in \{1, \ldots, n\}$ which is coprime to n such that

$$\mathfrak{P}_D = \pi_D \Delta = \Delta \pi_D , \quad \pi_D^n = \pi , \quad \pi_D \omega \pi_D^{-1} = \omega^{q^r} . \tag{12}$$

By nr: $D \to K$ we denote the reduced norm map.

2.2 Theorem. *For every $k \in \mathbb{N}_0$ and $t \in \{1, \dots, n\}$ one has*

$$\mathrm{nr}(1 + \mathfrak{P}_D^{kn+t}) = 1 + \mathfrak{p}^{k+1} .$$

Proof. The unramified extension $W := K(\omega)$ of K has degree n and is a splitting field for D. Moreover, by the paragraphs preceding [11, Theorem 14.5], the set $\{1, \pi_D, \dots, \pi_D^{n-1}\}$ is an $\mathcal{O}_W$-basis of Δ. We denote by θ the Galois automorphism of W over K with $\theta(\omega) = \omega^{q^r}$. Note that θ generates the Galois group $\mathrm{Gal}(W/K)$. As in the proof of [11, Theorem 14.6], we obtain an isomorphism $W \otimes_K D \to \mathrm{Mat}_n(W)$ of W-algebras such that

$$1 \otimes \pi_D \mapsto \begin{pmatrix} 0 & 1 & 0 & \cdots & 0 \\ 0 & 0 & 1 & \cdots & 0 \\ \vdots & & \ddots & \ddots & \vdots \\ 0 & \cdots & & 0 & 1 \\ \pi & 0 & \cdots & 0 & 0 \end{pmatrix}$$

and

$$1 \otimes w \mapsto \begin{pmatrix} w & & & \\ & \theta(w) & & \\ & & \ddots & \\ & & & \theta^{n-1}(w) \end{pmatrix}$$

for $w \in W$. If $\delta = a_0 + a_1 \pi_D + \dots + a_{n-1} \pi_D^{n-1}$, $a_i \in \mathcal{O}_W$, is an arbitrary element in Δ, then $1 \otimes \delta \mapsto A(\delta)$ with

$$A(\delta) = \begin{pmatrix} a_0 & a_1 & a_2 & \cdots & a_{n-1} \\ \theta(a_{n-1})\pi & \theta(a_0) & \theta(a_1) & \cdots & \theta(a_{n-2}) \\ \theta^2(a_{n-2})\pi & \theta^2(a_{n-1})\pi & \theta^2(a_0) & \cdots & \theta^2(a_{n-3}) \\ \vdots & \vdots & \vdots & & \vdots \\ \theta^{n-1}(a_1)\pi & \theta^{n-1}(a_2)\pi & \theta^{n-1}(a_3)\pi & \cdots & \theta^{n-1}(a_{n-1})\pi & \theta^{n-1}(a_0) \end{pmatrix}$$

An elementary computation using $\pi_D^n = \pi$ shows that

$$A(\delta \pi_D^{kn+t}) = \begin{pmatrix} a_{n-t}\pi^{k+1} & & * \cdot \pi^k \\ & \ddots & \\ * \cdot \pi^{k+2} & & \theta^{n-1}(a_{n-t})\pi^{k+1} \end{pmatrix}$$

for any $k \in \mathbb{N}_0$ and $t \in \{1, \dots, n\}$. There are always n consecutive diagonals involving a factor π^{k+1}, including the main diagonal. For $t = 1$ this block of n diagonals extends from the bottom left corner to the main diagonal. For $t = n$ it extends from the main diagonal to the top right corner. While t moves from 1 to n

this block moves diagonally from the lower left to the upper right corner. Every entry to the left and below this block is divisible by π^{k+2} and every entry to the right and above this block is divisible by π^k. Since $\mathrm{nr}(1+\delta) = \det(1+A(\delta))$, it follows immediately that

$$\mathrm{nr}(1 + \mathfrak{P}_D^{kn+t}) \subseteq 1 + \mathfrak{p}^{k+1},$$

for $k \in \mathbb{N}_0$ and $t \in \{1, \ldots, n\}$.

Next we will show that

$$\frac{1 + \mathfrak{P}_D^{kn+t}}{1 + \mathfrak{P}_D^{(k+1)n+t}} \xrightarrow{\ \mathrm{nr}\ } \frac{1 + \mathfrak{p}^{k+1}}{1 + \mathfrak{p}^{k+2}} \qquad (13)$$

is surjective for $k \geqslant 0$ and $t \in \{1, \ldots, n\}$. Without loss of generality we may assume $t = n$. For $b \in \mathcal{O}$ we have to find $\delta \in \Delta$ such that

$$\mathrm{nr}(1 + \delta\pi_D^{(k+1)n}) \equiv 1 + b\pi^{k+1} \mod 1 + \mathfrak{p}^{k+2}.$$

Since W/K is unramified, there exists an element $a \in \mathcal{O}_W$ such that $\mathrm{Tr}_{W/K}(a) = b$. Setting $\delta = a$ we obtain

$$\mathrm{nr}(1 + a\pi_D^{(k+1)n}) \equiv \det \begin{pmatrix} 1 + a\pi^{k+1} & & & * \\ 0 & 1 + \theta(a)\pi^{k+1} & & \\ & & \ddots & \\ 0 & & & 1 + \theta^{n-1}(a)\pi^{k+1} \end{pmatrix}$$
$$\equiv 1 + \mathrm{Tr}_{W/K}(a)\pi^{k+1} \equiv 1 + b\pi^{k+1} \mod \mathfrak{p}^{k+2}.$$

By induction on l it follows easily that

$$\mathrm{nr}_l : \frac{1 + \mathfrak{P}_D^{kn+t}}{1 + \mathfrak{P}_D^{(k+l)n+t}} \longrightarrow \frac{1 + \mathfrak{p}^{k+1}}{1 + \mathfrak{p}^{k+l+1}}$$

is surjective for all $k \geqslant 0$ and $l \geqslant 1$. Hence we have a short exact sequence of projective systems (indexed by l) of finite abelian groups

$$0 \to (\ker(\mathrm{nr}_l))_l \longrightarrow \left(\frac{1 + \mathfrak{P}_D^{kn+t}}{1 + \mathfrak{P}_D^{(k+l)n+t}} \right)_l \xrightarrow{\ \mathrm{nr}\ } \left(\frac{1 + \mathfrak{p}^{k+1}}{1 + \mathfrak{p}^{k+l+1}} \right)_l \to 0.$$

Since $\ker(\mathrm{nr}_l)$ is a finite abelian group for all l, it satisfies clearly the Mittag-Leffler condition, so that

$$\varprojlim_l \frac{1 + \mathfrak{P}_D^{kn+t}}{1 + \mathfrak{P}_D^{(k+l)n+t}} \xrightarrow{\ \mathrm{nr}\ } \varprojlim_l \frac{1 + \mathfrak{p}^{k+1}}{1 + \mathfrak{p}^{k+l+1}} \qquad (14)$$

is surjective by [8, Proposition II.9.1]. Since Δ, resp. $\mathcal{O}$, is complete relative to the $\mathfrak{P}_D$-adic, resp. $\mathfrak{p}$-adic, valuation, we derive from (14) immediately the assertion of the theorem. $\qquad\square$

2.3 Corollary. *Let $K/\mathbb{Q}_p$ be a finite field extension and let $A \cong \operatorname{Mat}_m(D)$ be a finite dimensional central simple K-algebra, where D is a division ring with $Z(D) = K$ and $[D : K] = n^2$. Furthermore, let $\mathfrak{M}$ be a maximal $\mathcal{O}$-order in A and let $\mathfrak{P}_{\mathfrak{M}}$ be the maximal ideal of $\mathfrak{M}$. Then*

$$\operatorname{nr}_{A/K}(1 + \mathfrak{P}_{\mathfrak{M}}^{kn+t}) = 1 + \mathfrak{p}^{k+1}$$

for all $k \in \mathbb{N}_0$ and all $t \in \{1, \ldots, n\}$.

Proof. This is an immediate consequence of Theorem 2.2, [11, Theorem 17.3] and the formula
$$\operatorname{nr}_{A/K}(X) = \operatorname{nr}_{D/K}(\operatorname{Ddet}(X))$$
for any $X \in \operatorname{GL}_m(D)$ (see [4, Equation (7.42)]). Here $\operatorname{Ddet}\colon \operatorname{GL}_m(D) \to D_{\mathrm{ab}}^{\times}$ denotes the Dieudonné determinant. $\qquad\square$

2.4 Corollary. *Let $K/\mathbb{Q}_p$ be a finite field extension and let A be a finite-dimensional central simple K-algebra. Furthermore, let $\mathfrak{M}$ be a maximal $\mathcal{O}_K$-order of A, let $\mathfrak{f}$ be a proper (two-sided) ideal of $\mathfrak{M}$ and set $\mathfrak{g} := K \cap \mathfrak{f}$. Then*

$$\operatorname{nr}_{A/K}(1 + \mathfrak{f}) = 1 + \mathfrak{g}.$$

Moreover, $\operatorname{nr}_{A/K}(\mathfrak{M}^{\times}) = \mathcal{O}_K^{\times}$.

Proof. The last statement is included for the sake of completeness and can be found in [4, Proposition 45.8]. In order to prove the first statement, we can assume that $A = \operatorname{Mat}_m(D)$ for some $m \in \mathbb{N}$ and some division ring D with $Z(D) = K$. Using the notation from 2.1 and applying [11, Theorem 17.3], we can also assume that $\mathfrak{M} = \operatorname{Mat}_m(\Delta)$ and that $\mathfrak{f} = \mathfrak{P}_{\mathfrak{M}}^{kn+t}$ for some $k \in \mathbb{N}_0$ and $t \in \{1, \ldots, n\}$ with $\mathfrak{P}_{\mathfrak{M}}$ denoting the maximal ideal of $\mathfrak{M}$. Now the first statement follows from Corollary 2.3 and from $\mathfrak{g} = \mathfrak{p}^{k+1}$. $\qquad\square$

3   An Algorithm to Compute cl($\mathcal{A}$) in the Group Algebra Case

In this section we present an algorithm which computes the locally free class group $\operatorname{cl}(\mathcal{A})$ of any $\mathcal{O}_K$-order $\mathcal{A}$ in the group algebra KG of a finite group G. Moreover, the algorithm computes the so-called kernel group $D(\mathcal{A})$, namely the kernel of the canonical map $\operatorname{cl}(\mathcal{A}) \to \operatorname{cl}(\mathfrak{M})$, where $\mathfrak{M}$ is a maximal $\mathcal{O}_K$-order of KG containing $\mathcal{A}$. It is well-known that $D(\mathcal{A})$ does not depend on the choice of $\mathfrak{M}$. The algorithm presented here has been implemented in Magma, cf. [9], however only for $K = \mathbb{Q}$. We expect it to be straightforward to extend it to arbitrary K.

3.1. *Input:* We assume that we are given a number field K, its ring of integers $R := \mathcal{O}_K$, a finite group G and an R-order $\mathcal{A}$ of KG.

3.2. *Computation of $I_{\mathfrak{g}}/P_{\mathfrak{g}}^+$:*

(a) Compute the order n of G, the exponent e of G, and the character table of G. The character table comes with a set $\mathcal{C}$ of representatives of conjugacy classes of G.

(b) Define $L := K(\zeta)$, where $\zeta \in \mathbb{C}$ is a root of unity of order e and compute $\Omega := \mathrm{Gal}(L/K)$.

(c) Compute representatives $\chi_1, \ldots, \chi_r$ of the Ω-orbits of the set $\mathrm{Irr}(G)$ of irreducible characters of G. Denote by X_i the Ω-orbit of χ_i. Furthermore, for $i = 1, \ldots, r$, compute the number field $K_i := K(\{\chi_i(g) \mid g \in \mathcal{C}\})$ and its ring of integers $R_i := \mathcal{O}_{K_i}$.

(d) For $\chi \in \mathrm{Irr}(G)$ let $e_\chi := \frac{\chi(1)}{|G|} \sum_{g \in G} \chi(g^{-1})g$ be the corresponding primitive central idempotent of the group algebra LG. Then, $e_i := \sum_{\chi \in X_i} e_\chi$, $i = 1, \ldots, r$, are the primitive central idempotents of $A := KG$.

(e) Using the Round 2-algorithm described in [7, Kapitel 3 and 4] compute a maximal R-order $\mathcal{M}$ in A that contains $\mathcal{A}$.

(f) Applying [7, Algorithmus (2.16)] compute the left conductor $\mathfrak{c}_l := \{x \in A \mid x\mathcal{M} \subset \mathcal{A}\}$, the right conductor $\mathfrak{c}_r := \{x \in A \mid \mathcal{M}x \subset \mathcal{A}\}$ and $\mathfrak{f} := \mathfrak{c}_r \cdot \mathfrak{c}_l$. Then $\mathfrak{f}$ is an ideal of $\mathcal{M}$ that is contained in $\mathcal{A}$. Compute $\mathfrak{f}_i := \mathfrak{f} \cdot e_i$ for $i = 1, \ldots, r$.

(g) Compute the unique K-algebra map $K_i \to A_i$ with the property that $\chi_i(g) \mapsto \sum_{\chi \in X_i} \chi(g)e_\chi$. This map identifies K_i with $Z(A_i)$. Using this identification, compute the ideal $\mathfrak{g}_i := R_i \cap \mathfrak{f}_i$ of R_i for $i = 1, \ldots, r$.

(h) For $i = 1, \ldots, r$ compute the Frobenius-Schur indicator $c(\chi_i) := |G|^{-1} \sum_{g \in G} \chi_i(g^2)$ of χ_i. It is known that $c(\chi_i) \in \{-1, 0, 1\}$, cf. [12, Section 13.2]. Obviously, $c(\chi) = c(\chi_i)$ for every $\chi \in X_i$. A Galois automorphism $\tau \in \Omega$ is a real embedding such that $A_i \otimes_{K_i, \tau} \mathbb{R}$ is a matrix algebra over the quaternions if and only if the Schur-Frobenius indicator equals -1, cf. [12, Section 13.2]. This allows to compute ∞_i for $i = 1, \ldots, r$.

(i) Using Algorithm 4.3.1 of [3] component-wise compute the ray class group $I_{\mathfrak{g}}/P_{\mathfrak{g}}^+$.

3.3 Remark. For computational reasons we wish to choose $\mathfrak{f}$ as large as possible. For special orders $\mathcal{A}$ there may be better ways to compute an ideal $\mathfrak{f}$ than the one described in (f). For example, if $\mathcal{A} = \mathcal{O}_K G$ is the integral group ring, then $\mathfrak{c}_r = \mathfrak{c}_l$, cf. [4, Theorem (27.8)], so that we can take $\mathfrak{f} = \mathfrak{c}_r = \mathfrak{c}_l$. Then $\mathfrak{f}$ is the largest ideal of $\mathcal{M}$ that is contained in $\mathcal{A}$. We are grateful to the referee to point out that, in many cases, the ideal $\mathfrak{f} := \mathfrak{c}\mathcal{M}$ with $\mathfrak{c} := \mathfrak{c}_r \cap C = \mathfrak{c}_l \cap C = \{c \in C \mid c\mathcal{M} \subseteq \mathcal{A}\}$ is a better choice.

3.4. Before we turn to the algorithm for the computation of generators of $K_1(\overline{\mathcal{A}})$ we state two preparatory lemmas.

Let $\mathfrak{g} = \prod_{\mathfrak{P} \in \mathcal{P}} \mathfrak{P}^{e_\mathfrak{P}}$ be the prime ideal decomposition of $\mathfrak{g}$ in $\mathcal{O}_C$, and set $\mathcal{P}' := \{\mathfrak{P} \cap \mathcal{A} \mid \mathfrak{P} \in \mathcal{P}\}$, a set of prime ideals of $\mathcal{A} \cap \mathcal{O}_C$. For every $\mathfrak{p} \in \mathcal{P}'$ consider the ideal

$$\mathfrak{q} := \bigcap_{\substack{\mathfrak{P} \in \mathcal{P} \\ \mathfrak{P} \cap \mathcal{A} = \mathfrak{p}}} (\mathfrak{P}^{e_\mathfrak{P}} \cap \mathcal{A}).$$

These are precisely the factors in the primary decomposition of $\mathfrak{g}$, cf. [1, Prop. 3.2]. We write $\mathfrak{Q}$ for the set of ideals $\mathfrak{q}$.

3.5 Lemma. *Assume the above notations. Then one has*

$$\mathfrak{f} = \bigcap_{\mathfrak{q} \in \mathfrak{Q}} (\mathfrak{q}\mathcal{A} + \mathfrak{f}) = \prod_{\mathfrak{q} \in \mathfrak{Q}} (\mathfrak{q}\mathcal{A} + \mathfrak{f}).$$

Proof. For $\mathfrak{q}_1, \mathfrak{q}_2 \in \mathfrak{Q}$, $\mathfrak{q}_1 \neq \mathfrak{q}_2$, one has

$$(\mathfrak{q}_1\mathcal{A} + \mathfrak{f}) + (\mathfrak{q}_2\mathcal{A} + \mathfrak{f}) = \mathcal{A}, \quad (\mathfrak{q}_1\mathcal{A} + \mathfrak{f})(\mathfrak{q}_2\mathcal{A} + \mathfrak{f}) = (\mathfrak{q}_2\mathcal{A} + \mathfrak{f})(\mathfrak{q}_1\mathcal{A} + \mathfrak{f}).$$

It follows easily that $\bigcap_{\mathfrak{q} \in \mathfrak{Q}} (\mathfrak{q}\mathcal{A} + \mathfrak{f}) = \prod_{\mathfrak{q} \in \mathfrak{Q}} (\mathfrak{q}\mathcal{A} + \mathfrak{f})$. Since $\mathfrak{g} = \prod_{\mathfrak{q} \in \mathfrak{Q}} \mathfrak{q}$ we conclude $\mathfrak{f} \subseteq \bigcap_{\mathfrak{q} \in \mathfrak{Q}} (\mathfrak{q}\mathcal{A} + \mathfrak{f}) = \prod_{\mathfrak{q} \in \mathfrak{Q}} (\mathfrak{q}\mathcal{A} + \mathfrak{f}) \subseteq \mathfrak{f}$. $\square$

3.6 Lemma. *Let $\pi : T \to S$ be an epimorphism of rings. Suppose that $\ker(\pi)$ is contained in the Jacobson radical $J(T)$ of T. If $s \in S^\times$, then every preimage t of s is a unit in T.*

Proof. This follows immediately from $tT + \ker(\pi) = T = Tt + \ker(\pi)$ and Nakayama's lemma. $\square$

3.7. *Computation of generators of $K_1(\overline{\mathcal{A}})$:*
  (a) Applying the Chinese remainder theorem, cf. [13, Theorem A10], to the decomposition of Lemma 3.5, we obtain

$$\mathcal{A}/\mathfrak{f} \cong \prod_{\mathfrak{q} \in \mathfrak{Q}} \mathcal{A}/(\mathfrak{q}\mathcal{A} + \mathfrak{f}), \tag{15}$$

This induces a decomposition $K_1(\overline{\mathcal{A}}) \cong \prod_{\mathfrak{q} \in \mathfrak{Q}} K_1(\mathcal{A}/\mathfrak{q}\mathcal{A} + \mathfrak{f})$ and our task is reduced to finding generators of the group $K_1(\mathcal{A}/\mathfrak{q}\mathcal{A} + \mathfrak{f})$ for every $\mathfrak{q} \in \mathfrak{Q}$.
  Let $\mathfrak{q} \in \mathfrak{Q}$ and let $\mathfrak{p} \in \mathcal{P}'$ be the associated prime ideal of $\mathcal{A} \cap \mathcal{O}_C$. Then we have a natural surjective ring homomorphism $\mathcal{A}/\mathfrak{q}\mathcal{A} + \mathfrak{f} \to \mathcal{A}/\mathfrak{p}\mathcal{A} + \mathfrak{f}$ and there exists a unique prime number p such that $p \in \mathfrak{p}$. Thus, the latter ring is an algebra over the field $\mathbb{F}_p$ with p elements. We have a commutative diagram

$$
\begin{array}{ccccccc}
1 & \longrightarrow & \dfrac{1 + \mathfrak{p}\mathcal{A} + \mathfrak{f}}{1 + \mathfrak{q}\mathcal{A} + \mathfrak{f}} & \longrightarrow & (\mathcal{A}/\mathfrak{q}\mathcal{A} + \mathfrak{f})^\times & \longrightarrow & (\mathcal{A}/\mathfrak{p}\mathcal{A} + \mathfrak{f})^\times & \longrightarrow & 1 \\
 & & \downarrow & & \downarrow & & \downarrow & & \\
1 & \longrightarrow & U & \longrightarrow & K_1(\mathcal{A}/\mathfrak{q}\mathcal{A} + \mathfrak{f}) & \longrightarrow & K_1(\mathcal{A}/\mathfrak{p}\mathcal{A} + \mathfrak{f}) & \longrightarrow & 1
\end{array}
$$

with natural maps, where U is defined as the kernel of the bottom right horizontal map. Since $\mathcal{A}/\mathfrak{q}\mathcal{A} + \mathfrak{f}$ and $\mathcal{A}/\mathfrak{p}\mathcal{A} + \mathfrak{f}$ are semilocal rings, the middle and right vertical maps are surjective, cf. [4, Theorem 40.31]. By a result of Vaserstein, cf. [4, Remark 40.32(ii)], the kernel of the right vertical map is generated by all elements of the form $(1+xy)(1+yx)^{-1}$ with $x,y \in \mathcal{A}/\mathfrak{p}\mathcal{A}+\mathfrak{f}$ such that $(1+xy)$ and $(1+yx)$ are units in $\mathcal{A}/\mathfrak{p}\mathcal{A}+\mathfrak{f}$. The same statement holds for $\mathcal{A}/\mathfrak{q}\mathcal{A}+\mathfrak{f}$. Since $\mathfrak{p}^l \subseteq \mathfrak{q}$ for some $l \in \mathbb{N}$, the ideal $\mathfrak{p}\mathcal{A}+\mathfrak{f}/\mathfrak{q}\mathcal{A}+\mathfrak{f}$ is contained in the Jacobson radical $J(\mathcal{A}/\mathfrak{q}\mathcal{A}+\mathfrak{f})$. Thus, every lift of a unit of $\mathcal{A}/\mathfrak{p}\mathcal{A}+\mathfrak{f}$ to $\mathcal{A}/\mathfrak{q}\mathcal{A}+\mathfrak{f}$ is a unit, cf. Lemma 3.6. This implies that the top right horizontal map is surjective, as well as the induced map between the kernels of the middle and right vertical maps. The snake lemma now implies that the left vertical map is surjective. Thus, we obtain an exact sequence

$$\frac{1+\mathfrak{p}\mathcal{A}+\mathfrak{f}}{1+\mathfrak{q}\mathcal{A}+\mathfrak{f}} \to K_1(\mathcal{A}/\mathfrak{q}\mathcal{A}+\mathfrak{f}) \to K_1(\mathcal{A}/\mathfrak{p}\mathcal{A}+\mathfrak{f}) \to 1.$$

So our task is reduced to finding generators of $\frac{1+\mathfrak{p}\mathcal{A}+\mathfrak{f}}{1+\mathfrak{q}\mathcal{A}+\mathfrak{f}}$ and $K_1(\mathcal{A}/\mathfrak{p}\mathcal{A}+\mathfrak{f})$.

(b) In order to compute generators of the multiplicative group $\frac{1+\mathfrak{p}\mathcal{A}+\mathfrak{f}}{1+\mathfrak{q}\mathcal{A}+\mathfrak{f}}$ we use the filtration

$$\mathfrak{p}\mathcal{A}+\mathfrak{f} \supseteq (\mathfrak{q}+\mathfrak{p}^2)\mathcal{A}+\mathfrak{f} \supseteq (\mathfrak{q}+\mathfrak{p}^4)\mathcal{A}+\mathfrak{f} \supseteq \cdots \supseteq (\mathfrak{q}+\mathfrak{p}^{2^{l-1}})\mathcal{A}+\mathfrak{f} \supseteq \mathfrak{q}\mathcal{A}+\mathfrak{f}$$

with l minimal such that $\mathfrak{p}^{2^l} \subseteq \mathfrak{q}$. For each integer $m \geqslant 0$, the map $x \mapsto x-1$ induces an isomorphism

$$\frac{1+(\mathfrak{q}+\mathfrak{p}^{2^m})\mathcal{A}+\mathfrak{f}}{1+(\mathfrak{q}+\mathfrak{p}^{2^{m+1}})\mathcal{A}+\mathfrak{f}} \to \frac{(\mathfrak{q}+\mathfrak{p}^{2^m})\mathcal{A}+\mathfrak{f}}{(\mathfrak{q}+\mathfrak{p}^{2^{m+1}})\mathcal{A}+\mathfrak{f}}$$

of abelian groups. Assuming that each of the modules can be represented by a $\mathbb{Z}$-basis, we apply Hermite normal form techniques to compute a $\mathbb{Z}$-basis for the right hand side. Lifting generators of $\frac{1+(\mathfrak{q}+\mathfrak{p}^{2^m})\mathcal{A}+\mathfrak{f}}{1+(\mathfrak{q}+\mathfrak{p}^{2^{m+1}})\mathcal{A}+\mathfrak{f}}$ to $1+(\mathfrak{q}+\mathfrak{p}^{2^m})\mathcal{A}+\mathfrak{f}$ and collecting these elements for $m=0,\ldots,l-1$ yields a set of elements of $\mathcal{A}$, whose classes modulo $1+\mathfrak{q}\mathcal{A}+\mathfrak{f}$ generate $\frac{1+\mathfrak{p}\mathcal{A}+\mathfrak{f}}{1+\mathfrak{q}\mathcal{A}+\mathfrak{f}}$.

(c) We put $B := \mathcal{A}/\mathfrak{p}\mathcal{A}+\mathfrak{f}$ and note that B is a finite $\mathbb{F}_p$-algebra. In order to compute generators of $K_1(B)$ we use the same arguments as above to obtain an exact sequence

$$1+J \to K_1(B) \to K_1(B/J) \to 1,$$

where J denotes the Jacobson radical of B. Algorithms for the computation of the Jacobson radical of associative algebras over $\mathbb{F}_p$ are, for example, discussed in [5, Sec. 2.3] or [6].

This reduces the problem to the computation of generators of $K_1(B/J(B))$ and of $1+J(B)$. The finite ring B/J is semisimple and thus isomorphic to a direct product of matrix rings $\mathrm{Mat}_s(F)$ over finite fields. In order to compute these simple components one can adapt the algorithms described in [5, Sec. 2.4] (see also [7, Sec. 5.2.1]). This leads to a probabilistic algorithm, which performs very well in practice.

Let now $\mathrm{Mat}_s(F)$ be a simple component of B. Using the fact that the canonical maps $F^\times \to K_1(F) \to K_1(\mathrm{Mat}_s(F))$ are isomorphisms, leaves us with the problem of finding a generator of $F^\times$, which we solve by trial and error.

Finally, we still have to find generators of $1 + J$. For that purpose we consider again a filtration, namely

$$1 + J \supseteq 1 + J^2 \supseteq 1 + J^4 \supseteq \cdots \supseteq 1 + J^{2^{l-1}} \supseteq 1,$$

with l minimal such that $J^{2^l} = 0$. Then we use the isomorphisms $1 + J^{2^m}/1 + J^{2^{m+1}} \to J^{2^m}/J^{2^{m+1}}$ induced by $x \mapsto x - 1$, for $m = 0, \ldots, l - 1$. The latter groups are $\mathbb{F}_p$-vector spaces and we compute a basis and proceed similar as in part (b).

3.8. *Computation of the image of $\hat{\partial}$: $K_1(\overline{\mathcal{A}}) \to I_{\mathfrak{g}}/P_{\mathfrak{g}}^+$:* In the previous subsection, generators of $K_1(\overline{\mathcal{A}})$ of the form (u) with $u \in \overline{\mathcal{A}}^{\times}$ were computed. Each such u we lift to an element $a \in \mathcal{A}$. Then, by Proposition 1.8, it suffices to compute the ideal $\mathrm{nr}(a)\mathcal{O}_C$. Instead of computing $\mathrm{nr}(a)\mathcal{O}_C$ we write $a = (a_i)_i$ with $a_i \in A_i$ and compute the norm $\alpha_i := N_{A_i/K_i}(a_i) \in R_i$. If $\dim_{K_i} A_i = n_i$, then one has $N_{A_i/K_i}(a_i) = \mathrm{nr}(a_i)^{n_i}$. But knowing the ideal $N_{A_i/K_i}(a_i)R_i$ allows us to compute the ideal $\mathrm{nr}(a_i)R_i$ in the free abelian group $I(K_i)$ of fractional ideals. Now we only have to compute the representative of $(\mathrm{nr}(a_i)R_i)_i$ in $I_{\mathfrak{g}}/P_{\mathfrak{g}}^+$ by component-wise application of [3, Algorithm 4.3.2].

3.9. *Computation of $D(G)$:* Consider the commutative diagram

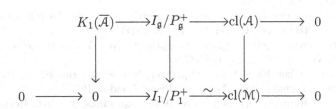

with exact rows as in (10), where we write I_1 for $I_{\mathcal{O}_C}$ and P_1^+ for $P_{\mathcal{O}_C}^+$. Since the vertical maps are surjective with respective kernels $K_1(\overline{\mathcal{A}})$, $(I_{\mathfrak{g}} \cap P_1^+)/P_{\mathfrak{g}}^+$, and $D(\mathcal{A})$, the snake lemma yields an exact sequence

$$K_1(\overline{\mathcal{A}}) \xrightarrow{\hat{\partial}} (I_{\mathfrak{g}} \cap P_1^+)/P_{\mathfrak{g}}^+ \longrightarrow D(\mathcal{A}) \longrightarrow 0.$$

Generators of $K_1(\overline{\mathcal{A}})$ have already been computed in 3.7, and the image of these generators under $\hat{\partial}$ is computed as in 3.8.

3.10 Remark. We conclude the paper with a remark on our implentation. We decided to choose the algebra system MAGMA because it includes both algorithms for group and representation theory and number theory. Many of the features that we need are already implemented in MAGMA, most importantly the computation of character tables and the computation of ray class groups in number fields. Moreover, we use many of the MAGMA functions which deal with associative algebras over finite fields. Here we should at least mention the computation of Jacobson radicals.

We computed a large number of locally free class groups for integral group rings $\mathbb{Z}G$. Our implementation performs well as long as the character fields $K_i, i = 1, \ldots, r$, are small, say $[K_i : \mathbb{Q}] < 20$. This is explained by the fact, that from the algorithmic point of view the computation of ray class groups is a very hard problem. It seems to be the most difficult and time-consuming part of the algorithm.

References

1. W. BLEY, M. ENDRES: Picard groups and refined discrete logarithms. *LMS J. Comput. Math.* **8** (2005), 1–16.
2. H. COHEN: A course in computational algebraic number theory. Springer GTM **138**, New York - Heidelberg 1995.
3. H. COHEN: Advanced topics in computational number theory. Springer GTM **193**, New York - Heidelberg, 2000.
4. C. CURTIS, I. REINER: Methods of representation theory, volume I and II. Wiley, 1981 and 1987.
5. W. EBERLY: Computations for Algebras and Group Representations. Doctoral Thesis, University of Toronto, 1989.
6. K. FRIEDL, L. RÓNYAI: Polynomial time solutions for some problems in computational algebra. Proceedings, 17th ACM Symposium on Theory of Computing, Providence, 1985, 153–162.
7. C. FRIEDRICHS: Berechnung von Maximalordnungen über Dedekindringen. Doctoral Thesis, Technische Universität Berlin, 2000.
8. R. HARTSHORNE: Algebraic geometry. Springer, New York - Heidelberg 1977.
9. MAGMA, Version V2.12, Sydney 2005.
10. J. NEUKIRCH: Class Field Theory. Springer, New York - Heidelberg 1986.
11. I. REINER: Maximal orders. Academic Press, London 1975.
12. J. P. SERRE: Représentations linéaires des groupes finis. 3ème éd., Hermann, Paris 1978.
13. R. G. SWAN: K-theory of finite groups and orders. Lecture Notes in Math. 149, Springer 1970.
14. S. M. J. WILSON: Reduced norms in the K-theory of orders, *J. Algebra* **46** (1977), 1–11.

Numerical Results on Class Groups of Imaginary Quadratic Fields

Michael J. Jacobson Jr., Shantha Ramachandran, and Hugh C. Williams*

Department of Computer Science, University of Calgary
2500 University Drive NW, Calgary, Alberta, Canada T2N 1N4
{jacobs, sramacha}@cpsc.ucalgary.ca, williams@math.ucalgary.ca

Abstract. Using techniques described in [3], we have computed the class number and class group structure of all imaginary quadratic fields with discriminant Δ for $0 < |\Delta| < 10^{11}$. A novel verification algorithm based on the Eichler Selberg Trace Formula [15] was used to ensure that the correctness of our results does not rely on any unproved hypothesis. We present the results of our computations, and remark on specific evidence that was found pertaining to a number of heuristics. In particular, we present data which supports some of the Cohen-Lenstra heuristics [8], Littlewood's bounds on $L(1, \chi)$ [14], and Bach's bound on the maximum norm of the prime ideals required to generate the class group [1].

1 Introduction

To increase the range of numerical evidence supporting various heuristics and asymptotic conjectures, we have computed the class number and class group structure for all imaginary quadratic fields with discriminant Δ for $0 < |\Delta| < 10^{11}$. Currently, the most comprehensive experimental computation for imaginary quadratic fields was performed by Buell [6], who looked at discriminants Δ for $0 < |\Delta| < 2.2 \cdot 10^9$. Buell called his work the "last" computation of such class numbers due to a number of reasons, most notably of which was feasibility.

This issue of feasibility has been overcome by using a more efficient algorithm for computing the class number. Buell's method for unconditionally computing class numbers is based on counting reduced binary quadratic forms for each discriminant, and requires $O(|\Delta|^{1/2})$ steps for each discriminant Δ. Our approach is to compute class numbers using a $O(|\Delta|^{1/4})$ algorithm whose correctness is conditional on the Extended Riemann Hypothesis (ERH), and then to verify the results unconditionally using a novel procedure derived from the Eichler Selberg Trace Formula [15]. This verification procedure is very efficient, so the combined algorithm still requires only $O(|\Delta|^{1/4})$ steps per discriminant, allowing us to increase the upper bound to 10^{11}. The computations required just over 1561 days of CPU time, or about 6 days of real time using a cluster. The verification required just over 2242 days of CPU time, or about 8 days of real time on the same cluster.

---

* All three authors are supported in part by NSERC of Canada.

F. Hess, S. Pauli, and M. Pohst (Eds.): ANTS 2006, LNCS 4076, pp. 87–101, 2006.

In this paper, we present the results of our computation with respect to a number of heuristics. We have tested some of the Cohen-Lenstra heuristics [8] regarding the class number h_Δ by calculating frequencies of various events. We have also tested Littlewood's bounds on the $L(1, \chi_\Delta)$ function [14] and Bach's bound on the maximum norm of the prime ideals required to generate the class group [1].

2 Computing the Structure of the Class Group

The computation was performed using two algorithms; one for computing the structure of the class group, and one for verifying the class numbers. The verification was necessary in order for our results to be unconditional.

2.1 Computation

To compute the class number h_Δ, the first step is to compute an estimate h^* such that $h^* \leq h_\Delta \leq 2h^*$. This is accomplished using methods found in [10], under the assumption of the ERH. Once h^* is found, both the structure of the class group and the class number are computed using an algorithm given by Buchmann, Jacobson and Teske (BJT) [3], which is an improvement on Shanks's well-known baby-step giant-step algorithm [16]. Let Cl_Δ be the class group of the quadratic order $\mathcal{O}_\Delta \in \mathbb{Q}(\sqrt{\Delta})$. We produce prime ideals $\mathfrak{p}_i$ of increasing norm, and iteratively compute subgroups $G_i = \langle \mathfrak{p}_1, \mathfrak{p}_2, ..., \mathfrak{p}_i \rangle$ of Cl_Δ. We continue to add generators until the order of G_i is greater than or equal to h^*. Then, we know that $Cl_\Delta = G_i$. In this way, if $h_\Delta = |G_i|$ is incorrect, it will always be a divisor of the real class number, a fact that is important for our verification algorithm. This method is quite efficient, and it also allows us to collect data on Bach's bounds [1], such as the number and size of prime ideals which are required to generate the class group.

This method of computation requires $O(i2^{\frac{l}{2}}\sqrt{|G|})$ steps, where i is the number of prime ideals used to generate G, and l is the number of elementary divisors of G. Therefore, this algorithm is more efficient than Shanks' algorithm, but is exponential in the rank of the class group. A more recent algorithm for computing the structure of the class group given by Buchmann and Schmidt (BS) [4] is not exponential in the rank of the class group, requiring $O(\sqrt{|G|})$ steps. We considered this algorithm as an option for computing the class numbers, but found that most groups in our computation range required less than 4 prime ideals to generate the group. In practice, the BS algorithm required more than 8 prime ideals before it became more efficient than the BJT algorithm, due to a larger big-O constant. Consequently, we chose to use the BJT algorithm for our computations.

2.2 Verification

In order to remove the dependence of our results on the ERH, we implemented a second algorithm for verifying the class numbers unconditionally. Our verification algorithm is based on the Eichler Selberg Trace Formula [15, Theorem 2.2]. It relates the trace of the Hecke operator T_n acting on the space $S_k(\Gamma_0(N), \chi)$ of cusp forms of weight k, level N and character χ to a sum of class numbers of imaginary quadratic fields. It is well known that for $N = 1, k = 2$ and $\chi = 1$, the space $S_2(\Gamma_0(1), 1)$ has dimension zero. Therefore, the traces of all Hecke operators are zero in this case. We use this equality to verify the correctness of our class numbers, as described below.

For $\Delta = e(\Delta)f(\Delta)^2$, let $e(\Delta)$ be a fundamental discriminant, that is, $e(\Delta)$ is square free, and $e(\Delta) \equiv 1 \pmod 4$ or $e(\Delta) \equiv 8, 12 \pmod{16}$. Let $H(\Delta) = h_w(e(\Delta))K(\Delta)$ denote the Kronecker class number of the quadratic order $\mathcal{O}_\Delta$, where

$$h_w(\Delta) = \begin{cases} h_\Delta & \text{if } |\Delta| > 4 \\ \frac{1}{2} & \text{if } \Delta = -4 \\ \frac{1}{3} & \text{if } \Delta = -3 \end{cases},$$

and

$$K(\Delta) = \sum_{t | f(\Delta)} t \prod_{q|t} \left(1 - \frac{\left(\frac{e(\Delta)}{q}\right)}{q} \right).$$

Using Theorem 2.2 and 2.5 from [15] with $k = 2$, we have the equality

$$Tr(T_n) = A_1 + A_2 + A_3 + A_4 = 0 , \tag{2.1}$$

where

$$A_1 = \frac{1}{12}\chi(n) , \qquad A_2 = -\frac{1}{2}H(-4n) - \sum_{t=1}^{\lceil\sqrt{4n}\rceil - 1} H(t^2 - 4n) ,$$

$$A_3 = -\left(\sum_{\substack{d|n \\ d<\sqrt{n}}} d \right) - \frac{1}{2}\chi(n)\sqrt{n} , \qquad A_4 = \sum_{d|n} d ,$$

and $\chi(n) = 1$ if n is a square, and 0 otherwise. Rearranging (2.1) gives us

$$H(-4n) + 2 \sum_{t=1}^{\lceil\sqrt{4n}\rceil - 1} H(t^2 - 4n)$$

$$= 2 \left(\sum_{d|n} d \right) - 2 \left(\sum_{\substack{d|n \\ d<\sqrt{n}}} d \right) - \chi(n)\sqrt{n} + \frac{1}{6}\chi(n) . \tag{2.2}$$

We verify all the class numbers for discriminants Δ for $0 < |\Delta| < 10^{11}$ by verifying that (2.2) is satisfied for a certain series of values of n. This is done in such a way that each discriminant in our range is incorporated into at least one equation. We compute the required number of n values using a simple brute force preprocessing approach. Starting with a maximum value of $n = \lceil \frac{|\Delta|}{4} \rceil$ where Δ is the discriminant with the smallest absolute value, we mark each discriminant that is incorporated into the sum $\sum_{t=1}^{\lceil \sqrt{4n} \rceil - 1} H(t^2 - 4n)$. Then, we reduce n by 1 and repeat until all discriminants have been incorporated into at least one of the equations.

The verification process is carried out by computing the left hand side of (2.2) for each value of n produced by the preprocessing step. This is done in parallel on a cluster, with different intervals of discriminants Δ in the range of $0 < |\Delta| < 10^{11}$ being computed by separate nodes. For each interval, the values of $t^2 - 4n$ that lie above the lower bound of the interval are computed for all values of n. Then, for each value of n, a sieve is used to factor out all square parts of the discriminants to give us a list of fundamental discriminants in the given interval. Finally, the sums of the class number of the discriminants whose fundamental parts lie in the given interval are all computed.

Once the left hand side of the equations in all intervals are computed, the master node sums them up to produce one value for each n. We then compute the right hand side of each equation and compare the two values. The equality will be violated if we discover a class number which was calculated erroneously. Due to the nature of our algorithm, if the number we computed is not equal to the class number, it is always a divisor of h_Δ. Therefore it is smaller than or equal to h_Δ. Thus, if all the equalities hold, then we are able to unconditionally verify our results.

3   Numerical Results

We implemented the algorithms for computation and verification using the C++ programming language, coupled with the number theory library NTL [19]. We used the 64-bit **long long** data type to represent discriminants and ideals, and NUCOMP [18,7] for ideal arithmetic. The class groups of 30396355052 imaginary quadratic fields were computed using a cluster with 256 2.4 GHz Xeon processors running Linux, each with 1 GB of RAM.

The entire computations required approximately 15 days of real time. The verification was successful, and all the results agreed with Buell's [6]. The computation itself took 1561 days of CPU time, or about 6 days of real time using the cluster. The verification took 2242 days of CPU time, or about 8 days of real time using the cluster. A separate program was used to tabulate the data and compare it to the heuristic results in question. This program ran in just under 237 days of CPU time, or approximately 22 hours of real time. The results are outlined in the following sections.

3.1   Bounds on $L(1, \chi)$

There has been significant interest [5,6,9,17] in the extreme values of $L(1, \chi_\Delta)$ due to the relationship between it and the class number h_Δ. This can be seen in the analytic class number formula,

$$L(1, \chi_\Delta) = \frac{h_\Delta \pi}{\sqrt{|\Delta|}} \ ,$$

where extreme values of $L(1, \chi_\Delta)$ correspond to extreme values of h_Δ.

In [14], Littlewood developed bounds on $L(1, \chi_\Delta)$, namely that under the ERH,

$$\{1 + o(1)\}(c_1 \log \log \Delta)^{-1} < L(1, \chi_\Delta) < \{1 + o(1)\}c_2 \log \log(\Delta) \ , \qquad (3.1)$$

where c_1 and c_2 are defined as follows:

$$c_1 = 12e^\gamma / \pi^2 \text{ and } c_2 = 2e^\gamma \text{ when } 2 \nmid \Delta$$
$$c_1 = 8e^\gamma / \pi^2 \text{ and } c_2 = e^\gamma \text{ when } 2 \mid \Delta \ .$$

In [17], Shanks investigated Littlewood's bounds, and defined two values he termed the *upper* and *lower Littlewood indices*

$$ULI = L(1, \chi_\Delta)/(c_2 \log \log \Delta)$$
$$LLI = L(1, \chi_\Delta)c_1 \log \log \Delta \ .$$

These indices effectively ignore the $o(1)$ given in Littlewood's bounds. We would expect extreme values of the LLI and the ULI to approach 1.

In order to test the validity of these conditional bounds, we recorded successive minimum and maximum values, and corresponding ULI and LLI values, of $L(1, \chi_\Delta)$ for discriminants Δ, with $\Delta \equiv 0 \pmod 4$, $\Delta \equiv 1 \pmod 8$ and $\Delta \equiv 5 \pmod 8$. The maximum $L(1, \chi_\Delta)$ found was 8.09414... ($ULI = 0.70996$) for $\Delta = -45716419031$. The maximum ULI value was 0.73202... ($L(1, \chi_\Delta) = 4.14624...$) for $\Delta = -27867502724$. The minimum $L(1, \chi_\Delta)$ found was 0.17448... ($LLI = 1.2188...$) for $\Delta = -8570250280$. The minimum LLI value was 1.10314... ($L(1, \chi_\Delta) = 0.39502...$) for $\Delta = -1012$.

In Table 1 we list successive maximum $L(1, \chi_\Delta)$ and corresponding ULI values with $\Delta \equiv 1 \pmod 8$, as the values in this congruence class are the overall maximum. We also list successive minimum $L(1, \chi_\Delta)$ and corresponding LLI values with $\Delta \equiv 5 \pmod 8$, as the values in this congruence class are the overall minimum. The $L(1, \chi_\Delta)$ values correspond to Buell's previous tabulations [6] and so we only display the maximum and minimum values which follow after Buell's data.

Following Buell, we also calculated the mean values of $L(1, \chi_\Delta)$ for discriminants $\Delta \equiv 0 \pmod 4$ and $\Delta \equiv 1 \pmod 4$. These values, 1.18639... and 1.58185... are similar to Buell's findings [6].

3.2 The Cohen-Lenstra Heuristics

In [8], Cohen and Lenstra presented a number of heuristics regarding class groups of quadratic number fields. During our computations, we tested the frequency with which odd primes p divide the class number h_Δ, the frequency that the odd part of the class group is non-cyclic, and the number of non-cyclic factors of the p-Sylow subgroups.

Divisibility of h_Δ by Odd Primes. For an imaginary quadratic field with discriminant Δ, the probability that an odd prime p divides the class number h_Δ is conjectured in [8] as

$$\text{prob}(p \mid h_\Delta) = 1 - \eta_\infty(p) , \tag{3.2}$$

where $\eta_\infty(p) = \prod_{k \geq 1} 1 - p^{-k}$. As observed by Buell [6], under the same heuristic assumptions, p^2 divides the class number h_Δ with probability $1 - \frac{p\eta_\infty(p)}{p-1}$ and p^3 divides the class number with probability $1 - \frac{p^3 \eta_\infty(p)}{(p-1)^2(p+1)}$. We define the value $p_l(x)$ as the observed ratio of discriminants less than x with $l \mid h_\Delta$ divided by the conjectured probability shown in (3.2). As x increases, we would expect the value of $p_l(x)$ to approach 1. Similarly, we define the ratios $p_{l^2}(x)$ for l^2 dividing the class number, and $p_{l^3}(x)$ for l^3 dividing the class number.

In Table 2, we present the values of $p_l(x)$ for small primes l. The values appear to approach 1 from below. The values of $p_{l^2}(x)$ and $p_{l^3}(x)$ approach 1 from below in a similar fashion, and so are not presented here. It should be noted that the ratios approach 1 at a slower rate for l^2 and an even slower rate for l^3.

Cyclic Cl_Δ^*. Define Cl_Δ^* to be the odd part of Cl_Δ. The heuristics given in [8] state that the probability that Cl_Δ^* is cyclic is equal to

$$\text{prob}(Cl_\Delta^* \text{ cyclic}) = \frac{\zeta(2)\zeta(3)}{3\zeta(6)C_\infty \eta_\infty(2)} , \tag{3.3}$$

where $C_\infty = \prod_{i>2} \zeta(i)$. This value is roughly 97.7575%. We define $c(x)$ as the observed ratio of discriminants less than x with Cl_Δ^* cyclic divided by the conjectured probability shown in (3.3). As x increases, we would expect the value of $c(x)$ to approach 1.

In Table 3, we present values of $c(x)$, along with the total number of discriminants less than x with Cl_Δ^* non-cyclic. As expected, the values of $c(x)$ approach 1 from above.

Non-Cyclic Factors of p-Sylow Subgroups. For an odd prime p, define the p-rank of Cl_Δ as the number of non-cyclic factors of the p-Sylow subgroup of Cl_Δ. The heuristics given in [8] state that the probability that the p-rank is equal to r is

$$\text{prob}(p\text{-rank of } Cl_\Delta = r) = \frac{\eta_\infty(p)}{p^{r^2} \eta_r(p)^2} . \tag{3.4}$$

We define $p_{l,r}(x)$ as the observed ratio of discriminants less than x with l-rank equal to r divided by the conjectured probability shown in (3.4). As x increases, we would expect the value of $p_{l,r}(x)$ to approach 1.

In Table 4, we present values of $p_{l,r}(x)$ for various values of small primes l and $r = 2, 3, 4$. As expected, the values tend to approach 1 from below fairly smoothly, but slowly.

3.3 First Occurrences of Non-cyclic p-Sylow Subgroups

In [6], Buell looked at what he called "exotic" groups, particular non-cyclic p-Sylow subgroups for various odd primes p. Following Buell, we have recorded both the first occurrence and the total number of discriminants for which a specific p-Sylow subgroup is "exotic". When dealing with the prime $p = 2$, we consider only the 2-Sylow subgroup of the principal genus (the subgroup of squares) of the class group, as was done in [9] and [6]. For brevity, data for $p = 2$ is not presented here but is available at [11]. In the following tables, only new occurrences of specific p-Sylow subgroups for odd primes p which were not discovered by Buell in [6] are listed. Full tables are available at [11].

In Table 5, we present the discriminants Δ with the smallest absolute value for which Cl_Δ has a rank 2 p-Sylow subgroup of the form $C(p^{e_1}) \times C(p^{e_2})$ for an odd prime p. We have tabulated and displayed those discriminants where $\Delta \equiv 0$ (mod 4) and those where $\Delta \equiv 1$ (mod 4) separately. We also list the number of discriminants $|\Delta| < 10^{11}$ for which each p-Sylow subgroup has the specified structure. We found several fields for which the p-Sylow subgroup has rank 2 for all odd primes $p \leq 173$.

In Table 6, we present the discriminants Δ with the smallest absolute value for which Cl_Δ has a rank 3 p-Sylow subgroup of the form $C(p^{e_1}) \times C(p^{e_2}) \times C(p^{e_3})$ for an odd prime p. Once again, we list discriminants in different congruence classes separately, and also the number of discriminants for which each p-Sylow subgroup has the specified structure. We found fields with p-Sylow subgroups of rank 3 for all odd primes $p \leq 13$. Although fields with 11 and 13-Sylow subgroups of rank 3 were already known [12,13], the discriminants we found are unconditionally the smallest in absolute value of any fields with these properties.

We found numerous examples of fields with rank 4 3-Sylow subgroups. This data is not included here, as Belabas [2] has listed all of these fields and minimal discriminants with 3-rank ≤ 5. We did not observe any fields with p-rank equal to 4 for $p > 3$.

In Table 7 we present the first occurrences of doubly non-cyclic class groups, and in Table 8 we present the first occurrences of trebly non-cyclic class groups. The most "exotic" of these class groups, for $\Delta = -61164913211$, is isomorphic to $C(3 \cdot 7 \cdot 19) \times C(3 \cdot 7 \cdot 19)$. In addition, we were able to find 4 discriminants for which the corresponding class groups are quadruply non-cyclic with respect to the primes 2, 3, 5 and 7. The smallest of these discriminants is $\Delta = -20777253551$ with $Cl_\Delta \cong C(4 \cdot 3 \cdot 5 \cdot 7) \times C(4 \cdot 3 \cdot 5 \cdot 7)$.

3.4 Number of Generators

In [1], Bach proved a theorem stating that under the ERH, prime ideals of norm less than $6 \log^2 |\Delta|$ are sufficient to generate the class group of a quadratic field. However, in [7], a tighter bound of $O(\log^{1+\epsilon} |\Delta|)$ was conjectured. Other authors, such as [6] and [9], have observed that in practice, Bach's bound seems to be excessive and attempt to find a constant c for which the tighter bound could hold.

We define $\max_p(\Delta)$ as the maximum norm of the prime ideals required to generate the class group of $\mathbb{Q}(\sqrt{\Delta})$. If Bach's theorem is true, we would expect that $\max_p(\Delta)/\log^2 |\Delta| \leq 6$. To test this theorem, we maintained values of $\max_p(\Delta)$ for all discriminants Δ with $0 < |\Delta| < 10^{11}$. In order to test the tighter bound given in [7], we tried to find the magnitude of the constant c for which $\max_p(\Delta) \leq c \log |\Delta|$. To do this, we looked at the ratio of $\max_p(\Delta)/\log |\Delta|$.

Throughout our computations, the maximum value of $\max_p(\Delta)$ we found was 353 for $\Delta = -42930759883$ and $\Delta = -88460711448$. The maximum value of $\max_p(\Delta)/\log^2 |\Delta|$ was $0.780042...$ for the discriminant $\Delta = -424708$, and the average value was $0.02481....$ The maximum value of $\max_p(\Delta)/\log |\Delta|$ was $14.41825...$ for the discriminant $\Delta = -42930759883$, and the average value was $0.60191....$ The maximum value of $\max_p(\Delta)/\log^2 |\Delta|$ remained constant for most of the computation, whereas the maximum of $\max_p(\Delta)/\log |\Delta|$ increased very slowly, suggesting that a bound of $O(\log^{1+\epsilon} |\Delta|)$ may indeed be the truth. Complete data for $\max_p(\Delta)$ and both ratios can be found in [11].

Following Buell [6], we also kept track of the first occurrences and total number of discriminants for which all prime ideals of norm up to a certain bound were necessary, with the maximum norm found being 353. We found that the total number of discriminants requiring all prime ideals of norm up to a prime p tended to decrease as p increased, except for when $p = 181$, where the number increased by over 60 times.

We also looked at the number of prime ideals that were required to generate the class group. The maximum number of prime ideals required to generate all discriminants Δ for $0 < |\Delta| < 10^{11}$ was 25 for the discriminant $\Delta = -75948116920$, but on average only $3.31359...$ were required, justifying our use of [3] to compute class groups as opposed to [4]. The complete table containing first occurrences of discriminants requiring k prime ideals and totals can be found in [11].

4 Conclusions and Future Work

We intend to extend our computations to include all class groups of imaginary quadratic fields with discriminant Δ for $0 < |\Delta| < 10^{12}$, in order to provide stronger evidence towards the heuristics mentioned in this paper. Additionally, we plan on modifying our verification algorithm to work with real quadratic fields, and carry out new computations in that setting.

Acknowledgements

We wish to thank Andrew Booker for suggesting the use of traces of Hecke operators on cusp forms as a verification procedure.

References

1. E. Bach, *Explicit bounds for primality testing and related problems*, Math. Comp. **55** (1990), no. 191, 355–380.
2. K. Belabas, *On quadratic fields with large 3-rank*, Math. Comp. **73** (2004), 2061–2074.
3. J. Buchmann, M.J. Jacobson, Jr., and E. Teske, *On some computational problems in finite abelian groups*, Math. Comp. **66** (1997), no. 220, 1663–1687.
4. J. Buchmann and A. Schmidt, *Computing the structure of a finite abelian group*, Math. Comp. **74** (2005), 2017–2026.
5. D.A. Buell, *Small class numbers and extreme values of L-functions of quadratic fields*, Math. Comp. **31** (1977), no. 139, 786–796.
6. _____, *The last exhaustive computation of class groups of complex quadratic number fields*, Number theory (Ottawa, ON, 1996) (Providence, RI), CRM Proceedings and Lecture Notes, vol. 19, Amer. Math. Soc., 1999, pp. 35–53.
7. H. Cohen, *A course in computational algebraic number theory*, Springer-Verlag, Berlin, 1993.
8. H. Cohen and H.W. Lenstra, Jr., *Heuristics on class groups of number fields*, Number Theory, Lecture notes in Math., vol. 1068, Springer-Verlag, New York, 1983, pp. 33–62.
9. M.J. Jacobson, Jr., *Experimental results on class groups of real quadratic fields (extended abstract)*, Algorithmic Number Theory - ANTS-III (Portland, Oregon), Lecture Notes in Computer Science, vol. 1423, Springer-Verlag, Berlin, 1998, pp. 463–474.
10. _____, *Applying sieving to the computation of quadratic class groups*, Math. Comp. **68** (1999), no. 226, 859–867.
11. M.J. Jacobson, Jr., S. Ramachandran, H.C. Williams, Supplementary tables for "Numerical results on class groups of imaginary quadratic fields," http://www.math.tu-berlin.de/~kant/ants/proceedings.html, 2006.
12. F. Leprévost, *Courbes modulaires et 11-rang de corps quadratiques*, Experimental Mathematics **2** (1993), no. 2, 137–146.
13. _____, *The modular points of a genus 2 quotient of $X_0(67)$*, Proceedings of the 1997 Finite Field Conference of the AMS, Contemporary Mathematics, vol. 245, 1999, pp. 181–187.
14. J.E. Littlewood, *On the class number of the corpus $P(\sqrt{-k})$*, Proc. London Math. Soc. **27** (1928), 358–372.
15. R. Schoof and M. van der Vlugt, *Hecke operators and the weight distributions of certain codes*, Journal of Combinatorial Theory, Series A **57** (1991), 163–186.
16. D. Shanks, *Class number, a theory of factorization, and genera*, Proc. Sympos. Pure Math, AMS, Providence, R.I., 1971, pp. 415–440.
17. _____, *Systematic examination of Littlewood's bounds on $L(1,\chi)$*, Proc. Sympos. Pure Math, AMS, Providence, R.I., 1973, pp. 267–283.
18. _____, *On Gauss and composition I, II*, Proc. NATO ASI on Number Theory and Applications (R.A. Mollin, ed.), Kluwer Academic Press, 1989, pp. 163–179.
19. V. Shoup, *NTL: A library for doing number theory*, Software, 2001, See http://www.shoup.net/ntl.

A   Appendix

Table 1. Successive $L(1, \chi)$ maxima and minima

Successive maxima ($\Delta \equiv 1 \pmod 8$)			Successive minima ($\Delta \equiv 5 \pmod 8$)		
Δ	$L(1,\chi)$	ULI	Δ	$L(1,\chi)$	LLI
1514970551	7.49759	0.68985	1930143763	0.18764	1.24439
2438526191	7.52739	0.68757	2426489587	0.18655	1.24146
2570169839	7.56669	0.69062	2562211723	0.18470	1.23020
2772244991	7.58892	0.69186	3030266803	0.18445	1.23160
3555265271	7.59038	0.68945	3416131987	0.18152	1.21415
5111994359	7.64749	0.69097	6465681643	0.18082	1.22069
6194583071	7.69307	0.69318	6623767483	0.17973	1.21375
7462642151	7.70257	0.69221	15442196323	0.17843	1.21922
7979490791	7.70933	0.69217	21538327507	0.17609	1.20857
8462822759	7.77325	0.69733	45640185427	0.17604	1.22007
12123145319	7.80183	0.69642	84291143203	0.17599	1.22914
13005495359	7.82594	0.69790	85702502803	0.17448	1.21885
17833071959	7.89105	0.70071			
29414861999	7.89941	0.69683			
35535649679	7.94608	0.69923			
42775233959	7.99504	0.70187			
45716419031	8.09414	0.70996			

Table 2. Values of $p_l(x)$

x	$p_3(x)$	$p_5(x)$	$p_7(x)$	$p_{11}(x)$	$p_{13}(x)$	$p_{17}(x)$	$p_{19}(x)$
1000000000	0.97327	0.99348	0.99609	0.99576	0.99489	0.99474	0.99347
2000000000	0.97624	0.99453	0.99687	0.99664	0.99621	0.99585	0.99522
3000000000	0.97783	0.99515	0.99737	0.99701	0.99666	0.99646	0.99601
4000000000	0.97888	0.99558	0.99760	0.99724	0.99708	0.99672	0.99657
5000000000	0.97966	0.99586	0.99778	0.99751	0.99738	0.99698	0.99698
6000000000	0.98029	0.99610	0.99786	0.99769	0.99757	0.99724	0.99718
7000000000	0.98080	0.99622	0.99791	0.99780	0.99771	0.99742	0.99737
8000000000	0.98122	0.99635	0.99800	0.99795	0.99787	0.99753	0.99745
9000000000	0.98159	0.99645	0.99808	0.99803	0.99799	0.99763	0.99757
10000000000	0.98191	0.99653	0.99818	0.99810	0.99812	0.99771	0.99770
20000000000	0.98391	0.99712	0.99861	0.99852	0.99853	0.99823	0.99824
30000000000	0.98496	0.99744	0.99876	0.99875	0.99876	0.99850	0.99852
40000000000	0.98567	0.99761	0.99887	0.99890	0.99889	0.99871	0.99868
50000000000	0.98619	0.99776	0.99896	0.99901	0.99900	0.99884	0.99880
60000000000	0.98661	0.99786	0.99901	0.99904	0.99906	0.99891	0.99888
70000000000	0.98695	0.99796	0.99905	0.99908	0.99912	0.99902	0.99893
80000000000	0.98723	0.99804	0.99909	0.99912	0.99916	0.99907	0.99902
90000000000	0.98748	0.99810	0.99913	0.99917	0.99920	0.99911	0.99906
100000000000	0.98770	0.99815	0.99915	0.99919	0.99924	0.99914	0.99910

Table 3. Number of noncyclic odd parts of class groups

x	total	non-cyclic	percent	c(x)
1000000000	303963510	5585092	1.83742	1.00414
2000000000	607927095	11356654	1.86809	1.00383
3000000000	911890759	17182389	1.88426	1.00366
4000000000	1215854223	23041817	1.89511	1.00355
5000000000	1519817699	28923395	1.90308	1.00347
6000000000	1823781240	34822620	1.90936	1.00341
7000000000	2127745010	40736296	1.91453	1.00336
8000000000	2431708386	46659753	1.91881	1.00331
9000000000	2735672001	52600902	1.92278	1.00327
10000000000	3039635443	58544601	1.92604	1.00324
20000000000	6079271092	118313612	1.94618	1.00303
30000000000	9118906425	178447518	1.95690	1.00292
40000000000	12158541989	238793386	1.96400	1.00285
50000000000	15198177465	299290965	1.96926	1.00280
60000000000	18237813070	359892824	1.97333	1.00275
70000000000	21277448334	420584966	1.97667	1.00272
80000000000	24317083860	481364092	1.97953	1.00269
90000000000	27356719791	542201863	1.98197	1.00267
100000000000	30396355052	603101904	1.98413	1.00264

Table 4. Values of $p_{l,r}(x)$

x	$pr_{3,2}(x)$	$pr_{5,2}(x)$	$pr_{7,2}(x)$	$pr_{11,2}(x)$	$pr_{13,2}(x)$	$pr_{3,3}(x)$	$pr_{5,3}(x)$	$pr_{7,3}(x)$	$pr_{3,4}(x)$
1000000000	0.84360	0.95708	0.96014	0.94248	0.92552	0.44305	0.92707	0.33360	0.08031
2000000000	0.86065	0.96351	0.96570	0.95530	0.94096	0.49026	0.90254	0.38920	0.08031
3000000000	0.86959	0.96705	0.97262	0.96803	0.94671	0.51636	0.89927	0.59306	0.10708
4000000000	0.87560	0.96977	0.97550	0.97268	0.95343	0.53591	0.91726	0.63940	0.12047
5000000000	0.88013	0.97125	0.97670	0.97199	0.95713	0.55085	0.93099	0.73392	0.11244
6000000000	0.88365	0.97289	0.97885	0.97394	0.95893	0.56132	0.93770	0.83400	0.13385
7000000000	0.88658	0.97382	0.98039	0.97483	0.96086	0.57142	0.93968	0.82605	0.16062
8000000000	0.88904	0.97467	0.98048	0.97631	0.96206	0.58047	0.93627	0.83400	0.19074
9000000000	0.89126	0.97566	0.98150	0.97711	0.96586	0.58722	0.93415	0.79075	0.20524
10000000000	0.89309	0.97642	0.98224	0.97917	0.96762	0.59382	0.93394	0.73392	0.20881
20000000000	0.90470	0.98042	0.98737	0.98407	0.97719	0.63193	0.93614	0.76172	0.24896
30000000000	0.91096	0.98264	0.98868	0.98628	0.98298	0.65288	0.94064	0.78581	0.24361
40000000000	0.91516	0.98384	0.98918	0.98709	0.98517	0.66798	0.95491	0.83956	0.25298
50000000000	0.91828	0.98481	0.98996	0.98755	0.98684	0.67905	0.95385	0.84956	0.26503
60000000000	0.92072	0.98541	0.99060	0.98830	0.98749	0.68821	0.95707	0.87106	0.27707
70000000000	0.92272	0.98605	0.99119	0.98845	0.98738	0.69563	0.96477	0.87530	0.29141
80000000000	0.92444	0.98653	0.99168	0.98917	0.98795	0.70175	0.96398	0.89098	0.29916
90000000000	0.92591	0.98704	0.99198	0.98966	0.98824	0.70727	0.96729	0.91060	0.31143
100000000000	0.92721	0.98743	0.99223	0.99025	0.98900	0.71201	0.96636	0.91628	0.31803

Table 5. Non-cyclic rank 2 p-Sylow subgroups

p	e_1	e_2	first even Δ	# even Δ	first odd Δ	# odd Δ
3	5	4	16887409796	45	4301015239	115
3	5	5	*	*	6743415071	4
3	6	3	2671485416	443	636617543	1279
3	6	4	49547047976	1	7274282423	32
3	7	3	17668343384	42	3541241903	269
3	7	4	*	*	47649110911	3
3	8	2	17082145064	241	1173834359	2289
3	8	3	*	*	37703425007	18
3	9	1	11132690456	2153	1106108639	20187
3	9	2	93287426216	1	11901791639	221
3	9	3	*	*	60543925679	1
3	10	1	98284577816	2	8795475911	2039
3	10	2	*	*	65798421911	4
3	11	1	*	*	52623967679	21
5	3	3	13603495364	16	1068156239	70
5	4	3	*	*	10036313687	8
5	5	2	26611903016	9	5180829911	129
5	6	1	25411429364	205	1614153239	3578
5	6	2	*	*	75913193999	1
5	7	1	*	*	48662190359	51
7	3	2	5468598824	115	528784319	397
7	3	3	*	*	40111506371	1
7	4	2	75003362216	1	16336216607	14
7	5	1	64461971636	18	5800676279	672
11	2	2	8124316712	19	4536377039	69
11	3	2	*	*	91355041631	1
11	4	1	89983172564	1	7219509359	95
13	2	2	15290030216	2	10692322055	12
13	3	1	5247449576	493	781846103	2375
13	4	1	*	*	55385334839	10
17	2	2	*	*	94733724779	1
17	3	1	28205334296	10	5767994839	201
19	3	1	*	*	5862529559	69
23	3	1	*	*	74447537447	2
29	2	1	5614832984	137	296873471	534
31	2	1	14560212776	62	362103671	367
37	2	1	33184320308	6	2793641999	99
41	2	1	29030848244	6	12558317543	49
43	2	1	*	*	28602441479	26
47	2	1	65816894324	2	20751947191	18
53	2	1	*	*	34862413351	3
59	2	1	*	*	65887828631	2
79	1	1	5114393428	154	888934163	445
83	1	1	2390420804	136	884989055	354

Table 5. (*continued*)

p	e_1	e_2	first even Δ	# even Δ	first odd Δ	# odd Δ
89	1	1	2339707096	99	1941485183	259
97	1	1	9388308724	70	2179032511	177
101	1	1	4293806984	45	758562611	164
103	1	1	19084053944	45	787024943	132
107	1	1	4576627816	40	4041299887	125
109	1	1	2202664232	30	4903396807	97
113	1	1	3422486836	30	1047199379	71
127	1	1	7127111912	21	3482629127	46
131	1	1	16018714472	12	2884161823	45
137	1	1	37914915092	4	4549823483	38
139	1	1	50553654520	4	8396560295	29
149	1	1	56336668888	4	15233330011	20
151	1	1	42941394424	4	13310472899	19
157	1	1	19416052676	5	15661511531	24
163	1	1	46586000024	3	10302820679	15
167	1	1	20926233044	2	22669688623	13
173	1	1	84419230376	4	14602373903	14

Table 6. Non-cyclic rank 3 p-Sylow subgroups

p	e_1	e_2	e_3	first even Δ	# even Δ	first odd Δ	# odd Δ
3	2	2	2	3457439416	18	364435991	35
3	3	2	2	18741973496	9	11037391871	8
3	3	3	2	*	*	20687610651	1
3	4	2	2	12251300788	4	9766538987	7
3	4	3	1	2245873412	29	522302531	67
3	4	4	1	*	*	26320580987	1
3	5	2	1	3130903236	272	413771887	625
3	5	2	2	*	*	45248632247	2
3	5	3	1	43721231572	5	2232519167	15
3	6	2	1	19996254456	61	376424303	165
3	6	2	2	*	*	9483757583	1
3	6	3	1	27291040424	1	53192765699	3
3	7	1	1	6382094504	373	461309711	1183
3	7	2	1	33828950744	4	4163792239	35
3	8	1	1	20594835764	24	5347129751	255
3	8	2	1	*	*	59714529551	3
3	9	1	1	*	*	12792023879	22
5	2	2	1	7095550408	9	6896149079	14
5	3	2	1	49468612564	1	29867315295	2
5	4	1	1	5871738932	32	3511272455	75
5	5	1	1	*	*	25384593659	5
7	2	1	1	2760876184	34	648153647	70
7	3	1	1	32727392168	4	19379510159	9

Table 6. (*continued*)

p	e_1	e_2	e_3	first even Δ	# even Δ	first odd Δ	# odd Δ
11	1	1	1	3035884424	6	23235125867	5
13	1	1	1	*	*	38630907167	2

Table 7. Doubly non-cyclic class groups

p_1	p_2	first even Δ	# even Δ	first odd Δ	# odd Δ
3	29	3395393108	167	557577743	446
3	31	3792995864	122	386659943	314
3	37	6112785556	55	1455428855	146
3	41	17658330596	26	1166119039	80
3	43	3286197848	30	4075192859	71
3	47	16964359736	14	485163311	44
3	53	41696300984	5	457096511	23
3	59	49943038232	5	10227491279	19
3	61	32515774996	4	8522929927	17
3	67	84253538216	2	26792580191	8
3	71	*	*	17614533947	8
3	73	*	*	11752995103	9
3	79	85480238756	1	51762875627	6
3	83	*	*	50476998239	4
3	97	*	*	43344787079	2
3	103	*	*	93069031703	1
3	109	*	*	35029686023	1
3	113	*	*	56428950647	1
5	19	3925533652	102	965381231	271
5	23	14260068616	33	336603767	108
5	29	15541379720	12	10138338695	29
5	31	11788579624	8	17205833747	23
5	37	10719968216	3	16249120831	8
5	41	*	*	26948199679	8
5	43	51986729896	1	71114945339	1
5	47	*	*	8182208159	4
5	53	*	*	22759605719	2
5	71	*	*	14917874303	1
5	73	*	*	63515115611	1
7	17	6198957812	37	1851928807	87
7	19	24082268968	7	5166049215	53
7	23	22198579640	10	2591136407	20
7	29	*	*	21164450935	8
7	31	18704562356	1	68200813691	1
7	37	*	*	49918973471	1
7	43	*	*	57006644887	1
7	47	*	*	98533572251	1
11	13	31664474564	11	13609279311	31

Table 7. (*continued*)

p_1	p_2	first even Δ	# even Δ	first odd Δ	# odd Δ
11	17	50159859416	2	41219120419	10
11	19	*	*	19439678123	2
11	23	*	*	94266055451	1
13	17	72831993636	1	41507696303	4
13	19	*	*	75779342435	2
17	23	*	*	54134972891	1

Table 8. Trebly non-cyclic class groups

p_1	p_2	p_3	first even Δ	# even Δ	first odd Δ	# odd Δ
3	5	7	6890424056	78	1475373743	264
3	5	11	49957566964	6	4643885759	30
3	5	13	84831842696	2	13308756863	14
3	5	17	*	*	60235736039	5
3	7	11	42843308072	2	*	*
3	7	13	*	*	38986878143	4
3	7	19	*	*	61164913211	1

Cyclic Polynomials Arising from Kummer Theory of Norm Algebraic Tori

Masanari Kida*

University of Electro-Communications, Chofu Tokyo, Japan
kida@sugaku.e-one.uec.ac.jp

Abstract. We study cyclic extensions arising from Kummer theory of norm algebraic tori. In particular, we compute quintic cyclic polynomials defining 'Kummer extension'. The polynomials do not only give all the quintic cyclic extensions over the rationals by choosing the parameters but also classify all such extensions. Some arithmetic properties of the polynomials are also derived.

1 Introduction

The aim of this paper is to study cyclic extensions arising naturally from Kummer theory of norm algebraic tori. In particular, we compute defining polynomials of the quintic cyclic extensions over the field of rational numbers. We can expect that they have nice algebraic and arithmetic properties like the classical Kummer polynomials. We will see that our expectation is considerably met.

In the second section, we briefly review the Kummer theory for norm algebraic tori, which is a basis of the discussion in the subsequent sections. In the third section, we search fields admitting the Kummer theory. By this consideration, we find that our quintic case is a seemingly rare case where $\mathbb{Q}$ can be taken as a base field. In the fourth section, we explain how to parameterize these fields by the points in a projective space. In the fifth section we compute the quintic cyclic polynomials defining 'Kummer extensions' and study some arithmetic properties.

2 Kummer Theory for Norm Tori

In this section, we explain the Kummer theory for the norm tori. For the proof, we refer to [1].

The notation introduced in this section will be used throughout the paper.

Let k be a field containing a prime field k_0. We fix an separable closure $\bar{k}$ of k and assume that all separable extensions of k are contained in the fixed separable closure $\bar{k}$. For a positive integer n, we choose a primitive n-th root of unity ζ_n such that $\zeta_d = \zeta_n^{n/d}$ holds for any divisor d of n. Let $m > 1$ be an integer prime to $\text{char}(k)$ and $K = k(\zeta_m)$ and $n = [K : k]$.

---

* This research is supported in part by Grant-in-Aid for Scientific Research (C) (No. 16540014), Ministry of Education, Science, Sports and Culture, Japan.

F. Hess, S. Pauli, and M. Pohst (Eds.): ANTS 2006, LNCS 4076, pp. 102–113, 2006.

The norm torus associated to K/k is defined by the exact sequence of algebraic k-tori:

$$1 \to R_{K/k}^{(1)} \mathbb{G}_m \to R_{K/k} \mathbb{G}_m \xrightarrow{N_{K/k}} \mathbb{G}_m \to 1,$$

where the last map $N_{K/k}$ is the induced norm map. The algebraic torus $R_{K/k}^{(1)} \mathbb{G}_m$ is of dimension $n - 1$ and splits over K, namely we have an isomorphism $R_{K/k}^{(1)} \mathbb{G}_m \cong \mathbb{G}_m^{n-1}$ defined over K. As an algebraic variety it is defined as a norm hyper-surface:

$$R_{K/k}^{(1)} \mathbb{G}_m = \operatorname{Spec}\left(k[x_1, \ldots, x_n]/(N_{K/k}(x_1, \ldots, x_n) - 1) \right)$$

and the points on it can be identified with matrices of determinant 1 in the regular representation of K over k for some fixed basis.

From now on we make the following assumptions:

(I) K/k is a non-trivial cyclic extension;
(II) m is square-free. Moreover if we write $m = p_1 \ldots p_r$ with distinct primes p_j $(j = 1, \ldots, r)$, we have $p_j \equiv 1 \pmod{n}$;
(III) $[k_0(\zeta_{p_j}) : k \cap k_0(\zeta_{p_j})] = n$ for each prime divisor p_j of m;
(IV) every prime ideal of $\mathbb{Z}[\zeta_n]$ lying above every p_j is principal.

By (IV) there exists an element $\lambda \in \mathbb{Z}[\zeta_n]$ of norm m. Let τ_j denote the element of $\operatorname{Gal}(k_0(\zeta_m)/k_0)$ satisfying $\tau_j(\zeta_m) = \zeta_m^j$. We identify $G = \operatorname{Gal}(K/k)$ as a subgroup of $\operatorname{Gal}(k_0(\zeta_m)/k_0)$ by the canonical homomorphism $\operatorname{Gal}(K/k) \cong \operatorname{Gal}(k_0(\zeta_m)/k \cap k_0(\zeta_m)) \hookrightarrow \operatorname{Gal}(k_0(\zeta_m)/k_0)$. Under this identification, the group G is generated by τ_s where s is an element of order n in $(\mathbb{Z}/m\mathbb{Z})^*$. The assumption (III) is equivalent to saying that $s \bmod p_j$ is also of order n. Using (III) we can show that, by twisting λ by the action of $\operatorname{Gal}(\mathbb{Q}(\zeta_n)/\mathbb{Q})$ if necessary, λ satisfies

$$s \equiv \zeta_n \pmod{\lambda}. \tag{1}$$

We further assume that the following system of Diophantine equations has an integral solution $(t_0, t_1, \ldots, t_{n-2})$:

$$\text{(V)} \quad \begin{cases} t_0 + t_1 \zeta_d + \cdots + t_{n-2} \zeta_d^{n-2} = 1 & \text{for all proper divisors } d \text{ of } n, \\ t_0 + t_1 \zeta_n + \cdots + t_{n-2} \zeta_n^{n-2} = \lambda, \end{cases}$$

Under these assumptions, we can show that the element λ of $\mathbb{Z}[\zeta_n]$ of norm m induces an endomorphism, which we denote by the same symbol λ, of $R_{K/k}^{(1)} \mathbb{G}_m$ of degree m in a natural way and that λ fits for the Kummer duality in the following theorem.

Theorem 1. *If K/k satisfies these five assumptions, then there exists a cyclic endomorphism λ of $R_{K/k}^{(1)} \mathbb{G}_m$ of degree m which induces the Kummer duality:*

$$R_{K/k}^{(1)} \mathbb{G}_m(k)/\lambda(R_{K/k}^{(1)} \mathbb{G}_m(k)) \cong \operatorname{Hom}_{\mathrm{cont}}(\operatorname{Gal}(\bar{k}/k), R_{K/k}^{(1)} \mathbb{G}_m[\lambda](\bar{k})). \tag{2}$$

Therefore if K/k satisfies our assumptions, then all the cyclic extensions of k of degree m are obtained by adjoining the coordinates of the inverse image of a point in $R^{(1)}_{K/k}\mathbb{G}_m(k)$ by λ:

$$k(\lambda^{-1}(P))\ \ (P \in R^{(1)}_{K/k}\mathbb{G}_m(k)).$$

This description is remakable in the theoretical sense, but is not really helpful for concrete studies of cyclic extension because it is usually given by a system of algebraic equations with many variables. A single equation defining the extension is more desirable. We shall compute polynomials defining these extensions in Section 5 for a particular case.

As we will see in Section 4, the points on $R^{(1)}_{K/k}\mathbb{G}_m(k)$ are rationally parameterized by the points on a projective space. Thus Noether's problem in Galois theory (see [2] and [3, pp. 115–124]) is completely solved in this case.

3   Fields Admitting Kummer Theory

In this section, we investigate which fields admits the descent Kummer theory in the previous section. The first four assumptions (I),(II),(III) and (IV) are rather easier to deal with. In [1, Section 5] we showed the following:

- If m is a prime number, the first three assumptions are satisfied.
- The assumption (IV) is satisfied particularly if $\mathbb{Z}[\zeta_n]$ is a principal ideal domain. There are 29 such $n > 1$ [4, Chapter 11].
- If n is a prime number and $k \not\ni \zeta_{p_j}$ for all p_j dividing m, then (III) is satisfied.

Here we discuss the solubility of the system of Diophantine equations (V):

$$\begin{cases} t_0 + t_1\zeta_d + \cdots + t_{n-2}\zeta_d^{n-2} = 1 & \text{for all proper divisors } d > 1 \text{ of } n, \\ t_0 + t_1\zeta_n + \cdots + t_{n-2}\zeta_n^{n-2} = \lambda. \end{cases}$$

The determinant of the coefficient matrix of this linear system is of the form $\prod_{i>j}(\zeta_n^i - \zeta_n^j)$ and, in particular, is non-zero. Thus the equation always has a unique rational solution.

If n is a prime, then $1, \zeta_n, \ldots, \zeta_n^{n-2}$ are independent over $\mathbb{Z}$ and therefore the solution of (V) is integral. But, in general, the determinant is a divisor of a power of n and it seems difficult to determine precisely when it cancels. One easy case is settled by the following lemma.

Lemma 1. The linear equations (V) has an integral solution if $n = 4$ for any $\lambda \in \mathbb{Z}[\zeta_4]$ with odd norm.

Proof. We write $\lambda = a + b\zeta_4$ with integers $a, b \in \mathbb{Z}$. Since its norm is odd, we have $a \not\equiv b \pmod 2$.

On the other hand, the equation (V) for $n = 4$ is

$$\begin{cases} t_0 - t_1 + t_2 = 1, \\ (t_0 - t_2) + t_1\zeta_4 = \lambda. \end{cases}$$

It is easy to observe that it has an integral solution if and only if there exists an integer t_2 satisfying $2t_2 = 1 - a + b$. This is equivalent to that $b - a$ is odd.   □

From the above consideration, it follows that typical examples of the fields admitting the descent Kummer theory are given by relatively large subfields of prime cyclotomic fields.

Now we make some numerical investigation for the general case. We shall search the subfields of $K = \mathbb{Q}(\zeta_m)$ $(3 \leq m \leq 97)$ admitting the Kummer theory. For each non-trivial subfield $k = K^{\langle \tau_s \rangle}$ of K satisfying (I),(II),(III) and (IV), we determine whether (V) has an integral solution or not. In the following table, we list m and $n = [K : k]$ and s and λ for which there exist an integral solution of the equation (V) (we omit the cases where n is a prime number or 4):

m	n	s	λ
13	6	4	$[3, -4]$
31	6	26	$[1, -6]$
37	9	16	$[1, 0, -1, 0, 0, 1]$
73	8	63	$[-1, -2, 2, 0]$
79	6	24	$[3, -10]$
89	8	52	$[-3, 0, 2, -2]$
97	8	47	$[-2, 3, -1, 1]$

Here λ's are given with respect to the basis $1, \zeta_n, \ldots, \zeta_n^{\varphi(n)-2}$. Note that the choice of λ depends on the choice of the primitive n-th root of unity (see (1)).

4   Parameterization of the Rational Points on Norm Tori

By the Kummer duality (2), the cyclic extensions of the base field k are parameterized by the group $R_{K/k}^{(1)}\mathbb{G}_m(k)$ of k-rational points. Thus it is important to understand this group of rational points. Recall first that the points in $R_{K/k}^{(1)}\mathbb{G}_m(k)$ can be identified with elements of K^* whose norm to k^* are 1:

$$R_{K/k}^{(1)}\mathbb{G}_m(k) \cong \ker(N_{K/k} : K^* \longrightarrow k^*).$$

Under this identification, we consider the following map:

$$\phi : K^* \longrightarrow R_{K/k}^{(1)}\mathbb{G}_m(k), \quad \beta \mapsto \frac{\beta}{\beta^\tau}$$

where $\tau = \tau_s$ is the previously fixed generator of the Galois group $\mathrm{Gal}(K/k)$. By Hilbert 90, this map is surjective. Furthermore, for a given element in $R_{K/k}^{(1)}\mathbb{G}_m(k)$, this β is uniquely determined up to multiplication with elements of k^* (see [5, Remark (3) after Lemma 10.2.4]). Regarding K as an n-dimensional vector space over k, we obtain a map

$$\phi : \mathbb{P}^{n-1}(k) \longrightarrow R_{K/k}^{(1)}\mathbb{G}_m(k).$$

Namely if $u = (u_1 : \cdots : u_n) \in \mathbb{P}^{n-1}(k)$, then

$$\phi(u) = \frac{\beta}{\beta^\tau}$$

with $\beta = u_1 \xi_1 + \cdots + u_n \xi_n$ where $\xi_1, \ldots, \xi_n$ is a fixed basis of K/k. We can write

$$\frac{\beta}{\beta^\tau} = v_1(u_1 : \cdots : u_n)\xi_1 + \cdots + v_n(u_1 : \cdots : u_n)\xi_n$$

where all $v_i(u_1 : \cdots : u_n)$ are rational functions of $u_1, \ldots, u_n$.

For given $\alpha \in R^{(1)}_{K/k}\mathbb{G}_m(k)$, it is also possible to compute β satisfying $\phi(\beta) = \alpha$ explicitly. In fact we can take

$$\beta = \sum_{0 \le i < n} \tau^i(x) \prod_{0 \le k < i} \tau^k(\alpha),$$

where $x \in K$ should be taken so that $\beta \ne 0$ holds (see [5, Remark (1) after Lemma 10.2.4]).

Remark 1. In [1] we took a parameterization coming from the geometry. Namely since $R^{(1)}_{K/k}\mathbb{G}_m$ is a rational variety if K/k is a cyclic extension, we have a morphism from a projective space to the torus. But, in general, this morphism does not induce a surjection on the rational points. Therefore the parameterization used in [1, Example 5.3] is not complete. On the other hand, the parameterization in [1, Example 5.2] is correct since the geometric parameterization is surjective on the rational points.

5   Cyclic Quintic Polynomials over the Rationals

By the discussion in the end of Section 3, it seems rare that our Kummer duality holds over the field of rational numbers. These seemingly rare cases happen when $(m, n) = (3, 2)$ and $(5, 4)$. A cyclic cubic polynomial corresponding to the former case is computed in [1, Example 5.2]. See also [6] for a different but equivalent expression of this cyclic cubic polynomial. We deal with the latter case $(m, n) = (5, 4)$ here. Let $k = \mathbb{Q}$ and $K = \mathbb{Q}(\zeta_5)$. The aim is computing a quintic Kummer polynomial parameterizing all cyclic quintic extensions over $\mathbb{Q}$. Throughout this section we simply write ζ for ζ_5. The Galois group of K/k is generated by $\tau = \tau_2 : \zeta \mapsto \zeta^2$. Hence we have $s = 2$. We take $\{\zeta, \zeta^2, \zeta^4, \zeta^3\}$ for a basis of K/k (they form a normal basis of K/k) and consider the norm equation $N(x_1, x_2, x_3, x_4)$ with respect to this basis:

$$
\begin{aligned}
N &= N(x_1, x_2, x_3, x_4)\\
&= N_{K/k}(x_1\zeta + x_2\zeta^2 + x_3\zeta^4 + x_4\zeta^3)\\
&= x_1^4 - x_1^3 x_2 - x_1^3 x_3 - x_1^3 x_4 + x_1^2 x_2^2 + 2x_1^2 x_2 x_3 + 2x_1^2 x_2 x_4 + x_1^2 x_3^2 - 3x_1^2 x_3 x_4\\
&\quad + x_1^2 x_4^2 - x_1 x_2^3 + 2x_1 x_2^2 x_3 - 3x_1 x_2^2 x_4 - 3x_1 x_2 x_3^2 - x_1 x_2 x_3 x_4 + 2x_1 x_2 x_4^2\\
&\quad - x_1 x_3^3 + 2x_1 x_3^2 x_4 + 2x_1 x_3 x_4^2 - x_1 x_4^3 + x_2^4 - x_2^3 x_3 - x_2^3 x_4 + x_2^2 x_3^2 + 2x_2^2 x_3 x_4\\
&\quad + x_2^2 x_4^2 - x_2 x_3^3 + 2x_2 x_3^2 x_4 - 3x_2 x_3 x_4^2 - x_2 x_4^3 + x_3^4 - x_3^3 x_4 + x_3^2 x_4^2 - x_3 x_4^3 + x_4^4.
\end{aligned}
$$

Our model of $R_{K/k}^{(1)}\mathbb{G}_m$ is

$$\text{Spec}\left(k[x_1, x_2, x_3, x_4]/(N(x_1, x_2, x_3, x_4) - 1)\right).$$

Let $\zeta_4 = i$. There are two prime ideals $(2+i)$ and $(2-i)$ in $\mathbb{Z}[i]$ lying above 5. In view of (1), we should choose $\lambda = 2 - i$. We solve the system of Diophantine equations (V) and obtain $(t_0, t_1, t_2) = (1, -1, -1)$. The endomorphism of the character module $\widehat{R_{K/k}^{(1)}\mathbb{G}_m}$ can be computed from (t_0, t_1, t_2) (see [1, Proposition 2.1]), which induces an endomorphism of the split torus $\mathbb{G}_m^3$ defined by

$$(X_1, X_2, X_3) \mapsto (X_1 X_2^{-1} X_3^{-1}, X_1 X_2^2, X_2 X_3^2). \tag{3}$$

We compute the corresponding endomorphism λ on the above model of $R_{K/k}^{(1)}\mathbb{G}_m$:

$$\lambda : (x_1, x_2, x_3, x_4) \mapsto (x_1', x_2', x_3', x_4')$$

where

$$
\begin{aligned}
x_1' =& - x_1^3 + 2x_1^2 x_2 + x_1^2 x_4 - 2x_1 x_2 x_4 + x_1 x_3^2 - x_2^3 \\
&+ x_2^2 x_3 + 2x_2^2 x_4 - 2x_2 x_3 x_4 - x_3^3 + 2x_3 x_4^2 - x_4^3, \\
x_2' =& - x_1^3 + 2x_1^2 x_4 + x_1 x_2^2 - 2x_1 x_2 x_3 + 2x_1 x_3^2 - 2x_1 x_3 x_4 \\
&- x_2^3 + 2x_2^2 x_3 + x_2 x_4^2 - x_3^3 + x_3^2 x_4 - x_4^3, \\
x_3' =& - x_1^3 + x_1^2 x_3 + 2x_1 x_2^2 - 2x_1 x_2 x_4 + x_1 x_4^2 - x_2^3 \\
&+ x_2 x_3^2 - 2x_2 x_3 x_4 + 2x_2 x_4^2 - x_3^3 + 2x_3^2 x_4 - x_4^3, \\
x_4' =& - x_1^3 + x_1^2 x_2 + 2x_1^2 x_3 - 2x_1 x_2 x_3 - 2x_1 x_3 x_4 \\
&+ 2x_1 x_4^2 - x_2^3 + x_2^2 x_4 + 2x_2 x_3^2 - x_3^3 + x_3 x_4^2 - x_4^3.
\end{aligned}
$$

On the other hand, a rational parameterization $\mathbb{P}^3(k) \longrightarrow R_{K/k}^{(1)}\mathbb{G}_m(k)$ is induced by

$$\beta = u_1 \zeta + u_2 \zeta^2 + u_3 \zeta^4 + u_4 \zeta^3 \mapsto \beta/\beta^\tau = v_1 \zeta + v_2 \zeta^2 + v_3 \zeta^4 + v_4 \zeta^3.$$

Explicitly we have

$$
\begin{aligned}
v_1 =& \frac{1}{N}(- 2u_1^3 u_2 + u_1^3 u_3 + u_1^3 u_4 + 2u_1^2 u_2^2 - u_1^2 u_2 u_4 - 3u_1^2 u_3 u_4 - u_1 u_2^3 - u_1 u_2^2 u_3 \\
&- 2u_1 u_2 u_3^2 + 6u_1 u_2 u_3 u_4 + u_1 u_2 u_4^2 + u_1 u_3^2 u_4 - u_1 u_3 u_4^2 - u_1 u_4^3 - u_2^3 u_3 + u_2^3 u_4 \\
&+ 2u_2^2 u_3^2 - 2u_2^2 u_4^2 - u_2 u_3^3 - 3u_2 u_3 u_4^2 + 2u_2 u_4^3 + u_3^4 - 2u_3^3 u_4 + 2u_3^2 u_4^2), \\
v_2 =& \frac{1}{N}(- u_1^3 u_2 + 2u_1^3 u_3 + u_1^2 u_2 u_3 - u_1^2 u_2 u_4 - 2u_1^2 u_3^2 - 3u_1^2 u_3 u_4 + 2u_1^2 u_4^2 + u_1 u_2^3 \\
&- u_1 u_2^2 u_3 - 3u_1 u_2^2 u_4 + 6u_1 u_2 u_3 u_4 + u_1 u_2 u_4^2 + u_1 u_3^3 - 2u_1 u_3^3 - 2u_2^3 u_3 + u_2^3 u_4 \\
&+ 2u_2^2 u_3^2 - u_2 u_3^3 - u_2 u_3^2 u_4 - 2u_2 u_3 u_4^2 - u_3^3 u_4 + 2u_3^2 u_4^2 - u_3 u_4^3 + u_4^4),
\end{aligned}
$$

$$v_3 = \frac{1}{N}(u_1^4 - 2u_1^3 u_2 - u_1^3 u_4 + 2u_1^2 u_2^2 + u_1^2 u_2 u_3 - 2u_1^2 u_3 u_4 + 2u_1^2 u_4^2 - u_1 u_2^2 u_3 - 3u_1 u_2^2 u_4$$
$$- 3u_1 u_2 u_3^2 + 6u_1 u_2 u_3 u_4 + u_1 u_3^3 - u_1 u_3 u_4^2 - u_1 u_4^3 - u_2^3 u_3 + 2u_2^3 u_4 + u_2^2 u_3 u_4$$
$$- 2u_2^2 u_4^2 + u_2 u_3^3 - u_2 u_3^2 u_4 + u_2 u_4^3 - 2u_3^3 u_4 + 2u_3^2 u_4^2 - u_3 u_4^3),$$

$$v_4 = \frac{1}{N}(-u_1^3 u_2 + u_1^3 u_3 - u_1^3 u_4 + 2u_1^2 u_2^2 - u_1^2 u_2 u_4 - 2u_1^2 u_3^2 + 2u_1^2 u_4^2 - u_1 u_2^3 - 2u_1 u_2^2 u_4$$
$$- 3u_1 u_2 u_3^2 + 6u_1 u_2 u_3 u_4 + 2u_1 u_3^3 + u_1 u_3^2 u_4 - u_1 u_3 u_4^2 - 2u_1 u_4^3 + u_2^4 - 2u_2^3 u_3$$
$$+ 2u_2^2 u_3^2 + u_2^2 u_3 u_4 - u_2 u_3^2 u_4 - 3u_2 u_3 u_4^2 + u_2 u_4^3 - u_3^3 u_4 + u_3 u_4^3)$$

where $N = N(\beta) = N(u_1, u_2, u_3, u_4)$. Now the system of equations

$$x_1' = v_1, \quad x_2' = v_2, \quad x_3' = v_3, \quad x_4' = v_4$$

defines our cyclic extension $L = k(x_1, x_2, x_3, x_4)$. Unfortunately it is difficult to solve this system of equations for one particular variable. This difficulty can be overcome by the method developed in [1]. The key observation is that, since our extension is a Kummer extension, the Galois action on (x_1, x_2, x_3, x_4) is given by the multiplication of an element in $\ker(\lambda)(\bar{k})$. Moreover the group $\ker(\lambda)(\bar{k})$ is isomorphic to the group of m-th roots of unity in K^* over K. Hence by

$$\zeta(x_1\zeta + x_2\zeta^2 + x_3\zeta^4 + x_4\zeta^3) = -x_3\zeta + (x_1 - x_3)\zeta^2 + (x_4 - x_3)\zeta^4 + (x_2 - x_3)\zeta^3,$$

we find that the action of a generator σ of $\mathrm{Gal}(L/k)$ on (x_1, x_2, x_3, x_4) is given by

$$(x_1, x_2, x_3, x_4) \mapsto (-x_3, x_1 - x_3, x_4 - x_3, x_2 - x_3).$$

Therefore the conjugates of x_1 over k are

$$x_1, \quad -x_3, \quad x_3 - x_4, \quad x_4 - x_2, \quad x_2 - x_1.$$

From this it follows that L is generated only by x_1 over k: $L = k(x_1)$. At this stage, we do not know how to express these by the parameters u_1, u_2, u_3 and u_4. For that purpose, we write x_1, x_2, x_3, x_4 in terms of X_1, X_2, X_3. Since X_i's are the coordinates on the split torus, they relate by

$$X_i = x_1\zeta^{\tau^{i-1}} + x_2(\zeta^2)^{\tau^{i-1}} + x_3(\zeta^4)^{\tau^{i-1}} + x_4(\zeta^3)^{\tau^{i-1}} \quad (i = 1, 2, 3).$$

It is also natural to define X_4 by this formula with $i = 4$. Then the equation satisfied by x_1 is

$$\begin{aligned}
F(T) =& F(u_1 : u_2 : u_3 : u_4; T)\\
=& (T - x_1)(T - (-x_3))(T - (x_3 - x_4))(T - (x_4 - x_2))(T - (x_2 - x_1))\\
=& T^5 + (\xi_3 X_1 X_3 + \xi_3^{\tau} X_2 X_4)T^3\\
& + (\xi_2 X_1^2 X_4 + \xi_2^{\tau} X_1 X_2^2 + \xi_2^{\tau^2} X_2 X_3^2 + \xi_2^{\tau^3} X_3 X_4^2)T^2\\
& + (-X_1 X_2 X_3 X_4/25 + \xi_{11} X_1^3 X_2 + \xi_{11}^{\tau} X_2^3 X_3 + \xi_{11}^{\tau^2} X_3^3 X_4 + \xi_{11}^{\tau^3} X_4^3 X_1\\
& + \xi_{12} X_1^2 X_3^2 + \xi_{12}^{\tau} X_2^2 X_4^2)T
\end{aligned}$$

$$+ (\xi_{01} X_1^5 + \xi_{01}^\tau X_2^5 + \xi_{01}^{\tau^2} X_3^5 + \xi_{01}^{\tau^3} X_4^5$$
$$+ \xi_{02} X_1^3 X_3 X_4 + \xi_{02}^\tau X_1 X_2^3 X_4 + \xi_{02}^{\tau^2} X_1 X_2 X_3^3 + \xi_{02}^{\tau^3} X_2 X_3 X_4^3$$
$$+ \xi_{03} X_1^2 X_2^2 X_3 + \xi_{03}^\tau X_2^2 X_3^2 X_4 + \xi_{03}^{\tau^2} X_1 X_3^2 X_4^2 + \xi_{03}^{\tau^3} X_1^2 X_2 X_4^2),$$

where

$$\xi_3 = \frac{1}{5}(3\zeta + 2\zeta^2 + 3\zeta^4 + 2\zeta^3), \qquad \xi_2 = \frac{1}{25}(2\zeta - \zeta^2 - 2\zeta^4 + \zeta^3),$$

$$\xi_{11} = \frac{1}{25}(\zeta + \zeta^4), \qquad \xi_{12} = \frac{1}{25}(-2\zeta - \zeta^2 - 2\zeta^4 - \zeta^3), \qquad (4)$$

$$\xi_{01} = \frac{1}{625}(\zeta - 2\zeta^2 - \zeta^4 + 2\zeta^3), \qquad \xi_{02} = \frac{1}{125}(\zeta^2 - \zeta^3)$$

$$\xi_{03} = \frac{1}{125}(-\zeta - \zeta^2 + \zeta^4 + \zeta^3).$$

Setting

$$\alpha_i = v_1 \zeta^{\tau^{i-1}} + v_2 (\zeta^2)^{\tau^{i-1}} + v_3 (\zeta^4)^{\tau^{i-1}} + v_4 (\zeta^3)^{\tau^{i-1}} \quad (i = 1, 2, 3, 4),$$

we shall write the coefficients of $F(T)$ in terms of these α_i's. By (3) we have

$$X_1 X_2^{-1} X_3^{-1} = \alpha_1, \ X_1 X_2^2 = \alpha_2, \ X_2 X_3^2 = \alpha_3. \qquad (5)$$

We solve these for X_1^5 and obtain

$$X_1^5 = \alpha_1^4 \alpha_2 \alpha_3^2. \qquad (6)$$

At this point, it is easy to see

$$L(\zeta) = K(X_1, X_2, X_3, X_4) = K \left(\sqrt[5]{\alpha_1^4 \alpha_2 \alpha_3^2} \right). \qquad (7)$$

Now starting from the second equation of (5), we can express X_2, X_3, X_4 using only X_1:

$$X_2 = \alpha_1^{-2} \alpha_3^{-1} X_1^2, \ X_3 = \alpha_1^{-3} \alpha_2^{-1} \alpha_3^{-1} X_1^4, \ X_4 = \alpha_1 X_1^{-2}. \qquad (8)$$

Substituting these formulas, we obtain

$$F(T) = T^5 + \frac{1}{2} \operatorname{Tr}(\xi_3 \alpha_1 \alpha_3) T^3 + \operatorname{Tr}(\xi_2 \alpha_1) T^2$$
$$+ \left(-\frac{1}{25} + \operatorname{Tr}(\xi_{11} \alpha_1^2 \alpha_2 \alpha_3) + \frac{1}{2} \operatorname{Tr}(\xi_{12} \alpha_1^2 \alpha_2^2) \right) T$$
$$+ \left(\operatorname{Tr}(\xi_{01} \alpha_1^4 \alpha_2 \alpha_3^2) + \operatorname{Tr}(\xi_{02} \alpha_1 \alpha_2 \alpha_3) + \operatorname{Tr}(\xi_{03} \alpha_2^2 \alpha_4) \right),$$

where Tr is the trace map from $K(u_1, \ldots, u_4)$ to $k(u_1, \ldots, u_4)$. Similarly we can compute a polynomial F_i for $x_i (i = 2, 3, 4)$. We state this as a theorem.

Theorem 2. *A polynomial having* x_i $(i = 1, 2, 3, 4)$ *as its zero is given by*

$$F_i(T) = T^5 + \frac{1}{2}\mathrm{Tr}(\xi_3^{\tau^{i-1}}\alpha_1\alpha_3)T^3 + \mathrm{Tr}(\xi_2^{\tau^{i-1}}\alpha_1)T^2$$
$$+ \left(-\frac{1}{25} + \mathrm{Tr}(\xi_{11}^{\tau^{i-1}}\alpha_1^2\alpha_2\alpha_3) + \frac{1}{2}\mathrm{Tr}(\xi_{12}^{\tau^{i-1}}\alpha_1^2\alpha_2^2)\right)T$$
$$+ \left(\mathrm{Tr}(\xi_{01}^{\tau^{i-1}}\alpha_1^4\alpha_2\alpha_3^2) + \mathrm{Tr}(\xi_{02}^{\tau^{i-1}}\alpha_1\alpha_2\alpha_3) + \mathrm{Tr}(\xi_{03}^{\tau^{i-1}}\alpha_2^2\alpha_4)\right),$$

where the constants ξ's are given by (4). In particular, every cyclic quintic extension of $\mathbb{Q}$ is obtained as the splitting field of F_1 with some parameter $u_1, \ldots, u_4 \in \mathbb{Q}$.

Of course, we can write down the equation using the original parameters u_1, u_2, u_3 and u_4. But it would require a few hundred lines.

Other generic families of cyclic quintic polynomials with less parameters are known (for example [7]). But our family has several advantages over previously known ones, because it inherits good properties from the classical Kummer theory. To tell a few, the Galois action is, as we have seen, easily deduced. Also we can tell when two fields with different parameters are isomorphic. Moreover the decomposition law of the fields can also be deduced from the classical framework of Kummer theory. We shall see these properties in the following.

Example 1. We compute some numerical examples of quintic cyclic polynomials for $(u_1 : u_2 : u_3 : u_4) \in \mathbb{P}^3(\mathbb{Q})$.

(i) The point $(u_1 : u_2 : u_3 : u_4) = (0 : 0 : -1 : 0)$ corresponds to $\zeta = (1, 0, 0, 0) \in R_{K/k}^{(1)}\mathbb{G}_m(\mathbb{Q})$ and the corresponding polynomial is

$$F(0 : 0 : -1 : 0; T) = T^5 - T^3 - \frac{2}{5}T^2 + \frac{1}{125}.$$

This polynomial defines a cyclic quintic extension L over $\mathbb{Q}$ with discriminant 5^8.

(ii) The point $(u_1 : u_2 : u_3 : u_4) = (1 : -1 : 0 : 0)$ corresponds to $\zeta + \zeta^3 + \zeta^4 = (1, 0, 1, 1) \in R_{K/k}^{(1)}\mathbb{G}_m(\mathbb{Q})$ and the corresponding polynomial is

$$F(1 : -1 : 0 : 0; T) = T^5 - 2T^3 - \frac{1}{5}T^2 + \frac{4}{5}T + \frac{7}{125}.$$

We shall show that the corresponding two fields are isomorphic. We have

$$(1, 0, 0, 0)(-1, 0, -1, 0) = (1, 0, 1, 1) \text{ in } R_{K/k}^{(1)}\mathbb{G}_m(\mathbb{Q}).$$

Hence if $(-1, 0, -1, 0)$ is contained in $\lambda(R_{K/k}^{(1)}\mathbb{G}_m(\mathbb{Q}))$, then it follows

$$(1, 0, 0, 0) \equiv (1, 0, 1, 1) \pmod{\lambda(R_{K/k}^{(1)}\mathbb{G}_m(\mathbb{Q}))}.$$

From our Kummer duality (2) we can conclude that two fields defined by $F(0 : 0 : -1 : 0; T)$ and $F(1 : -1 : 0 : 0; T)$ are isomorphic.

Now we show $(-1,0,-1,0) \in \lambda(R^{(1)}_{K/k}\mathbb{G}_m(\mathbb{Q}))$ by checking the corresponding point in the split torus is in the image of (3). The first coordinate of the point on the split torus corresponding to $(-1,0,-1,0)$ is $\alpha_1 = -\zeta - \zeta^4$. We also have $\alpha_2 = -\zeta^2 - \zeta^3$, $\alpha_3 = -\zeta^4 - \zeta$. In view of (8), the point $(\alpha_1, \alpha_2, \alpha_3) \in \mathbb{G}^3_m(K)$ is in the image of the map (3) if and only if (6) has a solution in K:

$$X_1^5 = -5\zeta^3 - 5\zeta^2 - 8.$$

In fact, we can find a solution $X_1 = \zeta^2 + \zeta^4$ corresponding to $(0,1,1,0) \in R^{(1)}_{K/k}\mathbb{G}_m(\mathbb{Q})$. Therefore we have $\lambda((0,1,1,0)) = (-1,0,-1,0)$ as desired.

Describing the decomposition law in $L/\mathbb{Q}$ is also an easy task since it is a Kummer extension. Indeed, we have the following theorem.

Theorem 3. *Let L be a cyclic quintic field over $\mathbb{Q}$ corresponding to the parameter $(u_1 : u_2 : u_3 : u_4) \in \mathbb{P}^3(\mathbb{Q})$. Let $\beta = u_1\zeta + u_2\zeta^2 + u_3\zeta^4 + u_4\zeta^3$. We may and do assume that β is an algebraic integer in $\mathbb{Q}(\zeta)$. Also let $\beta_i = \beta^{\tau^{i-1}}$ $(i = 1,2,3,4)$ and $B = \beta_1^4\beta_2^2\beta_3\beta_4^3$. Let p be a prime number and*

$$p\mathbb{Z}[\zeta] = (\mathfrak{p}_1 \ldots \mathfrak{p}_g)^e, \quad f = \deg(\mathfrak{p}_i)$$

the decomposition of p in $\mathbb{Q}(\zeta)$. For each $\mathfrak{p}_i$, let $v_{\mathfrak{p}_i}$ be the normalized valuation. Set $v_i = v_{\mathfrak{p}_i}(\beta)$. We define an integer $u = u_p$ by the following formula:

$$u = \begin{cases} 10v_1 & \text{if } p \equiv 2 \text{ or } 3 \pmod 5; \\ 5(v_1 + v_2) & \text{if } p \equiv 4 \pmod 5; \\ 4v_1 + 3v_2 + v_3 + 2v_4 & \text{if } p \equiv 1 \pmod 5; \\ 10v_1 & \text{if } p = 5. \end{cases}$$

(i) *The case where $p \neq 5$. A prime number p is unramified in $L/\mathbb{Q}$ if and only if 5 divides u. If it is the case, then p splits completely in $L/\mathbb{Q}$ if and only if the congruence $x^5 \equiv B \pmod{\mathfrak{p}^{1+u}}$ is soluble in $x \in K$.*
If $(5,u) = 1$, then p ramifies totally and tamely in $L/\mathbb{Q}$.
(ii) *The case where $p = 5$. Let a be the maximal integer in the set*

$$\{k \mid x^5 \equiv B \pmod{\mathfrak{p}^{k+u}} \text{ is soluble in } K\}.$$

Then we have
(a) 5 splits completely if and only if $a \geq 6$;
(b) 5 is inert if and only if $a = 5$;
(c) 5 ramifies totally if and only if $a \leq 4$.
When p ramifies, the exponent of the different of $L/\mathbb{Q}$ at a prime ideal of L lying above p is given by $9 - a$.

Proof. This theorem is a consequence of Hecke's theory describing the decomposition law in Kummer extensions of prime degree [5, 10.2.3]. In fact the decomposition in $L/\mathbb{Q}$ is completely determined by that of the classical Kummer extension $L(\zeta)/K$. On the other hand, if we write $\alpha_1 = \beta/\beta^\tau$, then it follows from (7) that $L(\zeta) = K(\sqrt[5]{\alpha_1^4\alpha_2\alpha_3^2}) = K(\sqrt[5]{B})$, where we have $B \in \mathbb{Z}[\zeta]$ by our assumption.

On the other hand, the decomposition types of p in $\mathbb{Q}(\zeta)$ are given by

$$p \equiv 2 \text{ or } 3 \pmod 5 \Longrightarrow e = g = 1, f = 4$$
$$p \equiv 4 \pmod 5 \Longrightarrow e = 1, f = g = 2$$
$$p \equiv 1 \pmod 5 \Longrightarrow e = f = 1, g = 4$$
$$p = 5 \Longrightarrow e = 4, f = g = 1.$$

From this it is easy to show that $u = v_{\mathfrak{p}_1}(B)$ holds in each case and that the 5-divisibility of u does not depend on the order of $\mathfrak{p}_i$'s.

Now there is no difficulty to deduce the result by applying Hecke's theorem to $K(\sqrt[5]{B})/K$.

The claim for the different also follows from Hecke's theorem and the chain relation of the differents.   □

The important point here is that we can choose a Kummer generator of $L(\zeta)/K$ canonically from the parameters $(u_1 : u_2 : u_3 : u_4)$. In the case of other generic cyclic polynomials such as Hashimoto-Tsunogai's polynomial in [7], Kummer generators are constructed usually by means of Lagrange resolvent (see below). This form of the generator does not readily imply the ramification property and the decomposition law.

Finally we study a relation between our polynomial and a well-known quintic family of Emma Lehmer. For $n \in \mathbb{Q}$, let

$$G_L^n(T) = T^5 + n^2 T^4 - (2n^3 + 6n^2 + 10n + 10)T^3$$
$$+ (n^4 + 5n^3 + 11n^2 + 15n + 5)T^2 + (n^3 + 4n^2 + 10n + 10)T + 1$$

be Lehmer's quintic polynomial. It is known that it defines a cyclic quintic extension over $\mathbb{Q}$ if the parameter n is chosen in $\mathbb{Q}$. We embed Lehmer's family into our family. By doing so, we can tell how large Lehmer's family is and when two Lehmer type polynomials define the same field. Also the decomposition law of the fields defined by G_L^n should be easily deduced from the theorem above.

Let us write $n = u/v$ with $u, v \in \mathbb{Z}$ and $(u, v) = 1, v > 0$ and let $G(u, v; T) = G_L^n(T)$ (cf. [8]).

Theorem 4. *The polynomials $F(u - v : -2v : -4v - u : -3v; T)$ and $G(u, v; T)$ define the same field.*

Proof. Let L_F and L_G be the quintic cyclic fields defined by $F(u-v, -2v, -4v-u, -3v; T)$ and $G(u, v; T)$ respectively. It is enough to show $L_F(\zeta) = L_G(\zeta)$ since L_F and L_G are the unique quintic subfields of $L_F(\zeta)$ and $L_G(\zeta)$ respectively. Let $\beta = (u - v)\zeta - 2v\zeta^2 + (-4v - u)\zeta^4 - 3v\zeta^3$. Writing $\beta_i = \beta^{\tau^{i-1}}$ $(i = 1, 2, 3, 4)$, we have $L_F(\zeta) = \mathbb{Q}(\zeta)(\sqrt[5]{\beta_1^4\beta_2^2\beta_3\beta_4^3})$ by (7). We now compute a Kummer generator of $L_G(\zeta)$. Let η be a root of G_L^n. In their paper [9], Schoof and Washington show that a generator σ of $\mathrm{Gal}(\mathbb{Q}(\eta)/\mathbb{Q})$ acts by

$$\sigma : \eta \mapsto \frac{(n+2) + n\eta - \eta^2}{1 + (n+2)\eta}.$$

Using this formula, we form a Lagrange resolvent ([5, Theorem 5.3.5 (4)]):

$$\vartheta = \frac{1}{5} \sum_{j=0}^{4} \zeta^{-j} \sigma^j(\eta).$$

Then ϑ is a Kummer generator of $L_G(\zeta)$. Namely we have $L_G(\zeta) = \mathbb{Q}(\zeta)(\vartheta)$ and $\vartheta^5 \in \mathbb{Q}(\zeta)$. We compute

$$B\vartheta^5 = -(1 + 2\zeta^2 + 2\zeta^3)^5 (5v - 2u\zeta^3 - u\zeta^2 - 2u\zeta)^5 (u\zeta^3 + 2u + 2u\zeta + 5v)^5$$
$$\times (2u + 2u\zeta^2 + u\zeta + 5v)^5 (u\zeta^3 + u + 5v - u\zeta^2 - u\zeta)^5 5^{-15} v^{-10} \in (\mathbb{Q}(\zeta)^*)^5.$$

This implies $L_F(\zeta) = L_G(\zeta)$ as we desired.   □

Hence Lehmer's quintics occupy the locus given by $(u - v : -2v : -4v - u : -3v)$ in the full parameter space $\mathbb{P}^3(k)$.

Since the polynomial $F(u - v : -2v : -4v - u : -3v; T)$ is of moderate length, we write it down here:

$$F(u - v, -2v, -4v - u, -3v; T) = T^5 + \frac{-u^4 - 6u^3v - 20u^2v^2 - 30uv^3 - 25v^4}{u^4 + 5u^3v + 15u^2v^2 + 25uv^3 + 25v^4} T^3$$
$$+ \frac{-u^3v - 4u^2v^2 - 10uv^3 - 10v^4}{u^4 + 5u^3v + 15u^2v^2 + 25uv^3 + 25v^4} T^2 + \frac{u^6v^2 + 7u^5v^3 + 24u^4v^4 + 45u^3v^5 + 50u^2v^6 + 25uv^7}{(u^4 + 5u^3v + 15u^2v^2 + 25uv^3 + 25v^4)^2} T$$
$$+ \frac{v^3(u^9 + 10u^8v + 53u^7v^2 + 184u^3u^6 + 454u^5v^4 + 815v^5u^4 + 1050v^6u^3 + 925v^7u^2 + 500v^8u + 125v^9)}{(u^4 + 5u^3v + 15u^2v^2 + 25uv^3 + 25v^4)^3}.$$

References

1. Kida, M.: Kummer theory for norm algebraic tori. J. Algebra **293** (2005) 427–447
2. Swan, R.G.: Noether's problem in Galois theory. In: Emmy Noether in Bryn Mawr (Bryn Mawr, Pa., 1982). Springer, New York (1983) 21–40
3. Brewer, J.W., Smith, M.K., eds.: Emmy Noether. Volume 69 of Monographs and Textbooks in Pure and Applied Mathematics. Marcel Dekker Inc., New York (1981)
4. Washington, L.C.: Introduction to cyclotomic fields. Second edn. Volume 83 of Graduate Texts in Mathematics. Springer-Verlag, New York (1997)
5. Cohen, H.: Advanced topics in computational number theory. Springer-Verlag, New York (2000)
6. Komatsu, T.: Arithmetic of Rikuna's generic cyclic polynomial and generalization of Kummer theory. Manuscripta Math. **114** (2004) 265–279
7. Hashimoto, K.i., Tsunogai, H.: Generic polynomials over $\mathbb{Q}$ with two parameters for the transitive groups of degree five. Proc. Japan Acad. Ser. A Math. Sci. **79** (2003) 142–145
8. Spearman, B.K., Williams, K.S.: The discriminant of a cyclic field of odd prime degree. Rocky Mountain J. Math. **33** (2003) 1101–1122
9. Schoof, R., Washington, L.C.: Quintic polynomials and real cyclotomic fields with large class numbers. Math. Comp. **50** (1988) 543–556

The Totally Real Primitive Number Fields of Discriminant at Most 10^9

Gunter Malle

Fachbereich Mathematik, Universität Kaiserslautern
Postfach 3049, D–67653 Kaiserslautern, Germany
malle@mathematik.uni-kl.de

Abstract. In this note we report on the enumeration of totally real number fields of discriminant at most 10^9 with no proper subfield and give some statistics on their properties.

1 Introduction

The enumeration of number fields with bounded discriminant (and fixed degree) has a long history. Often the aim was the determination of the field(s) with minimal possible absolute value of the discriminant for given degree and possibly given Galois group of the Galois closure. See for example the article [10]. Recently, the focus has shifted towards investigating the asymptotic behaviour of the counting function

$$N(G, X) := \#\{K/\mathbb{Q} \mid \mathrm{Gal}(K/\mathbb{Q}) = G \text{ and } |d(K/\mathbb{Q})| \leq X\}$$

of fields with Galois group G and discriminant bounded by X, as X goes to infinity, see for example [13,6]. The author has put forward in [13,14] a precise conjecture predicting the main term in the asymptotic of $N(G, X)$, which is known to hold for abelian groups G as well as for all groups of degree at most 4 except the alternating group $\mathfrak{A}_4$.

In this note we report on the computer calculation of all primitive totally real number fields of discriminant at most 10^9 (and in some cases even with larger discriminant). Here, a number field $K/\mathbb{Q}$ is called primitive if it contains no proper subfield except $\mathbb{Q}$. Previously, such lists were often only computed for imprimitive fields, since for those the existence of intermediate fields provides a natural reduction to two (easier) subproblems. Moreover, imprimitive fields of degree less than 10 are solvable and thus lend themselves to methods from class field theory. Thus our results complement earlier extensive computations for imprimitive extensions.

By the unconditional Odlyzko bound [16, p.223], a totally real number field of discriminant at most 10^9 has degree $n = (K/\mathbb{Q})$ at most 9. Assuming the generalized Riemann hypothesis, this can be improved to $n \leq 8$. By the result of Minkowski, for each fixed degree there exist only finitely many fields with bounded discriminant. Thus there is a finite number of totally real number fields of discriminant at most 10^9.

F. Hess, S. Pauli, and M. Pohst (Eds.): ANTS 2006, LNCS 4076, pp. 114–123, 2006.

Theorem 1. *There exist exactly 389 013 654 totally real primitive number fields of discriminant at most 10^9, as given in Table 1.*

Table 1. Totally real primitive number fields

n	number of fields
1	1
2	303,963,559
3	64,659,361
4	17,897,739
5	2,341,960
6	147,600
7	3,432
8	2
total	389,013,654

For degrees 2 and 3, as well as for degree 4 fields with Galois group $\mathfrak{A}_4$, all such fields were known before, see for example [7]. They can be found by using methods from class field theory. For degree 4 extensions with group $\mathfrak{S}_4$ as well as for extensions of degree at least 5, our results are new. In particular this seems to be the first time that large numbers of degree 5 and degree 6 fields with primitive Galois group are enumerated.

The method used to find these fields is an adaptation to the totally real case of the theorem of Hunter (see [4, Thm. 6.4.2] for example) which has already often been used before. This theorem states that in order to enumerate primitive elements of all primitive fields of fixed degree and bounded discriminant, an explicitly given finite set of possible minimal polynomials has to be searched. We have implemented this algorithm in the computer algebra system KANT [9], suitably tailored to just produce totally real fields. The resulting lists of polynomials were reduced using the Pari-gp command `polredabs`. The total computations took several years of CPU-time on a SUN-workstation.

We use our results to study the distribution of class numbers, the maximal number of extensions with fixed discriminant, and possible asymptotic expansions of the counting functions $N(G, X)$.

The leading term in the asymptotic behaviour of $N(\mathfrak{S}_n, X)$ is known for $n \leq 5$, and there is a prediction by Bhargava for $n \geq 6$. Our results show the following phenomenon, which had previously already been observed in degrees $n = 3, 4$: In the range of our tables, the actual number of $\mathfrak{S}_n$-fields is always below the proven respectively expected leading term, and consequently the first error term has a negative sign. Thus, for small discriminants, we find less $\mathfrak{S}_n$-extensions than expected. Furthermore, compared to the Cohen-Martinet heuristic for class numbers [8], we observe less fields than expected with class number larger than 1 in the range of our data. Nevertheless the proportion of such fields increases as the discriminant increases. Finally, in the range of our data the proportion of pairs, triples, ... of fields with equal discriminant increases with the size of the discriminant, so again there seem to be too few multiple discriminants at first.

2 Extensions of Degree 2 and 3

The only (primitive) group in degree 2 is the cyclic group C_2 of order 2. The number of real quadratic fields of discriminant at most 10^9 equals 303,963,559 by [7, Table 2.2], for example.

Both groups in degree 3, the cyclic group C_3 and the symmetric group $\mathfrak{S}_3$ are primitive. The number of totally real extensions of discriminant at most 10^9 can easily be computed using class field theory respectively the method of Belabas [1], and it equals 5008 for C_3-extensions respectively 64,654,353 for $\mathfrak{S}_3$-extensions by [7, Table 3.2 and 4.2].

3 Extensions of Degree 4

The primitive groups in degree 4 are the alternating group $\mathfrak{A}_4$ and the symmetric group $\mathfrak{S}_4$.

According to [7, Table 8.3] there are 2037 totally real $\mathfrak{A}_4$-extensions of $\mathbb{Q}$ of discriminant at most 10^9. It is easy to reproduce this result, obtained by using Kummer theory, with the Hunter method. The asymptotic of $\mathfrak{A}_4$-extensions of $\mathbb{Q}$ is expected to be given by

$$N(\mathfrak{A}_4, X) = c_1 X^{\frac{1}{2}} \log X + c_2 X^{\frac{1}{2}} + O(X^\alpha)$$

for positive constants c_1, c_2 and some $1/4 < \alpha < 1/2$ by [7, 8.1] (see also [14] for a prediction of the leading term). This, however, has not yet been proved (see [5] for partial results).

Table 9.3 in [7] gives the number of totally real $\mathfrak{S}_4$-fields of discriminant at most 10^7. Computations using Hunter's method yield the results in Table 2 for discriminant at most 10^9. (Note that the estimate for totally real $\mathfrak{S}_4$-fields up to 10^8 made in [7, 9.2] is within 0.5 percent of the correct value.)

Table 2. Totally real $\mathfrak{S}_4$-fields

X	$N(\mathfrak{S}_4, X)$	$E(\mathfrak{S}_4, X)$	α_1	α_2	$P(\mathfrak{S}_4, X)$	$\Delta(\mathfrak{S}_4, X)$
10^4	13	240	0.595	0.952	19	0.19
10^5	449	2086	0.664	0.923	462	0.22
10^6	8301	17047	0.705	0.908	8384	0.32
10^7	120622	132855	0.732	0.892	120877	0.34
10^8	1529634	1005137	0.750	0.876	1529780	0.27
10^9	17895702	7464718	0.764	0.870	17896723	0.33

By the theorem of Bhargava [3] the number of totally real $\mathfrak{S}_4$-fields grows linearly with the discriminant, with proportionality constant

$$c_0 := \frac{1}{48} \prod_{p \geq 2} \left(1 + \frac{1}{p^2} - \frac{1}{p^3} - \frac{1}{p^4}\right) = 0.0253477143104\ldots$$

where the product ranges over all primes p. It seems interesting to obtain information on the error term in this asymptotic behaviour. For this assume that $E(\mathfrak{S}_4, X) := c_0 X - N(\mathfrak{S}_4, X) \sim \lambda X^\alpha$ for some $\alpha < 1$, $\lambda > 0$. Then

$$\alpha_1(\mathfrak{S}_4, X) := \log E(\mathfrak{S}_4, X) / \log X$$

and

$$\alpha_2(\mathfrak{S}_4, X) := \log\left(E(\mathfrak{S}_4, X)/E(\mathfrak{S}_4, X/2)\right)/\log(2)$$

both converge to α when $X \to \infty$. In the fourth and fifth column of Table 2 we give $\alpha_1(\mathfrak{S}_4, X)$ and $\alpha_2(\mathfrak{S}_4, X)$. In the range of our data, the first increases monotonically, while the second decreases, suggesting that the exponent α of the error term should satisfy $0.76 \leq \alpha \leq 0.87$.

In fact, the authors of [7] expect, following a communication by Yukie, that $\alpha = 5/6$, and more precisely an asymptotic behaviour of the form

$$N(\mathfrak{S}_4, X) = c_0 X + c_1 X^{\frac{5}{6}} + c_2 X^{\frac{3}{4}} \log X + c_3 X^{\frac{3}{4}} + O(X^\beta)$$

with some $\beta < 3/4$. Using their count up to discriminant 10^7 they present least squares approximations to the constants $c_1, \ldots, c_3$. Assuming this form of the asymptotic expansion, using a least square method with our more extensive results we get the following approximations

$$c_1 = -0.354, \qquad c_2 = 0.012, \qquad c_3 = 0.418,$$

to the constants appearing in the conjectured expansion for $N(\mathfrak{S}_4, X)$. The value of

$$P(\mathfrak{S}_4, X) := c_0 X + c_1 X^{\frac{5}{6}} + c_2 X^{\frac{3}{4}} \log X + c_3 X^{\frac{3}{4}},$$

with c_1, c_2, c_3 as before, is given in column 6 of Table 2. The last column gives the quantity

$$\Delta(\mathfrak{S}_4, X) := \log |N(\mathfrak{S}_4, X) - P(\mathfrak{S}_4, X)| / \log(X).$$

The predicted number of totally real $\mathfrak{S}_4$-fields of discriminant at most 10^{10} equals $199\,133\,067$, which seems out of the range of the Hunter method at present.

We have used our data for $\mathfrak{S}_4$ to count multiplicities of discriminants. This question has aroused some interest recently. For example, Klüners [11] has shown that the number of $\mathfrak{S}_4$-fields with discriminant D is at most equal to $O_\epsilon(D^{1/2+\epsilon})$ for all $\epsilon > 0$. It is conjectured, though, that this number should rather be of the order $O_\epsilon(D^\epsilon)$, an expectation consistent with our data in Table 3.

It may be remarked from the table that the number of 7-tuples of discriminants is more than twice the number of 6-tuples. Most of these 7-tuples arise in the

Table 3. k-tuples of $\mathfrak{S}_4$-discriminants

k	1	2	3	4	5	6	7	8	9	10	11	12	13	14	15
$N(\mathfrak{S}_4)$															
$5 \cdot 10^6$	4117874	191762	150463	7913	1122	388	1016	52	5	5					
10^7	8128708	400282	319821	18453	2588	973	2485	124	19	8	1	1		1	
$1.5 \cdot 10^7$	12103274	614573	494938	29990	4288	1623	4198	197	41	21	2	3		1	1

following way. Take an $\mathfrak{S}_3$-field L of degree 3 whose class group contains an elementary abelian subgroup E of order 8. For any of the seven subgroups of E of order 2, there exists an unramified $\mathfrak{S}_4$-extension of L, so this gives rise to seven $\mathfrak{S}_4$-fields with the same discriminant, all containing L in their Galois closures. Similarly, one may obtain $2^4 - 1 = 15$-tuples of equal discriminants starting from $\mathfrak{S}_3$-fields with 2-rank four.

We have also calculated the class numbers of all $\mathfrak{S}_4$-fields in the range using the `bnfclgp`-command in Pari-gp (so the correctness of the results relies on a heuristic strengthening of the generalized Riemann hypothesis, as described in the Pari-manual). The total number of fields with given class number is displayed in Table 4.

Table 4. Class numbers of $\mathfrak{S}_4$-fields

h	1	2	3	4	5	6	7	8	9	10	11	12
$n(h)$	15354301	2074924	187536	206241	22469	21858	5271	13859	1750	2362	722	1839

h	13	14	15	16	17	18	19	20	21	22	23	24	25	26	27	28	29	30	31	32	33	34	35	36
$n(h)$	322	541	176	683	103	143	51	166	36	70	17	84	16	27	13	35	2	9	2	17	6	4	6	8

h	38	39	40	42	43	44	45	46	48	51	52	53	59
$n(h)$	6	3	5	2	4	2	1	2	3	1	2	1	1

According to the heuristic of Cohen and Martinet [8, (8.3)] the probability that the odd part h'_K of the class number of a totally real $\mathfrak{S}_4$-extension $K/\mathbb{Q}$ of degree 4 equals the odd integer h should be given by

$$\mathrm{pr}(h'_K = h) = c_4 \left(h^4 \prod_{p^j | h} (1 - p^{-j}) \right)^{-1}$$

for some explicit constant $c_4 = .978989...$, where the product runs over all (odd) prime powers dividing h.

In Table 5 we give the relative proportions of the first few class numbers for chunks of 4 500 000 consecutive $\mathfrak{S}_4$-fields, as well as the proportion predicted in [8]. One observes that all class numbers bigger than 1 appear less often than

Table 5. Relative proportions of $\mathfrak{S}_4$-class numbers

h'	1	3	5	7	9	11	13
$1 - 4500000$	.9881	.1025E$-$1	.121E$-$2	.273E$-$3	.08E$-$3	.32E$-$4	.13E$-$4
$4500001 - 9000000$	.9863	.1183E$-$1	.142E$-$2	.316E$-$3	.09E$-$3	.43E$-$4	.22E$-$4
$9000001 - 13500000$	.9855	.1244E$-$1	.147E$-$2	.359E$-$3	.12E$-$3	.50E$-$4	.22E$-$4
$13500001 - 17895702$	.9852	.1274E$-$1	.150E$-$2	.360E$-$3	.13E$-$3	.53E$-$4	.22E$-$4
Cohen $-$ Martinet	.9790	.1813E$-$1	.196E$-$2	.476E$-$3	.25E$-$3	.74E$-$4	.37E$-$4

expected, but the proportion slowly increases with the discriminant. In particular, the data do not seem to contradict the Cohen–Martinet-heuristic.

4 Extensions of Degree 5

All transitive subgroups of $\mathfrak{S}_5$ are primitive. There exist nine C_5-extensions of $\mathbb{Q}$ with discriminant at most 10^9, 302 totally real extensions with the dihedral group D_5 and 196 totally real extensions with the Frobenius group F_{20} of order 20. The results for the alternating and the symmetric group are collected in Table 6. All these can be obtained with Hunter's method. For square discriminants, the ideas presented in [14, Sect. 5.1] allow to get up to discriminant 10^{11}.

Table 6. Totally real $\mathfrak{A}_5$- and $\mathfrak{S}_5$-fields

X	$N(\mathfrak{A}_5, X)$
10^7	7
10^8	80
10^9	398
10^{10}	1874
10^{11}	8121

X	$N(\mathfrak{S}_5, X)$	$E(\mathfrak{S}_5, X)$	α_1	α_2
10^5	8	568	0.551	0.995
10^6	409	5348	0.621	0.970
10^7	9461	48106	0.668	0.947
10^8	162022	413648	0.702	0.927
10^9	2341055	3415647	0.726	0.912
2.10^9	5109739	6403665	0.732	0.907

By an as yet unpublished result of Bhargava (see [2]) the number of $\mathfrak{S}_5$-extensions of $\mathbb{Q}$ should grow linearly with the discriminant, with proportionality factor

$$c_0 := \frac{1}{240} \prod_{p \geq 2} \left(1 + \frac{1}{p^2} - \frac{1}{p^4} - \frac{1}{p^5}\right) = 0.005756702\ldots$$

where again p ranges over all primes. Let $E(\mathfrak{S}_5, X) := c_0 X - N(\mathfrak{S}_5, X)$. In Table 6 for $\mathfrak{S}_5$ we give

$$\alpha_1(\mathfrak{S}_5, X) := \log E(\mathfrak{S}_5, X)/\log(X), \quad \alpha_2(\mathfrak{S}_5, X) := \log \frac{E(\mathfrak{S}_5, X)}{E(\mathfrak{S}_5, X/2)}/\log(2),$$

which seem to indicate that the exponent α of the error term in the asymptotic

$$N(\mathfrak{S}_5, X) = c_0 X + O(X^\alpha)$$

Table 7. k-tuples of $\mathfrak{S}_5$-discriminants

k	1	2	3	4
$N(\mathfrak{S}_5)$				
500000	496190	1881	16	0
1000000	991675	4114	31	1
1500000	1486819	6503	57	1
2000000	1981818	8970	78	2
2500000	2476933	11377	99	4

and also the position of the second right-most pole of the associated ζ-function should satisfy $0.732 \le \alpha \le 0.907$. Note, though, that in the range of our table, the error is still (slightly) larger than the main term.

In Table 7 we have recorded the numbers of multiplets of $\mathfrak{S}_5$-discriminants in the range up to 10^9. Again the growth of the maximal multiplicity is very slow, consistent with the expectation that this maximal multiplicity should be $O_\epsilon(X^\epsilon)$.

Again using the Pari-gp command `bnfclgp` we have calculated the class numbers of the $\mathfrak{S}_5$-fields in the range. The total number of fields with given class number is displayed in Table 8.

By the heuristic of Cohen and Martinet the probability that the part h'_K prime to 5 of the class number of a totally real $\mathfrak{S}_5$-extension $K/\mathbb{Q}$ equals h should be given by

$$\mathrm{pr}(h'_K = h) = c_5 \left(h^5 \prod_{p^j \mid h} (1 - p^{-j}) \right)^{-1}$$

for some explicit constant $c_5 = 0.932929...$, where the product runs over all prime powers p^j, with $p \ne 5$, dividing h. In Table 9 we give the relative proportions of the first few class numbers for chunks of $780\,000$ consecutive $\mathfrak{S}_5$-fields, as well as the proportion predicted in [8]. The same observations as in the $\mathfrak{S}_4$-case apply.

Table 8. Class numbers of $\mathfrak{S}_5$-fields

h	1	2	3	4	5	6	7	8	9	10	11	12	13
$n(h)$	$2,292,467$	$41,944$	$5,094$	901	508	72	43	13	7	3	1	1	1

Table 9. Relative proportions of $\mathfrak{S}_5$-class numbers

h'	1	2	3	4	6	7	8
$1 - 780000$	.9837	.01433	.00166	.00024	.000018	.000009	0
$780001 - 1560000$	.9785	.01874	.00232	.00042	.000027	.000019	.000006
$1560001 - 2340000$	.9762	.02066	.00253	.00049	.000047	.000027	.000010
Cohen $-$ Martinet	.9329	.05831	.00576	.00243	.000360	.000065	.000086

5   Extensions of Degree 6

The primitive groups in degree 6 are the four non-solvable groups $PSL_2(5) \cong \mathfrak{A}_5$, $PGL_2(5) \cong \mathfrak{S}_5$, $\mathfrak{A}_6$ and $\mathfrak{S}_6$. There are precisely two $PGL_2(5)$-fields of discriminant at most 10^9. The data for the other three groups are collected in Table 10.

Table 10. Totally real $PSL_2(5)$-, $\mathfrak{A}_6$- and $\mathfrak{S}_6$-fields

X	$N(PSL_2(5), X)$
10^8	3
10^9	23
10^{10}	139
10^{11}	720

X	$N(\mathfrak{A}_6, X)$
10^9	22
10^{10}	159

X	$N(\mathfrak{S}_6, X)$	α_1
10^6	1	0.509
10^7	177	0.578
10^8	6513	0.629
10^9	147553	0.666
$4 \cdot 10^9$	847116	0.684

The results for $\mathfrak{A}_6$ and $\mathfrak{S}_6$ were obtained by Hunter's method, while the $\mathfrak{A}_5$- and $\mathfrak{S}_5$-fields were converted from the corresponding lists in degree 5 using a KANT-routine. According to a conjecture of Bhargava (see [2]), the number of totally real $\mathfrak{S}_6$-extensions of $\mathbb{Q}$ should grow linearly, with proportionality constant

$$c_1 := \frac{1}{1440} \prod_{p \geq 2} (1 + \frac{1}{p^2} + \frac{1}{p^3} - \frac{2}{p^5} - \frac{1}{p^6}) = 0.0011350200\ldots$$

In Table 10 we also give the value of $\alpha_1(\mathfrak{S}_6, X)$, defined as in the case of $\mathfrak{S}_4$ and $\mathfrak{S}_5$.

In the range considered here, there are at most two $\mathfrak{S}_6$-fields with the same discriminant. The number of pairs of fields with the same discriminant among the first $N(\mathfrak{S}_6)$ fields is given in Table 11.

In Table 12 we present the number $n(h)$ of totally teal $\mathfrak{S}_6$-fields of discriminant at most 10^9 with class number h, again calculated with the Pari-gp command bnfclgp. It turns out that all fields have class number at most 3.

Table 11. k-tuples of $\mathfrak{S}_6$-discriminants

$N(\mathfrak{S}_6)$	singles	pairs
200000	199822	89
400000	399540	230
600000	599292	354
800000	798978	511

Table 12. Class numbers of $\mathfrak{S}_6$-fields

h	1	2	3
$n(h)$	146, 960	578	15

6   Extensions of Degree 7

All transitive subgroups of $\mathfrak{S}_7$ are primitive. The totally real number fields of discriminant at most 10^9 can be enumerated by Hunter's method. One finds that there is one C_7-extension, one extension with the dihedral group D_7, and none with the Frobenius groups of order 21 and 42 (see also [12, Thm. 12]). There are two Kronecker-equivalent fields with group $\mathrm{PSL}_2(7)$ and one with group $\mathfrak{A}_7$. The data for the symmetric group $\mathfrak{S}_7$ are given in Table 13.

Table 13. Totally real $\mathfrak{S}_7$-fields

X	$N(\mathfrak{S}_7, X)$
10^8	71
10^9	3427

According to Pari-gp, all $\mathfrak{S}_7$-fields in the range have class number 1.

7   Extensions of Degree 8 and Higher

There are just two totally real primitive number fields of degree 8 with discriminant at most 10^9, both with Galois group $\mathfrak{S}_8$, with discriminants 483,345,053 respectively 707,295,133, as can be checked by Hunter's method (see also [12, Thm. 13]).

For degree 9, the bounds given by Hunter's theorem are very small, and it turns out that no primitive totally real fields of discriminant at most 10^9 exist. (According to [15] the smallest discriminant equals $9\,685\,993\,193$.) Totally real fields of degree $n \geq 10$ have discriminant larger than 10^{10} by the unconditional Odlyzko bound [16, p.223].

Acknowledgement. I thank Jürgen Klüners for useful conversations on the topic of this paper.

References

1. K. BELABAS, A fast algorithm to compute cubic fields. Math. Comp. **66** (1997), 1213–1237.
2. K. BELABAS, Paramétrisation de structures algébriques et densité de discriminants [d'après M. Bhargava]. Astérisque No. 299 (2005), 267-299.
3. M. BHARGAVA, The density of discriminants of quartic rings and fields. Ann. of Math. **162** (2005), 1031–1063.
4. H. COHEN, *A course in computational algebraic number theory.* Springer-Verlag, Berlin, 1993.
5. H. COHEN, Counting A_4 and S_4 number fields with given resolvent cubic. Pp. 159–168 in: Fields Inst. Commun., 41, Amer. Math. Soc., Providence, RI, 2004

6. H. Cohen, F. Diaz y Diaz, M. Olivier, A survey of discriminant counting. Pp. 80–94 in: Lecture Notes in Comput. Sci., 2369, Springer, Berlin, 2002.

7. H. Cohen, F. Diaz y Diaz, M. Olivier, Counting discriminants of number fields. To appear in J. Th. Nombres Bordeaux.

8. H. Cohen, J. Martinet, Class groups of number fields: numerical heuristics. Math. Comp. **48** (1987), 123–137.

9. M. Daberkow, C. Fieker, J. Klüners, M. Pohst, K. Roegner, K. Wildanger, KANT V4. J. Symbolic Comp. **24** (1997), 267–283.

10. D. Ford, M. Pohst, The totally real A_5 extension of degree 6 with minimum discriminant. Experiment. Math. **1** (1992), 231–235.

11. J. Klüners, The number of S_4-fields with given discriminant. To appear in Acta Arithmetica.

12. J. Klüners, G. Malle, A database for field extensions of the rationals. LMS J. Comput. Math. **4** (2001), 182–196.

13. G. Malle, On the distribution of Galois groups. J. Number Theory **92** (2002), 315–329.

14. G. Malle, On the distribution of Galois groups. II. Experiment. Math. **13** (2004), 129–135.

15. K. Takeuchi, Totally real algebraic number fields of degree 9 with small discriminant. Saitama Math. J. **17** (1999), 63–85.

16. L. C. Washington, Introduction to cyclotomic fields. 2nd ed. Graduate Texts in Mathematics, 83. Springer-Verlag, New York, 1997.

A Modular Method for Computing the Splitting Field of a Polynomial

Guénaël Renault[1,*] and Kazuhiro Yokoyama[2]

[1] LIP6-SPIRAL - Université Paris 6, 4, place Jussieu, F-75005 Paris, France
guenael.renault@lip6.fr
[2] Department of Mathematics, Rikkyo University
3-34-1 Nishi Ikebukuro, Toshima-ku, Tokyo, 171-8501, Japan
yokoyama@rkmath.rikkyo.ac.jp

Abstract. We provide a modular method for computing the splitting field K_f of an integral polynomial f by suitable use of the byproduct of computation of its Galois group G_f by p-adic Stauduhar's method. This method uses the knowledge of G_f with its action on the roots of f over a p-adic number field, and it reduces the computation of K_f to solving systems of linear equations modulo some powers of p and Hensel liftings. We provide a careful treatment on reducing computational difficulty. We examine the ability/practicality of the method by experiments on a real computer and study its complexity.

1   Introduction

This paper is a continuation of Section 5.3 in [21], where, in order to compute the splitting field of an integral polynomial f, the use of the approximations of its roots was suggested. Here we give its details, show its practicality by experiments and provide its complexity study. Moreover we give some techniques in order to increase the feasibility of this new method.

To compute the Galois group G_f of a monic integral polynomial f, the approach of p-adic approximation is very practical (see [21,9,8]). In this approach, one used the approximation of roots of f in a p-adic number field $\mathbb{Q}_p$ (or one of its extensions) in order to find integral roots of the relative resolvents used in Stauduhar's method (see [18]).

For computing the splitting field K_f, there are two approaches: one is constructing this field as a *simple extension* and the other, which is ours, as a *successive extension* given by the *splitting ideal*. Constructing the splitting field as a simple extension can be done by rather simpler computation, where the minimal polynomial of a primitive element of K_f is constructed. (Using p-adic approximations of all its conjugates, it can be computed efficiently.) But, in this setting, if one wants to compute products and sums of several roots of f, i.e.

---

* We wish to acknowledge the Japanese Ministry of Education, Science and Culture which supported the invitation of the first author in the University of Kyushu during September 2004 where this collaboration has been initiated.

F. Hess, S. Pauli, and M. Pohst (Eds.): ANTS 2006, LNCS 4076, pp. 124–140, 2006.

one wants to do arithmetic operations in $K_f \cong \mathbb{Q}[x_1, ..., x_n]/\mathcal{M}$, where each variable corresponds each root of f and $\mathcal{M}$ is the *splitting ideal* generated by all algebraic relations of roots of f, one have to compute the expressions of roots with respect to the primitive element. On the other hand, in our approach, we compute a Gröbner basis $\mathcal{G}$ of the splitting ideal $\mathcal{M}$ and hence, it is easy to perform arithmetic operations in $\mathbb{Q}[x_1, ..., x_n]/\mathcal{M}$. Moreover, in general, expressions by primitive elements tend to be suffered "expression swell", that is, huge coefficients appear and those harm the efficiency. So, for our purpose, simple extension does not seem suited.

In order to compute the splitting ideal $\mathcal{M}$ of a polynomial, there is a classical approach due to Kronecker using algebraic factoring algorithms. But, as shown in [2], it does not seem practical for polynomials having large Galois groups. Here, to overcome the difficulty, we use the knowledge of certain algebraic structures: the p-adic approximation of roots and the explicit action of the Galois group G_f. For the computation of a Gröbner basis of $\mathcal{M}$ we compute a theoretical form of our output with indeterminate coefficients representing the polynomials generating the basis. Then, we compute these polynomials by solving linear systems modulo a power of p and Hensel liftings. For the theoretical form, there is a well known dense generic one based on the knowledge of the degrees of the polynomials (see [21,4]). In Section 3, we show how a *careful study* on the symmetric representation of G_f allows to produce a sparser theoretical form and how to avoid the computation of polynomials in the basis. From this study we obtain, for a given symmetric representation of G_f, a *scheme* for the computation of $\mathcal{G}$. In Section 4, we show how to compute the polynomials of $\mathcal{G}$ with linear algebra and Hensel lifting and provide an effective test for an *early detection* strategy. We emphasise that one can combine other methods for the computation of $\mathcal{G}$ with the proposed scheme. For example we could combine sparse interpolations strategy effectively (dense interpolation formulas are given in [6,12]), this will be study in a future work. We also note that it is possible to translate the results presented in this article to polynomials over global fields.

2   Preliminaries

We provide necessary notions and summarize some results of [21].

2.1   Splitting Field and Galois Group over $\mathbb{Q}$

Let $f(x)$ be a monic square-free integral polynomial of degree n and $\underline{\alpha}$ the set of all its roots in an algebraic closure $\bar{\mathbb{Q}}$ of $\mathbb{Q}$. The splitting field K_f of f is the extension field $\mathbb{Q}(\underline{\alpha})$ obtained by adjoining $\underline{\alpha}$ to $\mathbb{Q}$. The group G_f of $\mathbb{Q}$-automorphisms of K_f acts faithfully on $\underline{\alpha}$, thus one can consider the permutation representation G_f of this group. Fixing a numbering of the roots $\underline{\alpha} = \{\alpha_1, ..., \alpha_n\}$ of f, G_f is viewed as a subgroup of S_n. The group G_f is called the Galois group of f.

To express K_f symbolically, the following epimorphism ϕ of $\mathbb{Q}$-algebra is considered:

$$\mathbb{Q}[x_1, ..., x_n] \ni x_i \longmapsto \alpha_i \in K_f$$

For simplicity, we write $X = \{x_1, \ldots, x_n\}$ and, more generally, for a subset E of $\{1, \ldots, n\}$ we write $X_E = \{x_i : i \in E\}$. Then K_f is represented by the residue class ring $\mathcal{A}$ of the polynomial ring $\mathbb{Q}[X]$ factored by the kernel $\mathcal{M}$ of ϕ. We call $\mathcal{M}$ *the splitting ideal of f associated with the assignment of the roots* $\alpha_1, \ldots, \alpha_n$. In this setting, computing K_f means to compute a *Gröbner basis* $\mathcal{G}$ of $\mathcal{M}$ (see [5]). If we choose the lexicographic order $\prec$ on terms with $x_1 \prec \cdots \prec x_n$, then the reduced Gröbner basis of $\mathcal{M}$ coincides with the generating set $\{g_1, g_2, \ldots, g_n\}$ obtained by *successive extensions*, that is, for each i,

1. g_i is a polynomial in $x_1, \ldots, x_i$ and monic with respect to x_i, and
2. $\mathbb{Q}(\alpha_1, \ldots, \alpha_i) \cong \mathbb{Q}[X_{\{1,\ldots,i\}}]/\langle g_1, \ldots, g_i \rangle$, where $\langle F \rangle$ denotes the ideal generated by an element or a set F. This implies that g_i is an irreducible factor of $f(x_i)$ over $\mathbb{Q}[X_{\{1,\ldots,i-1\}}]/\langle g_1, \ldots, g_{i-1} \rangle$ such that $g_i(\alpha_1, \ldots, \alpha_i) = 0$.

Thus this reduced Gröbner basis can be obtained by "algebraic factoring methods" (see [2]) and is said to be a *triangular basis* (see [11,6]). For a Gröbner basis $\mathcal{G} \subset \mathbb{Q}[X]$ and a polynomial P, let $\mathrm{NF}(P, \mathcal{G})$ denote the normal form of P in $\mathbb{Q}[X]$ with respect to $\mathcal{G}$ (see [5]).

The group S_n acts naturally on $\mathbb{Q}[X]$ with $x_i^\sigma = x_{i^\sigma}$ for $1 \leqslant i \leqslant n$ and $\sigma \in S_n$. Thus G_f is the $\mathbb{Q}$-automorphisms group of $\mathcal{A}$ denoted by $\mathrm{Aut}_\mathbb{Q}(\mathcal{A})$ (see [2,1]). We use the following notation for groups: for a group G acting on a set $\mathcal{S}$, the stabilizer in G of an element or a subset A of $\mathcal{S}$ is denoted by $\mathrm{Stab}_G(A)$, i.e. $\mathrm{Stab}_G(A) = \{\sigma \in G : A^\sigma = A\}$. If G is the full symmetric group on $\mathcal{S}$, we simply write $\mathrm{Stab}(A)$ for $\mathrm{Stab}_G(A)$. We denote by $\mathrm{Stab}_G([a_1, \ldots, a_k])$ the pointwise stabilizer of a subset $A = \{a_1, \ldots, a_k\}$ of $\mathcal{S}$, i.e. $\mathrm{Stab}_G([a_1, \ldots, a_k]) = \{\sigma \in G \mid a_i^\sigma = a_i, \forall i \in \{1, \ldots, k\}\}$. The set of right cosets of H in G is denoted by $H\backslash G$ and the set of all representatives of $H\backslash G$ by $H\backslash\backslash G$.

Definition 1. *We call the ideal generated by* $t_1 + a_1, \ldots, t_n + (-1)^{n-1}a_n$, *where* t_i *is the i-th elementary symmetric function on X and* $f(x) = x^n + a_1 x^{n-1} + \cdots + a_n$, *the* universal splitting ideal *of f and denote it by $\mathcal{M}_0$. We call the residue class ring $\mathbb{Q}[X]/\mathcal{M}_0$ the* universal splitting ring *of f over $\mathbb{Q}$ and denote it by $\mathcal{A}_0$.*

The reduced Gröbner basis of $\mathcal{M}_0$ is composed of the n *Cauchy's modules* of f (see [16]). Since S_n stabilizes $\mathcal{M}_0$, S_n also acts faithfully on $\mathcal{A}_0$, i.e. $S_n \subset \mathrm{Aut}_\mathbb{Q}(\mathcal{A}_0)$. We have the following theorem (see [14,3,21] for details and other references).

Theorem 1. *There is a one-to-one correspondence between the set of all primitive idempotents of $\mathcal{A}_0$ and the set of all prime divisors of $\mathcal{M}_0$. Let e be the primitive element corresponding to the fixed prime divisor $\mathcal{M}$. Then, $G_f = \mathrm{Stab}(\mathcal{M}) = \mathrm{Stab}(e)$ and $\mathcal{M}^\sigma = \{g \in \mathbb{Q}[X] \mid ge^\sigma = 0 \in \mathcal{A}_0\}$. Moreover, we have $\mathcal{M}_0 = \cap_{\sigma \in G_f \backslash\backslash S_n} \mathcal{M}^\sigma$ and $\mathcal{A}_0 = \oplus_{\sigma \in G_f \backslash\backslash S_n} e^\sigma \mathcal{A}_0 = \oplus_{\sigma \in G_f \backslash\backslash S_n} \mathbb{Q}[X]/\mathcal{M}^\sigma$.*

2.2 Splitting Field over p-adic Number Field

Now we consider the relation between the splitting ring over $\mathbb{Q}$ and that over a p-adic field $\mathbb{Q}_p$. The n-tuple $\underline{\alpha} = \{\alpha_1, \ldots, \alpha_n\}$ and the splitting ideal $\mathcal{M}$

associated with the assignment x_i to α_i are fixed. The primitive idempotent of $\mathcal{A}$ corresponding to $\mathcal{M}$ is denoted by e. For a prime integer p, we denote by $\mathbf{Z}_p^0$ (resp. $\mathbf{Z}_p$) the localization of $\mathbf{Z}$ at p (resp. the completion of $\mathbf{Z}_p^0$). We denote by π_p the projection from $\mathbf{Z}_p[X]$ to $\mathbb{F}_p[X]$ (the natural extension of the projection from $\mathbf{Z}$ to $\mathbb{F}_p$).

From now on, we fix a prime number p such that $\pi_p(f)$ is square-free. Let $\bar{\mathcal{M}}_0$ denote the ideal $\pi_p(\mathcal{M}_0 \cap \mathbf{Z}_p^0[X])$ in $\mathbb{F}_p[X]$ and $\mathcal{G}_0$ denotes the standard generating set of $\mathcal{M}_0$. By construction, the Cauchy's modules of f are polynomials with integral coefficients and monic in their greatest monomial. Thus, the set $\pi_p(\mathcal{G}_0)$ is a Gröbner basis of $\pi_p(\mathcal{M}_0 \cap \mathbf{Z}_p^0[X])$. Moreover, $\mathcal{G}_0$ is a Gröbner basis of the universal splitting ideal $\mathbb{Q}_p \otimes_{\mathbb{Q}} \mathcal{M}_0$ of f as a polynomial with coefficients in $\mathbb{Q}_p$ and that of $\mathbf{Z}_p[X] \otimes_{\mathbf{Z}_p^0} (\mathcal{M}_0 \cap \mathbf{Z}_p^0[X])$ over $\mathbf{Z}_p$. The ideal $\mathbb{Q}_p \otimes_{\mathbb{Q}} \mathcal{M}_0$ is denoted by $\mathcal{M}_0^{(\infty)}$. We denote $\mathbb{F}_p[X]/\bar{\mathcal{M}}_0$ by $\bar{\mathcal{A}}_0$ and $\mathbb{Q}_p[X]/\mathcal{M}_0^{(\infty)}$ by $\mathcal{A}_0^{(\infty)}$.

Theorem 2. *We have the following assertions:*
1. The projection π_p gives a one-to-one correspondence between the set of all primitive idempotents of $\mathcal{A}_0^{(\infty)}$ and that of $\bar{\mathcal{A}}_0$. Moreover, for each pair $(\bar{e}, e^{(\infty)})$ of corresponding primitive idempotents, $\mathrm{Stab}(\bar{e}) = \mathrm{Stab}(e^{(\infty)})$.
2. The idempotent e of $\mathcal{A}_0$ is also an idempotent of $\mathcal{A}_0^{(\infty)}$. Let $\bar{e}$ be a component of $\pi_p(e)$ and $e^{(\infty)}$ the primitive idempotent of $\mathcal{A}_0^{(\infty)}$ corresponding to $\bar{e}$. Then $\mathrm{Stab}(e)$ contains $\mathrm{Stab}(\bar{e})(=\mathrm{Stab}(e^{(\infty)}))$ and $\mathrm{Stab}(\pi_p(e)) = \mathrm{Stab}(e)$. Moreover, by letting $\mathcal{S} = \mathrm{Stab}(\bar{e})\backslash\backslash \mathrm{Stab}(e)$, $\pi_p(e) = \sum_{\sigma \in \mathcal{S}} \bar{e}^\sigma$ and $e = \sum_{\sigma \in \mathcal{S}} e^{(\infty)\sigma}$.

Now we fix a component $\bar{e}$ of $\pi_p(e)$ and its corresponding idempotent $e^{(\infty)}$ of $\mathcal{A}_0^{(\infty)}$. Let $\bar{\mathcal{M}}$ be the maximal ideal of $\mathbb{F}_p[X]$ corresponding to $\bar{e}$ and $\mathcal{M}^{(\infty)}$ the maximal ideal of $\mathbb{Q}_p[X]$ corresponding to $e^{(\infty)}$. Moreover, let $\mathcal{G}^{(\infty)}$ and $\bar{\mathcal{G}}$ be the reduced Gröbner basis of $\mathcal{M}^{(\infty)}$ and that of $\bar{\mathcal{M}}$ respectively.

Definition 2. *Let $\mathcal{G}^{(\infty)} = \{g_1^{(\infty)}, \ldots, g_n^{(\infty)}\}$. For a positive integer k, we call the set $\{g_1^{(\infty)} \bmod p^{k+1}, \ldots, g_n^{(\infty)} \bmod p^{k+1}\}$ the k-th approximation to $\mathcal{G}^{(\infty)}$ and denote it by $\mathcal{G}^{(k)}$. We note that $\mathcal{G}^{(0)} = \bar{\mathcal{G}}$.*

We can lift $\bar{\mathcal{G}}$ to $\mathcal{G}^{(\infty)}$ by Hensel construction. More precisely we have:

Theorem 3. *The reduced Gröbner basis $\mathcal{G}^{(\infty)}$ of $\mathcal{M}^{(\infty)}$ with respect to $\prec$ is contained in $\mathbf{Z}_p[X]$, and $\bar{\mathcal{G}}$ is lifted uniquely to $\mathcal{G}^{(\infty)}$ by Hensel construction.*

Proof. Theorem 21 in [21] gives the result and a construction based on a *linear iteration Hensel lifting*. Actually, its *quadratic iteration* version can be restated for this construction (see [15]). □

3   The Computation Scheme

In this section, we propose a framework for the computation of a Gröbner basis $\mathcal{G} = \{g_1, \ldots, g_n\}$ of the splitting ideal $\mathcal{M}$ of f with indeterminate coefficients strategy. We now assume the Galois group G_f of f is already computed as a subgroup of S_n. We show how the knowledge of the symmetric representation G_f can give a *good* theoretical form of $\mathcal{G}$, and then we provide some *techniques* which permit us to avoid computations of some g_i.

3.1   The Form of $\mathcal{G}$

Since we compute polynomials g_i with indeterminate coefficients strategy, we need to know the *potential terms* which may appear in g_i. The following allows to deduce $\deg_i(g_i)$, the degree in x_i of g_i, from $G_f = \mathrm{Stab}(\mathcal{M})$.

Proposition 1 (Theorem 5.3 [4]). *The degree d_i of g_i in x_i is given by*

$$d_i = |\mathrm{Stab}_{G_f}([1,\dots,i-1])|/|\mathrm{Stab}_{G_f}([1,\dots,i])|.$$

Reciprocally, the next proposition gives the characterization of all the Gröbner bases of $\mathcal{M}$. Its proof is immediate (see [5]).

Proposition 2. *Let $\mathcal{G} = \{g_1,\dots,g_n\}$ be a triangular set of polynomials of $\mathcal{M}$ such that $\deg_i(g_i) = d_i$. Then, $\mathcal{G}$ is a Gröbner basis of $\mathcal{M}$. Note that $\mathcal{G}$ is not necessarily reduced but it is minimal (see [5]).*

Thus, we want to compute such a triangular set $\mathcal{G}$. A generic form for such a Gröbner basis $\mathcal{G}$ can be retrieved from this: the terms of g_i's monomials are potentially $x_i^{k_i} x_{i-1}^{k_{i-1}} \cdots x_1^{k_1}$ with $0 \leqslant k_j < d_j$. In this case, the number of indeterminate coefficients is of the order of G_f which may be very large (this dense form is considered in [12]). Clearly, the sparser the basis $\mathcal{G}$ is, the most efficient the computation is, thus we are interested in finding a sparse one. For this task we introduce a definition.

Definition 3. *Let i be an integer in $\{1,\dots,n\}$. A subset E of $\{1,\dots,i\}$ containing i is said to be an i-relation if there exists a polynomial r_i in $\mathbb{Q}[X_E]$ such that*

$$\alpha_i^{d_i} + r_i(\underline{\alpha}) = 0 \ \text{and} \ \deg_i(r_i) < d_i.$$

An i-relation corresponds to a potential g_i in any $\mathcal{G}$, for example, the sets $\{1,\dots,i\}$, for $i = 1,\dots,n$, are the i-relations corresponding to the generic form of $\mathcal{G}$. The following proposition permits us to easily find an i-relation which may be smaller. Its proof is immediate by considering a minimal polynomial of α_i (see [15]).

Proposition 3. *Let i be an integer in $\{1,\dots,n\}$ and m be the minimal integer in $\{1,\dots,i-1\}$ such that $|\mathrm{Stab}_{G_f}([1,\dots,m])|/|\mathrm{Stab}_{G_f}([1,\dots,m,i])| = d_i$. Then, there exists an i-relation in $\{1,\dots,m,i\}$.*

If E_i is the maximal i-relation $\{1,\dots,i\}$ then, as one can see above, it is easy to identify the potential terms of the corresponding polynomial. The following result, which is a consequence of classical Galois theory, gives us the way of doing the same for more general i-relations:

Proposition 4. *Let $E = \{e_1 < e_2 < \cdots < e_s = i\}$ be an i-relation. Then, there exists a polynomial r_i as in Definition 3 such that*

$$deg_j(r_i) < |\mathrm{Stab}_{G_f}([e_1,\ldots,e_{j-1}])|/|\mathrm{Stab}_{G_f}([e_1,\ldots,e_j])|, \; \forall j \in \{1,\ldots,s\}\,.$$

The preceding proposition provides a relation between an i-relation and the maximal degree of each variable of the corresponding polynomial g_i. We now want to know the size of g_i.

Definition 4. *Let $E_i = \{e_1 < e_2 < \cdots < e_s = i\}$ be an i-relation. We define the finite sequence $d(E_i)_{e_1},\ldots,d(E_i)_{e_s}$ by*

$$d(E_i)_{e_j} = |\mathrm{Stab}_{G_f}([e_1,\ldots,e_{j-1}])|/|\mathrm{Stab}_{G_f}([e_1,\ldots,e_j])|, \; \forall j \in \{1,\ldots,s\}\,.$$

The degree of E_i is defined by $\prod_{j=1}^s d(E_i)_{e_j}$ and is denoted by $D(E_i)$.

Given an i-relation $E_i = \{a < b < \cdots < l = i\}$, then the number of terms $x_a^{k_a} x_b^{k_b} \cdots x_l^{k_l}$ which potentially appear in the corresponding g_i is $D(E_i)$. There might be different i-relations, so we give a *partial order* among all the i-relations.

Definition 5. *Let i be an integer in $\{1,\ldots,n\}$. An i-relation E_i is said to be minimal if $D(E_i)$ is minimal (among all the i-relation) and not any proper subset of E_i is an i-relation.*

We note that a minimal i-relation $E_i = \{e_1 < e_2 < \cdots < e_s = i\}$ verifies $d(E_i)_{e_j} \geqslant 2$ for all j, $1 \leqslant j < s$. Minimal i-relations for each $i = 1,\ldots,n$ correspond to polynomials g_i with a minimal number of coefficients and thus to a Gröbner basis $\mathcal{G}$ which have a sparse form. Note that an i-relation satisfying conditions of Proposition 3 may not be minimal.

3.2 Reducing the Number of Polynomials to Compute

We assume that the symmetric representation of G_f and an i-relation E_i for each i in $\{1,\ldots,n\}$ are known. Here we give techniques to avoid some computations of elements of $\mathcal{G}$. These techniques were already used in [13] with a partial knowledge of G_f. However, since we know the exact symmetric representation of G_f, we make use here of the whole power of these techniques.

Cauchy modules technique. Let $\mathcal{G} = \{g_1,\ldots,g_n\}$ be a triangular Gröbner basis of the ideal $\mathcal{M}$ with $deg_i(g_i) = d_i$. Let $\mathcal{O} = \{i_1 < i_2 < \cdots < i_k\}$ be the orbit of i under the action of $\mathrm{Stab}_{G_f}([1,\ldots,i-1])$. Then $i_1 = i$ and $k = d_i$. For a multivariate polynomial g, we denote by $\mathrm{E}(g,u)$ the multivariate polynomial obtained by replacing the greatest variable in g by a newly introduced indeterminate u. Then, the d_i (generalised) *Cauchy modules* of g_i are defined by: $c_1(g_i) = g_i$,

$$c_2(g_i) = \frac{\mathrm{E}(c_1, x_{i_2}) - \mathrm{E}(c_1, x_{i_1})}{(x_{i_2} - x_{i_1})}, \quad \cdots, \quad c_{d_i}(g_i) = \frac{\mathrm{E}(c_{d_i-1}, x_{i_{d_i}}) - \mathrm{E}(c_{d_i-1}, x_{i_{d_i-1}})}{(x_{i_{d_i}} - x_{i_{d_i-1}})}\,.$$

By construction, the following holds:

Lemma 1. *The Cauchy module $c_j(g_i)$ is a polynomial of $\mathbb{Q}[X_{\{1,\ldots,i_j\}}]$ which is monic as a polynomial in x_{i_j} with $\deg_{i_j}(c_j(g_i)) = d_i - j + 1$. Moreover, the polynomial $c_j(g_i)$ is in $\mathcal{M}$.*

As we know the symmetric representation of G_f we can know in advance if $c_j(g_i)$ has the same degree, in x_{i_j}, as g_{i_j}. In this case, in $\mathcal{G}$, g_{i_j} can be replaced by $c_j(g_i)$ and this set is still a Gröbner basis of $\mathcal{M}$ (see Proposition 2). So, in the construction of $\mathcal{G}$ we avoid the computation of g_{i_j}.

Transporters technique. Here we use the fact that the group G_f is the stabilizer of the ideal $\mathcal{M}$. Let $E_i = \{e_1 < e_2 < \cdots < e_s = i\}$ be an i-relation and $j \in \{i+1, \ldots, n\}$. A permutation $\sigma \in G_f$ is said to be an (i, j)-*transporter* if it satisfies:

$$\sigma(i) = j \text{ and } j = \max(\{\sigma(e) : e \in E_i\})$$

Proposition 5. *Let σ be an (i, j)-transporter and g_i the polynomial corresponding to E_i. Then, $\mathrm{NF}(g_i^\sigma, \{g_1, \ldots, g_{j-1}\})$ is a multiple of g_j as polynomials in $\mathcal{A} = (\mathbb{Q}[X_{\{1,\ldots,j-1\}}]/\langle g_1, \ldots, g_{j-1}\rangle)[x_j]$.*

Proof. Since σ is an (i, j)-transporter, the polynomial $\mathrm{NF}(g_i^\sigma, \{g_1, \ldots, g_{j-1}\})$ can be viewed as a univariate polynomial h in x_j over $\mathbb{Q}(\alpha_1, \ldots, \alpha_{j-1})$. Moreover, since $g_i^\sigma \in \mathcal{M}$, we have $h(\alpha_j) = 0$. Thus h is a multiple of the minimal polynomial of α_j over $\mathbb{Q}(\alpha_1, \ldots, \alpha_{j-1})$, hence h is a multiple of g_i as a polynomial of $\mathcal{A}$. $\square$

Corollary 1. *With the same notations as in Proposition 5, if the degree d_j is equal to d_i then g_i^σ can take the place of g_j in $\mathcal{G}$.*

As for the Cauchy's techniques, from the knowledge of an (i, j)-transporter σ satisfying conditions of Corollary 1, we can avoid the computation of the polynomial g_j since it can be replaced by g_i^σ.

4   Computing Splitting Fields by Linear Systems Solving

In this section, we assume the knowledge of G_f with its action over approximations of the roots of f in $\bar{\mathbb{Q}}_p$. Moreover, we assume that the *computation scheme* attached to G_f is known, in particular we know a corresponding i-relation E_i for each polynomial g_i of $\mathcal{G}$. We show how these knowledges can be used for the computation of $\mathcal{G}$ by linear systems solving. We denote by $Z(I)$ the algebraic variety associated to an ideal I of $\mathbb{Q}[X]$ or $\mathbb{F}_p[X]$.

4.1   Computation by Solving Systems of Linear Equations

Here we compute $g_1, \ldots, g_n$ by a *method of indeterminate coefficients*. Assume that the n-tuple $\underline{\alpha} = (\alpha_1, \ldots, \alpha_n)$ of roots of f lie in $Z(\mathcal{M})$. Recall that G_f is already presented as a sub-group of S_n and $\mathrm{Stab}(\mathcal{M}^{(\infty)}) = \mathrm{Aut}_{\mathbb{F}_p}(\mathbb{F}_p[X]/\bar{\mathcal{M}}) = G_{\pi_p(f)} \subset G_f$. We denote $|G_f|$ and $|G_{\pi_p(f)}|$ by N and $\bar{N}$, respectively.

Systems over the rationals. We fix an integer $i \in \{1, \ldots, n\}$. Each coefficient of g_i is replaced with an indeterminate, for simplicity, the terms $\prod_{e \in E_i} x_e^{m_e}$, where $0 \leqslant m_e < d(E_i)_e$, are sorted with respect to the lexicographic order and denoted by $t_1, \ldots, t_{D(E_i)}$. Then, with indeterminates $a_j^{(i)}$, we have $g_i = x_i^{d_i} + \sum_{j=1}^{D(E_i)} a_j^{(i)} t_j$. Since $\mathcal{G}$ is supposed to be a Gröbner basis of $\mathcal{M}$, the following equation holds for i.

$$g_i(\gamma) = 0 \text{ for every } \gamma \in Z(\mathcal{M}). \tag{1}$$

Let $E_i = \{e_1 < e_2 < \cdots < e_s\}$ and $\gamma = (\gamma_1, \ldots, \gamma_n)$ be an element of $Z(\mathcal{M})$. We denote by $\gamma(E_i)$ the projection of γ on the indexes given by E_i (i.e. $(\gamma_{e_1}, \ldots, \gamma_{e_s})$) and $Z(\mathcal{M})(E_i) = \{\gamma(E_i) : \gamma \in Z(\mathcal{M})\}$. Thus, we have $|Z(\mathcal{M})(E_i)| = D(E_i)$. Let G_{E_i} be the group $\mathrm{Stab}_{G_f}([e_1, \ldots, e_s])$ and $G_{E_i} \backslash\backslash G_f = \{\sigma_1, \ldots, \sigma_{D(E_i)}\}$. Then, we have $Z(\mathcal{M})(E_i) = \{\alpha(E_i)^{\sigma_1}, \ldots, \alpha(E_i)^{\sigma_{D(E_i)}}\}$ and

$$g_i(\gamma) = 0 \text{ for every } \gamma \in Z(\mathcal{M})(E_i). \tag{2}$$

The system (2) of equations becomes a linear system of $D(E_i)$ equations and $D(E_i)$ variables with matrix representation $-V_i = M_i A_i$, where $A_i = (a_j^{(i)})$, $V_i = ((\alpha_i^{d_i})^{\sigma_r})$ and $M_i = (t_c(\alpha(E_i)^{\sigma_r}))_{r,c}$ with $(r, c) \in \{1, \ldots, D(E_i)\}^2$. Since the set $\{t_1(\alpha(E_i)), \ldots, t_{D(E_i)}(\alpha(E_i))\}$ is a $\mathbb{Q}$-linear basis of $\mathbb{Q}(\{\alpha_e : e \in E_i\})$, this system has a unique solution. Thus we can compute g_i by solving the system of linear equations if we already know the *exact value of each root* α_i *of* f.

Systems over p-adic numbers. As we do not know the exact value of each α_i, we use the approximate value of roots of f in $\bar{\mathbb{Q}}_p$. In the sequel we use the same notations as Section 2. The ideal $\mathcal{M}$ may not be maximal if it is considered as an ideal in $\mathbb{Q}_p[X]$, more precisely we have:

Proposition 6. *Let $\mathcal{S}$ be the transversal* $\mathrm{Stab}(\bar{e})\backslash\backslash\mathrm{Stab}(e)$. *Then* $\mathbb{Q}_p \otimes_{\mathbb{Q}} \mathcal{M} = \cap_{\sigma \in \mathcal{S}} (\mathcal{M}^{(\infty)})^{\sigma}$, *and* $\pi_p(\mathcal{M} \cap \mathbf{Z}_p^0) = \cap_{\sigma \in \mathcal{S}} (\bar{\mathcal{M}})^{\sigma}$.

Proof. Let e be the idempotent of $\mathcal{A}_0$ corresponding to $\mathcal{M}$. As $\mathcal{M} = \{h \in \mathbb{Q}[X] \mid eg = 0 \in \mathbb{Q}[X]/\mathcal{M}_0\}$, the first equation can be derived directly from Theorem 1 (2) and Theorem 2 (2). The second equation can be also derived by considering the projection π_p. $\square$

By Proposition 6, we can reduce the system (2) to the following.

$$g_i(\gamma) = 0 \text{ for every } \gamma \in \cup_{\sigma} Z((\mathcal{M}^{(\infty)})^{\sigma})(E_i), \tag{3}$$

where σ ranges in $\mathcal{S} = G_{\pi_p(f)} \backslash\backslash G_f$. The system (3) consists of $D(E_i)$ variables and $D(E_i)$ linear equations over $\mathbb{Q}_p[X]/\mathcal{M}^{(\infty)}$ and it is equivalent to

$$\mathrm{NF}(g_i, (\mathcal{G}^{(\infty)})^{\sigma}) = 0 \text{ for every } \sigma \in G_{E_i} \backslash\backslash G_f. \tag{4}$$

Moreover, replacing $\mathcal{G}^{(\infty)}$ with $\mathcal{G}^{(k)}$, we have the following system which g_i mod p^{k+1} must satisfy.

$$\mathrm{NF}(g_i, (\mathcal{G}^{(k)})^{\sigma}) \equiv 0 \pmod{p^{k+1}} \text{ for every } \sigma \in G_{E_i} \backslash\backslash G_f. \tag{5}$$

The system (5) is considered as a system of $D(E_i)$ variables and $D(E_i)$ linear equations with coefficients in $(\mathbf{Z}/p^{k+1}\mathbf{Z})[X]/\mathcal{M}^{(k)}$. Especially, for the case $k = 0$, the system (5) is translated to the following system which $\pi_p(g_i)$ must satisfy: Fix a zero $\bar{\alpha} = (\bar{\alpha}_1, \ldots, \bar{\alpha}_n)$ in $Z(\bar{\mathcal{M}})$, and set $\pi_p(g_i) = x_i^{d_i} + \sum_{j=1}^{D(E_i)} \bar{a}_j^{(i)} t_j$. Let $\bar{A}_i = (\bar{a}_j^{(i)})$, $\bar{V}_i = ((\alpha_i^{d_i})^{\sigma_r})$ and $\bar{M}_i = (t_c(\bar{\alpha}(E_i)^{\sigma_r}))_{r,c}$ with $(r,c) \in \{1, \ldots, D(E_i)\}^2$. Then we have the identity $-\bar{V}_i = \bar{M}_i \bar{A}_i$.

Theorem 4. *For each i, $1 \leqslant i \leqslant n$, the following holds.*
1. The linear system corresponding to $-\bar{V}_i = \bar{M}_i \bar{A}_i$ has a unique solution over $\mathbb{F}_p$ which gives $\pi_p(g_i)$.
2. For a positive integer k, the system (5) has a unique solution which gives the approximation $g_i \bmod p^{k+1}$. Moreover, we can construct $g_i \bmod p^{k+1}$ from $\pi_p(g_i)$ by Hensel lifting.

Proof. Consider the expansion of $\det(M_i)$ and that of $\mathrm{disc}(f)$, where we consider each root α_i as an indeterminate y_i. Then, it can be shown that $\mathrm{disc}(f) = \prod_{j \neq k}(y_j - y_k)$ and by *discriminant composition formula* (see [14]) there exist integers $e_{j,k}$ such that $\det(M_i) = \prod_{1 \leqslant j < k \leqslant n}(y_j - y_k)^{e_{j,k}}$. As $\pi_p(f)$ is square-free, we conclude that $\det(\bar{M}_i) \neq 0$ and so the linear system corresponding to $-\bar{V}_i = \bar{M}_i \bar{A}_i$ has a unique solution and thus, the unique solution gives $\pi_p(g_i)$. We can show the second statement by the same argument and the fact that $\det(\bar{M}_i) \neq 0$. For the Hensel lifting we would like to apply the same construction as in Theorem 3. Since the ring $A = \mathbb{F}_p[X]/\pi_p(\mathcal{M} \cap \mathbf{Z}_p^0)$ is not a field, two cases are possible when we compute the *Bézout relation* with the *Extended Euclidean Algorithm* (EEA) with pseudo division in the first step of this lifting: At the end of the EEA a gcd is computed and it is invertible, in this case the lifting can continue; when EEA does not work, we can compute the Bézout relation by other methods. In this second case, we may use combination of EEA over A and *Chinese Remainder Theorem* or solving a system of linear equations derived from this relation. One can see also [17] for a general study about *Newton-Hensel* operator for general triangular sets. □

Remark 1. At each step k, the Hensel lifting of a polynomial g_i which corresponds to an i-relation $E_i = \{e_1 < \cdots < e_s = i\}$ can be done with two different points of view. The first one is to considerate g_i as a univariate polynomial with coefficients in the ring $R_{2k} = (\mathbb{Z}/p^{2k}\mathbb{Z})[X_{\{1,\ldots,i-1\}}]/\langle g_1, \ldots, g_{i-1}\rangle$. The second one is to see g_i in the univariate polynomial ring with coefficients in $R'_{2k} = (\mathbb{Z}/p^{2k}\mathbb{Z})[X_{E_i \setminus \{x_i\}}]/\langle g_1^*, \ldots, g_{s-1}^*\rangle$ where the polynomials g_j^* lying in $(\mathbb{Z}/p^{2k}\mathbb{Z})[X_{\{e_1,\ldots,e_j\}}]$ are the approximations of the polynomials which defines the extensions $\mathbb{Q}(\alpha_{e_1})$, $\mathbb{Q}(\alpha_{e_1}, \alpha_{e_2})$, ..., $\mathbb{Q}(\alpha_{e_1}, \alpha_{e_2}, \ldots, \alpha_{e_s})$. ($\{g_1^*, \ldots, g_{s-1}^*, g_i\}$ is the reduced Gröbner basis of the elimination ideal $\mathcal{M} \cap \mathbb{Q}[X_{E_i}]$.) In the latter case, we compute each g_j^* by solving linear system and Hensel lifting in the same manner as computation of g_i, recursively from g_1^* to g_{s-1}^*. We may also obtain g_j^* by transporter techniques by inspecting the action of G_f. In the former case, at the end of the lifting procedure the Gröbner basis $\mathcal{G}$ is necessarily reduced, but not in the latter case.

Theorem 4 gives two possible strategies (which can be mixed) for the computation of $\mathcal{G}_k = \{g_1 \bmod p^{k+1}, \ldots, g_n \bmod p^{k+1}\}$ a k-approximation of a triangular Gröbner basis $\mathcal{G}$ of $\mathcal{M}$:

1: By Hensel lifting, $\mathcal{G}^{(k)}$ is constructed from $\overline{\mathcal{G}}$ (see Theorem 3). From $\mathcal{G}^{(k)}$ we construct and solve the system 5 for each i, $1 \leqslant i \leqslant n$, the solutions are then $\mathcal{G}_k$.
2: From $\overline{\mathcal{G}}$ we construct and solve the systems 5 for each i, $1 \leqslant i \leqslant n$. The solutions are $\mathcal{G}_0$ and we can construct $\mathcal{G}_k$ by Hensel lifting.

Now, assume $\mathcal{G}_k = \{g_1 \bmod p^{k+1}, \ldots, g_n \bmod p^{k+1}\}$ is computed. Then we convert each $g_i \bmod p^{k+1}$ to a polynomial over $\mathbb{Q}$ by the well-known *rational reconstruction* technique. Let B_i be a bound on all absolute values of the numerators and denominators of coefficients of g_i. Then, as soon as $2B_i^2 < p^{k+1}$, the polynomial converted from $g_i \bmod p^{k+1}$ coincides with g_i (see [7]).

4.2 Estimation of the Bound B_i

Here we give details on the bound B_i for the rational reconstruction. Since coefficients of g_i correspond to the solution of the system (2), by Cramer's rule, the denominator of each coefficient of g_i divides $\det(M_i)$ and the numerator of the j-th coefficient of g_i divides $\det(M_i^{(j)})$, where $M_i^{(j)}$ is the matrix obtained by replacing the j-th column with V_i.

Lemma 2. *Let B_0 be the maximum of the absolute values of roots α_i's of f in $\mathbb{C}$. Then, for each i, B_i can be computed from $\{d(E_i)_e : e \in E_i\}$ and B_0.*

Proof. We assume w.l.o.g. that the bound B_0 is greater than 1. For each row of $M_i^{(j)}$ and each row of M_i, by replacing each α_k with B_0 and by denoting $d(E_i)_e$ by d_e we can bound the square-norm of these rows by the integer $\mathbb{B}_i^2 = \prod_{e \in E_i}(1 + B_0^2 + \cdots + B_0^{2(d_e-1)}) + B_0^{2d_i} = \prod_{e \in E_i} \frac{B_0^{2d_e}-1}{B_0^2-1} + B_0^{2d_i}$. Thus, as the determinant of a matrix is bounded by the product of square-norms of its rows (by the inequality of Hadamard), we can set $B_i = \mathbb{B}_i^{D(E_i)}$. □

If $B_0 > 2$, then we can set B_i as $B_0^{D(E_i)(\sum_{e \in E_i} d(E_i)_e)}$ and, since $\sum_{e \in E_i} d(E_i)_e \leqslant \sum_{1 \leqslant k \leqslant i} d_k \leqslant \sum_{1 \leqslant k \leqslant i} k$, the bit size of B_i is bounded by $O(n^2 D(E_i) \log(B_0))$. For the denominator, we can give a precise bound (see [10]).

Lemma 3. *For each i, there is a positive integer C_i computed from the set of degrees $\{d(E_i)_e : e \in E_i\}$ such that each $d(f)^{C_i} g_i$ belongs to $\mathbb{Z}[X]$.*

Proof. By the discriminant identity given in the proof of Theorem 4, $\det(M_i)$ is considered as a polynomial in each α_i. Then estimating the degree of $\det(M_i)$ in each α_j, we can obtain a bound on the denominators of coefficients of g_i. In fact, the degree of $\det(M_i)$ in α_j is bounded by $D_i = \frac{D(E_i)(\sum_{e \in E_i} d_e)}{n_0}$, where $n_0 = n$ if f is irreducible over $\mathbb{Q}$, and $n_0 = 1$ otherwise. Then, from the shape of $\mathrm{disc}(f)$, it can be shown easily that $C_i = \frac{D_i}{2}$ satisfies the statement. Moreover, if f is irreducible, we can set $C_i = \frac{D_i}{2(n-1)}$. □

The bound B_i given in Lemma 2 is in general very pessimistic. We will see in Section 4.3 how the problem of pessimistic theoretical bound can be avoided.

4.3 Check of Correctness and Early Detection

To improve the efficiency of the method, we can incorporate "early detection strategy" which is widely used in computer algebra. As the bound B_i tends to be large compared to the exact value, the technique is supposed to work very well in our case.

Conversion at Early Stage. Assume that we have computed $\mathcal{G}_k$, even though p^{k+1} does not exceed the theoretical bound. Suppose that we have obtained the first $j-1$ polynomials $\{g_1, \ldots, g_{j-1}\}$ of $\mathcal{G}$. We want to test if the Hensel lifting is enough for $g_j \mod p^{k+1}$. Thus, we try to convert it to a candidate polynomial over $\mathbb{Q}$ by rational reconstruction. Then we first check the following:

1. The conversion is done successfully for every coefficients of $g_i \mod p^{k+1}$.
2. The denominator of each coefficient of a candidate polynomial divides a certain power of $\mathrm{disc}(f)$ (See Lemma 3).

If the conversion does not satisfy the criteria above then p^{k+1} is not sufficient to afford the correct g_j. Thus, we continue the lifting process again. If, in the contrary, the conversion, say h_j, satisfy the criteria we have to prove that $h_j = g_j$ this is what we do now.

Correctness of Solution. Assume that we have a candidate polynomial h_j for the polynomial g_j corresponding to the j-relation E_j. We can check if $h_j = g_j$ by the following theorem.

Theorem 5. *We have* $h_j = g_j$ *if and only if* $\mathrm{NF}(c_j(f), \{g_1, \ldots, g_{j-1}, h_j\}) = 0$.

Proof. The *only-if-part* is clear, we have only to show the *if-part*. Let H be the triangular set $\{g_1, \ldots, g_{j-1}, h_j\}$. By the hypothesis, the ideal $\langle H \rangle$ contains $\{c_1(f), \ldots, c_j(f)\}$ which is the reduced Gröbner basis of the elimination ideal $\mathcal{M}_0 \cap \mathbb{Q}[X_{\{1,\ldots,j\}}]$. Thus, $\langle H \rangle$ is contained in a maximal ideal $\mathcal{M}'$ of $\mathbb{Q}[X_{\{1,\ldots,j\}}]$, which coincides with $\mathcal{M}^\sigma \cap \mathbb{Q}[X_{\{1,\ldots,j\}}]$ for some $\sigma \in S_n$. But, comparing the dimensions of the residue class rings, it follows that $\langle H \rangle = \mathcal{M}' = \mathcal{M}^\sigma \cap \mathbb{Q}[X_{\{1,\ldots,j\}}]$. Seeing their stabilizers, σ is the identity and $h_j = f_j$. □

By the similar manner and considering $\mathbb{Q}[X_{E_j}]$, we have alternative test for h_j in the case where Hensel liftings are done over R'_{2k}, see Remark 1.

Theorem 6. *Let* $h_j^*, \ldots, h_{s-1}^*, h_j$ *be constructed polynomials by Hensel lifting using* R'_{2k}, *where* $E_j = \{e_1, \ldots, e_{s-1}, e_s = j\}$. *We have* $h_j = g_j$ *if and only if* $\mathrm{NF}(f(x_{e_m}), \{g_1^*, \ldots, g_{s-1}^*, h_j\}) = 0$ *for all* m, $1 \leqslant m \leqslant s$.

5 Algorithms

Here we give a brief survey on the algorithms underlying of this method. We first give an algorithm for the construction of a *computation scheme*, then we give an algorithm for the computation of *splitting ideals*.

5.1   A Database of Computation Schemes

Given a subgroup G of S_n the following algorithm computes a corresponding *computation scheme*.

Algorithm 1: COMPUTATIONSCHEME(G)

> **Step 1** Compute the degrees $\deg_i(g_i)$ for $i = 1, \ldots, n$ (see Proposition 1).
> **Step 2** Apply the *Cauchy's technique* (see Lemma 1). Let $\mathcal{I}$ be the set of integers corresponding to the indexes of the g_i which cannot be obtained with this technique.
> **Step 3** For each integer i in $\mathcal{I}$, compute a minimal i-relation and store it in $\mathcal{E}$.
> **Step 4** Apply *transporter technique* on the i-relations in $\mathcal{E}$. Let $\mathcal{E}$ be the set of i-relations corresponding to the g_i which must be computed.
> **Return** $\mathcal{E}$ with the techniques for retrieving the other polynomials.

The set $\mathcal{E}$ depends only on the choice of the representative for G and on the chosen i-relations in **Step 3**. This set represents all the linear systems which are solved in our method. Thus, a measure of complexity is given by $|\mathcal{E}| = \sum_{E \in \mathcal{E}} D(E)$ and, in Algorithm 1, we compute $\mathcal{E}$ with minimal $|\mathcal{E}|$.

Definition 6. *For a given sub-group G of S_n, the minimal value of $|\mathcal{E}|$ is called the c-size of G and is denoted by $c(G)$.*

A conjugate of G with minimal c-size is called *c-minimal*. In a conjugacy class there may be a big difference, in term of c-size, between two of its representatives. For example, in the conjugacy class of $[2^4]S_4$ there are two representatives G_1 and G_2 with $c(G_1) = 8$ and $c(G_2) = 632$.

5.2   Algorithm for the Computation of Splitting Fields

Assume that the *computation scheme* of G_f is pre-computed (w.l.o.g. we can choose a representative of G_f which is c-minimal). We also suppose that all transversals of groups needed in our algorithm are pre-computed.

Given the polynomial f of degree n, our method for computing a Gröbner basis $\mathcal{G} = \{g_1, g_2, \ldots, g_n\}$ is describe with the following algorithm. We give only

Algorithm 2: SPLITTINGIDEAL($\mathcal{G}^{(k_0)}, G_f, p$)

> Let $\mathcal{I}$ be the indexes of the g_i we have to compute with linear systems.
> **for** $i = 1$ to n **do**
> > **if** $i \in \mathcal{I}$ **then**
> > > Construct/Solve $\mathcal{S}$ the linear system mod p^{k_0+1} corresponding to E_i.
> > > **S1: try** to convert the solution s_i of $\mathcal{S}$ to a rational polynomial h_i
> > > **if** the conversion of s_i above succeed **and** h_i satisfies the correctness test
> > > **then** The polynomial h_i is g_i.
> > > **else** Apply an Hensel lifting to s_i and goto step **S1**.
> > **else**
> > > Apply a Cauchy/Transporter technique in order to obtain g_i from g_j with $j < i$
> > **end if**
> **end for**
> **Return** $\mathcal{G}, G_f$.

the algorithm where early detections are used. One could use the theoretical bounds by applying some minor modifications (fix the exponent of p, cancel the *early detection* tests). A variant of Algorithm 2 is presented in [15].

5.3   Complexity Analysis

In this section, we study the complexity of Algorithm 2 *focusing on effects of the quantity $c(G)$*. We assume that a database containing a computation scheme of a c-minimal representative of each conjugacy class is already known. For *simplifying our analysis and extracting its typical behavior related to $c(G)$*, we choose liftings over R'_{2k} (see Remark 1) and consider a case where $k_0 = 0$ in input and EEA with pseudo division always works in the first step of the lifting. Also we assume that $\overline{N} = 1$, as this property is desired in efficient Galois group computation [9,21], and $\log \log(B_0)$ is quite small compared with n for B_0 defined in Lemma 2.

Since we use the *early detection strategy*, the complexity of our algorithm also depends on the size of the coefficients of the output $\mathcal{G}$. Let B_{true} be the maximum of the absolute values of denominators and numerators of coefficients of g_j^* and g_i appearing in the computation. By Lemma 2, the theoretical bound B_i on the coefficients of g_i can be also on those of g_j^*. Thus, B_{true} is supposed very much smaller than $B_1, \ldots, B_n$. In the sequel, for each integer $k \geqslant 0$, we denote by $\mathbb{M}(k)$ the cost of arithmetic over $\mathbb{Z}/p^{k+1}\mathbb{Z}$ as number of word operations. As the size of necessary p^{k+1} tends to be huge, we may apply fast multiplication techniques over $\mathbb{Z}/p^{k+1}\mathbb{Z}$. On the other hand, as the size n which can be handled here is not so large, we use ordinary techniques for polynomial multiplication.

We now sketch the complexity of each step of Algorithm 2 for computing one polynomial g_i with respect to the pre-computed i-relation $E_i = \{e_1, \ldots, e_s\}$. We note that the number of iterations is bounded by $O(\log \log(B_{true}))$.

Linear algebra: To compute a polynomial $g_i \bmod p$ with respect to the i-relation E_i, we have to construct the matrix $\bar{M}_i$ and solve $-\bar{V}_i = \bar{M}_i \bar{A}_i$ for $\bar{A}_i$. Under the assumption, the matrix $\bar{M}_i$ is constructed directly as a matrix over $\mathbb{F}_p$, and its construction takes $O(D(E_i)^2 \mathbb{M}(0))$ word operations. Then we solve the resulted $D(E_i) \times D(E_i)$ linear system which requires $O(D(E_i)^\omega \mathbb{M}(0))$ word operations. (Here, ω represents a feasible matrix multiplication exponent and $2 \leqslant \omega \leqslant 3$, see [20].) Thus, in total, it takes $O(D(E_i)^\omega \mathbb{M}(0))$ word operations.

Hensel lifting: At each step k, $g_i \bmod p^k$ is lifted to $g_i \bmod p^{2k}$ and this computation is executed over $R'_{2k} = (\mathbb{Z}/p^{2k}\mathbb{Z})[X_{E_i} \setminus \{x_i\}]/\langle g_1^*, \ldots, g_{s-1}^* \rangle$ (see Remark 1). At this step, by using ordinary polynomial multiplication, it takes $O(n^2)$ arithmetic operations over R'_{2k}, and hence it takes $O(n^2 D(E_i)^2 \mathbb{M}(2k-1))$ word operations. At the first step of the lifting, we also compute s, t in $R_1[x_i]$ such that (Bézout relation) $s\pi_p(f(x_i)) + t(g_i \bmod p) = 1$ by EEA, which takes $O(n^2 D(E_i)^2 \mathbb{M}(0))$ word operations. As we use quadratic Hensel construction, the total cost is dominated by the same order for the final step, and thus, it takes $O(n^2 D(E_i)^2 \mathbb{M}(\log(B_{true})))$ word operations.

Rational reconstruction: As each coefficient $a_j^{(i)}$ of $g_i \bmod p^{2k}$ can be converted to a rational number by EEA of $a_j^{(i)}$ and p^{2k}. By applying fast GCD computation techniques [20], it takes $O(\mathbb{M}(\log(B_{true})) \log \log(B_{true}))$ word operations for each $a_j^{(i)}$, as we can use the same symbol $\mathbb{M}(\log(B_{true}))$ for the cost of one multiplication of integers of word size $O(\log(B_{true}))$. Then, in total, it takes $O(D(E_i)\mathbb{M}(\log(B_{true}))(\log \log(B_{true}))^2)$ word operations. From the computed bound in Lemma 2 for B_{true}, $\log \log(B_{true}) = O(n \log(n))$ and the total cost of rational reconstruction is dominated by the cost of Hensel construction.

Auxiliary computation: As the computation of g_i is executed over R'_{2k}, $g_1^*, \ldots, g_{i-1}^*$ must already computed. (Some can be easily converted from already constructed g_j, $j < i$.) Each g_j^* is constructed by linear algebra and Hensel construction in the same manner as g_i, and it takes $O(D_j^\omega \mathbb{M}(0) + n^2 D_j^2 \mathbb{M}(\log(B_{true})))$ word operations, where $D_j = \prod_{\ell=1}^j d(E_i)_{e_\ell}$. As E_i is set to be minimal, $d(E_i)_{e_j} \geqslant 2$ for each $j < s$ and it follows easily that $\sum_{\ell=1}^{s-1} n^2 D_j^2 = O(n^2 D(E_i)^2)$ and $\sum_{\ell=1}^{s-1} D_j^\omega = O(D(E_i)^\omega)$. Hence the cost of auxiliary computation is dominated by the cost of Hensel construction steps for g_i.

Normal form computation: We use the same notation as in Auxiliary computation. For the correctness of g_i, normal forms of $f(x_{e_1}), \ldots, f(x_{e_{s-1}}), f(x_i)$ with respect to $\{g_1^*, \ldots, g_{s-1}^*, g_i\}$ are computed. These computations can be executed via powers of x_{e_j} and so it takes $O(\log(n)D_1^2 + \cdots + \log(n)D_{s-1}^2) = O(\log(n)D(E_i)^2)$ arithmetic operations over $\mathbb{Z}$.

Thus, by summing the quantities above among all the polynomials g_i, we obtain the following result:

Theorem 7. *Algorithm 2 with $k_0 = 0$ takes*

$$O(c(G)^\omega \mathbb{M}(k_0) + n^2 c(G)^2 \mathbb{M}(\log(B_{true})) + L)$$

word operations, where L is the total cost of normal form computations in correctness tests. Letting B' be the maximum of absolute values of integers appearing in normal form computations, L can be bounded by $O(\log(n)c(G)^2 \mathbb{M}(\log(B')))$ $\log \log(B_{true})$). (When k_0 is general we have almost the same result.) Moreover, for cases where the word size of B' is the almost same order as that of B_{true}, the above estimation can be simplified to $O(c(G)^\omega \mathbb{M}(k_0) + n^2 c(G)^2 \mathbb{M}(\log(B_{true})))$.

As B_{true} is a bound on coefficients of g_j^* and g_i, it might be greater than the actual bound B on coefficients of g_i's. But, in many cases for computation of successive extensions, the final element has coefficients of the maximal absolute value. Thus, for representing actual behaviors of computation, it may be allowed to use B_{true} instead of B.

6   Experiments and Remarks

We have implemented Algorithm 2 with the *computer algebra system* MAGMA (version 2.11) in the case of an irreducible monic integral polynomial. We choose

MAGMA since it has all the functionalities needed (Galois group computation, multivariate polynomial ring, permutation group). We have computed a database of c-minimal representatives (with their *computation scheme*) of each conjugacy class of transitive groups of degree up to 11. The experiments we made show that this first implementation is already very efficient.

Choice of the prime p: By Tchebotarev's density theorem, it is possible to compute a prime p such that $\overline{N} = 1$ and it may be found among $\mathcal{O}(|G_f|)$ number of primes. In our implementation, we choose the smallest such prime. One can see in the table that the time taken by this procedure is not significant compared with the rest of the computation.

The power k_0: In our implementation we take $k_0 = 10$. In this case, none of the tests presented in table need to be lifted after the linear resolution: the *early detection tests* pass. We will investigate, in a future work, some other power k_0 and compare the efficiency with the case where the Hensel lifting is needed.

Experiments timings: We tested polynomials from the database `galpols` of MAGMA. We give, for each example, the name of the group G in Butler and McKay's nomenclature, the order of G and the integer $c(G)$ (as the sum of the i-relations degrees). The column Tcheb. shows the timings of computing a prime p such that $\overline{N} = 1$, the column p gives this prime. The column Galois shows the timings of computing the Galois group, Matrix/Solve those for constructions and resolutions of the matrices respectively, NF the timings for the normal forms computations and Total the total timing

group	$|G|$	$c(G)$	Tcheb.	p	Galois	Matrix/Solve	NF	Total
$6T_{12}$	60	60 + 60	0.13	929	0.06	0.22 / 0.17	0.04	0.66
$6T_{13}$	72	12	0.11	619	0.03	0.01 / 0.01	0	0.18
$6T_{14}$	120	120	0.15	1447	0.05	0.44 / 0.44	0.06	1.18
$6T_{15}$	360	360	0.22	2437	0.0	3.69 / 6.51	0.21	10.79
$7T_5$	168	42	0.19	1879	0.06	0.05 / 0.04	0.04	0.41
$8T_{32}$	96	8 + 96 + 96	0.34	3413	0.13	0.55 / 0.59	0.14	1.870
$8T_{33}$	96	96 + 32	0.23	2099	0.14	0.32 / 0.3	0.34	1.42
$8T_{34}$	96	24 + 24 + 95	0.09	229	0.14	0.34 / 0.24	0.09	0.99
$8T_{35}$	128	8 + 16	0.31	2909	0.06	0.01 / 0.01	0.01	0.45
$8T_{36}$	168	168 + 168	0.06	211	0.14	1.78 / 1.59	1.63	5.360
$8T_{37}$	168	168 + 168	0.31	2969	0.1	1.76 / 2.26	1.15	5.72
$8T_{38}$	192	96 + 8	0.26	2503	0.1	0.29 / 0.29	0.05	1.09
$8T_{39}$	192	8 + 192	0.16	947	0.06	1.14 / 1.44	0.2	3.11
$8T_{41}$	192	24 + 96	0.4	4271	0.13	0.33 / 0.32	0.06	1.32
$8T_{42}$	288	24 + 24	0.46	5051	0.1	0.05 / 0.02	0.02	0.71
$8T_{43}$	336	336	0.29	3209	0.12	3.48 / 6.09	3.84	14.0
$8T_{44}$	384	8	1.05	14071	0.06	0.01 / 0.01	0.05	1.24
$8T_{45}$	576	24 + 576	0.36	3719	0.06	10.21 / 22.87	1.18	35.1
$8T_{46}$	576	24 + 576	0.56	6269	0.1	10.25 / 23.72	1.1	36.14
$8T_{47}$	1152	24	1.27	17299	0.05	0.03 / 0.02	0.0	1.44
$8T_{48}$	1344	336	5.56	78497	0.08	3.56 / 8.56	20.33	38.3
$9T_{21}$	162	54 + 54	0.59	6047	1.08	0.2 / 0.16	0.54	2.72
$9T_{22}$	162	27 + 54	0.12	461	0.16	0.13 / 0.09	0.08	0.65
$9T_{23}$	297	216 + 72	0.16	727	0.31	3.13 / 5.17	1.37	10.4
$9T_{24}$	324	18 + 108	0.24	1801	1.07	0.4 / 0.38	2.23	4.45
$9T_{25}$	324	27 + 324	0.16	953	1.03	3.41 / 5.49	0.33	10.63
$9T_{26}$	432	72	0.98	10273	0.3	0.18 / 0.16	7.43	9.15
$9T_{27}$	504	504	0.79	10103	0.42	7.98 / 18.6	105.49	133.64
$9T_{28}$	648	27	0.33	3037	1.38	0.03 / 0.02	0.01	1.87
$9T_{29}$	648	18 + 648	0.75	7883	0.43	13.17 / 38.74	1.44	55.21
$9T_{31}$	1296	18	0.33	2801	1.0	0.01 / 0.01	0.03	1.53
$9T_{32}$	1512	1512 + 1512	0.46	5167	0.27	142.17 / 608.1	1761.84	2523

of the procedure. The measurements were made on a personal computer with a 1.5Ghtz Intel Pentium 4 and 512MB of memory running GNU/Linux. As one can see, the size of $c(G)$ and the size of p^{k_0} heavily influenced the timings of constructions and resolutions of matrices like Theorem 7 shows. When $c(G)$ is big, two cases are possible: few big matrices to compute or a lot of little matrices to compute. The first case is more time consuming than the second. This is why there are some differences between examples with same size of $c(G)$ and p^{k_0} (for example, see the lines $8T_{37}$ and $6T_{15}$).

7   Conclusion and Future Works

We have presented a new method, with theoretical and practical aspects, for the computation of the splitting field of a polynomial f where the knowledge of the action of the Galois group over p-adic approximations of its roots is used.

We have introduced the notion of *computation scheme*. This new approach seems a good way for efficient computation of splitting fields. This framework is not limited to be used with linear systems solving. For example, we will study the integration of sparse interpolation formulas (like the dense ones in [6,12]) in our algorithm. Also, it would be interesting to study the possibility of using this approach in a dynamical strategy like the one of MAGMA (see [19]).

References

1. I. Abdeljaouad, S. Orange, G. Renault, and A. Valibouze. Computation of the decomposition group of a triangular ideal. *AAECC*, 15(3-4):279–294, 2004.
2. H. Anai, M. Noro, and K. Yokoyama. Computation of the splitting fields and the Galois groups of polynomials. In *Algorithms in algebraic geometry and applications (Santander, 1994)*, volume 143 of *Progr. Math.*, 29–50. Birkhäuser, Basel, 1996.
3. J.-M. Arnaudiès and A. Valibouze. Lagrange resolvents. *J. Pure Appl. Algebra*, 117/118:23–40, 1997. Algorithms for algebra (Eindhoven, 1996).
4. Ph. Aubry and A. Valibouze. Using Galois ideals for computing relative resolvents. *J. Symbolic Comput.*, 30(6):635–651, 2000. Algorithmic methods in Galois theory.
5. T. Becker and V. Weispfenning. *Gröbner bases*, volume 141 of *Graduate Texts in Mathematics*. Springer-Verlag, New York, 1993. In cooperation with H. Kredel.
6. X. Dahan and É. Schost. Sharp estimates for triangular sets. In *Proceedings of ISSAC '04*, pages 103–110, New York, NY, USA, 2004. ACM Press.
7. J. H. Davenport, Y. Siret, and E. Tournier. *Computer algebra*. Academic Press Ltd., London, second edition, 1993.
8. K. Geissler. *Berechnung von Galoisgruppen über Zhal- und Funktionenkörpern*. PhD thesis, Universität Berlin, 2003.
9. K. Geissler and J. Klüners. Galois group computation for rational polynomials. *J. Symbolic Comput.*, 30(6):653–674, 2000. Algorithmic methods in Galois theory.
10. L. Langemyr. Algorithms for a multiple algebraic extension. II. In *Proceedings of AAECC-9*, volume 539 of *LNCS*, pages 224–233. Springer, Berlin, 1991.
11. D. Lazard. Solving zero-dimensional algebraic systems. *J. Symbolic Comput.*, 13(2):117–131, 1992.
12. M. Lederer. Explicit constructions in splitting fields of polynomials. *Riv. Mat. Univ. Parma (7)*, 3*:233–244, 2004.
13. S. Orange, G. Renault, and A. Valibouze. Calcul efficace de corps de décomposition. LIP6 Research Report 005, Laboratoire d'Informatique de Paris 6, 2003.
14. M. Pohst and H. Zassenhaus. *Algorithmic Algebraic Number Theory*. Cambridge Univ. Press, Cambridge, 1989.
15. G. Renault. *Calcul efficace de corps de décomposition*. PhD thesis, Université Paris 6, 2005.
16. N. Rennert and A. Valibouze. Calcul de résolvantes avec les modules de Cauchy. *Experiment. Math.*, 8(4):351–366, 1999.
17. É. Schost. *Sur la résolution des systèmes polynomiaux à paramètres*. PhD thesis, École polytechnique, 2000.

18. R. Stauduhar. The determination of galois groups. *Math. Comp.*, 27:981–996, 1973.

19. A. K. Steel. A new scheme for computing with algebraically closed fields. In *ANTS-V*, Volume 2369 of *LNCS*, pages 491–505. Springer, Berlin, 2002.

20. J. von zur Gathen and J. Gerhard. *Modern computer algebra*. Cambridge University Press, Cambridge, second edition, 2003.

21. K. Yokoyama. A modular method for computing the Galois groups of polynomials. *J. Pure Appl. Algebra*, 117/118:617–636, 1997. Algorithms for algebra (Eindhoven, 1996).

On the Density of Sums of Three Cubes

Jean-Marc Deshouillers[1], François Hennecart[2], and Bernard Landreau[3]

[1] IMB, CNRS - Université Bordeaux 1 - Université Victor Segalen Bordeaux 2,
F-33076 Bordeaux Cedex, France
`jean-marc.deshouillers@math.u-bordeaux1.fr`
[2] LAMUSE, Université Jean-Monnet, F-42023 Saint-Etienne Cedex, France
`francois.hennecart@univ-st-etienne.fr`
[3] LAREMA, CNRS - Université d'Angers, F-49045 Angers Cedex, France
`bernard.landreau@univ-angers.fr`

Abstract. We give evidence that sums of 3 cubes have a positive density, the value of which, $0.0999425\ldots$, is that given by a probabilistic model we developed earlier.

1 Introduction

We are addressing a question concerning the statistical behaviour of the integers which can be expressed as a sum of s integers which are s^{th}-powers, and more specifically of sums of 3 cubes.

In general, the number $r_s(n)$ of representations of the integer n as a sum of s s^{th}-powers is bounded on average, since one readily checks the following relation

$$\frac{x}{s} \sim \left(\sum_{m \le (x/s)^{1/s}} 1\right)^s \le \sum_{n \le x} r_s(n) \le \left(\sum_{m \le x^{1/s}} 1\right)^s \sim x. \tag{1}$$

One may thus expect that $r_s(n)$ should behave according to some sort of a Poisson law and thus that the set of those integers n for which $r_s(n) > 0$ has a positive density. This approach was first considered by Erdős and Rényi in 1960 (cf. [4]), who built the first probabilistic model for the study of sums of s s^{th}-powers.

The case of squares ($s = 2$) has now been known for almost one century: Landau (cf. [6]) showed that the number of integers which are sums of 2 squares and less than x is equivalent to $Cx/\sqrt{\log x}$ for some explicit positive constant C, which implies that sums of 2 squares have a zero density. The inadequacy of the probabilistic model in this case can be related to the multiplicative structure of $r_2(.)$ or to the irregular distribution of sums of 2 squares in arithmetic progressions.

No multiplicative relation for $r_s(.)$ is known when $s \ge 3$ nor is likely to exist. As regards the irregularity of the distribution of s s^{th}-powers in arithmetic progressions, we built a probabilistic model taking it into account and leading to an almost sure positive density for the sums of s *quasi* s^{th}-powers when $s \ge 3$ and to an almost sure zero density when $s = 2$ (cf. [2]).

F. Hess, S. Pauli, and M. Pohst (Eds.): ANTS 2006, LNCS 4076, pp. 141–155, 2006.

In a previous ANTS meeting (cf. [3]), we presented computations concerning sums of 4 biquadrates (4th-powers) and showed their adequacy with our probabilistic model. More precisely, if we denote by $\nu^4(x)$ (resp. $\nu_d^4(x)$) the frequency of the sums of 4 biquadrates up to x, taking into account all the representations (resp. only the representations with distinct summands), we saw that for $x \leq 10^{14}$, the function $\nu^4(x)$ (resp. $\nu_d^4(x)$) essentially behaves as a decreasing (resp. increasing) function that would converge to the limit value given by our probabilistic model.

The case of the cubes is surprisingly different. First, the functions $\nu^3(x)$ and $\nu_d^3(x)$ (that will be respectively denoted by $\nu(x)$ and $\nu_d(x)$ in the sequel) start to be decreasing (when suitably smoothed); moreover, for x around 10^{12}, both functions fall below the probabilistic value[1]. After a few weeks of computation, reaching 10^{13}, we saw that ν_d was changing its behaviour and was starting to increase. However, the function ν was still decreasing; only after a few months, we saw, around 10^{14}, the function ν change its behaviour: computing up to $5 \cdot 10^{14}$ led us to think that we could possibly not predict whether our probabilistic model is adequate or not before reaching 10^{18}, a value which cannot be attained by actual computation. We thus changed our strategy, and stopped computing $\nu(x)$ and $\nu_d(x)$ for all values of x, but only for values of x belonging to some arithmetic progressions we thought of being "generic"; the different results match perfectly well one with the other: **We are now in a position to give strong evidence that sums of 3 cubes have a positive density, the value of which, 0.0999425..., is that given by our probabilistic model.**

2   The Probabilistic Density

For any modulus $K \geq 1$, and any integer k_0, we denote by $\rho(k_0, K)$ the number of integral triples $(k_1, k_2, k_3) \in [1, K]^3$ such that

$$k_0 \equiv k_1^3 + k_2^3 + k_3^3 \pmod{K}. \tag{2}$$

We put $\gamma = \Gamma(4/3)^3/6 = 0.1186\ldots$, and introduce

$$\delta(K) = \frac{1}{K} \sum_{k=1}^{K} \exp\left(-\gamma\rho(k, K)/K^2\right). \tag{3}$$

It is shown in [2] that $\delta(K)$ is multiplicatively increasing, in the sense that $K|K'$ implies $\delta(K) \leq \delta(K')$. Thus by putting

$$K_B = \prod_{p^\alpha \leq B} p^\alpha, \tag{4}$$

we deduce that $\delta = \lim_{B \to +\infty} \delta(K_B)$ exists. It is also proved in [2] that $\delta < 1$.

---

[1] To check this point, it was convenient to compute this expected limit to seven decimal places; we are indebted to Philippe Flajolet who suggested the use of Mellin's transform.

Our aim is now to give arguments supporting the conjecture that $1 - \delta$ is the asymptotic density of the sequence of sums of three cubes.

In order to compute the actual value of δ with a given precision, we shall use the bounds given in [3, Lemma, p. 197] that we recall here: for each K, one has

$$0 \le \delta - \delta(K) \le \frac{\gamma^2}{2}\left(\mathfrak{S} - \mathfrak{S}_2(K)\right), \tag{5}$$

where

$$\mathfrak{S}_2(K) = \frac{1}{K}\sum_{k=1}^{K}\left(\frac{\rho(k,K)}{K^2}\right)^2 \quad \text{and} \quad \mathfrak{S} = \lim_{B\to\infty}\mathfrak{S}_2(K_B).$$

3   Expressing δ in Terms of Gauss Sums

As in [3], we proceed in two steps: we first compute $\delta(K)$ for a given suitable value of K (as large as possible), which gives a lower bound for δ. We then deduce an upper bound by computing the corresponding $\mathfrak{S}_2(K)$, and also the limit value $\mathfrak{S}$. For this we write $\mathfrak{S}_2(K)$ as a product on prime divisors of K, and thus $\mathfrak{S}$ as an Eulerian product.

In the next section, we show that we have essentially two methods for computing $\delta(K)$, either by expanding it into a power series, or by expressing it as an integral over a complex line using inverse Mellin formula. We denote

$$\tau(k,K) = \frac{\rho(k,K)}{K^2}.$$

In both approaches, we are led to compute mean values

$$\mathfrak{S}_t(K) = \frac{1}{K}\sum_{k=1}^{K}\left(\tau(k,K)\right)^t$$

which are well defined if $t > 0$. We extend the definition to the whole complex plane by

$$\mathfrak{S}_z(K) = \frac{1}{K}\sum_{\substack{k=1\\\rho(k,K)\neq 0}}^{K}\left(\tau(k,K)\right)^z \tag{6}$$

Fortunately, $\mathfrak{S}_z(K)$ is a multiplicative function of the modulus K, thus

$$\mathfrak{S}_z(K) = \prod_{p^\alpha\|K}\mathfrak{S}_z(p^\alpha). \tag{7}$$

This property will be crucial in our computations.

3.1   Evaluation of $\mathfrak{S}_z(p^\alpha)$

For any positive integer q and any integer h, let us consider the Gaussian cubic sum

$$S(q,h) = \sum_{x=1}^{q}e\left(\frac{hx^3}{q}\right),$$

where $e(u) = \exp(2i\pi u)$. Then

$$\tau(k, p^\alpha) = \sum_{h=1}^{p^\alpha} \left(\frac{S(p^\alpha, h)}{p^\alpha} \right)^3 e\left(-\frac{hk}{p^\alpha} \right) = 1 + \Omega_k(p) + \Omega_k(p^2) + \cdots + \Omega_k(p^\alpha),$$

where

$$\Omega_k(q) = \sum_{\substack{h=1 \\ (h,q)=1}}^{q} \left(\frac{S(q, h)}{q} \right)^3 e\left(-\frac{hk}{q} \right).$$

We now recall some standard properties of the Gaussian sums (see for example [7]). For $(p, h) = 1$ we have

$$S(p^\ell, h) = \begin{cases} p & \text{if } \ell = 2 \text{ and } p \neq 3, \\ p^2 & \text{if } \ell = 3, \\ p^2 S(p^{\ell-3}, h) & \text{if } 3 < \ell. \end{cases} \tag{8}$$

Thus for $\ell \geq 4$ and any prime p, we get $\Omega_k(p^\ell) = \Omega_{k/p^3}(p^{\ell-3})$ if $p^3 | k$, and 0 otherwise. It thus suffices to compute $\Omega_k(p^\ell)$ for $\ell = 1, 2, 3$. For $p \neq 3$, we have

$$\Omega_k(p^2) = \begin{cases} 0 & \text{if } (p, k) = 1, \\ -1/p^2 & \text{if } p\|k, \\ 1/p - 1/p^2 & \text{if } p^2 | k, \end{cases} \qquad \Omega_k(p^3) = \begin{cases} 0 & \text{if } p^2 \nmid k, \\ -1/p & \text{if } p^2\|k, \\ 1 - 1/p & \text{if } p^3 | k. \end{cases}$$

If $p \equiv 2 \bmod 3$, any residue is a cubic residue modulo p, thus for $(p, h) = 1$, we have $S(p, h) = 0$, whence $\Omega_k(p) = 0$. For $1 \leq k \leq p^\alpha - 1$, we write $k = p^{3u+v} k'$ with $(p, k') = 1$ and $0 \leq v \leq 2$. This gives

$$\tau(k, p^\alpha) = \begin{cases} 1 + u(1 - 1/p^2) & \text{if } v = 0, \\ (u + 1)(1 - 1/p^2) & \text{otherwise.} \end{cases}$$

Moreover for $k = p^\alpha$, then $k' = 1$ and $\alpha = 3u + v$, we have

$$\tau(k, p^\alpha) = \begin{cases} 1 + u(1 - 1/p^2) & \text{if } v = 0 \text{ or } 1, \\ (u + 1)(1 - 1/p^2) + 1/p & \text{if } v = 2. \end{cases}$$

We denote by $\lfloor u \rfloor$ the integral part of u. For any complex number z, we obtain

$$\mathfrak{S}_z(p^\alpha) = \left(1 - \frac{1}{p} \right) \sum_{\beta=0}^{\alpha-1} \frac{1}{p^\beta} \left(\left\lfloor \frac{\beta}{3} \right\rfloor \left(1 - \frac{1}{p^2} \right) + 1 - \frac{\epsilon_\beta}{p^2} \right)^z$$

$$+ \frac{1}{p^\alpha} \left(\left\lfloor \frac{\alpha}{3} \right\rfloor \left(1 - \frac{1}{p^2} \right) + 1 + \epsilon'_\alpha \left(\frac{1}{p} - \frac{1}{p^2} \right) \right)^z, \tag{9}$$

where $\epsilon_\beta = 0$ if $3 | \beta$ and 1 otherwise, and $\epsilon'_\alpha = 1$ if $3 | (\alpha - 2)$ and 0 otherwise.

We now consider $p \equiv 1 \bmod 3$. Let g be a primitive root modulo p. We write $\mathrm{ind}_g(h)$ for the smallest nonnegative integers m such that $g^m \equiv h \bmod p$. We put

$$H_i = \sum_{x=1}^{p} e\left(\frac{g^i x^3}{p}\right), \qquad i = 0, 1, 2.$$

Let h be such that $(p, h) = 1$. Then $\mathrm{ind}_g(h) \equiv i$ modulo 3 for some $i = 0, 1$ or 2. It follows that $S(p, h) = H_i$. It is not difficult to see that H_0, H_1, H_2 are the solutions of the cubic equation $H^3 - 3pH - pa = 0$, where a denotes the unique integer $\equiv 1 \bmod 3$ such that $4p = a^2 + 27b^2$, for some b. We obtain

$$\Omega_k(p) = \begin{cases} A_p := a(p-1)/p^2 & \text{if } p\,|\,k, \\ B_p := (6p - a)/p^2 & \text{if } \mathrm{ind}_g(k) \equiv 0 \bmod 3, \\ C_p := (-3p - a)/p^2 & \text{if } \mathrm{ind}_g(k) \equiv 1 \text{ or } 2 \bmod 3. \end{cases} \tag{10}$$

For $1 \le k \le p^\alpha - 1$, we write $k = p^{3u+v} k'$ with $(p, k') = 1$ and $0 \le v \le 2$, and we get

$$\tau(k, p^\alpha) = \begin{cases} 1 + u(A_p + 1 - 1/p^2) + B_p & \text{if } v = 0 \text{ and } k' \text{ is a cube mod } p, \\ 1 + u(A_p + 1 - 1/p^2) + C_p & \text{if } v = 0 \text{ and } k' \text{ is not a cube mod } p, \\ (u+1)(A_p + 1 - 1/p^2) & \text{otherwise.} \end{cases}$$

If $k = p^\alpha$, whence $\alpha = 3u + v$ and $k' = 1$, we have

$$\tau(k, p^\alpha) = \begin{cases} 1 + u(A_p + 1 - 1/p^2) & \text{if } v = 0, \\ (u+1)(A_p + 1 - 1/p^2) + 1/p^2 & \text{if } v = 1, \\ (u+1)(A_p + 1 - 1/p^2) + 1/p & \text{if } v = 2. \end{cases}$$

Thus for any complex number z, we get

$$\mathfrak{S}_z(p^\alpha) = \left(1 - \frac{1}{p}\right) \left(\sum_{\substack{\beta=0 \\ 3\nmid\beta}}^{\alpha-1} \frac{1}{p^\beta} \left(\left\lfloor 1 + \frac{\beta}{3}\right\rfloor \left(A_p + 1 - \frac{1}{p^2}\right)\right)^z \right.$$

$$+ \sum_{\substack{\beta=0 \\ 3|\beta}}^{\alpha-1} \frac{1}{3p^\beta} \left(\left(\frac{\beta}{3}\left(A_p + 1 - \frac{1}{p^2}\right) + 1 + B_p\right)^z + 2\left(\frac{\beta}{3}\left(A_p + 1 - \frac{1}{p^2}\right) + 1 + C_p\right)^z\right)$$

$$\left. + \frac{1}{p^\alpha}\left(\left\lfloor \frac{\alpha}{3}\right\rfloor\left(A_p + 1 - \frac{1}{p^2}\right) + 1 + \epsilon_\alpha A_p + \epsilon'_\alpha\left(\frac{1}{p} - \frac{1}{p^2}\right)\right)^z\right), \tag{11}$$

where $\epsilon_\alpha = 0$ if $3|\alpha$ and 1 otherwise, and $\epsilon'_\alpha = 1$ if $3|(\alpha - 2)$ and 0 otherwise.

We finally treat the case $p = 3$. To easy the exposition, we restrict our attention to the case $3|\alpha$. For $1 \le k \le 3^\alpha$ we put $k = 3^{3u+v} k'$ with $0 \le v \le 2$ and $(p, k') = 1$. We have

$$\tau(k, 3^\alpha) = \sum_{t=0}^{\min(3u+3, \alpha)} \Omega_k(p^t) = \begin{cases} u(\tau(0, 27) - 1) + \tau(3^v k', 27) & \text{if } 1 \le k < 3^\alpha, \\ 1 + \frac{\alpha}{3}(\tau(0, 27) - 1) & \text{if } k = 3^\alpha. \end{cases}$$

By a direct computation, we easily check that $\tau(k, 27) = 0$, if $k \equiv \pm 4 \bmod 9$, 1 if $k \equiv \pm 2 \bmod 9$, $1/3$ if $3\|k$, 2 if $k \equiv \pm 1 \bmod 9$ or $9\|k$ and 3 if $27|k$, whence for $3|\alpha$ and $3u + v \leq \alpha$ we have for $(k', p) = 1$,

$$\tau(3^{3u+v}k', 3^\alpha) = \begin{cases} 2u & \text{if } v = 0 \text{ and } k' \equiv \pm 4 \bmod 9, \\ 2u + 1 & \text{if } v = 0 \text{ and } k' \equiv \pm 2 \bmod 9, \text{ or } \alpha = 3u, \\ 2u + 2 & \text{if } v = 0 \text{ and } k' \equiv \pm 1 \bmod 9, \text{ or } v = 2, \\ 2u + 1/3 & \text{if } v = 1. \end{cases}$$

Thus for any complex number z, we get for $3|\alpha$,

$$\mathfrak{S}_z(3^\alpha) \sum_{t=0}^{\alpha/3-1} \frac{1}{27^{t+1}} \left(6(2t)^z + 6(2t+1)^z + 6\left(2t + \frac{1}{3}\right)^z + 8(2t+2)^z \right)$$

$$+ \frac{1}{3^\alpha} \left(\frac{2\alpha}{3} + 1 \right)^z. \qquad (12)$$

3.2   Evaluation of $\mathfrak{S}$

In order to compute $\mathfrak{S}$, it is suitable to rewrite it as an Eulerian product. For this, we follow section 3.3 of [2], and we first get

$$\mathfrak{S}_2(K) = \prod_{p^\alpha \| K} \sum_{\beta=0}^{\alpha} \Omega(p^\beta), \quad \text{where} \quad \Omega(q) = \sum_{\substack{h=1 \\ (h,q)=1}}^{q} \left| \frac{S(q, h)}{q} \right|^6.$$

We now observe that $\Omega(p) = 0$ for $p \equiv 2 \bmod 3$ (since $S(p, h) = 0$ for $(p, h) = 1$), and when $p \equiv 1 \bmod 3$, we get

$$\Omega(p) = \frac{p-1}{3p^6}(H_0^6 + H_1^6 + H_2^6) = \frac{(p-1)(18p + a^2)}{p^4},$$

by Newton identities (recall that $4p = a^2 + 27b^2$ and $a \equiv 1 \bmod 3$). Moreover we obtain from (8) that for $p \neq 3$

$$\Omega(p^2) = \frac{p-1}{p^5}, \qquad \Omega(p^3) = \frac{p-1}{p^4},$$

and that for $\ell \geq 4$, and any prime p, $\Omega(p^\ell) = \Omega(p^{\ell-3})/p^3$. We also compute $\Omega(3) = 0$, $\Omega(9) = 20/27$ and $\Omega(27) = 2/81$. We thus deduce that

$$\mathfrak{S} = \frac{70}{39} \prod_{p \neq 3} (1 + T(p)), \quad \text{where} \quad T(p) = \frac{\Omega(p) + 1/p^3 - 1/p^5}{1 - 1/p^3}.$$

For a convenient bound P, we get by a basic C-program a numerical value of the partial Eulerian product $\prod_{p \neq 3, p \leq P}(1 + T(p))$. For large p, $p > P$, we use for $p \equiv 1 \bmod 3$ the upper bound $a \leq 2\sqrt{p}$ deduced from the identity $4p = a^2 + 27b^2$ to obtain the bound $T(p) \leq 22/p^2$ and for $p \equiv 2 \bmod 3$ the upper bound $1 + T(p) \leq 1/(1 - 1/p^3)$. By this way, this yields the upper bound for $P = 10^9$

$$\mathfrak{S} < 3.35141544. \qquad (13)$$

4   Computing δ - The Inverse Mellin Transform

The value of $\delta(K)$ for given modulus K of the type K_B, with B as large as possible, may be obtained by summing the series

$$\sum_{i=1}^{\infty}(-1)^{i+1}\frac{\gamma^i}{i!}\mathfrak{S}_i(K),$$

The general term of the series is, as in the biquadrates case (cf. [3]), at first decreasing for small values of i, then it quickly increases and finally goes to zero for large values of i. In case of cubes, no sufficiently good estimation of $\delta(K)$ are obtained in reasonable time by considering the bounds

$$\sum_{i=1}^{2n}(-1)^{i+1}\frac{\gamma^i}{i!}\mathfrak{S}_i(K) \le 1 - \delta(K) \le \sum_{i=1}^{2n-1}(-1)^{i+1}\frac{\gamma^i}{i!}\mathfrak{S}_i(K).$$

Indeed for large $K = K_B$ given by (4), we have to compute $\mathfrak{S}_i(K)$ for $i \le I(K)$ which is also very large. For $K = \prod_{p^\alpha<6600}p^\alpha$, we have $I(K) \simeq 18000$, and it only leads to $0.099919 < 1 - \delta < 0.099963$. The computation of these bounds by this method implies numbers with 2800 digits, and has needed 100 hours of cpu time on a standard PC. We used for this Pari which enables to compute with multiprecision.

To go further, as pointed out to us by Philippe Flajolet, the inverse Mellin transform can be used for the computation of $\delta(K)$. Indeed, we know that for any real numbers $x > 0$ and $c > 0$, we have

$$e^{-x} = \frac{1}{2\pi i}\int_{c-i\infty}^{c+i\infty} x^{-s}\Gamma(s)ds.$$

By (3) and (6) we thus deduce that when $9|K$ we have

$$\delta(K) = \frac{2}{9} + \frac{1}{2\pi i}\int_{c-i\infty}^{c+i\infty} \gamma^{-s}\mathfrak{S}_{-s}(K)\Gamma(s)ds, \qquad (14)$$

by considering separately the classes k for which $\rho(k,K) = 0$ (there are $2K/9$ such residues) from the others. We now recall the Stirling's formula: for any $z \in \mathbb{C}\setminus\mathbb{R}_-$, we have

$$\log\Gamma(z) = \left(z-\frac{1}{2}\right)\log z - z + \frac{1}{2}\log 2\pi + E(z), \qquad (15)$$

where

$$E(z) = \int_0^{+\infty}\frac{\theta(u)}{u+z}du, \quad\text{and}\quad \theta(u) = \lfloor u\rfloor - u + 1/2.$$

We put $\Theta(u) = \int_0^u \theta(t)dt$ and observe that $0 \le \Theta(u) \le 1/8$ for any $u \ge 0$. By partial summation, we thus obtain

$$|E(z)| \le \frac{1}{8}\int_0^{+\infty}\frac{du}{|u+z|^2}.$$

For $z = c + it$ where $c > 0$, this gives the bound

$$|E(c+it)| \leq \frac{1}{8} \int_0^{+\infty} \frac{du}{(u+c)^2 + t^2} = \frac{1}{8|t|} \int_{c/|t|}^{+\infty} \frac{dv}{v^2+1} \leq \frac{\pi}{16|t|},$$

for any $t \neq 0$. We deduce from this

$$\mathrm{Re}\log \Gamma(c+it) = \left(c - \frac{1}{2}\right) \log|t| - \frac{\pi|t|}{2} + \frac{\log 2\pi}{2} + E_c(t), \quad (t \neq 0),$$

where

$$|E_c(t)| \leq \frac{\pi}{16|t|}.$$

We take $c = \frac{1}{2}$. By (14), this yields

$$\left| e^{-x} - \frac{1}{2\pi i} \int_{c-iT}^{c+iT} x^{-s} \Gamma(s) ds \right| \leq x^{-\frac{1}{2}} \left(\frac{2}{\pi}\right)^{\frac{3}{2}} \exp(E_{1/2}(T)) \, e^{-\pi T/2}.$$

We thus obtain

$$\left| \delta(K) - \frac{2}{9} - \frac{1}{2\pi i} \int_{c-iT}^{c+iT} \gamma^{-s} \mathfrak{S}_{-s}(K) \Gamma(s) ds \right|$$

$$\leq \gamma^{-c} \mathfrak{S}_{-c}(K) \left(\frac{2}{\pi}\right)^{3/2} e^{-\pi T/2 + \pi/16T}. \quad (16)$$

We then compute for some large fixed T and $c = \frac{1}{2}$ a numerical approximating value of

$$\frac{1}{2\pi i} \int_{c-iT}^{c+iT} \gamma^{-s} \mathfrak{S}_{-s}(K) \Gamma(s) ds = \frac{1}{\pi} \mathrm{Re} \int_0^T \gamma^{-c-it} \mathfrak{S}_{-c-it}(K) \Gamma(c+it) dt$$

In order to use a numerical integration method, namely Simpson's rule, we need to control the size of the derivatives of the function $\Lambda(s) = \gamma^{-s} \mathfrak{S}_{-s}(K)\Gamma(s)$. For $s = c + it$ the j-th derivative of Γ satisfies $|\Gamma^{(j)}(s)| \leq G_j(c)$ where $G_j(c) = \int_0^\infty |\log(x)|^j x^{c-1} e^{-x} \, dx$. Now we denote by $\mathfrak{S}'_{-s}(K)$, $\mathfrak{S}''_{-s}(K)$ and so on the successive derivatives of $\mathfrak{S}_{-s}(K)$ according to the variable s.

We have

$$|\mathfrak{S}'_{-s}(K)| \leq \frac{1}{K} \sum_{\substack{k=1 \\ \rho(k,K) \neq 0}}^{K} |\log(\tau(k,K))|(\tau(k,K))^{-c}.$$

Using the fact that $\log t \leq t^{1/e}$ for $t > 1$ and $|\log t| \leq t^{-1/e}$ for $t < 1$ we get

$$|\mathfrak{S}'_{-s}(K)| \leq \mathfrak{S}_{-c+1/e}(K) + \mathfrak{S}_{-c-1/e}(K).$$

For the successive derivatives, we have for $j \geq 1$.

$$|\mathfrak{S}^{(j)}_{-s}(K)| \leq S_j(K,c) \quad (17)$$

where $S_j(K,c) = \mathfrak{S}_{-c+j/e}(K) + \mathfrak{S}_{-c-j/e}(K)$.

From (17), we get for instance for the second derivative of Λ the upper bound

$$|\Lambda''(c+it)| \leq \gamma^{-c} \left((\ln\gamma)^2 \mathfrak{S}_{-c}(K)\Gamma(c)+2|\ln\gamma|\left(\mathfrak{S}_{-c}(K)G_1(c)+S_1(K,c)\Gamma(c)\right)\right.$$
$$\left. +\mathfrak{S}_{-c}(K)G_2(c)+2G_1(c)S_1(K,c)+\Gamma(c)S_2(K,c)\right). \quad (18)$$

Similar bounds can be obtained for $\Lambda^{(j)}(c+it)$ for any $j \geq 1$.

Choosing $K = K_B$ with B around 10^8 in (4), our computation consists in estimating $\Lambda(c+it)$ for N values of t in the interval $[0,T]$. For such t and for each prime p such that $p^\alpha \| K$, we calculate, using (9), (11), (12) and Pari, $\mathfrak{S}_{-c-it}(p^\alpha)$ giving by mulitplicativity $\mathfrak{S}_{-c-it}(K)$, and we obtain by quadrature an approximation of $\delta(K)$. In order to bound the error term, we also calculate for $c = \frac{1}{2}$, $S_j(K,c)$ and $G_j(c)$ (by numerical integration), $1 \leq j \leq 4$. We thus obtain an upper bound $M_{K,c}^{(4)}$ for $|\Lambda^{(4)}(c+it)|$ for $0 \leq t \leq T$ and an upper bound for the error e_T in the quadrature by Simpson's rule which is given by

$$|e_T| \leq \frac{T^5 M_{K,c}^{(4)}}{2880 N^4}. \quad (19)$$

We deduce from this in view of (5), (13) and (16) (with $T = 15$), and also taking into account of the error term (19) entailed in the numerical integrating process (with $N = 2^{12}$), the bounds

$$0.09994250 < 1 - \delta < 0.09994254.$$

5 Comparing $1 - \delta$ with the Experimental Densities - First Observations

For $x \leq x_0 = 5 \cdot 10^{14}$, we have computed, on the one hand, the number $N(x)$ of integers up to x which are the sum of three cubes, and on the other hand, the number $N_d(x)$ of such integers restricted to sums of three distinct cubes (recall that our model concerns sums of distinct terms of random sequences).

The difference $N(x) - N_d(x)$ plainly is less than the number of integers n up to x which are sums of the type $n = 2n_1^3 + n_2^3$ which is less than $x^{2/3}$. Thus the two functions $\nu(x) = N(x)/x$ and $\nu_d(x) = N_d(x)/x$ join together when x tends to infinity, i.e. $\nu(x) - \nu_d(x) = o(1)$.

To compute $N(x)$ and $N_d(x)$, we proceed by intervals: each interval $I = [z, z+ \Delta]$ corresponds to a large array of bits initialized to 0, which are switched to 1, whenever the corresponding integer in I is a sum of three cubes: the computation of the sums of three cubes in I is done by a three stage DO loops indexed by (j, k, l) such that $j \geq k \geq l$ or $j > k > l$ according to the density we are dealing with. It turns out that the order in programming the loops on respectively j, k and l, have a strong influence on the cpu time: we notice that the loop indexed by j is the shortest one compared to the loops indexed on k and l. It follows that we save a lot of cpu time in programming it first. Here is the algorithm we used for covering all the sums of three cubes in the interval $(a, b]$.

```
Initialize to 0 an array with b − a bits
Initialize j to ⌊b^(1/3)⌋
While j³ > a
        Initialize k to min(j, ⌊(b − j³)^(1/3)⌋)
        While j³ + k³ > a and k ≥ 0
                Initialize l to min(k, ⌊(b − j³)^(1/3)⌋)
                While s = j³ + k³ + l³ − a > 0 and l ≥ 0
                        Switch the bit s to 1
                l ← l − 1
        k ← k − 1
j ← j − 1
```

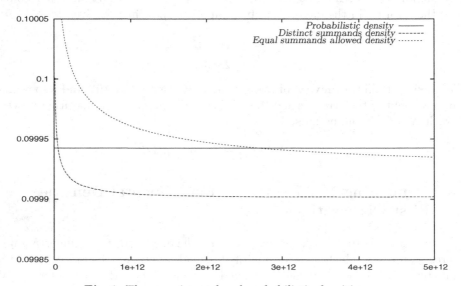

Fig. 1. The experimental and probabilistic densities

We were first disappointed to discover in Figure 1 that for small values of x the functions ν and ν_d are both decreasing, contrary to what we observed in the biquadrates case (cf. [3]). Moreover they pass below the probabilistic density $1 - \delta$ and continue to be globally decreasing.

Then around $x \simeq 5 \cdot 10^{12}$, the distinct summands density function $\nu_d(x)$ begins to grow up while $\nu(x)$ still decreases (see Figure 2).

We have to wait till $x \simeq 2 \cdot 10^{14}$ to finally observe $\nu(x)$ going up as well (see Figure 3).

We also computed for much larger x some values of $(N(x + t) - N(x))/t$ for $t = 10^{12}$. In view of these random experimentations, the fact that sums of three cubes has density $1 - \delta$ remains plausible.

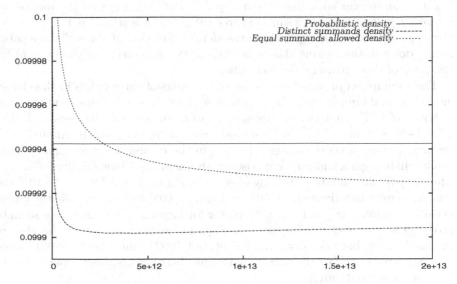

Fig. 2. The experimental and probabilistic densities

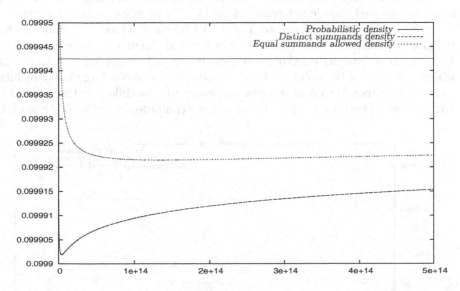

Fig. 3. The experimental and probabilistic densities

6   Comparing $1 - \delta$ with the Experimental Densities - A Modular Approach

Let p be a prime number $p \equiv 2 \bmod 3$. Since the set of the cubes modulo p coincides with the whole set of residues modulo p, we may reasonably expect that sums of three cubes are well-distributed in the residue classes modulo p.

Let a be an integer such that $0 \le a \le p-1$. Put $N(x; a, p)$ be the number of integers n up to x which are sums of three cubes and $n \equiv a$ mod p. For $x > 0$, let $\nu(x; a, p) = pN(x; a, p)/x$ be the relative density function of sums of three cubes falling down in the residue class a modulo p. We define similarly $\nu_d(x; a, p)$ for the sums of three pairwise distinct cubes.

The assumption of well-distribution of the sums of three cubes in the classes modulo p would imply that $\nu(x; a, p)$ is close to $\nu(x)$, as x becomes enough large in terms of p. The number of triples (u, v, w) of nonnegative integers such that $u^3 + v^3 + w^3 \le x$ and $u^3 + v^3 + w^3 \equiv a$ mod p is $\sim \gamma x/p$, as x tends to infinity. Since we need to have a good averaging process in the computation of $\nu(x; a, p)$, our result will be significiant for comparison with the probabilistic density $1 - \delta$, only when x/p is enough large. We may observe in Figure 4 that for $x \le 3 \cdot 10^{14}$ the relative density functions $\nu(x; 1, 503)$ and $\nu_d(x; 1, 503)$ look closely like the global density functions $\nu(x)$ and $\nu_d(x)$. Moreover for larger x, the former ones seem to prolong the later ones (see Figure 5). In Figure 6, we start to observe the effect of the modulus on both density functions $\nu(x; 43, 50021)$ and $\nu(x; 43, 50021)$ when x is not large enough. Afterwards, both lines become smoother, continuing to increase slowly but surely.

In Figure 7, the general outline is slightly similar than in Figure 6. However, a more pronounced chaotic behaviour of the density functions occurs for small x.

Let us explain how to compute $N(x; a, p)$ where $p \equiv 2$ mod 3 is prime. As for the global densities, we proceed interval by interval. Assume we want to compute the number of integers congruent to a modulo p in the interval $I = [z, z + \Delta]$ which are sums of three cubes. We first initialize to zero a large array of bits: each bit corresponds to some integer congruent to a modulo p in the interval I. For any sum of two cubes y less than $z + \Delta$, we compute the cubic root c modulo

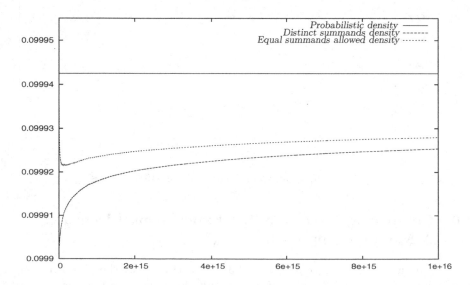

Fig. 4. The relative experimental and probabilistic densities in the class 1 mod 503

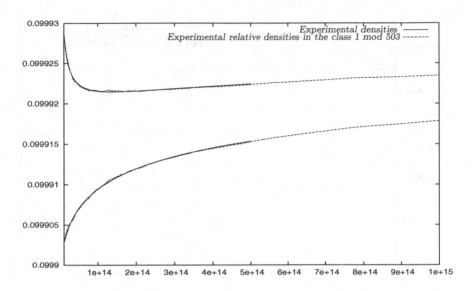

Fig. 5. Comparison between relative densities class 1 mod 503 and global densities

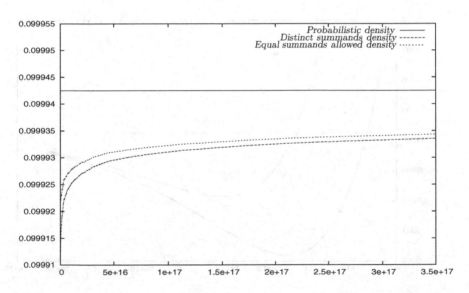

Fig. 6. The relative experimental and probabilistic densities in the class 43 mod 50021

p of $a - y$. Then to each sum $y + (c + mp)^3$ in the given interval, we obtain a sum of three cubes congruent to a modulo p, whose associated bit is switched if necessary to 1.

Notice that for each interval $I = [z, z + \Delta]$, we need to compute all sums of two cubes y, since they are too numerous to be stored once and for all in an

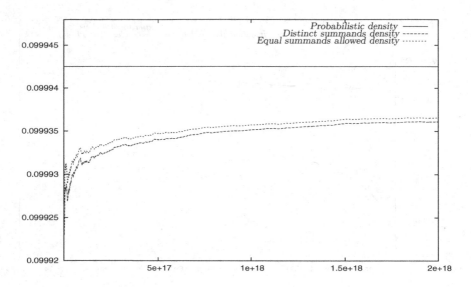

Fig. 7. The relative experimental and probabilistic densities in the class 2 mod 1000037

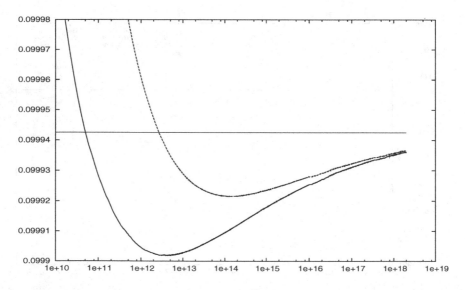

Fig. 8. The extrapolated experimental densities and the probabilistic density viewed in logarithmic scale

array. Moreover, when p is large compared to z, say $p^3 \geq x$, there is at most one value of m for which $y + (c + mp)^3$ is in I, and the most part of sums of two cubes less than $z + \Delta$ do not provide any sum of three cubes in the required class and in the given interval I.

By gathering all these numerical results, putting the computed density functions end to end on suitably chosen intervals, we obtain an extrapolated representation of the global density functions up to $x = 2 \cdot 10^{18}$. It is given in Figure 8 using a logarithmic scale in abscissa, according to the following interval division:

modulus	class	interval
1	0	$[0; 5 \cdot 10^{14}]$
503	1	$[5 \cdot 10^{14}; 10^{16}]$
50021	43	$[10^{16}; 3.5 \cdot 10^{17}]$
1000037	2	$[3.5 \cdot 10^{17}; 2 \cdot 10^{18}]$

Similar computations have been done with others moduli and lead to very close extrapolated densities.

References

1. Barrucand, P.: Sur la distribution empirique des sommes de trois cubes ou de quatre bicarrés, Note aux C. R. Acad. Sc. Paris **267** (1968), 409–411.
2. Deshouillers, J-M., Hennecart, F., Landreau, B.: Sums of powers: an arithmetic refinement to the probabilistic model of Erdős and Rényi. Acta Arith **85** (1998), 13–33
3. Deshouillers, J-M., Hennecart, F., Landreau, B.: Do sums of 4 biquadrates have a positive density? Proceedings ANTS 3, Lecture Notes in Comp. Science, Vol. 1423. Springer-Verlag, Berlin Heidelberg New York (1998), 196–203
4. Erdős, P., Rényi, A.: Additive properties of random sequences of positive integers. Acta Arith. **6** (1960), 83–110
5. Hooley, C.: On some topics connected with Waring's problem. J. Reine Angew. Math. 369 (1986), 110–153.
6. Landau, E.: Über die Einteilung der positiven ganzen Zahlen in vier Klassen nach der Mindestzahl der zu ihrer additiven Zusammensetzung erforderlichen Quadrate. Archiv der Math. und Phys. (3) **13** (1908), 305-312; Collected Works, v. 4, ed. L. Mirsky, I. J. Schoenberg, W. Schwarz and H. Wefelscheid, Thales Verlag, 1985, pp. 59–66.
7. Vaughan, R.C.: The Hardy-Littlewood method. Cambridge Tracts in Mathematics, 80. Cambridge University Press, Cambridge-New York, 1981.

The Mertens Conjecture Revisited

Tadej Kotnik[1] and Herman te Riele[2]

[1] Faculty of Electrical Engineering, University of Ljubljana, Slovenia
tadej.kotnik@fe.uni-lj.si
[2] CWI, P.O. Box 94079, 1090 GB Amsterdam, The Netherlands
herman@cwi.nl

Abstract. Let $M(x) = \sum_{1 \leq n \leq x} \mu(n)$ where $\mu(n)$ is the Möbius function. The Mertens conjecture that $|M(x)|/\sqrt{x} < 1$ for all $x > 1$ was disproved in 1985 by Odlyzko and te Riele [13]. In the present paper, the known lower bound 1.06 for $\limsup M(x)/\sqrt{x}$ is raised to 1.218, and the known upper bound -1.009 for $\liminf M(x)/\sqrt{x}$ is lowered to -1.229. In addition, the explicit upper bound of Pintz [14] on the smallest number for which the Mertens conjecture is false, is reduced from $\exp(3.21 \times 10^{64})$ to $\exp(1.59 \times 10^{40})$. Finally, new numerical evidence is presented for the conjecture that $M(x)/\sqrt{x} = \Omega_{\pm}(\sqrt{\log \log \log x})$.

1 Introduction

The Möbius function $\mu(n)$ is defined as follows

$$\mu(n) := \begin{cases} 1 & \text{if } n = 1, \\ 0 & \text{if } n \text{ is divisible by the square of a prime number,} \\ (-1)^k & \text{if } n \text{ is the product of } k \text{ distinct primes.} \end{cases}$$

Taking the sum of the values of $\mu(n)$ for all $1 \leq n \leq x$, we obtain the function

$$M(x) := \sum_{1 \leq n \leq x} \mu(n),$$

which is the difference between the number of squarefree positive integers $n \leq x$ with an *even* number of prime factors and those with an *odd* number of prime factors. The *Mertens conjecture* [11] states that

$$|M(x)|/\sqrt{x} < 1 \quad \text{for } x > 1.$$

This, but also the weaker assumption

$$|M(x)|/\sqrt{x} < C \quad \text{for } x > 1 \text{ and some } C > 1 \tag{1}$$

would imply the truth of the Riemann hypothesis. The Mertens conjecture was shown to be false by Odlyzko and te Riele in 1985 [13]. They proved the existence of some x for which $M(x)/\sqrt{x} > 1.06$, and some other x for which $M(x)/\sqrt{x} < -1.009$. In 1987, Pintz [14] gave an *effective* disproof of the Mertens conjecture

F. Hess, S. Pauli, and M. Pohst (Eds.): ANTS 2006, LNCS 4076, pp. 156–167, 2006.
© Springer-Verlag Berlin Heidelberg 2006

by showing that $|M(x)|/\sqrt{x} > 1$ for some $x \le \exp(3.21 \times 10^{64})$. Nowadays, it is generally believed that the function $M(x)/\sqrt{x}$ is *unbounded*, both in the positive and in the negative direction. In [8], for example, it is conjectured that

$$M(x)/\sqrt{x} = \Omega_{\pm}(\sqrt{\log\log\log x}). \tag{2}$$

In this paper, we improve the above results by showing that there exists an x for which $M(x)/\sqrt{x} > 1.218$ and an x for which $M(x)/\sqrt{x} < -1.229$ (Section 2), and that there exists an $x < \exp(1.59 \times 10^{40})$ for which the Mertens conjecture is false (Section 3). In addition, we provide new numerical evidence to support (2) (Section 4).

Notation The complex zeros of the Riemann zeta function are denoted by $\rho_j = \frac{1}{2} + i\gamma_j$ (we work in the range where the Riemann hypothesis is known to be true) with $\gamma_1 = 14.1347...$ and $\gamma_j < \gamma_{j+1}$, $j = 1, 2, \ldots$. Furthermore, we write $\psi_j = \arg \rho_j \zeta'(\rho_j)$ and $\alpha_j = |\rho_j \zeta'(\rho_j)|^{-1}$. We also consider the zeros ρ_j ordered according to *non-increasing* values of α_j, and denote them by $\rho_j^* = \frac{1}{2} + i\gamma_j^*$ with the corresponding quantities ψ_j^*, α_j^*, $j = 1, 2, \ldots$. For example, the first five ρ_j^*'s coincide with the first five ρ_j's, but $\rho_6^* = \rho_7$, $\rho_7^* = \rho_{10}$, and $\rho_8^* = \rho_6$ (with $\alpha_6^* = \alpha_7 = 0.0163\ldots, \alpha_7^* = \alpha_{10} = 0.0141\ldots$ and $\alpha_8^* = \alpha_6 = 0.0137\ldots$).

2 Improvement of the Upper and Lower Bounds for $M(x)/\sqrt{x}$

2.1 Background

We describe the approach which led to the disproof of the Mertens conjecture and which is the basis of the experiments which we have carried out to extend the results of Odlyzko and te Riele concerning the function $M(x)/\sqrt{x}$.

For large x, computational results on $M(x)$ are generally based on the following result due to Titchmarsh [18, Theorem 14.27].

Theorem 1. *If all the zeros of the Riemann zeta-function are simple, then there is an increasing sequence $\{T_n\}$ such that*

$$M(x) = \lim_{n \to \infty} \sum_{|\gamma| < T_n} \frac{x^\rho}{\rho \zeta'(\rho)} - R(x) + \sum_{n=1}^{\infty} \frac{(-1)^{n-1}(2\pi/x)^{2n}}{(2n)! n \, \zeta(2n+1)} \tag{3}$$

where $R(x) = 2 - \frac{\mu(x)}{2}$ if x is an integer, and $R(x) = 2$ otherwise.

On the Riemann hypothesis, we have $\rho = \frac{1}{2} + i\gamma$, so that (3) can be rewritten as

$$\frac{M(x)}{\sqrt{x}} = 2 \lim_{n \to \infty} \sum_{0 < \gamma < T_n} \frac{\cos(\gamma \log x - \psi_\gamma)}{|\rho \zeta'(\rho)|} + O(x^{-1/2}), \tag{4}$$

with $\psi_\gamma = \arg \rho \, \zeta'(\rho)$, and where we have also taken into account that in Theorem 1, $R(x) = O(1)$ and the second series is $O(x^{-2})$. Hence, as n increases, the sum in (4) will eventually converge to $M(x)/\sqrt{x}$, with the remaining error on the order of magnitude of $1/\sqrt{x}$. However, very little is known about the rate of this convergence, as the coefficients $|\rho\zeta'(\rho)|^{-1}$ do not form a monotonically decreasing sequence, but instead behave quite irregularly. For some values of x up to 10^{14}, this rate of convergence has been studied computationally, and several thousands of terms generally suffice to bring the error below 1% [8], but for much larger x this approach is not feasible.

If instead of an isolated value of $M(x)/\sqrt{x}$, a weighted average of this function in some x-range is of interest, the problem becomes somewhat more tractable. Namely, in these cases the terms of the above sum are multiplied by a function of bounded support, and the series in (3) is transformed into a finite sum. Two such cases will appear in Theorems 2 and 3 below.

We write $x = e^y$, $-\infty < y < \infty$, and define

$$m(y) := M(x)x^{-1/2} = M(e^y)e^{-y/2},$$

and

$$\overline{m} := \limsup_{y\to\infty} m(y), \quad \underline{m} := \liminf_{y\to\infty} m(y).$$

Then we have the following [5,6,7,13]

Theorem 2. *Let*

$$h(y,T) := 2 \sum_{0<\gamma<T} \left[\left(1 - \frac{\gamma}{T}\right)\cos(\pi\frac{\gamma}{T}) + \pi^{-1}\sin(\pi\frac{\gamma}{T}) \right] \frac{\cos(\gamma y - \psi_\gamma)}{|\rho\zeta'(\rho)|} \quad (5)$$

where $\rho = \beta + i\gamma$ are the complex zeros of the Riemann zeta function which satisfy $\beta = \frac{1}{2}$ and which are simple. Then for any y_0,

$$\underline{m} \le h(y_0, T) \le \overline{m}$$

and any value $h(y,T)$ is approximated arbitrarily closely, and infinitely often, by $M(x)/\sqrt{x}$.

Since

$$(1-t)\cos(\pi t) + \pi^{-1}\sin(\pi t) > 0 \text{ for } 0 < t < 1$$

and since it is known that $\sum_\rho |\rho\zeta'(\rho)|^{-1}$ diverges [18, Section 14.27], the sum of the *coefficients* of $\cos(\gamma y - \psi_\gamma)$ in (5) can be made arbitrarily large by choosing T large enough. Consequently, if we could find a value of y such that *all* of the $\gamma y - \psi_\gamma$ are close to integer multiples of 2π, then we could make $h(y,T)$ arbitrarily large. This would contradict, by Theorem 1, any conjecture of the form (1). If the γ's were linearly independent over the rationals, then by Kronecker's theorem (see, e.g., [4, Theorem 442]) there would indeed exist, for any $\epsilon > 0$, integer values of y satisfying

$$|\gamma y - \psi_\gamma - 2\pi m_\gamma| < \epsilon$$

for all $\gamma \in (0,T)$ and certain integers m_γ. This would show that $h(y,T)$, and hence $M(x)/\sqrt{x}$, can be made arbitrarily large. On the same assumptions, a similar argument can be given to imply that $h(y,T)$, and hence $M(x)/\sqrt{x}$, can be made arbitrarily large on the negative side. No good reason is known why among the γ's there should exist any linear dependencies over the rationals (see, e.g., [1]).

The approach which actually led to a disproof of the Mertens conjecture was based on the now well-known lattice basis reduction (L^3-) algorithm of Lenstra, Lenstra and Lovász [9] for finding short vectors in lattices. With this algorithm, the above mentioned inhomogeneous Diophantine approximation problem could be solved for a much larger number of terms in (5) than before. Any value of y that would come out was likely to be quite large, viz., of the order of 10^{70} in size. Therefore, it was necessary to compute the first 2000 γ's with an accuracy of about 75 decimal digits (actually, 100 decimal digits were used). The best lower and upper bounds found for $\overline{m}$ and $\underline{m}$ were 1.06 and -1.009, respectively.

2.2   Computation of New Lower and Upper Bounds for $M(x)/\sqrt{x}$

In order to find a y such that each of the numbers

$$\eta_j := (\gamma_j^* y - \psi_j^*) \bmod 2\pi, \ 1 \le j \le n, \tag{6}$$

is small, Odlyzko transformed this problem into a problem about short vectors in lattices, as described in [13]. The lattice L used is generated by the columns $\underline{v}_1, \underline{v}_2, \ldots, \underline{v}_{n+2}$ of the following $(n+2) \times (n+2)$ matrix (here $[x]$ means the greatest integer $\le x$):

$$
\begin{array}{ccccccc}
-[\sqrt{\alpha_1^*}\psi_1^* 2^\nu] & [\sqrt{\alpha_1^*}\gamma_1^* 2^{\nu-10}] & [2\pi\sqrt{\alpha_1^*}2^\nu] & 0 & \cdots & 0 \\
-[\sqrt{\alpha_2^*}\psi_2^* 2^\nu] & [\sqrt{\alpha_2^*}\gamma_2^* 2^{\nu-10}] & 0 & [2\pi\sqrt{\alpha_2^*}2^\nu] & \cdots & 0 \\
\vdots & & & & & \vdots \\
-[\sqrt{\alpha_n^*}\psi_n^* 2^\nu] & [\sqrt{\alpha_n^*}\gamma_n^* 2^{\nu-10}] & 0 & 0 & \cdots & [2\pi\sqrt{\alpha_n^*}2^\nu] \\
2^\nu n^4 & 0 & 0 & 0 & \cdots & 0 \\
0 & 1 & 0 & 0 & \cdots & 0 \\
\end{array}
\tag{7}
$$

where ν is an integer satisfying $2n \le \nu \le 4n$. The L^3 algorithm produces a reduced basis $\underline{v}_1', \underline{v}_2', \ldots, \underline{v}_{n+2}'$ for the lattice L, where each new basis vector is a linear combination of the $n+2$ given basis vectors. Now the $(n+1)$-st coordinate of $\underline{v}_1$, which has value $2^\nu n^4$, is very large compared to all the other entries of the original basis. Since the reduced basis is a basis for the lattice L, it should contain precisely one vector $\underline{w}$ which has a nonzero coordinate in the $(n+1)$-st position and that coordinate should be $\pm 2^\nu n^4$. Without loss of generality this may be taken to be $2^\nu n^4$. Given the original lattice basis, the j-th coordinate of this vector $\underline{w}$ equals, for $1 \leq j \leq n$:

$$z\left[\sqrt{\alpha_j^*}\gamma_j^* 2^{\nu-10}\right] - \left[\sqrt{\alpha_j^*}\psi_j^* 2^\nu\right] - m_j\left[2\pi\sqrt{\alpha_j^*}2^\nu\right]$$

and the $(n+2)$-nd coordinate is z, for some integers $z, m_1, m_2, \ldots, m_n$. If the length of $\underline{w}$ is small, all of the

$$z\sqrt{\alpha_j^*}\gamma_j^* 2^{\nu-10} - \sqrt{\alpha_j^*}\psi_j^* 2^\nu - m_j 2\pi\sqrt{\alpha_j^*}2^\nu$$

will be small, i.e., all of the

$$\beta_j = \sqrt{\alpha_j^*}(y\gamma_j^* - \psi_j^* - 2\pi m_j)$$

will be very small, where $y = z/1024$. The reason for the presence of the numbers α_j^* in the lattice basis is that we want to make the sum

$$\sum_{j=1}^{n} \alpha_j^* \cos(\gamma_j^* y - \psi_j^* - 2\pi m_j)$$

large. If the cos-arguments are all close to zero, this sum will be approximately

$$\sum_{j=1}^{n} \alpha_j^* - \frac{1}{2}\sum_{j=1}^{n}[\sqrt{\alpha_j^*}(\gamma_j^* y - \psi_j^* - 2\pi m_j)]^2,$$

and therefore we want the second sum to be small. This corresponds to minimizing the euclidean norm of the vector $(\beta_1, \beta_2, \ldots, \beta_n)$ which is what the L^3 algorithm attempts to do.

In order to obtain values of y for which $h(y, T)$ will be negative, similar lattices can be used with only one change, namely that ψ_j^* is replaced by $\psi_j^* + \pi$, so that the cosine-arguments mod 2π will be close to π and the cosine-values close to -1.

We have applied the L^3 algorithm with the matrix (7) as input, for all the combinations (ν, n) in the range $\nu = 8, 9, \ldots, 400$, $n = [\nu/4], [\nu/4]+1, \ldots, [\nu/2]$. To this end we used the function *qflll* from the PARI/GP package [15]. For a given ν, the precision by which the computations were carried out was chosen to be $\log_{10}(2^{2\nu})$ decimal digits. For each combination of ν and n a number $z = z(\nu, n)$ was generated as described above and we computed the local maximum of $h(y, T)$ as defined in (5) with y in the neighborhood of $z/1024$, and $T = \gamma_{10000}$. The γ_j's were computed to an accuracy of about 250 decimal digits using the Mathematica package [10], and, as a check, using the PARI/GP package. The computing time was about 600 CPU hours on the SGI Altix 3700 Aster system of the Academic Computing Centre Amsterdam (SARA).

Figure 1 gives for each $\nu = 8, 9, \ldots, 400$ and for each value of $z(\nu, n)$ which was found by the L^3 algorithm, a scatter plot of the positive values of

$$h(z(\nu, n)/1024, \gamma_{2000}).$$

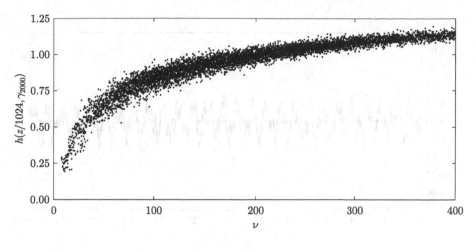

Fig. 1.

For increasing values of ν, the corresponding h-values are increasing on average, but at a rate that seems to decrease. For the negative values of h the pattern is very similar. Reaching 1.3 and -1.3 would likely require a value of ν in the neighborhood of 800.

For the most promising values of h obtained, we computed the local maximum resp. minimum of $h(y, \gamma_{10000})$ in the neighborhood of $y = z/1024$. On the positive side, our champion (found with $\nu = 379, n = 98$) is

$$y = -233029271\,5134531215\,0140181996\,7723401020\,4456785091\backslash$$
$$6681557518\,6743434036\,9240230890\,8933261706\,9029233958\,2730162362.807965$$

with

$$h(y, \gamma_{10000}) = 1.218429$$

and on the negative side, our champion (found with $\nu = 396, n = 102$) is

$$y = -1608\,7349754400\,0919817483\,9640165505\,4685212472\,2284778177\backslash$$
$$5539303027\,5350690810\,7957194829\,6433602695\,1442102295\,3212754000.679958$$

with

$$h(y, \gamma_{10000}) = -1.229385.$$

Figure 2 compares the typical behaviour of $M(e^y)/e^{y/2}$ (top) with the behaviour of $h(y, \gamma_{10000})$ around the 1.218–spike (middle) and around the -1.229–spike (bottom). Notice the four large negative spikes to the left and to the right of the champion positive spike, and the four large positive spikes to the left and to the right of the champion negative spike. This suggests that a very large spike in one direction is usually accompanied by several large spikes in the opposite direction. Notice also that the bottom graph, when inverted with respect to the

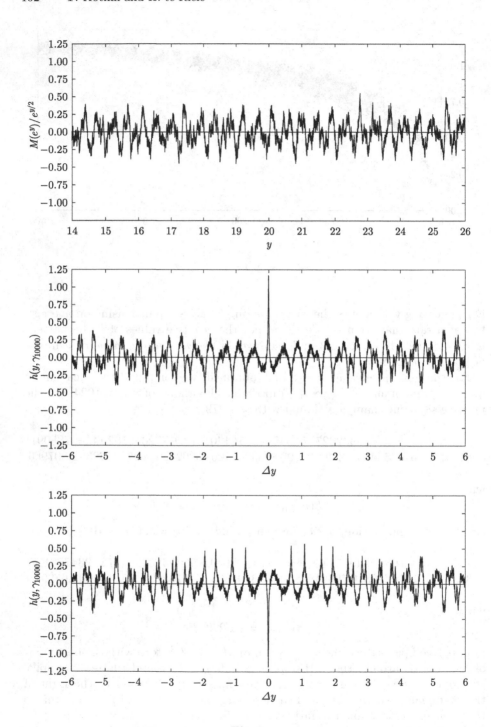

Fig. 2.

horizontal axis, very much resembles the middle graph. This is explained by the fact that the two functions plotted there are sums of cosines of which the first 98 main terms are aligned very well in $\Delta y = 0$.

Part of the L^3–output was also used to reduce the upper bound for which the Mertens conjecture is known to be false (Section 3) and for the computations concerning the growth rate of $M(x)/\sqrt{x}$ (Section 4).

3   Reduction of the Smallest Known x for which $|M(x)/\sqrt{x}| > 1$

Two years after Odlyzko and te Riele disproved the Mertens conjecture, Pintz [14] published a theorem which gives an *explicit* upper bound for the smallest x for which the Mertens conjecture is false:

Theorem 3. *Let*

$$h_P(y,T) := 2 \sum_{0 < \gamma < T} e^{-1.5 \times 10^{-6} \gamma^2} \frac{\cos(\gamma y - \psi_\gamma)}{|\rho \zeta'(\rho)|}. \tag{8}$$

If there exists a $y \in [e^7, e^{5 \times 10^4}]$ with $|h_P(y,T)| > 1 + e^{-40}$ for $T = 1.4 \times 10^4$, then $|M(x)|/\sqrt{x} > 1$ for some $x < e^{y+\sqrt{y}}$.

For the number $y = y_0 \approx 3.2097 \times 10^{64}$ as given by Odlyzko and te Riele in their Table 3 (line $i = 21$) in [13], on request of Pintz, te Riele computed $h_P(y_0, T)$ and found the value -1.00223, which implies, by Pintz's Theorem, that the Mertens conjecture is false for some $x < \exp(3.21 \times 10^{64})$.

We have computed $h_P(y, T)$ for many *smaller* values of y, resulting from our application of the L^3 algorithm in Section 2.2, in order to attempt to further reduce the upper bound for the smallest x for which the Mertens conjecture is false. The smallest y for which we found a value of $|h_P(y, T)| > 1 + e^{-40}$ is:

$$y = 1\,5853191167\,3595000428\,9014722171\,6268116204.984802$$

with $h_P(y, T) = -1.00819$. This shows that there exists an

$$x < \exp(1.59 \times 10^{40})$$

for which the Mertens conjecture is false. It is very likely that there is still substantial room for improvement of this result. For example, in [8] it is suggested that the first violation of the Mertens conjecture should occur not too far from $x \approx \exp(5 \times 10^{23})$.

4   Estimation of the Order of Magnitude of $M(x)/\sqrt{x}$

4.1   Existing Results and Conjectures

The strongest unconditional results on the order of magnitude of $M(x)$ are of the general form $M(x) = o(x)$. Thus Walfisz [19] proved that

$$M(x) = O\left(x\exp\left(-A\frac{(\log x)^{3/5}}{(\log\log x)^{1/5}}\right)\right) \quad \text{for some } A > 0$$

and Ford [2] has recently shown that we may take $A = 0.2098$. A proof of the Riemann hypothesis would strengthen this to

$$M(x) = O(x^{1/2+\varepsilon}) \quad \text{for every } \varepsilon > 0.$$

It is also known that

$$M(x) = \Omega_\pm(x^{1/2})$$

and from the disproof of the Mertens conjecture by Odlyzko and te Riele it follows that the multiplicative constant is larger than 1 into both the positive and the negative direction. Nonetheless, the question whether $M(x)/\sqrt{x}$ is unbounded remains open, although many experts suppose that this is the case, and some arguments in favor of this have been presented in Section 2.1. During the last decades, several conjectures on the order of magnitude of $M(x)/\sqrt{x}$ have been set forth. Good and Churchhouse [3], as well as Lévy in a comment to Saffari [17], have conjectured that

$$\limsup_{x\to\infty}\frac{|M(x)|}{\sqrt{x}\log\log x} = C \tag{9}$$

with $C = \frac{\sqrt{12}}{\pi} = 1.1026\ldots$ according to Good and Churchhouse, whereas $C = \frac{6\sqrt{2}}{\pi^2} = 0.8597\ldots$ according to Lévy. Conjectures of the type (9) seem questionable, however, both on theoretical grounds and because of very poor agreement with experimental observations (see e.g. [13, p. 140] and [8, pp. 479–480]). More recently, Ng [12], partly building on unpublished work by Gonek, conjectured that

$$\limsup_{x\to\infty}\frac{|M(x)|}{\sqrt{x}(\log\log\log x)^{5/4}} = B \tag{10}$$

for some $B > 0$, while Kotnik and van de Lune [8] observed experimentally that estimates of the largest positive and negative values of $M(x)/\sqrt{x}$ in the range $10^4 \le x \le 10^{10^{10}}$ are quite close to $\frac{1}{2}\sqrt{\log\log\log x}$ and $-\frac{1}{2}\sqrt{\log\log\log x}$, respectively. If this would also hold asymptotically, it would contradict both (9) and (10). In a somewhat more conservative spirit, Kotnik and van de Lune finally conjectured that

$$M(x)/\sqrt{x} = \Omega_\pm(\sqrt{\log\log\log x}),$$

which is weaker than both (9) and (10), as these correspond to $M(x)/\sqrt{x} = \Omega(\sqrt{\log\log x})$ and $M(x)/\sqrt{x} = \Omega((\log\log\log x)^{5/4})$, respectively.

4.2   New Results

Theorem 1 suggests that sums of the type

$$h_1(y,T) := 2\sum_{0<\gamma<T}\frac{\cos(\gamma y - \psi_\gamma)}{|\rho\zeta'(\rho)|}$$

should, as T increases, eventually converge to the respective values of $M(e^y)/\sqrt{e^y}$. As mentioned in Section 2.1, the essential problem of the estimation of $M(e^y)/\sqrt{e^y}$ by means of such sums is that their convergence can only be estimated empirically. Observing that the sums obtained with the first ten thousand ζ-zeros ($T = \gamma_{10^4} = 9877.782...$) and those obtained with the first million ζ-zeros ($T = \gamma_{10^6} = 600269.677...$) consistently differ by less than 1% (see e.g. Tables 3 and 4 in [8], as well as Figure 4 in the same source), Kotnik and van de Lune used these sums in estimating the largest positive and negative values of $M(x)/\sqrt{x}$ in the range $10^4 \le x \le 10^{10^{10}}$ (see previous subsection), and thereby came to the observations mentioned in the preceding subsection. In terms of y, the range they investigated is $9.210... \le y \le 23025850929.940...$, and the estimates were obtained by sampling the sums in y-increments of 0.0005, and then searching for the local extrema if an exceptionally large value was found. While this approach practically eliminated the possibility of missing an exceptionally large positive or negative value, due to uniform y-increments a significant extension of this approach (e.g., to the smallest values of y for which $|M(e^y)|/\sqrt{e^y}$ is known to exceed 1; see Section 3) would be impossible.

As the L^3 algorithm can be used to generate isolated y-values for which either $M(e^y)/\sqrt{e^y}$ or $-M(e^y)/\sqrt{e^y}$ is likely to be very large, this offered a possibility to considerably extend the estimation of the order of magnitude of $M(e^y)/\sqrt{e^y}$ by means of the sums $h_1(y, T)$.

In Figure 3, we extend the study of Kotnik and van de Lune [8] with the estimates $h_1(y, \gamma_{10^4})$ obtained in this manner. The hollow squares and circles give the increasingly large values of $M(e^y)/\sqrt{e^y}$ and $h_1(y, \gamma_{10^6})$, respectively, obtained in [8], and the solid circles give the values of $h_1(y, \gamma_{10^4})$ found by the L^3 algorithm. We observe that also with y up to $\approx 10^{110}$ (i.e., $\sqrt{\log \log y}$ up to ≈ 2.35), the estimates of the largest positive and negative values of $M(x)/\sqrt{x}$ are quite close to $\frac{1}{2}\sqrt{\log \log \log x}$ and $-\frac{1}{2}\sqrt{\log \log \log x}$, respectively. Nevertheless,

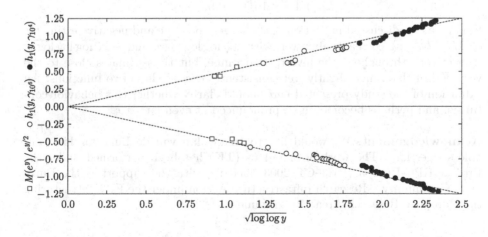

Fig. 3.

at the very largest y-values the positive and negative estimates appear to be systematically somewhat above the first and somewhat below the second of these two functions, respectively. This suggests that a further extension, perhaps to $y \approx 10^{500}$, could provide some additional insight into these observations.

5   Conclusions and Discussion

We have presented improvements of $\limsup M(x)/\sqrt{x}$ and $\liminf M(x)/\sqrt{x}$ with respect to the results obtained by Odlyzko and te Riele in 1985, and an improvement of the upper bound of the smallest x for which $|M(x)| > \sqrt{x}$ with respect to the result obtained by Pintz in 1987. The approach that led to our results was based on systematic application of the L^3 algorithm to a relatively extensive set of combinations (ν, n), and also on considerable increase in computing power with respect to what was available to the researchers twenty years ago.

There are several directions in which our results could be improved further. Concerning the limsup and liminf results, the methods described in Section 2 could be extended in an elementary manner to larger values of ν and n, and as we state in that section, values of ν close to 800 would probably lead to limsup and liminf values close to 1.3 and -1.3, respectively. In contrast, a further improvement of the upper bound of the first violation of the Mertens conjecture would very likely require a strengthening of Theorem 3, in the sense of reducing the value of the constant 1.5×10^{-6} in the exponent occurring in (8). Namely, there are values of y smaller than the upper bound presented in Section 3 for which the sum in (8) only falls short of exceeding 1 by a small amount. An example is

$$y = 35499\,1618091406\,4844654619\,4090311725.687444$$

for which

$$h_P(y, 1.4 \times 10^4) = 0.991549.$$

Finally, as we discussed in Section 4, the largest positive and negative estimates of $M(x)/\sqrt{x}$ agree reasonably well with $\frac{1}{2}\sqrt{\log\log\log x}$ and $-\frac{1}{2}\sqrt{\log\log\log x}$, respectively, throughout the investigated range, but the estimates close to the very end of this range slightly, yet consistently exceed these two functions. An extension of the study presented here should clarify whether this behavior continues, and perhaps becomes more pronounced at even larger x.

Acknowledgements. We would like to thank Jan van de Lune for his valuable suggestions. The work of one of us (TK) has been performed under the Project HPC-Europa (RII3–CT–2003–506079), with the support of the European Community – Research Infrastructure Action under the FP6 "Structuring the European Research Area" Programme.

References

1. P.T. Bateman, J.W. Brown, R.S. Hall, K.E. Kloss and R.M. Stemmler. Linear relations connecting the imaginary parts of the zeros of the zeta function. In: A.O.L. Atkin and B.J. Birch (editors). *Computers in Number Theory*. New York, 1971, 11–19.
2. K. Ford. Vinogradov's integral and bounds for the Riemann zeta function. *Proc. Lond. Math. Soc.*, 85:565–633, 2002.
3. I.J. Good and R.F. Churchhouse. The Riemann hypothesis and pseudorandom features of the Möbius sequence. *Math. Comp.*, 22:857–861, 1968.
4. G.H. Hardy and E.M. Wright. *An introduction to the theory of numbers*. Oxford at the Clarendon Press, Fourth Edition, 1975.
5. A.E. Ingham. On two conjectures in the theory of numbers. *Amer. J. Math.*, 64:313–319, 1942.
6. W.B. Jurkat. On the Mertens conjecture and related general Ω-theorems. In H. Diamond, editor, *Analytic Number Theory*, pages 147–158, Providence, 1973.
7. W. Jurkat and A. Peyerimhoff. A constructive approach to Kronecker approximations and its application to the Mertens conjecture. *J. reine angew. Math.*, 286/287:322–340, 1976.
8. T. Kotnik and J. van de Lune. On the order of the Mertens function. *Exp. Math.*, 13:473–481, 2004.
9. A.K. Lenstra, H.W. Lenstra, Jr., and L. Lovász. Factoring polynomials with rational coefficients. *Math. Ann.*, 261:515–534, 1982.
10. Mathematica, version 5.0, Wolfram Research, Inc., Champaign, IL, 2003. http://www.wolfram.com/
11. F. Mertens. Über eine zahlentheoretische Funktion. *Sitzungsberichte Akad. Wiss. Wien IIa*, 106:761–830, 1897.
12. N. Ng. The distribution of the summatory function of the Möbius function. *Proc. Lond. Math. Soc.*, 89:361–389, 2004.
13. A.M. Odlyzko and H.J.J. te Riele. Disproof of the Mertens conjecture. *J. reine angew. Math.*, 357:138–160, 1985.
14. J. Pintz. An effective disproof of the Mertens conjecture. *Astérisque*, 147–148: 325–333, 1987.
15. PARI/GP, version 2.2.11 (alpha), Bordeaux, 2005. http://pari.math.u-bordeaux.fr/
16. H.J.J. te Riele. Computations concerning the conjecture of Mertens. *J. reine angew. Math.*, 311/312:356–360, 1979.
17. B. Saffari. Sur la fausseté de la conjecture de Mertens. Avec une observation par Paul Lévy. *C. R. Acad. Sci.*, 271(A):1097–1101, 1970.
18. E.C. Titchmarsh. *The Theory of the Riemann Zeta-Function*. Oxford, 1951.
19. A. Walfisz. *Weylsche Exponentialsummen in der neueren Zahlentheorie*. VEB Deutscher Verlag, 1963.

Fast Bounds on the Distribution of Smooth Numbers*

Scott T. Parsell[1] and Jonathan P. Sorenson[2]

[1] Mathematics and Actuarial Science, Butler University, Indianapolis, IN 46208 USA
sparsell@butler.edu
http://blue.butler.edu/~sparsell
[2] Computer Science and Software Engineering, Butler University
Indianapolis, IN 46208 USA
sorenson@butler.edu
http://www.butler.edu/~sorenson

Abstract. Let $P(n)$ denote the largest prime divisor of n, and let $\Psi(x,y)$ be the number of integers $n \leq x$ with $P(n) \leq y$. In this paper we present improvements to Bernstein's algorithm, which finds rigorous upper and lower bounds for $\Psi(x,y)$. Bernstein's original algorithm runs in time roughly linear in y. Our first, easy improvement runs in time roughly $y^{2/3}$. Then, assuming the Riemann Hypothesis, we show how to drastically improve this. In particular, if $\log y$ is a fractional power of $\log x$, which is true in applications to factoring and cryptography, then our new algorithm has a running time that is polynomial in $\log y$, and gives bounds as tight as, and often tighter than, Bernstein's algorithm.

1 Introduction

For a positive integer n, let $P(n)$ denote the largest prime divisor of n. If $P(n) \leq y$, then n is said to be *y-smooth*. Smooth numbers are utilized by many integer factoring and discrete logarithm algorithms, and hence they are of interest in cryptography [19,22]. Define $\Psi(x,y)$ to be the number of integers $n \leq x$ that are *y*-smooth. In this paper, we present improvements to an algorithm of Bernstein[4,5], based on discrete generalized power series, which gives rigorous upper and lower bounds for $\Psi(x,y)$.

1.1 Previous Work

To compute the exact value of $\Psi(x,y)$, one could simply factor all the integers up to x using a sieve. The Buchstab identity

$$\Psi(x,y) = \Psi(x,2) + \sum_{2 < p \leq y} \Psi(x/p,p)$$

---

* This work was supported by a grant from the Holcomb Research Institute. We wish to thank the referee, whose comments helped improve this paper.

F. Hess, S. Pauli, and M. Pohst (Eds.): ANTS 2006, LNCS 4076, pp. 168–181, 2006.

leads to a simple recursive algorithm. Bernstein presents several algorithms in his thesis [3]. See [17] for several more. All of these algorithms are far too slow for use in applications related to factoring and cryptography.

There are a number of asymptotic estimates for $\Psi(x, y)$ in the literature [8,10,11,13,14,15,18,20,21], many of which lead to algorithms.

Dickman's function, $\rho(u)$, is defined as the unique continuous solution to

$$\rho(u) = 1 \quad \text{(for } 0 \leq u \leq 1\text{)},$$
$$\rho(u - 1) + u\rho'(u) = 0 \quad \text{(for } u > 1\text{)}.$$

It is well-known that the estimate $\Psi(x, y) \approx x\rho(\log x/\log y)$ holds; for example Hildebrand [13] proved that for $\varepsilon > 0$, we have

$$\Psi(x, y) = x\rho(u)\left(1 + O_\varepsilon\left(\frac{\log(u + 1)}{\log y}\right)\right)$$

where $y \geq 2$ and $u := u(x, y) = \log x/\log y$ satisfies $1 \leq u \leq \exp[(\log y)^{3/5-\varepsilon}]$. This range can be extended if we assume the Riemann Hypothesis. Highly accurate estimates for $\rho(u)$ can be computed quickly using numerical integration; see for example [27].

Hildebrand and Tenenbaum [14] gave a more complicated estimate for $\Psi(x, y)$ using the saddle-point method. Define

$$\zeta(s, y) := \prod_{p \leq y}(1 - p^{-s})^{-1},$$
$$\phi(s, y) := \log \zeta(s, y),$$
$$\phi_k(s, y) := \frac{d^k}{ds^k}\phi(s, y) \quad (k \geq 1).$$

Let a be the unique solution to $\phi_1(a, y) + \log x = 0$. Then

$$\Psi(x, y) = \frac{x^a \zeta(a, y)}{a\sqrt{2\pi\phi_2(a, y)}}\left(1 + O\left(\frac{1}{u} + \frac{(\log y)}{y}\right)\right)$$

uniformly for $2 \leq y \leq x$. This theorem has led to a string of algorithms that, in practice, appear to give significantly better estimates to $\Psi(x, y)$ than those based on Dickman's function [17,24,25]. Recently, Suzuki [26] showed how to estimate $\Psi(x, y)$ quite nicely in only $O(\sqrt{\log x \log y})$ operations using this approach.

Bernstein's algorithm [4,6] provides a very nice compromise between computing an exact value of $\Psi(x, y)$ (which is very slow) and computing an estimate (which is fast, but not as reliably accurate): compute rigorous upper and lower bounds for $\Psi(x, y)$. Bernstein's algorithm introduces an accuracy parameter α, and his algorithm creates upper and lower bounds for $\Psi(x, y)$ that are off by at most a factor of $1 + O(\alpha^{-1}\log x)$, implying a choice of, say, $\alpha \asymp \log x \log \log y$. As we will show in the next section, Bernstein's algorithm has a running time of

$$O\left(\frac{y}{\log \log y} + \frac{y \log x}{(\log y)^2} + \alpha \log x \log \alpha\right)$$

arithmetic operations, which is roughly linear in y. It also generates, for free, rigorous bounds on $\Psi(x', y)$ for certain values of $x' < x$.

1.2   New Results

We present two improvements to Bernstein's algorithm.

Our first improvement is a simple one that Bernstein mentioned but did not analyze. In essence, the idea is to use an algorithm to compute $\pi(t)$, the number of primes up to t, for many values of t with $2 \leq t \leq y$, rather than use a prime number sieve that finds all primes up to y. The result, Algorithm 3.1, has the same accuracy as the original, with a running time of

$$O\left(\alpha \frac{y^{2/3}}{\log y} + \alpha \log x \log \alpha\right)$$

operations.

Our second improvement is to choose a parameter z, with $1451 \leq z < y$ and $z \asymp \alpha^4 (\log \alpha)^2$, and then use the $\pi(t)$ algorithm for $t \leq z$, but use the fast-to-compute estimate

$$|\pi(t) - \mathrm{li}(t)| < \frac{\sqrt{t} \log t}{8\pi} \qquad (t \geq 1451)$$

for $t > z$, where $\mathrm{li}(t)$ is the logarithmic integral. The above inequality follows from work of Schoenfeld [23] under the assumption of the Riemann Hypothesis (see also [9, Exercise 1.36]). This improvement, Algorithm 4.1, leads to a running time of

$$O\left(\alpha \frac{z^{2/3}}{\log z} + \alpha \log x \log \alpha y\right)$$

operations, with a relative error of at most $O(\alpha^{-1} \log x)$. In particular, if we take $\alpha \asymp \log x (\log \log y)^2$, say, resulting in $z \asymp (\log x)^4 (\log \log x)^2 (\log \log y)^8$, we obtain the running time of

$$O((\log x)^{11/3} (\log \log x)^{1/3} (\log \log y)^{22/3})$$

operations. In applications related to factoring and discrete logarithms, we have $\log x \approx (\log y)^3$, so that our algorithm runs in time polynomial in $\log y$. With such a small running time, we can choose to make α larger, resulting in more accurate upper and lower bounds for $\Psi(x, y)$, in less time.

1.3   A Comparison

Below we compare the relative error and running times (with big-Oh understood) for several different algorithms.

For $\log x = (\log y)^2$ so that $u = \log y$ we have:

Relative Error	Algorithm	Running Time
$\log\log y/\log y$	$x\rho(u)$	$(\log y)^2$
$(\log y)^{-1}$	Suzuki [26]	$(\log y)^{3/2}$
$(\log y)^{-2}$	Bernstein [4,6]	y
$(\log y)^{-2}$	Algorithm 4.1	$(\log y)^{44/3+o(1)}$
$(\log y)^{-3}$	Bernstein [4,6]	y
$(\log y)^{-3}$	Algorithm 4.1	$(\log y)^{55/3+o(1)}$
y^{-1}	Bernstein [4,6]	$y(\log y)^3$
y^{-1}	Algorithm 4.1	$y(\log y)^3$

For $\log x = (\log y)^3$ so that $u = (\log y)^2$ we have:

Relative Error	Algorithm	Running Time
$\log\log y/\log y$	$x\rho(u)$	$(\log y)^4$
$(\log y)^{-1}$	Suzuki [26]	$(\log y)^2$
$(\log y)^{-2}$	Bernstein [4,6]	y
$(\log y)^{-2}$	Algorithm 4.1	$(\log y)^{55/3+o(1)}$
$(\log y)^{-3}$	Bernstein [4,6]	y
$(\log y)^{-3}$	Algorithm 4.1	$(\log y)^{22+o(1)}$
y^{-1}	Bernstein [4,6]	$y(\log y)^4$
y^{-1}	Algorithm 4.1	$y(\log y)^4$

1.4 Organization

The rest of this paper is organized as follows. In §2 we review Bernstein's algorithm and provide a running time analysis. In §3 we present and analyze our first improved algorithm. In §4 we present the second improved algorithm, along with a running time analysis. In §5 we perform an accuracy analysis of the algorithm from §4. Finally in §6 we present some timing results.

2 Bernstein's Algorithm

In this section, we review Bernstein's algorithm [4,6] that gives rigorous upper and lower bounds for $\Psi(x,y)$. We also give a running time analysis.

Consider a discrete generalized power series

$$F(X) = \sum_r a_r X^r,$$

where r ranges over the real numbers. The a_r may lie in any fixed ring or field, although we will limit our interest to the reals. We require that, for any real h, the set $\{r \le h : a_r \ne 0\}$ is finite. We write

$$\mathrm{distr}_h F := \sum_{r \le h} a_r,$$

the sum of the coefficients of F on powers of X below h.

We make the reasonable restriction that x be a power of 2. Define $\lg x := \log_2 x$, and let $h := \lg x$ so that $2^h = x$. Then for $|X| < 1$ we have

$$\Psi(2^h, y) = \mathrm{distr}_h \sum_{P(n) \le y} X^{\lg n}$$

$$= \mathrm{distr}_h \prod_{p \le y} \left(1 + X^{\lg p} + X^{2 \lg p} + \cdots\right)$$

$$= \mathrm{distr}_h \prod_{p \le y} \left(1 - X^{\lg p}\right)^{-1}$$

$$= \mathrm{distr}_h \exp \sum_{p \le y} \log \left(1 - X^{\lg p}\right)^{-1}$$

$$= \mathrm{distr}_h \exp \left(\sum_{p \le y} \sum_{k \ge 1} \frac{1}{k} X^{k \lg p}\right).$$

Here we used the identity $\log(1 - t)^{-1} = \sum_{k \ge 1} t^k / k$ for $|t| < 1$.

To reduce the number of terms in this power series, we approximate each prime p using a fractional power of 2. Define $\underline{p} \le p$ and $\overline{p} \ge p$ as such.

Replacing p with $\underline{p}$ in the series above, we denote the resulting series by $B^+(x, y)$, which overestimates Ψ:

$$\Psi(2^h, y) \le B^+(x, y) := \mathrm{distr}_h \exp \left(\sum_{p \le y} \sum_{k \ge 1} \frac{1}{k} X^{k \lg \underline{p}}\right).$$

Replacing p with $\overline{p}$, we denote the resulting series by $B^-(x, y)$ which underestimates Ψ:

$$\Psi(2^h, y) \ge B^-(x, y) := \mathrm{distr}_h \exp \left(\sum_{p \le y} \sum_{k \ge 1} \frac{1}{k} X^{k \lg \overline{p}}\right).$$

We now present the algorithm for computing a lower bound for $\Psi(x, y)$. Computing the upper bound is similar.

Algorithm 2.1. Recall that $x = 2^h$. WLOG we are computing $B^-(x, y)$, the lower bound.

1. Choose an *accuracy parameter* α, an integer, that satisfies $2 \log x < \alpha \lg 3 < (\log x) e^{\sqrt{\log y}}$.

2. Find the primes up to y, and for each p, compute $\overline{p}$ such that

$$\alpha \lg \overline{p} = \lceil \alpha \lg p \rceil \qquad (1)$$

(and similarly $\alpha \lg \underline{p} = \lfloor \alpha \lg p \rfloor$ for the upper bound).
For example, if $\alpha = 10$, then $\overline{2} = 2$, $\overline{3} := 2^{16/10} \approx 3.03$, $\overline{5} := 2^{24/10} \approx 5.28$, and $\overline{7} := 2^{29/10} \approx 7.46$.

3. Compute $\overline{G}(X) := \displaystyle\sum_{p \leq y} \sum_{k=1}^{\lfloor h/\lg \overline{p} \rfloor} \frac{1}{k} X^{k \lg \overline{p}}$.

4. Compute $\exp \overline{G}(X)$ using an FFT-based algorithm.

5. Compute $\mathrm{distr}_h \exp \overline{G}(X)$ by summing the coefficients.

Note that one can compute $\mathrm{distr}_{h'} \exp \overline{G}(X)$ for any $h' \leq h$ along the way, giving a lower bound for $\Psi(2^{h'}, y)$ as well, essentially for free.

Theorem 2.2. *When y is sufficiently large, Algorithm 2.1 computes upper and lower bounds, $B^+(x,y)$ and $B^-(x,y)$, for $\Psi(x,y)$ satisfying*

$$\frac{B^-(x,y)}{\Psi(x,y)} \geq 1 - \frac{\log x}{\alpha \lg 3} \qquad and \qquad \frac{B^+(x,y)}{\Psi(x,y)} \leq 1 + \frac{2 \log x}{\alpha \lg 3}$$

using at most

$$O\left(\frac{y}{\log \log y} + \frac{y \log x}{(\log y)^2} + \alpha \log x \log \alpha \right)$$

arithmetic operations.

Proof. If we set

$$\varepsilon_1 = \max_{p \leq y} \left(\frac{\lg \overline{p}}{\lg p} - 1 \right) \qquad and \qquad \varepsilon_2 = \max_{p \leq y} \left(1 - \frac{\lg \underline{p}}{\lg p} \right) \quad .$$

and take $\varepsilon \geq \max\{\varepsilon_1, \varepsilon_2\}$, then one has

$$\Psi(x^{1/(1+\varepsilon)}, y) = \mathrm{distr}_h \prod_{p \leq y} (1 - X^{(1+\varepsilon)\lg p})^{-1} \leq B^-(x,y)$$

and

$$\Psi(x^{1/(1-\varepsilon)}, y) = \mathrm{distr}_h \prod_{p \leq y} (1 - X^{(1-\varepsilon)\lg p})^{-1} \geq B^+(x,y).$$

Hildebrand [16] shows that $\Psi(cx, y) \leq c\Psi(x,y)$ when y is sufficiently large and $c \geq 1 + \exp(-\sqrt{\log y})$. Taking $c = x^{\varepsilon/(1 \pm \varepsilon)}$, we find that

$$\frac{B^-(x,y)}{\Psi(x,y)} \geq x^{-\varepsilon/(1+\varepsilon)} \geq 1 - \varepsilon \log x \quad and \quad \frac{B^+(x,y)}{\Psi(x,y)} \leq x^{\varepsilon/(1-\varepsilon)} \leq 1 + 2\varepsilon \log x,$$

provided that x is sufficiently large and

$$\exp(-\sqrt{\log y}) < \varepsilon \log x < 1/2.$$

In view of (1), we can take $\varepsilon = 1/(\alpha \lg 3)$.

As for the running time, Step 2 can be done with a prime sieve [2], taking $O(y/\log\log y)$ operations. In Step 3, $\overline{G}(X)$ will have $O(\alpha h)$ nonzero terms, and so takes $O(hy/(\log y)^2)$ time to construct. The FFT-based exponentiation algorithm in Step 4 takes only $O(\alpha h \log(\alpha h))$ operations [7]. Finally, Step 5 takes only $O(\alpha h)$ time. Adding this up gives the stated runtime bound. □

In practice, likely one of the first two terms will dominate the running time.

3   The First Improvement

Define $n_i := \pi(2^{i/\alpha}) - \pi(2^{(i-1)/\alpha})$, the number of primes p such that $\alpha \lg \overline{p} = i$, or equivalently $\alpha \lg p = i - 1$.

We improve Bernstein's algorithm by first computing the n_i values, and then use them to compute $\overline{G}(X)$.

Algorithm 3.1. WLOG we are computing $B^-(x,y)$, the lower bound.

1. Choose an *accuracy parameter* α, an integer, that satisfies $2\log x < \alpha \lg 3 < (\log x)e^{\sqrt{\log y}}$.
2. Compute the n_i values for $\alpha \le i \le \alpha \lg y$.
3. Compute $\overline{G}(X) := \displaystyle\sum_{i=\alpha}^{\lfloor \alpha \lg y \rfloor} n_i \sum_{k=1}^{\lfloor h\alpha/i \rfloor} \frac{1}{k} X^{ki/\alpha}$.
4. Compute $\exp \overline{G}(X)$ using an FFT-based algorithm.
5. Compute $\mathrm{distr}_h \exp \overline{G}(X)$ by summing the coefficients.

Similarly, for the upper bound we have

$$\underline{G}(X) := \sum_{i=\alpha-1}^{\lfloor \alpha \lg y \rfloor - 1} n_{i+1} \sum_{k=1}^{\lfloor h\alpha/i \rfloor} \frac{1}{k} X^{ki/\alpha}.$$

Bernstein mentions this improvement in his paper [6], but gives no analysis, and his code (downloadable from cr.yp.to) does not use it.

Theorem 3.2. *When y is sufficiently large, Algorithm 3.1 computes upper and lower bounds, $B^+(x,y)$ and $B^-(x,y)$, for $\Psi(x,y)$ satisfying*

$$\frac{B^-(x,y)}{\Psi(x,y)} \ge 1 - \frac{\log x}{\alpha \lg 3} \quad and \quad \frac{B^+(x,y)}{\Psi(x,y)} \le 1 + \frac{2\log x}{\alpha \lg 3}$$

using at most

$$O\left(\alpha \frac{y^{2/3}}{\log y} + \alpha \log x \log \alpha\right)$$

arithmetic operations.

Again, we expect the first term to dominate the running time.

Proof. The accuracy analysis of Algorithm 3.1 is identical to that of Algorithm 2.1, so we only need to perform a runtime analysis. We can use the algorithm of Deléglise and Rivat[12] to compute $\pi(t)$ in time $O(t^{2/3}/(\log t)^2)$. This means that it takes

$$O\left(\alpha \log y \cdot \frac{y^{2/3}}{(\log y)^2}\right)$$

operations to compute all the n_i values (Step 2). The time to construct $\overline{G}(X)$ or $\underline{G}(X)$ (Step 3) is then proportional to

$$\sum_{i=\alpha}^{\lfloor \alpha \lg y \rfloor} \frac{\alpha \log x}{i} = O(\alpha \log x \log \alpha).$$

The remaining steps have the same complexity as Algorithm 2.1.   □

4   The Second Improvement

Next we show how to make Bernstein's algorithm faster and tighter, especially when y is large. The idea is to choose a parameter $z < y$, and only compute the n_i values for $i \leq \alpha \lg z$. For larger i, we estimate n_i using the prime number theorem and the Riemann Hypothesis. This introduces more error, but the greatly improved running time allows us to choose a larger α to more than compensate.

Assuming the Riemann Hypothesis, we have

$$|\pi(t) - \text{li}(t)| < \frac{\sqrt{t}\log t}{8\pi} \tag{2}$$

when $t \geq 1451$ (see [23,9]), so we require that $z > 1451$. We note that a very good estimate for $\text{li}(t)$ can be computed in $O(\log t)$ time (see equations 5.1.3 and 5.1.10, or even 5.1.56, in [1]).

Define $n_i^{\pm}$, our upper and lower bound estimates for n_i, as follows:

– For $i \leq \alpha \lg z$, $n_i^- := n_i^+ := n_i$.

– For $i > \alpha \lg z$, $n_i^- := \max\left\{0, \left(\text{li}(2^{i/\alpha}) - \frac{\sqrt{2^{i/\alpha}}\log(2^{i/\alpha})}{8\pi}\right) - \sum_{j<i} n_j^-\right\},$

and $n_i^+ := \max\left\{0, \left(\text{li}(2^{i/\alpha}) + \frac{\sqrt{2^{i/\alpha}}\log(2^{i/\alpha})}{8\pi}\right) - \sum_{j<i} n_j^+\right\}.$

We define $G^-(X)$ by replacing n_i with n_i^- in the definition of $\overline{G}(X)$:

$$G^-(X) := \sum_{i=\alpha}^{\lfloor \alpha \lg y \rfloor} n_i^- \sum_{k=1}^{\lfloor h\alpha/i \rfloor} \frac{1}{k} X^{ki/\alpha},$$

and define

$$A^-(2^h, y) := \text{distr}_h \exp G^-(X).$$

We define $G^+(X)$ and $A^+(x,y)$ in a similar way for the upper bound.

Note that, for $A^-(x,y)$ to be a rigorous lower bound on $\Psi(x,y)$, it is not necessary for $n_i^- \le n_i$, but merely that, for every i,

$$\sum_{j\le i} n_j^- \le \sum_{j\le i} n_j = \pi(2^{i/\alpha}).$$

Similarly, for $A^+(x,y)$ to be a rigorous upper bound it suffices that, for every i,

$$\sum_{j\le i} n_j^+ \ge \sum_{j\le i} n_j = \pi(2^{i/\alpha}).$$

We achieve this assuming the Riemann Hypothesis. This leads us to the following algorithm.

Algorithm 4.1. WLOG we are computing $A^-(x,y)$.

1. Choose an *accuracy parameter* α, an integer, that satisfies $2\log x < \alpha \lg 3 < (\log x)e^{\sqrt{\log y}}$, and choose a parameter $z < y$ with $z \asymp \alpha^4 (\log \alpha)^2$.
2. Compute the n_i^- values as defined above.
3. Compute $G^-(X) := \displaystyle\sum_{i=\alpha}^{\lfloor \alpha \lg y \rfloor} n_i^- \sum_{k=1}^{\lfloor h\alpha/i \rfloor} \frac{1}{k} X^{ki/\alpha}$.
4. Compute $\exp G^-(X)$ using the FFT.
5. Compute $\mathrm{distr}_h \exp G^-(X)$ by summing the coefficients.

In the next section we prove the following:

Theorem 4.2 (RH). *When y is sufficiently large, Algorithm 4.1 computes upper and lower bounds, $A^+(x,y)$ and $A^-(x,y)$, for $\Psi(x,y)$ satisfying*

$$\frac{A^-(x,y)}{\Psi(x,y)} \ge 1 - \frac{\alpha \log x \log z}{6\sqrt{z}} - \frac{\log x}{\alpha \lg 3} + \frac{(\log x)^2 \log z}{6\sqrt{z}\lg 3}$$

and

$$\frac{A^+(x,y)}{\Psi(x,y)} \le 1 + \frac{\alpha \log x \log z}{3\sqrt{z}} + \frac{2\log x}{\alpha \lg 3} + \frac{2(\log x)^2 \log z}{3\sqrt{z}\lg 3}.$$

Because $\alpha \gg \log x$, asymptotically we can ignore the last term in each case. The other two terms balance when α is asymptotic to $z^{1/4}/\sqrt{\log z}$. This justifies our choosing z proportional to $\alpha^4 (\log \alpha)^2$ in Step 1 of the algorithm, and this implies that

$$\frac{\Psi(x,y)}{A^\pm(x,y)} = 1 + O\left(\frac{\log x}{\alpha}\right).$$

To achieve a tighter bound with $A^\pm(x,y)$ than is obtained with $B^\pm(x,y)$ in Algorithm 3.1, we will simply choose α larger. For example, if in Algorithm 3.1 we used $\alpha \asymp \log x \log \log y$, then in our improved algorithm we might use $\alpha \asymp \log x (\log \log y)^2$. As we will see in §6, we can tolerate a larger α and still get a faster running time.

Theorem 4.3. *Algorithm 4.1 computes* $A^+(x,y)$ *and* $A^-(x,y)$ *in*

$$O\left(\alpha \frac{z^{2/3}}{\log z} + \alpha \log x \log \alpha y\right)$$

operations.

Proof. We have the following:

- It takes $O(\alpha z^{2/3}/\log z)$ time to compute the n_i^- for $i \leq \alpha \lg z$ in Step 2.
- It takes $O(\alpha \log x \log y)$ time to compute the n_i^- for $i > \alpha \lg z$ in Step 2.
- The remaining steps take at most $O(\alpha \log x \log \alpha)$ steps, the same as in Algorithm 3.1.

Adding this up completes the proof. ☐

If we choose $\alpha \asymp \log x (\log \log y)^2$, say, making $z \asymp (\log x)^4 (\log \log x)^2 (\log \log y)^8$, then the running time is

$$O((\log x)^{11/3} (\log \log x)^{1/3} (\log \log y)^{22/3}).$$

In applications to factoring, we have, roughly, $\log x \approx (\log y)^3$, so in this case our running time is $(\log y)^{11+o(1)}$, which, asymptotically, is significantly better than $y^{2/3+o(1)}$.

5 An Accuracy Analysis

In this section, we present the proof of Theorem 4.2.

For the purposes of accuracy analysis, we will redefine n_i^- and n_i^+ for $i > \alpha \lg z$ as

$$n_i^- := \text{li}(2^{i/\alpha}) - \frac{\sqrt{2^{i/\alpha}} \log(2^{i/\alpha})}{8\pi} - \left(\text{li}(2^{(i-1)/\alpha}) + \frac{\sqrt{2^{(i-1)/\alpha}} \log(2^{(i-1)/\alpha})}{8\pi}\right)$$

and

$$n_i^+ := \text{li}(2^{i/\alpha}) + \frac{\sqrt{2^{i/\alpha}} \log(2^{i/\alpha})}{8\pi} - \left(\text{li}(2^{(i-1)/\alpha}) - \frac{\sqrt{2^{(i-1)/\alpha}} \log(2^{(i-1)/\alpha})}{8\pi}\right).$$

On recalling (2), we may rewrite this as

$$n_i^- = L_i - \Delta_i \leq n_i \leq L_i + \Delta_i = n_i^+, \tag{3}$$

where

$$L_i := \text{li}(2^{i/\alpha}) - \text{li}(2^{(i-1)/\alpha})$$

and

$$\Delta_i := \frac{2^{i/(2\alpha)} \log 2}{8\pi\alpha} \left(i + \frac{i-1}{2^{1/(2\alpha)}}\right) \leq \frac{i 2^{i/(2\alpha)} \log 2}{4\pi\alpha}. \tag{4}$$

These $n_i^{\pm}$ values lead to weaker bounds on $\Psi(x,y)$ than those used in Algorithm 4.1, but they are much easier to work with, and the results we obtain still apply to Algorithm 4.1.

It follows easily from (3) that

$$n_i^- \geq n_i(1-\delta_i) \quad \text{and} \quad n_i^+ \leq n_i(1+\delta_i), \tag{5}$$

where $\delta_i := 2\Delta_i/n_i$. Moreover, it follows from (3) and (4) after some computation that

$$\pi(w)-\pi(w/c) \geq \mathrm{li}(w)-\mathrm{li}(w/c)-\frac{\sqrt{w}\log w}{4\pi} \geq \left(1-\frac{1}{c}\right)\mathrm{li}(w)-\frac{w\log c}{c(\log w)^2}-\sqrt{w}\log w.$$

Taking $c = 2^{1/\alpha}$ and noting that

$$1-\frac{1}{c} = \sum_{k=1}^{\infty}\frac{(-1)^{k+1}(\log 2)^k}{k!\alpha^k} \geq \frac{0.9\log 2}{\alpha}$$

for $\alpha \geq 4$, we find that

$$\pi(w)-\pi(2^{-1/\alpha}w) \geq \frac{0.9w\log 2}{\alpha\log w}-\frac{w}{\alpha(\log w)^2} \geq \frac{(\log 2)^2 w}{\alpha\log w},$$

provided that w is sufficiently large and $\alpha \leq w^{1/4}$. Thus on taking $w = 2^{i/\alpha}$, we obtain

$$n_i \geq \frac{2^{i/\alpha}\log 2}{i}$$

for $i > \alpha\lg z$, provided that $\alpha \leq z^{1/4}$ and z is sufficiently large. Thus by (4) we have

$$\delta_i \leq \frac{i^2}{4\pi\alpha 2^{i/(2\alpha)}} \leq \frac{\alpha(\lg z)^2}{4\pi\sqrt{z}} \leq \frac{\alpha(\log z)^2}{6\sqrt{z}} := \delta \tag{6}$$

for $i > \alpha\lg z$, since the expression $i^2/2^{i/(2\alpha)}$ is a decreasing function of i for $i > 4\alpha/(\log 2)$. Write

$$g_i(X) = \sum_{k=1}^{\infty}\frac{X^{ki/\alpha}}{k},$$

and let $t = h/\lg z = \log x/\log z$. Since the smallest power of X in $g_i(X)$ is at least $X^{\lg z}$ when $i > \alpha\lg z$, we have

$$\mathrm{distr}_h \exp G^-(X) = \mathrm{distr}_h\left[\exp\left(\sum_{p\leq z}\sum_{k=1}^{\infty}\frac{X^{k\lg \overline{p}}}{k}\right)\exp\left(\sum_{i=\lfloor\alpha\lg z\rfloor+1}^{\lfloor\alpha\lg y\rfloor}n_i^- g_i(X)\right)\right]$$

$$= \mathrm{distr}_h\left[\exp\left(\sum_{i=\alpha}^{\lfloor\alpha\lg z\rfloor}n_i g_i(X)\right)\sum_{j=0}^{t}\frac{1}{j!}\left(\sum_{i=\lfloor\alpha\lg z\rfloor+1}^{\alpha\lg y}n_i^- g_i(X)\right)^j\right]$$

$$\geq (1-\delta)^t\mathrm{distr}_h\exp\overline{G}(X),$$

on recalling (5). It therefore follows from (6) that

$$\frac{A^-(x,y)}{B^-(x,y)} = \frac{\mathrm{distr}_h \exp G^-(X)}{\mathrm{distr}_h \exp \overline{G}(X)} \geq (1-\delta)^t \geq 1 - t\delta \geq 1 - \frac{\alpha \log x \log z}{6\sqrt{z}}.$$

Similarly, since $(1+\delta)^t \leq 1 + 2t\delta$ whenever $2t\delta \leq 1$, one has

$$\frac{A^+(x,y)}{B^+(x,y)} \leq (1+\delta)^t \leq 1 + \frac{\alpha \log x \log z}{3\sqrt{z}},$$

provided that

$$\alpha \leq \frac{3\sqrt{z}}{\log z \log x}.$$

On combining these bounds with the conclusion of Theorem (2.2), we find that

$$\frac{A^-(x,y)}{\Psi(x,y)} \geq 1 - \frac{\alpha \log x \log z}{6\sqrt{z}} - \frac{\log x}{\alpha \lg 3} + \frac{(\log x)^2 \log z}{6\sqrt{z}\lg 3}$$

and

$$\frac{A^+(x,y)}{\Psi(x,y)} \leq 1 + \frac{\alpha \log x \log z}{3\sqrt{z}} + \frac{2\log x}{\alpha \lg 3} + \frac{2(\log x)^2 \log z}{3\sqrt{z}\lg 3}.$$

Thus we start to obtain reasonably accurate upper and lower bounds as soon as

$$2\log x < \min\left(\frac{6\sqrt{z}}{\alpha \log z}, \alpha \lg 3\right),$$

and one can optimize the error terms by taking $\alpha \asymp z^{1/4}(\log z)^{-1/2}$, as suggested in Algorithm 4.1. This completes the proof of Theorem 4.2.

6 Timing Results

We estimated $\Psi(2^{255}, 2^{28})$ using Algorithm 3.1 with $\alpha = 32$ and using Algorithm 4.1 with $\alpha = 64$. We used $z = 23216$.

We obtained the following:

$$B^-(x,y) \approx 39235936 \times 10^{60}$$
$$A^-(x,y) \approx 39259233 \times 10^{60}$$
$$A^+(x,y) \approx 43345488 \times 10^{60}$$
$$B^+(x,y) \approx 51166381 \times 10^{60}$$

Algorithm 3.1 took 12.6 seconds, and Algorithm 4.1 took 2.1 seconds.

Note that we used a prime sieve in place of a $\pi(t)$ algorithm to compute the n_i values for Algorithm 3.1 and to compute the n_i values with $i \leq \alpha \lg z$ for Algorithm 4.1.

This experiment was done on a Pentium IV 1.3 GHz running Fedora Core v.4; we used the Gnu C++ compiler and Bernstein's code (psibound-0.50 from cr.yp.to) with modifications. (The code is available from the second author via e-mail.)

Notes

- If the FFT exponentiation algorithm is the runtime bottleneck (Step 4), then Algorithm 3.1 will perform better in practice; Algorithm 4.1 only does better when the bottleneck is finding the primes up to y (Step 2).
- Unless y is quite large, finding the primes up to y (or z) and using them to compute the n_i values is more efficient in practice than using an algorithm for $\pi(t)$.
- As with all timing experiments, the results depend on the platform, the compiler, and the programmer.

References

1. M. Abramowitz and I. A. Stegun. *Handbook of Mathematical Functions*. Dover, 1970.
2. A. O. L. Atkin and D. J. Bernstein. Prime sieves using binary quadratic forms. *Mathematics of Computation*, 73:1023–1030, 2004.
3. Daniel J. Bernstein. Enumerating and counting smooth integers. Chapter 2, PhD Thesis, University of California at Berkeley, May 1995.
4. Daniel J. Bernstein. Bounding smooth integers. In J. P. Buhler, editor, *Third International Algorithmic Number Theory Symposium*, pages 128–130, Portland, Oregon, June 1998. Springer. LNCS 1423.
5. Daniel J. Bernstein. Arbitrarily tight bounds on the distribution of smooth integers. In Bennett, Berndt, Boston, Diamond, Hildebrand, and Philipp, editors, *Proceedings of the Millennial Conference on Number Theory*, volume 1, pages 49–66. A. K. Peters, 2002.
6. Daniel J. Bernstein. Proving primality in essentially quartic time. To appear in *Mathematics of Computation*; http://cr.yp.to/papers.html#quartic, 2006.
7. R. P. Brent. Multiple precision zero-finding methods and the complexity of elementary function evaluation. In J. F. Traub, editor, *Analytic Computational Complexity*, pages 151–176. Academic Press, 1976.
8. E. R. Canfield, P. Erdős, and C. Pomerance. On a problem of Oppenheim concerning "Factorisatio Numerorum". *Journal of Number Theory*, 17:1–28, 1983.
9. R. Crandall and C. Pomerance. *Prime Numbers, a Computational Perspective*. Springer, 2001.
10. N. G. de Bruijn. On the number of positive integers $\leq x$ and free of prime factors $> y$. *Indag. Math.*, 13:50–60, 1951.
11. N. G. de Bruijn. On the number of positive integers $\leq x$ and free of prime factors $> y$, II. *Indag. Math.*, 28:239–247, 1966.
12. M. Deléglise and J. Rivat. Computing $\pi(x)$: the Meissel, Lehmer, Lagarias, Miller, Odlyzko method. *Math. Comp.*, 65(213):235–245, 1996.
13. A. Hildebrand. On the number of positive integers $\leq x$ and free of prime factors $> y$. *Journal of Number Theory*, 22:289–307, 1986.
14. A. Hildebrand and G. Tenenbaum. On integers free of large prime factors. *Trans. AMS*, 296(1):265–290, 1986.
15. A. Hildebrand and G. Tenenbaum. Integers without large prime factors. *Journal de Théorie des Nombres de Bordeaux*, 5:411–484, 1993.
16. Adolf Hildebrand. On the local behavior of $\Psi(x,y)$. *Trans. Amer. Math. Soc.*, 297(2):729–751, 1986.

17. Simon Hunter and Jonathan P. Sorenson. Approximating the number of integers free of large prime factors. *Mathematics of Computation*, 66(220):1729–1741, 1997.

18. D. E. Knuth and L. Trabb Pardo. Analysis of a simple factorization algorithm. *Theoretical Computer Science*, 3:321–348, 1976.

19. A. J. Menezes, P. C. van Oorschot, and S. A. Vanstone. *Handbook of Applied Cryptography*. CRC Press, Boca Raton, 1997.

20. Pieter Moree. *Psixyology and Diophantine Equations*. PhD thesis, Rijksuniversiteit Leiden, 1993.

21. Karl K. Norton. *Numbers with Small Prime Factors, and the Least kth Power Non-Residue*, volume 106 of *Memoirs of the American Mathematical Society*. American Mathematical Society, Providence, Rhode Island, 1971.

22. C. Pomerance, editor. *Cryptology and Computational Number Theory*, volume 42 of *Proceedings of Symposia in Applied Mathematics*. American Mathematical Society, Providence, Rhode Island, 1990.

23. L. Schoenfeld. Sharper bounds for the Chebyshev functions $\theta(x)$ and $\psi(x)$. II. *Mathematics of Computation*, 30(134):337–360, 1976.

24. Jonathan P. Sorenson. A fast algorithm for approximately counting smooth numbers. In W. Bosma, editor, *Proceedings of the Fourth International Algorithmic Number Theory Symposium (ANTS IV)*, pages 539–549, Leiden, The Netherlands, 2000. LNCS 1838.

25. K. Suzuki. An estimate for the number of integers without large prime factors. *Mathematics of Computation*, 73:1013–1022, 2004. MR 2031422 (2005a:11142).

26. K. Suzuki. Approximating the number of integers without large prime factors. *Mathematics of Computation*, 75:1015–1024, 2006.

27. J. van de Lune and E. Wattel. On the numerical solution of a differential-difference equation arising in analytic number theory. *Mathematics of Computation*, 23:417–421, 1969.

Use of Extended Euclidean Algorithm in Solving a System of Linear Diophantine Equations with Bounded Variables

Parthasarathy Ramachandran

Indian Institute of Science, Bangalore KA 560 012, India
parthar@mgmt.iisc.ernet.in

Abstract. We develop an algorithm to generate the set of all solutions to a system of linear Diophantine equations with lower and upper bounds on the variables. The algorithm is based on the Euclid's algorithm for computing the GCD of rational numbers. We make use of the ability to parametrise the set of all solutions to a linear Diophantine equation in two variables with a single parameter. The bounds on the variables are translated to bounds on the parameter. This is used progressively by reducing a n variable problem into a two variable problem. Computational experiments indicate that for a given number of variables the running times decreases with the increase in the number of equations in the system.

1 Introduction

Consider the following problem:

$$\text{Find all } \boldsymbol{x} \in \mathbb{Z}^n \text{ such that } \boldsymbol{Ax} = \boldsymbol{b}, \ \boldsymbol{l} \leq \boldsymbol{x} \leq \boldsymbol{u} \tag{1}$$

where m is the number of equations, n is the number of variables, $\boldsymbol{A} \in \mathbb{N}^{m \times n}$, $\boldsymbol{b} \in \mathbb{N}^m$, $\boldsymbol{l}$ and $\boldsymbol{u} \in \mathbb{N}^n$. Without the bounds on the variables the problem can be solved in polynomial time. The variable bounds make its NP-complete. This problem has important applications in discrete optimisation. Literature pertaining to the Frobenius problem and solution to a single linear Diophantine equation is of relevance to the topic discussed here.

The Frobenius problem seeks the largest integer b' such that $\boldsymbol{ax} = b'$ does not have a non-negative integer solution. Rödseth and Selmer and Beyer solves the Frobenius problem in 3 variables [9,11]. The more general problem for an arbitrary number of variables is known to be NP-hard, the objective of the current research being to refine bounds [4,12,2]. The solution to the Frobenius problem has been used to solve a linear Diophantine equation in the field of non–negative integers. Greenberg solve a linear Diophantine equation in three variables for non–negative integers using the ability to solve the three variable Frobenius problem [6]. Filgueiras and Tomás have given a complete characterisation of the set of minimal solutions over non–negative integers for a Diophantine equation

F. Hess, S. Pauli, and M. Pohst (Eds.): ANTS 2006, LNCS 4076, pp. 182–192, 2006.

with three unknowns using the known properties of congruence [5]. These are some very special cases of the problem that is being addressed in this paper.

The procedures for finding the general solution of a single linear Diophantine equation has been studied by Bond [3], Morito and Salkin [8], and Kertzner [7]. The general solution has been represented as a function of $n - 1$ independent integer parameters for a problem in n variables. Though these methods have represented the set of all solutions in a compact manner, the problem of locating one or more solutions that satisfies the bounds on the variables is open.

Aardal et al. [1] have studied a very closely related problem that is being discussed in this paper. The problem that they address seeks to identify if there exists a $x \in \mathbb{Z}^n$ such that $Ax = b$, $l \le x \le u$. They propose a method that is based on lattice basis reduction. The algorithm finds a short vector x_d such that $Ax_d = b$ and the basis of the null space X_0 such that $AX_0 = 0$. The problem is now recast as finding an integer multiplier λ such that $l \le (x_d + \lambda X) \le u$. The algorithm branches on different linear combinations of columns of X_0 in order to satisfy the variable bounds. This algorithm has no guidance for the branching algorithm. This make it very unreliable for generating the set of all solutions, which is the topic of this paper.

In the approach that we take in this paper, we make use of the ability to characterise the set of all solutions to a linear Diophantine equation in two variables using a single parameter. The bounds on the variables are translated into bounds on the parameter, which solves the two variable problem. For the more general problem with n variables, the problem dimension is progressively reduced to two, in which one of the variables is the actual variable and the other an artificial one. Then we roll back the problem by solving for one variable at a time using a branching scheme. The advantage of this approach is that it will generate the set of all solutions and also provides a guide to the branching scheme in terms on the bounds on the free parameter.

The rest of this paper is organised as follows. In Sect. 2 we discuss the problem of a linear Diophantine equation in two variables. Specifically, the bounds on the variables are translated to bounds on the parameter used to describe the set of all solutions. Specifically we prove that the conversion of the bounds on the variables into bounds on the parameter used to describe the set of all solutions leads to a computationally efficient algorithm. Next in Sect. 3, we use these results in developing an algorithm for the general problem. The algorithm reduces the problem recursively into a two variable problem with progressive updates on the bounds. Finally in Sect. 4, we discuss the results from our computational experiments.

2   Two Variable Problem

We first consider the two variable single equation problem to develop bounds on the parameter used to represent the set of all solutions. The algorithm to solve a linear Diophantine equation without any constraints on the variables is an extension of the Euclidean algorithm for finding the greatest common divisor (GCD) of two integers. This is a logarithmic time algorithm. Given two natural numbers a_1 and a_2, the Euclidean algorithm determines the GCD of these numbers $a' = \gcd\{a_1, a_2\}$, and also find integers γ and ϵ such that $\gamma a_1 + \epsilon a_2 = a'$. The Euclidean algorithm can be summarised as follows [10]:

– Determine a series of 3×2 matrices $A_0, A_1, A_2, \ldots$ as follows:

$$A_0 = \begin{bmatrix} a_1 & a_2 \\ 1 & 0 \\ 0 & 1 \end{bmatrix} \tag{2}$$

$$A_k = \begin{bmatrix} a_{1,k} & a_{2,k} \\ \gamma_k & \delta_k \\ \epsilon_k & \zeta_k \end{bmatrix} \tag{3}$$

– A_{k+1} is generated from A_k as follows:
  • If k is even and $a_{2,k} > 0$, subtract $\lfloor a_{1,k}/a_{2,k} \rfloor$ times the second column of A_k from the first column
  • If k is odd and $a_{1,k} > 0$, subtract $\lfloor a_{2,k}/a_{1,k} \rfloor$ times the first column of A_k from the second column
  This step is performed until $a_{1,k} = 0$ or $a_{2,k} = 0$. If $a_{1,k}$ is zero, then $a_{2,k}$ is $a' = \mathrm{g.c.d}\{a_1, a_2\}$, and vice-versa.

To solve a Diophantine equation in two variables $a_1 x_1 + a_2 x_2 = b$, there exists an integer solution only if a' evenly divides b, i.e, $a' \mid b$. When this condition is satisfied, an integral solution for the two variable Diophantine equation is given by $x_1^0 = \gamma b/a'$ and $x_2^0 = \epsilon b/a'$. WLOG, all other solutions can be generated by [7],

$$x_1 = x_1^0 + \left(\frac{a_2}{a'}\right) t, \quad x_2 = x_2^0 - \left(\frac{a_1}{a'}\right) t \quad \forall t \in \mathbb{Z}. \tag{4}$$

However, there are no guarantees that any of these solutions will abide by the bounding constraints on the variables.

If the variables have to abide by the binding constraints, then this has to be achieved by way of choosing an appropriate value for the parameter t if such a one exists. This leads to the following lemma on the bounded solution to a linear Diophantine equation is two variables.

Lemma 1. *Given the general solution to a linear Diophantine equation in two variables $a_1 x_1 + a_2 x_2 = b$ by equation 4, there exists an integral solution satisfying the boundary conditions $l \leq x \leq u$ iff there exists a $t \in \mathbb{Z}$ such that*

$$Max\left\{\frac{(l_1 - x_1^0)a'}{a_2}, \frac{(x_2^0 - u_2)a'}{a_1}\right\} \leq t \leq Min\left\{\frac{(u_1 - x_1^0)a'}{a_2}, \frac{(x_2^0 - l_2)a'}{a_1}\right\}. \tag{5}$$

Proof. The proof is rather direct. By applying the limits on the first variable x_1, the parameter t can be bounded by

$$\frac{(l_1 - x_1^0)a'}{a_2} \leq t \leq \frac{(u_1 - x_1^0)a'}{a_2}, \tag{6}$$

and the bounds on the second variable x_2 leads to

$$\frac{(l_2 - x_2^0)a'}{a_1} \leq -t \leq \frac{(u_2 - x_2^0)a'}{a_1}. \tag{7}$$

Since there are two bounds, the most restrictive bound dominates and hence the bound on the parameter t given by eqn. 5.   □

Example 1. Consider the following instance of single equation two variable problem: find $\boldsymbol{x} \in \mathbb{Z}^2$ such that:

$$17689x_1 + 8345x_2 = 15856125 \qquad 0 \leq x_1 \leq 543; 0 \leq x_2 \leq 1267 \tag{8}$$

By applying the Euclid's algorithm the general solution to this problem can be given by

$$x_1 = 37483879500 + 8345t \qquad x_2 = -79455042375 - 17689t \qquad \forall t \in \mathbb{Z} \tag{9}$$

Use of bounds (5) the parameter t can be restricted to $-4491777.01635 \leq t \leq -4491776.98705$. The only integer value of t with in this range leads to a solution of $x_1 = 435$ and $x_2 = 978$.

2.1 Analysis of the Parameter Bounds

A direct approach to the single equation two variable problem will be to branch on the different combinations of permissible values for the two variables. Given that $l_1 \leq x_1 \leq u_1$, the number of possible values for x_1 is given by $w_1 = u_1 - l_1 + 1$ and similarly $w_2 = u_2 - l_2 + 1$. Hence the direct branching algorithm will have to consider $w_1 w_2$ possible solutions. However, as we show next, the bounds on the parameter helps in reducing this computational effort.

Lemma 2. *Let t^l and t^u be the lower and upper bounds on the parameter t as defined by (5). Then $t^u - t^l \leq w_1 w_2$.*

Proof. Let $f_1 = a'/a_1$ and $f_2 = a'/a_2$. Since $a' \leq \mathrm{Min}\{a_1, a_2\}$, $f_1 \leq 1$ and $f_2 \leq 1$. Given the bounds on the parameter t by (5) the following four cases arise.

Case 1: Suppose that the limiting constraint on the parameter t is given by

$$\frac{(l_1 - x_1^0)a'}{a_2} \leq t \leq \frac{(u_1 - x_1^0)a'}{a_2}. \tag{10}$$

The number of permissible values of the parameter t is given by $(u_1 - x_1^0)f_2 - (l_1 - x_1^0)f_2$, which gets simplified to $(u_1 - l_1)f_2$, i.e., $(w_1 - 1)f_2$. Since $f_2 \leq 1$, we obtain that $(w_1 - 1)f_2 < w_1 w_2$.

Case 2: Suppose that the limiting constraint on the parameter t is given by

$$\frac{(x_2^0 - u_2)a'}{a_1} \leq t \leq \frac{(x_2^0 - l_2)a'}{a_1}. \tag{11}$$

Then we deduce as in Case 1 that $(w_2 - 1)f_1 < w_1 w_2$.

Case 3: Suppose that the limiting constraint on the parameter t is given by

$$\frac{(l_1 - x_1^0)a'}{a_2} \le t \le \frac{(x_2^0 - l_2)a'}{a_1}. \tag{12}$$

Since,

$$\frac{(x_2^0 - u_2)a'}{\alpha_1} \le \frac{(l_1 - x_l^0)a'}{a_2}, \tag{13}$$

Case 2 provides the result $(x_2^0 - l_2)f_1 - (l_1 - x_1^0)f_2 \le \text{Max}\{(w_2 - 1)f_1, (w_1 - 1)f_2\} < w_1w_2$.

Case 4: Suppose that the limiting constraint on the parameter t is given by

$$\frac{(x_2^0 - u_2)a'}{a_1} \le t \le \frac{(u_1 - x_1^0)a'}{a_2}. \tag{14}$$

then as in Case 3, we obtain $(u_1 - x_1^0)f_2 - (x_2^0 - u_2)f_1) \le \text{Max}\{(w_2 - 1)f_1, (w_1 - 1)f_2\} < w_1w_2$.   □

As the above Lemma 2 establishes, conversion of the bounds on the two variables into bounds on the parameter t helps in reducing the computational effort.

2.2   System of Equations in Two Variables

A system of equations in two variables can be solved using Lemma 1, the use of which will give raise a set of lower and upper bounds for the corresponding parameters t. Let $t^l \in \mathbb{R}^m$ be the set of lower bounds for the m equations, and $t^u \in \mathbb{R}^m$ be the set of upper bounds as specified by (5). Also, let a_{ij} be the coefficient of the jth variable in ith equation. Now the two variable linear Diophantine system problem can be restated as one to find all $t \in \mathbb{Z}^m : t^l \le t \le t^u$ that satisfies the following two equations.

$$x_{11}^0 + \left(\frac{a_{12}}{a_1'}\right)t_1 = x_{21}^0 + \left(\frac{a_{22}}{a_2'}\right)t_2 = \cdots = x_{m1}^0 + \left(\frac{a_{m2}}{a_m'}\right)t_m \tag{15}$$

$$x_{12}^0 - \left(\frac{a_{11}}{a_1'}\right)t_1 = x_{22}^0 - \left(\frac{a_{21}}{a_2'}\right)t_2 = \cdots = x_{m2}^0 - \left(\frac{a_{m1}}{a_m'}\right)t_m \tag{16}$$

We now give an algorithm in Fig. 1 to solve this problem.

Lemma 3. *The algorithm* **SLDS-2V** *for computing the set of all solutions to a linear Diophantine system in two bounded variables is correct. Given $w = \text{Max}(t_i^u - t_i^l)$, $i = 1, 2, \cdots, m$, the algorithm finishes in time $O(mw^2)$.*

Proof. If $m = 1$, the algorithm iterates over the set of all the integers in between t_1^l and t_1^u and populates the solution set S. The IF clause at step 9 will never be satisfied and hence the solution set will keep growing. If $m \ge 2$ the IF clause at step 9 might be satisfied for some of the solutions, and the corresponding count gets incremented. Step 13 checks to see if a solutions satisfies all the m

---

Algorithm SLDS-2V
Input: $A \in \mathbb{N}^{m \times 2}$, $b \in \mathbb{N}^m$, $l, u \in \mathbb{N}^2$
Output: All $x : Ax = b$, $l \leq x \leq u$
Initialise: Set of solutions $S = \emptyset$
Notation: $S(x_1, x_2).c$ is an integer used to track the number of equations that (x_1, x_2) solves

1. For $i = 1$ to m
2.   Find $\{a'_i, \gamma_i, \epsilon_i\}$ such that $a'_i = \gcd\{a_{i1}, a_{i2}\}$ and $\gamma_i a_{i1} + \epsilon_i a_{i2} = a'_i$
3. $x^0_{i1} = \gamma_i b_i / a'_i$ and $x^0_{i2} = \epsilon_i b_i / a'_i$
4. $t^l_i = \text{Max} \left\{ (l_1 - x^0_{i1}) a'_i / a_{i2}, (x^0_{i2} - u_2) a'_i / a_{i1} \right\}$
5. $t^u_i = \text{Min} \left\{ (u_1 - x^0_{i1}) a'_i / a_{i2}, (x^0_{i2} - l_2) a'_i / a_{i1} \right\}$
6.   For all $t \in \mathbb{Z} : t^l_i \leq t \leq t^u_i$
7. $x_1 = x^0_{i1} + (a_{i2}/a'_i)t$
8. $x_2 = x^0_{i2} - (a_{i1}/a'_i)t$
9.   If $(x_1, x_2) \in S$
10.   then $S(x_1, x_2).c + +$
11.   Else
12. $S = S \cup (x_1, x_2)$
13. Remove all (x_1, x_2) from S if $S(x_1, x_2).c \neq m$

Fig. 1. Algorithm **SLDS-2V**

equations and removes the ones that do not satisfy all the equations. Note that all the candidate solutions in S will satisfy the lower and upper bounds on the variables. The remaining solution in the set S will satisfy all the m equations and also the variable bounds.

From Lemma 2, we know that a single Diophantine equation in two variables has a running time of the order $O(w)$ (since $f \leq 1$). Steps 9 through 12 of the algorithm searches to see if the solution has been already generated. In the worst case it will finish in $O(w)$. The **SLDS-2V** algorithm iterates over m single Diophantine equations. Finally it filters out those solutions which are not generated by all the m equations. Hence the total running time of this algorithm will be $O(mw^2)$. $\qquad\qquad\square$

3   The General Problem

To solve a linear Diophantine system with n bounded variables we reduce the problem recursively into a two variable problem with progressive update on the bounds. For the kth equation, the Euclidean algorithm described earlier yields a'_{k1}, γ_{k1}, and ϵ_{k1} satisfying $a'_{k1} = \text{g.c.d.}\{a_{k1}, a_{k2}\}$ and $\gamma_{k1} a_{k1} + \epsilon_{k1} a_{k2} = a'_{k1}$. This is used to reduce a n variable equation into the following equation with $n - 1$ variables

$$a'_{k1}\xi_{k1} + a_{k3}x_3 + \cdots + a_{kn}x_n = b_k \qquad l \leq x \leq u, \; l'_{k1} \leq \xi_{k1} \leq u'_{k1} \qquad (17)$$

where $\xi_{k1} = c_{k1}x_1 + c_{k2}x_2$, c_{k1} and c_{k2} are integers such that $a_{k1} = a'_{k1}c_{k1}$ and $a_{k2} = a'_{k1}c_{k2}$. Furthermore, $l'_{k1} = c_{k1}l_1 + c_{k2}l_2$ and $u'_{k1} = c_{k1}u_1 + c_{k2}u_2$. This process of combining the variables is continued until it reduces to a two variables,

$$a'_{kn-2}\xi_{kn-2} + a_{kn}x_n = b_k \qquad l \le x \le u, \ l'_{kn-2} \le \xi_{kn-2} \le u'_{kn-2} \qquad (18)$$

where $l'_{kn-2} = c'_{kn-3}l'_{kn-3} + c_{kn-1}l_{kn-1}$ and $u'_{kn-2} = c'_{kn-3}u'_{kn-3} + c_{kn-1}u_{kn-1}$. Lemma 1 provides the bounds on the free parameter and algorithm SLDS-2V will generate the set of all solutions with in the bounds. In this two variable problem one of the variables x_n is a given variable, and the other ξ_{kn-2} is an artificial variable constructed for the kth equation in the system. ξ_{kn-2} will have to be rolled back to the other variables. However not all the solutions generated by SLDA-2V need to be considered. This is because if a solution needs to be considered and rolled back the x_n must have been generated for all the other $m-1$ equations. In other words for each equation only the common solutions for the actual variable will have to be considered for the next stage. Having identified the common solutions, x_n will be used in back tracking into a new two dimensional problem as follows:

$$a'_{kn-3}\xi_{kn-3} + a_{kn-1}x_{n-1} = b_k - a_{kn}x_n \qquad 1 \le x \le u, \ l'_{kn-3} \le \xi_{kn-3} \le u'_{kn-3} \tag{19}$$

Next we formally present the algorithm **SLDS-nV** in Fig. 3 that uses the above described principle in deriving the set of all solutions for a linear Diophantine system with variable bounds. This algorithm uses a pre-processing procedure that rolls up the n variable problem into a 2 variable problem. This is presented in the algorithm **SLDA-Pre-nV** in Fig. 2.

The pre-processing algorithm **SLDA-Pre-nV** works as follows. For each of the m equations, it rolls up the actual variables two at a time except for the first iteration. In the first iteration for equation i, the Euclidean algorithm is used to compute γ_{i1}, ϵ_{i1}, and $a'_{i1} = \gcd\{a_{i1}, a_{i2}\}$. For the next $n-2$ iterations it rolls up the gcd from the previous iteration with the co-efficient of the next variable. For instance in the second iteration for the ith equation the Euclid's algorithm is used to compute γ_{i2}, ϵ_{i2}, and $a'_{i2} = \gcd\{a'_{i1}, a_{i3}\}$. Also the lower and upper bounds on the artificial variables are computed at each stage.

Lemma 4. *The* **SLDA-Pre-nV** *algorithm is correct. For any input* $A \in \mathbb{N}^{m \times n}$, $b \in \mathbb{N}^m$, $l, u \in \mathbb{N}^n$ *it finishes in time* $O(mn)$.

The **SLDS-nV** algorithm in Fig. 3 uses the pre-processed data from **SLDA-Pre-nV** as input. It is a *depth-first* search algorithm. The algorithm builds a tree where each node corresponds to a variable. The *root* node is artificial that ties together all the branches. There will be a total m levels to this *depth-first* search tree with each level corresponding to a variable.

The **SLDS-nV** algorithm works as follows. The algorithm is invoked with the *root* node as one of the arguments. Any node has the following attributes:

Algorithm SLDS-Pre-nV
Input: $A \in \mathbb{N}^{m \times n}$, $b \in \mathbb{N}^m$, $l, u \in \mathbb{N}^n$
Output: $A' \in \mathbb{N}^{m \times n-1}$, $\gamma \in \mathbb{Z}^{m \times n-1}$, $\epsilon \in \mathbb{Z}^{m \times n-1}$,
$\qquad\quad L' \in \mathbb{R}^{m \times n-2}$, $U' \in \mathbb{R}^{m \times n-2}$

1. For $i = 1$ to m
2.   Find $\{a'_{i1}, \gamma_{i1}, \epsilon_{i1}\}$ such that $a'_{i1} = \gcd\{a_{i1}, a_{i2}\}$ and
       $\gamma_{i1} a_{i1} + \epsilon_{i1} a_{i2} = a'_{i1}$
3.   Find c_1 and c_2 such that $a_{i1} = a'_{i1} c_1$ and $a_{i2} = a'_{i1} c_2$
4. $l'_{i1} = c_1 l_1 + c_2 l_2$ and $u'_{i1} = c_1 u_1 + c_2 u_2$
5.   For $j = 2$ to $n - 1$
6.   Find $\{a'_{ij}, \gamma_{ij}, \epsilon_{ij}\}$ such that $a'_{ij} = \gcd\{a'_{ij-1}, a_{ij+1}\}$ and
           $\gamma_{ij} a'_{ij-1} + \epsilon_{ij} a_{ij+1} = a'_{ij}$
7.   Find c_{i1} and c_{i2} such that $a'_{ij-1} = a'_{ij} c_1$ and $a_{ij+1} = a'_{ij} c_2$
8. $l'_{ij} = c_1 l'_{j-1} + c_2 l_{j+1}$ and $u'_{ij} = c_1 u'_{j-1} + c_2 u_{j+1}$

Fig. 2. Algorithm **SLDS-Pre-nV**

1. *node.b* is the current right hand side vector for the linear Diophantine system. After solving for a variable the right hand side is updated.
2. *node.variable* is the attribute that indicates the variable that need to be solved next. For example, the *root.variable* will be set to m while invoking the routine for the first time.
3. *root.child* is a list that maintains the list of its children
4. *root.parent* points to the parent of a node

The **SLDS-nV** procedure is invoked with the root node as its attribute, and the recursive algorithm branches for all the possible values that solves the system at all the stages.

3.1 Analysis of the SLDS-nV Algorithm

We next show that the **SLDS-nV** algorithm is correct. Next we give a bound regarding the running time of this algorithm.

Lemma 5. *The* **SLDS-nV** *algorithm that computes the set of all solutions to a linear Diophantine system with bounded variables is correct.*

Proof. We prove the correctness of the algorithm by induction on the number of variables. If $n = 2$, the algorithm iterates through the set of equations in steps 2 through six and constructs the set of candidate solutions for the variable x_2. Step 7 filters them to only those that are generated by all the equations. Steps 9 through 13 creates a node for each of the remaining solutions to variable x_2. Since $n = v = 2$, the recursive call at step 15 is skipped. In steps 17 through 25 values for the variable x_1 is computed and the program terminates.

Algorithm SLDS-nV
Input: *node*, $A \in \mathbb{N}^{m \times n}$, $b \in \mathbb{N}^m$, $l, u \in \mathbb{N}^n$, $A' \in \mathbb{N}^{m \times n-1}$,
$\qquad \gamma \in \mathbb{Z}^{m \times n-1}$, $\epsilon \in \mathbb{Z}^{m \times n-1}$, $L' \in \mathbb{R}^{m \times n-2}$, $U' \in \mathbb{R}^{m \times n-2}$
Output: All $x : Ax = b, l \leq x \leq u$
Initialise: Set of solutions $S = \emptyset$, *root.b* = b, *root.variable* = n
Notation: A_i refers to the ith column of matrix A

1. $v = node.variable$, $y = \emptyset$
2. For $i = 1$ to m
3. $\quad$ Solve the two variable problem corresponding to variables
   $\quad\quad \xi_{iv-2}$ and x_v to generate the solution set for the given variable x_v^t
4. $\quad$ For all $x_v \in x_v^t$
5. $\quad\quad$ If $x_v \in y$ then $y(x_v).c{+}{+}$
6. $\quad\quad$ Else $y = y \cup x_v$
7. Filter the solution set y to retain only those in which $y(x_v) = m$
8. For $j = 1$ to $|y|$
9. $\quad$ *newnode.variable*= $v - 1$
10. $\quad$ *newnode.value*= y_j
11. $\quad$ *newnode.b* = *node.b*$-A_v y_j$
12. $\quad$ *newnode.parent* = *node*
13. $\quad$ *node.child*.add(*newnode*)
14. $\quad$ If $v > 2$
15. $\quad\quad$ then SLDS-nV(*newnode*)
16. $\quad$ Else
17. $\quad\quad$ $z = \emptyset$
18. $\quad\quad$ For $i = 1$ to m
19. $\quad\quad\quad$ $x_0^t =$*newnode.*b$_i$$/A_{i1}$
20. $\quad\quad\quad$ $z = z \cup x_o^t$
21. $\quad\quad$ If z is a set of identical numbers
22. $\quad\quad\quad$ *newnode1.variable*= $v - 2$
23. $\quad\quad\quad$ *newnode1.value*= z_1
24. $\quad\quad\quad$ *newnode1.parent* = *newnode*
25. $\quad\quad\quad$ *newnode.child*.add(*newnode1*)

Fig. 3. Algorithm **SLDS-nV**

If $n = 3$, the recursive step is invoked for each of the possible solutions for the variable x_3. The recursive calls are made with nodes with the attribute *variable*= 2, which takes us to the original case. $\qquad\qquad\square$

Lemma 6. *The worst case running time of the* **SLDS-nV** *algorithm is* $O(mw^n)$.

Proof. Each node first solves a two variable linear Diophantine system problem. By Lemma 3, the running time of steps 2 through 7 is $O(mw^2)$. For a two variable problem, the number of possible solutions is $O(w)$. For each one of these possible solutions a recursive call is made to **SLDS-nV**. This gives us the following recursive relationship,

$$T(n) = wT(n-1) + O(mw^2) \qquad\qquad (20)$$

and $T(2) = O(mw^2)$. Solving this recursive relationship yields a worst case time running time $O(mw^n)$.   □

4 Computational Experience

The performance of the algorithm was tested using a sequential implementation of the algorithm in Java on a workstation with AMD64 1.8 ghz CPU and 2 gb RAM. Random problem instances were generated with bounds on the co-efficnet matrix A and the vector b.

A set of experiments was conducted to test the effect of number of equations on a given number of variables. Problem instances were randomly generated such that A co-efficients were generated from the uniform distribution $U[0, 100]$, and the variable upper bound was fixed at 10 for all the variables. The results are tabulated in Table 1 that are the average running times of 10 experiments each. An interesting pattern emerges from the computational experiments. For given number of variables, the running times decreases with the number of equations. This can possibly be attributed to the decrease in the size of the set of all solutions as the number of equations increases. The worst case complexity of the algorithm also point in that dircction with the number of variables alone contributing significantly to the running times.

Table 1. Impact of number of equations

No of equations	No. of variables		
	5	7	9
2	0.0264151	0.1362856	11.9158421
3	0.0265163	0.1236470	5.4392386
5	0.0254278	0.1436144	4.4304221
10	0.0247979	0.0926015	2.2889441
50	0.0297468	0.0450302	0.3642787
100	0.0375549	0.0531305	0.3501732

References

1. K. Aardal, C.A.J. Hurkens, and A.K. Lenstra. Solving a system of Diophantine equation with lower and upper bounds on the variables. *Mathematics of Operations Research*, 25:427–442, 2000.
2. Matthias Beck and Shelemyahu Zacks. Refined upper bounds for the linear Diophantine problem of Frobenius. *Advances in Applied Mathematics*, 32:454–467, 2004.
3. James Bond. Calculating the general solution of a linear Diophantine equation. *American Mathematical Monthly*, 74:955–957, 1967.
4. P. Erdös and R. L. Graham. On a linear Diophantine problem of Frobenius. *Acta Arithmetica*, 21:399–408, 1972.

5. Miguel Filgueiras and Ana Paula Tomás. A fast method for finding the basis of non–negative solutions to a linear Diophantine equation. *Journal of Symbolic Computation*, 19:507–526, 1995.
6. Harold Greenberg. Solution to a linear Diophantine equation for nonnegative integers. *Journal of Algorithms*, 9:343–353, 1988.
7. Stanley Kertzner. The linear Diophantine equation. *American Mathematical Monthly*, 88:200–203, 1981.
8. Susumu Morito and Harvey M. Salkin. Using the Blankinship algorithm to find the general solution of a linear Diophantine equation. *Acta Informatica*, 13:379–382, 1980.
9. Ö. J. Rödseth. On a linear Diophantine problem of Frobenius. *Journal für die reine und angewandte Mathematik*, 301:431–440, 1978.
10. Alexander Schrijver. *Theory of Linear and Integer Programming*. John Wiley & Sons, 1986.
11. E. S. Selmer and Ö. Beyer. On a linear Diophantine problem of Frobenius in three variables. *Journal für die reine und angewandte Mathematik*, 301:161–170, 1978.
12. Y. Vitek. Bounds for a linear Diophantine problem of Frobenius. *Journal of the London Mathematical Society*, 10:79–85, 1975.

The Pseudosquares Prime Sieve

Jonathan P. Sorenson[*]

Computer Science and Software Engineering,
Butler University
Indianapolis, IN 46208 USA
sorenson@butler.edu
http://www.butler.edu/~sorenson

Abstract. We present the pseudosquares prime sieve, which finds all primes up to n. Define p to be the smallest prime such that the pseudosquare $L_p > n/(\pi(p)(\log n)^2)$; here $\pi(x)$ is the prime counting function. Our algorithm requires only $O(\pi(p)n)$ arithmetic operations and $O(\pi(p) \log n)$ space. It uses the pseudosquares primality test of Lukes, Patterson, and Williams.

Under the assumption of the Extended Riemann Hypothesis, we have $p \leq 2(\log n)^2$, but it is conjectured that $p \sim \frac{1}{\log 2} \log n \log \log n$. Thus, the conjectured complexity of our prime sieve is $O(n \log n)$ arithmetic operations in $O((\log n)^2)$ space. The primes generated by our algorithm are proven prime unconditionally. The best current unconditional bound known is $p \leq n^{1/(4\sqrt{e}-\epsilon)}$, implying a running time of roughly $n^{1.132}$ using roughly $n^{0.132}$ space.

Existing prime sieves are generally faster but take much more space, greatly limiting their range ($O(n/\log \log n)$ operations with $n^{1/3+\epsilon}$ space, or $O(n)$ operations with $n^{1/4}$ conjectured space). Our algorithm found all 13284 primes in the interval $[10^{33}, 10^{33} + 10^6]$ in about 4 minutes on a 1.3GHz Pentium IV.

We also present an algorithm to find all pseudosquares L_p up to n in sublinear time using very little space. Our innovation here is a new, space-efficient implementation of the wheel datastructure.

1 Introduction

A *prime number sieve* is an algorithm that finds all prime numbers up to a bound n. The fastest known sieves take $O(n/\log \log n)$ arithmetic operations [2,10,18,23], which is quite fast, considering there are $\pi(n) \sim n/\log n$ primes to find. However in practice, the utility of a prime number sieve is often limited by how much memory space it needs. For example, a sieve that uses $O(\sqrt{n})$ space [2,19,23] cannot, on current hardware, generate primes larger than about 10^{18}.

---

[*] This work was supported by a grant from the Holcomb Research Institute. Special thanks to Hugh Williams for helpful discussions and for hosting me at the University of Calgary in March of 2005, and thanks to the referees, whose comments improved this paper.

F. Hess, S. Pauli, and M. Pohst (Eds.): ANTS 2006, LNCS 4076, pp. 193–207, 2006.

Even with Galway's clever improvements [11] to the Atkin-Bernstein sieve [2], the space requirement is still $n^{1/3+\epsilon}$, giving an effective limit of roughly 10^{27}. Galway also has a sieve with conjectured space use of $O(n^{1/4})$ that runs linear time [12, ch. 6]. (Note that the space needed to write down the output, the primes up to n, is not included.)

If we applied trial division to each integer up to n separately, we would only need $O(\log n)$ space, but the time of $O(n\sqrt{n}/\log n)$ would be prohibitive. We could sieve by a few primes, then apply a quick base-2 pseudoprime test to remove most composites [17] and then use a prime test. If we used the AKS test [1] with Bernstein's complexity improvements [8], the result would be a sieve that takes $n(\log n)^{2+o(1)}$ operations. (The modified AKS test is $(\log n)^{4+o(1)}$ bit operations; we save a $\log n$ factor with the 2-psp test, and another $\log n$ factor because in this paper we count arithmetic operations instead of bit operations.) We can improve the time to $O(n(\log n)^2)$ by using Miller's prime test [16], but then our output is correct only if the ERH is true. And of course if we are willing to accept probable primes, the Miller-Rabin [20] or Solovay-Strassen [22] tests could give us $O(n\log n)$ operations. But in most applications for prime sieves, we need to be certain of our output.

In this paper, we present a new prime number sieve, the *pseudosquares prime sieve* (Algorithm PSSPS), that uses very little space and yet is fast enough to be practical. It uses an Eratosthenes-like sieve followed by the pseudosquares prime test of Lukes, Patterson, and Williams [15] (which effectively includes a base-2 pseudoprime test). Our sieve has a conjectured running time of $O(n\log n)$ arithmetic operations and $O((\log n)^2)$ bits of space. This is the complexity we observed in practice, and is as fast as using one of the probabilistic tests mentioned above. Assuming the ERH, we obtain $O(n(\log n)^2/\log\log n)$ operations and $O((\log n)^3/\log\log n)$ space. But in any case, the primes generated by our sieve are unconditionally proven prime.

Often, the user actually needs to find all primes in a short interval. On average, assuming reasonable conjectures, and after precomputation, our new algorithm will find all primes in an interval containing n of length at least $(\log n)^2$ at a cost of $O(\log n)$ operations per integer. In particular, we found all the primes in the interval $[10^{33}, 10^{33} + 10^6]$ in just over 4 minutes on a 1.3 GHz Pentium IV running Linux.

We also present a new, space-efficient implementation of the wheel data structure that leads to an algorithm for finding all pseudosquares $L_p \leq n$ in time $O(n \cdot \exp[-c\log n/\log\log n])$ for a constant $c > 0$. This data structure may prove to be useful in other areas of computational number theory.

For more on finding pseudosquares, see Wooding and Williams [28] in this volume. For recent work on prime number sieves, see [2,11,12,23].

The rest of this paper is organized as follows. In §2 we discuss some preliminaries, including pseudosquares, followed by a description of our algorithm in §3. In §4 we present our new wheel data structure and give our algorithm for finding pseudosquares. We conclude in §5 with some timings.

2   Preliminaries

2.1   Model of Computation

Our model of computation is a RAM with a potentially infinite, direct access memory. If n is the input, then all arithmetic and memory access operations on integers of $O(\log n)$ bits are assigned unit cost. Memory may be addressed either at the bit level or at the word level, where each machine word is composed of $O(\log n)$ bits.

When we present code fragments, we use a C++ style that should be familiar to most readers [25]. We occasionally declare integer variables with an INT datatype instead of the int datatype. This indicates that these integers typically exceed 32 bits in practice and may require special implementation (we used the Gnu-MP mpz_t datatype and associated functions [13]). We still limit INTs to $O(\log n)$ bits.

The space used by an algorithm under our model is counted in bits. The space used by the output of a prime number sieve (the list of primes up to n) is not counted against the algorithm. For further discussion, see [10].

2.2   Some Definitions

p always denotes a prime, with p_i denoting the ith prime, so that $p_1 = 2$. For integers a, b let $\gcd(a, b)$ denote the greatest common divisor of a and b. We say a and b are *relatively prime* if $\gcd(a, b) = 1$. For a positive integer m let $\phi(m)$ be the number of positive integers up to m that are relatively prime to m, with $\phi(1) = 1$. The number of primes up to x is given by $\pi(x)$. An integer x is a square, or *quadratic residue*, modulo p if the Legendre symbol $(x/p) = 1$.

We say $f(n) = O(g(n))$ if there exists a constant $c > 0$ such that $f(n) \leq c \cdot g(n)$ for all sufficiently large n. We write $f(n) = \Theta(g(n))$ if $f(n) = O(g(n))$ and $g(n) = O(f(n))$. We say $f(n) = o(g(n))$ if $\lim_{n \to \infty} f(n)/g(n) = 0$.

2.3   Pseudosquares

The *pseudosquare* L_p is the least non-square positive integer satisfying these two properties:

1. $L_p \equiv 1 \pmod 8$, and
2. L_p is a quadratic residue modulo every odd prime $q \leq p$.

Thus $L_3 = 73$ and $L_5 = 241$. See Williams [26, §16.2].

Lemma 2.1 (Lukes, Patterson, and Williams[15]). *Let n and s be positive integers. If*

1. *All prime divisors of n exceed s,*
2. *$n/s < L_p$ for some prime p,*

3. $p_i^{(n-1)/2} \equiv \pm 1$ (mod n) *for all primes* $p_i \leq p$,
4. $2^{(n-1)/2} \equiv -1$ (mod n) *when* $n \equiv 5$ (mod 8),
   $p_i^{(n-1)/2} \equiv -1$ (mod n) *for some* $p_i \leq p$ *when* $n \equiv 1$ (mod 8),

then n *is a prime or a prime power.*

Note that if n is prime, then the conditions of the lemma hold with $s = 1$ and $n < L_p$.

2.4   Useful Estimates

Here $x, x_1, x_2 > 0$, and except for (5), all sums and products are only over primes.

$$\sum_{p \leq x} \frac{1}{p} = \log\log x + O(1); \tag{1}$$

$$\sum_{p \leq x} \log p = x(1 + o(1)); \tag{2}$$

$$\sum_{p \leq x} 1 = \pi(x) = \frac{x}{\log x}(1 + o(1)); \tag{3}$$

$$\prod_{p \leq x} \frac{p-1}{p} = O\left(\frac{1}{\log x}\right); \tag{4}$$

$$\sum_{\substack{x_1 < d \leq x_2 \\ \gcd(d,m)=1}} \frac{1}{d} = \frac{\phi(m)}{m}\log(x_2/x_1)(1 + o(1)). \tag{5}$$

For proofs of (1)–(4), see Hardy and Wright [14]. For a proof of (5), see [23, Lemma 1].

2.5   The Wheel

A *wheel*, as we will use it, is a data structure that encapsulates information about the integers relatively prime to the first k primes. Generally speaking, a wheel can often be used to reduce the running time of a prime number sieve by a factor proportional to $\log p_k$. Pritchard was the first to show how to use a wheel in this way [18,19]. We begin with the following definitions:

$$M_k := \prod_{i=1}^{k} p_i;$$
$$W_k(y) := \{x \leq y : \gcd(x, M_k) = 1\};$$
$$W_k := W_k(M_k).$$

Let $\#S$ denote the cardinality of the set S. We have (see (2) and (4)):

$$\log M_k = p_k(1 + o(1));$$

$$\#W_k = \phi(M_k) = M_k \prod_{i=1}^{k} \frac{p-1}{p} = O\left(\frac{M_k}{\log\log M_k}\right);$$

$$\#W_k(n) = O\left(\frac{n}{\log\log M_k}\right).$$

Our data structure, then, is an array $W[]$ of records or structs, indexed by $0 \ldots (M_k - 1)$, defined as follows:

- $W[x]$.rp is 1 if $x \in W_k$, and 0 otherwise.
- $W[x]$.dist is $d = y - x$, where $y > x$ is minimal with $\gcd(y, M_k) = 1$.

We say that W is the kth wheel, with size M_k. For our C++ notation, we will declare W to be of class type $\texttt{Wheel}(k)$, where k is an integer parameter. We can construct a wheel of size M_k in $O(M_k)$ operations.

For examples of the wheel data structure, see [19,24].

3   Algorithm PSSPS

3.1   Precomputations and Main Loop

We first construct a table of pseudosquares up to $n/(\log n)^2$ using the algorithm we describe later in §4. In the code fragment below, this is stored in an array pss[] of structs or records:

- pss[i].prime is the smallest prime p (an int) such that
- pss[i].pss is L_p (an INT).

So pss[1].prime$=3$ with pss[1].pss$=73$, and pss[2].prime$=5$ with pss[2]. pss$=241$. In practice, we can use the table from Wooding [27, pp. 92–93], which has 49 entries, with the largest being pss[49].pss $= 2953634874009003108800401$, pss[49].prime $= 353$. Storing this table requires $O(\pi(p)\log n)$ space.

Next we specify the parameters p, segment size Δ, and sieve limit s:

- Let p be the smallest prime such that the pseudosquare $L_p > n/(\pi(p)(\log n)^2)$.
- $\Delta := \Theta(\pi(p)\log n)$. Note $\Delta \gg p$.
- $s := \lfloor n/L_p \rfloor + 1 = \Theta(\Delta \log n)$.

We conjecture $p \sim (1/\log 2)\log n \log\log n$ (see below). Making Δ larger improves overall performance; we choose here to give it roughly the same size as the pseudosquares table so it does not dominate overall space use. Our choice for s will balance the time spent in sieving versus the time applying the pseudosquares prime test.

In practice, we might choose Δ first. One normally chooses Δ to be as large as possible yet small enough to fit in cache memory, say around 2^{20}. Then choose $s = \Theta(\Delta \log n)$, and pick the smallest prime p so that $L_p > n/s$. If this choice for p is larger than our largest pseudosquares table entry, we simply set p to the largest entry (353) and set $s := \lfloor n/L_p \rfloor + 1$. Once p and s are set, the pseudosquares table is no longer needed.

We wrap up precomputation by building a wheel of size $\Theta(\log n)$. In practice, a wheel of size $30 = 2 \cdot 3 \cdot 5$ ($k = 3$) works fine. We must have $p_k \leq p$.

Our main loop iterates over segments of size Δ.

```
int p,s,delta;
INT l,r,n;                    // declare multiprec. ints
input(n);
pssentry pss[];               // Pseudosquare table
PssBuild(pss,n);              // Builds the pseudosq. tbl.
Initialize();                 // Compute p,delta,s etc.
Wheel W(pi(log(n)));          // Wheel of size O(log n)

//** Main Loop
Primelist PL(p);              // Find primes up to p
output(PL);                   // Output primes up to p
for(l=p; l<n; l=l+delta)      // Loop over segments
{
  r=min(l+delta,n);
  sieve(l,r,p,s,PL,W);        // Sieve the inverval [l+1,r]
}
```

Precomputation is dominated by the time to build the pseudosquares table; this is $o(n)$ operations and $O(\pi(p) \log n)$ space (we "cripple" the running time of our algorithm from §4 to meet the space bound). Constructing the wheel takes $O(\log n)$ operations and space. The list of primes up to p takes at most $O(p)$ operations and space. We will analyze the cost of the main loop at the end of this section.

3.2 Finding Primes in a Segment

Here we implement the `sieve()` function called in the main loop above. We begin by sieving, then we perform the pseudosquares prime test, and we finish by removing perfect powers.

Sieving. We sieve by the primes up to p, and then we sieve by integers from p to s using the wheel.

Here our `BitVector` class is created with left and right endpoints (ℓ and r), of length Δ, that supports functions to set and clear bits. Also, the member function `first(x)` will return the first integer larger than ℓ divisible by x.

```
BitVector B(delta,l,r);    // bit vector for the interval
B.setall();                // assume all are prime to start

//** Sieve by primes up to p
int i; INT x;
for(i=1; i<=PL.length(); i++)
```

```
// Loop through multiples of PL[i]:
for(x=B.first(PL[i]); x<=r; x=x+PL[i])
    B.clear(x);

//** Sieve by integers d up to s, gcd(d,m)=1
int d, m=W.size();            // m is the size of the wheel
for(d=W[p%m].next; d<=min(s,sqrt(l)); d=d+W[d%m].next)
    // Loop through multiples of d:
    for(x=B.first(d); x<=r; x=x+d)
        B.clear(x);
```

At this point, B represents only those integers from the interval $[\ell+1, r]$, with no prime divisors smaller than $\min\{s, \sqrt{\ell}\}$. In practice, one can implement this so that all the work done in these inner loops requires only single-precision integers: simply work with $x - \ell$ rather than x.

The time to sieve by primes is proportional to

$$\sum_{p_i \le p} \left(1 + \frac{\Delta}{p_i}\right) = O(\pi(p) + \Delta \log\log p) = o(\Delta \log n).$$

Sieving by integers generated by the wheel between p and s takes time proportional to

$$\sum_{\substack{p < d \le s \\ \gcd(d,m)=1}} \left(1 + \frac{\Delta}{d}\right) = O\left(\frac{\phi(m)}{m}(s + \Delta \log(s/p))\right)$$

using (5). This simplifies to $O((s + \Delta \log(s/p))/\log\log\log n) = o(\Delta \log n)$ using (4). In total, this phase requires $o(\Delta \log n)$ operations and $O(\Delta)$ space.

The Pseudosquares Prime Test. The next phase of our algorithm is based on Lemma 2.1, due to Lukes, Patterson, and Williams [15]. We code this prime test as function psspt(), which tests conditions (3) and (4) of the lemma. We make sure to perform the $2^{(n-1)/2} \mod n$ test first, for this has the effect of performing a base-2 pseudoprime test [17].

```
INT x;
for(x=l+1; x<=r; x++)  // loop over the interval
    if(B[x]==1)            // x meets conditions (1) & (2)
        if(!psspt(x,p))    // if x fails the test
            B.clear(x);    // x is not prime
```

Because of our earlier sieving, only $O(\Delta/\log s) = O(\Delta/\log\log n)$ integers remain that pass conditions (1) and (2) for our prime test. (Recall that $s = \Theta(\Delta \log n)$.) Function psspt() will first effectively perform a base-2 pseudoprime test. This takes $O(\log n)$ arithmetic operations per test, for a total time to this point of $O(\Delta(\log n)/\log\log n) = o(\Delta \log n)$. From [17] and elsewhere in the literature, we know that only $O(n/\log n)$ integers up to n pass the base-2 pseudoprime test, or an average of $O(\Delta/\log n)$ per interval. (A particular interval could conceivably have more than this.) The `psspt()` function performs $\pi(p) - 1$ more modular exponentiations, at a cost of $O(\log n)$ arithmetic operations each, on each remaining integer for an overall average cost of $O(\pi(p)\Delta)$ operations.

Removing Perfect Powers. At this point, the only remaining integers represented by B are either prime or the power of a prime. Note that if $n \leq 6.4 \cdot 10^{37}$, only primes remain and we are done [26, p. 417].

To remove the prime powers, in theory we use a perfect power testing algorithm [6,7,9] which, in our model of computation, requires sublinear time per integer on average, making the cost negligible ($o(\Delta)$ operations on average, since we only perform the tests on the remaining $O(\Delta/\log n)$ integers). In practice, one can very efficiently enumerate perfect powers using a priority queue data structure; we leave the details to the reader in the interest of space.

3.3 Complexity

Let us summarize what we have from above:

- Precomputation takes $o(n)$ operations and $O(\pi(p)\log n)$ space (dominated by building the pseudosquares table).
- Sieving a segment takes $o(\Delta \log n)$ operations; a segment takes $O(\Delta) = O(\pi(p)\log n)$ space.
- Performing base-2 pseudoprime tests and the pseudosquares prime test takes, on average, $O(\pi(p)\Delta)$ operations per interval.
- Removing perfect powers takes $o(\Delta)$ operations on average.

By multiplying the average cost per segment by n/Δ, the number of segments, we prove the following.

Theorem 3.1. *Let p be defined as above. Algorithm PSSPS finds all primes up to n using $O(\pi(p)n) + o(n \log n)$ arithmetic operations and $O(\pi(p) \log n)$ space.*

The work of Bach and Huelsbergen [4] implies the following conjecture.

Conjecture 3.2. $\log L_p \sim \log 2 \frac{p}{\log p}$, or equivalently, $p \sim \frac{1}{\log 2} \log L_p \log\log L_p$.

Lukes, Patterson, and Williams [15] studied the relationship between L_p and p for all known pseudosquares, and their data supports the conjecture. See also [28].

Corollary 3.3. *If Conjecture 3.2 is true, then Algorithm PSSPS finds all primes up to n in $O(n \log n)$ arithmetic operations and $O((\log n)^2)$ space.*

Fortunately in practice, Conjecture 3.2 appears to hold.

Corollary 3.4. *If the ERH is true, then Algorithm PSSPS finds all primes up to n in $O(n(\log n)^2/\log\log n)$ arithmetic operations and $O((\log n)^3/\log\log n)$ space.*

This follows from Bach's Theorem [3], which implies $p < 2(\log n)^2$, or asymptotically $p < (1 + o(1))(\log n)^2$. Note that this weaker result still outperforms the use of Miller's prime test [16] or AKS [1,8] in a prime sieve.

Currently the best unconditional result is $p \leq L_p^{1/(4\sqrt{e}-\epsilon)} \approx L_p^{0.1516...}$, due to Schinzel [21]. Since we use $L_p \approx n/p$, we obtain that $p \approx n^{1/(4\sqrt{e}+1-\epsilon)} \approx n^{0.132}$. This implies the following much weaker result:

Corollary 3.5. *Let $\epsilon > 0$. Algorithm PSSPS finds all primes up to n in $O(n^{1+1/(4\sqrt{e}+1-\epsilon)}) \approx n^{1.132}$ arithmetic operations and $O(n^{1/(4\sqrt{e}+1-\epsilon)}) \approx n^{0.132}$ space.*

Algorithm 3.1 from [23] would require a running time of roughly $n^{1.368}$ to stay within the same space bound. Of course, an AKS-based sieve would give the best unconditional result.

4   Finding Pseudosquares

In this section we present a sublinear-time algorithm to find all pseudosquares $L_p \leq n$. It makes use of a new way to implement a wheel-like datastructure that uses significantly less space.

We begin by presenting our new wheel datastructure, after which we show how to adapt it to find pseudosquares.

4.1   A New Wheel

As mentioned in §2.5, the wheel datastructure is used primarily to enumerate integers relatively prime to M_k, like this:

```
for(x=1; x<n; x=x+W[x%m].dist)
    output(x);
```

Here we present a new implementation of the wheel, which has the following differences:

- The space used by the wheel is proportional to the sum of the moduli instead of their product: $O(\log M_k \sum_{i=1}^{k} p_i) = O((\log M_k)^3/\log\log M_k)$ bits instead of $O(\log p_k \prod_{i=1}^{k} p_i) = O(M_k \log\log M_k)$ bits. This is a huge savings.
- The integers relatively prime to M_k are not enumerated in ascending order.

An Example - Enumerating Primes up to 100. Sometimes it is best to introduce a new datastructure with an example. We construct our new wheel with moduli $2, 3, 5, 7$ to enumerate 1, plus the primes p_i with $7 < p_i \leq 100$.

For each prime modulus p_i except for 2, we create an array of structs or records, indexed from $0 \ldots p_i - 1$, each of which has 2 fields. Let m_i be the *input modulus*, which is $m_i := 2 \cdot 3 \cdot \cdots \cdot p_{i-1}$. For our example, $m_2 = 2, m_3 = 6$, and $m_4 = 30$. Here $0 \leq x < p_i$.

- $W[i][x]$.rp is 1 if $\gcd(x, p_i) = 1$, 0 otherwise (int),
- $W[i][x]$.jump is the smallest multiple $j > 0$ of the input modulus m_i such that $\gcd(x + j, p_i) = 1$ (INT).

This gives us the following three datastructures:

$$p_2 = 3: \begin{array}{|c c c} 0 & 1 & 2 \\ \hline \text{rp} \; 0 & 1 & 1 \\ \text{jump} \; 2 & 4 & 2 \end{array} (m_2 = 2, \phi(3) = 2)$$

$$p_3 = 5: \begin{array}{|c c c c c} 0 & 1 & 2 & 3 & 4 \\ \hline \text{rp} \; 0 & 1 & 1 & 1 & 1 \\ \text{jump} \; 6 & 6 & 6 & 6 & 12 \end{array} (m_3 = 6, \phi(5) = 4)$$

$$p_4 = 7: \begin{array}{|c c c c c c c} 0 & 1 & 2 & 3 & 4 & 5 & 6 \\ \hline \text{rp} \; 0 & 1 & 1 & 1 & 1 & 1 & 1 \\ \text{jump} \; 30 & 30 & 30 & 30 & 30 & 60 & 30 \end{array} (m_4 = 30, \phi(7) = 6)$$

We will explain how to compute the jump fields below.

Using Recursion. To enumerate the primes (and 1) we use the recursive function below.

Let k denote the number of prime moduli in the wheel; recall that 2 is not given a datastructure, so there will be $k - 1$ levels to the recursion. To enumerate integers relatively prime to M_k up to n, we call enumerate(2,1,n). For our example, we use enumerate(2,1,100) and $k = 4$.

```
function enumerate(int i,INT x,INT n)
{
  //** make sure gcd(x,p[i])=1
  if(!W[i][x%p[i]].rp)  x=x+W[i][x%p[i]].jump;

  if(i==k) // base case for the recursion
     for( ; x<n; x=x+W[k][x%p[k]].jump)  output(x);
  else // recursive case for the recursion
  {
    for(int cnt=0; cnt<p[i]-1; cnt++)
    {
      enumerate(i+1,x,n); // recursive call
      x=x+W[i][x%p[i]].jump;
    }
  }
}
```

Note that we are assuming pass-by-value here, so that changes to x in recursive calls are not reflected in the calling function. Let x_i denote the value of x during the recursive call with input i.

So x_2 will take the values 1 and 5.

When x_2 is 1, x_3 loops through 1, 7, 13, and 19. When x_2 is 5, x_3 loops through 11, 17, 23, and 29. ($x_2 = 5$ is not relatively prime to $p_3 = 5$, so the first if-statement is triggered, adding 6 to get 11.)

The values x_4 loops through are listed in the table below, giving the primes from 11 to 100, plus 1.

```
 1 31 61 91|11 41 71
37 67 97   |17 47
13 43 73   |23 53 83
19 79      |29 59 89
```

The values to the left of the line arise from $x_2 = 1$; they are $\equiv 1 \bmod 6$. The values to the right of the line arise from $x_2 = 5$; they are $\equiv 5 \bmod 6$.

Computing jumps. Computing the rp field for each prime is the same as for the basic wheel, and takes time linear in p_i. Computing the jump fields is a bit trickier, and takes $O(p_i^2 \log p_i)$ operations.

For each column $x = 0 \ldots p_i - 1$, we do the following:

1. Compute a list of distances to *all* other residue classes that are relatively prime to p_i.
   For $p_i = 7$ and $x = 5$, we get the list $1, 3, 4, 5, 6$.
2. For each distance d in the list, use the extended Euclidean algorithm [5, §4.3] to find a multiple of the modulus, $a \cdot p_i$, such that $d + ap_i$ is divisible by m_i, the input modulus.
   Continuing our example, for $d = 1$ we must use $a = 17$ to get $1 + 17 \cdot 7 = 120$. Repeating this for the entire list gives $120 = 17 \cdot 7 + 1, 150 = 21 \cdot 7 + 3, 60 = 8 \cdot 7 + 4, 180 = 25 \cdot 7 + 5, 90 = 12 \cdot 7 + 6$.
3. Write down the smallest number from the list computed in the last step.
   In our example, it is 60.

The value of jump entries will not exceed $p_i m_i$.

The total time to build a datastructure for the first k primes is proportional to $k \cdot p_k^2 \log p_k = O(p_k^3)$ operations. The space needed is proportional to $k \cdot p_k \log M_k = O(p_k^3 / \log p_k)$.

If $n > M_k$, then analyzing the running time reduces to counting the number of times output(x) is called, which gives us $O((\phi(M_k)/M_k)n)$ operations.

Theorem 4.1. *Let $n > M_k$. Using our new implementation of the wheel datastructure, we can enumerate integers up to n relatively prime to M_k in $O((\phi(M_k)/M_k)n)$ operations. Precomputing the datastructure requires $O(p_k^3)$ operations and $O(p_k^3 / \log p_k)$ space.*

4.2   Enumerating Pseudosquares

To search for pseudosquares $L_p \leq n$, we simply make a few minor changes to our new wheel datastructure and enumerate() function from above:

1. We choose k so that $M_k \leq n$, but as large as possible. We assume that all pseudosquares L_{p_i} with $p_i \leq p_k$ are already known. (If not, find them recursively with a smaller n.)
2. Our first prime is $p_3 = 5$, with input modulus $m_3 = 24$; we know $L_p \equiv 1 \pmod{24}$ for $p \geq 3$. Each successive input modulus satisfies $m_i := p_{i-1}m_{i-1}$.

3. We change the rp field to a qr field, set to 1 if x is a quadratic residue modulo p_i, and 0 otherwise. This can be computed in linear time by setting all the qr bits to 0, then square each integer $1 \ldots p_i - 1$ modulo p_i and mark the corresponding qr field with a 1.
4. Compute the jump field from the qr field as if it were the rp field.
5. Replace $p_i - 1$ with $(p_i - 1)/2$ in the loop control for the recursive case of the enumerate() function.
6. Integers x that are output by the enumerate() function are checked to see if they are quadratic residues modulo the primes p_i with $p_k < p_i \le p$. If they pass, then we check to see if they are squares. The average cost for this is $O(1)$ operations per x value if we precompute a table of quadratic residues modulo several primes $p_i > p_k$.
7. An integer x that passes all these tests is L_p; output it, and find the next prime to serve as p to begin the search for the next pseudosquare.

The algorithm described above can find all pseudosquares $L_p \le n$ in $O(n2^{-k}/\log p_k)$ operations, as the output() function will be called roughly

$$\frac{1}{4} \cdot \prod_{i=1}^{k} \frac{(p_i - 1)/2}{p_i} \cdot n$$

times. By our choice for k, we have $k = \Theta(\log n / \log\log n)$. We have proven the following.

Theorem 4.2. *Our algorithm will find all pseudosquares $L_p \le n$ in*

$$O(n \exp[-c \log n / \log\log n])$$

operations, for $c > 0$ fixed, using $O(p + (\log n)^3/\log\log n)$ space.

Conjecture 3.2 implies only $O((\log n)^3/\log\log n)$ space is needed; assuming the ERH instead does not increase this bound.

We use this algorithm in our prime sieve with $p_k \approx (\log n)^{2/3}$ to keep our space usage under control, yet maintain a $o(n)$ running time.

Our crude implementation of this algorithm found $L_{223} \approx 1.16 \times 10^{16}$ in about 17 hours on a single 1.3GHz Pentium IV processor.

Robert Threlfal observed that this wheel can be used to factor integers of the form $n = p^2 q$, p, q prime, by using $(-1/n)$, $(2/n)$, $(3/n)$, $(5/n)$, etc. to initialize the datastructures to search for q.

5   Timing Results

In our first set of results (Table 1), we compared our new sieve to the sieve of Eratosthenes and the Atkin-Bernstein sieve to find the primes up to 10^9.

We used a 1.3 GHz Pentium IV running Linux, with the Gnu g++ compiler. The code for the Atkin-Bernstein and Eratosthenes sieves came, unmodified, from Dan Bernstein's website (http://cr.yp.to). Our code for Algorithm

Table 1. Sieve Algorithm Comparison

Algorithm	Time in Seconds	Δ
Atkin-Bernstein	7.2	—
Eratosthenes	5.9	—
PSSPS	58.1	25000
PSSPS	103.5	10000
PSSPS	183.0	5000
PSSPS	367.2	2500

Table 2. Finding all primes between n and $n + 10^6$

n	Rem.	$Time_s$	s	Primes	$Time_p$	p	L_p	$Time_{tot}$
10^{15}	28845	1.25	31622776	28845	0	0	0	1.25
10^{16}	28774	2.17	57063204	27168	17.51	67	$1.75 \cdot 10^8$	19.68
10^{17}	28286	4.19	111269821	25463	19.49	79	$8.98 \cdot 10^8$	23.68
10^{18}	31717	2.09	42343580	24280	24.55	103	$2.36 \cdot 10^{10}$	26.64
10^{19}	31628	2.48	50951495	23069	27.36	113	$1.96 \cdot 10^{11}$	29.84
10^{20}	27342	23.29	509514950	21632	39.73	113	$1.96 \cdot 10^{11}$	63.02
10^{21}	28668	15.95	348208470	20832	44.59	131	$2.87 \cdot 10^{12}$	60.54
10^{22}	31814	2.7	55885834	19757	55.63	173	$1.78 \cdot 10^{14}$	58.33
10^{23}	30253	6.64	143644910	18939	55.84	181	$6.96 \cdot 10^{14}$	62.48
10^{24}	30879	4.06	85900327	18149	63.90	211	$1.16 \cdot 10^{16}$	67.96
10^{25}	27748	39.06	859003269	17549	63.43	211	$1.16 \cdot 10^{16}$	102.49
10^{26}	29965	6.54	140326390	16587	67.00	233	$7.12 \cdot 10^{17}$	73.54
10^{27}	30512	5.18	98057476	16139	74.55	263	$1.01 \cdot 10^{19}$	79.73
10^{28}	29863	8.14	143167432	15606	80.97	277	$6.98 \cdot 10^{19}$	89.11
10^{29}	30368	6.11	106761861	15002	107.82	293	$9.36 \cdot 10^{20}$	113.93
10^{30}	30944	4.24	73264612	14496	117.25	331	$1.36 \cdot 10^{22}$	121.49
10^{31}	30616	5.74	100509639	13955	116.26	347	$9.94 \cdot 10^{22}$	122.00
10^{32}	28542	18.95	338566228	13653	122.46	353	$2.95 \cdot 10^{23}$	141.41
10^{33}	26244	121.22	3385662272	13284	124.56	353	$2.95 \cdot 10^{23}$	245.78

PSSPS was not optimized for single-precision use; it used functions from the GnuMP package for arithmetic, and in particular, to perform modular exponentiations for the pseudosquares prime tests.

In Table 1, for each sieve we give the time to find the primes to 10^9 in seconds. For our new algorithm, we also show different times for various choices for Δ, the size of our interval. In every case, Algorithm PSSPS sieved up to $s = 31622$ and used $p = 0$, the largest entry from the pseudosquares table used for prime tests (a value of 0 indicates no such tests were performed).

Our goal here was to verify our results and to see how bad the $\log n \log \log n$ factor in the running time affects Algorithm PSSPS. When we used $\Delta = 500$, we were able to force the algorithm to use $p = 17$, but the running time became quite large. Simply put, our algorithm is not appropriate for inputs this small; it ends up performing what is, essentially, the sieve of Eratosthenes in a non-efficient way.

Next, in Table 2 we show how our sieve performs when finding all primes in an interval of length 10^6 for much larger values for n. The first column gives n, the starting point of the interval searched for primes. The next three columns report the performance of the sieving stage, giving the number of integers that are free of factors below s (the remainder), $Time_s$, the time sieving took in seconds, and the value of s used. The next four columns present the results from the pseudosquares prime tests, with the number of primes found first, followed by the time in seconds ($Time_p$), the value of p, and the approximate value of L_p used by the prime test. The last column gives the total time, $Time_{tot}$ (sieving plus prime tests). The number of tests performed (beginning with a base-2 psp test) matches the number in column 2 (the remainder), with the $Primes$ column giving the number of integers that pass the test. Note that $s \cdot L_p$ should match or exceed $n + 10^6$. When $p = 0$, sieving only was used, in which case $s \geq \lfloor \sqrt{n + 10^6} \rfloor$ must hold.

Since L_{353} is currently the largest pseudosquare known, any increase in size beyond 33 decimal digits would have to be absorbed entirely by using a larger value for s, which will greatly degrade performance unless a correspondingly longer interval is used.

Notice that the pseudosquares prime test was not even used until the input was 16 digits in length.

References

1. Manindra Agrawal, Neeraj Kayal, and Nitin Saxena. PRIMES is in P. http://www.cse.iitk.ac.in/news/primality_v3.pdf, 2003.
2. A. O. L. Atkin and D. J. Bernstein. Prime sieves using binary quadratic forms. *Mathematics of Computation*, 73:1023–1030, 2004.
3. E. Bach. *Analytic Methods in the Analysis and Design of Number-Theoretic Algorithms*. MIT Press, Cambridge, 1985.
4. Eric Bach and Lorenz Huelsbergen. Statistical evidence for small generating sets. *Math. Comp.*, 61(203):69–82, 1993.
5. Eric Bach and Jeffrey O. Shallit. *Algorithmic Number Theory*, volume 1. MIT Press, 1996.
6. Eric Bach and Jonathan P. Sorenson. Sieve algorithms for perfect power testing. *Algorithmica*, 9(4):313–328, 1993.
7. Daniel J. Bernstein. Detecting perfect powers in essentially linear time. *Math. Comp.*, 67(223):1253–1283, 1998.
8. Daniel J. Bernstein. Proving primality in essentially quartic time. To appear in *Mathematics of Computation*; http://cr.yp.to/papers.html#quartic, 2006.
9. Daniel J. Bernstein, Hendrik W. Lenstra Jr., and Jonathan Pila. Detecting perfect powers by factoring into coprimes. To appear in *Mathematics of Computation*, 2006.
10. Brian Dunten, Julie Jones, and Jonathan P. Sorenson. A space-efficient fast prime number sieve. *Information Processing Letters*, 59:79–84, 1996.

11. William F. Galway. Dissecting a sieve to cut its need for space. In *Algorithmic number theory (Leiden, 2000)*, volume 1838 of *Lecture Notes in Comput. Sci.*, pages 297–312. Springer, Berlin, 2000.

12. William F. Galway. *Analytic Computation of the Prime-Counting Function*. PhD thesis, University of Illinois at Urbana-Champaign, 2004. Available at http://www.math.uiuc.edu/~galway/PhD_Thesis/.

13. T. Granlund. The Gnu multiple precision arithmetic library, edition 4.1.3. http://www.swox.se/gmp/manual, 2004.

14. G. H. Hardy and E. M. Wright. *An Introduction to the Theory of Numbers*. Oxford University Press, 5th edition, 1979.

15. R. F. Lukes, C. D. Patterson, and H. C. Williams. Some results on pseudosquares. *Math. Comp.*, 65(213):361–372, S25–S27, 1996.

16. G. Miller. Riemann's hypothesis and tests for primality. *Journal of Computer and System Sciences*, 13:300–317, 1976.

17. Carl Pomerance, J. L. Selfridge, and Samuel S. Wagstaff, Jr. The pseudoprimes to $25 \cdot 10^9$. *Math. Comp.*, 35(151):1003–1026, 1980.

18. P. Pritchard. A sublinear additive sieve for finding prime numbers. *Communications of the ACM*, 24(1):18–23,772, 1981.

19. P. Pritchard. Fast compact prime number sieves (among others). *Journal of Algorithms*, 4:332–344, 1983.

20. M. O. Rabin. Probabilistic algorithm for testing primality. *Journal of Number Theory*, 12:128–138, 1980.

21. A. Schinzel. On pseudosquares. *New Trends in Prob. and Stat.*, 4:213–220, 1997.

22. R. Solovay and V. Strassen. A fast Monte Carlo test for primality. *SIAM Journal on Computing*, 6:84–85, 1977. Erratum in vol. 7, p. 118, 1978.

23. Jonathan P. Sorenson. Trading time for space in prime number sieves. In Joe Buhler, editor, *Proceedings of the Third International Algorithmic Number Theory Symposium (ANTS III)*, pages 179–195, Portland, Oregon, 1998. LNCS 1423.

24. Jonathan P. Sorenson and Ian Parberry. Two fast parallel prime number sieves. *Information and Computation*, 144(1):115–130, 1994.

25. B. Stroustrup. *The C++ Programming Language*. Addison-Wesley, 2nd edition, 1991.

26. Hugh C. Williams. *Édouard Lucas and primality testing*. Canadian Mathematical Society Series of Monographs and Advanced Texts, 22. John Wiley & Sons Inc., New York, 1998. A Wiley-Interscience Publication.

27. Kjell Wooding. Development of a high-speed hybrid sieve architecture. Master's thesis, The University of Calgary, Calgary, Alberta, November 2003.

28. Kjell Wooding and H. C. Williams. Doubly-focused enumeration of pseudosquares and pseudocubes. In F. Hess, S. Pauli, and M. Pohst (Eds.), *Proceedings of the 7th International Algorithmic Number Theory Symposium (ANTS VII)*, pages 208–221, Berlin, Germany, 2006. LNCS 4076.

Doubly-Focused Enumeration of Pseudosquares and Pseudocubes

Kjell Wooding and Hugh C. Williams

Centre for Information Security and Cryptography, University of Calgary
2500 University Dr. NW, Calgary, Alberta, T2N 1N4, Canada
{kjell, williams}@math.ucalgary.ca

Abstract. This paper offers numerical evidence for a conjecture that primality proving may be done in $(\log N)^{3+o(1)}$ operations by examining the growth rate of quantities known as pseudosquares and pseudocubes. In the process, a novel method of solving simultaneous congruences— doubly-focused enumeration— is examined. This technique, first described by D. J. Bernstein, allowed us to obtain record-setting sieve computations in software on general purpose computers.

1 Motivation

In August 2002, Agrawal, Kayal, and Saxena [1] described an unconditional, deterministic algorithm for proving primality with time complexity $(\log N)^{10.5+o(1)}$. This result was later improved by Lenstra and Pomerance (described in [5]) to $(\log N)^{6+o(1)}$. Bernstein [6] (and independently Cheng [10]) then generalized an argument of Berrizbeitia [7] to produce a random-time primality provining algorithm with complexity $(\log N)^{4+0(1)}$. Given these results, an obvious question to ask may be: "how far can the time complexity of unconditional, deterministic primality proving be improved"? This paper offers numerical evidence for a conjecture that primality may be proved with complexity $(\log N)^{3+o(1)}$.

2 The Generalized Sieve Problem

Definition 1. *Define a* sieve ring, ρ_i, *to be a modulus,* M_i, *together with a set of j acceptable residues,* $\mathcal{R}_i = \{r_{i,j} \mid 0 \leq r_{i,j} < M_i\}$. *Given*

1. *$A, B \in \mathbb{Z}$ with $B > A$ (the sieve bounds);*
2. *$k \geq 1$ sieve rings $\rho_i, \ldots, \rho_k$ whose moduli $M_1, \ldots, M_k$ are relatively prime in pairs[1].*

The Generalized Sieve Problem (GSP) is the problem of finding all $x \in \mathbb{Z}$ such that $A \leq x < B$ and

$$x \pmod{M_i} \in \mathcal{R}_i \text{ for all } i = 1, \ldots, k$$

---

[1] The requirement that sieve moduli be pairwise relatively prime is not strictly required. Since it greatly simplifies the discussion, however, relative primality of sieve moduli will be assumed throughout this paper.

F. Hess, S. Pauli, and M. Pohst (Eds.): ANTS 2006, LNCS 4076, pp. 208–221, 2006.

These solutions, x, are called the solutions admitted *by the sieve problem. k is called the* width *of the sieve problem.*

Equivalently, the sieve problem $\mathcal{S}$ may be expressed in terms of sets:

$$\mathcal{S} = \bigcap_{i=1}^{k} \{x \in \mathbb{Z} \mid x \pmod{M_i} \in \mathcal{R}_i, \quad A \leq x < B\}$$

Definition 2. *The sieve problems*

$$\mathcal{S}_1 = \bigcap_{i=1}^{r} \{x \in \mathbb{Z} \mid x \pmod{M_i} \in \mathcal{R}_i, \quad A \leq x < B\}$$
$$\mathcal{S}_2 = \bigcap_{j=1}^{s} \{x \in \mathbb{Z} \mid x \pmod{M_j} \in \mathcal{R}_j, \quad A \leq x < B\}$$

are equivalent *if and only if $\mathcal{S}_1 = \mathcal{S}_2$ for all $A, B \in \mathbb{Z}$; i.e. for any choice of bounds the set of solutions admitted by each of the sieve problems is the same.*

With this notion of equivalence, the following theorem may now be demonstrated.

Theorem 1. *Given a sieve problem,*

$$\mathcal{S} = \bigcap_{i=1}^{k} \{x \in \mathbb{Z} \mid x \pmod{M_i} \in \mathcal{R}_i, \quad A \leq x < B\}$$

of width $k > 1$ an equivalent sieve problem of width $k - 1$ can be formed.

Proof. Consider any two sieve rings, $\rho_1 = \{M_1, \mathcal{R}_1\}$ and $\rho_2 = \{M_2, \mathcal{R}_2\}$. Clearly, any solution $x \in \mathcal{S}$ satisfies both $x \pmod{M_1} \in \mathcal{R}_1$ and $x \pmod{M_2} \in \mathcal{R}_2$.

Define the set $\mathcal{R}$ to be the Chinese Remainder Theorem (CRT) combination of all residues from the sets $\mathcal{R}_1$ and $\mathcal{R}_2$; *i.e.* let $M = M_1 \cdot M_2$, $N_i = \frac{M}{M_i}$, $\xi_i \equiv N_i^{-1} \pmod{M_i}$. Then

$$\mathcal{R} = \{r \mid r \equiv \xi_1 N_1 r_1 + \xi_2 N_2 r_2 \pmod{M}, \quad r_1 \in \mathcal{R}_1, r_2 \in \mathcal{R}_2\}$$

Now, form a new sieve problem, replacing the sieve rings $\{M_1, \mathcal{R}_1\}$ and $\{M_2, \mathcal{R}_2\}$ with the newly constructed ring $\{M, \mathcal{R}\}$. The width of this new sieve problem is $k - 1$. By the CRT, $x \pmod{M_1 M_2} \in \mathcal{R}$ if and only if $x \pmod{M_1} \in \mathcal{R}_1$ and $x \pmod{M_2} \in \mathcal{R}_2$. Equivalence of the sieve problems follows from Definition 2. □

Corollary 1. *A sieve problem, $\mathcal{S} = \bigcap_{i=1}^{k}\{x \in \mathbb{Z} \mid x \pmod{M_i} \in \mathcal{R}_i, \quad A \leq x < B\}$ consisting of k sieve rings can be replaced by an equivalent sieve problem consisting of a single sieve ring:*

$$\mathcal{S} = \{x \in \mathbb{Z} \mid x \pmod{M} \in \mathcal{R}, \quad A \leq x < B\}$$

where $M = \prod_{i=1}^{k} M_i$, and $\mathcal{R}$ consists of the CRT combinations of the sets $\mathcal{R}_i$ for $i = 1, 2, \ldots, k$.

3   Transforming and Parallelizing Sieve Problems

3.1   Sieve Rotation

Recall that start (A) and end (B) values for a sieve problem are specified such that any solution, x, lies in the interval $A \leq x < B$. To offer the largest possible sieve range to the underlying implementation[2], sieve problems with a nonzero start value are often translated to equivalent problems in the interval $0 \leq x < H$ by adjusting the acceptable residues, $r_{i,j} \in \mathcal{R}_i$; *i.e.* taking $s = A$ in the (bijective) map

$$\sigma[s] : r_{i,j} \mapsto (r_{i,j} - s) \pmod{M_i}$$
$$(A, B) \mapsto (A - s, B - s).$$

This is called a *rotation* of the sieve problem.

3.2   Normalization

When a sieve problem contains a modulus for which there is but a single acceptable residue, even greater savings may be achieved. In [15], D. H. Lehmer described a technique for eliminating single-valued congruences from sieve problems. He called this technique *normalization*, and it works as follows. First, all single-valued residue conditions are combined (via the CRT) into the single congruence

$$x \equiv r \pmod{m}. \tag{1}$$

Thus, any solution to this sieve problem must be of the form $x = r + ym$ for some y. Rather than searching for solutions x that satisfy all sieve congruences, it is more efficient to search for acceptable values of y by using the relationship in Equation 1 to translate the remaining sieve criteria; *i.e.* for all $r_{i,j} \in \mathcal{R}_i$, and all $\mathcal{R}_i \in \mathcal{S}$, apply the (bijective) map:

$$\eta[m, r] : r_{i,j} \mapsto (r_{i,j} - r)m^{-1} \pmod{M_i} \tag{2}$$
$$(0, H) \mapsto \left(0, \left\lceil \frac{(H - r)}{m} \right\rceil\right)$$

In practice, most sieve problems are transformed both by rotation—to eliminate all nonzero start values—and normalization—to eliminate any single-residue congruence conditions. Since rotation affects all sieve rings including those which give rise to the normalization map, the result of applying rotation to a sieve problem, $\mathcal{S}$, with acceptable residues $r_{i,j} \in \mathcal{R}_i$ and normaliztion $\eta[m, r]$ is given by the map:

$$\sigma[s] : \eta[m, r] \mapsto \eta[m, (r - s) \bmod m] \tag{3}$$
$$r_{i,j} \mapsto (r_{i,j} - s) \pmod{M_i}$$

---

[2] To ensure that intermediate values fit into a machine word, for instance.

Definition 3. *Given a sieve problem $\mathcal{S}$ operating over the interval $A \leq x < B$ for which the solutions satisfy $x = my + r$, we define its* normalized *form as the (equivalent) sieve problem $\eta[m, r](\mathcal{S})$ operating over the interval $0 \leq y < \left\lceil \frac{H-r}{m} \right\rceil$, where $H = B - A$.*

4   Parallel Implementation

An additional application of the normalization transformation arises if multiple sieve units are employed in parallel. Consider a sieve ring $\rho = \{M, \mathcal{R}\}$ with $|\mathcal{R}|$ acceptable residues. The sieve problem containing this ring may be partitioned into $|\mathcal{R}|$ parallel problems by applying a normalization, $\eta[M, r_j]$, for each of the $r_j \in \mathcal{R}$ to the remaining sieve rings, and running one normalized problem on each of the available sieve devices. The solution set $\mathcal{S}$ is simply the union of the results for each of the $|\mathcal{R}|$ parallelized sieve problems.

This optimization can be useful even if the normalized sieve problems are solved consecutively. As demonstrated by Lehmer [16], an effective speedup of $\frac{M_i}{|\mathcal{R}_i|}$ is achieved by executing each of the normalized sieve problems in series. Lukes [18] called this technique *multiple residue optimization*. Bernstein [4] called it *singly-focused enumeration*.

5   Doubly-Focused Enumeration

Doubly-focused enumeration, first described by Bernstein [4] makes use of an explicit form of the Chinese Remainder Theorem to map a sieve problem, $\mathcal{S}$ into two smaller sieve problems whose solutions may then be combined to retrieve all $x \in \mathcal{S}$. To illustrate this technique, we first require a lemma.

Lemma 1. *Every x in the range $0 \leq x < H$ may be expressed as the difference $t_p M_n - t_n M_p$, where M_p, M_n are relatively prime, $0 \leq t_p < \left\lceil \frac{H + M_n M_p}{M_n} \right\rceil$, and $0 \leq t_n < M_n$.*

Proof. Consider the arithmetic progression obtained by fixing t_n and varying t_p in the expression[3]:

$$x = t_p M_n - t_n M_p$$

This progression is capable of producing any $x \equiv -t_n M_p \pmod{M_n}$. Thus, if $t_n M_p \pmod{M_n}$ is made to range over all residue classes $\{0, 1, 2, \ldots, M_n - 1\}$, the resulting arithmetic progressions can be used to produce all possible integers x in an interval $[0, H)$ by varying t_p.

Consider $t_n \in \{0, 1, 2, \ldots, M_n - 1\}$. Since $\gcd(M_n, M_p) = 1$, it is straightforward to show that the set $\{t \mid t \equiv t_n M_p \pmod{M_n}, \ 0 \leq t_n < M_n\}$ forms a complete reduced residue system. If not, then for some $0 \leq i, j < M_n, i \neq j$, the

---

[3] Subtraction is used in the CRT decomposition of x instead of the more traditional addition to allow both sieve problems in the doubly-focused enumeration to operate in the same direction. See Appendix A for a description of the algorithm.

congruence: $i \cdot M_p \equiv j \cdot M_p \pmod{M_n}$ would hold. Multiplying both sides by $M_p^{-1} \pmod{M_n}$, however, gives $i \equiv j \pmod{M_n}$, a contradiction.

Hence, it is sufficient to consider $0 \leq t_n < M_n$ to produce the necessary arithmetic progressions. Since t_n is always nonnegative, choose $t_p \geq 0$, and as $t_n < M_n$, it follows that $t_p < \left\lceil \frac{H + M_n M_p}{M_n} \right\rceil$.   □

5.1 The Doubly-Focused Transformation

The doubly-focused transformation $\delta[M_n, M_p]$ works as follows. Consider a sieve problem:

$$\mathcal{S} = \bigcap_{i=1}^{k} \{x \in \mathbb{Z} \mid x \pmod{M_i} \in \mathcal{R}_i, \quad 0 \leq x < H \}. \tag{4}$$

Partition the moduli $M_1, M_2, \ldots, M_k$ into two distinct sets, $\mathcal{M}_n$ and $\mathcal{M}_p$, and define the quantities M_n and M_p as the products of the moduli in these sets: $M_n = \prod_{i=1}^{s} M_i$ and $M_p = \prod_{j=s+1}^{k} M_j$ respectively.

From Lemma 1, a solution $x \in \mathcal{S}$ can be written $x = t_p M_n - t_n M_p$. Reducing this expression modulo M_n and M_p respectively, it is clear that any solution x must satisfy the congruences:

$$x \equiv -t_n M_p \pmod{M_n} \in \mathcal{R}_n$$
$$x \equiv t_p M_n \pmod{M_p} \in \mathcal{R}_p.$$

Thus, rather than sieving for solutions x in the original problem, we can search instead for solutions t_n, t_p in two translated sieve problems, and recombine these solutions to obtain solutions in the original interval as follows. Set $\mathcal{T}_p = \eta[M_n^*, 0](\mathcal{R}_p)$, $\mathcal{T}_n = \eta[M_p^*, 0](\mathcal{R}_n)$, $M_n^* = M_n^{-1} \bmod M_p$ and $M_p^* = (-M_p)^{-1} \bmod M_n$. The doubly-focused map is given by:

$$\delta[M_n, M_p] : \mathcal{S} \to (\mathcal{S}_p, \mathcal{S}_n),$$

where

$$\mathcal{S}_p = \bigcap_{i=1}^{s} \left\{ t_p \in \mathbb{Z} \mid t_p \pmod{M_p} \in \mathcal{T}_p, \quad 0 \leq t_p < \left\lceil \frac{H + M_n M_p}{M_n} \right\rceil \right\} \tag{5}$$

$$\mathcal{S}_n = \bigcap_{i=s+1}^{k} \{t_n \in \mathbb{Z} \mid t_n \pmod{M_n} \in \mathcal{T}_n, \quad 0 \leq t_n < M_n \}. \tag{6}$$

We call these new problems the *positive* and *negative* sieve problems, respectively. The solutions of these new problems may be converted back into solutions to the original sieve problem (4) using only a moderate amount of storage; *i.e.* by considering one t_n output at a time, and maintaining an array of solutions for t_p, where $0 \leq t_p M_n - t_n M_p < H$. See Appendix A for a more detailed description of this algorithm.

6   The Calgary Scalable Sieve

CASSIE, the CAlgary Scalable SIEve, is a software-based tookit for representing and solving congruential sieve problems. The toolkit implements the transformations of Sections 3-5 as a set of extension to the scripting language Tcl [20]. For parallel sieving, CASSIE was implemented on the University of Calgary's Advanced Cryptography Lab (ACL); a Beowulf cluster consisting of 152 dual-Xeon Pentium IV processors running at 2.4 GHz.

CASSIE is written in portable C and was compiled using GCC (2.96 and later) under Red Hat Linux 7.3 (kernel 2.4.18-27.7) on the Intel Pentium IV architecture. It has since been ported to the AMD 64-bit Opteron architecture, and the OpenBSD 3.8 operating system.

7   Pseudosquares and Primality Testing

The pseudosquare problem, first considered by Kraitchik [14] is characterized in the following manner:

Definition 4. *Given an integer x, a pseudosquare $M_{2,x}$ is defined as the least positive integer satisfying:*

1. $M_{2,x} \equiv 1 \pmod 8$
2. The Legendre symbol $\left(\frac{M_{2,x}}{q}\right) = 1$ for all odd primes $q \leq x$
3. $M_{2,x}$ is not a perfect square.

In other words, the pseudosquare, $M_{2,x}$ behaves (locally) like a perfect square modulo all small primes $q \leq x$,

Perhaps the most interesting application of pseudosquares is in the area of primality testing. In [19], Lukes *et al.* indicated that a sufficiently rapid growth rate of pseudosquares would lead to a deterministic polynomial-time algorithm for determining the prime character of an integer N.

Theorem 2. *If*

1. All prime divisors $q|N$ exceed the bound $B \in \mathbb{Z}^+$,
2. $\frac{N}{B} < M_{2,x}$ for some integer, x,
3. $p_i^{\frac{N-1}{2}} \equiv \pm 1 \pmod N$ for all primes p_i, $2 \leq p_i \leq x$,
4. $p_j^{\frac{N-1}{2}} \equiv -1 \pmod N$ for some odd $p_j \leq x$ when $N \equiv 1 \pmod 8$, or
   $2^{\frac{N-1}{2}} \equiv -1 \pmod N$ when $N \equiv 5 \pmod 8$

then N is a prime or a power of a prime.

Note that if N is prime, the conditions of Theorem 2 hold with $B = 1$ and $N < M_{2,x}$. The main consequence of this result is that if $M_{2,x}$ grows sufficiently quickly—*i.e.* if $p < c(\log M_{2,x})^k$ for fixed constants c, k—then Theorem 2 offers an unconditional, deterministic polynomial-time primality test.

One interesting applictaion of Theorem 2 was noted by D.J Bernstein [3]. He observed that, when combined with the Pollard rho technique, Theorem 2

integers offers the fastest known method for proving the primality of 100-bit integers.

7.1 Pseudosquare Growth

In [19], Lukes *et al.* appealed to a result of Bach [2] which, under the assumtions of the Extended Riemann Hypothesis (ERH), give a lower bound for the growth of pseudosquares:

$$\log M_{2,x} > \sqrt{x/2}. \tag{7}$$

In [21], Schinzel refined the bounds on $M_{2,x}$ to:[4]

$$(1-\epsilon)\sqrt{x} < \log M_{2,x} < (2\log 2 + \epsilon)\frac{x}{\log x}$$

for any $\epsilon > 0$ with $x > x_0(\epsilon)$. Thus, under the conditions of the ERH, Theorem 2 offers a deterministic polynomial-time primality test.

Lukes offers an alternate prediction for the growth rate of $M_{2,x}$ [18, pp. 111]. Assume solutions for $M_{2,x}$ are equidistributed in the range $0 < x < 8p_2p_3 \cdots p_n$, and so $M_{2,x} \approx \frac{8p_2p_3\cdots p_n}{\prod_{i=1}^{n}(p_i-1)/2}$. By Merten's Theorem [12] and the Prime Number Theorem, $M_{2,x} \approx c2^n \log x$ for $c = 2e^\gamma$ where $\gamma = 0.57721$ is Euler's constant. Under these stated assumptions, $M_{2,x}$ would have a growth rate of the form $2^{(x/\log x)(1+o(1))}$; *i.e.*

$$\log M_{2,x} \approx \frac{x \cdot \log 2}{\log x} \tag{8}$$

If these predictions hold, then via Theorem 2, primality proving may be done using $(\log N)^{1+o(1)}$ modular exponentiations. Since performing modular exponentiation incurs a complexity of $(\log N)^{2+o(1)}$ (using, for instance, the techniques of Schönhage and Strassen [22]) we may conjecture that the primality of an arbitrary integer N may be proved with $(\log N)^{3+o(1)}$ operations.

8 Pseudosquare Results

In [18], Lukes offered empirical evidence to support the growth estimates in (7) and (8) by computing the pseudosquares for all primes $q \le 277$, and comparing the results to (7) and (8). In [4], Bernstein extended this result to $q \le 281$. Using CASSIE, the table of pseudosquares was extended to include all primes $q \le 359$ and it was shown that the predictions of (7) and (8) still hold. These results are given in Section 8.1.

To achieve these new results, two separate computations were performed. The first computation was a doubly-focused enumeration implemented on two AMD Athlon MP 2000+ processors. To partition the problem over these processors,

---

[4] Assuming the Extended Riemann Hypothesis (ERH). In the same paper, an unconditional result is also given.

the acceptable pseudosquare residues for 3, 5, and 8 were combined to produce the normalizations $x = 1 + 120y$, and $x = 49 + 120y$. The primes from 7 to 73 were used as doubly-focused moduli, arranged in the following manner:

$$\mathcal{M}_n = \{7, 13, 29, 31, 71, 41, 43, 59, 61\}$$
$$\mathcal{M}_p = \{11, 17, 19, 23, 73, 37, 47, 53, 67\}.$$

Values emerging from the doubly-focused sieve were further filtered by examining their quadratic character modulo the primes 79–127. Remaining results were filtered to remove the perfect squares, and then tested against the primes up to 400 to determine where their pseudosquare behaviour broke down.

The run was completed on April 6, 2003, and achieved an effective canvass rate of 2.06×10^{15} trials per second. In addition to verifying the previous results of [19] and [4], we were able to find 6 previously unknown pseudosquare values: $M_{2,293}$ through $M_{2,317}$. These results are summarized in Table 1.

Table 1. Pseudosquare Results (2-processors)

p	$M_{2,p}$	Source
283	533 552 663 339 828 203 681	Wooding, Williams (2003)
293,307	936 664 079 266 714 697 089	*CASSIE*
311,313,317	2 142 202 860 370 269 916 129	(2 processors)

Once CASSIE had proved successful in the initial run, the pseudosquare computation was retooled for implementation in software over 180 processors. To partition the problem in this manner, the acceptable pseudosquare residues for 3, 5, 8, 11, 13, and 17 were combined to produce 180 acceptable residue classes (mod 120120). Using the MPI library [11], the CASSIE software running on each of the ACL nodes was able to determine which normalization to use.

The rest of the problem setup was identical for each of the processors. The primes from 17 to 83 were arranged into two sets:

$$\mathcal{M}_n = \{17, 23, 29, 31, 37, 41, 47, 53, 71\}$$
$$\mathcal{M}_p = \{19, 43, 59, 61, 67, 73, 79, 83\}.$$

Values emerging from the doubly-focused sieve were further filtered by examining their quadratic character modulo the primes 89–127. Remaining results were filtered to remove the perfect squares, and then tested against the primes up to 400 to determine where their pseudosquare behaviour broke down.

The run was completed on July 26, 2003 and achieved an effective canvass rate of 1.05×10^{18} trials per second. 6 previously unknown pseudosquare values were obtained: $M_{2,331}$ to $M_{2,359}$. These results are summarized in Table 2.

Table 2. Pseudosquare Results (180 processors)

p	$M_{2,p}$	Source
331	13 649 154 491 558 298 803 281	
337	34 594 858 801 670 127 778 801	Wooding, Williams (2003)
347, 349	99 492 945 930 479 213 334 049	*CASSIE / ACL*
353, 359	295 363 487 400 900 310 880 401	(180 processors)
367	$> 120120 \times 2^{64}$	

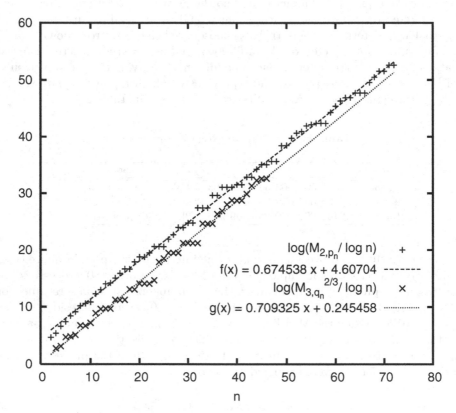

Fig. 1. Pseudosquare and pseudocube growth vs. n

8.1   Numerical Confirmation of Growth Preditions

In Figure 1, pseudosquare growth is shown as a function of n, where p_n is the nth prime. The straight line represents the least squares line fitted to this data, and is given by:

$$y = 0.67454x + 4.60704$$

Even with the relatively small number of data points, the slope of the least squares fit in Figure 1 appears to be approaching the predicted value of $\log 2 = 0.69315$, *i.e.* M_{2,p_n} has a growth rate of the form $2^{n(1+o(1))}$.

To date, the pseudosquare results obtained using CASSIE support the predictions of Equations (7) and (8). This is at least empirical evidence that the polynomial-time nature of the primality test of Theorem 2 holds even in the absence of the ERH.

9 Pseudocubes

The pseudocube problem may be defined in a manner analagous to that of the pseudosquares.

Definition 5. *Given an integer, x, the pseudocube $M_{3,x}$ is defined as the least positive integer satisfying:*

1. $M_{3,x} \equiv \pm 1 \pmod 9$
2. $M_{3,x}^{\frac{q-1}{3}} \equiv 1 \pmod q$ *for all primes $q \leq x$, $q \equiv 1 \pmod 3$*
3. $q \nmid M_{3,x}$ *for all primes $q \leq x$, $q \not\equiv 1 \pmod 3$.*
4. $M_{3,x}$ *is not a perfect cube.*

In [8], Berrizbeitia *et al.* generalized the primality test of Theorem 2 (with $B = 1$) to involve pseudocubes. This definition makes use of the ring of Eisenstein integers, $\mathbb{Z}[\omega]$ where $\omega = e^{2\pi i/3}$. The general properties of these integers are described in [13].

Theorem 3. *Let N be odd, $3 \nmid N$. Define $N^* = N$ if $N \equiv 1 \pmod 3$ and $N^* = -N$ if $N \equiv -1 \pmod 3$. Furthermore, if $q \equiv 1 \pmod 3$ and prime, define $\alpha_q \in \mathbb{Z}[\omega]$ by $\alpha_q \overline{\alpha_q} = q$. If*

1. $N < (M_{3,x})^{2/3}$.
2. $\left(\frac{\alpha_q}{N}\right)_3 \equiv \lambda_q^{\frac{(N^*-1)}{3}} \pmod N$ *for all primes $q \equiv 1 \pmod 3$, $q \leq x$.*
3. $q \nmid N$ *for all primes $q \equiv 2 \pmod 3$, $q \leq x$.*

where $\alpha \in \mathbb{Z}[\omega]$, $\lambda_q = \alpha_q/\overline{\alpha_q}$, then N is a prime or a power of a prime.   □

9.1 Pseudocube Growth

In [8], Berrizbeitia *et al.* conjectured that, under the same assumptions that were used to forumulate (8), pseudocube growth should eventually outpace the pseudosquares; *i.e.*

$$\frac{(M_{q_n})^{2/3}}{M_{2,p_n}} \approx \frac{c_2^{(}2/3)3^{2n/3}(\log q_n)^{4/3}}{c_1 2^n \log p_n} \geq c \left(\frac{3^{2/3}}{2}\right)^n > 1 \qquad (9)$$

where p_n is the nth prime, and q_n is the nth prime for which $q \equiv 1 \pmod 3$. Thus, the primality test of Theorem 3 should be more efficient than that of Theorem 2 (where $B = 1$) for sufficiently large n.

9.2 Pseudocube Results

To extend the table of known pseudocubes, the following computation was performed. First, the acceptable residues for 2, 7, 9, 13 and 31 were combined to produce 160 residue classes (modulo 50778). These classes were then used as normalizations on each of 160 processors of the ACL cluster.

The remaining primes congruent to 1 (mod 3) between 19 and 127 were then arranged into two sets:

$$\mathcal{M}_p = \{19, 43, 79, 103, 37, 73\}$$
$$\mathcal{M}_n = \{127, 61, 67, 97, 109\}.$$

and used in a doubly-focused enumeration

Solutions emerging from the sieve were further filtered to exlude perfect cubes, and all solutions divisible by the primes $q \equiv 2$ (mod 3).

The computation took (on average) 626102 seconds on each of 160 nodes to test all solutions up to $H = 1.45152 \times 10^{22}$, achieving an effective canvass rate of 2.31×10^{16} trials per second. In the end, 11 new pseudocubes were obtained, $M_{3,367}$ through $M_{3,487}$. These results are summarized in Table 3.

9.3 Comparison with Growth Predictions

In Figure 1, pseudocube growth is shown as a function of n, where p_n is the nth prime. The straight line represents the least squares line fitted to this data, and is given by:

$$y = 0.709325x + 0.245458$$

Unfortunately, even with the new pseudocube results of Section 9.2, primality proving via pseudocubes is not yet more efficient than the pseudosquare method.

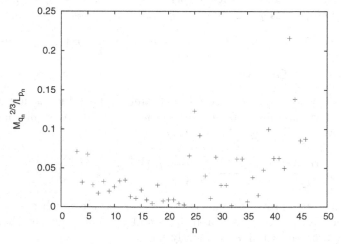

Fig. 2. Ratio of pseudocube to pseudosquare growth

Table 3. Least Pseudocubes for primes $q \equiv 1 \pmod 3$

q	$M_{3,q}$	Source
367	996 438 651 365 898 469	
373	2 152 984 914 389 968 651	
379	12 403 284 862 819 956 587	
397, 209, 421	37 605 274 105 479 228 611	
433	205 830 039 006 337 114 403	Wooding, Williams (2005)
439	1 845 193 818 928 603 436 441	CASSIE/ACL
457	7 854 338 425 385 225 902 393	(160 processors)
463	12 904 554 928 068 268 848 739	
487	13 384 809 548 521 227 517 303	
499	$> 1.45152 \times 10^{22}$	

However, if the trend illustrated in Figure 1 continues (and to a lesser degree, Figure 2) , it would appear that the conjecture underlying Equation 9 is sound; *i.e.* $0.709325 - 0.674538 = 0.034787$ is approaching the predicted $\log \frac{3^{2/3}}{2}$, and hence, the efficiency of the pseudocube primality test should eventually exceed that of the pseudosquares.

10 Summary

In this paper, we have offered additional numerical evidence for a conjecture that primality proving may be done in $(\log N)^{3+o(1)}$ operations. We have done this by extending the tables of known pseudosquares and pseudocubes using CASSIE, a software-based congruential sieve employing D. J. Bernstein's technique of doubly-focused enumeration.

References

1. Agrawal, Manindra, Kayal, Neeral and Saxena, Nitin: PRIMES is in P. Annals of Mathematics. 160(2). (2004) 781–793
2. Bach, E.: Explicit bounds for primality testing and related problems. Mathematics of Computation. 55(191). (1990) 355–380
3. Bernstein, D. J.: More news from the Rabin-Williams front. Conderence slides from Mathematics of Public Key Cryptography (MPKC). University of Illinois at Chicago. Available from http://cr.yp.to/talks/2003.11.08-2/slides.ps (2003)
4. Bernstein, D. J.: Doubly Focused Enumeration Of Locally Square Polynomial Values. High Primes and Misdemeanors—Lectures in honour of the 60th birthday of Hugh Cowie Williams. Alf van der Poorten, Andreas Stein, eds. (2004) 69–76
5. Bernstein, D. J.: Proving primality after Agrawal-Kayal-Saxena, (Unpublished). Available from http://cr.yp.to/papers.html#aks. (2003)
6. Bernstein, D. J.: Proving primality in Essentially Quartic Random Time. Mathematics of Computation (to appear). Available from http://cr.yp.to/papers.html#quartic (2004)

7. Berrizbeitia, P.: Sharpening *PRIMES is in P* for a large family of numbers, (Preprint). Available from http://arxiv.org/ps/math.NT/0211334 (2002)
8. Berrizbeitia, P., Müller, S. and Williams, H. C.: Pseudocubes and Primality Testing, Algorithmic Number Theory: 6th International Symposium, ANTS-VI, Burlington, VT, USA, June 13-18, 2004. Proceedings. Duncan Buell, ed. LNCS 3076. Springer-Verlag GmbH. (2004)
9. Bronson, N. D. and Buell, D. A.: Congruential sieves on FPGA computers, Mathematics of computation, 1943–1993: a half-century of computational mathematics: Mathematics of Computation 50th Anniversary Symposium, August 9–13, 1993, Vancouver, British Columbia. Walter Gautschi, ed. (1994) 547–551
10. Cheng, Qi: Primality Proving via One Round in ECPP and One Iteration in AKS Advances in Cryptology—Crypto 2003, Santa Barbara, LNCS 2729. pp 338 – 348 (2003)
11. Gropp, W., Lusk, E. and Skjellum, A.: Using MPI: Portable Parallel Programming with the Message Passing Interface. MIT Press. (1994)
12. Hardy, G. H. and Wright, E. M.: An Introduction to the Theory of Numbers, Oxford University Press, 5ed. (1979)
13. Ireland, K. and Rosen, M.: A Classical Introduction to Modern Number Theory. Volume 84 of Graduate Texts in Mathematics. 2ed. Springer-Verlag (1990)
14. Kraitchik, Maurice: Récherches sur la Théorie des Nombres. Tome I Gauthier-Villars. (1924)
15. Lehmer, D. H.: The Sieve Problem for All-Purpose Computers. Mathematical Tables and Other Aids to Computation. Vol 7, no 41. (1953) 6–14
16. Lehmer, D. H.: The mechanical combination of linear forms. American Mathematical Monthly. Vol 35, no 4. (1928) 114–121
17. Lehmer, D. H., Lehmer, E. and Shanks, D.: Integer Sequence having Prescribed Quadratic Character, Mathematics of Computation. 24(110). (1970) 433–451
18. Lukes, Richard F.: A Very Fast Electronic Number Sieve. Ph.D. Thesis. University of Manitoba. (1995)
19. Lukes, Richard F., Patterson, Cameron D. and Williams, H. C.: Some results on pseudosquares, Mathematics of Computation. 65(213). (1996) 361–372
20. Ousterhout, John K.: Tcl and the Tk Toolkit. Addison Wesley. (1994)
21. Schinzel, A.: On Pseudosquares, New Trends in Probability and Statistics. Vol. 4.(1997) 213–220
22. Schönhage A. and Strassen, V.: Schnelle Multiplikation grosser Zahlen, Computing. Vol. 7. (1971) 281–292

A   Appendix—Enumeration Algorithms

Given a normalized sieve problem $\mathcal{S}$, define an accessor function next($\mathcal{S}$) which returns the next sieve output. Now, fix M_n, M_p and consider the doubly-focused sieve problem given by $\delta [M_n, M_p] (\mathcal{S})$. This consists of an upper bound, H, and positive and negative sieve problems $\mathcal{S}_p, \mathcal{S}_n$. Given the current output of positive and negative sieves $(t_p, t_n$ respectively), Algorithm 1 (HUNT) describes a method for obtaining an initial solution $x \in \mathcal{S}$ such that $x = t_p M_n - t_n M_p$, $0 \leq x < H$. Algorithm 2 (DFSIEVE) describes a method for obtaining all such solutions $x \in \mathcal{S}$ in the given interval.

Algorithm 1. HUNT: Find an initial solution

INPUTS: $H, \mathcal{S}_p := \{\mathcal{R}_p, M_p\}, \mathcal{S}_n := \{\mathcal{R}_n, M_n\}, t_n, t_p$
OUTPUT: $t_p \in \mathcal{S}_p, t_n \in \mathcal{S}_n$ such that $x = t_p M_n - t_n M_p, 0 \le x < H$.

1: $x \leftarrow t_p M_n - t_n M_p$
2: **repeat**
3:   **while** $x < 0$ **do**
4:     $t_p \leftarrow \textbf{next}(\mathcal{S}_p)$
5:     $x \leftarrow (t_p M_n - t_n M_p)$
6:   **end while**
7:   **while** $(x \ge H)$ **do**
8:     $t_n \leftarrow \textbf{next}(\mathcal{S}_n)$
9:     $x \leftarrow (t_p M_n - t_n M_p)$
10:   **end while**
11: **until** $(x \ge 0)$
12: Return (t_p, t_n)

Algorithm 2. DFSIEVE: Doubly-focused enumeration of sieve solutions

INPUTS: $H, \mathcal{S}_p := \{\mathcal{R}_p, M_p\}, \mathcal{S}_n := \{\mathcal{R}_n, M_n\}$
OUTPUTS: All solutions x such that $x = t_p M_n - t_n M_p, t_p \in \mathcal{S}_p, t_n \in \mathcal{S}_n, 0 \le x < H$.

1: $t_p \leftarrow \textbf{next}(\mathcal{S}_p)$
2: $t_n \leftarrow \textbf{next}(\mathcal{S}_n)$
3: $(t_p, t_n) \leftarrow HUNT(H, \mathcal{S}_p, \mathcal{S}_n, t_p, t_n)$
4: **while** not DONE **do**
5:   Append $x_{\text{last}} = t_p M_n - t_n M_p$
6:   $t_p \leftarrow \textbf{next}(\mathcal{S}_p); a_p \leftarrow t_p M_n$
7:   **while** $(a_p - a_n < H)$ **do**
8:     Extend x array with $x_{\text{last}} \leftarrow a_p - a_n$
9:   **end while**
10:   **for all** x_i in $x_{\text{first}}, \ldots, x_{\text{last}}$ **do**
11:     Filter and/or print x_i
12:   **end for**
13:   $\delta_n \leftarrow \textbf{next}(\mathcal{S}_n) - t_n; \Delta_n \leftarrow \delta_n M_p; a_n \leftarrow a_n + \delta_n$
14:   **if** $t_n \ge M_n$ **then**
15:     **return** (DONE)
16:   **end if**
17:   **for all** x_i in $x_{\text{first}}, \ldots, x_{\text{last}}$ **do**
18:     $x_i = x_i - \Delta_n M_p$
19:     **if** $x_i < 0$ **then**
20:       Delete x_{first}
21:       **if** x array empty **then**
22:         $t_n \leftarrow \textbf{next}(\mathcal{S}_n)$
23:         $(t_p, t_n) \leftarrow HUNT(H, \mathcal{S}_p, \mathcal{S}_n, t_p, t_n)$
24:       **end if**
25:     **end if**
26:   **end for**
27: **end while**

Practical Lattice Basis Sampling Reduction

Johannes Buchmann and Christoph Ludwig

Technische Universität Darmstadt, Fachbereich Informatik
Hochschulstr. 10, 64289 Darmstadt, Germany
{buchmann, cludwig}@cdc.informatik.tu-darmstadt.de

Abstract. We propose Simple Sampling Reduction (SSR) that makes Schnorr's Random Sampling Reduction (RSR) practical. We also introduce generalizations of SSR that yield bases with several short basis vectors and that, in combination, generate shorter basis vectors than SSR alone. Furthermore, we give a formula for $\Pr[\,\|\mathbf{v}\|^2 \leq x\,]$ provided $\mathbf{v}$ is randomly sampled from SSR's search space. We describe two algorithms that estimate the probability that a further SSR iteration will find an even shorter vector, one algorithm based on our formula for $\Pr[\,\|\mathbf{v}\|^2 \leq x\,]$, the other based on the approach of Schnorr's RSR analysis. Finally, we report on some cryptographic applications.

1 Introduction

Lattice basis reduction aims to efficiently compute lattice bases that consist of vectors as short as possible. After the renowned LLL algorithm was published [1], lattice reduction was no longer a topic for number theorists only, but it was also applied in integer programming, coding theory, and cryptography. Over the last twenty years, many algorithms were proposed that advance on LLL, trading shorter base vectors for longer computing times.

Recently, Schnorr [2] proposed Random Sampling Reduction (RSR), a new lattice reduction technique that combines LLL-like algorithms with the exhaustive search of a lattice point set that is likely to contain short vectors. He concluded that his new algorithms improves on the previous most efficient algorithms by a fourth root. Unfortunately, the RSR algorithm as well as its analysis depend on two assumptions on the Gram-Schmidt decomposition of LLL reduced bases. It is clear that, in practice, LLL reduced lattice bases satisfy these assumptions only in some approximate sense, if at all. It turns out that in particular one assumption, the so called (GSA), fails regularly, which renders RSR impractical.

We propose a modification of Schnorr's RSR– named Simple Sampling Reduction (SSR) – that makes Sampling Reduction practical. We also describe several generalizations of SSR that yield more short basis vectors and that proceed even when SSR could not find shorter vectors anymore. SSR has to estimate the probability that it will find a vector shorter than the first basis vector. We propose two algorithms that compute such estimates: The first one is very efficient, but tends to return a rather pessimistic bound; the second algorithm gives a more

F. Hess, S. Pauli, and M. Pohst (Eds.): ANTS 2006, LNCS 4076, pp. 222–237, 2006.

exact result based on the convolution theorem, but is more expensive. Due to the lack of space this paper omits some details and the proofs; they are presented in full length in [3].

The rest of this paper is structured as follows: Sect. 2 introduces some definitions and notations. Sect. 3 describes SSR, points out the differences between SSR and RSR, and reports on the typical behavior of SSR. Sect. 4 describes the algorithms called by SSR that estimate the further success probability. Sect. 5 proposes several generalizations of SSR and describes their impact on the reduction result. Finally, Sect. 6 summarizes some empirical results if Sampling Reduction is used to run lattice based attacks on some cryptosystems.

2 Notation

We assume the Euclidean metric on $\mathbb{R}^d$. A lattice L is a discrete subgroup of $\mathbb{R}^d$, its dimension is $\dim(L) := \dim(L \otimes_{\mathbb{R}} \mathbb{R})$. The first minimum of L is $\lambda_1(L) := \min\{\, \|\mathbf{x}\| \mid 0 \neq \mathbf{x} \in L \,\}$.

Every n-dimensional lattice L, $n \geq 1$, has infinitely many (ordered) bases $B = [\mathbf{b}_1, \ldots, \mathbf{b}_n] \in \mathbb{R}^{d \times n}$ such that $L = L(B) := \{\, B\mathbf{x} \mid \mathbf{x} \in \mathbb{Z}^n \,\}$. Two lattice bases B, B' generate the same lattice L if and only if there is a unimodular matrix $U \in \mathbb{Z}^{n \times n}$ (i. e., $\det U = \pm 1$) such that $B' = BU$. We consider integer coefficient lattices only, whence $B \in \mathbb{Z}^{d \times n}$.

Throughout this paper, $B = \hat{B}R$ is the Gram-Schmidt decomposition of the lattice basis B, i. e. the columns $\hat{\mathbf{b}}_j$ of $\hat{B} \in \mathbb{Q}^{d \times n}$ are pairwise perpendicular and $R = (r_{i,j}) \in \mathbb{Q}^{n \times n}$ is unit upper triangular. $\pi_j : \mathbb{R}^d \to \lin\{\, \mathbf{b}_1, \ldots, \mathbf{b}_{j-1} \,\}^{\perp}$ denominates the orthogonal projection onto the orthogonal space of the first $j-1$ base vectors.

A basis $B = \hat{B}R$ is said to be δ-LLL reduced ($\delta \in (1/4, 1)$) if and only if

$$|r_{i,j}| \leq 1/2 \qquad \text{for all } 1 \leq i < j \leq n,$$
$$\|\hat{\mathbf{b}}_{j+1}\|^2 \geq (\delta - r_{j,j+1}^2)\|\hat{\mathbf{b}}_j\|^2 \quad \text{for all } 1 \leq j \leq n.$$

An LLL variant due to Schnorr and Euchner [4] computes δ-LLL reduced bases in, heuristically, $O(dn^4)$ arithmetic steps on $O(n)$ bit integers, provided the input basis vectors $\mathbf{b}$ satisfy $\log\|\mathbf{b}\| \in O(n)$. (Nguyen and Stehlé [5] recently proposed a floating-point LLL variant that guarantees this complexity.) We have $\|\mathbf{b}_1\| \leq (\delta - 1/4)^{-(n-1)/2}\lambda_1(L(B))$ for any δ-LLL reduced basis B. δ is $3/4$ in the original LLL algorithm, which leads to the norm bound $\|\mathbf{b}_1\| \leq 2^{(n-1)/2}\lambda_1(L(B))$.

3 Simple Sampling Reduction

We outline in this section our Simple Sampling Reduction algorithm that makes Sampling Reduction practical, point out some differences to Schnorr's Random Sampling Reduction, and describe its empirical behavior.

Algorithm 1. SSR

Input: − generating system $G \in \mathbb{Z}^{d \times m}$ of lattice L, $\dim(L) = n$
      − search space bound $u_{\max} \in \mathbb{N}$
Output: LLL reduced basis $B = \hat{B}R$ of L such that
      − $0.99\|\mathbf{b}_1\|^2 < \min\{\|\mathbf{v}\|^2 \mid \mathbf{v} \in S_{u_{\max},B}\}$ or
      − $CSSS((\|\hat{\mathbf{b}}_1\|^2, \dots, \|\hat{\mathbf{b}}_n\|^2), u_{\max}) = false$.

  procedure SSR$(G, u_{\max})$
    $(B, \flat, R) \leftarrow$ LLL(G) /* $B = \hat{B}R$, $\flat = (\|\hat{\mathbf{b}}_1\|^2, \dots, \|\hat{\mathbf{b}}_n\|^2)$ */
    while CSSS$(\flat, u_{\max}) = true$ **do**

      for x **from** 0 **to** $2^{u_{\max}} - 1$ **do**
        $\mathbf{v} \leftarrow$ Sample(B, R, x)
        if $\|\mathbf{v}\|^2 \leq 0.99\|\mathbf{b}_1\|^2$ **then** break
        if $x = 2^{u_{\max}} - 1$ **then** terminate("no short vector")
      end for

      $(B, \flat, R) \leftarrow$ LLL$([\mathbf{v}, \mathbf{b}_1, \dots, \mathbf{b}_n])$

    end while

    terminate("further progress unlikely")
  end procedure

3.1 The Simple Sampling Reduction Algorithm

The overall structure of Simple Sampling Reduction (SSR, Alg. 1) is as follows: The algorithm operates on generating systems G of an n-dimensional lattice L. SSR first applies some LLL-type reduction to G and obtains the basis B together with its Gram-Schmidt decomposition. It then iterates the main loop as long as the subalgorithm Check Search Space Size (CSSS) deems the probability of finding a vector shorter than $\mathbf{b}_1$ sufficient; we defer the discussion of possible CSSS implementations to Sect. 4.

In each iteration of the main loop, SSR enumerates the $2^{u_{\max}}$ elements of some search space $S_{u_{\max},B} \subset L$ – that we describe below – by means of the function Sample. The enumeration loop is left as soon as the length of a sampled vector $\mathbf{v}$ is at most $\sqrt{0.99}\|\mathbf{b}_1\|$. If there is no such vector in $S_{u_{\max},B}$, then SSR terminates. The last step in each iteration of the main loop is that SSR applies again the LLL-type reduction to the new generating system G formed by prepending $\mathbf{v}$ to the basis B.

In each iteration of the outer loop, $\mathbf{b}_1$ becomes shorter by a factor at most $\sqrt{0.99}$. Before the first iteration, $\|\mathbf{b}_1\| \leq 2^{(n-1)/2}\lambda_1(L)$ because B is LLL reduced. So SSR terminates after $O(n)$ main loop iterations since then $S_{u_{\max},B}$ will not contain a sufficiently short vector anymore.

We can express the length of a lattice vector $\mathbf{v}$ in terms of its Gram-Schmidt vector representation; i. e., $\|\mathbf{v}\|^2 = \sum_{j=1}^n \nu_j^2 \|\hat{\mathbf{b}}_j\|^2$ for $\mathbf{v} = \sum_{j=1}^n \nu_j \hat{\mathbf{b}}_j$. Therefore, if a vector $\mathbf{v}$ is short then we expect its Gram-Schmidt coefficients ν_j to be small as well. However, the Gram-Schmidt vectors contribute to the length of $\mathbf{v}$ to different degrees; the smaller $\|\hat{\mathbf{b}}_j\|$, the more leeway there is for ν_j. In an LLL reduced basis, the length of the Gram-Schmidt vectors $\hat{\mathbf{b}}_j$ typically decreases with j. This observation motivates the following definition of the search space $S_{u,\mathsf{B}}$.

Definition 1. *Let* $\mathsf{B} \in \mathbb{Z}^{d \times n}$ *be a lattice basis with Gram-Schmidt decomposition* $\mathsf{B} = \hat{\mathsf{B}}\mathsf{R}$ *and* $1 \leq u < n$. *The search space* $S_{u,\mathsf{B}}$ *is the set of all lattice vectors* $\mathbf{v} = \sum_{j=1}^{n} \nu_j \hat{\mathbf{b}}_j \in L(B)$ *such that*

$$\nu_j \in \begin{cases} (-1/2, 1/2] & \text{if } 1 \leq j < n-u, \\ (-1, 1] & \text{if } n-u \leq j < n, \\ \{1\} & \text{if } j = n. \end{cases} \tag{1}$$

If $\mathbf{v} = \sum_{j=1}^{n} \nu_j \hat{\mathbf{b}}_j, \mathbf{v}' = \sum_{j=1}^{n} \nu_j' \hat{\mathbf{b}}_j \in L(\mathsf{B})$ and $\nu_j = \nu_j'$ for $j = j_0 + 1, \ldots, n$, then $\nu_{j_0} - \nu_{j_0}' \in \mathbb{Z}$. This implies that $S_{u,\mathsf{B}}$ has 2^u elements. Using a floating point representation of the Gram-Schmidt coefficient matrix $\mathsf{R} \in \mathbb{Q}^{n \times n}$, we can construct an algorithm Sample that implements a bijection $\{0, \ldots, 2^u - 1\} \to S_{u,\mathsf{B}}$ in $O(n^2)$ integer and floating point operations [3].

3.2   Simple vs. Random Sampling Reduction

Schnorr's RSR requires two assumptions. The first guarantees that RSR eventually terminates, the second determines a coefficient in a formula that RSR evaluates in each outer loop iteration. Unfortunately, neither assumption holds in practice, whence RSR as described by Schnorr is impractical. We describe in the following both assumptions and point out how SSR works around them.

The *Randomness Assumption* (RA) states that the Gram-Schmidt coefficients of the sampled lattice vectors behave like pairwise independent uniform random variables on the intervals defined by (1). If this were strictly true, then sampling would eventually produce a sufficiently short vector. However, $S_{u,\mathsf{B}}$ is a finite set. In practice, we will therefore encounter search spaces that do not contain such a short vector; in such a situation the inner RSR loop becomes infinite.

Since RSR does indeed randomly sample the elements of the search space, it cannot determine for sure that there is no sufficiently short vector in $S_{u,\mathsf{B}}$. SSR systematically enumerates the search space elements and can terminate as soon as it exhausted the search space. As a side effect, SSR searches $S_{u,\mathsf{B}}$ faster than RSR: The latter will sample some vectors repeatedly; if $u > 8$, the sampling loop of RSR thus misses, on average, in $|S_{u,\mathsf{B}}|$ iterations $(1 - 1/|S_{u,\mathsf{B}}|)^{|S_{u,\mathsf{B}}|}|S_{u,\mathsf{B}}| \geq (1 - 2^{-8})^{2^8}|S_{u,\mathsf{B}}| \geq \frac{1}{3}|S_{u,\mathsf{B}}|$ lattice points.

According to the *Geometric Series Assumption* (GSA), there is for every LLL reduced lattice basis B some (GSA) coefficient $q_{\mathsf{B}} \in (0,1)$ such that $\|\hat{\mathbf{b}}_j\|^2 = q_{\mathsf{B}}^{j-1}\|\mathbf{b}_1\|^2$ for all $1 \leq j \leq n$. RSR relies on (GSA) in several ways. First, it computes the search space size exponent u by a formula that explicitly refers to q_{B}. This leaves the question how to determine a suitable value for q_{B}, since, in practice, we can only expect lattice bases to approximate the (GSA) if at all. In fact, we observe that the lattice bases computed in the course of Sampling Reduction

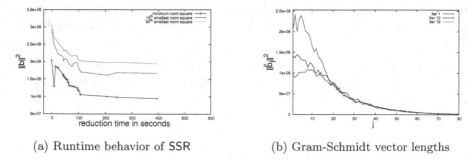

(a) Runtime behavior of SSR   (b) Gram-Schmidt vector lengths

Fig. 1. Typical reduction by SSR

approximate (GSA) worse and worse, even if the Gram-Schmidt vector lengths of the initial basis indeed resemble a geometric sequence. SSR, in contrast, relies on a user defined search space size, so it does not need to reference q_B.

Second, (GSA) guarantees that $\min_j \|\hat{\mathbf{b}}_j\|^2$ will eventually become larger than $(6/k)^{(n-1)/k}\|\mathbf{b}_1\|^2$, whereupon RSR terminates. Since we cannot rely on (GSA) in practice, SSR delegates the decision when to leave the main loop to CSSS. Unfortunately, without (GSA) we are no longer able to give an a priori bound on the length of $\mathbf{b}_1$ that is better than the postcondition of the LLL-type reduction used. We can only provide empirical data.

3.3   Typical Behavior of SSR

In the following, we give an example of typical SSR behavior.

Fig. 1 shows the results of SSR applied to a lattice basis in dimension $n = 180$ that were obtained as in Micciancio's variant of the GGH cryptosystem [6,7]. The original lattice basis O was uniformly chosen from $\{-180, \ldots, 180\}^{180 \times 180}$ and then transformed into Hermite normal form H = HNF(O). H was finally BKZ reduced with BKZ parameters $(\delta, \beta) = (0.99, 5)$.

We see in Fig. 1(a) the squared length of the nth shortest vector in the course of SSR. The intermittent LLL-type reduction was always BKZ with parameters $(\delta, \beta) = (0.99, 5)$. The squared length of the shortest vector in the SSR output is reduced by a factor $\approx 1/2$ compared to the shortest vector in the original BKZ-reduced basis. The remaining base vectors also became shorter but less so; only the first ten base vectors are particularly small. We therefore observe a much larger gap between the length of the shortest and the 15th shortest vector than between the 15th and the 30th shortest vector.

It stands out that most of the reduction happened in the first 15 iterations within 108 seconds. The following iterations did not significantly improve the minimum norm square but required much more sampled vectors (at a rate of ≈ 5200 samples per second on a 2.4 GHz Intel Pentium 4 machine).

Fig. 1(b) exhibits the (GSA) behavior of the basis generated by sampling. We see that the input basis B more or less approximates (GSA). However, inserting the short vectors found by sampling and re-reducing the generating system with

BKZ, $(\delta, \beta) = (0.99, 5)$ makes only the very first Gram-Schmidt vectors shorter. In the end, the graph of $\|\hat{\mathbf{b}}_j\|^2$ forms a distinct hump around $j = 10$. But SSR does not affect the lengths of the Gram-Schmidt vectors $\hat{\mathbf{b}}_j$ with, say, $j > 20$ at all. Put differently, the length relative to $\|\mathbf{b}_1\|$ of the vast majority of Gram-Schmidt vectors grows with every iteration, whence the probability to find an even shorter lattice vector by sampling drops.

4 Check Search Space Size

In this section we propose two non-trivial implementations of CSSS and discuss their trade-offs.

A trivial implementation of CSSS always returns true. This implies that $S_{u_{\max},\mathsf{B}}$ is always exhaustively searched, no matter how small the success probability; in particular, in the last SSR iteration the search space will always be enumerated once in vain. This may be acceptable for small $u_{\max}$, but it is not practical if the enumeration of $S_{u_{\max},\mathsf{B}}$ takes many hours, if not days. We propose in the following two CSSS implementations that estimate the success probability of the sampling loop and return true if and only if they deem the probability $\geq 1/2$. The correctness proofs of both algorithms require (RA), but the failure of (RA) in practice is likely to cause only a small error. Neither algorithm needs (GSA), though.

4.1 Event Based CSSS

We describe $\mathsf{CSSS_{event}}$, that is based on the approach of Schnorr's analysis of RSR. $\mathsf{CSSS_{event}}$ (Alg. 2) estimates the sampling loop's success probability by maximizing the probability of some probability event $(\mathcal{E}_{k,u,r})$ over all parameters that do not yield a too large conditional mean value for $\|\mathbf{v}\|^2$.

Definition 2. *Let* $\mathsf{B} = \hat{\mathsf{B}}\mathsf{R} \in \mathbb{Z}^{d \times n}$ *be a lattice basis,* $1 \leq u < n$, $1 < k < n - u$, *and* $r \in [0, 1]$. *Let* $\mathbf{v} = \sum_{j=1}^{n} \nu_j \hat{\mathbf{b}}_j = \mathsf{Sample}(\mathsf{B}, \mathsf{R}, x)$ *for random* $x \in_R \{0, \ldots, 2^u - 1\}$. *The permutation* $\sigma_{u,\mathsf{B}} \in \mathsf{Sym}(n)$ *sorts the first* $n - u - 1$ *Gram-Schmidt vectors by non-increasing length, i. e.,*

$$
\begin{aligned}
\|\hat{\mathbf{b}}_{\sigma_{u,\mathsf{B}}(j)}\|^2 &\geq \|\hat{\mathbf{b}}_{\sigma_{u,\mathsf{B}}(j+1)}\|^2 && \text{for } j \in \{1, \ldots, n - u - 2\}, \\
\sigma_{u,\mathsf{B}}(j) &= j && \text{for } j \in \{n - u, \ldots, n\}.
\end{aligned}
\tag{2}
$$

The event $(\mathcal{E}_{k,u,r})$ *is defined by*

$$
\nu^2_{\sigma_{u,\mathsf{B}}(j)} \leq \tfrac{1}{4} r^{k-j} \qquad \text{for all } j \in \{1, \ldots, k\}. \tag{$\mathcal{E}_{k,u,r}$}
$$

Lemma 1. *Assume (RA) and let* $\mathsf{B} = \hat{\mathsf{B}}\mathsf{R}$, k, u, *and* r *as in Def. 2. Set*

$$
s_j(k, u, r) = \begin{cases} \frac{1}{12} r^{k-j} & \text{if } 1 \leq j < k, \\ \frac{1}{12} & \text{if } k \leq j < n - u, \end{cases} \qquad s_j(k, u, r) = \begin{cases} \frac{1}{3} & \text{if } n - u \leq j < n, \\ 1 & \text{if } j = n. \end{cases}
$$

Algorithm 2. CSSS$_{\text{event}}$

Input: – $\mathfrak{b}$: vector of squared Gram-Schmidt vector lengths $\mathfrak{b} = (\|\hat{\mathfrak{b}}_1\|^2, \dots, \|\hat{\mathfrak{b}}_n\|^2)$
       where $\mathsf{B} = \hat{\mathsf{B}}\mathsf{R} \in \mathbb{Z}^{d \times n}$, $n > 3$
    – $u_{\max}$: $\log_2$ of maximum search space size
Output: $\exists k \in \{1, \dots, n - u_{\max} - 1\}, u \in \{1, \dots, u_{\max}\}, r \in [0,1]$:
     $E(\mathsf{B}; k, u, r) \le 0.99\|\mathbf{b}_1\|^2$ and $2^{u_{\max}} \ge 1/\left(\frac{1}{2}\Pr[(\mathcal{E}_{k,u,r})]\right)$

    procedure LogProbBound$(\ell, k, u, \|\mathbf{b}_1\|^2)$ /* $E(\mathsf{B}; k, u, r)$ as in Lemma 1 */
      if $E(\mathsf{B}; k, u, 1) \le 0.99\|\mathbf{b}_1\|^2$ **then return** -1
      if $E(\mathsf{B}; k, u, 0) \ge 0.99\|\mathbf{b}_1\|^2$ **then return** $-\infty$
      $r_{\max} \leftarrow$ Solve$(E(\mathsf{B}; k, u, q) = 0.99\|\mathbf{b}_1\|^2, r \in [0,1])$ /* e.g., RegulaFalsi */
      return $\lfloor \frac{k(k-1)}{4}\log_2(r_{max}) - 1\rfloor$
    end procedure

    procedure CSSS$_{\text{event}}(\mathfrak{b}, u_{\max})$
      for u **from** $u_{\max}$ **downto** 1 **do**
        $\ell \leftarrow (\|\hat{\mathbf{b}}_{\sigma_{u,\mathsf{B}}(1)}\|^2, \dots, \|\hat{\mathbf{b}}_{\sigma_{u,\mathsf{B}}(n)}\|^2)$ /* $\sigma_{u,\mathsf{B}}$ as in Def. 2 */
        if $u_{\max} \ge -\max_{1 \le k < n - u_{\max}}$ LogProbBound$(\ell, k, u, \|\mathbf{b}_1\|^2)$ **then return** true
      end for
      return false
    end procedure

Then we have $E(\mathsf{B}; k, u, r) := \sum_{j=1}^{n} s_j(k, u, r)\|\hat{\mathbf{b}}_{\sigma_{u,\mathsf{B}}(j)}\|^2 = E\left[\|\mathbf{v}\|^2 \mid (\mathcal{E}_{k,u,r})\right]$ as well as $\Pr[(\mathcal{E}_{k,u,r})] = r^{k(k-1)/4}$.
   If $E\left[\|\mathbf{v}\|^2 \mid (\mathcal{E}_{k,u,r})\right] \le 0.99\|\mathbf{b}_1\|^2$, then $\Pr[\|\mathbf{v}\|^2 \le 0.99\|\mathbf{b}_1\|^2] \ge \frac{1}{2}\Pr[(\mathcal{E}_{k,u,r})]$.

Since both $E(\mathsf{B}; k, u, r)$ and $\Pr[(\mathcal{E}_{k,u,r})]$ are continuous and strictly increasing in r, LogProbBound is able to determine for given k and u the maximum $r_{\max} \in [0,1]$ such that $E\left[\|\mathbf{v}\|^2 \mid (\mathcal{E}_{k,u,r})\right] \le 0.99\|\mathbf{b}_1\|^2$ by textbook root finding algorithms. Lemma 1 implies $\Pr[\|\mathbf{v}\|^2 \le \|\mathbf{b}_1\|^2] \ge \frac{1}{2}\Pr[(\mathcal{E}_{k,u,r_{\max}})]$. LogProbBound returns the logarithm of this bound, rounded towards $-\infty$.

   CSSS$_{\text{event}}$ checks whether there are admissible parameters k, u such that the search space size is larger than the reciprocal of the probability bound computed by LogProbBound. Then we can expect to find a sufficiently short vector in $S_{u,\mathsf{B}}$ by the usual probability enhancement argument.

   In practice, the success probability estimate by CSSS$_{\text{event}}$ is rather pessimistic. It takes only those short vectors into account that also satisfy $(\mathcal{E}_{k,u,r_{\max}})$. We found that the number of samples required is typically by a factor 2^{10} smaller than implied by the maximum result of LogProbBound.

4.2   Convolution Based CSSS

In the following, we explain CSSS$_{\text{Fourier}}$, that computes $\Pr[\|\mathbf{v}\|^2 \le \|\mathbf{b}_1\|^2]$ by means of the convolution of the distribution functions of the sampled vectors' Gram-Schmidt coefficients.

Algorithm 3. CSSS$_{\text{Fourier}}$

Input: − $\mathfrak{b}$: vector of squared Gram-Schmidt vector lengths $\mathfrak{b} = (\|\hat{\mathbf{b}}_1\|^2, \ldots, \|\hat{\mathbf{b}}_n\|^2)$
where $\mathsf{B} = \hat{\mathsf{B}}\mathsf{R} \in \mathbb{Z}^{d \times n}$, $n > 3$
− u: $\log_2$ of maximum search space size
Output: true if and only if $\Pr\left[\|\mathbf{v}\|^2 \leq 0.99\|\mathbf{b}_1\|^2\right] > 2^{1-u_{\max}}$

    procedure Density $((t_1,\ldots,t_m), T, N)$ /* $T > 2\sqrt{t_1^2 + \cdots + t_m^2}, N > 1$ */
      for j from 0 to $N/2$ **do**
        $(\Phi_j, k) \leftarrow (\varphi_{t_1,\ldots,t_m}(j/T), 0)$
        repeat
          $\delta_\Phi \leftarrow \varphi_{t_1,\ldots,t_m}((j+kN)/T) + \varphi_{t_1,\ldots,t_m}((j-kN)/T)$
          $(\Phi_j, k) \leftarrow (\Phi_j + \delta_\Phi, k+1)$
        until $\delta_\Phi < \varepsilon_{\text{FPA}}$ /* equality up to floating point precision */
      end for
      return $\frac{N}{4T}\text{DFT}^{-1}((\Phi_0,\ldots,\Phi_{N/2}, \overline{\Phi_{N/2-1}},\ldots,\overline{\Phi_1})^t)$
    end procedure

    procedure CSSS$_{\text{Fourier}}$ $(\mathfrak{b}, u)$
      $(t_1,\ldots,t_{n-1}) \leftarrow (\frac{1}{2}\|\hat{\mathbf{b}}_1\|, \ldots, \frac{1}{2}\|\hat{\mathbf{b}}_{n-u-1}\|, \|\hat{\mathbf{b}}_{n-u}\|, \ldots, \|\hat{\mathbf{b}}_{n-1}\|)/\|\hat{\mathbf{b}}_1\|$
      $N \leftarrow 2^D$ /* D global parameter, 2^D DFT sample points */
      $T \leftarrow \min\{2^\tau \mid \tau \in \mathbb{Z} \text{ and } 2^\tau > 2\sqrt{t_1^2 + \cdots + t_{n-1}^2}\}$
      $(d_0,\ldots,d_{N-1})^t \leftarrow \text{Density}((t_1,\ldots,t_{n-1}), T, N)$
      $p \leftarrow \text{Quadrature}(0.99 - \|\hat{\mathbf{b}}_n\|^2/\|\mathbf{b}_1\|^2, (d_0,\ldots,d_{N-1}), T, N)$
      return $\log_2(1-p) \leq -2^{-u}$
    end procedure

Definition 3. *The function*

$$\text{Fr} : \mathbb{R} \to \mathbb{C} : t \mapsto \int_0^t e^{2\pi i \frac{x^2}{4}}\, dx$$

is named Fresnel integral function. For positive real numbers $t_1,\ldots,t_n$, $n > 0$, define $\varphi_{t_1,\ldots,t_n} : \mathbb{R} \to \mathbb{C}$ as the unique continuous function such that, for $y > 0$,

$$\varphi_{t_1,\ldots,t_n}(y) = \prod_{j=1}^n \frac{\text{Fr}(2t_j\sqrt{y})}{2t_j\sqrt{y}} \quad and \quad \varphi_{t_1,\ldots,t_n}(-y) = \overline{\varphi_{t_1,\ldots,t_n}(y)}.$$

Theorem 1. *Let $\mathsf{B} = \hat{\mathsf{B}}\mathsf{R} \in \mathbb{Z}^{d\times n}$, $n > 2$, be a lattice basis and $1 \leq u < n$. Let $\mathbf{v} \in_R S_{u,\mathsf{B}}$ be uniformly sampled. For $1 \leq j < n$ set $t_j = \|\hat{\mathbf{b}}_j\|$ if $n-u \leq j < n$ and $t_j = \frac{1}{2}\|\hat{\mathbf{b}}_j\|$ else. Then, under (RA),*

$$\Pr\left[\|\mathbf{v}\|^2 \leq x\right] = \int_0^{x-\|\hat{\mathbf{b}}_n\|^2} (\mathcal{F}^{-1}\varphi_{t_1,\ldots,t_{n-1}})(s)\, ds, \tag{3}$$

where $\mathcal{F}^{-1}\varphi_{t_1,\ldots,t_{n-1}} : \mathbb{R} \to \mathbb{C} : x \mapsto \int_\mathbb{R} \varphi_{t_1,\ldots,t_{n-1}}(y)e^{-2\pi i x y}\, dy$ is the inverse Fourier transform of $\varphi_{t_1,\ldots,t_{n-1}}$.

Sketch of Proof. By (RA), the probability $\Pr[\,\nu_j^2\|\hat{\mathbf{b}}_j\|^2 \le x\,]$ is $\sqrt{x}/t_j$ if $0 < x < t_j^2$. Its distribution is $d_j(x) = 1/(2t_j\sqrt{x})$ if $0 < x < t_j^2$ and $d_j(x) = 0$ else. Since d_j is a real function, $\mathcal{F}d_j$ has the Hermitian property $\mathcal{F}d_j(-y) = \overline{\mathcal{F}d_j(y)}$. Assume $y > 0$. Then

$$(\mathcal{F}d_t)(y) = \int_{\mathbb{R}} d_t(z)e^{2\pi i yz}\,dz = \int_0^{t^2} \frac{1}{2t\sqrt{z}}e^{2\pi i yz}\,dz$$

$$= \int_0^{2\sqrt{y}t} \frac{2\sqrt{y}}{2tx}e^{2\pi i \frac{x^2}{4}}\frac{x}{2y}\,dx = \frac{1}{2t\sqrt{y}}\int_0^{2t\sqrt{y}} e^{2\pi i \frac{x^2}{4}}\,dx = \frac{\mathrm{Fr}(2t\sqrt{y})}{2t\sqrt{y}}$$

by means of the substitution $x(z) = 2\sqrt{y}\sqrt{z}$. Therefore, $\varphi_{t_j} = \mathcal{F}d_j$.

Let $D : \mathbb{R} \to \mathbb{R}$ be the distribution function of $\Pr[\,\|\mathbf{v}\|^2 \le x\,]$. Then D is the convolution product $d_1 * \cdots * d_{n-1}$ by the sum distribution formula and (RA). Therefore, (3) follows by the convolution theorem.   □

CSSS$_{\text{Fourier}}$ (Alg. 3) is a mostly straightforward numerical implementation of (3) by means of the inverse discrete Fourier transform DFT^{-1} and integration by some textbook quadrature formula (e. g., Simpson's rule). To counteract the so called aliasing effect [8], though, the subroutine **Density** determines the sample values of the DFT^{-1} input as $\Phi_j = \sum_{k \in \mathbb{Z}} \varphi_{t_1,\ldots,t_{n-1}}((j + kN)/T)$. This series converges sufficiently fast, since $|\varphi_{t_1,\ldots,t_{n-1}}| \in O(|y|^{-(n-1)/2})$ and we are interested in highly dimensional lattices only.

The Fresnel integral can be efficiently evaluated up to arbitrary precision [8]. There are freely available implementations that compute the (inverse) discrete Fourier transform in $O(N \log N)$ time and achieve an L_2 error as small as $O(\sqrt{\log N})$ [9]. Therefore, CSSS$_{\text{Fourier}}$ is sufficiently efficient.

CSSS$_{\text{Fourier}}$ determines the sample loop's success probability much more precisely than CSSS$_{\text{event}}$, but it is also much more expensive. Thus, CSSS$_{\text{Fourier}}$ offers an advantage over CSSS$_{\text{event}}$ if the search space is very large; then the sampling loop outweighs the runtime of CSSS$_{\text{Fourier}}$ and its higher precision is more important.

5   Generalizations

We described the behavior of SSR in Sect. 3.3. Better results require either that SSR enumerates larger search spaces or that the intermittent LLL-type reduction has to return shorter basis vectors – the latter typically means that BKZ is run with a larger block size. In this section we describe some Sampling Reduction variants that yield more short vectors and, in combination, a better SVP approximation than SSR alone.

5.1   Pool Sampling Reduction

SSR concentrates on reducing the length of the first base vector. Typically, it significantly reduces the length of the very first base vector only. The *Pool Sampling Reduction* (PoolSR) that we describe in the following generates more short vectors in the result basis.

Algorithm 4. PoolSR

Input: – generating system $G \in \mathbb{Z}^{d \times m}$ of lattice L, $\dim(L) = n$
        – search space bound $u_{max} \in \mathbb{N}$
        – pool size $s \geq 1$
        – (randomized) acceptor algorithm $\mathcal{A} : L(B) \rightarrow \{\text{true}, \text{false}\}$
Output: LLL reduced basis $B = \hat{B}R$ of L such that
        – $0.99\|\mathbf{b}_1\|^2 < \min\{\|\mathbf{v}\|^2 \mid \mathbf{v} \in S_{u_{max}, B}\}$ or
        – $CSSS(B, R, u_{max}) = \text{false}$.

  procedure PoolSR$(G, u_{max}, s, \mathcal{A})$
    $(B, \flat, R) \leftarrow$ LLL(G) /∗ $B = \hat{B}R$, $\flat = (\|\hat{\mathbf{b}}_1\|^2, \ldots, \|\hat{\mathbf{b}}_n\|^2)$ ∗/

    while CSSS$(\flat, u_{max}) = \text{true}$ **do**
      $t \leftarrow 0$ /∗ Invariant: $\|\mathbf{p}_1\| \leq \cdots \leq \|\mathbf{p}_t\|$, $t \leq s$ ∗/
      for x **from** 0 **to** $2^{u_{max}} - 1$ **do**
        $\mathbf{v} \leftarrow$ Sample(B, R, x)
        if $\|\mathbf{v}\|^2 \leq 0.99\|\mathbf{b}_1\|^2$ or $\mathcal{A}(\mathbf{v}) = \text{true}$ **then**
          $j_{\mathbf{v}} \leftarrow \min\{j \in \{1, \ldots, t + 1\} \mid j = t + 1$ or $\|\mathbf{v}\| \leq \|\mathbf{p}_j\|\}$
          $[\mathbf{p}_1, \ldots, \mathbf{p}_{t+1}] \leftarrow [\mathbf{p}_1, \ldots, \mathbf{p}_{j_{\mathbf{v}} - 1}, \mathbf{v}, \mathbf{p}_{j_{\mathbf{v}}}, \ldots, \mathbf{p}_t]$
          $t \leftarrow \min\{s, t + 1\}$
        end if
        if $t > 0$ and $\|\mathbf{p}_1\|^2 \leq 0.99\|\mathbf{b}_1\|^2$ **then** break
        if $x = 2^{u_{max}} - 1$ **then** terminate("no short vector")
      end for

      $(B, \flat, R) \leftarrow$ LLL$([\mathbf{p}_1, \ldots, \mathbf{p}_t, \mathbf{b}_1, \ldots, \mathbf{b}_n])$
    end while
    terminate("further progress unlikely")
  end procedure

Before the sampling loop in SSR finds a sufficiently short vector, it is likely to generate lattice points $\mathbf{v}$ that are about the same length as $\mathbf{b}_1$. SSR discards them even though they are shorter than most of the base vectors. PoolSR (Alg. 4) includes some of the shortest sampled vectors in the generating system the LLL reduction of which becomes the lattice basis in the next outer loop iteration.

Besides the input generating system and the maximum search space size, PoolSR expects a pool size $s \geq 1$ and some (possibly randomized) acceptor algorithm $\mathcal{A}$. We suggest a reasonable acceptor below. PoolSR maintains a "pool" $[\mathbf{p}_1, \ldots, \mathbf{p}_t]$, $t \leq s$, of sampled lattice vectors, sorted in non-decreasing length order. A sampled vector $\mathbf{v}$ is is added to the pool if its length is at most $\sqrt{0.99}\|\mathbf{b}_1\|$ or if the acceptor $\mathcal{A}$ returns true on input $\mathbf{v}$. The longest pool vector is dropped if otherwise the pool grew larger than s elements. PoolSR leaves the sampling loop as soon as the length of the shortest pool vector falls below $\sqrt{0.99}\|\mathbf{b}_1\|$. The generating system that is used to compute the next lattice basis B is formed by prepending the complete pool to the current basis B. (SSR prepends $\mathbf{p}_1$ only.)

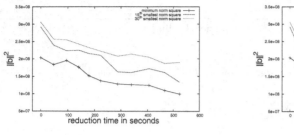
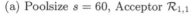

(a) Poolsize $s = 60$, Acceptor $\mathcal{R}_{1,1}$

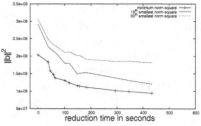

(b) Poolsize $s = 60$, Acceptor $\mathcal{R}_{1,25}$

Fig. 2. Typical runtime behavior of PoolSR

Vectors slightly longer than $\mathbf{b}_1$ are still shorter than most base vectors, so, for our experiments, we wanted to included them in the pool. However, the optimal threshold length is not obvious. To address this we used an acceptor $\mathcal{R}_{\tau,\alpha}$ that selects pool vector candidates $\mathbf{v}$ with a chance that decreases as $\|\mathbf{v}\|^2$ increases; more precisely, every invocation of $\mathcal{R}_{\tau,\alpha}$ is a random experiment such that $\Pr[\,\mathcal{R}_{\tau,\alpha}(\mathbf{v}) = \mathrm{true}\,] = \min\{\,1, e^{\alpha(\tau - \|\mathbf{v}\|^2/\|\mathbf{b}_1\|^2)}\,\}$ This ensures that all vectors shorter than $\tau\mathbf{b}_1$ are always selected for the pool and the probability for longer vectors to be included decreases quickly.

Fig. 2 shows the typical progress of PoolSR on the same input basis as in Sect. 3.3 for poolsize $s = 60$ and with acceptors $\mathcal{R}_{1,1}$ as well as $\mathcal{R}_{1,25}$. PoolSR does not significantly improve the SVP approximation compared to SSR. This is because the lengths of the Gram-Schmidt vectors $\hat{\mathbf{b}}_j$ do not differ much when comparing PoolSR with SSR. Therefore, The probability that the sampling loop will succeed in finding a significantly shorter vector is about the same in PoolSR and SSR. In contrast to SSR, however, we observe in the result basis a block of 15 to 30 vectors at the beginning that are about the same length as $\mathbf{b}_1$, independent from the value of α. After this block the remaining base vectors suddenly become longer; the gap in the squared length of the leading block vectors and the shortest remaining base vectors is typically 30 % to 50 % of $\|\mathbf{b}_1\|^2$.

If we run PoolSR with $\mathcal{R}_{1,1}$, then the pool grows to its maximum size $s = 60$ in almost all iterations, even if the inner loop is left after less than 2^{10} samples. The mean squared length of the pool vectors varies from $1.2\|\mathbf{b}_1\|^2$ to $1.6\|\mathbf{b}_1\|^2$, so the bulk of the pool vectors is not much shorter than the average base vector and does not contribute much towards our objective to generate bases with many short vectors. However, the reduction of that large generating systems is expensive; In the experiment shown in Fig. 2(a), only 43 seconds were spent in the sampling loop, but 481 seconds in the BKZ reduction.

The situation is different with large acceptor parameter $\alpha = 25$: Then the pool holds rarely more than 5 vectors and their average squared length is $1.1\|\mathbf{b}_1\|^2$. In consequence, PoolSR spends much less time in the intermittent BKZ reduction and the overall runtime is not much longer than the runtime of SSR. (149 seconds spend in BKZ in the example depicted in Fig. 2(b), compared with 285 seconds

---

Algorithm 5. ShortProjectionSR

Input: – generating system $G \in \mathbb{Z}^{d \times m}$ of lattice L, $\dim(L) = n$
 – search space bound $u_{\max} \in \mathbb{N}$
 – reduction factor $\gamma \in (0,1)$
 – set T of target indices, $\emptyset \neq T \subseteq \{1, \ldots, n - u_{\max}\}$
Output: LLL reduced basis $B = \hat{B}R$ of L such that $\forall j \in T$:
 – $\gamma \|\mathbf{b}_j\|^2 < \min\{\pi_j(\|\mathbf{v}\|^2) \mid \mathbf{v} \in S_{u_{\max},B}\}$ or
 – $\mathsf{CSSS}((\|\hat{\mathbf{b}}_j\|^2, \ldots, \|\hat{\mathbf{b}}_n\|^2), u_{\max}) = \text{false}$.

  procedure ShortProjectionSR($B, u_{\max}, \gamma, T, \delta$)
    $(B, \flat, R) \leftarrow \mathsf{LLL}(G)$ /* $B = \hat{B}R$, $\flat = (\|\hat{\mathbf{b}}_1\|^2, \ldots, \|\hat{\mathbf{b}}_n\|^2)$ */

    while $\mathsf{CSSS}'(\flat, \gamma, T, u_{\max}) = \text{true}$ **do**

        for x from 0 to $2^{u_{\max}} - 1$ **do**
          $\mathbf{v} \leftarrow$ Sample(B, R, x)
          $t \leftarrow \min\{j \in T \mid \|\pi_j(\mathbf{v})\|^2 \leq \gamma \|\hat{\mathbf{b}}_j\|^2\} \cup \{\infty\}$
          if $t \in T$ **then** break
          if $x = 2^{u_{\max}} - 1$ **then** terminate("no short vector")
        end for

        $(B, \flat, R) \leftarrow \mathsf{LLL}([\mathbf{b}_1, \ldots, \mathbf{b}_{t-1}, \mathbf{v}, \mathbf{b}_t, \ldots, \mathbf{b}_n])$
    end while

    terminate("further progress unlikely")
  end procedure

---

spent on sampling.) The base vector lengths in the end result is similar to the experiments with $\alpha = 1$, though.

5.2 Short Projection Sampling Reduction

BKZ reduction of lattice bases that arise from the NTRU cryptosystem [10] yields bases where $\|\mathbf{b}_1\|^2 \ll \max_j \|\hat{\mathbf{b}}_j\|^2$. It is therefore very unlikely that sampling will find a vector shorter than $\mathbf{b}_1$ unless the search space is unfeasibly huge. Even if we reduce some lattice basis that initially approximates (GSA) reasonably well, the intermittent LLL-type reduction causes in practice a "bump" in the sequence $(\|\hat{\mathbf{b}}_j\|^2)_{j=1,\ldots,n}$ that brings the Sampling Reduction soon to a halt. *Short Projection Sampling Reduction* (ShortProjectionSR) addresses both problems by inserting sampled vectors in between the base vectors rather than to prepend them.

ShortProjectionSR (Alg. 5) takes two parameters more than SSR, the reduction factor γ and the set T of target indices. T specifies at which positions a sample vector may be inserted into the lattice basis to form the new generating system. The reduction factor determines when a sample vector projection is considered short enough; ShortProjectionSR leaves the sampling loop if there is an index $j \in T$ such that the sample vector $\mathbf{v}$ satisfies $\|\pi_j(\mathbf{v})\|^2 \leq \gamma \|\hat{\mathbf{b}}_j\|$. The minimum index for which this condition holds is stored in the variable t. Then $\mathbf{v}$ is inserted in B at column t and this generating system is reduced again with an LLL-type reduction as in SSR.

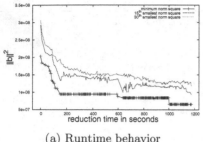

stage	algorithm	parameters	u_{max}
1	PoolSR	$\gamma = 0.99, s = 60, \mathcal{A} = \mathcal{R}_{1,25}$	25
2	ShortProjectionSR	$\gamma = 0.75, T = \{1, 21, \ldots, 25\}$	24
3	ShortProjectionSR	$\gamma = 0.75, T = \{1, 41, \ldots, 45\}$	24
4	ShortProjectionSR	$\gamma = 0.75, T = \{1, 61, \ldots, 65\}$	24
5	PoolSR	$\gamma = 0.99, s = 60, \mathcal{A} = \mathcal{R}_{1,25}$	25
6	ShortProjectionSR	$\gamma = 0.85, T = \{1, 21, \ldots, 25\}$	24
7	ShortProjectionSR	$\gamma = 0.85, T = \{1, 41, \ldots, 45\}$	24
8	ShortProjectionSR	$\gamma = 0.85, T = \{1, 61, \ldots, 65\}$	24
9	ShortProjectionSR	$\gamma = 0.99, T = \{1, \ldots, 5\}$	24

(a) Runtime behavior   (b) Algorithms and Parameters used

Fig. 3. Behavior of combined PoolSR and ShortProjectionSR

The CSSS implementations discussed in Sect. 4 assume $\gamma = 0.99$ and $T = \{1\}$. It is very easy to modify them into implementations CSSS' that take γ and T into account as well.

ShortProjectionSR can be used to tackle the peak in the Gram-Schmidt vector lengths of, e. g., NTRU bases. Here ShortProjectionSR is used to make Sampling Reduction ignore the initial Gram-Schmidt vectors that impede the success of the sampling loop. But also if the reduction of a basis that initially approximated (GSA) well stops due to the mentioned bump in $(\|\hat{\mathbf{b}}_j\|^2)$, ShortProjectionSR can be used to reduce this bump and therefore make Sampling Reduction proceed. Thus ShortProjectionSR is best used in turns with PoolSR. For optimal results, these turns and the ShortProjectionSR parameters need to be interactively controlled.

Fig. 3 exhibits the result of PoolSR and ShortProjectionSR applied to the example used before. Subtable (b) shows the parameters used in this experiment; with $(t_j, N_j) = (500, 75)$ if $j \in \{1, 5\}$ and $(t_j, N_j) = (150, 60)$ else, stage j ran until t_j seconds were spent in the current stage, the outer reduction loop was iterated N_j times, or $\text{CSSS}'_{\text{event}}$ returned false, whatever happened first.

In the end, $\|\mathbf{b}_1\|^2$ was reduced to 6.6×10^7, i. e., a third of its original value. Even more notably, the combination of PoolSR and ShortProjectionSR improved the basis globally such that the average squared base vector length was finally only 1.5×10^8, a reduction by 25 % to 40 % compared with the results of the previous algorithms. We can thus achieve approximately the same length reduction for all base vectors, not only for the very first ones as with SSR. We pay for this improvement with a runtime factor 2 to 3, compared with PoolSR.

5.3   Further Generalizations

Generalized Search Space. The Sample algorithm computes lattice vectors $\mathbf{b} = \sum_j \nu_j \hat{\mathbf{b}}_j$ such that $\nu_j^2 > 1/4$ only if $n - u \leq j$. This is motivated by (GSA), that implies $\hat{\mathbf{b}}_{n-u}, \ldots, \hat{\mathbf{b}}_n$ are the shortest Gram-Schmidt vectors of the LLL reduced basis B. In practice, that is not always the case. If we choose the search space better adapted to the actual Gram-Schmidt vectors than we can expect that the sampled vectors will be shorter on average.

Definition 4. *Let* $\mathsf{B} \in \mathbb{Z}^{d \times n}$ *be a lattice basis with Gram-Schmidt decomposition* $\mathsf{B} = \hat{\mathsf{B}}\mathsf{R}$. *Let* $\mathbf{c} = (c_1, \ldots, c_n) \in \mathbb{N}^n$ *and assume there is* $j_0 \in \{1, \ldots, n\}$ *such that* $c_j = 0$ *if and only if* $j > j_0$. *Then the generalized search space* $V_{\mathbf{c},\mathsf{B}}$ *is the set of all lattice vectors* $\mathbf{v} = \sum_{j=1}^{n} \nu_j \hat{\mathbf{b}}_j \in L(\mathsf{B})$ *subject to* $\nu_j \in (-c_j/2, c_j/2]$ *if* $1 \leq j < j_0$, $\nu_j \in \{1, \ldots, c_j\}$ *if* $j = j_0$, *and* $\nu_j = 0$ *else.*

The cardinality of $V_{\mathbf{c},\mathsf{B}}$ is $N = \prod_{j=1}^{j_0} c_j$. We have $S_{u,\mathsf{B}} = V_{\mathbf{c},\mathsf{B}}$ if and only if $\mathbf{c} = (1, \ldots, 1, 2, \ldots, 2, 1)$ with exactly u entries equal 2. The necessary modifications of the Sample algorithm and the CSSS implementations are covered in [3] in more detail.

Distributed Sampling. Sampling Reduction does not modify the basis during the sampling phase at all. It runs the same algorithm Sample again and again for only slightly different input data. This is an algorithm pattern for which SIMD-parallelization (single instruction, multiple data) excels. To implement such a distributed Sampling Reduction, we partition the search space in disjunct parts $V_{\mathbf{c},\mathsf{B}} = \bigcup_{j=1}^{N} V_j$ where N is typically a small multiple of the number of available computing nodes. Then we schedule the search space parts on the computing nodes; for this we need to transfer only once the basis and the specification of the V_j assigned. As soon as a node has exhausted its section of the search space, the next available V_j is scheduled on this node. This continues until either a node returns a sufficiently short vector – in which case all other nodes are notified and abort their task – or all V_j were searched without success.

If N is large enough to keep all nodes busy but the search space parts are not too small to avoid unnecessary communication overhead between the nodes and the scheduler, then software packages for distributed computations in, say, department LANs yield a near optimal efficiency, i. e., the attainable speedup scales almost linearly in the number of nodes. For instance, our primary test hardware computes 2^{16} samples in dimension 180 in about 10 to 15 seconds. So if each search packet covers at least 2^{16} vectors, then we expect that a distributed implementation results in a very high efficiency.

6   Cryptographic Applications

We tried SSR and its generalizations on some classes of lattice bases that stem from cryptographic applications, namely public keys in Micciancio's GGH variant [7], NTRU lattices [10], and Knapsack lattices [11]. For this experiments we developed a C++ framework for lattice reduction and (re-)implemented both the Schnorr-Euchner variant of LLL and BKZ based on the implementation found in Shoups NTL [12] as well as the proposed algorithms. An interface to the scripting language Python made it possible to evaluate the reduction results and reset arbitrary parameters between iterations. We will make this framework and its Python binding available on our homepage.

In our experiments we consistently found that we could break many instances with a significantly smaller block size of the intermittent BKZ reduction compared with BKZ only attacks.

The Sampling Reduction attacks were in some cases (mostly GGH and Knapsack lattices) up to three times faster, in other cases the overall runtime of Sampling Reduction was a small multiple of the runtime of BKZ attacks with larger block size. For example, a knapsack instance with 100 weights of maximum bitlength 102 (i. e., density 0.9804) and a solution vector of Hamming weight 10 was solved by ShortProjectionSR in 1004 seconds. The intermittent BKZ reduction was done with block size $\beta = 5$. In contrast, BKZ alone needed 2500 seconds with a block size of $\beta = 30$ to solve the same instance.

Sampling Reduction was in many, but not in every experiment faster than BKZ alone. However, most of the time was spent on sampling; the accumulated time spent in the BKZ algorithm when called from Sampling Reduction was always significantly shorter than the runtime of the BKZ only attacks. (This was also true for the GGH and NTRU experiments.) Since the sampling loop can efficiently be distributed on cheap hardware – which is not the case for LLL-type reduction – we expect that an attacker can benefit from Sampling Reduction. [3] describes these experiments and their results in detail.

7   Conclusion

We demonstrated that Sampling Reduction can be made practical and proposed several Sampling Reduction variants. We proposed alternative methods to estimate the success probability of further sampling iterations; on of these methods is based on a formula for the probability function of $\|\mathbf{v}\|^2$ introduced in this paper. Experiments have shown that lattice based attacks on cryptosystems can succeed in less time than if run with BKZ alone, in particular if the attacker takes advantage of Sampling Reduction's potential for distributed computation.

References

1. Lenstra, A.K., Lenstra, H.W., Lovász, L.: Factoring polynomials with rational coefficients. Math. Ann. **261** (1982) 515–534
2. Schnorr, C.P.: Lattice reduction by random sampling and birthday methods. In Alt, H., Habib, M., eds.: STACS 2003: 20th Annual Symposium on Theoretical Aspects of Computer Science. Volume 2607 of LNCS., Springer (2003) 146–156
3. Ludwig, C.: Practical Lattice Basis Sampling Reduction. PhD thesis, TU Darmstadt (2005) Available at http://elib.tu-darmstadt.de/diss/000640/.
4. Schnorr, C.P., Euchner, M.: Lattice basis reduction: Improved practical algorithms and solving subset sum problems. Math. Programming **66** (1994) 181–199
5. Nguyen, P.Q., Stehlé, D.: Floating-point LLL revisited. In Cramer, R., ed.: Proceedings of Eurocrypt'05, Springer (2005) Extended Abstract available at ftp://ftp.di.ens.fr/pub/users/pnguyen/EC05.pdf.
6. Goldreich, O., Goldwasser, S., Halevi, S.: Public-key cryptosystems from lattice reduction problems. In Kaliski, Jr., B.S., ed.: Advances in Cryptology – Crypto'97. Volume 1294 of LNCS., Springer-Verlag (1997) 112–131

7. Micciancio, D.: Improving lattice based cryptosystems using the Hermite normal form. In Silverman, J.H., ed.: Cryptography and Lattices. Volume 2146 of LNCS., Springer-Verlag (2001) 126–145
8. Press, W.H., Teukolsky, S.A., Vetterling, W.T., Flannery, B.P.: Numerical Recipes in C. 2nd edn. Cambridge University Press (1992)
9. Frigo, M., Johnson, S.G.: The design and implementation of FFTW3. Proceedings of the IEEE **93** (2005) 216–231 special issue on "Program Generation, Optimization, and Platform Adaptation".
10. Hoffstein, J., Pipher, J., Silverman, J.H.: NTRU: A ring-based public key cryptosystem. In Buhler, J.P., ed.: Algorithmic Number Theory (ANTS III). Volume 1423 of LNCS., Springer-Verlag (1998)
11. Coster, M.J., Joux, A., LaMacchia, B.A., Odlyzko, A.M., Schnorr, C.P., Stern, J.: Improved low-density subset sum algorithms. Comput. Complexity **2** (1992) 111–128
12. Shoup, V.: NTL – a library for doing number theory (2005) Release 5.4.

LLL on the Average

Phong Q. Nguyen[1,*] and Damien Stehlé[2]

[1] CNRS & École normale supérieure, DI, 45 rue d'Ulm, 75005 Paris, France
http://www.di.ens.fr/~pnguyen
[2] LORIA / Université Nancy 1
615 rue du J. Botanique, 54602 Villers-lès-Nancy, France
stehle@maths.usyd.edu.au
http://www.loria.fr/~stehle

Abstract. Despite their popularity, lattice reduction algorithms remain mysterious in many ways. It has been widely reported that they behave much more nicely than what was expected from the worst-case proved bounds, both in terms of the running time and the output quality. In this article, we investigate this puzzling statement by trying to model the average case of lattice reduction algorithms, starting with the celebrated Lenstra-Lenstra-Lovász algorithm (L^3). We discuss what is meant by lattice reduction on the average, and we present extensive experiments on the average case behavior of L^3, in order to give a clearer picture of the differences/similarities between the average and worst cases. Our work is intended to clarify the practical behavior of L^3 and to raise theoretical questions on its average behavior.

1 Introduction

Lattices are discrete subgroups of $\mathbb{R}^n$. A basis of a lattice L is a set of $d \leq n$ linearly independent vectors $\mathbf{b}_1, \ldots, \mathbf{b}_d$ in $\mathbb{R}^n$ such that L is the set $L[\mathbf{b}_1, \ldots, \mathbf{b}_d] = \left\{ \sum_{i=1}^{d} x_i \mathbf{b}_i, x_i \in \mathbb{Z} \right\}$ of all integer linear combinations of the $\mathbf{b}_i$'s. The integer d matches the dimension of the linear span of L: it is called the dimension of the lattice L. A lattice has infinitely many bases (except in trivial dimension ≤ 1), but some are more useful than others. The goal of *lattice reduction* is to find interesting lattice bases, such as bases consisting of reasonably short and almost orthogonal vectors. Finding good reduced bases has proved invaluable in many fields of computer science and mathematics (see [9,14]), particularly in cryptology (see [22,24]).

The first lattice reduction algorithm in arbitrary dimension is due to Hermite [15]. It was introduced to show the existence of Hermite's constant and of lattice bases with bounded orthogonality defect. Very little is known on the complexity of Hermite's algorithm: the algorithm terminates, but its polynomial-time

---

* The work described in this article has in part been supported by the Commission of the European Communities through the IST program under contract IST-2002-507932 ECRYPT.

complexity remains an open question. The subject had a revival with Lenstra's celebrated work on integer programming [19,20], which used an approximate variant of Hermite's algorithm. Lenstra's variant was only polynomial-time for fixed dimension, which was however sufficient in [19]. This inspired Lovász to develop a polynomial-time variant of the algorithm, which reached a final form in [18] where Lenstra, Lenstra and Lovász applied it to factor rational polynomials in polynomial time, from whom the name L^3 comes. Further refinements of L^3 were later proposed, notably by Schnorr [27,28]. Currently, the most efficient provable variant of L^3 known in case of large entries, called L^2, is due to Nguyen and Stehlé [23], and is based on floating-point arithmetic. Like L^3, it can be viewed as a relaxed version of Hermite's algorithm.

OUR CONTRIBUTION. One of the main reasons why lattice reduction has proved invaluable in many fields is the widely reported experimental fact that lattice reduction algorithms, L^3 in particular, behave much more nicely than what could be expected from the worst-case proved bounds, both in terms of the running time and the output quality. However, to our knowledge, this mysterious phenomenon has never been described in much detail. In this article, we try to give a clearer picture and to give heuristic arguments that explain the situation. We start by discussing what is meant by the average case of lattice reduction, which is related to notions of random lattices and random bases. We then focus on L^3. Regarding the output quality, it seems as if the only difference between the average and worst cases of L^3 in high dimension is a change of constants: while the worst-case behavior of L^3 is closely related to Hermite's constant in dimension two $\gamma_2 = \sqrt{4/3}$, the average case involves a smaller constant whose value is only known experimentally: ≈ 1.04. So while L^3 behaves better than expected, it does not behave that much better: the approximation factors seem to remain exponential in d. Regarding the running time, there is no surprise for the so-called integer version of L^3, except when the input lattice has a special shape such as knapsack-type lattices. However, there can be significant changes with the floating-point variants of L^3. We give a family of bases for which the average running time should be asymptotically close to the worst-case bound, and explain why for reasonable input sizes the executions are faster.

APPLICATIONS. Guessing the quality of the bases output by L^3 is very important for several reasons. First, all lattice reduction algorithms known rely on L^3 at some stage and their behavior is therefore strongly related to that of L^3. A better understanding of their behavior should provide a better understanding of stronger reduction algorithms such as Schnorr's BKZ [27] and is thus useful to estimate the hardness of lattice problems (which is used in several public-key cryptosystems, such as NTRU [16] and GGH [11]). Besides, if after running L^3, one obtains a basis which is worse than expected, then one should randomize the basis and run L^3 again. Another application comes from the so-called floating-point (*fp* for short) versions of L^3. These are very popular in practice because they are usually much faster. They can however prove tricky to use because they require tuning: if the precision used in *fp*-arithmetic is not chosen carefully, the

algorithm may no longer terminate, and if it terminates, it may not give an
L^3-reduced basis. On the other hand, the higher the precision, the slower the
execution. Choosing the right precision for *fp*-arithmetic is thus important in
practice and it turns out to be closely related to the average-case quality of the
bases output by the L^3 algorithm.

The table below sums up our results, for d-dimensional lattices whose initial
basis vectors are of lengths smaller than B, with $n = \Theta(d)$ and $d = O(\log B)$.

	$\frac{\|\mathbf{b}_1\|}{(\det L)^{1/d}}$	Running time of L^2	Required prec. for L^2
Worst-case bound	$(4/3)^{d/4}$	$O(d^5 \log^2 B)$	$\approx 1.58d + o(d)$
Average-case estim.	$(1.02)^d$	$O(d^4 \log^2 B) \to O(d^5 \log^2 B)$	$0.18d + o(d)$

ROAD MAP. In Section 2 we provide necessary background on L^3. We discuss
random lattices and random bases in Section 3. Then we describe our experimental observations on the quality of the computed bases (Section 4), the running
time (Section 5) and the numerical behavior (Section 6).

ADDITIONAL MATERIAL. All experiments were performed with `fplll-1.2`, available at `http://www.loria.fr/~stehle/practLLL.html`. The data used to
draw the figures of the paper and some others are also available at this URL.

2   Background

NOTATION. All logarithms are in base 2. Let $\| \cdot \|$ and $\langle \cdot, \cdot \rangle$ be the Euclidean
norm and inner product of $\mathbb{R}^n$. The notation $\lceil x \rfloor$ denotes a closest integer to x.
Bold variables are vectors. All the lattices we consider are integer lattices, as
usual. All our complexity results are given for the bit complexity model, without
fast integer arithmetic. Our *fpa*-model is a smooth extension of the IEEE-754
standard, as provided by NTL [30] (RR class) and MPFR [26].

We recall basic notions from algorithmic geometry of numbers (see [22]).

First Minimum. If L is a lattice, we denote by $\lambda(L)$ its *first minimum*.

Gram Matrix. Let $\mathbf{b}_1, \ldots, \mathbf{b}_d$ be vectors. Their *Gram matrix* $G(\mathbf{b}_1, \ldots, \mathbf{b}_d)$ is
the $d \times d$ symmetric matrix $(\langle \mathbf{b}_i, \mathbf{b}_j \rangle)_{1 \le i,j \le d}$ formed by all the inner products.

Gram-Schmidt Orthogonalization. Let $\mathbf{b}_1, \ldots, \mathbf{b}_d$ be linearly independent
vectors. The *Gram-Schmidt orthogonalization* (GSO) $[\mathbf{b}_1^*, \ldots, \mathbf{b}_d^*]$ is the orthogonal family defined as follows: $\mathbf{b}_i^*$ is the component of $\mathbf{b}_i$ orthogonal to the
linear span of $\mathbf{b}_1, \ldots, \mathbf{b}_{i-1}$. We have $\mathbf{b}_i^* = \mathbf{b}_i - \sum_{j=1}^{i-1} \mu_{i,j} \mathbf{b}_j^*$ where $\mu_{i,j} = \langle \mathbf{b}_i, \mathbf{b}_j^* \rangle / \|\mathbf{b}_j^*\|^2$. For $i \le d$, we let $\mu_{i,i} = 1$. The lattice L spanned by the $\mathbf{b}_i$'s
satisfies $\det L = \prod_{i=1}^{d} \|\mathbf{b}_i^*\|$. The GSO family depends on the order of the vec-
tors. If the $\mathbf{b}_i$'s are integer vectors, the $\mathbf{b}_i^*$'s and the $\mu_{i,j}$'s are rational. In what
follows, the *GSO family* denotes the $\mu_{i,j}$'s, together with the quantities $r_{i,j}$'s
defined as: $r_{i,i} = \|\mathbf{b}_i^*\|^2$ and $r_{i,j} = \mu_{i,j} r_{j,j}$ for $j < i$.

Size-Reduction. A basis $[\mathbf{b}_1, \dots, \mathbf{b}_d]$ is *size-reduced* with factor $\eta \geq 1/2$ if its GSO family satisfies $|\mu_{i,j}| \leq \eta$ for all $j < i$. The i-th vector $\mathbf{b}_i$ is *size-reduced* if $|\mu_{i,j}| \leq \eta$ for all $j < i$. Size-reduction usually refers to $\eta = 1/2$, but it is essential for *fp* variants of L^3 to allow larger η.

L^3-Reduction. A basis $[\mathbf{b}_1, \dots, \mathbf{b}_d]$ is L^3-reduced with factor (δ, η) with $1/4 < \delta \leq 1$ and $1/2 \leq \eta < \sqrt{\delta}$ if the basis is η-size-reduced and if its GSO satisfies the $(d-1)$ Lovász conditions $(\delta - \mu_{\kappa,\kappa-1}^2)r_{\kappa-1,\kappa-1} \leq r_{\kappa,\kappa}$ (or equivalently $\delta\|\mathbf{b}_{\kappa-1}^*\|^2 \leq \|\mathbf{b}_\kappa^* + \mu_{\kappa,\kappa-1}\mathbf{b}_{\kappa-1}^*\|^2$), which implies that the $\|\mathbf{b}_\kappa^*\|$'s never drop too much. Such bases have useful properties (see [18]), like providing approximations to the shortest and closest vector problems. In particular, the first vector is relatively short: $\|\mathbf{b}_1\| \leq \beta^{(d-1)/4}(\det L)^{1/d}$, where $\beta = 1/(\delta - \eta^2)$. And the first basis vector is at most exponentially far away from the first minimum: $\|\mathbf{b}_1\| \leq \beta^{(d-1)/2}\lambda(L)$. L^3-reduction usually refers to the factor $(3/4, 1/2)$ initially chosen in [18], in which case $\beta = 2$. But the closer (δ, η) is to $(1, 1/2)$, the shorter $\mathbf{b}_1$ should be. In practice, one usually selects $\delta \approx 1$ and $\eta \approx 1/2$, so that $\beta \approx 4/3$ and therefore $\|\mathbf{b}_1\| \lesssim (4/3)^{(d-1)/4}(\det L)^{1/d}$. The L^3 algorithm obtains in polynomial time a $(\delta, 1/2)$-L^3-reduced basis where $\delta < 1$ can be chosen arbitrarily close to 1. The L^2 algorithm achieves a factor (δ, η), where $\delta < 1$ can be arbitrarily close to 1 and $\eta > 1/2$ arbitrarily close to $1/2$.

The L^3 Algorithm. The L^3 algorithm [18] is described in Figure 1. It computes an L^3-reduced basis in an iterative fashion: the index κ is such that at any stage of the algorithm, the truncated basis $[\mathbf{b}_1, \dots, \mathbf{b}_{\kappa-1}]$ is L^3-reduced. At each loop

---

Input: A basis $[\mathbf{b}_1, \dots, \mathbf{b}_d]$ and $\delta \in (1/4, 1)$.
Output: An L^3-reduced basis with factor $(\delta, 1/2)$.
1. Compute the rational GSO, i.e., all the $\mu_{i,j}$'s and $r_{i,i}$'s.
2. $\kappa{:=}2$. While $\kappa \leq d$ do
3.   Size-reduce $\mathbf{b}_\kappa$ using the algorithm of Figure 2, that updates the GSO.
4. $\kappa'{:=}\kappa$. While $\kappa \geq 2$ and $\delta r_{\kappa-1,\kappa-1} \geq r_{\kappa',\kappa'} + \sum_{i=\kappa-1}^{\kappa'-1}\mu_{\kappa',i}^2 r_{i,i}$, do $\kappa{:=}\kappa - 1$.
5.   For $i = 1$ to $\kappa - 1$, $\mu_{\kappa,i}{:=}\mu_{\kappa',i}$. Insert $\mathbf{b}_{\kappa'}$ right before $\mathbf{b}_\kappa$.
6. $\kappa{:=}\kappa + 1$.
7. Output $[\mathbf{b}_1, \dots, \mathbf{b}_d]$.

Fig. 1. The L^3 Algorithm

Input: A basis $[\mathbf{b}_1, \dots, \mathbf{b}_d]$, its GSO and an index κ.
Output: The basis with $\mathbf{b}_\kappa$ size-reduced and the updated GSO.
1. For $i = \kappa - 1$ down to 1 do
2. $\mathbf{b}_\kappa{:=}\mathbf{b}_\kappa - \lceil\mu_{\kappa,i}\rfloor\mathbf{b}_i$.
3.   For $j = 1$ to i do $\mu_{\kappa,j}{:=}\mu_{\kappa,j} - \lceil\mu_{\kappa,i}\rfloor\mu_{i,j}$.
4. Update the GSO accordingly.

Fig. 2. The size-reduction algorithm

242   P.Q. Nguyen and D. Stehlé

iteration, κ is either incremented or decremented: the loop stops when κ reaches the value $d+1$, in which case the entire basis $[\mathbf{b}_1, \ldots, \mathbf{b}_d]$ is L^3-reduced.

If L^3 terminates, it is clear that the output basis is L^3-reduced. What is less clear *a priori* is why L^3 has a polynomial-time complexity. A standard argument shows that each swap decreases the quantity $\Delta = \prod_{i=1}^{d} \|\mathbf{b}_i^*\|^{2(d-i+1)}$ by at least a factor $\delta < 1$, while $\Delta \geq 1$ because the $\mathbf{b}_i$'s are integer vectors and Δ can be viewed as a product of squared volumes of lattices spanned by some of the $\mathbf{b}_i$'s. This proves that there are $O(d^2 \log B)$ swaps, and therefore loop iterations, where B is an upper bound on the norms of the input basis vectors. It remains to estimate the cost of each loop iteration. This cost turns out to be dominated by $O(dn)$ arithmetic operations on the basis matrix and GSO coefficients $\mu_{i,j}$ and $r_{i,i}$ which are rational numbers of bit-length $O(d \log B)$. Thus, the overall complexity of L^3 is $O((d^2 \log B) \cdot dn \cdot (d \log B)^2) = O(d^5 n \log^3 B)$.

L^3 with *fpa*. The cost of L^3 is dominated by the operations on the GSO coefficients which are rationals with huge numerators and denominators. It is therefore tempting to replace the exact GSO coefficients by *fp* approximations. But doing so in a straightforward manner leads to numerical anomalies. The algorithm is no longer guaranteed to be polynomial-time: it may not even terminate. And if ever it terminates, the output basis may not be L^3-reduced. The main number theory computer packages [7,21,30] contain heuristic *fp*-variants of L^3 *à la* Schnorr-Euchner [29] suffering from stability problems. On the theoretic side, the fastest provable *fp* variant of L^3 is Nguyen-Stehlé's L^2 [23], whose running time is $O(d^4 n(d + \log B) \log B)$. The main algorithmic differences with Schnorr-Euchner's *fp* L^3 are that the integer Gram matrix is updated during the execution (thus avoiding cancellations while computing scalar products with *fpa*), and that the size-reduction algorithm is replaced by a lazy variant (this idea was already in Victor Shoup's NTL code). In L^2, the worst-case required precision for *fpa* is $\leq 1.59d + o(d)$. The proved variant of `fplll-1.2` implements L^2.

3 Random Lattices

In this section, we give the main methods known to generate random lattices and random bases, and describe the random bases we use in our experiments.

3.1 Random Lattices

When experimenting with L^3, it seems natural to work with random lattices, but what is a random lattice? From a practical point of view, one could just select randomly generated lattices of interest, such as lattices used in cryptography or in algorithmic number theory. This would already be useful but one might argue that it would be insufficient to draw conclusions, because such lattices may not be considered random in a mathematical sense. For instance, in many cryptanalyses, one applies reduction algorithms to lattices whose first minimum is much shorter than all the other minima.

From a mathematical point of view, there is a natural notion of random lattice, which follows from a measure on n-dimensional lattices with determinant 1 introduced by Siegel [31] back in 1945, to provide an alternative proof of the Minkowski-Hlwaka theorem. Let $X_n = SL_n(\mathbb{R})/SL_n(\mathbb{Z})$ be the space of (full-rank) lattices in $\mathbb{R}^n$ modulo scale. The group $G = SL_n(\mathbb{R})$ possesses a unique (up to scale) bi-invariant Haar measure, which can be thought of as the measure it inherits as a hypersurface in $\mathbb{R}^{n^2}$. When mapping G to the quotient $X_n = G/SL_n(\mathbb{Z})$, the Haar measure projects to a finite measure μ on the space X_n which we can normalize to have total volume 1. This measure μ is G-invariant: if $A \subseteq X_n$ is measurable and $g \in G$, then $\mu(A) = \mu(gA)$. In fact, μ can be characterized as the unique G invariant Borel probability measure on X_n. This gives rise to a natural notion of random lattices. The recent articles [2,4,12] propose efficient ways to generate lattices which are random in this sense. For instance, Goldstein and Mayer [12] show that for large N, the (finite) set $\mathcal{L}_{n,N}$ of n-dimensional integer lattices of determinant N is uniformly distributed in X_n in the following sense: given any measurable subset $A \subseteq X_n$ whose boundary has zero measure with respect to μ, the fraction of lattices in $\mathcal{L}_{n,N}/N^{1/n}$ that lie in A tends to $\mu(A)$ as N tends to infinity.

Thus, to generate lattices that are random in a natural sense, it suffices to generate uniformly at random a lattice in $\mathcal{L}_{n,N}$ for large N. This is particularly easy when N is prime. Indeed, when p is a large prime, the vast majority of lattices in $\mathcal{L}_{n,p}$ are lattices spanned by row matrices of the following form:

$$
R_p^n = \begin{pmatrix}
p & 0 & 0 & \cdots & 0 \\
x_1 & 1 & 0 & \cdots & 0 \\
x_2 & 0 & 1 & \ddots & \vdots \\
\vdots & \vdots & \ddots & \ddots & 0 \\
x_{n-1} & 0 & \cdots & 0 & 1
\end{pmatrix},
$$

where the x_i's are chosen independently and uniformly in $\{0, \ldots, p-1\}$.

3.2 Random Bases

Once a lattice has been selected, it would be useful to select a random basis, among the infinitely many bases. This time however, there is no clear definition of what is a random basis, since there is no finite measure on $SL_n(\mathbb{Z})$. Since we mostly deal with integer lattices, one could consider the Hermite normal form (HNF) of the lattice, and argue that this is the basis which gives the least information on the lattice, because it can be computed in polynomial time from any basis. However, it could also be argued that the HNF may have special properties, depending on the lattice. For instance, the HNF of NTRU lattices [16] is already reduced in some sense, and does not look like a random basis at all. A random basis should consist of long vectors: the orthogonality defect should not be bounded, since the number of bases with bounded orthogonality defect is bounded. In other words, a random basis should not be reduced at all.

A heuristic approach was used for the GGH cryptosystem [11]. Namely, a secret basis was transformed into a large public basis of the same lattice by multiplying generators of $SL_n(\mathbb{Z})$ in a random manner. However, it is difficult to control the size of the entries, and it looks hard to obtain theoretical results.

One can devise a less heuristic method as follows. Consider a full-rank integer lattice $L \subseteq \mathbb{Z}^n$. If B is much bigger than $(\det L)^{1/n}$, it is possible to sample efficiently and uniformly points in $L \cap [-B, B]^n$ (see [1]). For instance, if $B = (\det L)/2$, one can simply take an integer linear combination $x_1 b_1 + \cdots + x_n b_n$ of a basis, with large coefficients x_i, and reduce the coordinates modulo $\det L = [\mathbb{Z}^n : L]$. That is easy for lattices in the previous set $\mathcal{L}_{n,p}$ where p is prime. Once we have such a sampling procedure, we note that n vectors randomly chosen in such a way will with overwhelming probability be linearly independent. Though they are unlikely to form a lattice basis (rather, they will span a sublattice), one can easily lift such a full-rank set of linearly independent vectors of norm $\leq B$ to a basis made of vectors of norm $\leq B\sqrt{n}/2$ using Babai's nearest plane algorithm [5] (see [1] or [22, Lemma 7.1]). In particular, if one considers the lattices of the class $\mathcal{L}_{n,p}$, it is easy to generate plenty of bases in a random manner in such a way that all the coefficients of the basis vectors are $\leq p\sqrt{n}/2$.

3.3 Random L^3-Reduced Bases

There are two natural notions of random L^3 bases. One is derived from the mathematical definition. An L^3-reduced basis is necessarily Siegel-reduced (following the definition of [8]), that is, its $\mu_{i,j}$'s and $\|b_i^*\|/\|b_{i+1}^*\|$'s are bounded. This implies [8] that the number of L^3-reduced bases of a given lattice is finite (for any reduction parameters), and can be bounded independently of the lattice. Thus, one could define a random L^3 basis as follows: select a random lattice, and among all the finitely many L^3-reduced bases of that lattice, select one uniformly at random. Unfortunately, the latter process is impractical, but it might be interesting to prove probabilistic statements on such bases. Instead, one could try the following in practice: select a random lattice, then select a random basis, and eventually apply the L^3 algorithm. The output basis will not necessarily be random in the first sense, since the L^3 algorithm may bias the distribution. However, intuitively, it could also be viewed as some kind of random L^3 basis. In the previous process, it is crucial to select a random-looking basis (unlike the HNF of NTRU lattices). For instance, if we run the L^3 algorithm on already reduced (or almost reduced) bases, the output basis will differ from a typical L^3-reduced basis.

3.4 Random Bases in Our Experiments

In our experiments, besides the Goldstein-Mayer [12] bases of random lattices, we considered two other types of random bases. The Ajtai-type bases of dimension d and factor α are given by the rows of a lower triangular random matrix B with:

$$B_{i,i} = \lfloor 2^{(2d-i+1)^\alpha} \rfloor \quad \text{and} \quad B_{j,i} = \text{rand}(-B_{i,i}/2, B_{i,i}/2) \text{ for all } j > i.$$

Similar bases have been used by Ajtai in [3] to show the tightness of worst-case bounds of [27]. The bases used in Coppersmith's root-finding method [10] bear some similarities with what we call Ajtai-type bases.

We define the knapsack-type bases as the rows of the $d \times (d+1)$ matrices:

$$\begin{pmatrix} A_1 & 1 & 0 & \ldots & 0 \\ A_2 & 0 & 1 & \ldots & 0 \\ \vdots & \vdots & \vdots & \ddots & \vdots \\ A_d & 0 & 0 & \ldots & 1 \end{pmatrix},$$

where the A_i's are sampled independently and uniformly in $[-B, B]$, for some given bound B. Such bases often occur in practice, e..g., in cryptanalyses of knapsack-based cryptosystems, reconstructions of minimal polynomials and detections of integer relations between real numbers. The behavior of L^3 on this type of bases and on the above R_p^{d+1}'s look alike.

Interestingly, we did not notice any significant change in the output quality or in the geometry of reduced bases between all three types of random bases.

4   The Output Quality of L^3

For fixed parameters δ and η, the L^3 and L^2 algorithms output bases $\mathbf{b}_1, \ldots, \mathbf{b}_d$ such that $\|\mathbf{b}_{i+1}^*\|^2/\|\mathbf{b}_i^*\|^2 \geq \beta = 1/(\delta - \eta^2)$ for all $i < d$, which implies that:

$$\|\mathbf{b}_1\| \leq \beta^{(d-1)/4}(\det L)^{1/d} \quad \text{and} \quad \|\mathbf{b}_1\| \leq \beta^{(d-1)/2}\lambda(L).$$

It is easy to prove that these bounds are tight in the worst case: both are reached for some reduced bases of some particular lattices. However, there is a common belief that they are not tight in practice. For instance, Odlyzko wrote in [25]:

This algorithm [...] usually finds a reduced basis in which the first vector is much shorter than guaranteed [theoretically]. (In low dimensions, it has been observed empirically that it usually finds the shortest non-zero vector in a lattice.)

We argue that the quantity $\|\mathbf{b}_1\|/(\det L)^{1/d}$ remains exponential on the average, but is indeed far smaller than the worst-case bound: for δ close to 1 and η close to $1/2$, one should replace $\beta^{1/4} \approx (4/3)^{1/4}$ by ≈ 1.02, so that the approximation factor $\beta^{(d-1)/4}$ becomes $\approx 1..02^d$. As opposed to the worst-case bounds, the ratio $\|\mathbf{b}_1\|/\lambda(L)$ should also be $\approx 1.02^d$ on the average, rather than being the square of $\|\mathbf{b}_1\|/(\det L)^{1/d}$. Indeed, if the Gaussian heuristic holds for a lattice L, then $\lambda(L) \approx \sqrt{\frac{d}{2\pi e}}(\det L)^{1/d}$. The Gaussian heuristic is only a heuristic in general, but it can be proved for random lattices (see [2,4]), and it is unlikely to be wrong by an exponential factor, unless the lattice is very special.

Heuristic 1. *Let δ be close to 1 and η be close to $1/2$. Given as input a random basis of almost any lattice L of sufficiently high dimension (e.g., larger than 40), L^3 and L^2 with parameters δ and η output a basis whose first vector $\mathbf{b}_1$ satisfies $\|\mathbf{b}_1\|/(\det L)^{1/d} \approx (1.02)^d$.*

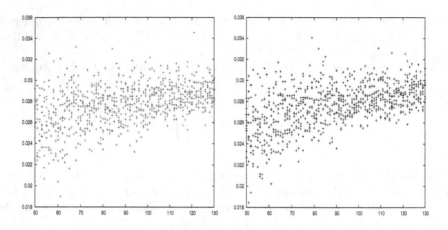

Fig. 3. Variation of $\frac{1}{d} \log \frac{\|\mathbf{b}_1\|}{(\det L)^{1/d}}$ as a function of d

4.1 A Few Experiments

In Figure 3, we consider the variations of the quantity $\frac{1}{d} \log \frac{\|\mathbf{b}_1\|}{(\det L)^{1/d}}$ as the dimension d increases. On the left side of the figure, each point is a sample of the following experiment: generate a random knapsack-type basis with $B = 2^{100 \cdot d}$ and reduce it with L^2 (the fast variant of $\texttt{fplll-1.2}$ with $(\delta, \eta) = (0.999, 0.501)$). The points on the right side correspond to the same experiments, but starting with Ajtai-type bases, with $\alpha = 1.2$. The two sides of Figure 3 are similar and the quantity $\frac{1}{d} \log \frac{\|\mathbf{b}_1\|}{(\det L)^{1/d}}$ seems to converge slightly below 0.03 (the corresponding worst-case constant is ≈ 0.10). This means that the first output vector $\mathbf{b}_1$ usually satisfies $\|\mathbf{b}_1\| \approx (1.02)^d (\det L)^{1/d}$. The exponential quantity $(1.02)^d$ remains tiny even in moderate dimensions: e.g., $(1.02)^{50} \approx 2.7$ and $(1.02)^{100} \approx 7.2$. These data may explain why in the 80's, cryptanalysts used to believe that L^3 returns vectors surprisingly small compared to the worst-case bound.

4.2 The Configuration of Local Bases

To understand the shape of the bases that are computed by L^3, it is tempting to consider the local bases of the output bases, i.e.., the pairs $(\mathbf{b}_i^*, \mu_{i+1,i}\mathbf{b}_i^* + \mathbf{b}_{i+1}^*)$ for $i < d$. These pairs are the components of $\mathbf{b}_i$ and $\mathbf{b}_{i+1}$ which are orthogonal to $\mathbf{b}_1, \ldots, \mathbf{b}_{i-1}$.. We experimentally observe that after the reduction, local bases seem to share a common configuration, independently of the index i. In Figure 4, a point corresponds to a local basis (its coordinates are $\mu_{i+1,i}$ and $\|\mathbf{b}_{i+1}^*\|/\|\mathbf{b}_i^*\|$) of a basis returned by the fast variant of $\texttt{fplll-1.2}$ with parameters $\delta = 0.999$ and $\eta = 0.501$, starting from a knapsack-type basis with $B = 2^{100 \cdot d}$. The 2100 points correspond to 30 reduced bases of 71-dimensional lattices. This distribution seems to stabilize between the dimensions 40 and 50.

Figure 4 is puzzling. First of all, the $\mu_{i+1,i}$'s are not uniformly distributed in $[-\eta, \eta]$, as one may have thought *a priori*. As an example, the uniform

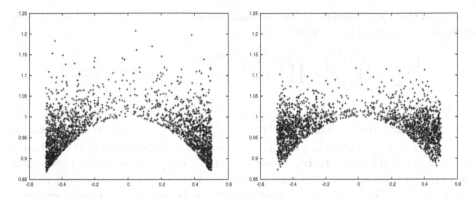

Fig. 4. Distribution of the local bases after L^3 (left) and deep-L^3 (right)

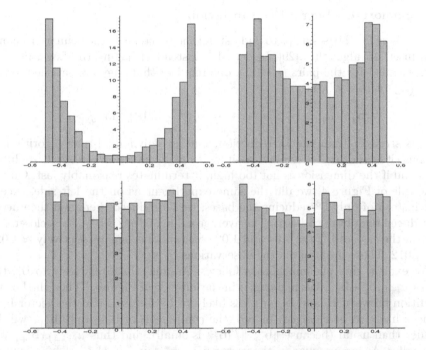

Fig. 5. Distribution of $\mu_{i,i-1}$ (top left), $\mu_{i,i-2}$ (top right), $\mu_{i,i-5}$ (bottom left) and $\mu_{i,i-10}$ (bottom right) for 71-dimensional lattices after L^3

distribution was used as an hypothesis Theorem 2 in [17]. Our observation therefore invalidates this result. This non-uniformity is surprising because the other $\mu_{i,j}$'s seem to be uniformly distributed in $[-\eta, \eta]$, in particular when $i - j$ becomes larger, as it is illustrated by Figure 5. The mean value of the $|\mu_{i+1,i}|$'s is close to 0.38. Besides, the mean value of $\|\mathbf{b}_i^*\|/\|\mathbf{b}_{i+1}^*\|$ is close to 1.04, which matches the 1.02 constant of the previous subsection. Indeed, if the local bases behave independently, we have:

$$\frac{\|\mathbf{b}_1\|^d}{\det L} = \prod_{i=1}^{d} \frac{\|\mathbf{b}_1\|}{\|\mathbf{b}_i^*\|} = \prod_{i=1}^{d-1} \left(\frac{\|\mathbf{b}_i^*\|}{\|\mathbf{b}_{i+1}^*\|}\right)^{d-i+1} \approx (1.04)^{d^2/2} \approx (1.02)^{d^2}.$$

A possible explanation of the shape of the pairs $(\mathbf{b}_i^*, \mu_{i+1,i}\mathbf{b}_i^* + \mathbf{b}_{i+1}^*)$ is as follows. During the execution of L^3, the ratios $\|\mathbf{b}_i^*\|/\|\mathbf{b}_{i+1}^*\|$ are decreasing steadily. At some moment, the ratio $\|\mathbf{b}_i^*\|/\|\mathbf{b}_{i+1}^*\|$ becomes smaller than $\sqrt{4/3}$. When it does happen, relatively to $\mathbf{b}_i^*$, either $\mu_{i+1,i}\mathbf{b}_i^* + \mathbf{b}_{i+1}^*$ lies in one of the corners of Figure 4 or is close to the vertical axis. In the first case, it does not change since $(\mathbf{b}_i^*, \mu_{i+1,i}\mathbf{b}_i^* + \mathbf{b}_{i+1}^*)$ is reduced. Otherwise $\mathbf{b}_i$ and $\mathbf{b}_{i+1}$ are to be swapped since $\mu_{i+1,i}\mathbf{b}_i^* + \mathbf{b}_{i+1}^*$ is not in the fundamental domain.

4.3   Schnorr-Euchner's Deep Insertion

The study of local bases helps to understand the behavior of the Schnorr-Euchner deep insertion algorithm [29]. In deep-L^3, instead of having the Lovász conditions satisfied for the pairs $(i, i+1)$, one requires that they are satisfied for all pairs (i, j) with $i < j$, i.e.:

$$\|\mathbf{b}_j^* + \mu_{j,j-1}\mathbf{b}_{j-1}^* + \ldots + \mu_{j,i}\mathbf{b}_i^*\|^2 \geq \delta\|\mathbf{b}_i^*\|^2 \text{ for } j > i.$$

This is stronger than the L^3-reduction, but no polynomial-time algorithm to compute it is known. Yet in practice, if we deep-L^3-reduce an already L^3-reduced basis and if the dimension is not too high, it terminates reasonably fast. On the right side of Figure 4, we did the same experiment as on the left side, except that instead of only L^3-reducing the bases, we L^3-reduced them and then deep-L^3-reduced the obtained bases. The average value of $\|\mathbf{b}_i^*\|/\|\mathbf{b}_{i+1}^*\|$'s is closer to 1 than in the case of L^3: the 1.04 and 1.02 constants become respectively ≈ 1.025 and 1.012. These data match the observations of [6].

We explain this phenomenon as follows. Assume that, relatively to $\mathbf{b}_i^*$, the vector $\mu_{i+1,i}\mathbf{b}_i^* + \mathbf{b}_{i+1}^*$ lies in a corner in the left side of Figure 4. Then the Lovász condition between $\mathbf{b}_i$ and $\mathbf{b}_{i+2}$ is less likely to be fulfilled, and the vector $\mathbf{b}_{i+2}$ is more likely to be changed. Indeed, the component of $\mathbf{b}_{i+2}$ onto $\mathbf{b}_{i+1}^*$ will be smaller than usual (because $\|\mathbf{b}_{i+1}^*\|/\|\mathbf{b}_i^*\|$ is small), and thus $\mu_{i+2,i+1}\mathbf{b}_{i+1}^*$ will be smaller. As a consequence, the vector $\mu_{i+2,i}\mathbf{b}_i^* + \mu_{i+2,i+1}\mathbf{b}_{i+1}^* + \mathbf{b}_{i+2}^*$ is more likely to be shorter than $\mathbf{b}_i^*$, and thus $\mathbf{b}_{i+2}$ is more likely to change. Since the corner local bases arise with high frequency, deep-L^3 often performs insertions of depths higher than 2 that would not be performed by L^3.

5   The Practical Running Time of L^3

In this section, we argue that the worst case complexity bound $O(d^4(d+n)(d+\log B)\log B)$ is asymptotically reached for some classes of random bases, and explain how and why the running time is better in practice. Here we consider bases for which $n = \Theta(d) = O(\log B)$, so that the bound above becomes $O(d^5 \log^2 B)$. Notice that the heuristic codes do not have any asymptotic meaning since they do not terminate when the dimension increases too much (in particular, the working precision must increase with the dimension). Therefore, all the experiments described in this section were performed using the proved variant of fplll-1.2.

We draw below a heuristic worst-case complexity analysis of L^2 that will help us to explain the difference between the worst case and the practical behavior:

- There are $O(d^2 \log B)$ loop iterations.
- In a given loop iteration, there are usually two iterations within the lazy size-reduction: the first one makes the $|\mu_{\kappa,i}|$'s smaller than η and the second one recomputes the $\mu_{\kappa,i}$'s and $r_{\kappa,i}$'s with better accuracy. This is incorrect in full generality (in particular for knapsack-type bases as we will see below), but is the case most often.
- In each iteration of the size-reduction, there are $O(d^2)$ arithmetic operations.
- Among these, the most expensive ones are those related to the coefficients of the basis and Gram matrices: these are essentially multiplications between integers of lengths $O(\log B)$ and the x_i's, of lengths $O(d)$.

We argue that the analysis above is tight for Ajtai-type random bases.

Heuristic 2. *Let $\alpha > 1$. When d grows to infinity, the average cost of the L^2 algorithm given as input a randomly and uniformly chosen d-dimensional Ajtai-type basis with parameter α is $\Theta(d^{5+2\alpha})$.*

In this section, we also claim that the bounds of the heuristic worst-case analysis are tight in practice for Ajtai-type random bases, except the $O(d)$ bound on size of the x_i's. Finally, we detail the case of knapsack-type random bases.

5.1 L^2 on Ajtai-Type Random Bases

Firstly, the $O(d^2 \log B)$ bound on the loop iterations seems to be tight in practice, as suggested by Figure 6. The left side corresponds to Ajtai-type random bases with $\alpha = 1.2$: the points are the experimental data and the continuous line is the gnuplot interpolation of the type $f(d) = a \cdot d^{3.2}$ (we have $\log B = O(d^{1.2})$). The right side has been obtained similarly, for $\alpha = 1.5$, and $g(d) = b \cdot d^{3.5}$. With Ajtai-type bases, size-reductions contain extremely rarely more than two iterations. For example, for $d \leq 75$ and $\alpha = 1.5$, fewer than 0.01% of the size-reductions involve more than two iterations. The third bound of the heuristic worst case analysis is also reached.

These similarities between the worst and average cases do not go on for the size of the integers involved in the arithmetic operations. The x_i's computed during the size-reductions are most often shorter than a machine word, which makes it difficult to observe the $O(d)$ factor in the complexity bound coming from them. For an Ajtai-type basis with $d \leq 75$ and $\alpha = 1.5$, fewer than 0.2% of the non-zero x_i's are longer than 64 bits. In the worst case [23], we have $|x_i| \lesssim (3/2)^{\kappa-i} M$, where M is the maximum of the $\mu_{\kappa,j}$'s before the lazy size-reduction starts, and κ is the current L^3 index. In practice, M happens to be small most of the time.

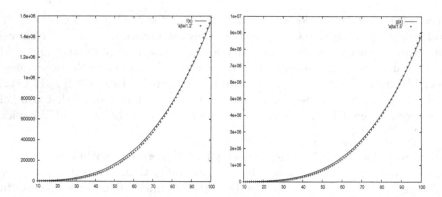

Fig. 6. Number of loop iterations of L^2 as a function of d, for Ajtai-type random bases

We argue that the average situation is $|x_i| \approx (1.04)^{\kappa-i} M$. This bound remains exponential, but for a small M, x_i becomes larger than a machine word only in dimensions higher than several hundreds. We define:

$$\mu_{\kappa,i}^{(\text{final})} = \mu_{\kappa,i}^{(\text{initial})} - \sum_{j=i+1}^{\kappa-1} x_j \mu_{j,i} = \mu_{\kappa,i}^{(\text{initial})} - \sum_{j=i+1}^{\kappa-1} \left\lfloor \mu_{\kappa,j}^{(\text{final})} \right\rceil \mu_{j,i}.$$

We model the $\left(\mu_{\kappa,\kappa-i}^{(\text{final})} \right)_i$'s by the random variables U_i defined as follows:

$$U_0 = R_0 \quad \text{and} \quad U_i = R_i + \sum_{j=1}^{i-1} U_j R'_{i,j} \quad \text{if } i \geq 1,$$

where the R_i's and $R'_{i,j}$'s are uniformly distributed respectively in $[-a, a]$ for some constant a and in $[-\eta, \eta]$. We assume that the R_i's and $R'_{i,j}$'s are pairwise independent. These hypotheses on the $\mu_{i,j}$'s are strong. In particular we saw in Section 4 that the $\mu_{i,i-1}$'s are not uniformly distributed in $[-\eta, \eta]$. Nevertheless, this simplification does not significantly change the asymptotic behavior of the sequence (U_i) and simplifies the technicalities. Besides, to make the model closer to the reality, we could have rounded the U_j's, but since these quantities are growing to infinity, this should not change much the asymptotic behavior. The independence of the R_i's and $R'_{i,j}$'s and their symmetry give:

$$\mathbb{E}\left[U_i\right] = 0, \ \mathbb{E}\left[U_i^2\right] = \mathbb{E}\left[R_i^2\right] + \sum_{j=1}^{i-1} \mathbb{E}\left[U_j^2\right] \cdot \mathbb{E}\left[R'^2_{i,j}\right] = \frac{2a^3}{3} + \frac{2\eta^3}{3} \sum_{j=1}^{i-1} \mathbb{E}\left[U_j^2\right].$$

As a consequence, for i growing to infinity, we have $\mathbb{E}[U_i^2] \approx \left(\frac{2\eta^3}{3} + 1\right)^i$. If we choose $\eta \approx 1/2$, we get $\frac{2\eta^3}{3} + 1 \approx \frac{13}{12} \approx 1.08$. We thus expect the $|x_i|$'s to be of length $\lesssim (\log_2 1.04) \cdot d \approx 0.057 \cdot d$. To sum up, the x_i's should have length $O(d)$

in practice, but the $O(\cdot)$-constant is tiny. For example, the quantity $(1.04)^d$ becomes larger than 2^{64} for $d \geq 1100$. Since we cannot reduce lattice bases which simultaneously have this dimension and reach the other bounds of the heuristic worst-case complexity analysis, it is at the moment impossible to observe the asymptotic behavior. The practical running time is rather $O(d^4 \log^2 B)$.

5.2 L² on Knapsack-Type Bases

In the case of knapsack-type bases there are fewer loop iterations than in the worst case: the quantity $\Delta = \prod_{i=1}^{d} \| \mathbf{b}_i^* \|^{d-i+1}$ of L³'s analysis satisfies $\Delta = B^{O(d)}$ instead of $\Delta = B^{O(d^2)}$. This ensures that there are $O(d \log B)$ loop iterations, so that the overall cost of L² for these lattice bases is $O(d^4 \log^2 B)$. Here we argue that asymptotically one should expect a better complexity bound.

Heuristic 3. *When d and $\log B$ grow to infinity with $\log B = \Omega(d^2)$, the average cost of the L^2 algorithm given as input a randomly and uniformly chosen d-dimensional knapsack-type basis with entries of length $\leq \log B$ is $\Theta(d^3 \log^2 B)$.*

In practice for moderate dimensions, the phenomenon described in the previous subsection makes the cost even lower: close to $O(d^2 \log^2 B)$ when $\log B$ is significantly larger than d.

First, there are $\Theta(d \log B)$ main loop iterations. These iterations are equally distributed among the different values of $\kappa_{\max}$: we define κ_{max} as the maximum of the indices κ since the beginning of the execution of the algorithm, i.e., the number of basis vectors that have been considered so far. We have $\kappa_{\max} = 2$ at the beginning, then $\kappa_{\max}$ is gradually incremented up to $d + 1$, when the execution of L² is over. The number of iterations for each $\kappa_{\max}$ is roughly the same, approximately $\Theta(\log B)$. We divide the execution into $d - 1$ phases, according to the value of $\kappa_{\max}$. We observe experimentally that at the end of the phase of a given $\kappa_{\max}$, the current basis has the following shape:

$$
\begin{pmatrix}
a_{1,1} & a_{1,2} & \cdots & a_{1,\kappa_{\max}+1} & 0 & 0 & \cdots & 0 \\
a_{2,1} & a_{2,2} & \cdots & a_{2,\kappa_{\max}+1} & 0 & 0 & \cdots & 0 \\
\vdots & \vdots & \ddots & \vdots & \vdots & \vdots & \ddots & \vdots \\
a_{\kappa_{\max},1} & a_{\kappa_{\max},2} & \cdots & a_{\kappa_{\max},\kappa_{\max}+1} & 0 & 0 & \cdots & 0 \\
A_{\kappa_{\max}+1} & 0 & \cdots & 0 & 1 & 0 & \cdots & 0 \\
A_{\kappa_{\max}+2} & 0 & \cdots & 0 & 0 & 1 & \cdots & 0 \\
\vdots & \vdots & \ddots & \vdots & \vdots & \vdots & \ddots & \vdots \\
A_d & 0 & \cdots & 0 & 0 & 0 & \cdots & 1
\end{pmatrix},
$$

where the top left $a_{i,j}$'s satisfy: $|a_{i,j}| = O\left(B^{\frac{1}{\kappa_{\max}}}\right)$.

We subdivide each $\kappa_{\max}$-phase in two subphases: the first subphase is the first loop iteration of L² for which $\kappa = \kappa_{\max}$, and the second one is made of the other iterations with the same $\kappa_{\max}$. The first subphase shortens the vector $\mathbf{b}_{\kappa_{\max}}$: its length decreases from $\approx B$ to $\leq \sqrt{\kappa_{\max}(\max_{i<\kappa_{\max}} \|\mathbf{b}_i\|^2) + 1} \lesssim B^{\frac{1}{\kappa_{\max}-1}}$. This subphase costs $O(d \log^2 B)$ bit operations (see [23]): there are $O(\log B/d)$

loop iterations in the lazy size-reduction; each one involves $O(d^2)$ arithmetic operations; among them, the most costly are the integer multiplications between the x_i's (that are $O(d)$-bit long) and the coefficients of the basis and Gram matrices (their lengths are $O(\log B/d)$, except the $\langle \mathbf{b}_\kappa, \mathbf{b}_i \rangle$'s which occur with frequency $1/O(\kappa)$). The global cost of the first subphases is $O(d^2 \log^2 B)$. This is negligible in comparison to the overall cost of the second subphases.

Let $\mathbf{b}'_i$ be the vector obtained from $\mathbf{b}_i$ after the first subphase of the phase for which $\kappa_{\max} = i$, that is, right after its first size-reduction. Let $C(d, B)$ be the overall cost of the second subphases in dimension d and for input A_i's satisfying $|A_i| \le B$. We divide the execution of the algorithm as follows: it starts by reducing a knapsack-type basis of dimension $\lfloor d/2 \rfloor$; let $\left(\mathbf{b}''_1, \dots, \mathbf{b}''_{\lfloor d/2 \rfloor} \right)$ be the corresponding L^3-reduced vectors; if we exclude the $\lceil d/2 \rceil$ remaining first subphases, then L^2 reduces the basis $\left(\mathbf{b}''_1, \dots, \mathbf{b}''_{\lfloor d/2 \rfloor}, \mathbf{b}'_{\lfloor d/2+1 \rfloor}, \dots, \mathbf{b}'_d \right)$, where all the lengths of the vectors are bounded by $\approx B^{2/d}$. As a consequence, we have:

$$C(d, B) = C(d/2, B) + O(d^5 (\log B/d)^2) = C(d/2, B) + O(d^3 \log^2 B),$$

from which one easily gets $C(d, B) = O(d^3 \log^2 B)$, as long as $d^2 = O(\log B)$.

5.3   Many Parameters Can Influence the Running Time

We list below a few tunings that should be performed if one wants to optimize L^3 and L^2 for particular instances:

- Firstly, use as less multiprecision arithmetic as possible. If you are in a medium dimension (e.g., less than 170), you may avoid multiprecision *fpa* (see Section 6). If your input basis is made of short vectors, like for NTRU lattices, try using chip integers instead of multiprecision integers.
- Detect if there are scalar products cancellations: if these cancellations happen very rarely, use a heuristic variant that does not require the Gram matrix. Otherwise, if such cancellations happen frequently, a proved variant using the Gram matrix may turn out to be cheaper than a heuristic one recomputing exactly many scalar products.
- It is sometimes recommended to weaken δ and η. Indeed, if you increase η and/or decrease δ, it will possibly decrease the number of iterations within the lazy size-reduction and the number of global iterations. However, relaxed L^3-factors require a higher precision: for a given precision, the dimension above which L^2 might loop forever decreases (see Section 6).

6   "Numerical Stability" of L^3

In this section, we discuss problems that may arise when one uses *fpa* within L^3. The motivation is to get a good understanding of the "standard" numerical behavior, in order to keep the double precision as long as possible with low chances of failure. Essentially, two different phenomena may be encountered: a lazy size-reduction or consecutive Lovász tests may be looping forever. The output may also be incorrect, but most often if something goes wrong, the execution loops within a size-reduction. We suppose here that either the Gram matrix is maintained exactly during the execution or that the problems arising from scalar product cancellations do not show up.

It is shown in [23] that for some given parameters δ and η, a precision of $\left(\log \frac{(1+\eta)^2}{\delta-\eta^2} + \varepsilon\right) \cdot d + o(d)$ is sufficient for L^2 to work correctly, for any constant $\varepsilon > 0$. For δ close to 1 and η close to $1/2$, it gives that a precision of $1.6 \cdot d + o(d)$ suffices. A family of lattice bases for which this bound seems to be tight is also given. Nevertheless, in practice the algorithm seems to work correctly with a much lower precision: for example, the double precision (53 bit-long mantissæ) seems sufficient most of the time up to dimension 180. We argue here that the average required precision grows linearly with the dimension, but with a significantly lower constant.

Heuristic 4. *Let δ be close to 1 and η be close to $1/2$. For almost every lattice, with a precision of $0.18 \cdot d + o(d)$ bits for the fp-calculations, the L^2 algorithm performs correctly when given almost any input basis.*

This heuristic has direct consequences for a practical implementation of L^2: it helps guessing what precision should be sufficient in a given dimension, and thus a significant constant factor can be saved for the running time.

We now give a justification for the heuristic above. For a fixed size of mantissa, we evaluate the dimension for which things should start going wrong. First, we evaluate the error made on the Gram-Schmidt coefficients and then we will use these results for the behavior of L^3: to do this, we will say that L^3 performs plenty of Gram-Schmidt calculations (during the successive loop iterations), and that things go wrong if at least one of these calculations is erroneous.

We consider the following random model, which is a simplified version of the one described in Section 4 (the simplification should not change the asymptotic results but helps for the analysis).

- The $\mu_{i,j}$'s for $i > j$ are chosen randomly and independently in $[-\eta, \eta]$. They share a distribution that is symmetric towards 0. This implies that $\mathbb{E}[\mu] = 0$. We define $\mu_2 = \mathbb{E}[\mu^2]$ and $\mu_{i,i} = 1$.
- The $\frac{r_{i,i}}{r_{i+1,i+1}}$'s are chosen randomly and independently in $(0, \beta]$. These choices are independent of those of the $\mu_{i,j}$'s. We define $\alpha = \mathbb{E}\left[\frac{r_{i,i}}{r_{i+1,i+1}}\right]$.
- The random variables $\mu_{i,j}$ and $\frac{r_{i,i}}{r_{i+1,i+1}}$ determine the Gram matrix of the initial basis. Let $r_{1,1}$ be an arbitrary constant. We define the following random variables, for $i \geq j$: $\langle \mathbf{b}_i, \mathbf{b}_j \rangle = r_{1,1} \sum_{k=1}^{j} \mu_{j,k}\mu_{i,k} \prod_{l=1}^{k-1}(r_{l,l}/r_{l+1,l+1})^{-1}$.
- We define the random variables $r_{i,j} = r_{1,1}\mu_{i,j} \prod_{l=1}^{j-1}(r_{l,l}/r_{l+1,l+1})^{-1}$ (for $i \geq j$).
- We assume that we do a relative error $\varepsilon = 2^{-\ell}$ (with ℓ the working precision) while translating the exact value $\|\mathbf{b}_1\|^2$ into a *fp* number: $\Delta\|\mathbf{b}_1\|^2 = \varepsilon\|\mathbf{b}_1\|^2$.

We have selected a way to randomly choose the Gram matrix and to perform a rounding error on $\|\mathbf{b}_1\|^2$. To simplify the analysis, we suppose that there is no rounding error performed on the other $\langle \mathbf{b}_i, \mathbf{b}_j \rangle$'s. Our goal is to estimate the amplification of the rounding error $\Delta\|\mathbf{b}_1\|^2$ during the calculations of approximations of the $r_{i,j}$'s and $\mu_{i,j}$'s. We neglect high-order error terms. More precisely, we study the following random variables, defined recursively:

$$\Delta r_{1,1} = \Delta\|\mathbf{b}_1\|^2 = \varepsilon\|\mathbf{b}_1\|^2,$$

$$\Delta r_{i,j} = -\sum_{k=1}^{j-1}(\Delta r_{i,k}\mu_{j,k} + \Delta r_{j,k}\mu_{i,k} - \Delta r_{k,k}\mu_{i,k}\mu_{j,k}) \text{ when } i \geq j,$$

The $\mu_{a,b}$'s and $\frac{r_{b,b}}{r_{b+1,b+1}}$'s that may not be independent with $\Delta r_{i,k}$ are those for which $b < k$. As a consequence, $\Delta r_{i,k}, \mu_{j,k}, \frac{r_{j-1,j-1}}{r_{j,j}}, \frac{r_{j-2,j-2}}{r_{j-1,j-1}}, \ldots, \frac{r_{k,k}}{r_{k+1,k+1}}$ are pairwise independent, $\Delta r_{j,k}, \mu_{i,k}, \frac{r_{j-1,j-1}}{r_{j,j}}, \frac{r_{j-2,j-2}}{r_{j-1,j-1}}, \ldots, \frac{r_{k,k}}{r_{k+1,k+1}}$ are pairwise independent, and $\Delta r_{k,k}, \mu_{i,k}, \mu_{j,k}, \frac{r_{j-1,j-1}}{r_{j,j}}, \frac{r_{j-2,j-2}}{r_{j-1,j-1}}, \ldots, \frac{r_{k,k}}{r_{k+1,k+1}}$ are pairwise independent, for all (i,j,k) satisfying $i > j > k$. Therefore, for any $i > j$:

$$\mathbb{E}\left[\frac{\Delta r_{i,j}}{r_{j,j}}\right] = -\sum_{k=1}^{j-1}\left(\prod_{l=k}^{j-1}\mathbb{E}\left[\frac{r_{l,l}}{r_{l+1,l+1}}\right]\right)\left(\mathbb{E}\left[\frac{\Delta r_{i,k}}{r_{k,k}}\right]\mathbb{E}[\mu_{j,k}] + \mathbb{E}\left[\frac{\Delta r_{j,k}}{r_{k,k}}\right]\mathbb{E}[\mu_{i,k}]\right.$$
$$\left. -\mathbb{E}\left[\frac{\Delta r_{k,k}}{r_{k,k}}\right]\mathbb{E}[\mu_{i,k}]\mathbb{E}[\mu_{j,k}]\right).$$

Because $\mathbb{E}[\mu_{j,k}] = \mathbb{E}[\mu_{i,k}] = 0$, we get $\mathbb{E}\left[\frac{\Delta r_{i,j}}{r_{j,j}}\right] = 0$, for all $i > j$. Similarly, we have, for $i > 1$:

$$\mathbb{E}\left[\frac{\Delta r_{i,i}}{r_{i,i}}\right] = \mu_2\sum_{k=1}^{j-1}\left(\prod_{l=k}^{i-1}\mathbb{E}\left[\frac{r_{l,l}}{r_{l+1,l+1}}\right]\right)\mathbb{E}\left[\frac{\Delta r_{k,k}}{r_{k,k}}\right] = \mu_2\sum_{k=1}^{j-1}\alpha^{i-k}\mathbb{E}\left[\frac{\Delta r_{k,k}}{r_{k,k}}\right].$$

We obtain that $\mathbb{E}\left[\frac{\Delta r_{i,i}}{r_{i,i}}\right]$ is close to $(\alpha(1+\mu_2))^i\varepsilon$. For example, if the $\mu_{i,j}$'s are uniformly chosen in $[-1/2, 1/2]$, if $\alpha = 1.04$ (as observed in Section 4), and if $\varepsilon \approx 2^{-53}$, we get $\mathbb{E}\left[\frac{\Delta r_{i,i}}{r_{i,i}}\right] \approx 1.13^i \cdot 2^{-53}$. For $i = 200$, this is close to 2^{-17}.

We have analyzed very roughly the influence of the rounding error made on $\|\mathbf{b}_1\|^2$, within the Gram-Schmidt orthogonalization for L^3-reduced bases. If we want to adapt this analysis to L^2, we must take into account the number of $r_{i,j}$'s and $\mu_{i,j}$'s that are computed during the execution. To simplify we consider only the $r_{d,d}$'s, which are *a priori* the less accurate. We suppose that all the computations of $r_{d,d}$ through the execution are independent. Let K be the number of iterations for which $\kappa = d$. We consider that an error on $r_{d,d}$ is significant if $\frac{\Delta r_{d,d}}{r_{d,d}}$ is at least 2^{-3}. If such an error occurs, the corresponding Lovász test is likely to be erroneous. Under such hypotheses, the probability of failure is of the order of $1 - (1 - 2^{-17+3})^K \approx K2^{-14}$. In case of several millions of Lovász tests, it is likely that there is one making L^3 behave unexpectedly.

The above analysis is completely heuristic and relies on very strong hypotheses, but it provides orders of magnitude that one seems to encounter in practice. For random bases, we observe infinite loops in double precision arising around dimensions 175 to 185, when there are a few millions Lovász tests.

Acknowledgments. We thank Guillaume Hanrot for helpful discussions. The writing of this paper was completed while the second author was visiting the University of Sydney, whose hospitality is gratefully acknowledged.

References

1. M. Ajtai. Generating hard instances of lattice problems (extended abstract). In *Proc. of STOC 1996*, pages 99–108. ACM, 1996.
2. M. Ajtai. Random lattices and a conjectured 0-1 law about their polynomial time computable properties. In *Proc. of FOCS 2002*, pages 13–39. IEEE, 2002.
3. M. Ajtai. The worst-case behavior of Schnorr's algorithm approximating the shortest nonzero vector in a lattice. In *Proc. of STOC 2003*, pages 396–406. ACM, 2003.
4. M. Ajtai. Generating Random Lattices According to the Invariant Distribution. Draft, 2006.
5. L. Babai. On Lovász lattice reduction and the nearest lattice point problem. *Combinatorica*, 6:1–13, 1986.
6. W. Backes and S. Wetzel. Heuristics on lattice reduction in practice. *ACM Journal of Experimental Algorithms*, 7:1, 2002.
7. C. Batut, K. Belabas, D.. Bernardi, H. Cohen, and M. Olivier. PARI/GP computer package version 2. Av. at http://pari.math.u-bordeaux.fr/.
8. J. W. S. Cassels. *Rational quadratic forms*, volume 13 of *London Mathematical Society Monographs*. Academic Press Inc. [Harcourt Brace Jovanovich Publishers], London, 1978.
9. H. Cohen. *A Course in Computational Algebraic Number Theory, 2nd edition*. Springer-V., 1995.
10. D. Coppersmith. Small solutions to polynomial equations, and low exponent RSA vulnerabilities. *Journal of Cryptology*, 10(4):233–260, 1997.
11. O. Goldreich, S. Goldwasser, and S. Halevi. Public-key cryptosystems from lattice reduction problems. In *Proc. of Crypto 1997*, volume 1294 of *LNCS*, pages 112–131. Springer-V., 1997.
12. D. Goldstein and A. Mayer. On the equidistribution of Hecke points. *Forum Mathematicum*, 15:165–189, 2003.
13. G. Golub and C. van Loan. *Matrix Computations*. J. Hopkins Univ. Press, 1996.
14. L. Groetschel, L. Lovász, and A. Schrijver. *Geometric Algorithms and Combinatorial Optimization*. Springer-V., 1988.
15. C. Hermite. Extraits de lettres de M. Hermite à M. Jacobi sur différents objets de la théorie des nombres, deuxième lettre. *Journal für die reine und angewandte Mathematik*, 40:279–290, 1850.
16. J. Hoffstein, J. Pipher, and J. H. Silverman. NTRU: a ring based public key cryptosystem. In *Proc. of ANTS III*, volume 1423 of *LNCS*, pages 267–288. Springer-V., 1998.
17. H.. Koy and C. P. Schnorr. Segment LLL-reduction of lattice bases with floating-point orthogonalization. In *Proc. of CALC '01*, volume 2146 of *LNCS*, pages 81–96. Springer-V., 2001.

18. A. K. Lenstra, H. W. Lenstra, Jr., and L. Lovász. Factoring polynomials with rational coefficients. *Mathematische Annalen*, 261:513–534, 1982.
19. H. W. Lenstra, Jr. Integer programming with a fixed number of variables. Technical report 81-03, Mathematisch Instituut, Universiteit van Amsterdam, 1981.
20. H. W. Lenstra, Jr. Integer programming with a fixed number of variables. *Mathematics of Operations Research*, 8(4):538–548, 1983.
21. Magma. The Magma computational algebra system for algebra, number theory and geometry. Av. at http://www.maths.usyd.edu.au:8000/u/magma/.
22. D. Micciancio and S. Goldwasser. *Complexity of lattice problems: a cryptographic perspective*. Kluwer Academic Press, 2002.
23. P. Q. Nguyen and D. Stehlé. Floating-point LLL revisited. In *Proc. of Eurocrypt 2005*, volume 3494 of *LNCS*, pages 215–233. Springer-V., 2005.
24. P. Q. Nguyen and J. Stern. The two faces of lattices in cryptology. In *Proc. of CALC '01*, volume 2146 of *LNCS*, pages 146–180. Springer-V., 2001.
25. A. M. Odlyzko. The rise and fall of knapsack cryptosystems. In *Proc. of Cryptology and Computational Number Theory*, volume 42 of *Proc. of Symposia in Applied Mathematics*, pages 75–88. AMS, 1989.
26. The SPACES Project. MPFR, a LGPL-library for multiple-precision floating-point computations with exact rounding. Av. at http://www.mpfr.org/.
27. C. P. Schnorr. A hierarchy of polynomial lattice basis reduction algorithms. *Theoretical Computer Science*, 53:201–224, 1987.
28. C. P. Schnorr. A more efficient algorithm for lattice basis reduction. *Journal of Algorithms*, 9(1):47–62, 1988.
29. C. P. Schnorr and M. Euchner. Lattice basis reduction: improved practical algorithms and solving subset sum problems. *Mathematics of Programming*, 66:181–199, 1994.
30. V. Shoup. NTL, Number Theory Library. Av. at http://www.shoup.net/ntl/.
31. C. L. Siegel. A mean value theorem in geometry of numbers. *Annals of Mathematics*, 46(2):340–347, 1945.

On the Randomness of Bits Generated
by Sufficiently Smooth Functions

Damien Stehlé

LORIA / Université Nancy 1, 615 rue du J. Botanique, 54602 Villers-lès-Nancy,
France
stehle@maths.usyd.edu.au
http://www.loria.fr/~stehle

Abstract. Elementary functions such as sin or exp may naively be considered as good generators of random bits: the bit-runs output by these functions are believed to be statistically random most of the time. Here we investigate their computational hardness: given a part of the binary expansion of $\exp x$, can one recover x? We describe a heuristic technique to address this type of questions. It relies upon Coppersmith's heuristic technique — itself based on lattice reduction — for finding the small roots of multivariate polynomials modulo an integer. For our needs, we improve the lattice construction step of Coppersmith's method: we describe a way to find a subset of a set of vectors that decreases the Minkowski theorem bound, in a rather general setup including Coppersmith-type lattices.

1   Introduction

Using expansions of real numbers is a natural idea to build pseudo-random number generators (PRNG). In the paper [2] in which Blum, Blum and Shub introduce the celebrated "$x^2 \bmod N$" PRNG, they first investigate the so-called "$1/N$" generator. A secret integer N is chosen, and the output bits are consecutive bits of the binary expansion of $1/N$, starting from a specified rank (the most significant bits are hidden, otherwise one would recover N by simply applying the inverse function). The PRNG is efficient but, unfortunately, it is cryptographically insecure: with a run of $2 \log_2 N + O(1)$ bits, one can recover N in time polynomial in the bit-size of N, and thus compute the bits that are to come next. This bit-run is not random since one can predict the remainder from the beginning of it. Instead of rationals, one could use algebraic numbers, i.e., roots of a degree d monic univariate integer polynomial $P(x)$, where d and a bound H on the magnitude of the coefficients are specified. This question was raised by Manuel Blum and answered negatively in [10]: if the first $O(d^2 + d \log H)$ bits of a root of P are known then one can recover P in polynomial time and thus compute the sequence himself.

In the present paper, we address a generalization of this type of questions to smooth mathematical functions like trigonometric functions, exponentials, logarithms, ... Let f be such a function over $[0, 1]$ and x an n-bit long secret integer.

F. Hess, S. Pauli, and M. Pohst (Eds.): ANTS 2006, LNCS 4076, pp. 257–274, 2006.

The output of the PRNG is the bit-run of the binary expansion of $f\left(\frac{x}{2^n}\right)$, starting from a specified rank: the first bits (or digits, the base being irrelevant) are kept hidden to make impossible the use of f^{-1} if it exists. These generated bits are believed to be statistically random. They would correspond to so-called normal numbers [1]. They were introduced by Borel [5] who showed they are overwhelming. Besides, statistical randomness is far weaker than unpredictability. Here we will consider the PRNG as weak if from a sequence of length polynomial in n one can recover the integer x in time polynomial in n.

The problem described above is connected to the so-called table maker's dilemma [11], a difficulty encountered while implementing an elementary mathematical function f in a given floating-point (fp for short) precision — for example in double precision (with 53-bit long mantissæ). In tight to all cases, the image $f(x)$ of the fp-number x cannot be represented exactly. The value $f(x)$ has to be rounded to $\diamond(f(x))$, where $\diamond(a)$ is a closest fp-number to a. Unfortunately, the exact value $f(x)$ can be extremely close to the middle of two consecutive representable numbers (or even worse, it could be exactly the middle) and thus many bits of $f(x)$, that is to say a very sharp approximation to $f(x)$, may be needed to compute $\diamond(f(x))$. The maximum of the number of needed bits, taken over the input fp-numbers, helps getting an efficient implementation of f: for any input x, one computes a close enough approximation to $f(x)$, and then round this approximation to the closest representable number. This quantity is usually computed by finding the fp-numbers x for which $f(x)$ has a long run of zeros, or a long run of ones, starting just after the rounding bit. Instead of "inverting" an arbitrary sequence of bits in the context of the PRNG described above, we "invert" a sequence of zeros or a sequence of ones. The overall approach we describe generalizes and improves the technique developed in [21] to find bad cases for the rounding of functions.

We tackle these issues with Coppersmith's lattice-based technique for calculating the small roots of multivariate polynomials modulo an integer [7]. It is heuristic, which is the only reason why our result is heuristic. For our needs, we improve the lattice construction step of this technique. In Coppersmith's technique, a family of polynomials is first derived from the polynomial whose roots are wanted. This family naturally gives a lattice basis and short vectors of this lattice possibly provide the wanted roots. The main difficulty is to choose cleverly the family: the goal is to find polynomials for which the Minkowski bound of the corresponding lattice is as low as possible, making possible the computation of larger roots. We present a general technique to choose a good subset of polynomials within a family of polynomials. Boneh and Durfee already presented such a technique in [3] but it applies only to very specific lattice bases. Our technique is more general, though slightly less powerful in the case of [3], and could be of interest wherever Coppersmith's method is used.

Road-Map of the Paper. In Section 2 we describe the problem we tackle and related issues. In Section 3 we give the minimal background on lattices and Coppersmith's method. We describe our algorithm in Section 4, give its complexity analysis in Section 5 and demonstrate experimentally its efficiency in Section 6. In Section 7, we discuss a few generalizations and open problems.

Notation. We define $[\![a, b]\!]$ as the set of integers in $[a, b]$. For any integer n, we let $[a, b)_n$ denote the $\frac{m}{2^n}$'s where $a \leq \frac{m}{2^n} < b$ and $m \in \mathbb{Z}$. For example $[1/2, 1)_{53}$ corresponds to the positive fp-numbers in double precision with exponent -1. For any real x, we let $\lfloor x \rfloor, \lceil x \rceil, \lfloor x \rceil$ and $(x \bmod 1)$ denote its floor, ceiling, closest integer and centered fractional part $x - \lfloor x \rceil$. For $\lceil x \rceil$ and $(x \bmod 1)$, if x is half an odd integer, we choose any of the two possibilities. In the following, vectors are written in bold and for a vector $\mathbf{x} \in \mathbb{R}^n$, $\|\mathbf{x}\|$ and $\|\mathbf{x}\|_1$ are its L_2 and L_1 norms, i.e., $\sqrt{\sum_{i=1}^n x_i^2}$ and $\sum_{i=1}^n |x_i|$. For the complexity results, we use the bit complexity model. The notation $\mathcal{P}(x_0, \ldots, x_k)$ shall be read as "polynomial in $x_0, \ldots, x_k$". Finally, all logarithms are in base 2.

2 Bits Generated by Mathematical Functions

In the present section, after describing the equation we tackle, we explain how it relates to the problems mentioned above: the PRNG and the search of bad cases for the rounding of functions. We describe our results afterwards.

2.1 The Equation to be Solved

Let $f : [0, 1] \to [\alpha, \beta]$ (for some reals α and β) be a function, and N_1, N_2, M be three integers. Let $c \in \mathbb{R}$. We are interested in solving the following equation:

$$\left| \left[N_2 \cdot f\left(\frac{x}{N_1}\right) - c \right] \bmod 1 \right| \leq \frac{1}{M}, \text{ for } x \in [\![0, N_1]\!]. \tag{1}$$

We are given an approximation c to the exact value $f(x/N_1)$, but the most significant bits are hidden. If the outputs of f behave sufficiently randomly, then as soon as $M = \Omega(N_1)$, the solution x, if there is one, should be unique.

The Pseudo-Random Number Generator. We study the following PRNG, based on a given function f. Fix two security parameters n_1 and n_2. Choose a secret seed $x_0 \in [\![0, 2^{n_1}]\!]$. Compute $f(x_0/2^{n_1})$, throw away the first bits up to the one of weight 2^{-n_2}, and then output as many as needed of the following ones. More precisely, the output corresponds to the bits of $(2^{n_2} \cdot f(x_0/2^{n_1}) \bmod 1)$, up to some rank. One must choose a large enough n_2 to make out of reach the guess of the hidden bits and the use of f^{-1} if it exists. To obtain an efficient PRNG one may choose an efficiently computable function f. For example $\sin, \log, \exp$ are computable in time quasi-linear in n_1, n_2 and the number of output bits [6]. In this context, n_1 and n_2 are thus polynomially related. Breaking this PRNG can be reduced to (1). Indeed, suppose that one has seen the first $m + 1$ output bits, giving some $y_0 \in [\![2^m, 2^{m+1} - 1]\!]$. The seed x_0 satisfies:

$$\left| \left[2^{n_2} \cdot f(x_0/2^{n_1}) - y_0/2^{m+1} \right] \bmod 1 \right| \leq 2^{-m}.$$

It is an instance of (1): take $N_1 = 2^{n_1}, N_2 = 2^{n_2}, M = 2^m$ and $c = y_0/2^{m+1}$.

The Table Maker's Dilemma. Let f be an elementary function, and n be the input-output precision for an implementation of f over $[1/2, 1)_n$. The input fp-numbers such that $\diamond(f(x))$ is hard to compute are the x's in $[1/2, 1)_n$ with:

$$|[2^{n+1} \cdot f(x) + 1/2] \text{ cmod } 1| \leq \epsilon \text{ in the case of a rounding to nearest,}$$
$$|2^n \cdot f(x) \text{ cmod } 1| \leq \epsilon \text{ in the case of a directed rounding,}$$

where directed roundings are the roundings towards $\infty, -\infty$ and 0. Obviously these equations are instances of (1). The smaller ϵ, the more difficult the computation of $\diamond(f(x))$, because a tighter approximation of the exact value needs to be computed to decide the last bit of $\diamond(f(x))$. If f behaves randomly enough and we want to find the worst input, then ϵ will be set to $\approx 2^{-n}$: we expect $O(1)$ solutions (the bits should be independent and uniformly distributed), containing the worst input. If this equation cannot be solved efficiently for this ϵ, it may be interesting to show that it has no solution for a much smaller ϵ, in order to ensure that f has no "exact output": such outputs (at the middle between two fp-numbers in the case of the rounding to nearest, and exactly a fp-number in the case of a directed rounding) are in full generality impossible to handle because even very accurate approximations do not help deciding the rounding direction.

Other Related Problems. Integer factorization can also be reduced to (1). Take an n-bit long integer $N = pq$ with $p \leq q$. Then p is a solution to:

$$\left| 2^{\lfloor n/2 \rfloor} \cdot \left(\frac{N}{2^n} \frac{1}{x/2^{\lceil n/2 \rceil}} \right) \text{ cmod } 1 \right| = 0 \text{ for } x \in \left[0, 2^{\lceil n/2 \rceil} \right].$$

Similarly, solving (1) could help obtaining integer points on curves. For example, take two integers a and b and suppose we want to find the pairs of integers (x, y) satisfying $y^2 = x^3 + ax + b$ and $0 \leq x \leq 2^n$ for an even n. Then we can consider (1) with $f(x) = \sqrt{x^3 + (a2^{-2n})x + (b2^{-3n})}$ and $N_2 = 2^{3n/2}$.

Unfortunately, our heuristic method seems to fail for algebraic functions and does not help solving the two problems above.

2.2   Description of the Results

From now on, we fix $f : [0, 1] \to [\alpha, \beta]$ and suppose that f is $\mathcal{C}^\infty$ and that its successive derivatives satisfy: $\forall i \geq 0, \forall x \in [0, 1], |f^{(i)}(x)| \leq i!K$ for some K. We suppose that the derivatives of f are efficiently computable (the first n_2 bits of $f^{(i)}(x)$ where x is n_1-bit long shall be computable in time $\mathcal{P}(i, n_1, n_2)$). We also suppose that the quantities $\max_{x \in [0,1]} |f^{(i)}(x)|$ are efficiently computable. For example, we can choose $f = \exp$ or $f = \sin$.

In the following sections, we describe and analyze an algorithm that, given as inputs N_1, N_2, M and c satisfying $MN_2 \geq N_1$, finds all the solutions to (1) in essentially (see Theorem 3 for the exact statement):

$$\mathcal{P}(\log(N_1 N_2 M)) \cdot 2^{\frac{\log^2(N_1 N_2)}{4 \log(M N_2)}} \text{ bit operations.}$$

Some comments need to be made on this statement. Firstly, the algorithm always finishes and gives the correct output but its running time is only heuristic: in the worst case, it might fall down to an exhaustive search. The heuristic assumption under which the result holds will be made explicit below. Secondly, notice that in the case of the table maker's dilemma, we get a (heuristic) $\mathcal{P}(n)2^{n/2}$ running time by choosing $N_1 = N_2 = M = 2^n$. This improves the best complexity bound previously known, i.e., a (heuristic) $\mathcal{P}(n)2^{4n/7}$ running time [21]. Finally, by choosing $M = 2^{\log^2(N_1 N_2)}$, we obtain a polynomial time algorithm that breaks the PRNG: a run of $(n_1+n_2)^2$ output bits suffices to recover the seed efficiently. Amazingly, this quadratic bound matches the result of Nesterenko and Waldschmidt [17] for exp when specialized to our context: their result implies that for $f = \exp, c = 0$ and $M \geq 2^{k(n_1+n_2)^2}$ for some constant k, there is no non-trivial solution to (1). Our work can be seen as a constructive variant of [17].

3 Preliminaries

We start this section by stating some algorithmic results on lattices (see [13] for more details) before describing Coppersmith's technique and our method for selecting a good subset of polynomials within a set of polynomials.

3.1 Lattices and the L^3 Algorithm

A lattice L is a set of all linear integer combinations of $d \leq n$ linearly independent vectors $\mathbf{b}_i$ over $\mathbb{R}$, that is $L = \{\sum_{i=1}^{d} x_i \mathbf{b}_i, x_i \in \mathbb{Z}\}$. The $\mathbf{b}_i$'s are called a basis of L. A given lattice has infinitely many bases (as soon as $d \geq 2$). The lattice dimension $\dim L = d$ does not depend on the choice of the basis, neither does the embedding dimension n. The determinant of the lattice L is defined by:

$$\det L = \prod_{i=1}^{d} \|\mathbf{b}_i^*\|, \tag{2}$$

where $[\mathbf{b}_1^*, \ldots, \mathbf{b}_d^*]$ is the Gram-Schmidt orthogonalization of $[\mathbf{b}_1, \ldots, \mathbf{b}_d]$, that is: $\mathbf{b}_1^* = \mathbf{b}_1$, and $\mathbf{b}_i^* = \mathbf{b}_i - \sum_{j=1}^{i-1} \frac{\langle \mathbf{b}_i, \mathbf{b}_j^* \rangle}{\|\mathbf{b}_j^*\|^2} \mathbf{b}_j^*$. This definition extends the usual definition of the determinant to non-square matrices (except for the sign). The determinant is a lattice invariant: it is independent of the chosen basis of L.

Most often, bases of interest are made of rather short vectors. Minkowski [16] showed that any lattice L contains a vector $\mathbf{b} \neq \mathbf{0}$ satisfying the so-called Minkowski bound:

$$\|\mathbf{b}\| \leq \sqrt{\dim L} \cdot (\det L)^{\frac{1}{\dim L}}.$$

Unfortunately, Minkowski's proof is not constructive and no efficient way to find such a short vector is known. In 1982, Lenstra, Lenstra and Lovász [12] gave a polynomial time algorithm computing a so-called L^3-reduced basis that, among others, contains a vector that satisfies a weakened version of Minkowski's bound.

Theorem 1 ([18]). *Let* $\mathbf{B}_1, \ldots, \mathbf{B}_d \in \mathbb{Z}^n$ *be independent vectors with lengths smaller than* B, *and* $d = O(\log B)$. *In* $O(d^4 n \log^2 B)$ *bit operations, one can find a basis* $\mathbf{b}_1, \ldots, \mathbf{b}_d$ *of the lattice spanned by the* $\mathbf{B}_i$'s *that satisfies:*

$$\|\mathbf{b}_1\| \leq 2^d (\det L)^{\frac{1}{d}} \quad and \quad \|\mathbf{b}_2\| \leq 2^d (\det L)^{\frac{1}{d-1}}.$$

This theorem covers up all we need to know about the L^3 algorithm for our current needs: lattice reduction is to be used as a black box. Of course, for practical issues, one should dismantle the black box and tune it for the application.

3.2 Small Roots of Bivariate Polynomials Modulo an Integer

Coppersmith's method [7] is a general technique to find all small roots of polynomial equations modulo an integer. It heavily relies upon the L^3 algorithm, that dominates the running time. It is provable in the case of univariate polynomials: if P is a degree d monic polynomial in $(\mathbb{Z}/N\mathbb{Z})[x]$, then one can find all its roots smaller than $N^{1/d}$ in time polynomial in d and $\log N$. It is only heuristic for multivariate polynomials. It has proved very powerful in public-key cryptography: the univariate variant [4,7,15] as well as the multivariate one [3,8,14].

Suppose we search the solutions to the equation:

$$P(x, y) = 0 \mod N, \tag{3}$$

where P is a bivariate polynomial with integer coefficients, the modulus N is an integer, and x and y are integer unknowns. Since in general solving such a polynomial equation is hard, we restrict ourselves to finding the small solutions: $|x| \leq X$ and $|y| \leq Y$, for some bounds X and Y that are as large as possible.

Coppersmith's technique depends on an integer parameter $\alpha \geq 1$ to be chosen to maximize the reachable bounds X and Y (most often, α growing to infinity is asymptotically the optimal choice). The method is made of four main steps.

1. First, a large set $\mathbb{P}$ of polynomial equations modulo N^α is derived from (3). We use powers of P shifted by powers of variables: $N^{\alpha-i} P(x, y)^i x^j y^k$ for $i \in [\![0, \alpha]\!]$ and $j, k \geq 0$. This is the *polynomials selection step*. If (x_0, y_0) is a solution to (3), then it must be a root modulo N^α of all the polynomials in $\mathbb{P}$.
2. In the *polynomials-to-lattice step*, we transform the family of polynomials $\mathbb{P}$ into a lattice $L_{X,Y}[\mathbb{P}]$. We list and sort (arbitrarily) the monomials $x^j y^k$ appearing in the polynomials of $\mathbb{P}$. Suppose there are n such monomials. If a polynomial $Q(x, y)$ has all its monomials belonging to the monomials appearing in the selected family, then we map it to an n-dimensional vector whose coefficient corresponding to the monomial $x^j y^k$ is the coefficient of $Q(x \cdot X, y \cdot Y)$ for this monomial. This map is obviously a bijection. The lattice $L_{X,Y}[\mathbb{P}]$ we consider is spanned by the vectors of $\mathbb{R}^n$ that are obtained from the selected polynomials *via* the map described above. Since any vector of this lattice is an integral linear combination of the vectors corresponding to the selected polynomials, any solution (x_0, y_0) to (3) is a root modulo N^α of all the polynomials corresponding to the vectors of $L_{X,Y}[\mathbb{P}]$.

3. If the polynomials of $\mathbb{P}$ were linearly independent, we got a lattice basis in the previous step, and we now run an L^3-type algorithm on it. If the polynomials are linearly dependent, this is not a problem since L^3 can be modified to manage generating vectors instead of basis vectors (but the analysis of the method becomes more intricate). After this *lattice-reduction step*, which is the computationally dominating step, we have a basis of L made of vectors whose lengths are related to $\det(L_{X,Y}[\mathbb{P}])$, as described in Theorem 1.

4. In the reduced basis of $L_{X,Y}[\mathbb{P}]$, we take all the vectors of L_1 norm $< N^\alpha$. Any solution (x_0, y_0) to (3) modulo N is a root over $\mathbb{Z}$ of all the polynomials corresponding to these vectors. It therefore remains to solve these equations over the integers. We call this the *root-finding step*. There are several ways to perform this step: with any variable elimination method (for example through a resultant computation) or with Hensel's lifting. In all cases, we need at least two lattice vectors with small L_1 norm to solve the system of equations. Furthermore these polynomials can share factors, in which case it may be impossible to recover any useful information. It is not known how to work around this difficulty. This step is the heuristic one in Coppersmith's method for bivariate polynomials. This is the reason why the running-time bound of our method for solving (1) is only heuristic. At the end, all the possible solutions to (3) need to be checked, since some might be spurious.

The Heuristic Assumption. In the present paper, we do the following heuristic assumption: if two polynomials correspond to the first two vectors of an L^3-reduced basis computed during any lattice reduction step of Coppersmith's bivariate method, then they do not share any factor.

Theorem 1 ensures we will obtain two sufficiently short vectors, as long as:

$$\sqrt{n} 2^{\dim L[\mathbb{P}]} \cdot (\det L[\mathbb{P}])^{\frac{1}{\dim L[\mathbb{P}] - 1}} < N^\alpha.$$

When expliciting the above inequality as a relation on X and Y, we obtain what we call the *Coppersmith equation*. The goal of the analysis of a particular use of Coppersmith's method is to find the family $\mathbb{P}$ providing the best Coppersmith equation, that is to say the one allowing the largest reachable X and Y. The target is thus to minimize Minkowski's quantity $(\det L[\mathbb{P}])^{\frac{1}{\dim L[\mathbb{P}] - 1}}$.

3.3 Finding a Good Family of Polynomials

As we have seen above, the strength of Coppersmith's method is determined by the polynomials selection. Often, the family $\mathbb{P}$ is chosen so that the corresponding lattice basis is square and triangular. This makes the determinant computation simple (the determinant being in this case the product of the absolute values of the diagonal entries) and this often gives the "good" bound, which means that after many trials, one could not find a better family of polynomials. Nevertheless, in some cases, one can improve Minkowski's bound by choosing polynomials giving a non-square matrix. This is the case for example in [3] and in our present situation. Here we give a mean to bound the determinant of lattices given by bases that are not necessarily square and triangular.

Suppose we have a $d \times n$ matrix B whose rows span a lattice L. Suppose further that the entries of B satisfy: $|B_{i,j}| \le wr_i \cdot wc_j$, for some wr_i's and wc_j's (with $i \le d$ and $j \le n$). This is the case for all Coppersmith-type lattice bases. We say that such a matrix is bounded by the products of the wr_i's and wc_j's.

Theorem 2. *Let B be a $d \times n$ matrix bounded by the product of some quantities wr_i's and wc_j's. Let $L[B]$ be the lattice spanned by the rows of the matrix B, and P the product of the d largest wc_j's. We have:*

$$\det L[B] \le 2^{\frac{d(d-1)}{2}} \cdot \sqrt{n} \cdot \left(\prod_{i=1}^{d} wr_i \right) \cdot P.$$

The result follows from basic row and column operations and Hadamard's bound $\det L[B] \le \prod_{i \le d} \|\mathbf{b}_i\|$, where the $\mathbf{b}_i$'s are the vectors corresponding to the rows of the matrix B. The proof is given in appendix. Theorem 2 can be used to find a subset of a given set of vectors that improves the term "$(\det L)^{\frac{1}{\dim L}}$" in Minkowski's bound. Suppose we have a $d \times n$ matrix B bounded by the product of some wr_i's and wc_j's. We consider the d vectors given by the rows of B. We begin by ordering the wc_j's decreasingly and the wr_i's increasingly. Then the subset of cardinality k that gives the best bound *via* Theorem 2 is the one corresponding to any k smallest wr_k's. We compute the quantities $\left(\prod_{i=1}^{k} wr_i \cdot wc_{n-i} \right)^{1/k}$ for all $k \le d$, and keep the vectors giving the smallest value.

Notice that our method does not necessarily gives the best subset of a given set of vectors: in particular, it makes no use of the possibly special shape (not even triangular) of the coefficients. E.g., it fails to give the 0..292 bound of [3].

4   The Algorithm and Its Correctness

The algorithm we study is described in Figure 1. It takes as input the quantities N_1, N_2, M and c, as well as three parameters T, d and α that will be chosen in order to improve the efficiency of the algorithm. The output are all the solutions to (1) for the given N_1, N_2, M and c. The overall architecture of the algorithm is as follows. The initial search interval $[\![0, N_1]\!]$ is divided into $\frac{N_1}{2T}$ subintervals of length $2T$. Each subinterval is considered independently (possibly on different machines): for each of them, we approximate the function f by a degree d polynomial P; we solve (1) for P instead of f with a smaller M (to take care of the distance between f and P); to perform this last step, we use Coppersmith's method with the bivariate polynomial $P(x) + y$.

At Step 1, we compute the embedding dimension n of the lattices we will reduce, and the list of the monomials that will appear in the polynomials generated during Coppersmith's method. During the execution of the algorithm, the integer t_0 increases: at any moment, all the solutions to (1) below t_0 have already been found and it remains to find those that are between t_0 and N_1. The set S contains all the solutions to (1) that are $\le t_0$. Finally, the value T' is half the

---

Input: $N_1, N_2, M \in \mathbb{Z}, c \in \mathbb{R}$ with finitely many bits. Three parameters $t, d, \alpha \in \mathbb{Z}$.

Output: All the x's in $[\![0, N_1]\!]$ that are solution to (1).

1. $n := \frac{(\alpha+1)(d\alpha+2)}{2}$, $\{e_1, \ldots, e_n\} := \{x^i y^j, i + dj \leq d\alpha\}$, $T := 2^t, T' := T, S := \emptyset$.

2. $t_0 := 0$. While $t_0 \leq N_1$, do

3.   If $t_0 + 2T' \geq N_1, T' := \lfloor \frac{N_1 - t_0}{2} \rfloor$.

4.   If $T' = 0$, then

5.   Add t_0 in S if it is solution to (1).

6. $t_0 := t_0 + 1, T' := T$.

7.   Else

8. $t_m := t_0 + T', P(x) := c + f\left(\frac{t_m}{N_1}\right) + f'\left(\frac{t_m}{N_1}\right)\frac{x}{N_1} + \ldots + \frac{1}{d!}f^{(d)}\left(\frac{t_m}{N_1}\right)\frac{x^d}{N_1^d}$.

9. $\epsilon := \left(\max_{x \in [0,1]}\left|f^{(d+1)}(x)\right|\right) \cdot \frac{N_2}{(d+1)!}\left(\frac{T'}{N_1}\right)^{d+1}$, $M' := \frac{1}{\epsilon + 1/M}$.

10. $\left\{g_1, \ldots, g_{\frac{\alpha(\alpha+1)}{2}}\right\} := \{x^i (N_2 P(x) + y)^j, i + j \leq \alpha\}$.

11.   Create the $\frac{\alpha(\alpha+1)}{2} \times n$ matrix B such that $B_{k,l}$ is the coefficient of the monomial e_l in the polynomial $g_k\left(xT', \frac{1}{M'}y\right)$.

12. L^3-reduce the rows of B. Let $\mathbf{b}_1, \mathbf{b}_2$ be the first two output vectors.

13. $test := 1$. If $\|\mathbf{b}_1\|_1 \geq 1$ or $\|\mathbf{b}_2\|_1 \geq 1, test := 0$.

14.   Let $Q_1(x,y), Q_2(x,y)$ be the polynomials corresponding to $\mathbf{b}_1$ and $\mathbf{b}_2$.

15. $R(x) := \mathrm{Res}_y(Q_1(x,y), Q_2(x,y))$. If $R(x) = 0$, then $test := 0$.

16.   If $test = 0$, then $T' := \lfloor T'/2 \rfloor$.

17.   Else, for any root x_0 of R belonging to $[\![-T', T']\!]$, add $t_m + x_0$ in S if it is a solution to (1), $t_0 := t_0 + 2T' + 1, T' := T$.

18. Return S.

<figure>**Fig. 1.** The algorithm solving (1)</figure>

size of the current subinterval in which we are searching solutions to (1): usually, we have $T' = T$, but if Coppersmith's method fails, then we halve T'. The quantity T' makes the algorithm valid even if Coppersmith's method repeatedly fails. As a drawback the running time bound only heuristic. If there are too many consecutive failures of Coppersmith's method, then we might have $T' = 0$: in this case we test if the current value t_0 is a solution to (1) (Step 5). At Step 8, we approximate the function f by its degree d Taylor expansion P at the center of the considered interval. At Step 9, we compute the error ϵ made by approximating f by P. We update M accordingly. At Step 10, we generate the family of polynomials that will be used in Coppersmith's method for the bivariate polynomial $N_2 P(x) + y$: we are searching the roots (x_0, y_0) of $N_2 P(x) + y$ modulo 1 such that $|x_0| \leq T'$ and $|y_0| \leq 1/M'$. Coppersmith's method can fail for two reasons: either we do not find two vectors of small enough L_1 norm (this is detected at Step 13), or the two bivariate polynomials corresponding to these vectors share a factor (this is detected at Step 15). If Coppersmith's method does not fail, we go to Step 17: all the solutions of (1) that are in the considered subinterval must be roots of the y-resultant of the polynomials corresponding to the two small vectors that we found. If this polynomial has no integer root in the considered subinterval, then it means there were no solution to (1) in this subinterval; if it has roots, we test them to avoid those that are not solution to (1).

In the algorithm of Figure 1, the calculations are described with real numbers (at Steps 8 to 15). It is possible to replace these real numbers by others with finitely many bits, or by integers. At Step 8, we can replace the polynomial P by a polynomial $\tilde{P}$ whose coefficients approximate those of P: it suffices that $\max_{x \in [-T', T']} |N_2 \cdot [P(x) - \tilde{P}(x)]| = O(1/M)$. This can be ensured by taking the $O(\log M + \log N_2 + d \log T) = O(\log M + \log N_2 + d \log N_1)$ most significant bits of each coefficient. At Step 9, we can take $M' := \left\lfloor \frac{1}{2\epsilon + 1/M} \right\rfloor$ instead of $M' := \frac{1}{\epsilon + 1/M}$. The computations of Step 10 and 11 are then performed exactly. At Step 12, we need integer entries to use Theorem 1. Since all the entries of the matrix B are reals with finitely many bits, it suffices to multiply them by a sufficiently large power of 2: since we took reals with $O(\log M + \log N_2 + d \log N_1)$ bits to construct the polynomial $\tilde{P}$, multiplying by 2^ℓ with $\ell = O(\alpha(\log M + \log N_2 + d \log N_1))$ is sufficient. Once these modifications are performed, the remaining steps of the algorithm compute over the integers. In the following, we will keep the initial description (of Figure 1) to avoid unnecessary technicalities.

The main result of the paper is the following:

Theorem 3. *Let $N_1, N_2, M \in \mathbb{Z}$ and $c \in \mathbb{R}$ with finitely many bits. Let $t, d, \alpha \geq 0$. Given $N_1, N_2, M, c, t, d, \alpha$ as input, the algorithm of Figure 1 outputs all the solutions $x \in [\![0, N_1]\!]$ to (1). If test is never set to 0 at Step 15, the algorithm finishes in time $\mathcal{P}(n_1, n_2, m, d, \alpha) \frac{N_1}{2^t}$, as long as $N_1 \leq M N_2, \log d = O(\alpha)$ and:*

$$t \leq \min\left(n_1 - \frac{m + n_2 + O(1)}{d+1}, n_1 - \frac{(n_1 + n_2)^2}{4(m+n_2)}(1 + \epsilon_1) + \epsilon_2\right),$$

with $n_1 = \log N_1, n_2 = \log N_2, m = \log M$ and, for α growing to ∞:

$$\epsilon_1 = O(1/\alpha) + dO(1/\alpha^2)$$
$$\epsilon_2 = \frac{1}{m + n_2} O(\alpha^2) + \frac{n_1}{m + n_2}(O(\alpha) + dO(1/\alpha))$$
$$+ (n_1 + n_2)(O(1/\alpha) + d(1/\alpha^3)) + mO(1/\alpha^2).$$

Corollary 4 (Table Maker's Dilemma). *Let $N_1 = 2^n$ and $N_2 = 2^{n+e}$ with $e \in \{0, 1\}$. Let $\epsilon > 0$. Suppose that $d = 3, M = 2^n$, and $t = \frac{n}{2}(1-\epsilon)$. Suppose also that test is never set to 0 at Step 15. Then one can choose α, d and t such that the algorithm of Figure 1 finds the solutions to (1) in time $\mathcal{P}(n) \cdot 2^{\frac{n}{2}(1+\epsilon)}$.*

Corollary 5 (Inverting the PRNG). *Let $N_1 = 2^{n_1}$ and $N_2 = 2^{n_2}$. Suppose that $M = 2^{(n_1 + n_2)^2}$. Suppose also that test is never set to 0 at Step 15. Then one can choose α, d and t such that the algorithm of Figure 1 finds the solutions to (1) in time polynomial in n_1 and n_2.*

The following table gives the parameters providing the two corollaries above.

	M	α	d	t
Corollary 4	2^n	$O(n)$	3	$n/2$
Corollary 5	$2^{(n_1+n_2)^2}$	$O(n_1 + n_2)$	$O((n_1 + n_2)^2)$	$n_1 + O(1)$

We can give a precise complexity estimate in Corollary 5 by using Theorem 1. The most expensive step of the algorithm is Step 12. The dimension of the lattice is $O(\alpha^2) = O((n_1 + n_2)^2)$, its embedding dimension is $O(d\alpha^2) = O((n_1 + n_2)^4)$, and the entries, when considered as integers (see the discussion above) are of length $O((n_1 + n_2)^4)$. Therefore, the overall cost is $O((n_1 + n_2)^{20})$.

Proof of Correctness of the Algorithm. Since we test any returned solution, it suffices to check that we do not miss any. Let $x_0 \in [\![0, N_1]\!]$ be such a solution. The integer x_0 belongs to at least one of the considered subintervals: let $[\![t_m - T', t_m + T']\!]$ be the smallest of them. For this subinterval, at Step 15 we have $test = 1$. Besides, for any $x \in [\![t_m - T', t_m + T']\!]$ we have, with $x' = x - t_m$:

$$|N_2 P(x') \text{ cmod } 1| \leq |N_2 P(x') - N_2 f(x/N_1) - c| + |(N_2 f(x/N_1) + c) \text{ cmod } 1|$$
$$\leq \epsilon + 1/M \leq 1/M'.$$

As a consequence, we have $|N_2 P(x_0 - t_m) \text{ cmod } 1| \leq 1/M'$. Let $y_0 = -N_2 P(x_0 - t_m) \text{ cmod } 1$. The pair $(x_0 - t_m, y_0)$ is a root of the bivariate polynomial $N_2 P(x) + y$ modulo 1. It is therefore a root of any of the g_k's modulo 1, and of their integer linear combinations. In particular, it is a root of Q_1 et Q_2 modulo 1.

Besides, we have the inequalities $|x_0 - t_m| \leq T'$ and $|y_0| \leq 1/M'$. Since the L_1 norms of the vectors $\mathbf{b}_1$ and $\mathbf{b}_2$ are smaller than 1, the pair $(x_0 - t_m, y_0)$ is a root of the polynomials Q_1 and Q_2 over $\mathbb{Z}$. It implies that if $R = \text{Res}_y(Q_1(x, y), Q_2(x, y))$ is non-zero, then we find x_0 at Step 17 of the algorithm.

5   Analysis of the Algorithm

This section is devoted to proving the complexity statement of Theorem 3. We suppose that $test$ is never set to 0 at any Step 15. It implies that with a correct choice of the input parameters we always have $T' = T$ (except possibly when t_0 is close to N_1). The definition of M' gives $M' = O(M)$. Besides, Taylor's theorem and the condition $t \leq n_1 - \frac{m + n_2 + O(1)}{d+1}$ of Theorem 3 ensure that for any $x \in [\![t_m - T, t_m + T]\!]$ the quantity ϵ computed at Step 9 satisfies $\epsilon = O(1/M)$. We thus have $M' = \Theta(M)$. To simplify, we identify M' with M and T' with T.

Our goal is to prove that if the conditions of Theorem 3 are fulfilled, then $test$ if never set to 0 at Step 13, which means that the L_1 norms of the first two vectors output by L^3 are smaller than 1. Theorem 1 ensures that if the following condition is satisfied, then $\mathbf{b}_1$ and $\mathbf{b}_2$ will be short enough:

$$\sqrt{\frac{d\alpha(\alpha + 1)}{2}} \cdot 2^{O\left(\frac{\log M + \log N_2 + d \log N_1}{\alpha - 1}\right)} \cdot 2^{\frac{\alpha(\alpha+1)}{2}} \cdot (\det L[\mathbb{P}])^{\frac{2}{\alpha(\alpha-1)}} < 1, \quad (4)$$

where we used the classical relation between the L_1 and L_2 norms, took care of the fact that we have to scale the lattice to the integers to use Theorem 1, and $\mathbb{P}$ is the family of polynomials $\{x^i (N_2 P(x) + y)^j, 0 \leq i + j \leq \alpha\}$.

Let B be the $\frac{\alpha(\alpha+1)}{2} \times \frac{d\alpha(\alpha+1)}{2}$ matrix where the entry $B[(i, j); (i', j')]$ is the coefficient of the polynomial $Q_{i,j}(x, y) = (xT)^i (N_2 P(xT) + y/M)^j$ corresponding to the monomial $x^{i'} y^{j'}$. We order the rows and columns of B by increasing value of $i + dj$, and by increasing value of j in case of equality. We can write:

$$B = \begin{pmatrix} B_1 & 0 \\ B_2 & B_3 \end{pmatrix},$$

where B_1 is square, lower-triangular and corresponds to the rows such that $0 \leq i + dj \leq \alpha$, and B_3 is rectangular and corresponds to the rows such that $\alpha < i + dj \leq d\alpha$ and $0 \leq i + j \leq \alpha$. We easily obtain $\det L[\mathbb{P}] = |\det B_1| \cdot |\det B_3|$.

Lemma 6. *We have* $|\det B_1| = T^{\frac{1}{6d}(\alpha^3 + O(\alpha^2))} M^{-\frac{1}{6d^2}(\alpha^3 + O(\alpha^2))}$.

We are to bound the determinant of the matrix B_3 by using Theorem 2.

Lemma 7. *The matrix B_3 is bounded by the product of the quantities:*

$$wr_{i,j} = 2^j (d+1)^j K^j N_1^i N_2^j \quad and \quad wc_{i',j'} = T^{i'} / (M^{j'} N_1^{i'} N_2^{j'}),$$

with $i + j \in [\![0, \alpha]\!]$, $i + dj \in [\![\alpha + 1, d\alpha]\!]$ *and* $i' + j' \in [\![\alpha + 1, d\alpha]\!]$.

Theorem 2 gives the inequality $|\det B_3| \leq 2^{O(\alpha^4)} \cdot \sqrt{d} \cdot R \cdot P$, where R is the product of the $wr_{i,j}$'s with $i + j \in [\![0, \alpha]\!]$ and $i + dj \in [\![\alpha + 1, d\alpha]\!]$, and P is the product of the $(\dim B_3)$ largest $wc_{i,j}$'s with $i + dj \in [\![\alpha + 1, d\alpha]\!]$.

Lemma 8. *We have the following relations, for α growing to infinity:*

$$\dim B_3 = \left(\tfrac{1}{2} - \tfrac{1}{2d}\right)(\alpha^2 + O(\alpha)),$$
$$R = 2^{O(\alpha^3)}(d+1)^{O(\alpha^3)} N_1^{\frac{d-1}{6d}(\alpha^3 + O(\alpha^2))} N_2^{\frac{d^2-1}{6d^2}(\alpha^3 + O(\alpha^2))}.$$

In order to bound P, we write $P \leq 2^{O(\alpha^3)} \prod_{k=\tau}^{k_{\max}} 2^{k \cdot c_k}$, with $c_k = \#\{(i,j), \alpha < i + dj \leq d\alpha$ and $k \leq (t - n_1)i - (m + n_2)j < k + 1\}$. We also define the parameter $\tau = -\sqrt{(n_1 - t)(m + n_2)}\alpha$. It is fixed so that $\sum_{k=\tau}^{k_{\max}} c_k \approx \dim B_3$, which means that we take sufficiently many columns. Finally we define:

$$k_{\max} = \begin{cases} \lfloor \alpha(t - n_1) \rfloor & \text{if } MN_2 \geq (N_1/T)^d, \\ \lfloor -(m + n_2)\alpha/d \rfloor & \text{if } MN_2 \leq (N_1/T)^d. \end{cases}$$

Figure 2 shows the relations between the variables i, j and k when $MN_2 \in [(N_1/T)^d, (N_1/T)^{d+1}]$. The dashed area corresponds to the submatrix B_1 of B and is not considered in the study of B_3. We split the set of valid pairs (i,j) (the largest triangle, without the dashed area) into three zones Z_1, Z_2 and Z_3 depending on the value of k. In the lemma below we consider only Z_1 and Z_2 since $\tau\alpha$ corresponds to an index k belonging to Z_2.

Lemma 9. *If $MN_2 \in [N_1/T, (N_1/T)^{d+1}]$ and α grows to infinity, then:*

$$\sum_{k=\tau}^{k_{\max}} c_k = \left(\frac{1}{2} - \frac{1}{2d}\right)(\alpha^2 + O(\alpha)) = (\dim B_3)(1 + O(1/\alpha))$$

$$\sum_{k=\tau}^{k_{\max}} kc_k = \frac{(n_1 - t)d + (m + n_2) - 2d^2\sqrt{(n_1 - t)(m + n_2)}}{6d^2}(\alpha^3 + O(\alpha^2)).$$

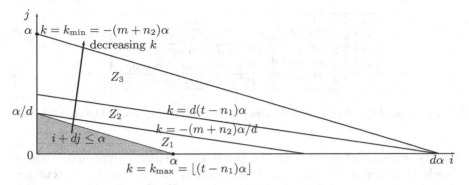

Fig. 2. Relations between i, j and k when $(N_1/T)^d \leq MN_2 \leq (N_1/T)^{d+1}$

We can now batch the partial results of Lemmata 6, 8 and 9.

Theorem 10. *If $N_1/T \leq MN_2 \leq (N_1/T)^{d+1}$ and α grows to infinity, we have:*

$$\det L[\mathbb{P}] \leq 2^{O(\alpha^4)} \cdot (N_1 N_2)^{\frac{\alpha^3}{6}+O(\alpha^2)} \cdot 2^{-\frac{1}{3}\sqrt{(n_1-t)(m+n_2)}(\alpha^3+O(\alpha^2))} \cdot (MT)^{O(\alpha^2)}.$$

By using (4) and Theorem 10, it is easy to end the proof of Theorem 3.

6   Experimental Data

A C implementation of the algorithm is available at the URL `http://www.loria.fr/~stehle/bacsel.html`. The code relies on GNU MP [9] for the integer arithmetic, on MPFR [19] for the *fp*-arithmetic and on a *fp*-L^3 available at the URL `http://www.loria.fr/~stehle/fpLLL-1.3.tar.gz`. The code is not meant to be efficient, but to be a proof of feasibility. The timings given in Figure 3 are thus overestimations of what may be possible with a more accurate implementation. Tuning the code is not an obvious task since the algorithm depends on many parameters. A previous implementation for the particular parameters $d = \alpha = 2$ was written by Paul Zimmermann and is available at the URL `http://www.loria.fr/~zimmerma/free/wclr-1.6.1.tar.gz`. This former implementation was used to find the worst cases for the correct rounding of the function 2^x over $[1/2, 1)$ in double extended precision (64-bits mantissæ) for all rounding modes. For example, the worst case for the rounding to nearest is:

$$2^{\frac{15741665561440311501}{2^{64}}} = \underbrace{1.110\ldots110}_{64}\underbrace{10\ldots0}_{63}11\ldots$$

The corresponding computation lasted a time equivalent to ≈ 7 years on a single Opteron 2.4 GhZ. With the new code and $d = 3$ and $\alpha = 2$, this computation should be speeded up significantly. Nevertheless, for the application to the table maker's dilemma, $n = 64$ seems to be the bound of feasibility. In particular, the quadruple precision (113-bit mantissæ) remains far out of reach.

n	d α M T	time
53	3 2 2^{53} $2^{20.45}$	7.7 days
64	2 2 2^{64} $2^{23.95}$	3.2 years (estimated time)
64	3 2 2^{64} $2^{24.60}$	2.6 years (estimated time)
113	3 2 2^{113} $2^{44.45}$	$3 \cdot 10^9$ years (estimated time)

Fig. 3. Estimated time to find a worst case of exp over $[1/2, 1)$, on an Opteron 2.4 GhZ

For the inversion of the PRNG, the complexity, though polynomial, remains too high for extensive computations with a growing value of n. As an example, we found the seed $x_0 = 17832265507654526358 \cdot 2^{-64} \in [1/2, 1)_{64}$ such that:

$$\exp x = b_{-1}b_0.b_1b_2 \ldots b_{64}c_1c_2c_3 \ldots c_{400} \ldots,$$

with the 400 bits c_i known. The computation was performed in time equivalent to less than one week on an Opteron 2.4 GhZ.

For the transcendental functions we tried, we did not encounter a single failure of Coppersmith's method. On the contrary, the method failed for all algebraic functions f that we tried: in this case the running-time is between what is expected and that of an exhaustive search, and it does not seem to decrease when we increase the parameter d. We have two heuristic explanations for this phenomenon: firstly, for some algebraic functions, there can be too many solutions to write them in polynomial time; secondly, as described in Section 2, it would give a polynomial time algorithm for integer factorization.

7   Generalizations and Open Problems

One can extend the algorithm and its analysis to the case of functions of several variables [20]. For the PRNG, if the input precision is of n bits for all variables, and the first n bits of the output are kept hidden, the number of bits needed to recover the multivariable seed is $O(n^{k+1})$, where k is the number of variables.

An open problem related to our algorithm is to prove Theorem 3 without the heuristic assumption on the resultant computation. Since the method fails in practice for some functions (in particular for algebraic functions), the task would be to give a sufficient condition on the function f for the method to work. An intermediate question is to determine under which conditions Coppersmith's method can be made provable for multivariate polynomials.

Another interesting problem is to determine if the $O(n^2)$ bound is the best possible: can we invert in polynomial time the PRNG with significantly fewer than n^2 bits? This bound matches the one of [17] and might thus be considered as somehow natural. The gap between the probabilistic injectivity of the PRNG ($m = O(n)$) and its polynomial-time invertibility ($m = O(n^2)$) is puzzling.

Acknowledgments. The author thanks Guillaume Hanrot, Vincent Lefèvre and Paul Zimmermann for their helpful comments. The writing of the present paper was completed while the author was visiting the University of Sydney, whose hospitality is gratefully acknowledged.

References

1. D. H. Bailey and R. E. Crandall. Random generators and normal numbers. *Experimental Mathematics*, 11(4):527–546, 2002.
2. L. Blum, M. Blum, and M. Shub. A simple unpredictable pseudo-random number generator. *SIAM Journal on Computing*, 15(2):364–383, 1986.
3. D. Boneh and G. Durfee. Cryptanalysis of RSA with private key d less than $N^{0.292}$. *IEEE Transactions on Information Theory*, 46(4):233–260, 2000.
4. D. Boneh, G. Durfee, and N. Howgrave-Graham. Factoring $n = pq^r$ for large r. In *Proc. of Eurocrypt 1999*, volume 1666 of *LNCS*, pages 326–337. Springer-V., 1999.
5. É. Borel. Les probabilités dénombrables et leurs applications arithmétiques. *Rendiconti del Circolo Matematico di Palermo*, 27:247–271, 1909.
6. R. Brent. Fast multiple precision zero-finding methods and the complexity of elementary function evaluation. *Journal of the ACM*, 23:242–251, 1976.
7. D. Coppersmith. Small solutions to polynomial equations, and low exponent RSA vulnerabilities. *Journal of Cryptology*, 10(4):233–260, 1997.
8. M. Ernst, E. Jochens, A. May, and B. de Weger. Partial key exposure attacks on RSA up to full size exponents. In *Proc. of Eurocrypt 2005*, number 3494 in LNCS, pages 371–386. Springer-V., 2005.
9. T. Granlund. The GNU MP Bignum Library. Available at http://www.swox.com/.
10. R. Kannan, A. K. Lenstra, and L. Lovász. Polynomial factorization and non-randomness of bits of algebraic and some transcendental numbers. In *Proc. of STOC 1984*, pages 191–200. ACM, 1984.
11. V. Lefèvre. *Moyens arithmétiques pour un calcul fiable*. PhD thesis, ÉNS Lyon, 2000.
12. A. K. Lenstra, H. W. Lenstra, Jr., and L. Lovász. Factoring polynomials with rational coefficients. *Mathematische Annalen*, 261:513–534, 1982.
13. L. Lovász. *An Algorithmic Theory of Numbers, Graphs and Convexity*. SIAM Publications, 1986. CBMS-NSF Regional Conference Series in Applied Mathematics.
14. A. May. *New RSA Vulnerabilities Using Lattice Reduction Methods*. PhD thesis, University of Paderborn, 2003.
15. A. May. Computing the RSA secret key is determinisitic polynomial time equivalent to factoring. In *Proc. of Crypto 2004*, volume 3152 of *LNCS*, pages 213–219. Springer-V., 2004.
16. H. Minkowski. *Geometrie der Zahlen*. Teubner-Verlag, 1896.
17. Yu. V. Nesterenko and M. Waldschmidt. On the approximation of the values of exponential function and logarithm by algebraic numbers. *Matematicheskie Zapiski*, 2:23–42, 1996.
18. P. Nguyen and D. Stehlé. Floating-point LLL revisited. In *Proc. of Eurocrypt 2005*, volume 3494 of *LNCS*, pages 215–233. Springer-V., 2005.
19. The SPACES Project. MPFR, a LGPL-library for multiple-precision floating-point computations with exact rounding. Available at http://www.mpfr.org/.
20. D. Stehlé. *Algorithmique de la réduction de réseaux et application à la recherche de pires cas pour l'arrondi de fonctions mathématiques*. PhD thesis, Université Nancy 1, 2005.
21. D. Stehlé, V. Lefèvre, and P. Zimmermann. Searching worst cases of a one-variable function. *IEEE Transactions on Computers*, 54(3):340–346, 2005.

Missing Proofs

Proof of Theorem 2. Recall that the determinant is defined by (2). If one of the wr_i's is zero, the determinant is obviously zero and the result holds. Suppose now that all wr_i's are positive. Let B' be the matrix B after having divided the i-th row by wr_i for all i. Wlog we suppose that $wc_1 \geq wc_2 \geq \ldots \geq wc_n$ (otherwise we perform a permutation of the columns, which does not change the determinant). It suffices to show that: $\det B' \leq 2^{\frac{d(d-1)}{2}} \sqrt{n} \cdot wc_1 \cdot \ldots \cdot wc_d$.

We prove it by induction on d and n. If $n < d$, this is obvious since the rows must be linearly dependent, so that $\det B' = 0$. If $d = 1$, the determinant is exactly the length of the unique vector. Suppose now that $n \geq d \geq 2$. We apply a permutation on the rows of the matrix B' in order to have $|B'_{1,1}| \geq |B'_{i,1}|$ for any i. This last operation does not change the determinant. For all $i \geq 2$, we perform the transformation $B'_i := B'_i - \frac{B'_{i,1}}{B'_{1,1}} B'_1$, which does not change the determinant either (B'_i is the i-th row of the matrix B'). Wlog we suppose that $B'_{1,1} \neq 0$: otherwise all the $B'_{i,1}$'s are zero, and we obtain the result by induction on n by removing the first column. The transformations we have performed on the matrix B' have given a new matrix B' for which $|B'_{1,1}| \leq wc_1$, $B'_{i,1} = 0$ for any $i \geq 2$ and $|B'_{i,j}| \leq 2wc_j$, for any pair (i,j). This last statement follows from the fact that for any $i \leq d$, we have $\left| \frac{B'_{i,1}}{B'_{1,1}} \right| \leq 1$. We apply the result inductively on the $(d-1) \times n$ matrix B'' at the bottom of the matrix B':

$$\det B'' \leq 2^{\frac{(d-1)(d-2)}{2}} \sqrt{n} \cdot (2wc_2) \cdot \ldots \cdot (2wc_d) \leq 2^{\frac{d(d-1)}{2}} \sqrt{n} \cdot wc_2 \cdot \ldots \cdot wc_d.$$

Proof of Corollary 4. Let $m = n_1 = n_2 = n$ and $d = 3$. If α grows to infinity, we have: $\epsilon_1 = O(1/\alpha), \epsilon_2 = 1/n \cdot O(\alpha^2) + O(\alpha) + nO(1/\alpha)$. We fix α sufficiently large to ensure that the terms "$O(1/\alpha)$" of ϵ_1 and ϵ_2 become smaller than ϵ (with absolute values). We get, for n growing to infinity: $|\epsilon_1| \leq \epsilon, |\epsilon_2| \leq O(1) + n\epsilon$. Finally, the equation to be satisfied becomes, for n growing to infinity:

$$t \leq \min\left(\frac{n}{2} - O(1), n - \frac{n}{2}(1+\epsilon) - n\epsilon + O(1) \right).$$

For n larger than some constant, this inequality is satisfied if $t = n(1 - 4\epsilon)/2$.

Proof of Corollary 5. Let $d = O((n_1 + n_2)^2)$ and $\alpha = O(n_1 + n_2)$. Then for $n_1 + n_2$ growing to infinity, we have $\epsilon_1 = O(1)$ and $\epsilon_2 = O(1)$. The equation to be satisfied becomes $t \leq n_1 - O(1)$, for $n_1 + n_2$ growing to infinity.

Proof of Lemma 6. The determinant of B_1 is:

$$\prod_{i+dj \leq \alpha} (xT)^i \cdot (P(xT) + \frac{y}{M})^j [x^i y^j] = \prod_{i+dj \leq \alpha} T^i M^{-j} = (T^{\frac{1}{6d}} M^{\frac{-1}{6d^2}})^{\alpha^3 + O(\alpha^2)}.$$

Proof of Lemma 7. The row (i,j) of B_3 with $i + j \in [\![0,\alpha]\!]$ and $i + dj \in [\![\alpha + 1, d\alpha]\!]$ corresponds to the polynomial $(xT)^i \left(P(xT) + \frac{y}{M} \right)^j$. Its column $(i' + j')$ with $i' + j' \in [\![\alpha + 1, d\alpha]\!]$ corresponds to the coefficient of this polynomial for the monomial $x^{i'} y^{j'}$. We have the following inequalities:

$$|B_3[(i,j);(i',j')]| \le (xT)^i \left(P(xT) + \frac{y}{M}\right)^j [x^{i'} y^{j'}]$$

$$\le T^i \left[\sum_{k=0}^{j} \binom{j}{k} (P(xT))^{j-k} M^{-k} y^k\right] [x^{i'-i} y^{j'}]$$

$$\le 2^j T^i M^{-j'} \left[\sum_{k=0}^{d} a_k T^k x^k\right]^{j-j'} [x^{i'-i}].$$

Besides, we have $a_k \le K \frac{N_2}{N_1^k}$, which gives that:

$$|B_3[(i,j);(i',j')]| \le 2^j K^{j-j'} T^i N_2^{j-j'} M^{-j'} \left[\sum_{k=0}^{d} \left(\frac{T}{N_1}\right)^k x^k\right]^{j-j'} [x^{i'-i}].$$

$$\le 2^j K^{j-j'} T^i N_2^{j-j'} M^{-j'} \left(\frac{T}{N_1}\right)^{i'-i} \left[\sum_{k=0}^{d} x^k\right]^{j-j'} [x^{i'-i}]$$

$$\le \left(2^j (d+1)^j K^j N_1^i N_2^j\right) \cdot \left(\frac{T^{i'}}{M^{j'} N_1^{i'} N_2^{j'}}\right).$$

Proof of Lemma 8. For the first relation, one can write:
$$\dim B_3 = \dim B - \dim B_1 = \frac{1}{2}(\alpha^2 + O(\alpha)) - \frac{1}{2d}(\alpha^2 + O(\alpha)).$$
The proof of the second relation is similar:

$$\prod_{\substack{i+j \in [\![0,\alpha]\!] \\ i+dj \in [\![\alpha+1,d\alpha]\!]}} N_1^i N_2^j = \left(\prod_{i+j \le \alpha} N_1^i N_2^j\right) \cdot \left(\prod_{i+dj \le \alpha} N_1^i N_2^j\right)^{-1}$$

$$= N_1^{\frac{d-1}{6d}(\alpha^3 + O(\alpha^2))} N_2^{\frac{d^2-1}{6d^2}(\alpha^3 + O(\alpha^2))}.$$

Proof of Lemma 9. We restrict ourselves to the first statement and to the case where $MN_2 \in [(N_1/T)^d, (N_1/T)^{d+1}]$. The other proofs are similar. We have:

$$\sum_{k=\tau}^{k_{max}} c_k = \sum_{i=0}^{\alpha-1} \# \left[\![\left\lfloor \frac{\alpha-i}{d}\right\rfloor, \left\lfloor \frac{(t-n_1)i-\tau}{m+n_2}\right\rfloor\right]\!] + \sum_{i=\alpha}^{\left\lfloor \frac{\tau}{t-n_1}\right\rfloor} \# \left[\![0, \left\lfloor \frac{(t-n_1)i-\tau}{m+n_2}\right\rfloor\right]\!]$$

$$= \sum_{i=0}^{\left\lfloor \frac{\tau}{t-n_1}\right\rfloor} \frac{(t-n_1)i-\tau}{m+n_2} - \sum_{i=0}^{\alpha-1} \frac{\alpha-i}{d} + O(\alpha)$$

$$= \frac{t-n_1}{2(m+n_2)} \left[\frac{\tau}{t-n_1} + O(1)\right]^2 - \frac{\tau}{m+n_2} \left[\frac{\tau}{t-n_1} + O(1)\right] - \frac{\alpha^2}{2d} + O(\alpha)$$

$$= \frac{\alpha^2}{2}(1 + O(1/\alpha)) - \frac{\alpha^2}{2d} + O(\alpha) = \left(\frac{1}{2} - \frac{1}{2d}\right)(\alpha^2 + O(\alpha))$$

End of the Proof of Theorem 3. Since $d = 2^{O(\alpha)}$, Theorem 10 gives that:

$$|\det L[\mathbb{P}]| \leq 2^{O(\alpha^4)+(n_1+n_2)\left(\frac{\alpha^3}{6}+O(\alpha^2)\right)-\frac{1}{3}\sqrt{(n_1-t)(m+n_2)}(\alpha^3+O(\alpha^2))+(m+t)O(\alpha^2)}.$$

As a consequence, to ensure that 4 is satisfied, it suffices that:

$$O(\alpha^2) + (m + n_2 + dn_1)O(1/\alpha) + (n_1 + n_2)(\alpha + O(1)) + (m + t)O(1)$$
$$< 2\sqrt{(n_1 - t)(m + n_2)}(\alpha + O(1)),$$

which is implied by the simpler equation:

$$O(\alpha)+(n_1+n_2)\left(1+O\left(\frac{1}{\alpha}\right)+dO\left(\frac{1}{\alpha^2}\right)\right)+mO\left(\frac{1}{\alpha}\right) < 2\sqrt{(n_1-t)(m+n_2)}.$$

This last equation is itself implied by the condition of Theorem 3.

Computing a Lower Bound for the Canonical Height on Elliptic Curves over $\mathbb{Q}$

John Cremona[1] and Samir Siksek[2]

[1] School of Mathematical Sciences, University of Nottingham
University Park, Nottingham NG7 2RD, UK
John.Cremona@nottingham.ac.uk
[2] Institute of Mathematics, University of Warwick, Coventry, CV4 7AL, UK
siksek@maths.warwick.ac.uk

Abstract. Let E be an elliptic curve over the rationals. A crucial step in determining a Mordell-Weil basis for E is to exhibit some positive lower bound $\lambda > 0$ for the canonical height $\hat{h}$ on non-torsion points.

We give a new method for determining such a lower bound, which does not involve any searching for points.

1 Introduction

Let E be an elliptic curve over the rationals $\mathbb{Q}$ given by a minimal Weierstrass model

$$E : \quad y^2 + a_1 xy + a_3 y = x^3 + a_2 x^2 + a_4 x + a_6. \qquad (1)$$

The task of explicitly computing a Mordell-Weil basis for $E(\mathbb{Q})$ can be divided into three steps (see [10]):

(i) A 2-descent (possibly combined with higher descents) is used to determine a basis $P_1, \ldots, P_r$ for a subgroup of $E(\mathbb{Q})$ of finite index.

(ii) A positive lower bound λ for the canonical height $\hat{h}(P)$ of non-torsion points is somehow determined. The geometry of numbers now gives an upper bound B on the index n of the subgroup of $E(\mathbb{Q})$ spanned by $P_1, \ldots, P_r$.

(iii) A sieving procedure is finally used to deduce a Mordell-Weil basis.

In step (ii) a rather indirect procedure has been used in the past to determine a lower bound $\lambda > 0$ for the canonical height $\hat{h}(P)$ of non-torsion points. The difference $h - \hat{h}$ between the logarithmic and canonical heights is known to be bounded on $E(\mathbb{Q})$; the best current bounds are to be found in [7]. Suppose that $h(P) - \hat{h}(P) \leq K$ for all non-zero rational points P. If $\hat{h}(P) < \lambda$ then $h(P) < K + \lambda$. To show that all non-torsion points P satisfy $\hat{h}(P) \geq \lambda$ one can search for all points satisfying $h(P) < K + \lambda$. More explicitly, write $x(P) = X/Z^2$ where X, Z are coprime integers and Z positive; then we must search for all points P satisfying

$$|X| < \exp(K + \lambda), \qquad Z < \exp\left((K + \lambda)/2\right).$$

F. Hess, S. Pauli, and M. Pohst (Eds.): ANTS 2006, LNCS 4076, pp. 275–286, 2006.

If the bound on the height difference K is large, then we are forced to search a huge region before achieving our goal; this is quite often impractical.

In this paper we propose a more direct method for determining a positive lower bound λ for the canonical height of non-torsion points.

For reasons to be explained later, it is convenient to work with the subgroup

$$E_{\mathrm{gr}}(\mathbb{Q}) = E(\mathbb{Q}) \cap E_0(\mathbb{R}) \cap \prod_{p|\Delta} E_0(\mathbb{Q}_p);$$

the subscript "gr" stands for good reduction, and Δ denotes the minimal discriminant of E, i.e. the discriminant of the minimal model (1). We give a method of determining a positive lower bound μ for the canonical height of non-torsion points P in $E_{\mathrm{gr}}(\mathbb{Q})$. Then, if c is the least common multiple of the Tamagawa indices $c_p = [E(\mathbb{Q}_p) : E_0(\mathbb{Q}_p)]$ (including $p = \infty$), we know that $\lambda = \mu/c^2$ is a lower bound for the canonical height of non-torsion points in $E(\mathbb{Q})$.

The basic idea of our approach is very simple: on $E_{\mathrm{gr}}(\mathbb{Q})$, the canonical height satisfies

$$\hat{\mathrm{h}}(P) \geq \log \max\{1, |x(P)|\} - b$$

where b is a constant that depends on the model for E and is typically small. Now if P is 'far from' the the point of order 2 on $E_0(\mathbb{R})$ then its x-coordinate is large and so the canonical height is large. If on the other hand P is 'close to' the point of order 2 on $E_0(\mathbb{R})$ then $x(2P)$ is large, and so $\hat{\mathrm{h}}(P) = \frac{1}{4}\hat{\mathrm{h}}(2P)$ is also large. We extend this idea as follows. Suppose that we want to prove that a certain $\mu > 0$ is a lower bound for the height for non-torsion points on $E_{\mathrm{gr}}(\mathbb{Q})$. We suppose that there is a non-torsion point $P \in E_{\mathrm{gr}}(\mathbb{Q})$ satisfying $\hat{\mathrm{h}}(P) \leq \mu$ and we use this to deduce a series of bounds $|x(nP)| \leq B_n(\mu)$ where the $B_n(\mu)$ are explicit constants. With the aid of the elliptic logarithm, we solve the simultaneous inequalities $|x(nP)| \leq B_n(\mu)$ with $n = 1, \ldots, k$ for some suitably chosen k. If there is no solution then we deduce that $\hat{\mathrm{h}}(P) > \mu$ for all non-torsion points on $E_{\mathrm{gr}}(\mathbb{Q})$. Otherwise we simply start again with a smaller value of μ, or a bigger value of k, or both.

We note that estimates for heights of points of infinite order on elliptic curves have previously been given by Silverman [11] and Hindry and Silverman [8]. Those estimates are theoretical, and too small for practical use; see the concluding remarks at the end of the paper.

2 Heights

In this section we gather some basic facts needed about local and canonical heights, with no claims of originality. A good reference is [13]. The reader is warned that there are several normalizations of local and canonical heights as explained in [7, section 4].

We define the usual constants associated to a Weierstrass model (1) as follows (see [12, page 46]):

$$b_2 = a_1^2 + 4a_2,$$
$$b_4 = 2a_4 + a_1 a_3,$$
$$b_6 = a_3^2 + 4a_6,$$
$$b_8 = a_1^2 a_6 + 4a_2 a_6 - a_1 a_3 a_4 + a_2 a_3^2 - a_4^2,$$
$$\Delta = -b_2^2 b_8 - 8b_4^3 - 27b_6^2 + 9b_2 b_4 b_6.$$

Let

$$f(P) = 4x(P)^3 + b_2 x(P)^2 + 2b_4 x(P) + b_6,$$
$$g(P) = x(P)^4 - b_4 x(P)^2 - 2b_6 x(P) - b_8;$$

so that $x(2P) = g(P)/f(P)$. We let $\mathcal{M}$ be the set of all primes of $\mathbb{Q}$ (including ∞). For $p \in \mathcal{M}$, define the function $\Phi_p : E(\mathbb{Q}_p) \to \mathbb{R}$ by

$$\Phi_p(P) = \begin{cases} 1 & \text{if } P = 0, \\ \dfrac{\max\{|f(P)|_p, |g(P)|_p\}}{\max\{1, |x(P)|_p\}^4} & \text{otherwise.} \end{cases} \tag{2}$$

It is straightforward to see that Φ_p is a continuous and hence bounded function on $E(\mathbb{Q}_p)$ (the boundedness follows immediately from the fact that $E(\mathbb{Q}_p)$ is compact with respect to the p-adic topology).

We define the local height $\lambda_p : E(\mathbb{Q}_p) \backslash \{0\} \to \mathbb{R}$ for all $p \in \mathcal{M}$ (including $p = \infty$) by

$$\lambda_p(P) = \log \max\{1, |x(P)|_p\} + \sum_{i=0}^{\infty} \frac{1}{4^{i+1}} \log \Phi_p(2^i P). \tag{3}$$

One definition of the canonical height $\hat{h} : E(\mathbb{Q}) \to \mathbb{R}$ is by the formula

$$\hat{h}(P) = \sum_{p \in \mathcal{M}} \lambda_p(P). \tag{4}$$

The canonical height extends to a quadratic form on $\mathbb{R} \otimes_{\mathbb{Z}} E(\mathbb{Q})$; in particular $\hat{h}(nP) = n^2 \hat{h}(P)$ for any integer n. Moreover, $P \in E(\mathbb{Q})$ is torsion if and only if $\hat{h}(P) = 0$. For non-torsion rational points the canonical height is strictly positive.

The following lemma is standard; see for example [14].

Lemma 1. *Let p be a finite prime and $P \in E_0(\mathbb{Q}_p) \backslash \{0\}$ (i.e. P is a point of good reduction). Then*

$$\lambda_p(P) = \log \max\{1, |x(P)|_p\}.$$

In particular, non-archimedean local heights are non-negative for points of good reduction; this is not true for points of bad reduction. We are interested in obtaining a positive lower bound for the canonical height, using its expression (4) as a sum of local heights, and thus it is sensible to restrict ourselves to points in $E_{\text{gr}}(\mathbb{Q}) \backslash \{0\}$.

Our next lemma is immediate from (4) and Lemma 1.

Lemma 2. *Suppose* $P \in E_{\mathrm{gr}}(\mathbb{Q}) \backslash \{0\}$. *Then*

$$\hat{\mathrm{h}}(P) = \lambda_\infty(P) + \log(\mathrm{denom}(x(P))).$$

2.1 The Archimedean Local Height Difference

Define $\alpha \in \mathbb{R}_+$ by

$$\alpha^{-3} = \inf_{P \in E_0(\mathbb{R})} \Phi_\infty(P), \tag{5}$$

where the exponent -3 has been chosen to simplify the formulae appearing later. This can be computed as in [7] or [10], with a slight adjustment since we are looking only at points on $E_0(\mathbb{R})$. The following lemma can be deduced easily from the definition of local heights (3).

Lemma 3. *If* $P \in E_0(\mathbb{R}) \backslash \{0\}$ *then*

$$\log \max\{1, |x(P)|\} - \lambda_\infty(P) \le \log \alpha.$$

In particular this inequality is true for all $P \in E_{\mathrm{gr}}(\mathbb{Q}) \backslash \{0\}$.

3 Multiplication by n

Let n be a positive integer. It is possible that multiplication by n annihilates some of the groups $E_0(\mathbb{Q}_p)/E_1(\mathbb{Q}_p)$: a non-torsion point $P \in E_{\mathrm{gr}}(\mathbb{Q})$ will be killed (mapped into $E_1(\mathbb{Q}_p)$) if p divides the denominator of $x(nP)$. In this section we give a lower estimate for the contribution that multiplication by n makes to the canonical height of nP.

For finite primes p, let e_p be the exponent of the group

$$E_{\mathrm{ns}}(\mathbb{F}_p) \cong E_0(\mathbb{Q}_p)/E_1(\mathbb{Q}_p).$$

Define

$$D_E(n) = \sum_{p < \infty,\, e_p | n} 2(1 + \mathrm{ord}_p(n/e_p)) \log p. \tag{6}$$

That this sum is finite follows from the following proposition; clearly, it is easily computable.

Proposition 1. *With notation as above, if* $e_p \mid n$ *then* $p \le (n+1)^2$. *Hence the sum defining* $D_E(n)$ *is finite. Moreover, if* P *is a non-torsion point in* $E_{\mathrm{gr}}(\mathbb{Q})$ *and* $n \ge 1$, *then*

$$\hat{\mathrm{h}}(nP) \ge \lambda_\infty(nP) + D_E(n).$$

Proof. Suppose that $e_p \mid n$. By definition e_p is the exponent of $E_{\mathrm{ns}}(\mathbb{F}_p)$. If p is a prime of singular reduction for E then $E_{\mathrm{ns}}(\mathbb{F}_p)$ is a cyclic group of order

$p - 1$, $p + 1$ or p depending on whether E has split multiplicative, non-split multiplicative or additive reduction at p. In either case we see that

$$n \geq e_p = |E_{\mathrm{ns}}(\mathbb{F}_p)| \geq p - 1$$

and so certainly $p \leq (n+1)^2$. Suppose now that p is a prime of good reduction. We know that

$$E_{\mathrm{ns}}(\mathbb{F}_p) = E(\mathbb{F}_p) \cong \mathbb{Z}/d_1 \times \mathbb{Z}/d_2$$

where $d_2 | d_1$ and $d_1 = e_p$. Hence

$$(\sqrt{p} - 1)^2 = p + 1 - 2\sqrt{p} \leq |E(\mathbb{F}_p)| = d_1 d_2 \leq d_1^2 = e_p^2 \leq n^2,$$

from which we deduce that $p \leq (n+1)^2$.

In view of Lemma 2, the proof is complete on showing

$$\log(\mathrm{denom}(x(nP))) \geq D_E(n)$$

for non-torsion $P \in E_{\mathrm{gr}}(\mathbb{Q})$. This easily follows from the structure of $E(\mathbb{Q}_p)$, since if $e = \mathrm{ord}_p(n/e_p)$ then $nP \in E_{e+1}(\mathbb{Q}_p)$, so $\mathrm{ord}_p(\mathrm{denom}(x(nP))) \geq 2(e+1)$.   □

4   A Bound for Multiples of Points of Good Reduction

Recall that our aim is to exhibit some positive μ such that $\hat{\mathrm{h}}(P) > \mu$ for all non-torsion points in $E_{\mathrm{gr}}(\mathbb{Q})$. In this section we first suppose that $\mu > 0$ is given, assume that P is a point in $E_{\mathrm{gr}}(\mathbb{Q})$ satisfying $\hat{\mathrm{h}}(P) \leq \mu$, and deduce a sequence of inequalities satisfied by the x-coordinates of the multiples nP for $n = 1, 2, 3, \ldots$. We then show that for sufficiently small positive μ and a suitable n (given explicitly in Corollary 1 below), there are no points P such that $x(nP)$ satisfies the inequality. In the following two sections we will explain how to combine the inequalities for several n, enabling us to obtain better values of μ such that $\hat{\mathrm{h}}(P) > \mu$ for non-torsion points in $E_{\mathrm{gr}}(\mathbb{Q})$.

Let α and D_E be defined as above in (5) and (6). For $\mu > 0$ and $n \in \mathbb{Z}^+$ define

$$B_n(\mu) = \exp\left(n^2\mu - D_E(n) + \log\alpha\right).$$

Proposition 2. *If $B_n(\mu) < 1$ then $\hat{\mathrm{h}}(P) > \mu$ for all non-torsion points on $E_{\mathrm{gr}}(\mathbb{Q})$. On the other hand, if $B_n(\mu) \geq 1$ then for all non-torsion points $P \in E_{\mathrm{gr}}(\mathbb{Q})$ with $\hat{\mathrm{h}}(P) \leq \mu$, we have*

$$-B_n(\mu) \leq x(nP) \leq B_n(\mu).$$

Proof. Suppose P is a non-torsion point on $E_{\mathrm{gr}}(\mathbb{Q})$ with $\hat{\mathrm{h}}(P) \leq \mu$. From the inequalities in Lemma 3 and Proposition 1 we see that

$$
\begin{aligned}
\log\max\{1, |x(P)|\} &\leq \lambda_\infty(P) + \log\alpha \\
&\leq \hat{\mathrm{h}}(nP) - D_E(n) + \log\alpha \\
&= n^2\,\hat{\mathrm{h}}(P) - D_E(n) + \log\alpha \\
&\leq n^2\mu - D_E(n) + \log\alpha.
\end{aligned}
$$

Thus
$$\max\{1, |x(nP)|\} \leq B_n(\mu).$$

If $B_n(\mu) < 1$ then we have a contradiction, and in this case we deduce that $\hat{h}(P) > \mu$ for all non-torsion points on $E_{\text{gr}}(\mathbb{Q})$.

If instead $B_n(\mu) \geq 1$, then $|x(nP)| \leq B_n(\mu)$ and the proposition follows.   □

Corollary 1. *Define α as in (5). Let p be a prime greater than $\sqrt{\alpha}$, and set $n = e_p$ and $\mu_0 = n^{-2}(D_E(n) - \log\alpha)$. Then $\mu_0 > 0$ and for all non-torsion $P \in E_{\text{gr}}(\mathbb{Q})$ we have*
$$\hat{h}(P) \geq \mu_0.$$

Proof. We have $D_E(n) \geq 2\log p > \log\alpha$, so certainly $\mu_0 > 0$. Now for all $\mu < \mu_0$ we have $n^2\mu - D_E(n) + \log\alpha < 0$, so that $B_n(\mu) < 1$ and hence $\hat{h}(P) > \mu$ by the Proposition. Since this holds for all $\mu < \mu_0$, we have $\hat{h}(P) \geq \mu_0$ as required.

As pointed out to us by the anonymous referee, we could use this Corollary by itself to provide a suitable positive lower bound for the height of non-torsion points in $E_{\text{gr}}(\mathbb{Q})$. However we can obtain a better bound (see Example 1 for an example) by combining the information from several different n simultaneously.

5   Solving Inequalities Involving the Multiples of Points

Proposition 2 gives a sequence of inequalities involving the multiples of non-torsion points P in $E_{\text{gr}}(\mathbb{Q})$ satisfying $\hat{h}(P) \leq \mu$. We would like to solve these inequalities. One approach is to use division polynomials. We have found this impractical as the degree and coefficients of division polynomials grow rapidly with the multiple considered. Instead we have found it convenient to use the elliptic logarithm.

For the reader's convenience, we give here a very brief description of the elliptic logarithm $\varphi : E_0(\mathbb{R}) \to \mathbb{R}/\mathbb{Z}$. We can rewrite the Weierstrass model (1) as
$$(2y + a_1x + a_3)^2 = 4x^3 + b_2x^2 + 2b_4x + b_6.$$

Let β be the largest real root of the right-hand side; thus β is the x-coordinate of the unique point of order 2 on $E_0(\mathbb{R})$. Let
$$\Omega = 2\int_\beta^\infty \frac{dx}{\sqrt{4x^3 + b_2x^2 + 2b_4x + b_6}}.$$

If $P = (\xi, \eta) \in E_0(\mathbb{R})$ with $2\eta + a_1\xi + a_3 \geq 0$ then let
$$\varphi(P) = \frac{1}{\Omega}\int_\xi^\infty \frac{dx}{\sqrt{4x^3 + b_2x^2 + 2b_4x + b_6}};$$

otherwise we let
$$\varphi(P) = 1 - \varphi(-P).$$

The elliptic logarithm can be very rapidly computed using arithmetic-geometric means; see Algorithm 7.4.8 in [2]. What matters most to us is that $\varphi : E_0(\mathbb{R}) \to \mathbb{R}/\mathbb{Z}$ is an isomorphism (of real Lie groups). We shall find it convenient to identify $\mathbb{R}/\mathbb{Z}$ with the interval $[0,1)$.

Suppose that ξ is a real number satisfying $\xi \geq \beta$. Then there exists η such that $2\eta + a_1\xi + a_3 \geq 0$ and $(\xi, \eta) \in E_0(\mathbb{R})$. Define

$$\psi(\xi) = \varphi\left((\xi, \eta)\right) \in [1/2, 1).$$

In words, $\psi(\xi)$ is the elliptic logarithm of the "higher" of the two points with x-coordinate ξ.

For real ξ_1, ξ_2 with $\xi_1 \leq \xi_2$ we define the subset $\mathcal{S}(\xi_1, \xi_2) \subset [0,1)$ as follows:

$$\mathcal{S}(\xi_1, \xi_2) = \begin{cases} \emptyset & \text{if } \xi_2 < \beta \\ [1 - \psi(\xi_2), \psi(\xi_2)] & \text{if } \xi_1 < \beta \leq \xi_2 \\ [1 - \psi(\xi_2), 1 - \psi(\xi_1)] \cup [\psi(\xi_1), \psi(\xi_2)] & \text{if } \xi_1 \geq \beta. \end{cases}$$

The following lemma is clear.

Lemma 4. *Suppose $\xi_1 < \xi_2$ are real numbers. Then $P \in E_0(\mathbb{R})$ satisfies $\xi_1 \leq x(P) \leq \xi_2$ if and only if $\varphi(P) \in \mathcal{S}(\xi_1, \xi_2)$.*

If $\bigcup[a_i, b_i]$ is a disjoint union of intervals and $t \in \mathbb{R}$, we define

$$t + \bigcup[a_i, b_i] = \bigcup[a_i + t, b_i + t]$$

and (for $t > 0$)

$$t \bigcup[a_i, b_i] = \bigcup[ta_i, tb_i].$$

Proposition 3. *Suppose $\xi_1 < \xi_2$ are real numbers and n a positive integer. Define*

$$\mathcal{S}_n(\xi_1, \xi_2) = \bigcup_{t=0}^{n-1} \left(\frac{t}{n} + \frac{1}{n}\mathcal{S}(\xi_1, \xi_2)\right).$$

Then $P \in E_0(\mathbb{R})$ satisfies $\xi_1 \leq x(nP) \leq \xi_2$ if and only if $\varphi(P) \in \mathcal{S}_n(\xi_1, \xi_2)$.

Proof. By Lemma 4 we know that $P \in E_0(\mathbb{R})$ satisfies $\xi_1 \leq x(nP) \leq \xi_2$ if and only if $\varphi(nP) \in \mathcal{S}(\xi_1, \xi_2)$.

Denote the multiplication by n map on $\mathbb{R}/\mathbb{Z}$ by ν_n. If $\delta \in [0,1)$ then

$$\nu_n^{-1}(\delta) = \left\{\frac{t}{n} + \frac{\delta}{n} \ : \ t = 0, 1, 2, \ldots, n-1\right\}.$$

However $\varphi(nP) = n\varphi(P) \pmod 1$. Therefore,

$$\varphi(nP) \in \mathcal{S}(\xi_1, \xi_2) \iff \varphi(P) \in \nu_n^{-1}\left(\mathcal{S}(\xi_1, \xi_2)\right) = \mathcal{S}_n(\xi_1, \xi_2).$$

$\square$

6   The Algorithm

Putting together Propositions 2 and 3 we deduce our main result.

Theorem 1. *Let $\mu > 0$. If $B_n(\mu) < 1$ for some positive integral n, then $\hat{h}(P) > \mu$ for all non-torsion P in $E_{\mathrm{gr}}(\mathbb{Q})$.*

On the other hand, if $B_n(\mu) \geq 1$ for $n = 1, \ldots, k$, then every non-torsion point $P \in E_{\mathrm{gr}}(\mathbb{Q})$ such that $\hat{h}(P) \leq \mu$ satisfies

$$\varphi(P) \in \bigcap_{n=1}^{k} \mathcal{S}_n\left(-B_n(\mu), B_n(\mu)\right).$$

In particular, if

$$\bigcap_{n=1}^{k} \mathcal{S}_n\left(-B_n(\mu), B_n(\mu)\right) = \emptyset \tag{7}$$

then $\hat{h}(P) > \mu$ for all non-torsion P in $E_{\mathrm{gr}}(\mathbb{Q})$.

In practice we have found the following procedure effective. We start with $\mu = 1$, $k = 5$ and compute $B_n(\mu)$ for $n = 1, \ldots k$. If any of these values of $B_n(\mu) < 1$ then we have succeeded in proving that $\mu = 1$ is a lower bound for the canonical height of non-torsion points of good reduction. Otherwise we compute $\bigcap_{n=1}^{k} \mathcal{S}_n\left(-B_n(\mu), B_n(\mu)\right)$ (as a union of intervals); if this is empty, then again we have succeeded in proving that $\mu = 1$ is a lower bound. Finally, if this intersection is non-empty we have failed to prove that $\mu = 1$ is a suitable lower bound.

Our course now proceeds differently according to whether we have succeeded or failed to show that $\mu = 1$ is a lower bound. If we succeed, we now repeatedly multiply μ by 1.1 and use the same method to try to prove that the new value of μ is still a lower bound. We return the last succeeding value of μ as the output to the algorithm.

If, on the other hand, we failed with $\mu = 1$ then we repeatedly multiply μ by 0.9 and increase k by 1 until we achieve success; the bound returned is then the first successful value of μ.

It is easy to use the proof of Corollary 1 to Proposition 2 to show that our algorithm will succeed in obtaining a positive lower bound μ, after a finite number of steps.

Alternative strategies are clearly possible here; instead of using scaling factors of $11/10$ and $9/10$ we could instead use a larger factor such as 2 or $1/2$ respectively, and then successively replace the scaling factor by its square root and apply a back-tracking method to converge to the optimal value of μ; the details may be left to the reader.

7   Reduced Models

The canonical height is independent of the model chosen for the elliptic curve. Our lower bound is however not model-independent. The constant α defined in (5) is dependent on b_2, b_4, b_6; all other constants and maps in the above discussion are model-independent. To improve our lower bound for the canonical height it is sensible to choose a model that minimises the value of α. We have no theoretical method for deciding on the best model here. The models for elliptic curves appearing in Cremona's tables [3], [5] (as well as those appearing in the earlier Antwerp IV tables [1]) are known as standardized models: we say that the model (1) for E is a standardized model if it is minimal with $a_1, a_3 \in \{0, 1\}$, and $a_2 \in \{-1, 0, 1\}$. Each elliptic curve has a unique standardized model. Practical experience shows that—for the purpose of obtaining a good lower bound for the canonical height—it is usually preferable to choose a model that reduces, in the sense of [4] but with respect to translations only, the cubic polynomial

$$f(X) = 4x^3 + b_2 x^2 + 2b_4 x + b_6.$$

We call this model the reduced model. For the convenience of the reader we give here the formulae, adapted from [4], for doing this.

If the discriminant $\Delta > 0$, then we let

$$P = b_2^2 - 24b_4, \qquad Q = 2b_2 b_4 - 36b_6.$$

Let r be the nearest integer to $-Q/(2P)$. The reduced model for E is given by replacing x by $x + r$ in (1).

If $\Delta < 0$ we let β be the unique real root of f. Let

$$h_0 = 144\beta^2 + 24b_2\beta + 48b_4 - b_2^2, \quad h_1 = 24b_2\beta^2 + 6(b_2^2 - 8b_4)\beta + 4b_2 b_4.$$

Let r be the nearest integer to $-h_1/(2h_0)$. Again, the reduced model for E is given by replacing x by $x + r$ in (1).

8   Examples

We have implemented our algorithm in `pari/gp`, and used the program to compute some examples.

Example 1. Consider the elliptic curve E (with code 60490d1 in [5]), given by the standardized model

$$y^2 + xy + y = x^3 + 421152067x + 105484554028056.$$

Our program shows that for non-torsion points in $E_{\mathrm{gr}}(\mathbb{Q})$

$$\hat{h}(P) > 1.9865.$$

If we apply the method used to prove Corollary 1 to Proposition 2 above, we do not obtain as good a bound. Here $\log(\alpha) = 3.3177\ldots$ and $\sqrt{\alpha} = 5.253\ldots,$

so we should use a prime $p \geq 7$. Rather than use $p = 7$ for which $e_p = 9$ we do better to take $p = 19$ with $n = e_p = 6$. Then $D_E(6) = 2\log 114$ and $\mu_0 = (2\log 114 - 3.317)/36 = 0.17$.

We note that the curve E has only one real component. Moreover, it has good reduction at all primes except 2, 5, 23 and 263 where the Tamagawa indices are 2, 21, 2 and 3 respectively. Hence if $P \in E(\mathbb{Q})$ then $42P \in E_{\mathrm{gr}}(\mathbb{Q})$. It follows that

$$\hat{h}(P) > 1.9865/42^2 = 0.001126$$

for non-torsion points in $E(\mathbb{Q})$.

This curve has rank 1, and a point of infinite order is

$$P = (3583035/169, 24435909174/2197)$$

with $\hat{h}(P) = 6.808233192$. It follows that the index of the subgroup $\langle P \rangle$ in $E(\mathbb{Q})$ is at most $\sqrt{6.808233192/0.001126} < 78$. We may check that $P \notin pE(\mathbb{Q})$ for all primes $p < 78$ (using the method of p-saturation introduced in [10]) and deduce that $E(\mathbb{Q}) = \langle P \rangle$.

For this curve the bound between logarithmic and canonical heights can be shown by the method of [7] to be at most 22.8, so finding a lower bound for $\hat{h}$ through searching would be prohibitive. However, if we apply the method of [7] to the subgroup $E_{\mathrm{gr}}(\mathbb{Q})$ we find that the height difference for points in the subgroup is only 3.3, so in fact we could have found a lower bound for the restriction of $\hat{h}$ to the subgroup by searching for points with small logarithmic height.

Finally, we can use our bound to prove that $E(\mathbb{Q}) = \langle P \rangle$ more simply as follows. First we apply p-saturation with $p = 2, 3$ and 7 to show that the index $[E(\mathbb{Q}) : \langle P \rangle]$ is not divisible by the primes dividing the Tamagawa numbers; then we observe that while $P \notin E_{\mathrm{gr}}(\mathbb{Q})$, we have $2P \in E_{\mathrm{gr}}(\mathbb{Q})$. It follows that $[E(\mathbb{Q}) : E_{\mathrm{gr}}(\mathbb{Q})] = 2$, when a priori this index could have been as large as 42. And moreover the index $m = [E(\mathbb{Q}) : \langle P \rangle] = [E_{\mathrm{gr}}(\mathbb{Q}) : \langle 2P \rangle]$ is coprime to $2, 3, 7$ and satisfies $m^2 \leq \hat{h}(2P)/1.9865 < 14$, so $m = 1$.

This method of saturating $E(\mathbb{Q})$, by first saturating $E_{\mathrm{gr}}(\mathbb{Q})$ and separately saturating at primes dividing the Tamagawa numbers, can also be used for curves of higher rank, though the details are more complicated. We will return to this in a future paper.

Example 2. (Statistics) We ran our program on all the 4081 optimal curves in our online tables [5] with conductors 7000–8000. The smallest lower bound we obtained for that range was 0.022 for curve 7042d1, and the largest was 11.879 for 7950r1. It took 941 seconds to compute the lower bounds for these curves, an average of 0.23 seconds for each curve.

Applications. Using this method we intend to show that the generators listed for the curves in the database [5] do generate the full Mordell-Weil group, modulo torsion, in every case. The present situation (January 2006) is that not all have been checked, the exceptions being those for which we have not yet obtained a lower bound for the height of non-torsion points, and hence do not have a bound in the index of saturation. Similarly, the algorithm described here will be incorporated into the first author's program `mwrank` (see [6]) for computing Mordell-Weil groups via 2-descent.

9 Concluding Remarks

As pointed out by the referee, it would be possible to extend this method to elliptic curves defined over any totally real number field; we leave the details to the interested reader. It would be rather harder, though, to extend our method to fields with a non-real complex embedding since we would then have to intersect subsets of the unit square instead of the unit interval.

Lastly, at the insistence of the referee, we conclude with a few words comparing our lower bound for the canonical height and earlier theoretical bounds, due to Silverman [11] and to Hindry and Silverman [8]. The bounds of [11] are not completely explicit. In [8], Hindry and Silverman give a lower bound for the canonical height of non-torsion points on elliptic curves over number fields (and function fields). For example, if E is an elliptic curve over $\mathbb{Q}$, write

$$\sigma = \frac{\log|\Delta|}{\log N}$$

where Δ is the minimal discriminant and N is the conductor of E. Specializing Theorem 0.3 of [8] we obtain that

$$\hat{h}(P) \geq \frac{2\log|\Delta|}{(20\sigma)^8 10^{1.1+4\sigma}}.$$

for non-torsion points $P \in E(\mathbb{Q})$. For example, for the elliptic curve E in our Example 1, this gives the lower bound for non-torsion points

$$\hat{h}(P) \geq 3.2\ldots \times 10^{-42},$$

as compared with our lower bound for non-torsion points $\hat{h}(P) > 0.001126$. However such a crude numerical comparison is not very useful, for two reasons:

- The bounds in [8] are much more general; undoubtedly the methods there could produce better bounds if specialised to elliptic curves over the rationals. It would be interesting to pursue this.
- A conjecture of Lang (mentioned in [8]) states that there is some absolute constant $c > 0$ such that $\hat{h}(P) \geq c\log|\Delta|$ for all elliptic curves E and non-torsion points $P \in E(\mathbb{Q})$. The objective of [8] seems to have been to prove a statement that is as close as possible to Lang's conjecture. Our aim is rather different.

References

1. B.J. Birch and W. Kuyk (eds.), *Modular Functions of One Variable IV*, Lecture Notes in Mathematics **476**, Springer-Verlag, 1975.
2. H. Cohen, *A Course in Computational Algebraic Number Theory (Third Corrected Printing)*. Graduate Texts in Mathematics **138**, Springer-Verlag, 1996.

3. J.E. Cremona, *Algorithms for modular elliptic curves*, second edition, Cambridge University Press, 1996.
4. J.E. Cremona, *Reduction of binary cubic and quartic forms*, LMS J. Comput. Math. **2** (1999), 62–92.
5. J.E. Cremona, *Elliptic Curve Data*,
   http://www.maths.nott.ac.uk/personal/jec/ftp/data/INDEX.html.
6. J.E. Cremona, mwrank, a program for computing Mordell-Weil groups of elliptic curves over $\mathbb{Q}$. Available from
   http://www.maths.nott.ac.uk/personal/jec/mwrank.
7. J.E. Cremona, M. Prickett and S. Siksek, *Height difference bounds for elliptic curves over number fields*, Journal of Number Theory **116** (2006), 42–68.
8. M. Hindry and J. H. Silverman, *The Canonical Height and integral points on elliptic curves*, Invent. Math. **93** (1988), 419–450.
9. PARI/GP, version 2.2.8, Bordeaux, 2004; available from
   http://pari.math.u-bordeaux.fr/.
10. S. Siksek, *Infinite descent on elliptic curves*, Rocky Mountain Journal of Mathematics **25**, number 4 (Fall 1995), 1501–1538.
11. J.H. Silverman, *Lower bound for the canonical height on elliptic curves*, Duke Mathematical Journal **48** (1981), 633–648.
12. J.H. Silverman, *The arithmetic of elliptic curves*, GTM 106, Springer-Verlag, 1986.
13. J. H. Silverman, *Advanced topics in the arithmetic of elliptic curves*, GTM 151, Springer-Verlag, 1994.
14. J.H. Silverman, *Computing heights on elliptic curves*, Math. Comp. **51** (1988), 339-358.

Points of Low Height on Elliptic Curves and Surfaces I: Elliptic Surfaces over $\mathbb{P}^1$ with Small d

Noam D. Elkies

Department of Mathematics, Harvard University, Cambridge, MA 02138 USA
elkies@math.harvard.edu

Abstract. For each of $n = 1, 2, 3$ we find the minimal height $\hat{h}(P)$ of a nontorsion point P of an elliptic curve E over $\mathbf{C}(T)$ of discriminant degree $d = 12n$ (equivalently, of arithmetic genus n), and exhibit all (E, P) attaining this minimum. The minimal $\hat{h}(P)$ was known to equal $1/30$ for $n = 1$ (Oguiso-Shioda) and $11/420$ for $n = 2$ (Nishiyama), but the formulas for the general (E, P) were not known, nor was the fact that these are also the minima for an elliptic curve of discriminant degree $12n$ over a function field of any genus. For $n = 3$ both the minimal height $(23/840)$ and the explicit curves are new. These (E, P) also have the property that that mP is an integral point (a point of naïve height zero) for each $m = 1, 2, \ldots, M$, where $M = 6, 8, 9$ for $n = 1, 2, 3$; this, too, is maximal in each of the three cases.

1 Introduction

1.1 Statement of Results

Let K be a function field of a curve C of genus g over a field k of characteristic zero,[1] and E a nonconstant elliptic curve over K. Let d be the degree of the discriminant of E (considered as a divisor on C), a natural measure of the complexity of E; and let $\hat{h} : E(K) \to \mathbf{Q}$ be the canonical height. Necessarily $12|d$; in fact it is known that $d = 12n$ where n is the arithmetic genus of the elliptic surface $\mathcal{E}$ associated with E. It is not hard to show that, given d, the set of numbers H that can occur as the canonical height of a rational point on E is discrete. In particular, for each $d = 12n$ there is a minimal positive height $\hat{h}_{\min}(d)$, and also a minimal positive height $\hat{h}_{\min}(g, d)$ for elliptic curves over function fields of genus g (except for $g = d = 0$, when E is a constant curve over $\mathbf{P}^1$ and thus has no points of positive height). It is thus a natural problem to compute or estimate these numbers $\hat{h}_{\min}(d)$ and $\hat{h}_{\min}(g, d)$. This paper is the first of a series concerned with different aspects of this problem.

In this paper we determine $\hat{h}_{\min}(12n)$ for $n = 1, 2$ and $\hat{h}_{\min}(0, 12n)$ for $n = 1, 2, 3$. Since we are working in characteristic zero, we may assume $k = \mathbf{C}$, when every genus-zero curve is isomorphic to $\mathbf{P}^1$ and its function field is isomorphic to $\mathbf{C}(T)$.

---

[1] One can also usefully define the canonical height etc. in positive characteristic, but we need to use the ABC conjecture for K and thus must assume that K has characteristic zero.

F. Hess, S. Pauli, and M. Pohst (Eds.): ANTS 2006, LNCS 4076, pp. 287–301, 2006.
© Springer-Verlag Berlin Heidelberg 2006

Theorem 1. *i)* *(Oguiso-Shioda [7])* $\hat{h}_{\min}(0,12) = 1/30$.
ii) $\hat{h}_{\min}(12) = 1/30$. *Moreover, let E be an elliptic curve with $d = 12$ over a complex function field K, and $P \in E(K)$. Then the following are equivalent: (a)* $\hat{h}(P) = 1/30$; *(b) Each of $P, 2P, 3P, 4P, 5P, 6P$ is an integral point on E; (c)* $K \cong \mathbf{C}(T)$, *and (E, P) is equivalent to the curve*

$$E_1(q) : Y^2 + (s' - (q+1)s)XY + qss'(s-s')Y = X^3 - qss'X^2 \qquad (1)$$

over the $(s : s')$ line with the rational point $P : (X,Y) = (0,0)$, for some $q \in \mathbf{C}$ other than 0 or 1.

Theorem 2. *i)* *(Nishiyama [6])* $\hat{h}_{\min}(0,24) = 11/420$.
ii) $\hat{h}_{\min}(24) = 11/420$. *Moreover, let E be an elliptic curve with $d = 24$ over a complex function field K, and $P \in E(K)$. Then the following are equivalent: (a)* $\hat{h}(P) = 11/420$; *(b) mP is an integral point on E for each $m = 1, 2, \ldots, 8$; (c)* $K \cong \mathbf{C}(T)$, *and (E, P) is equivalent to the curve*

$$
\begin{aligned}
E_2(u) : Y^2 &+ (r^2 - r'^2 + (u-2)rr')XY \\
&- r^2r'(r + r')(r + ur')(r + (u-1)r')Y \\
&= X^3 - rr'(r+r')(r+ur')X^2
\end{aligned} \qquad (2)
$$

over the $(r : r')$ line with the rational point $P : (X,Y) = (0,0)$, for some $u \in \mathbf{C}$ other than 0, 1.

Theorem 3. *i)* $\hat{h}_{\min}(0,36) = 23/840$.
ii) Let $E/\mathbf{C}(T)$ be an elliptic curve with $d = 36$, and P a rational point on E. Then the following are equivalent: (a) $\hat{h}(P) = 23/840$; *(b) mP is an integral point on E for each $m = 1, 2, \ldots, 9$; (c) (E, P) is equivalent to the curve*

$$
\begin{aligned}
E_3(A) : Y^2 &+ (At^3 + (1-2A)t^2t' - (A+1)tt'^2 - t'^3)XY \\
&- t^3t'(t + t')(At + t')(At^2 + tt' + t'^2)Y \\
&= X^3 - tt'(t + t')(At + t')(At^2 + tt' + t'^2)Y
\end{aligned} \qquad (3)
$$

over the $(t : t')$ line with the rational point $P : (X,Y) = (0,0)$, for some $A \in \mathbf{C}$ other than 0, 1.

The values of $\hat{h}_{\min}(12)$ and $\hat{h}_{\min}(24)$ are new. Note that we do not claim to determine $\hat{h}_{\min}(36)$. As indicated, the values of $\hat{h}_{\min}(0,12)$ and $\hat{h}_{\min}(0,24)$ (the first parts of Theorems 1 and 2) were already known, but were obtained using techniques that are specific to the geometry of rational and K3 elliptic surfaces and do not readily generalize past $n = 2$. Our approach lets us treat all three cases uniformly, and in principle lets us determine $\hat{h}_{\min}(0,12n)$ for any n, though the computations rapidly become infeasible as n grows beyond 3. The minimizing (E, P) had not been previously exhibited, except for a single case of a rational elliptic surface with a section of height 1/30 obtained by Shioda in a later paper [11], which we will identify with $E_1(4/5)$.

The connections with integral multiples of P (see statement (b) of part (ii) of each Theorem) are also new. We do not expect them to persist past $n = 3$, and in fact find that for $n = 4$ the largest number of consecutive integral multiples occurs for (E, P) with $\hat{h}(P) = 19/630$ or $13/360$, whereas $\hat{h}_{\min}(0, 48) \leq 41/1540 < 19/630 < 13/360$. We shall say more about integrality later; for now we content ourselves with the following remarks. A point on an elliptic curve over a function field $k(C)$ is said to be integral if it is a nonzero point whose naïve height vanishes. Geometrically, if we regard E as an elliptic surface $\mathcal{E}$ over C, and a rational point $P \in E(K)$ as a section s_P of $\mathcal{E}$, this means that s_P is disjoint from the zero-section s_0 of $\mathcal{E}$. Since $g = 0$ in our case, we can give an explicit algebraic characterization of integrality. Write E in extended Weierstrass form as

$$Y^2 + a_1 XY + a_3 Y = X^3 + a_2 X^2 + a_4 X + a_6 \tag{4}$$

where each a_i is a homogeneous polynomial of degree $i \cdot n$ in two variables. Then a rational point (X, Y) is integral if X, Y are homogeneous polynomials of degrees $2n, 3n$ respectively. The equation (4) depends on the choice of coordinates X, Y on E; replacing X, Y by

$$\delta^2(X + \alpha_2), \qquad \delta^3(Y + \alpha_1 X + \alpha_3) \tag{5}$$

(some α_i and nonzero δ) yields an isomorphic curve. If moreover $\delta \in \mathbf{C}^*$ and each α_i is a homogeneous polynomial of degree $i \cdot n$ then the new equation for E has the same discriminant degree and the same integral points.

1.2   Outline of this Paper

For each $n = 1, 2, 3$ we prove Theorem n, except for the implications (a),(b)$\Rightarrow$(c) of part (ii), which require different methods that we defer to a later paper. Our proofs use the following ingredients:

- $\hat{h}(mP) = m^2 \hat{h}(P)$ for all $m \in \mathbf{Z}$.
- If $mP \neq 0$ then

$$\hat{h}(mP) = h(mP) + \sum_v \lambda_v(mP), \tag{6}$$

  where $h(\cdot)$ is the naïve height and the sum extends over all places $v \in C(\mathbf{C})$ lying under singular fibers E_v of E. (All places of K are of degree 1 thanks to our use of the algebraically closed field $\mathbf{C}$ for k.) The local corrections $\lambda_v(mP)$ are described further below.
- The naïve height takes values in $\{0, 2, 4, 6, \ldots\}$, and satisfies $h(m'P) \leq h(mP)$ for any integers m, m' such that $m'|m$ and $mP \neq 0$.
- Each local correction $\lambda_v(mP)$ depends only on the Kodaira type of the fiber E_v and on the component of E_v meeting P. We shall call this component c_v. The values of $\lambda_v(\cdot)$ are known explicitly for all Kodaira types and each possible component, see for instance [13, Thm. 5.2].

- Finally, the condition that E have discriminant degree $d = 12n$ imposes two conditions on the Kodaira types of the singular fibers. The first condition is

$$d = \sum_v d_v, \tag{7}$$

where d_v is the local discriminant degree of E_v. This allows only finitely many collections of fiber types. The second condition follows from an inequality due to Shioda [9, Cor. 2.7 (p.30)], and eliminates some of these collections that have too few fibers. According to this condition, if a nonconstant elliptic curve of discriminant degree d over a function field $K = \mathbf{C}(C)$ has a nontorsion point then the conductor degree of the curve strictly exceeds $(d/6) + \chi(C)$. Here $\chi(C) = 2 - 2g$ is the Euler characteristic of C. The conductor degree may be defined as the number of multiplicative fibers plus twice the number of additive fibers; thus it is also a sum of invariants of the singular fibers. When $(g, d) = (0, 12n)$ we have $\chi(C) = 2$ and $d/6 = 2n$, so the conductor degree is at least $2n + 3$.

We shall refer to these constraints as the "combinatorial conditions" on $\hat{h}(P)$, $h(mP)$, and the collection of (E_v, c_v) that arise for (E, P). (For other uses of such conditions to obtain lower bounds on heights, see for instance [3,14] and work referenced in these sources.) In general the combinatorial conditions yield only a lower bound on $\hat{h}_{\min}(0, 12n)$, because they allow some possibilities that do not actually occur for any (E, P). But for each of $n = 1$, 2, and 3 this lower bound turns out to be attained by some (E, P) over $\mathbf{C}(T)$, namely those exhibited in statement (c) of part (ii) of Theorem n. (Note that we do not yet need to derive the formulas for these (E, P), nor to prove that they are the only ones possible.) Moreover, using (6) we can check that $\hat{h}(P) = \hat{h}_{\min}(0, 12n)$ if and only if the naïve height $h(mP)$ vanishes for all m up to 6, 8, or 9 respectively.

Still, already at $n = 1$ we see some redundancy. The combinatorial conditions allow $\hat{h}(P) = 1/30$ to be attained in any of five ways, four of which are realized by the curves $E_1(q)$ of Theorem 1 for suitable choices of q. Shioda's $E_1(4/5)$ has singular fibers of types I_5, I_3, I_2, and II. (We specify the components c_v later in the paper.) The fibers of $E_1(-1)$ have types I_5, IV, I_2, and I_1, while those of $E_1(4)$ have types I_5, I_3, III, and I_1. In all other cases, the fibers of $E_1(q)$ have types I_5, I_3, I_2, I_1, I_1: the first three at $s = 0$, $s' = 0$, $s' = s$, and the last two at the roots of the quadratic $(q + 1)^3 s^2 = (11q^2 - 14q + 2)ss' + (q - 1)s'^2$. When $q = 4/5$, these roots coincide and the two I_1 fibers merge to form a II; likewise at $q = -1$ or $q = 4$, one of the I_1 fibers merges with the I_3 or I_2 fiber to form a IV or III respectively. (The one merger that does not occur is $I_1 + I_1 \to I_2$.) But none of these degenerations changes $\hat{h}(P)$, nor any $h(mP)$, nor the conductor degree N. In fact a fiber of type II, III, or IV contributes as much to our formulas for $\hat{h}(P), h(mP), N$ as a pair of fibers of types I_1 and I_ν ($\nu = 1$, 2, or 3). Thus it is enough to minimize $\hat{h}(P)$ under the further assumption that no fibers of type II, III, or IV occur. We find similar replacements for all components of fibers of the remaining additive types I_ν^*, II^*, III^*, IV^*. See Proposition 2. This simplifies the computation of the combinatorial lower bound on $\hat{h}_{\min}(0, 12n)$: instead of an exhaustive search over all combinations of (E_v, c_v), we need only try those for which each E_v is multiplicative (of type I_ν for $\nu = d_v$).

We programmed the search over all partitions $\{d_v\}$ of $12n$ in GP [8] and ran it on a Sun Ultra 60. This took only a fraction of a second for $n = 1$, five seconds for $n = 2$, and five minutes for $n = 3$. It took about an hour to carry out the same computation for $n = 4$, and about 20 hours for $n = 5$; but the resulting bounds are probably not attained: as we shall see in a later paper, the required (E_v, c_v) data impose more conditions than the number of parameters needed to specify (E, P). We do produce explicit (E, P) that show $\hat{h}_{\min}(0, 48) \leq 41/1540$ and $\hat{h}_{\min}(0, 60) \leq 261/10010$, and conjecture that these are the correct values of $\hat{h}_{\min}(0, 12n)$ for $n = 4, 5$. We have not attempted to extend the computation past $n = 5$.

1.3   Coming Attractions

Happily, the computation of the surfaces (1,2,3) not only completes the proofs of Theorems 1 through 3 but also points the way to further results and connections. We outline these here, and defer detailed treatment to a later paper in this series. In each step of the computation we in effect obtain a new birational model for the moduli space, call it $\mathcal{X}$, of pairs (E, P) consisting of an elliptic curve and a point on it. Our new parametrizations of this rational surface $\mathcal{X}$ have several other applications. One is a geometric interpretation of Tate's method for exhibiting the generic elliptic curve with an N-torsion point: we readily locate the modular curves $X_1(N)$ ($N \leq 16$) on $\mathcal{X}$, together with nonconstant rational functions of minimal degree that realize each $X_1(N)$ as an algebraic curve of genus ≤ 2. Arithmetically, we can use our parametrizations of $\mathcal{X}$ to find (E, P) over $\mathbf{Q}$ (or over some other global field) such that P is a nontorsion point with small $\hat{h}(P)$, and/or with many integral multiples in the minimal model of E. For instance, we prove that there are infinitely many $(E, P)/\mathbf{Q}$ such that mP is integral for each $m = 1, 2, \ldots, 11, 12$. Our numerical results for a isolated curves (E, P) over $\mathbf{Q}$ may be found on the Web at http://www.math.harvard.edu/~elkies/low_height.html . They include new records for consecutive integral multiples and for the Lang ratio $\hat{h}(P)/\log|\Delta_E|$. We have mP integral for each $m = 1, 2, \ldots, 13, 14$ for

$$E : Y^2 + XY = X^3 - 139761580X + 1587303040400, \qquad (8)$$

an elliptic curve of conductor $1029210 = 2 \cdot 3 \cdot 5 \cdot 7 \cdot 13^2 \cdot 29$, and P the nontorsion point $(X, Y) = (11480, 1217300)$; and we find the curve

$$Y^2 + XY = X^3 - 161020013035359930X + 248692506247420690486 41252 \quad (9)$$

of conductor $3476880330 = 2 \cdot 3 \cdot 5 \cdot 7 \cdot 11 \cdot 23 \cdot 31 \cdot 2111$ with the nontorsion point $(-296994156, 6818852697078)$ of canonical height[2] $\hat{h}(P) = .0190117\ldots <$

---

[2] There are two standard normalizations, differing by a factor of 2, for the canonical height of a point on an elliptic curve over $\mathbf{Q}$. We use the larger one, which is the one consistent with our formulas for function fields.

$1.691732 \cdot 10^{-4} \log |\Delta_E|$. The curves (8,9) are the specializations of our formula (3) with $(A, t/t') = (35/32, -8/15), (33/23, 115/77)$.

Our simplified formula for $\hat{h}(mP)$ (Proposition 2) also bears on the asymptotic behavior of $\hat{h}_{\min}(g, 12n)$ for fixed g as $n \to \infty$. Hindry and Silverman [3] used the combinatorial conditions (except for the condition: $h(m'P) \leq h(mP)$ if $m'|m$) to show that there exists $C > 0$ such that

$$\hat{h}(g, 12n) \geq Cn - O_g(1), \tag{10}$$

This proved the function-field case of a conjecture of Lang [4, p.92]. The error terms $O_g(1)$ are effectively computed, and can be omitted entirely if $g \leq 1$. Hindry and Silverman also produce an explicit constant C, but it is quite small: about $7 \cdot 10^{-10}$. Their approach requires a point meeting every additive fiber in its identity component, which they achieved by working with $12P$ instead of P, at the cost of a factor of $1/12^2$ in C. Our results here let one apply the same methods directly to P, thus saving a factor of 12^2 and raising C to about 10^{-7}. In a later paper we show how to gain another factor of approximately 5000, raising the lower bound on $\liminf_n \hat{h}(g, 12n)/n$ to $1/2111$. This is within an order of magnitude of the correct value: for all $n \equiv 0 \mod 5$ we obtain $\hat{h}_{\min}(0, 12n) \leq 261n/50050$ via base change from our $n = 5$ example.

2   The Naïve and Canonical Heights

We collect here the facts we shall use about elliptic curves E over function fields K in characteristic zero, the associated elliptic surface $\mathcal{E}$, and the naïve and canonical height functions on $E(K)$.

2.1   The Naïve Height

The *naïve height* $h(P)$ of a nonzero $P \in E(K)$ can be defined using intersection theory on the elliptic surface $\mathcal{E}$ associated to some model of E. Let s_0 be the zero-section of the elliptic fibration $\mathcal{E} \to C$, and s_P the section corresponding to P. Then $h(P) := 2s_P \cdot s_0$. Since we assumed that $P \neq 0$, the sections s_0, s_P are distinct curves on $\mathcal{E}$. Hence their intersection number $s_P \cdot s_0$ is a nonnegative integer, and $h(P)$ is a nonnegative even integer. Moreover $h(P) = 0$ if and only if s_P is disjoint from s_0, in which case we say that P is an *integral point* on E.

When $C = \mathbf{P}^1$, we can give an equivalent algebraic definition of $h(P)$ in terms of a Weierstrass equation of E. This definition emphasizes the analogy with the canonical height in the more familiar case of an elliptic curve over $\mathbf{Q}$. Recall that each coefficient a_i in the Weierstrass equation (4) is a homogeneous polynomial of degree $i \cdot n$ in the projective coordinates on $\mathbf{P}^1$. Then the coordinates x, y of a nonzero $P \in E(K)$ are homogeneous rational functions of degrees $2n, 3n$. If x, y are written as fractions "in lowest terms", as quotients of coprime homogeneous polynomials, then the denominators are (up to scalar multiple) the square and cube of some polynomial ζ. The roots of ζ, with multiplicity, are the images on $\mathbf{P}^1$ of the intersection points of s_0 and s_P. Hence $s_P \cdot s_0 = \deg \zeta$. Therefore

$h(P)$ is the degree of the denominator ζ^2 of x, which is also the number of poles of x counted with multiplicity. An integral point is one for which ζ is a nonzero scalar and thus x, y are homogeneous polynomials of degrees $2n, 3n$.

For an arbitrary base curve C, the coefficients a_i are global sections of $\mathcal{L}^{\otimes i}$ for some line bundle $\mathcal{L}$ on C, and x, y are meromorphic sections of $\mathcal{L}^{\otimes 2}, \mathcal{L}^{\otimes 3}$. The pole divisors of x, y are $2Z, 3Z$ for some effective divisor Z on C, whose degree is $s_P \cdot s_0$; thus again $h(P)$ is the degree of the pole divisor $2Z$ of x, and P is integral iff $Z = 0$ iff x, y are global sections of $\mathcal{L}^{\otimes 2}, \mathcal{L}^{\otimes 3}$. A linear change of coordinates according to (5) yields the same notion of integrality if and only if $\delta \in \mathbf{C}^*$ and $\alpha_i \in \Gamma(\mathcal{L}^{\otimes i})$ for each i.

We shall need one more property of the naïve height beyond its relation with the canonical height and the fact that $h(mP) \in \{0, 2, 4, 6, \ldots\}$ $(mP \neq 0)$:

Lemma 1. *Let P be a point on an elliptic curve over $k(C)$, and let m, m' be any integers such that $m'|m$ and $mP \neq 0$. Then $h(m'P) \leq h(mP)$.*

Proof: Each point of $s_{m'P} \cap s_0$ is also a point of intersection of s_{mP} with s_0, to at least the same multiplicity. Hence $s_{m'P} \cdot s_0 \leq s_{mP} \cdot s_0$, so

$$h(m'P) = 2s_{m'P} \cdot s_0 \leq 2s_{mP} \cdot s_0 = h(mP)$$

as claimed.   □

Remarks:
1. We could also state the result as: The naïve height of a point is less than or equal to the naïve height of any of its multiples that is not the zero point. This is a more natural formulation (the first point does not have to be written as $m'P$), but less convenient for our purposes.
2. In the proof, "at least the same multiplicity" can be strengthened to "exactly the same multiplicity" in our characteristic-zero setting. In general $h(mP)$ may strictly exceed $h(m'P)$ because $s_{mP} \cap s_0$ may also contain points where $m'P$ reduces to a nontrivial (m/m')-torsion point.

The naïve height satisfies further inequalities along the lines of Lemma 1, for instance

$$h(6P) + h(P) \geq h(2P) + h(3P). \tag{11}$$

Lemma 1 suffices for the proofs of Theorems 1–3 in the genus-zero case, but inequalities such as (11) are sometimes needed to exclude possible configurations with positive g, as we shall see for $d = 24$. The strongest such inequality we found is:

Lemma 2. *Let P be a point on an elliptic curve over $k(C)$, and let m be any integer such that $mP \neq 0$. Then*

$$\sum_{m'|m} \mu(m/m')\, h(m'P) \geq 0. \tag{12}$$

Proof: The left-hand side can be interpreted as twice the number of points of C, counted with multiplicity, at which $mP = 0$ but $m'P \neq 0$ for each proper factor m' of m.   □

Inequality (11) is the special case $m = 6$ of this Lemma. The sum in (12) may be considered as an analogue of the formula $\prod_{m'|m}(x^{m'} - 1)^{\mu(m/m')}$ for the m-th cyclotomic polynomial. We recover Lemma 1 by summing the inequality (12) over all factors of m, including m itself but not 1, to obtain $h(mP) \geq h(P)$, which is equivalent to Lemma 1 by the first Remark above.

2.2 Local Invariants, and Shioda's Inequality

To go from the naïve to the canonical height we must use the minimal model of E for the elliptic surface $\mathcal{E}$. We next describe this model, collect some known facts on the singular fibers of $\mathcal{E}$, and give Shioda's lower bound on the conductor degree.

Whereas a naïve height could be defined for any model of E,[3] the canonical height requires the Néron minimal model. It is known that there exists a minimal line bundle $\mathcal{L}$ on C with the following property: let D be a divisor on C such that $O(D) \cong \mathcal{L}$; then E is isomorphic to a curve with an extended Weierstrass equation (4) whose coefficients a_i are global sections of iD. In characteristic zero we can easily obtain D and $\mathcal{L}$ by putting E in narrow Weierstrass form $Y^2 = X^3 + a_4 X + a_6$. Then D is the smallest divisor such that $(a_4) + 4D \geq 0$ and $(a_6) + 6D \geq 0$. In other words, we can regard a_4, a_6 as global sections of $\mathcal{L}^{\otimes 4}, \mathcal{L}^{\otimes 6}$ such that there is no point of C where a_4 and a_6 vanish to order at least 4 and 6 respectively. Once we have $a_i \in \Gamma(\mathcal{L}^{\otimes i})$, we can regard the Weierstrass equation (4) as a surface in the plane bundle $\mathcal{L}^{\otimes 2} \oplus \mathcal{L}^{\otimes 3}$ over C. If all the roots of the discriminant $\Delta \in \Gamma(\mathcal{L}^{\otimes 12})$ are distinct then this surface is smooth and is the minimal model of E. Otherwise it has isolated singularities, which we blow up as many times as needed (we may follow Tate's algorithm [16]) to obtain the minimal model $\mathcal{E}$. This is a smooth algebraic surface of arithmetic genus $n = \deg \mathcal{L}$, equipped with a map to C with generic fiber E and $\omega_{\mathcal{E}/C} \cong \mathcal{L}$. See for instance [1, pp.149ff.].

We shall need much information about the singular fibers that can arise for the elliptic fibration $\mathcal{E} \to C$. We extract from Tate's table [16, p.46] the following local data for each possible Kodaira type of a singular fiber E_v: the discriminant degree d_v, the conductor degree N_v, and the structure of the group $E_v/(E_v)_0$ of multiplicity-1 components. We also list in each case the root lattice L_v that E_v contributes to the Néron-Severi lattice $\mathrm{NS}(\mathcal{E})$ of $\mathcal{E}$. In each case, L_v has rank $d_v - N_v$, and $E_v/(E_v)_0 \cong L_v^*/L_v$ where $L_v^* \subset L_v \otimes \mathbf{Q}$ is the dual lattice. The lattice "A_0" that appears for Kodaira types I_1 and II is the trivial lattice of rank zero. For Kodaira type I_ν^*, the group $E_v/(E_v)_0$ always has order 4, and has exponent 2 or 4 according as ν is even or odd. For positive ν of either parity, a fiber of type

---

[3] Two models may yield different heights h, h', but $h' = h + O(1)$ holds for any pair of naïve heights on the same curve. It also follows that the property $\hat{h} = h + O(1)$ of the canonical height does not depend on the choice of naïve height h.

I_ν^* has a distinguished multiplicity-1 component of order 2 in $E_v/(E_v)_0$, namely the one closest to the identity component. In the L_v picture, the distinguished component corresponds to the nontrivial coset of $D_{4+\nu}$ in $\mathbf{Z}^{4+\nu}$. When $\nu = 0$ there is no distinguished component: all three non-identity components of multiplicity 1 are equivalent, as are all three nontrivial cosets due to the triality of D_4.

Kodaira type	$I_\nu (\nu > 0)$	II	III	IV	I_ν^*	IV*	III*	II*
d_v	ν	2	3	4	$6 + \nu$	8	9	10
N_v	1	2	2	2	2	2	2	2
$E_v/(E_v)_0$	$\mathbf{Z}/\nu\mathbf{Z}$	$\{0\}$	$\mathbf{Z}/2\mathbf{Z}$	$\mathbf{Z}/3\mathbf{Z}$	$D_{4+\nu}^*/D_{4+\nu}$	$\mathbf{Z}/3\mathbf{Z}$	$\mathbf{Z}/2\mathbf{Z}$	$\{0\}$
root lattice	$A_{\nu-1}$	A_0	A_1	A_2	$D_{4+\nu}$	E_6	E_7	E_8

The discriminant and conductor degrees d, N of $\mathcal{E}$ are sums of the discriminant and conductor degrees of the singular fibers:

$$12n = d = \sum_v d_v, \qquad N = \sum_v N_v. \tag{13}$$

Hence $d - N = \sum_v (d_v - N_v) = \sum_v \mathrm{rk}\, L_v$ is the rank of the subgroup $\oplus_v L_v$ of $\mathrm{NS}(\mathcal{E})$ due to the singular fibers. Shioda used this to prove [9, Cor. 2.7 (p.30)]:

Proposition 1. *Let E be a nonconstant elliptic curve over a function field $K = k(C)$ of genus g, with discriminant and conductor degrees $d = 12n$ and N. Then*

$$N \geq 2n + (2 - 2g) + r, \tag{14}$$

where r is the rank of the Mordell-Weil group $E(K)$.

Proof: Let $T \subseteq \mathrm{NS}(\mathcal{E})$ be the subgroup spanned by s_0, the generic fiber, and $\oplus_v L_v$. Then we have a short exact sequence (see for instance [10, Thm. 1.3]):

$$0 \to T \to \mathrm{NS}(\mathcal{E}) \to E(K) \to 0, \tag{15}$$

where the map $\mathrm{NS}(\mathcal{E}) \to E(K)$ is the sum on the generic fiber. Taking ranks, we find

$$\mathrm{rk}\,\mathrm{NS}(\mathcal{E}) = \mathrm{rk}\, T + \mathrm{rk}\, E(K) = 2 + (d - N) + r. \tag{16}$$

But $\mathrm{NS}(\mathcal{E})$ embeds into $H^{1,1}(\mathcal{E}, \mathbf{Z})$, a group of rank $h^{1,1}(\mathcal{E}) = 10n + 2g$. Hence $\mathrm{rk}\,\mathrm{NS}(\mathcal{E}) \leq 10n + 2g$. Therefore

$$N \geq (d + 2 + r) - (10n + 2g) = 2n + (2 - 2g) + r,$$

as claimed.   □

Remarks
  1. Since $r \geq 0$ it follows that

$$N \geq 2n + (2 - 2g) = (d/6) + \chi \tag{17}$$

for any nonconstant elliptic surface. This weaker inequality is sufficient for most of our purposes, even though we are interested in curves with a non-torsion point, for which the strict inequality $N > (d/6) + \chi$ holds because $r > 0$.

2. The inequality (17) is now usually known as the "Szpiro inequality", but Shioda's paper [9] predates Szpiro's [15] by almost two decades (see also [12, p.114]). It is by now well-known that (17) can be proved by elementary means via Mason's theorem [5] (the ABC inequality for function fields). Can one also give an elementary proof of Shioda's inequality, or even of its consequence that $r = 0$ if $N = (d/6) + \chi$?

3. The requirement that E not be a constant curve is essential. There is an analogous statement for constant curves but many details must change. Suppose E is such a curve, that is, $\mathcal{E} = C \times E_0$ for some elliptic curve E_0/k. Then $E(K)$ is not finitely generated, because it contains a copy of $E_0(k)$. Still, $E(K)/E_0(k)$ is finitely generated, and identified with the group $NS(\mathcal{E})/T$. Again we call the rank of this group r. Since $n = d = N = 0$ in this setting, we obtain the inequality $r + 2 \leq h^{1,1}(C \times E_0) - 2$. But for a constant curve, $h^{1,1}(C \times E_0) = 2g + 2$, instead of the $2g$ that one would expect from the $10n + 2g$ formula. Hence $r \leq 2g$. This can also be proved using the identification of $E(K)/E_0(k)$ with $\mathrm{End}(\mathrm{Jac}(C), E_0)$, an approach that also yields the equality condition: clearly $r = 2g$ if $g = 0$; if $g > 0$ then $r = 2g$ if and only if E_0 has complex multiplication and $\mathrm{Jac}(C)$ is isogenous with E_0^g. See for instance [2].

4. The hypothesis of characteristic zero, too, is essential here. In positive characteristic, one cannot decompose the second Betti number $b_2(\mathcal{E})$ as $h^{2,0} + h^{1,1} + h^{0,2}$, so one has only the weaker upper bound $b_2(\mathcal{E})$ on $\mathrm{rk}(NS(\mathcal{E}))$. This upper bound exceeds the characteristic-zero bound by $2g$ for a constant curve and $2(n + g - 1)$ for a nonconstant one. For instance, a constant curve $C \times E_0$ has $r \leq 4g$, with equality if and only if either $g = 0$ or E_0 and $\mathrm{Jac}(C)$ are both supersingular. In general $\mathcal{E}$ is said to be "supersingular" if $NS(\mathcal{E}) \cong \mathbf{Z}^{b_2(\mathcal{E})}$; such surfaces were studied and used in [10,2].

2.3   Local Height Corrections

We next list the local height corrections $\lambda_v(mP)$ for each of the Kodaira types. For convenience we abuse notation by using mP to refer also to the section s_{mP}.

– If mP is on the identity component of E_v then

$$\lambda_v(mP) = d_v/6. \tag{18}$$

In particular this covers fibers of type II or II*.

– If E_v is of type I_ν and P passes through component $a \in \mathbf{Z}/\nu\mathbf{Z}$, let $x = \bar{a}/\nu$ for any lift $\bar{a}$ of a to $\mathbf{Z}$; then

$$\lambda_v(mP) = \nu B(mx), \tag{19}$$

where $B(\cdot)$ is the second Bernoulli function $B(z) := \sum_{n=1}^{\infty} \cos(2\pi n)/(\pi n)^2$. Since B is $\mathbf{Z}$-periodic, the choice of $\bar{a}$ does not matter. Likewise, since $B(z) = B(-z)$ it does not matter that a cannot be canonically distinguished from $-a$. We have

$$B(z) = z^2 - z + \frac{1}{6} \tag{20}$$

for all $z \in [0, 1]$, so in particular $B(0) = 1/6$. Hence $\lambda_v(mP) = v/6$ if mP passes through the identity component of E_v, as also asserted by (18) in that case.

- If E_v is of type III, IV, I_0^*, III*, or IV*, and mP passes through a non-identity component of E_v, then $\lambda_v(mP) = 0$.
- Finally, suppose E_v is of type I_ν^* ($\nu > 0$) and that mP passes through a non-identity component. If that component is the distinguished one of order 2 then $\lambda_v(mP) = v/6$. Otherwise $\lambda_v(mP) = -v/12$. (We could have also allowed $\nu = 0$, when there is no distinction among the three non-identity components, but $\lambda_v(mP) = v/6 = -v/12 = 0$ for all of them.)

We record two applications of these formulas for future use:

Lemma 3. *Let E be an elliptic curve of discriminant degree $12n$ over a function field K, and P any nonzero point of $E(K)$. Then*

$$-n \leq \hat{h}(P) - h(P) \leq 2n. \tag{21}$$

Proof: For each v we have $-d_v/12 \leq \lambda_v \leq d_v/6$. Summing over v yields (21). □

Lemma 4. *Let E be an elliptic curve of discriminant degree $12n$ over a function field K, and P any point of $E(K)$. If for some integer m the multiple mP is a nonzero integral point then $\hat{h}(mP) \leq 2n/m^2$.*

Proof: By our formulas for λ_v we have $\lambda_v(mP) \leq d_v/6$ for all v. Hence

$$m^2 \hat{h}(P) = \hat{h}(mP) = h(mP) + \sum_v \lambda_v(mP) \leq h(mP) + \sum_v d_v/6. \tag{22}$$

But $h(mP) = 0$ since mP is integral, and $\sum_v d_v/6 = d/6 = 2n$. Hence $m^2 \hat{h}(P) \leq 2n$, and the Lemma follows. □

2.4   Reduction to the Semistable Case

Recall that an elliptic curve is said to be *semistable* if all its singular fibers are of type I_ν for some ν. Suppose E/K is semistable and P is a nontorsion point in $E(K)$. We associate to (E, P) an element γ of the abelian group $\mathbf{G}$ of formal $\mathbf{Z}$-linear combinations of orbits of $\mathbf{Q}$ under the infinite dihedral group D_∞ generated by $z \mapsto z+1$ and $z \leftrightarrow 1-z$. We denote by $[z]$ the generator of $\mathbf{G}$ corresponding to the orbit of z. Then γ is defined as a sum of local contributions

$\gamma_v \in \mathbf{G}$ that record the types $\nu(v)$ of the singular fibers E_v and the component $c_v = a(v) \in \mathbf{Z}/(\nu(v))\mathbf{Z}$ of each fiber that contains P, as follows:

$$\gamma_v := \sum_v \gcd(a(v), \nu(v)) \cdot \left[\frac{a(v)}{\nu(v)}\right]. \tag{23}$$

Then each of the height corrections $\hat{h}(mP) - h(mP)$, as well as the discriminant degree, are images of γ under homomorphisms $\boldsymbol{\lambda}_m, \mathbf{d}$ from $\mathbf{G}$ to $\mathbf{Q}$ or $\mathbf{Z}$, and the conductor is bounded above by the image of a homomorphism $\mathbf{N} : \mathbf{G} \to \mathbf{Z}$. We define these homomorphisms on the generators of $\mathbf{G}$ and extend by linearity. Suppose $\mathbf{Q} \ni z = a/b$ with $b > 0$ and $\gcd(a,b) = 1$. Note that b is an invariant of the action of D_∞. Then we set

$$\boldsymbol{\lambda}_m([z]) := b\, B_2(mz), \qquad \mathbf{d}([z]) := b, \qquad \mathbf{N}([z]) := 1. \tag{24}$$

Then our formulas (19,13) yield the identities

$$\hat{h}(mP) = h(mP) + \boldsymbol{\lambda}_m(\gamma) \quad (m = 1, 2, 3, \ldots), \qquad 12n = d = \mathbf{d}(\gamma) \tag{25}$$

and the estimate

$$N \leq \mathbf{N}(\gamma). \tag{26}$$

(This last is an upper bound rather than an identity because each v contributes 1 to N and $\gcd(a(v), \nu(v)) \geq 1$ to $\mathbf{N}(\gamma)$.) It follows that

$$\mathbf{N}(\gamma) \geq N \geq (d/6) + (2 - 2g) + r \geq \frac{1}{6}\mathbf{d}(\gamma) + 3 - 2g. \tag{27}$$

The second step is Shioda's inequality (Prop. 1), and the third step uses the positivity of r, which follows from our hypothesis that P is nontorsion.

To generalize these formulas to curves that may not be semistable, it might seem that we would have to extend $\mathbf{G}$ with generators that correspond to Kodaira types other than I_ν. But we can associate to any additive fiber E_v an element of $\mathbf{G}$ whose images under $\boldsymbol{\lambda}_m$ and $\mathbf{d}$ coincide with $\lambda_v(mP)$ and d_v, and whose image under $\mathbf{N}$ is $\geq N_v$. (Note that we already did this for multiplicative fibers with $f = \gcd(a(v), \nu(v)) > 1$, replacing them in effect by f fibers with a, ν coprime and the same value of a/ν.) As in the multiplicative case, this element is positive, in the sense that it is a nonzero formal linear combination of elements of $\mathbf{Q}/D_\infty$ with nonnegative coefficients. Specifically, we have:

Proposition 2. *Let E be an elliptic curve over a function field K of genus g, and $P \in E(K)$ a nontorsion point. Define for each singular fiber E_v a positive $\gamma_v \in \mathbf{G}$, depending on (E_v, c_v) as follows:*

- *If E_v is multiplicative, γ_v is defined by (23).*
- *If c_v is the identity component then $\gamma_v := d_v\,[0]$.*
- *If c_v is a non-identity component of a fiber E_v of type III, IV, IV*, or III* then γ_v is respectively*

$$[1/2] + [0], \quad [1/3] + [0], \quad 2 \cdot [1/2] + 2 \cdot [0], \quad 3 \cdot [1/3] + 3 \cdot [0].$$

– If c_v is a distinguished component of a fiber E_v of type I_ν^* then

$$\gamma_v := 2\,[1/2] + (\nu + 2)\,[0].$$

– If c_v is a non-distinguished, non-identity component of a fiber E_v of type I_ν^* then

$$\gamma_v := (\mu + 2)\,[1/2] + 2\,[0]$$

if $\nu = 2\mu$, and

$$\gamma_v := [1/4] + (\mu + 1)\,[1/2] + [0]$$

if $\nu = 2\mu + 1$ for some integer μ.

Then:
  i) $\lambda_v(mP) = \boldsymbol{\lambda}_m(\gamma_v)$ for each $m = 1, 2, 3, \ldots$;
  ii) $d_v = \mathbf{d}(\gamma_v)$; and
  iii) $N_v \leq \mathbf{N}(\gamma_v)$.
Thus (25,26,27) hold for $\gamma := \sum_v \gamma_v$. Equality in (iii) holds if and only if E_v is either a multiplicative fiber with $\gcd(a, \nu) = 1$, a fiber of type III or IV with c_v a non-identity component, or a fiber of type II.

[Note that, as was true for the λ_v formulas, the first two formulas in Prop. 2 overlap in the case of a multiplicative fiber with $a(v) = 0$, but give the same answer in this case. Here both prescriptions yield $\gamma_v = \nu(v) \cdot [0]$ for such v.]

Proof: The multiplicative case was seen already. For each of the other Kodaira types, it is straightforward to verify that $\lambda_v(mP) = \boldsymbol{\lambda}_m(\gamma_v)$ for each nonnegative m less than the exponent of the finite group $E_v/(E_v)_0$ (which is at most 4), and to check that $d_v = \mathbf{d}(\gamma_v)$, and that $N_v \leq \mathbf{N}(\gamma_v)$, with strict inequality except in the three cases listed. We recover (25,26,27) by summing over v.   □

3   The Values of $\hat{h}_{\min}(0, 12n)$ for $n = 1, 2, 3$, and Consecutive Integral Multiples

For each n we can use the formulas and results above to obtain a lower bound on $\hat{h}_{\min}(g, 12n)$. When $g = 0$ and $n = 1, 2, 3$ we also show that this bound is attained if and only if mP is integral for $m \leq M = 6, 8, 9$, and verify that the (E, P) exhibited in Theorem n satisfy those conditions.

Suppose E is an elliptic curve over $\mathbf{C}(T)$ with discriminant degree $12n$. Let P be a nontorsion rational point on E, and γ the associated element of $\mathbf{G}$. From γ and $\hat{h}(P)$ we can recover all the naïve heights $h(mP)$ from the first formula in (25): $h(mP) = m^2\hat{h}(P) - \boldsymbol{\lambda}_m(\gamma)$. Given n and an upper bound H on $\hat{h}(P)$, there are only finitely many candidates for the pair $(\gamma, \hat{h}(P))$: there are finitely many $\gamma > 0$ with $\mathbf{d}(\gamma) = 12n$, and for each one there are only finitely many possible choices for $h(P)$ consistent with $h(P) + \boldsymbol{\lambda}_1(\gamma) = \hat{h}(P) \in (0, H]$. For each candidate $(\gamma, \hat{h}(P))$ we can check the condition $m'|m \Rightarrow h(mP) \geq h(m'P) \geq 0$. Only finitely many m need be checked for each $(\gamma, \hat{h}(P))$: by Lemma 3 we know that $h(mP) \geq 0$ once $m^2\hat{h}(P) \geq n$, and $h(mP) \geq h(m'P)$ for each $m'|m$ once

$m^2 \hat{h}(P) \geq 4n$. The minimal $\hat{h}(P)$ among the $(\gamma, \hat{h}(P))$ that pass these tests is then our lower bound on $\hat{h}_{\min}(g, 12n)$. [We could also test the more complicated inequality of Lemma 2, which may further improve the bound; instead we checked that inequality after the fact when necessary.]

We wrote a GP program to compute this bound by exhaustive search, and ran it with $H = 2n/M^2$ for $n = 1, 2, 3$. We chose this upper bound H to ensure that, by Lemma 4, we would also find all feasible $(\gamma, \hat{h}(P))$ such that $h(mP) = 0$ for each $m = 1, 2, 3, \ldots, M$. For $n = 1$, we found that the minimum occurs for

$$\gamma = [1/5] + [1/3] + [1/2] + 2\,[0], \qquad \hat{h}(P) = 1/30, \tag{28}$$

and is the unique $(\gamma, \hat{h}(P))$ such that $h(mP) = 0$ for each $m \leq 6$. For $n = 2$, we found that the minimum occurs for

$$\gamma = [1/11] + 2\,[2/5] + [1/3], \qquad \hat{h}(P) = 4/165; \tag{29}$$

but this is not feasible because $h(mP) = 0, 2, 2, 2$ for $m = 2, 4, 6, 12$, so inequality (11) is violated when $m = 2$. Our lower bound on $\hat{h}_{\min}(g, 24)$ is thus the next-smallest value, which occurs for

$$\gamma = [1/7] + [2/5] + [1/4] + [1/3] + [1/2] + 3\,[0], \qquad \hat{h}(P) = 11/420, \tag{30}$$

and is the unique $(\gamma, \hat{h}(P))$ such that $h(mP) = 0$ for each $m \leq 8$.

On the other hand, the $(\gamma, \hat{h}(P))$ pairs of (28,30) are also those associated with the curves and points E, P exhibited in (1,2). Hence those E, P attain our lower bounds $1/30, 11/420$ on $\hat{h}_{\min}(12), \hat{h}_{\min}(24)$, as well as the upper bounds 6 and 8 on the number of consecutive integral multiples for $n = 1$ and $n = 2$. This proves all of Theorems 1 and 2 except for the claims that every (E, P) attaining those bounds is isomorphic with some $E_1(q)$ or $E_2(u)$.

For $n = 3$, we find that there is a unique $(\gamma, \hat{h}(P))$ such that $h(mP) = 0$ for each $m \leq 9$, namely

$$\gamma = [1/8] + [3/7] + [1/5] + [1/4] + 2\,[1/3] + [1/2] + 4\,[0], \quad \hat{h}(P) = 23/840. \tag{31}$$

Again these are the γ and $\hat{h}(P)$ for the (E, P) exhibited in the Introduction (formula (3)). But we do not claim that $\hat{h}_{\min}(36) = 23/840$: Lemma 2 eliminates the second-smallest pair

$$(\gamma, \hat{h}(P)) = ([1/13] + [3/8] + [3/7] + [1/5] + [1/3],\ 229/10920)$$

(which violates the inequality (11) in the same way that (29) did), but not several other possibilities with $\hat{h}(P) < 23/840$. We next list all these possibilities, in order of increasing $\hat{h}(P)$:

γ	$\hat{h}(P)$
$[1/13] + [3/11] + [3/8] + 2\,[1/2]$	$23/1144 \approx .02010$
$[1/13] + [3/8] + [2/7] + [1/4] + 2\,[1/2]$	$17/728 \approx .02335$
$[1/11] + [4/9] + [2/7] + [1/4] + [1/3] + 2\,[0]$	$65/2772 \approx .02345$
$[1/12] + [3/11] + [3/8] + 2\,[1/2] + [0]$	$7/264 \approx .02652$
$[1/11] + [3/7] + 2\,[1/5] + [1/4] + 2\,[1/2]$	$41/1540 \approx .02662$

$$\tag{32}$$

(For comparison, $229/10920 \approx .02097$ and $23/840 \approx .02738$.) We have $\mathbf{d}(\gamma) \leq 7$ for each entry in the table (32); therefore by Prop. 1 none of them can occur for an elliptic curve over $\mathbf{P}^1$. (Even the weaker inequality (17) would suffice here; either of those inequalities also excludes (29) for $n = 2$, and would thus be enough to obtain $\hat{h}_{\min}(0, 24)$, but the determination of $\hat{h}_{\min}(24)$ required a further argument.) Thus $\hat{h}_{\min}(0, 36) = 23/840$, proving Theorem 3 except for the claim that every (E, P) satisfying conditions (a) and (b) is of the form $E_3(A)$ for some A.

Acknowledgements. I thank J. Silverman and T. Shioda for helpful correspondence concerning their papers and related issues, and M. Watkins for carefully reading a draft of this paper. This work was made possible in part by funding from the Packard Foundation and the National Science Foundation.

References

1. Barth, W., Peters, C., Van de Ven, A.: *Compact Complex Surfaces*. Berlin: Springer, 1984.
2. Elkies, N.D.: Mordell-Weil lattices in characteristic 2, I: Construction and first properties. *International Math. Research Notices* 1994 #8, 343–361.
3. Hindry, M., Silverman, J.H.: The canonical height and integral points on elliptic curves, *Invent. Math.* **93** (1988), 419–450.
4. Lang, S.: *Elliptic Curves: Diophantine Analysis*. Berlin: Springer, 1978.
5. Mason, R.C.: *Diophantine Equations over Function Fields*, London Math. Soc. Lect. Note Ser. **96**, Cambridge Univ. Press 1984. See also pp.149–157 in Springer LNM **1068** (1984) [=proceedings of Journées Arithmétiques 1983 (Noordwijkerhout), H. Jager, ed.].
6. Nishiyama, K.-i.: The minimal height of Jacobian fibrations on K3 surfaces, *Tohoku Math. J.* (2) **48** (1996), 501–517.
7. Oguiso, K., Shioda, T.: The Mordell-Weil lattice of a rational elliptic surface, *Comment. Math. Univ. St. Pauli* **40** (1991), 83–99.
8. PARI/GP, versions 2.1.1–4, Bordeaux, 2000–4, http://pari.math.u-bordeaux.fr .
9. Shioda, T.: Elliptic Modular Surfaces, *J. Math. Soc. Japan* **24** (1972), 20–59.
10. Shioda, T.: On the Mordell-Weil lattices. *Comment. Math. Univ. St. Pauli* **39** (1990), 211–240.
11. Shioda, T.: Existence of a Rational Elliptic Surface with a Given Mordell-Weil Lattice, *Proc. Japan Acad. (Ser. A)* **68** (1992), 251–255.
12. Shioda, T.: Some remarks on elliptic curves over function fields, *Astérisque* **209** (1992) [=proceedings of Journées Arithmétiques 1991 (Genève), D.F. Coray and Y.-F. S. Pétermann, eds.], 99–114.
13. Silverman, J.H.: Computing Heights on Elliptic Curves, *Math. of Computation* **51** #183 (July 1988), 339–358.
14. Silverman, J.H.: A lower bound for the canonical height on elliptic curves over abelian extensions, *J. Number Theory* **104** (2005), 353–372.
15. Szpiro, L.: Discriminant et conducteur des courbes elliptiques. *Astérisque* **183** (1990) [=*Séminaire sur les Pinceaux de Courbes Elliptiques*, Paris 1988], 7–18.
16. Tate, J.: Algorithm for Determining the Type of a Singular Fiber in an Elliptic Pencil. Pages 33–52 in *Modular Functions of One Variable IV* (Lect. Notes in Math. **476** (1975); Birch, B.J., Kuyk, W., eds.).

Shimura Curves for Level-3 Subgroups of the $(2, 3, 7)$ Triangle Group, and Some Other Examples

Noam D. Elkies

Department of Mathematics, Harvard University, Cambridge, MA 02138 USA
elkies@math.harvard.edu

Abstract. The $(2,3,7)$ triangle group is known to be associated with a quaternion algebra A/K ramified at two of the three real places of $K = \mathbf{Q}(\cos 2\pi/7)$ and unramified at all other places of K. This triangle group and its congruence subgroups thus give rise to various Shimura curves and maps between them. We study the genus-1 curves $\mathcal{X}_0(3)$, $\mathcal{X}_1(3)$ associated with the congruence subgroups $\Gamma_0(3)$, $\Gamma_1(3)$. Since the rational prime 3 is inert in K, the covering $\mathcal{X}_0(3)/\mathcal{X}(1)$ has degree 28, and its Galois closure $\mathcal{X}(3)/\mathcal{X}(1)$ has geometric Galois group $\mathrm{PSL}_2(\mathbf{F}_{27})$. Since $\mathcal{X}(1)$ is rational, the covering $\mathcal{X}_0(3)/\mathcal{X}(1)$ amounts to a rational map of degree 28. We compute this rational map explicitly. We find that $\mathcal{X}_0(3)$ is an elliptic curve of conductor $147 = 3 \cdot 7^2$ over $\mathbf{Q}$, as is the Jacobian $\mathcal{J}_1(3)$ of $\mathcal{X}_1(3)$; that these curves are related by an isogeny of degree 13; and that the kernel of the 13-isogeny from $\mathcal{J}_1(3)$ to $\mathcal{X}_0(3)$ consists of K-rational points. We also use the map $\mathcal{X}_0(3) \to \mathcal{X}(1)$ to locate some complex multiplication (CM) points on $\mathcal{X}(1)$. We conclude by describing analogous behavior of a few Shimura curves associated with quaternion algebras over other cyclic cubic fields.

1 Introduction

1.1 Review: Quaternion Algebras over K, Shimura Modular Curves, and the $(2, 3, 7)$ Triangle Group

Let K be the field $\mathbf{Q}(\cos 2\pi/7)$, which is the totally real cubic field of minimal discriminant (namely, discriminant 49); let O_K be the ring $\mathbf{Z}[2\cos 2\pi/7]$ of algebraic integers in K; and let A be a quaternion algebra over K ramified at two of the three real places and at no other place of K. This algebra exists because the set of ramified places has even cardinality, and is determined uniquely up to K-algebra isomorphism. (See for instance [Vi] for this and other basic results on quaternion algebras and Shimura curves, and [E2] for our computational context.) All maximal orders in A are conjugate because A is indefinite and O_K has narrow class number 1; we fix one maximal order $\mathcal{O} \subset A$. Let $\Gamma(1)$ be the group of elements of $\mathcal{O}$ of reduced norm 1. Since A has exactly one unramified real place, $\Gamma(1)$ embeds into $\mathrm{SL}_2(\mathbf{R})$ as a discrete co-compact subgroup. Let $\mathcal{H}$ be the upper half-plane, with the usual action of $\mathrm{PSL}_2(\mathbf{R}) = \mathrm{SL}_2(\mathbf{R})/\{\pm 1\}$.

F. Hess, S. Pauli, and M. Pohst (Eds.): ANTS 2006, LNCS 4076, pp. 302–316, 2006.

The quotient of $\mathcal{H}$ by $\Gamma(1)/\{\pm 1\}$ then has the structure of a compact Riemann surface, and thus of an algebraic curve over $\mathbf{C}$. In fact this quotient has a natural structure as a curve over K, namely the Shimura curve associated to $\Gamma(1)$. We call this Shimura curve $\mathcal{X}(1)$, in analogy with the modular elliptic curve $X(1)$ of the classical theory of elliptic and modular curves: as $X(1)$ parametrizes elliptic curves, $\mathcal{X}(1)$ parametrizes certain abelian varieties which we shall call "$\mathcal{O}$-varieties".[1] By work of Shimura [Sh1], based on the classical work of Fricke [F1, F2], the group $\Gamma(1)/\{\pm 1\} \subset \mathrm{SL}_2(\mathbf{R})$ is the $(2,3,7)$ triangle group (the group generated by products of pairs of reflections in the sides of a hyperbolic triangle of angles $\pi/2$, $\pi/3$, $\pi/7$). Hence $\mathcal{X}(1)$ is a curve of genus 0 with elliptic points of orders 2, 3, 7. We choose the rational coordinate $t : \mathcal{X}(1) \xrightarrow{\sim} \mathbf{P}^1$ that takes the values $0, 1, \infty$ respectively at these three points; this determines t uniquely (as the classical modular invariant j is determined by its values $1728, 0, \infty$ at the elliptic points of orders 2, 3 and the cusp).

Suppose now that $\Gamma \subset \Gamma(1)/\{\pm 1\}$ is a subgroup of finite index d. Then $\mathcal{X} = \mathcal{H}/\Gamma$ is a compact Riemann surface with a degree-d map to $\mathcal{X}(1)$ that is unramified away from the elliptic points of $\mathcal{X}(1)$. Composing the map $\pi : \mathcal{X} \to \mathcal{X}(1)$ with our isomorphism $t : \mathcal{X}(1) \xrightarrow{\sim} \mathbf{P}^1$ yields a rational function $t \circ \pi$ of degree d on $\mathcal{X}$ that is unramified except above $t = 0$, $t = 1$, and $t = \infty$. Such a rational function is often called a "Belyi function" on $\mathcal{X}$, in tribute to Belyi's theorem [Bel] that a Riemann surface admits such a function if and only if it can be realized as an algebraic curve over $\overline{\mathbf{Q}}$. It will be convenient, and should cause no confusion, to use t also as the notation for this rational function on $\mathcal{X}$.

In particular, if Γ is a congruence subgroup of $\Gamma(1)/\{\pm 1\}$ — that is, if there exists a nonzero ideal N of O_K such that Γ contains the image in $\Gamma(1)/\{\pm 1\}$ of the normal subgroup

$$\Gamma(N) := \{g \in \Gamma(1) \mid g \equiv 1 \bmod N\mathcal{O}\}$$

— then $\mathcal{X}$ is also a Shimura modular curve, parametrizing $\mathcal{O}$-varieties with some level-N structure. For example, $\Gamma(N)$ itself is a congruence subgroup,

---

[1] Warning: unlike the case of Shimura curves associated with quaternion algebras over $\mathbf{Q}$, here an $\mathcal{O}$-variety is not simply a principally polarized abelian variety with endomorphisms by $\mathcal{O}$. Indeed there can be no abelian variety V with $\mathrm{End}(V) \otimes \mathbf{Q} = A$, because the set of ramified primes of A neither contains nor is disjoint from the set of real places of K. (See for instance [Mu], specifically Thm.1 on p.192 (positivity of the Rosati involution) together with the classification in Thm.2 on p.201.) The moduli description is quite complicated (see [Sh1, Ca]), and requires an auxiliary quadratic extension K'/K, with the field K' totally imaginary; for instance, for our case $K = \mathbf{Q}(\cos 2\pi/7)$ we may take $K' = K(\sqrt{-7}) = \mathbf{Q}(e^{2\pi i/7})$. The moduli description yields $\mathcal{X}(1)$ as a curve over K', but fortunately $\mathcal{X}(1)$ descends to a curve over K independent of the choice of K'. The same is true of the curves $\mathcal{X}(N), \mathcal{X}_1(N), \mathcal{X}_0(N)$ and the maps $\mathcal{X}(N) \to \mathcal{X}_1(N) \to \mathcal{X}_0(N) \to \mathcal{X}(1)$ (to be introduced soon) that concern us in this paper, at least provided that the extension K'/K satisfies some local conditions at the primes dividing N. I am grateful to Benedict Gross for explaining these subtleties and suggesting the references to [Ca, Mu]. Fortunately very little of this difficult theory is needed for our computations.

corresponding to a Shimura curve parametrizing $\mathcal{O}$-varieties with full level-N structure. We call this curve $\mathcal{X}(N)$, again in analogy with the classical case. If N is a prime ideal with residue field k, the ring $\mathcal{O}/N\mathcal{O}$ is isomorphic with the ring $M_2(k)$ of 2×2 matrices over k, because A is unramified at all finite places of k, including N. We choose one isomorphism $\iota : \mathcal{O}/N\mathcal{O} \xrightarrow{\sim} M_2(k)$; the choice does not matter because all are equivalent under conjugation by $\mathcal{O}^*$. We then obtain congruence subgroups $\Gamma_0(N) \rhd \Gamma_1(N) \supset \Gamma(N)$, where $\Gamma_0(N)$ consists of those $g \in \Gamma(1)$ for which $\iota(g)$ is upper triangular, and $\Gamma_1(N)$ is the kernel of the map from $\Gamma_0(N)$ to $k^*/\{\pm 1\}$ taking g to the top left entry of the matrix $\iota(g) \bmod N\mathcal{O}$. The corresponding Shimura curves are naturally denoted $\mathcal{X}_0(N)$, $\mathcal{X}_1(N)$ respectively. Once one has formulas for the cover $\mathcal{X}_0(N) \to \mathcal{X}(1)$, one can use them as in [Se] to obtain explicit extensions of K or $\mathbf{Q}$ (the latter when the residual characteristic of N does not split in O_K) whose normal closures have Galois groups $\mathrm{PSL}_2(k)$, $\mathrm{Aut}(\mathrm{PGL}_2(k))$, or other groups intermediate between these two.

Analogous to the Atkin-Lehner involution w_N of the classical modular curve $\mathrm{X}_0(N)$, we have an involution w_N of $\mathcal{X}_0(N)$, which can be constructed either from the normalizer of $\Gamma_0(N)$ in A^* or by invoking the dual isogeny of the "cyclic" isogeny between $\mathcal{O}$-varieties parametrized by a generic point of $\Gamma_0(N)$. Once explicit formulas are known for both the map $\mathcal{X}_0(N) \to \mathcal{X}(1)$ and the involution w_N, one can easily compute other interesting data. By eliminating $p \in \mathcal{X}_0(N)$ from the system $T_1 = t(p)$, $T_2 = t(w_N(p))$, we obtain the "modular polynomial" $\Phi_N(T_1, T_2)$, which is the irreducible polynomial such that $\Phi_N(t_1, t_2) = 0$ if and only if $t_1, t_2 \in \mathbf{C}$ are t-coordinates of "N-isogenous" points on $\mathcal{X}(1)$. The solutions of $\Phi_N(t, t) = 0$ are then coordinates of points of complex multiplication (CM) on $\mathcal{X}(1)$, that is, points that parametrize $\mathcal{O}$-varieties with CM. Finally, with some more effort we can obtain equations defining the recursive tower of curves $\mathcal{X}_0(N^n)$ ($n = 1, 2, 3, \ldots$), whose reduction modulo any prime ϖ of O_K not dividing N is asymptotically optimal over the quadratic extension of the residue field of ϖ (see [E1]).

If Γ is a proper normal subgroup, congruence or not, in $\Gamma(1)/\{\pm 1\}$, then the quotient group G acts on $\mathcal{X}$. By the Riemann-Hurwitz formula, $d = |G| = 84(g - 1)$, where g is the genus of $\mathcal{X}$. That is, $(\mathcal{X}, G)$ is a "Hurwitz curve": a curve of genus $g > 1$ that attains the Hurwitz bound ([H], see also [ACGH, Ch.I, Ex.F-3 ff., pp.45–47]) on the number of automorphisms of a curve of genus g over a field of characteristic zero. Conversely, all Hurwitz curves arise in this way for some proper normal subgroup of $\Gamma(1)$. For example, if N is a prime ideal with residue field k then $\mathcal{X}(N)$ is a curve with $G = \mathrm{PSL}_2(k)$ that attains the Hurwitz bound.

The first few possibilities for N yield k of size 7, 8, 13, and 27. The first of these, with N the prime of K above the totally ramified rational prime 7, yields the famed Klein quartic of genus 3, which also arises as the classical modular curve $\mathrm{X}(7)$ (see [E3]). The second, with N above the inert rational prime 2, yields the Fricke-Macbeath curve of genus 7 [F3, Ma]. If N is any of the three primes above the split rational prime 13 then $\mathcal{X}(N)$ is a curve of genus 14 that can be defined over K but not over $\mathbf{Q}$ [Str]. In each of those cases, $\mathcal{X}_0(N)$ has genus zero.

The next case, and the main focus of this paper, is the prime of residue field $\mathbf{F}_{27}$ above the inert rational prime 3. We call the resulting Shimura curves $\mathcal{X}_0(3)$, $\mathcal{X}_1(3)$, $\mathcal{X}(3)$. Since this prime is invariant under $\mathrm{Gal}(K/\mathbf{Q})$, these Shimura curves and their Belyi maps to $\mathcal{X}(1)$ (though not the action of $\mathrm{PSL}_2(\mathbf{F}_{27})$ on $\mathcal{X}(3)$) can be defined over $\mathbf{Q}$ [DN, Wo]. Here for the first time $\mathcal{X}_0(N)$ has positive genus; we calculate that $\mathcal{X}_0(3)$ and $\mathcal{X}_1(3)$ both have genus 1. The determination of the curve $\mathcal{X}_0(3)$, and of its degree-28 Belyi map to $\mathcal{X}(1)$ — that is, the degree-28 function t on $\mathcal{X}_0(3)$ — requires techniques beyond those of [E2].

In section 2 of this paper, we exhibit equations for this cover and show how we compute them. We locate the coefficients to high p-adic precision via Newton's method, using $p = 29$, a prime at which the cover has good reduction but some of the coefficients are known in advance modulo p and the rest can be determined algebraically. We then compute the simplest rational numbers consistent with the p-adic approximations, and prove that they are in fact the correct coefficients. The same method, possibly extended by exhaustive searching mod p, can be used to compute other such covers. (For instance, we used it to compute a previously unknown Shimura-Belyi function of degree 26 connected with the $(2,3,8)$ triangle group. We shall give the details of this computation, and other Shimura-Belyi maps for the $(2,3,8)$ triangle group, in a later paper.)

Having obtained equations for the map $\mathcal{X}_0(3) \to \mathcal{X}(1)$, we use those equations as explained above to locate several CM points on $\mathcal{X}(1)$. In particular, we confirm our conjecture from [E2] for the rational point of CM discriminant -11: we had computed its t-coordinate to high enough precision to convincingly guess it as an element of $\mathbf{Q}$, but could not prove this guess. It also follows immediately from our formulas that $\mathcal{X}_0(3)$ has a $\mathbf{Q}$-rational point, and is thus an elliptic curve. In section 3 of this paper, we note some arithmetic properties of this curve, and use the modular structure to explain them. In particular, $\mathcal{X}_0(3)$ admits a rational 13-isogeny to another elliptic curve, which we identify as the Jacobian of $\mathcal{X}_1(3)$. We then also explain why the kernel of the dual isogeny from this Jacobian to $\mathcal{X}_0(3)$ must consist of K-rational points, a fact first noted via direct computation by Mark Watkins [Wa].

In the final section we give some other examples suggested by these ideas. Watkins later found in the tables of elliptic curves two pairs of 7-isogenous curves, of conductors $162 = 2 \cdot 9^2$ and $338 = 2 \cdot 13^2$, with 7-torsion structure over the cyclic cubic fields of discriminant 9^2 and 13^2. He suggested that each of these pairs might thus be the Jacobians of the Shimura modular curves $\mathcal{X}_0(2), \mathcal{X}_1(2)$ for a quaternion algebra over the cyclic cubic field of discriminant 9^2, 13^2 respectively that is ramified at two of its three real places and at no finite primes. We verify that this suggestion is correct for the curves of conductor $2 \cdot 9^2$, corresponding to a subgroup of the $(2,3,9)$ triangle group. We cannot at present compute the analogous curves and maps for the Shimura curves of level 2 associated with a quaternion algebra over the cubic field of conductor 13^2, though we verify that both curves have genus 1. It might be possible to identify the curves using more refined arithmetical information as in [GR, K], but this would still leave open the problems of explicitly computing the maps $\mathcal{X}_0(2) \to \mathcal{X}(1)$ and w_2.

2   The Curve $\mathcal{X}_0(3)$

2.1   Preliminaries

We use the notations of the previous section. Let t be the degree-28 Belyi function on $\mathcal{X}_0(3)$. The elements of orders 2, 3, 7 in $\mathrm{PSL}_2(\mathbf{F}_{27})$ act on $\mathbf{P}^1(\mathbf{F}_{27})$ by permutations with cycle structure 2^{14}, $3^9 1$, 7^4. We denote by P_0 the unique simple zero of $(t-1)$ on $\mathcal{X}_0(3)$ (corresponding to the unique fixed point of an element of order 3 in $\mathrm{PSL}_2(\mathbf{F}_{27})$). Since $\mathcal{X}_0(3)$ and t are defined over $\mathbf{Q}$, so is P_0. The preimages other than P_0 of $0, 1, \infty$ under t constitute disjoint effective divisors D_{14}, D_9, D_4, which are the sum of 14, 9, 4 distinct points on $\mathcal{X}_0(3)$ respectively, such that

$$t^*(0) = 2D_{14}, \quad t^*(1) = 3D_9 + (P_0), \quad t^*(\infty) = 7D_4.$$

By the Riemann-Hurwitz formula, $\mathcal{X}_0(3)$ is a curve of genus 1. We give $\mathcal{X}_0(3)$ the structure of an elliptic curve by choosing P_0 as the origin of the group law. Since P_0 is the unique point of $\mathcal{X}_0(3)$ parametrizing a cyclic 3-isogeny between the order-3 elliptic point $t = 1$ of $\mathcal{X}(1)$ and itself, P_0 must be fixed under w_3. Therefore w_3 is the unique involution of $\mathcal{X}_0(3)$ as an elliptic curve, namely, multiplication by -1.

Proposition 1. *i) The differential dt has divisor $D_{14} + 2D_9 - 8D_4$. The divisors D_{14}, D_9, D_4 are linearly equivalent to $14(P_0), 9(P_0), 4(P_0)$, respectively.*

ii) There are nonzero rational functions F_{14}, F_9, F_4 on $\mathcal{X}_0(3)$ with divisors $D_{14} - 14(P_0), D_9 - 9(P_0), D_4 - 4(P_0)$. For each choice of F_{14}, F_9, F_4, there exist nonzero scalars α, β such that $F_{14}^2 = \alpha F_4^7 + \beta F_9^3$ and $t = F_{14}^2 / \alpha F_4^7$. If the F_n are defined over $\mathbf{Q}$ then α, β are rational as well.

Proof: i) Since $\mathcal{X}_0(3)$ has genus one, the divisor of dt is linearly equivalent to zero. This divisor is regular except for poles of order 8 at the four points of D_4. Moreover, dt has simple zeros at the points of D_{14} and double zeros at the points of D_9. This accounts for $14 + 2 \cdot 9 = 32$ zeros, same as the number $8 \cdot 4$ of poles, so dt has no further zeros (which we could also have deduced from the fact that the map t is unramified when $t \notin \{0, 1, \infty\}$).

It follows that $D_{14} + 2D_9 \sim 8D_4$. This together with the linear equivalence of the divisors $t^*(0), t^*(1), t^*(\infty)$ yields our claim that $D_n \sim n(P_0)$ for each $n = 14, 9, 4$; for instance

$$D_4 = 49D_4 - 48D_4 \sim 7(3D_9 + (P_0)) - 6(D_{14} + 2D_9)$$
$$= 3(3D_9 - 2D_{14}) + 7(P_0) \sim -3(P_0) + 7(P_0) = 4(P_0).$$

ii) Functions F_n with divisors $D_n - n(P_0)$ exist by part (i). The rational functions t and F_{14}^2 / F_4^7 have the same divisor, so $F_{14}^2 / F_4^7 = \alpha t$ for some nonzero scalar α.

Likewise $t-1$ and F_9^3/F_4^7 have the same divisor, so $F_9^3/F_4^7 = \alpha_0(t-1)$ for some nonzero scalar α_0. Eliminating t yields the desired identity $F_{14}^2 = \alpha F_4^7 + \beta F_9^3$ with $\beta = \alpha/\alpha_0$. It follows that α, and thus also β, is rational if each F_n is defined over $\mathbf{Q}$. This completes the proof of Proposition 1.   ◇

Now fix a nonzero holomorphic differential ω on $X_0(3)$ defined over $\mathbf{Q}$, and define a derivation $f \mapsto f'$ on the field of rational functions on $X_0(3)$ by $df = f'\omega$.

Proposition 2. *i) Let F_{14}, F_9, F_4 be as in Proposition 1(ii). Then F_{14}, F_9^2, F_4^6 are scalar multiples of*

$$3F_4 F_9' - 7F_4'F_9\,, \quad 2F_4 F_{14}' - 7F_4'F_{14}\,, \quad 2F_9 F_{14}' - 3F_9'F_{14}$$

respectively.
ii) The functions

$$\frac{14F_9 F_4'' - 13F_9'F_4'}{F_4}\,, \quad \frac{6F_4 F_9'' - 29F_4'F_9'}{F_9}\,, \quad \frac{6F_4 F_{14}'' - 43F_4'F_{14}'}{F_{14}}$$

on $X_0(3)$ are regular except at P_0.

Proof: i) By Proposition 1 we have $t = F_{14}^2/\alpha F_4^7$. Taking the logarithmic derivative, we find

$$\frac{dt}{t} = \left(2\frac{F_{14}'}{F_{14}} - 7\frac{F_4'}{F_4}\right)\omega = \frac{2F_4 F_{14}' - 7F_4'F_{14}}{F_4 F_{14}}\,\omega.$$

Since dt has divisor $D_{14}+2D_9-8D_4$, the divisor of dt/t is $2D_9 - D_{14} - D_4$. Thus $F_4 F_{14} dt/t$ is a differential with divisor $2D_9 - 18(P_0)$, same as the divisor of F_9^2. Since the divisor of a differential on a genus-one curve is linearly equivalent to zero, it follows that $2F_4 F_{14}' - 7F_4'F_{14}$ is a multiple of F_9^2, as claimed. The formulas for F_{14} and F_4^6 are obtained in the same way by computing the logarithmic derivatives of $t-1$ and $t/(t-1)$ respectively.

ii) Each of these is obtained by substituting one of the identities in (i) into another. Since the computations are similar and we shall use only the result concerning $(6F_4 F_9'' - 29F_4'F_9')/F_9$, we prove this result, and again leave the other two as exercises.

We have $F_{14} = \gamma_1(3F_4 F_9' - 7F_4'F_9)$ and $F_9^2 = \gamma_2(2F_4 F_{14}' - 7F_4'F_{14})$ for some nonzero scalars γ_1, γ_2. Hence

$$F_9^2 = \gamma_1\gamma_2\big(2F_4(3F_4 F_9' - 7F_4'F_9)' - 7F_4'(3F_4 F_9' - 7F_4'F_9)\big)$$
$$= \gamma_1\gamma_2\big((6F_4 F_9'' - 29F_4'F_9')F_4 + (49F_4'^2 - 14F_4 F_4'')F_9\big).$$

Dividing by F_9, we find that

$$(6F_4 F_9'' - 29F_4'F_9')\frac{F_4}{F_9} + (49F_4'^2 - 14F_4 F_4'') = \frac{1}{\gamma_1\gamma_2}F_9,$$

which is regular away from P_0. The same is true for $49F_4'^2 - 14F_4 F_4''$, and thus also for $(6F_4 F_9'' - 29F_4'F_9')F_4/F_9$. Since F_4 and F_9 have no common zeros, it follows that $(6F_4 F_9'' - 29F_4'F_9')/F_9$ has no poles except for a multiple pole at P_0, as claimed.   ◇

2.2   Computation

Theorem 1. *The curve $\mathcal{X}_0(3)$ is isomorphic with the elliptic curve*

$$y^2 + y = x^3 + x^2 - 44704x - 3655907.$$

The functions F_n on this curve may be given by

$$F_4 = x^2 - 1208x - 227671 + 91y,$$
$$F_9 = (8x^3 - 384x^2 - 13656232x - 678909344)y$$
$$\quad - (1015x^4 + 770584x^3 + 163098512x^2 + 29785004488x + 2319185361392),$$
$$F_{14} = 8x^7 + 400071x^6 - 343453068x^5 - 238003853192x^4$$
$$\quad - 116011622641292x^3 - 15704111899877744x^2$$
$$\quad - 95316727595264672x + 53553894620234333456$$
$$\quad - (8428x^5 + 19974360x^4 + 18880768004x^3 + 4128079708928x^2$$
$$\quad + 335969653582304x + 17681731246686360)y,$$

satisfying

$$64F_4^7 - 343F_9^3 = F_{14}^2,$$

and then

$$t = \frac{F_{14}^2}{64F_4^7} = 1 - \frac{343F_9^3}{64F_4^7}.$$

Proof: With a symbolic algebra package one may readily confirm the identity among the F_n, and might even feasibly (albeit not happily) verify the Galois group by following the 28 preimages of a point on the t-line as that point loops around $t = 0$ and $t = 1$; this would suffice to prove the theorem (since the cover $\mathcal{X}_0(3) \to \mathcal{X}(1)$ is determined by its Galois group and ramification behavior), but would not explain the provenance of the formulas. We thus devote most of the proof to the computation of the F_n.

We begin by observing that our proofs of Propositions 1 and 2 used the ramification behavior of the cover $\mathcal{X}_0(3) \to \mathcal{X}(1)$, but not its Galois group or the Shimura-curve structure. This will remain true in the rest of our computation.[2] We thus show that the ramification behavior uniquely determines the degree-28 cover. In particular this yields the following purely group-theoretical statement (that we could also have checked directly): any permutations $\sigma_2, \sigma_3, \sigma_7$ of 28 letters with cycle structure 2^{14}, $3^9 1$, 7^4 such that $\sigma_2\sigma_3\sigma_7 = 1$ are equivalent under conjugation in S_{28} with our generators of $PSL_2(\mathbf{F}_{27})$, and that we do not have to verify the Galois group as suggested in the previous paragraph.

Our strategy is to first find the cover modulo the prime 29, which occurs in one of the formulas in Proposition 2(ii), and then use Newton's method over $\mathbf{Q}_{29}$ to

---

[2] We shall retain the notation $\mathcal{X}_0(3)$ for the curve and w_3 for the involution, rather than introduce separate notations for an elliptic curve that we have not yet identified with $\mathcal{X}_0(3)$ and the multiplication-by-(-1) map on the curve.

compute a lift of the coefficients to sufficient 29-adic precision to recognize them as rational numbers. The prime 29 is large enough to guarantee good reduction of any Belyi cover of degree 28: if a Belyi cover has bad reduction at some prime p then the normal closure of the cover has a Galois group whose order is a multiple of p [Bec].

We may assume that each F_n $(n = 14, 9, 4)$ is scaled so that it has 29-adically integral coefficients and does not reduce to zero mod 29. In characteristic 29, the second function in Proposition 2(ii) simplifies to $6F_4 F_9''/F_9$. Again we use the fact that F_4, F_9 have no common zeros to conclude that $\xi := F_9''/F_9$ is regular away from P_0. At P_0, we know that F_9 has a pole of order 9; thus F_9'' has a pole of order 11, and ξ has a double pole. Since ξ is regular elsewhere, it follows that ξ is invariant under w_3.

We claim that F_9 is anti-invariant, that is, that $w_3(F_9) = -F_9$. Indeed the differential equation $f'' = \xi f$ satisfied by F_9 must also hold for the invariant and anti-invariant parts of F_9, call them F_9^+, F_9^-. Now F_9^- has a pole of order 9 at P_0. If F_9^+ is not identically zero then its valuation at P_0 is $-d$ for some $d \in \{0, 2, 4, 6, 8\}$; comparing leading coefficients in the local expansion about P_0 of $F_9^{+''} = \xi F_9^+$ and $F_9^{-''} = \xi F_9^-$, we obtain $9 \cdot 10 \equiv d(d+1) \bmod 29$, which is impossible. Hence $F_9^+ = 0$ as claimed. (We could also have obtained $F_9^+ = 0$ by arguing that $t = 1$ is the only supersingular point on $\mathcal{X}(1)$ mod 29, whence the zeros of F_9, which are the preimages of $t = 1$, must be permuted by w_3. But this would break our promise not to rely on the modular provenance of the cover.)

Now let $y^2 = x^3 + ax + b$ be a (narrow) Weierstrass equation for $\mathcal{X}_0(3)$, and choose $\omega = dx/y$. Then our derivation $f \mapsto f'$ is characterized by $x' = y$ and $y' = (3x^2 + a)/2$. For the rest of this paragraph we calculate modulo 29. We have seen that F_9 is a scalar multiple of $(x^3 + s_1 x^2 + s_2 x + s_3)y$ for some $s_1, s_2, s_3 \in \mathbf{F}_{29}$. Equating coefficients in $F_9'' = \xi F_9$, we find first that $\xi = 8x + 6s_1$, then that $s_2 = -12s_1^2 - 8a$ and $s_3 = 7b - 3s_1^3 - as_1$, and finally that $s_1 b + s_1^4 + 9as_1^2 + 9a^2 = 0$. We have also seen that F_4 is a scalar multiple of $x^2 + t_1 y + t_2 x + t_4$ for some $t_1, t_2, t_4 \in \overline{\mathbf{F}_{29}}$. Using Proposition 2(i) we compute F_{14} and F_4^6 up to scaling in terms of s_1, a, b and the t_i, and compare with $(x^2 + t_1 y + t_2 x + t_4)^6$. After matching the leading (degree-24) coefficients,[3] we find that the degree-23 coefficients agree identically, but in degree 22 we find $t_2 = 11(t_1^2 - s_1)$, and the degree 21 comparison yields $t_1 = 0$ or $t_1^2 = 5s_1$. We cannot have $t_1 = 0$, for then F_4 would be even, and then so would F_{14} (since F_{14} is proportional to $3F_4 F_9' - 7F_4' F_9$), which is impossible since the nonzero odd function F_9^3 is a linear combination of F_4^7 and F_{14}^2. Therefore $t_1^2 = 5s_1$. Comparing the next few coefficients in our two expressions for F_4^6, we learn that $t_4 = s_1^2 + 3a$ and $a = 9s_1^2$. This completes the determination of our cover mod 29 up to scaling. For instance, we may use the equation $y^2 = x^3 + 9x + 1$ for $\mathcal{X}_0(3)$ over $\overline{\mathbf{F}_{29}}$, and then check that

$$F_4 = x^2 + 11y - 14x - 1, \quad F_9 = (x^3 + x^2 + 3x - 5)y,$$

---

[3] The "degree" of a polynomial in x, y is the order of its pole at P_0. The vector space of rational functions on $\mathcal{X}_0(3)$ that are regular away from P_0 has basis $\{x^i | i = 0, 1, 2, \ldots\} \cup \{x^i y | i = 0, 1, 2, \ldots\}$; the monomials x^i, $x^i y$ have degrees $2i$, $2i + 3$.

$$F_{14} = (x^7 - 14x^6 - 5x^5 - 9x^4 - 10x^3 + 2x^2 + 10x - 7) - (8x^5 - x^3 - 3x^2 + x + 3)y$$

satisfy the identity $F_4^7 - 6F_9^3 = F_{14}^2$.

We now use Hensel's lemma to show that there is a unique such identity over $\mathbf{Z}_{29}$. To do this we regard $\alpha F_4^7 + \beta F_9^3 = F_{14}^2$ as an overdetermined system of equations in a, b, α, β, and the coefficients of F_4 and F_9. More precisely, we eliminate the ambiguity in the various choices of scalar multiples by requiring that F_4 and F_9 each have leading coefficient 1, setting $b = 9a$, and defining F_{14} by the formula $3F_4 F_9' - 7F_4'F_9$ from Proposition 2(i). (Any elements of $\mathbf{Z}_{29}$ that reduce to 1 mod 29 would do for the leading coefficients, as would any element of $\mathbf{Z}_{29}$ that reduces to 9 mod 29 for b/a.) We include in our equations the x^i coefficients of F_9 ($i = 0, 1, 2, 3, 4$), which were known to vanish mod 29 but not in $\mathbf{Z}_{29}$. By our analysis thus far, this system of equations has a unique solution mod 29. We compute the Jacobian matrix of our system of equations at that solution, and verify that this matrix has full rank. Therefore the solution lifts uniquely to $\mathbf{Z}_{29}$ by Hensel's lemma. This is also the unique solution over $\mathbf{Q}_{29}$, because 29 is a prime of good reduction as noted above.

It remains to recover the coefficients as rational numbers. We compute them mod 29^{128} by applying a Newton iteration 7 times 29-adically. (At each step, instead of computing the derivative mod $29^{2^{n-1}}$ symbolically, we approximate each partial derivative by evaluating the function mod 29^{2^n} at two points that differ by $29^{2^{n-1}}$ times the corresponding unit vector. We also streamline the computation by requiring at each step that only the x^i coefficients match, ignoring the $x^i y$ terms; this suffices because the resulting submatrix of the Jacobian matrix still has full rank mod 29.) We then use 2-dimensional lattice reduction to guess the rational numbers represented by those 29-adic approximations, and verify the guess by checking that the resulting F_4^7, F_9^3, F_{14}^2 are $\mathbf{Q}$-linearly dependent. This completes the determination of the cover $\mathcal{X}_0(3) \to \mathcal{X}(1)$ and the proof that it is characterized by its degree and ramification behavior. Finally we bring the curve $y^2 = x^3 + ax + b$ to minimal form and replace F_4, F_9, F_{14} by the smallest rational multiples that eliminate the denominators of those rational functions. This yields the formulas in Theorem 1. $\diamond$

In the same way that we proved F_9 is an odd function mod 29, we can use the other two formulas in Proposition 2(ii) to prove that F_4 and F_{14} are even mod 13 and 43 respectively. This is confirmed by Theorem 1: the terms containing y in F_4 and F_{14} are divisible by 13 and 43 respectively. The corresponding result for F_9 mod 29 is obscured by the minimal form of $\mathcal{X}_0(3)$, which makes it harder to recognize odd functions; but F_9 can be seen to be congruent mod $2y + 1$ to a multiple of 29, as expected.

2.3   Application: The CM Point of Discriminant −11

In [E2, 5.3], we noted that $\mathcal{X}(1)$ has a unique CM point of discriminant −11, and therefore that this point has rational t-coordinate. We then described high-precision numerical computations that strongly suggest this t-coordinate is

$$\frac{88983265401189332631297917}{45974167834557869095293} = \frac{7^3 43^2 127^2 139^2 307^2 659^2 11}{3^3 13^7 83^7},$$

with $t-1$ having numerator $2^9 29^3 41^3 167^3 281^3$. But we could not prove that this is correct. Our formulas for the cover $X_0(3) \to X(1)$ now let us do this:

Corollary to Theorem 1. *The CM-11 point on $X(1)$ has t-coordinate equal $7^3 43^2 127^2 139^2 307^2 659^2 11/3^3 13^7 83^7$, and lies under the two points of $X_0(3)$ with $(x,y) = (-10099/64, -1/2 \pm 109809\sqrt{-11}/512)$.*

Proof: The CM-11 point has two preimages in $X_0(3)$ that are switched by w_3, corresponding to a pair of "3-isogenies" between (the abelian variety parametrized by) that point and itself, namely the pair $(-1 \pm \sqrt{-11})/2$ of norm-3 elements of the endomorphism ring. A point of $X(1)$ is 3-isogenous to itself if and only if its t-coordinate satisfies $t = t(P) = t(w_3 P)$ for some point P on $X_0(3)$. Equivalently, P is either a common pole of the functions $t(P), t(w_3 P)$ on $X_0(3)$ or a zero of the rational function $t(P) - t(w_3 P)$ on that curve. The former cannot happen here because by our formulas F_4 and $F_4 \circ w_3$ have no common zeros. Since $t(P) - t(w_3 P)$ is an odd function on $X_0(3)$, we easily locate its zeros by writing it as $2y + 1$ times a rational function of x. We find that the latter function vanishes at $x = \infty$, $x = -1097/8$, $x = -10099/64$, and the roots of four irreducible polynomials of degree 3 and two irreducibles of degree 6. At $x = \infty$ we have $P = P_0$ and $t = 1$. At $x = -1097/8$ we have $y = -1/2 \pm 6615\sqrt{-2}/32$, and calculate that $t = 1092830632334/1694209959$, which we already identified as the CM-8 point by a similar computation with the curve $X_0(2)$ [E2, p.39]. The irrational values of x that are roots of the irreducible polynomials of degree 3 and 6 yield irrational t-values of the same degree. Hence the CM-11 point, being rational, must have $x = -10099/64$, for which we calculate the values of y and t given in the statement of the Corollary.   ◇

The CM-8 point also has an element of norm 3 in its endomorphism ring, namely $1 + \sqrt{-2}$. The fact that the CM points with $x = -10099/64$ have discriminant -11 also follows from the fact that their y-coordinates are conjugates over $\mathbf{Q}(\sqrt{-11})$. The irrational values of x have degrees 1 and 2 over K; in particular, the cubic irrationalities yield four $\mathrm{Gal}(K/\mathbf{Q})$-orbits of CM points. Taking $c = 2\cos(2\pi/7)$, we find that these CM points have x-coordinates $189c - 19$, $-(189c^2 + 567c + 397)$, $-2(756c^2 + 1701c + 671)$, $-2(3591c^2 + 8127c + 2939)$, and their $\mathrm{Gal}(K/\mathbf{Q})$ conjugates. We determine their CM fields as the fields of definition of the points' y-coordinates: they are the fields obtained by adjoining to K the square roots of the totally negative algebraic integers $c^2 + 2c - 7$, $c - 6$, $-(3c^2 + 2c + 3)$, and $c^2 - 2c - 11$, of norms -167, -239, -251, and -491 respectively.

3   The Curve $X_1(3)$, and Its Jacobian $J_1(3)$

The tables [BK, Cr] of elliptic curves of low conductor indicate that the curve 147-B2(J), which we identified with $X_0(3)$, is 13-isogenous with another elliptic curve over $\mathbf{Q}$, namely 147-B1(I): $[0, 1, 1, -114, 473]$. Now 13-isogenies between elliptic curves over $\mathbf{Q}$, though not hard to find (via a rational parametrization of the classical modular curve $X_0(13)$), are rare: this 13-isogeny, and its twist by $\mathbf{Q}(\sqrt{-7})$ [curves 147-C1(A) and 147-C2(B)], are the only examples up to conductor 200 [BK];[4] even up to conductor 1000, the only other examples are the twists of these 13-isogenies by $\mathbf{Q}(\sqrt{-3})$, which appear at conductor 441. The fact that $\mathcal{X}_0(3)$ admits a rational 13-isogeny thus seemed a remarkable coincidence. Mark Watkins [Wa] observes that this curve 147-B1(I) has an even more unusual property: not only is there a 13-isogeny from this curve to $\mathcal{X}_0(3)$, but the kernel of the isogeny consists of points rational over K. Whereas the classical modular curve $X_0(13)$, which parametrizes (generalized) elliptic curves with a rational 13-isogeny, has genus 0, the curve $X_1(13)$, which parametrizes (generalized) elliptic curves with a rational 13-torsion point, has genus 2. Thus by Mordell-Faltings $X_1(13)$ has only finitely many K-rational points. Hence there are only finitely many isomorphism classes of elliptic curves defined over K, let alone over $\mathbf{Q}$, with a K-rational 13-torsion point — and we have found one of them[5]

by computing the Jacobian of the Shimura modular curve $\mathcal{X}_1(3)$!

These observations are explained by considering $\mathcal{X}_1(3)$. This curve, like $\mathcal{X}_0(3)$, is defined over $\mathbf{Q}$, and the cyclic cover $\mathcal{X}_1(3) \to \mathcal{X}_0(3)$ has degree $(3^3-1)/2 = 13$. This cover must be unramified, because the only elliptic point on $\mathcal{X}_0(3)$ is of order 3, which is coprime to 13. Hence $\mathcal{X}_1(3)$ has genus 1. It does not quite follow that $\mathcal{X}_1(3)$ is isomorphic over $\mathbf{Q}$ with the 13-isogenous elliptic curve 147-B1(I), because $\mathcal{X}_1(3)$ need not have any $\mathbf{Q}$-rational point. However, the Jacobian $\mathcal{J}_1(3)$ of $\mathcal{X}_1(3)$ is an elliptic curve, and is also 13-isogenous with $\mathcal{X}_0(3)$ because the cover $\mathcal{X}_1(3) \to \mathcal{X}_0(3)$ induces a map of the same degree from $\mathcal{J}_1(3)$ to $\mathcal{X}_0(3)$. Since the elliptic curve 147-B1(I) is the only one 13-isogenous with $\mathcal{X}_0(3)$ over $\mathbf{Q}$, we conclude that it is isomorphic with $\mathcal{J}_1(3)$. Furthermore, the geometric Galois group of the cover $\mathcal{X}_1(3)/\mathcal{X}_0(3)$ is canonically $k^*/\{\pm1\}$, where k is the residue field of the prime above 3 in K. Working over K, we can choose a generator for this group, which acts on $\mathcal{X}_1(3)$ by translation by some element of $\mathcal{J}_1(3)$, and this element is a 13-torsion point that generates the kernel of the isogeny from $\mathcal{J}_1(3)$ to $\mathcal{X}_0(3)$.

This is all quite reminiscent of the situation for the classical modular curves $X_0(11)$, $X_1(11)$, which are 5-isogenous elliptic curves, with the kernel of the

---

[4] Table 4 in [BK], which gives all curves of conductor $2^a 3^b$, gives several other examples of conductor $20736 = 2^8 3^4$, all related by quadratic twists.

[5] Not, however, the unique one. The relevant twist of $X_1(13)$ by K is isomorphic with the curve $y^2 = 4x^6 + 8x^5 + 37x^4 + 74x^3 + 57x^2 + 16x + 4$, which inherits from $X_1(13)$ a 3-cycle generated by $(x,y) \mapsto (-x^{-1}-1, y/x^3)$. Rational points related by this 3-cycle and/or the hyperelliptic involution parametrize the same curve, with a different choice of generator of the torsion subgroup. The orbit of $\mathbf{Q}$-rational points with $x \in 0, -1, \infty$ yields our curve $\mathcal{J}_1(3)$. A computer search finds that there is at least one other orbit, represented by $(x,y) = (17/16, 31585/2048)$, which leads to a second elliptic curve over $\mathbf{Q}$ with a K-rational 13-torsion point. We calculate that this is the curve with coefficients

$$[0, 1, 1, -698193059962601758382254, 710086743590898842902587481252037]$$

and conductor $8480886141 = 3 \cdot 7^2 \cdot 13 \cdot 251 \cdot 17681$.

isogeny $X_1(11) \to X_0(11)$ canonically isomorphic with $\mathbf{F}_{11}^*/\{\pm 1\}$ and thus consisting of $\mathbf{Q}$-rational 5-torsion points. Also, $X_1(11)$ has considerably smaller height (visible in its smaller coefficients) and discriminant than $X_0(11)$, as does $\mathcal{J}_1(3)$ compared with $\mathcal{X}_0(3)$. The comparison becomes even clearer if we work with models of these curves minimal over K, where the additive reduction at the rational prime 7 becomes good reduction at the prime $(2 - c)$ of K above 7.

Now for the classical modular curves the fact that $X_1(11)$ has a simpler equation than $X_0(11)$ illustrates a general phenomenon noted in [Ste]: the minimal height in an isogeny class of elliptic curves over $\mathbf{Q}$ is conjecturally attained by the optimal quotient of the Jacobian of $X_1(N)$, not $X_0(N)$ (unless the X_1- and X_0-optimal quotients coincide). Does this always happen also for Shimura curves? We observed the same behavior in several other cases, one of which appears in the next section. But there are no extensive tables of optimal quotients of Jacobians of Shimura curves on which one might test such a conjecture. Although Vatsal [Va, Thm. 1.11] has proven Stevens' conjecture for curves with a rational ℓ-isogeny for prime $\ell \geq 7$, his methods cannot apply in our setting (even though all our curves have a suitable isogeny) because they rely on congruences between q-expansions of modular forms, a tool that is not available to us in the Shimura-curve setting.

4   Some Other Shimura Curves of Genus 1

Watkins notes several other examples of curves in [Cr] that behave similarly for other choices of cyclic cubic fields K, and asks whether they, too, can be explained as Shimura modular curves or Jacobians. We checked that this is the case for at least one of these examples, at conductor $162 = 2 \cdot 9^2$. In this section, we outline this computation and describe a 7-isogeny in conductor $338 = 2 \cdot 13^2$ that should also involve Shimura modular curves.

For the curves of conductor 162, we start with a quaternion algebra over $K_9 = \mathbf{Q}(\cos 2\pi/9)$ ramified at two of the three real places of K_9 and at none of its finite primes. The resulting modular group $\Gamma(1)$ is again a triangle group, this time with signature $(2,3,9)$ rather than $(2,3,7)$ (see [T], class XI). Since the rational prime 2 is inert in K_9, we have modular curves $\mathcal{X}_0(2)$ and $\mathcal{X}_1(2)$ with geometric Galois group $\mathrm{PSL}_2(\mathbf{F}_8)$ over the rational curve $\mathcal{X}(1)$. We calculate that here $\mathcal{X}_0(2)$ already has genus 1. Since the Belyi map $\mathcal{X}_0(2) \to \mathcal{X}(1)$ has degree as low as $8 + 1 = 9$, we can find its coefficients with little difficulty even without resorting to the methods we used to obtain the equations in Theorem 1. We place the elliptic points of $\mathcal{X}(1)$ at $t = 1, 0, \infty$. Then t is a rational function of degree 9 on $\mathcal{X}_0(2)$ with a ninth-order pole, which we use as the base point P_0 to give $\mathcal{X}_0(2)$ the structure of an elliptic curve. The zero divisor of t is then $3D_3$ for some divisor D_3 of degree 3. Hence $D_3 - 3(P_0)$ is a 3-torsion divisor on $\mathcal{X}_0(2)$. By computing the divisor of dt as in the proof of Proposition 1(i), we see that the corresponding 3-torsion point on $\mathcal{X}_0(2)$ is also the simple zero of $t - 1$. In particular, this 3-torsion point cannot be trivial. (This fact could also be deduced by noting that if the 3-torsion point were trivial then t would be the cube of a rational function on $\mathcal{X}_0(2)$, and so could not have Galois group $\mathrm{PSL}_2(\mathbf{F}_8)$.) The general elliptic curve with a rational 3-torsion point has the form $y^2+a_1xy+a_3y = x^3$, with the torsion point at $(x,y) = (0,0)$. Solving for $(a_3 : a_1^3)$ and the coefficients of t, we soon find a model of $\mathcal{X}_0(2)$ with $(a_1, a_3) = (15, 128)$ and $t = (y - x^2 - 17x)^3/(2^{14}y)$, and with the double roots of $t - 1$ occurring at the zeros of $(x + 9)y + 6x^2 + 71x$ other than $(0,0)$. Reducing $\mathcal{X}_0(2)$ to standard minimal form, we find the curve 162-B3(I): $[1, -1, 1, -95, -697]$. As expected, this curve attains its 3-adic potential good reduction over K_9. The involution w_2 is the unique involution of this elliptic curve that fixes the simple zero $(0, -128)$ of $t - 1$. For instance, w_2 sends the point at infinity to the other 3-torsion point $(0,0)$, at which $t = -17^3/2^7$. Hence $t = -17^3/2^7$ is a CM point on $\mathcal{X}(1)$, the unique point 2-isogenous with the elliptic point $t = \infty$. Solving $t(P) = t(w_2(P))$ yields the further CM point $t = 17^3 5^3/2^6$, which must have CM field $K_9(\sqrt{-7})$ because P and $w_2(P)$ are conjugate over $\mathbf{Q}(\sqrt{-7})$.

Besides the known 3-torsion point, we find that $\mathcal{X}_0(2)$ also has a 7-isogeny with the curve 162-B1(G): $[1, -1, 1, -5, 5]$. It is known (see for instance the "remarks on isogenies" in [BK]) that the isogeny class of this curve is the unique one, up to quadratic twist, with both a 3- and a 7-isogeny over $\mathbf{Q}$. We have already accounted for the 3-torsion point using the ramification behavior of the map $\mathcal{X}_0(2) \to \mathcal{X}(1)$. The 7-isogeny is again explained using $\mathcal{X}_1$. This time the cyclic cover $\mathcal{X}_1(2) \to \mathcal{X}_0(2)$ has degree $2^3 - 1 = 7$, and the only elliptic point of $\mathcal{X}_0(2)$ is the simple zero of $t - 1$, which has order 2. Since $\gcd(2, 7) = 1$, the cyclic cover is again unramified, and $\mathcal{J}_1(2)$ must be an elliptic curve 7-isogenous with $\mathcal{X}_0(2)$. Hence $\mathcal{J}_1(2)$ is isomorphic with the elliptic curve 162-B1(G): $[1, -1, 1, -5, 5]$. Also as before, the kernel of the isogeny must consist of points rational over K_9, and again Watkins confirms [Wa] that this is the case for this curve. This time it turns out $\mathcal{J}_1(2)$ is the unique elliptic curve over $\mathbf{Q}$ with the correct torsion structure: a 3-torsion point over $\mathbf{Q}$, and a 7-torsion point over K_9 that is proportional to its Galois conjugates. Again we note that it is $\mathcal{J}_1(2)$, not $\mathcal{X}_0(2)$, that has the smaller discriminant and height.

The next case is the cyclic cubic field K_{13} of discriminant $169 = 13^2$. Once more we use a quaternion algebra over this field that is ramified at two of its three real places and at none of the finite primes. By Shimizu's formula ([Sh2, Appendix], quoted in [T, p.207]), the resulting curve $\mathcal{X}(1)$ has hyperbolic area

$$\frac{1}{16\pi^6}d_{K_{13}}^{3/2}\zeta_{K_{13}}(2) = -\frac{1}{2}\zeta_{K_{13}}(-1).$$

Since K_{13} is cyclotomic, we can compute $\zeta_{K_{13}}(2)$ or $\zeta_{K_{13}}(-1)$ by factoring ζ_K as a product of the Riemann zeta function and two Dirichlet L-series. We find that $\mathcal{X}(1)$ has normalized hyperbolic area $1/6$. Since $\Gamma(1)$ is not on Takeuchi's list of triangle groups [T], the curve $\mathcal{X}(1)$ must have at least four elliptic points, or positive genus and at least one elliptic point. The only such configuration that attains an area as small as $1/6$ is genus zero, three elliptic points of order 2, and one of order 3. Again we use the prime of K above the inert rational prime 2 to construct modular curves $\mathcal{X}_0(2)$ and $\mathcal{X}_1(2)$ with geometric Galois group $\mathrm{PSL}_2(\mathbf{F}_8)$ over the rational curve $\mathcal{X}(1)$. In the degree-9 cover $\mathcal{X}_0(2) \to \mathcal{X}(1)$, the elliptic point of order 3 has 3 triple preimages, and each of the order-2 points has one simple and four double preimages. Hence $\mathcal{X}_0(2)$ has genus 1 by Riemann-Hurwitz. Again the cyclic cover $\mathcal{X}_1(2) \to \mathcal{X}_0(2)$ has degree 7, relatively prime to the orders of the elliptic points on $\mathcal{X}_0(2)$, so $\mathcal{X}_1(2)$ also has genus 1.

In this setting it is not clear that either $\mathcal{X}_0(2)$ or $\mathcal{X}_1(2)$ is an elliptic curve over $\mathbf{Q}$, since neither curve is forced to have a $\mathbf{Q}$-rational divisor of degree 1. (The elliptic points of order 2 may be Galois conjugates, not individually rational over $\mathbf{Q}$.) But we can still consider the Jacobians $\mathcal{J}_0(2)$ and $\mathcal{J}_1(2)$, which are elliptic curves 7-isogenous over $\mathbf{Q}$, with the kernel of the 7-isogeny $\mathcal{J}_1(2) \to \mathcal{J}_0(2)$ consisting of points rational over K_{13}. Watkins suggests, by analogy with the cases of conductor 147 and 162, that this 7-isogeny should connect curves of conductor $2 \cdot 13^2 = 338$. According to the tables of [Cr], there are three pairs of 7-isogenous curves of conductor 338. Watkins computes that of the 6 curves involved in these isogenies, only 338-B1: $[1, -1, 1, 137, 2643]$ has 7-torsion points rational over K_{13}, and thus proposes this curve as $\mathcal{J}_1(2)$, and the 7-isogenous curve 338-B2: $[1, -1, 1, -65773, -6478507]$ as $\mathcal{J}_0(2)$. This must be correct, since these Jacobians should again have multiplicative reduction at 2 and good reduction at all primes of K_{13} other than 2, whence their conductor must be $2 \cdot 13^2$ (there being no curve of conductor 2 over $\mathbf{Q}$). But here I have not obtained an explicit rational function or even determined the cross-ratio of the elliptic points on $\mathcal{X}(1)$, which would be quite a demanding computation using the methods of [E2].

Acknowledgements. This work was supported in part by NSF grants DMS-0200687 and DMS-0501029. Symbolic and numerical computations were assisted by the computer packages PARI/GP [P] and MAXIMA. I am also grateful to Mark Watkins for the reference to [Va] and other relevant correspondence [Wa], to Watkins and the referees for careful readings of and corrections to earlier drafts of this paper, to the referees for suggesting several other bibliographical sources, and to Benedict Gross for the remarks in footnote 1.

References

[ACGH]   Arbarello, E., Cornalba, M., Griffiths, P.A., Harris, J.: *Geometry of Algebraic Curves, Vol. I.* New York: Springer, 1985.

[Bec]   Beckmann, S.: Ramified primes in the field of moduli of branched coverings of curves, *J. of Algebra* **125** (1989), 236–255.

[Bel]   Belyi, G.V.: On the Galois extensions of the maximal cyclotomic field, *Izv. Akad. Nauk SSSR* **43** (1979), 267–76 (= *Math. USSR Izv.* **14** (1980), 247–256); see also: A new proof of the three-point theorem, *Mat. Sb.* **193** (2002) #3, 21–24 (= *Sb. Math.* **193** (2002) #3–4, 329–332).

[BK]   Birch, B.J., Kuyk, W., ed.: *Modular Functions of One Variable IV.* Lect. Notes in Math. **476**, 1975.

[Ca]   Carayol, H.: Sur la mauvaise réduction des courbes de Shimura, *Compositio Math.* **59** (1986) #2, 151–230.

[Cr]   Cremona, J.E.: *Algorithms for modular elliptic curves.* Cambridge University Press, 1992.

[DN]   Doi, K., Naganuma, H.: On the algebraic curves uniformized by arithmetical automorphic functions, *Annals of Math. (2)* **86** (1967), 449–460.

[E1]   Elkies, N.D.: Explicit modular towers, pages 23–32 in *Proceedings of the Thirty-Fifth [1997] Annual Allerton Conference on Communication, Control and Computing* (T. Başar and A. Vardy, eds.; Univ. of Illinois at Urbana-Champaign, 1998) = http://arXiv.org/abs/math.NT/0103107 .

[E2]   Elkies, N.D.: Shimura curve computations, pages 1–47 in *Algorithmic number theory* (Portland, OR, 1998; J. Buhler, ed.; Lect. Notes in Computer Sci. #1423; Berlin: Springer, 1998) = http://arXiv.org/abs/math.NT/0005160 .

[E3]   Elkies, N.D.: The Klein quartic in number theory, pages 51–102 in *The Eightfold Way: The Beauty of Klein's Quartic Curve* (S.Levy, ed.; Cambridge Univ. Press, 1999); online at http://www.msri.org/publications/books/Book35/ .

[F1]   Fricke, R.: Über den arithmetischen Charakter der zu den Verzweigungen $(2,3,7)$ und $(2,4,7)$ gehörenden Dreiecksfunctionen, *Math. Ann.* **41** (1893), 443–468.

[F2]   Fricke, R.: Entwicklungen zur Transformation fünfter und siebenter Ordnung einiger specieller automorpher Functionen, *Acta Math.* **17** (1893), 345–395.

[F3]   Fricke, R.: Ueber eine einfache Gruppe von 504 Operationen, *Math. Ann.* **52** (1899), 321–339.

[GR]   González, J., Rotger, V.: Equations of Shimura curves of genus 2 (http://arXiv.org/abs/math.NT/0312434), to appear in *International Math. Research Letters*.

[H]   Hurwitz, A.: Über algebraische Gebilde mit eindeutigen Transformationen in sich, *Math. Annalen* **41** (1893), 403–442.

[K]   Kurihara, A.: On some examples of equations defining Shimura curves and the Mumford uniformization, *J. Fac. Sci. Univ. Tkyo, Sec. IA* **25** (1979), 277–301.

[Ma]   Macbeath, A.M.: On a curve of genus 7, *Proc. LMS* **15** (1965), 527–542.

[Mu]   Mumford, D.: *Abelian Varieties*. London: Oxford Univ. Press, 1970.

[P]   PARI/GP, versions 2.1.1–4, Bordeaux, 2000–4, http://pari.math.u-bordeaux.fr .

[Se]   Serre, J.-P.: *Topics in Galois Theory*. Boston: Jones and Bartlett 1992.

[Sh1]   Shimura, G.: Construction of class fields and zeta functions of algebraic curves, *Ann. of Math.* **85** (1967), 58–159.

[Sh2]   Shimizu, H.: On zeta functions of quaternion algebras, *Ann. of Math.* **81** (1965), 166–193.

[Ste]   Stevens, G.: Stickelberger elements and modular parametrizations of elliptic curves, *Invent. Math.* **98** (1989) #1, 75–106.

[Str]   Streit, M.: Field of definition and Galois orbits for the Macbeath-Hurwitz curves, *Arch. Math.* **74** (2000) #5, 342–349

[T]   Takeuchi, K.: Commensurability classes of arithmetic triangle groups, *J. Fac. Sci. Univ. Tokyo* **24** (1977), 201–212.

[Va]   Vatsal, V.: Multiplicative subgroups of $J_0(N)$ and applications to elliptic curves, *J. Inst. Math. Jussieu* 4 (2005) #2, 281–316.

[Vi]   Vignéras, M.-F.: *Arithmétique des Algèbres de Quaternions*, Berlin: Springer, 1980 (SLN **800**).

[Wa]   Watkins, M.: e-mail communication, 2003–2005.

[Wo]   Wolfart, J.: Belyi Surfaces with Many Automorphisms, pages 97–112 in *Geometric Galois Actions, 1. Around Grothendieck's Esquisse d'un Programme* (L. Schneps, P. Lochak, eds.; *London Math. Soc. Lect. Note Series* **242**), Cambridge University Press, 1997.

The Asymptotics of Points of Bounded Height on Diagonal Cubic and Quartic Threefolds

Andreas-Stephan Elsenhans and Jörg Jahnel

Universität Göttingen, Mathematisches Institut
Bunsenstraße 3–5, D-37073 Göttingen, Germany[*]
{elsenhan, jahnel}@uni-math.gwdg.de

Abstract. For the families $ax^3 = by^3 + z^3 + v^3 + w^3$, $a, b = 1, \ldots, 100$, and $ax^4 = by^4 + z^4 + v^4 + w^4$, $a, b = 1, \ldots, 100$, of projective algebraic threefolds, we test numerically the conjecture of Manin (in the refined form due to Peyre) about the asymptotics of points of bounded height on Fano varieties.

1 Introduction — Manin's Conjecture

Let V be a projective algebraic variety over $\mathbb{Q}$. We fix an embedding $\iota\colon V \to \mathbf{P}^n_{\mathbb{Q}}$. In this situation, there is the well-known *naive height* $\mathrm{H}_{\mathrm{naive}}\colon V(\mathbb{Q}) \to \mathbb{R}$ which is given by $\mathrm{H}_{\mathrm{naive}}(P) := \max_{i=0,\ldots,n} |x_i|$. Here, $(x_0 : \ldots : x_n) := \iota(P) \in \mathbf{P}^n(\mathbb{Q})$ where the projective coordinates are integers satisfying $\gcd(x_0, \ldots, x_n) = 1$.

It is of interest to ask for the asymptotics of the number of $\mathbb{Q}$-rational points on V of bounded naive height. This applies particularly to the case V is a Fano variety as those are expected to have many rational points (at least after a finite extension of the ground-field). Simplest examples of Fano varieties are complete intersections in $\mathbf{P}^n_{\mathbb{Q}}$ of a multidegree $(d_1, \ldots, d_r)$ such that $d_1 + \ldots + d_r \leq n$. In this case, Manin's conjecture reads as follows.

Conjecture 1. *Let $V \subseteq \mathbf{P}^n_{\mathbb{Q}}$ be a non-singular complete intersection of multidegree $(d_1, \ldots, d_r)$. Assume $\dim V \geq 3$ and $k := n + 1 - d_1 - \ldots - d_r > 0$. Then, there exists a Zariski open subset $V^\circ \subseteq V$ such that*

$$\#\{x \in V^\circ(\mathbb{Q}) \mid \mathrm{H}_{\mathrm{naive}}(x)^k < B\} \sim CB \qquad (1)$$

for a constant C.

Example 2. Let $V \subset \mathbf{P}^4_{\mathbb{Q}}$ be a smooth hypersurface of degree 4. Conjecture 1 predicts $\sim CB$ rational points of height $< B$. However, the hypersurface

---

[*] The computer part of this work was executed on the Linux PCs of the Gauß Laboratory for Scientific Computing at the Göttingen Mathematisches Institut. Both authors are grateful to Prof. Y. Tschinkel for the permission to use these machines as well as to the system administrators for their support. Special thanks go to two anonymous referees whose valuable comments and suggestions helped to substancially improve this article.

F. Hess, S. Pauli, and M. Pohst (Eds.): ANTS 2006, LNCS 4076, pp. 317–332, 2006.
© Springer-Verlag Berlin Heidelberg 2006

$x^4 + y^4 = z^4 + v^4 + w^4$ contains the line given by $x = z$, $y = v$, and $w = 0$ on which there is quadratic growth, already. This explains the necessity of the restriction to a Zariski open subset $V^\circ \subseteq V$.

Remark 3. Conjecture 1 is proven for $\mathbf{P}^n$, linear subspaces, and quadrics. Further, it is established [Bi] in the case that the dimension of V is very large compared to $d_1, \dots, d_r$. Generalizations are known to be true in a number of further particular cases. A complete list may be found in the survey article [Pe2, sec. 4].

In this note, we report numerical evidence for Conjecture 1 in the case of the varieties $V_{a,b}^e$ given by $ax^e = by^e + z^e + v^e + w^e$ in $\mathbf{P}_{\mathbb{Q}}^4$ for $e = 3$ and 4.

Remark 4. By the Noether-Lefschetz Theorem, the assumptions made on V imply that $\mathrm{Pic}(V_{\overline{\mathbb{Q}}}) \cong \mathbb{Z}$ [Ha1, Corollary IV.3.2]. This is no longer true in dimension two. See Remark 6.6 for more details.

The Constant. Conjecture 1 is compatible with results obtained by the classical circle method (e.g. [Bi]). Motivated by this, E. Peyre [Pe1] provided a description of the constant C expected in (1). In the situation considered here, Peyre's constant is equal to the Tamagawa-type number

$$\tau_{\mathbf{H}}(V) := \prod_{p \in \mathbb{P} \cup \{\infty\}} (1 - \tfrac{1}{p}) \omega_{\mathbf{H},p}(V(\mathbb{Q}_p)).$$

In this formula, the Tamagawa measure $\omega_{\mathbf{H},p}$ is given in local p-adic analytic coordinates $x_1, \dots, x_d$ by $\|\frac{\partial}{\partial x_1} \wedge \dots \wedge \frac{\partial}{\partial x_d}\|_p \, dx_1 \dots dx_d$. Here, each dx_i denotes a Haar measure on $\mathbb{Q}_p$ which is normalized in the usual manner. $\frac{\partial}{\partial x_1} \wedge \dots \wedge \frac{\partial}{\partial x_d}$ is a section of $\mathcal{O}(-K) \cong \mathcal{O}(-k)$.

For p finite, one has a canonical model $\mathscr{V} \subseteq \mathbf{P}_{\mathbb{Z}_p}^n$ of V. This defines the norm $\|.\|_p$ on $\mathcal{O}(-k)$. It is almost immediate from the definition that

$$\omega_{\mathbf{H},p}(V(\mathbb{Q}_p)) = \lim_{n \to \infty} \frac{\#V(\mathbb{Z}/p^n\mathbb{Z})}{p^{dn}}.$$

The Place at Infinity. Here, $\|.\|_{\sup}$ on $\mathcal{O}(1)$ is the hermitian metric, in the sense of complex geometry, defined by $\|x_i\|_{\sup} := \inf_{j=0,\dots,n} |x_i/x_j|$. This induces the hermitian metric $\|.\|_\infty := \|.\|_{\sup}^{-k}$ on $\mathcal{O}(-k)$.

Lemma 5. *If $V \subset \mathbf{P}^n$ is a hypersurface defined by the equation $f = 0$ then*

$$\omega_\infty(V(\mathbb{R})) = \frac{1}{2} \int_{\substack{f(x_0,\dots,x_n)=0 \\ |x_0|,\dots,|x_n| \leq 1}} \omega_{\mathrm{Leray}}.$$

The Leray measure ω_{Leray} on $\{(x_0, \dots, x_n) \in \mathbb{R}^{n+1} \mid f(x_0, \dots, x_n) = 0\}$ is related to the usual hypersurface measure by the formula $\omega_{\mathrm{Leray}} = \frac{1}{\mathrm{grad}\, f} \omega_{\mathrm{hyp}}$. On the other hand, one may also write

$$\omega_{\mathrm{Leray}} = \frac{1}{|\frac{\partial f}{\partial x_i}(x_0, \dots, x_n)|} dx_0 \dots \widehat{dx_i} \dots dx_n.$$

Proof. The equivalence of the two descriptions of the Leray measure is a standard calculation. The main assertion is a particular case of [Pe1, Lemma 5.4.7]. 2 is the number of roots of unity in $\mathbb{Q}$.   □

Remark 6. There are several ways to generalize Conjecture 1.

i) One may consider more general heights corresponding to the tautological line bundle $\mathcal{O}(1)$. This includes to
   a) replace the maximum norm by an arbitrary continuous hermitian metric on $\mathcal{O}(1)$. This would affect the domain of integration for the factor at infinity.
   b) multiply $H_{\text{naive}}(x)$ with a function that depends on the reduction of x modulo some $N \in \mathbb{N}$. This augments Conjecture 1 by an equidistribution statement.
ii) Instead of complete intersections, one may consider arbitrary projective Fano varieties V. Then, H_{naive}^k needs to be replaced by a height defined by the anticanonical bundle $\mathcal{O}(-K_V)$.

If $\text{Pic}(V_{\overline{\mathbb{Q}}}) \not\cong \mathbb{Z}$ then the description of the constant C gets more complicated in several ways. First, there is an additional factor $\beta := \#H^1(\text{Gal}(\overline{\mathbb{Q}}/\mathbb{Q}), \text{Pic}(V_{\overline{\mathbb{Q}}}))$. Further, instead of the factors $(1 - \frac{1}{p})$ one has to write $1/L_p(1, \text{Pic}(V_{\overline{\mathbb{Q}}}))$, L_p being the local L-function corresponding to the Picard group. Finally, the Tamagawa measure has to be taken not of the full variety $V(\mathbb{A}_{\mathbb{Q}})$ but of the subset which is not affected by the Brauer-Manin obstruction.

If $\text{Pic}(V) \not\cong \mathbb{Z}$ already over $\mathbb{Q}$ then the right hand side of (1) has to be replaced by $CB \log^t B$. For the exponent of the log-term, there is the expectation that $t = \text{rk} \, \text{Pic}(V) - 1$. There are, however, examples [BT] in dimension three in which the exponent is larger. The constant C gets equipped with yet another additional factor α which depends on the structure of the effective cone in $\text{Pic}(V)$ and on the position of $-K_V$ in it [Pe1, Définition 2.4].

Finally, there is a generalization to arbitrary number fields [Pe1].

2   Computation of the Tamagawa Number

Counting Points over Finite Fields. We consider the projective varieties $V_{a,b}^e$ given by $ax^e = by^e + z^e + v^e + w^e$ in $\mathbf{P}_{\mathbb{F}_p}^4$. We assume $a, b \neq 0$ (and $p \nmid e$) in order to avoid singularities. Observe that, even for large p, these are at most e^2 varieties up to obvious $\mathbb{F}_p$-isomorphism as $\mathbb{F}_p^*$ consists of no more than e cosets modulo $(\mathbb{F}_p^*)^e$.

It follows from the Weil conjectures, proven by P. Deligne [De, Théorème (8.1)], that

$$\#V_{a,b}^e(\mathbb{F}_p) = p^3 + p^2 + p + 1 + E_{a,b}^e$$

where the error-term $E_{a,b}^e$ may be estimated by $|E_{a,b}^e| \leq C_e p^{3/2}$.

Here, $C_3 = 10$ and $C_4 = 60$ as $\dim H^3(\mathcal{V}^3, \mathbb{R}) = 10$ for every smooth cubic threefold and $\dim H^3(\mathcal{V}^4, \mathbb{R}) = 60$ for every smooth quartic threefold in $\mathbf{P}^4(\mathbb{C})$. These dimensions result from the Weak Lefschetz Theorem together with

F. Hirzebruch's formula [Hi, Satz 2.4] for the Euler characteristic which actually works in much more generality.

Remark 7. Suppose $e = 3$ and $p \equiv 2 \pmod 3$. Then, $\#V_{a,b}^3(\mathbb{F}_p) = \#V_{a,b}^1(\mathbb{F}_p)$ as $\gcd(p - 1, 3) = 1$. Similarly, for $e = 4$ and $p \equiv 3 \pmod 4$, one has $\gcd(p - 1, 4) = 2$ and $\#V_{a,b}^4(\mathbb{F}_p) = \#V_{a,b}^2(\mathbb{F}_p)$. In these cases, the error term vanishes and $\#V_{a,b}^e(\mathbb{F}_p) = p^3 + p^2 + p + 1$.

In the remaining cases $p \equiv 1 \pmod 3$ for $e = 3$ and $p \equiv 1 \pmod 4$ for $e = 4$, our goal is to compute the number of $\mathbb{F}_p$-rational points on $V_{a,b}^e$. As $V_{a,b}^e \subseteq \mathbf{P}^4$, there would be an obvious $O(p^4)$-algorithm. We can do significantly better than that.

Definition 8. *Let K be a field and let $x \in K^n$ and $y \in K^m$ be two vectors. Then, their* convolution $z := x * y \in K^{n+m-1}$ *is defined to be* $z_k := \sum_{i+j=k} x_i y_j.$

Theorem 9 (FFT convolution). *Let $n = 2^l$ and K be a field which contains the $2n$-th roots of unity. Then, the convolution $x * y$ of two vectors x, y of length $\leq n$ can be computed in $O(n \log n)$ steps.*

Proof. The idea is to apply the Fast Fourier Transform (FFT) [Fo, Satz 20.3]. The connection to the convolution is shown in [Fo, Satz 20.2, or CLR, Theorem 32.8]. □

Theorem 9 is the basis for the following algorithm.

Algorithm 10 (FFT point counting on $V_{a,b}^e$)

i) *Initialize a vector $x[0 \ldots p]$ with zeroes.*

ii) *Let r run from 0 to $p - 1$ and increase $x[r^e \bmod p]$ by 1.*

iii) *Calculate $\tilde{y} := x * x * x$ by FFT convolution.*

iv) *Normalize by putting $y[i] := \tilde{y}[i] + \tilde{y}[i + p] + \tilde{y}[i + 2p]$ for $i = 0, \ldots, p - 1$.*

v) *Initialize N as zero.*

vi) *(Counting points with first coordinate $\neq 0$)*
Let j run from 0 to $p - 1$ and increase N by $y[(a - bj^4) \bmod p]$.

vii) *(Counting points with first coordinate 0 and second coordinate $\neq 0$)*
Increase N by $y[(-b) \bmod p]$.

viii) *(Counting points with first and second coordinate 0)*
Increase N by $(y[0] - 1)/(p - 1)$.

ix) *Return N as the number of all $\mathbb{F}_p$-valued points on $V_{a,b}^e$.*

Remark 11. For the running-time, step iii) is dominant. Therefore, the running-time of Algorithm 10 is $O(p \log p)$.

To count, for fixed e and p, $\mathbb{F}_p$-rational points on $V_{a,b}^e$ with varying a and b, one needs to execute the first four steps only once. Afterwards, one may perform steps v) through ix) for all pairs (a, b) of elements from a system of representatives for $\mathbb{F}_p^*/(\mathbb{F}_p^*)^e$. Note that steps v) through ix) alone are of complexity $O(p)$.

We ran this algorithm for all primes $p \leq 10^6$ and stored the cardinalities in a file. This took several days of CPU time.

Remark 12. There is a formula for $\#V_{a,b}^e(\mathbb{F}_p)$ in terms of Jacobi sums. A skilful manipulation of these sums should lead to another efficient algorithm which serves the same purpose as Algorithm 10.

The Local Factors at Finite Places. We are interested in the Euler product

$$\tau_{a,b,\mathrm{fin}}^e := \prod_{p \in \mathbb{P}} \left(1 - \frac{1}{p}\right) \lim_{n \to \infty} \frac{\#V_{a,b}^e(\mathbb{Z}/p^n\mathbb{Z})}{p^{3n}}.$$

Lemma 13. a) (Good reduction)
If $p \nmid abe$ then the sequence $(\#V_{a,b}^e(\mathbb{Z}/p^n\mathbb{Z})/p^{3n})_{n \in \mathbb{N}}$ is constant.
b) (Bad reduction)
i) *If p divides ab but not e then the sequence $(\#V_{a,b}^e(\mathbb{Z}/p^n\mathbb{Z})/p^{3n})_{n \in \mathbb{N}}$ becomes stationary as soon as p^n divides neither a nor b.*
ii) *If $p = 2$ and $e = 4$ then the sequence $(\#V_{a,b}^e(\mathbb{Z}/p^n\mathbb{Z})/p^{3n})_{n \in \mathbb{N}}$ becomes stationary as soon as 2^n does not divide $8a$ or $8b$.*
iii) *If $p = 3$ and $e = 3$ then the sequence $(\#V_{a,b}^e(\mathbb{Z}/p^n\mathbb{Z})/p^{3n})_{n \in \mathbb{N}}$ becomes stationary as soon as 3^n divides neither $3a$ nor $3b$.*

Theorem 14. *For every pair (a, b) of integers such that $a, b \neq 0$, the Euler product $\tau_{a,b,\mathrm{fin}}^e$ is convergent.*
Proof. Let p be a prime bigger than $|a|$, $|b|$, and e. Then, the factor at p is $\tau_p := (1 - \frac{1}{p})(1 + p + p^2 + p^3 + D_p p^{3/2})/p^3$ where $|D_p| \leq C_e$ for $C_3 = 10$ and $C_4 = 60$, respectively.

Taking the logarithm, we consider $\sum_p \log \tau_p$. In the case $e = 3$, the sum is effectively over the primes $p \equiv 1 \pmod 3$. If $e = 4$ then summation extends over all primes $p \equiv 1 \pmod 4$. In either case, we take a sum over one-half of all primes. This leads to the following estimate,

$$\sum_{p \geq N} |\log \tau_p| \leq \sum_{p \geq N} \left[\frac{C_e}{p^{3/2}} + O(p^{-5/2})\right] \sim \frac{C_e}{2} \int_N^\infty \frac{1}{t^{3/2} \log t} \, dt$$

$$\leq \frac{C_e}{2 \log N} \int_N^\infty t^{-3/2} dt = \frac{C_e}{\sqrt{N} \log N}. \qquad \square$$

Remark 15. We are interested in an explicit upper bound for $|\sum_{p \geq 10^6} \log \tau_p|$. Using Taylor's formula, one gets

$$\left| \sum_{p \geq 10^6} \log \tau_p - \sum_{p \geq 10^6} \frac{D_p}{p^{3/2}} \right| \leq 10^{-8}.$$

Since $\frac{D_p}{p^{3/2}}$ is zero for $p \equiv 3 \pmod 4$ (or $p \equiv 2 \pmod 3$), the sum should be compared with $\log(\zeta_K(3/2))$. Here, ζ_K is the Dedekind Zeta function of $K = \mathbb{Q}(i)$ or $\mathbb{Q}(\zeta_3)$, respectively. This yields

$$\sum_{\substack{p \geq 10^6 \\ p \equiv 1 \; (\mathrm{mod}\; 4)}} \frac{1}{p^{3/2}} \leq \log \frac{\sqrt{\zeta_{\mathbb{Q}(i)}(3/2)}}{(1 - 2^{-3/2})^{-1/2} \cdot \prod_{p \equiv 3 \; (\mathrm{mod}\; 4)} (1 - p^{-3})^{-1/2} \cdot \prod_{\substack{p < 10^6 \\ p \equiv 1 \; (\mathrm{mod}\; 4)}} (1 - p^{-3/2})^{-1}}$$

and, for the other case, a similar estimate containing $\zeta_{\mathbb{Q}(\zeta_3)}(3/2)$. Note that the infinite product in the denominator converges a lot faster than the left hand side.

Using **Pari**, we evaluated the right hand side numerically. We found 0.39% for the quartic and 0.065% for the cubic as upper bounds for the error of approximation.

Remark 16. In practice, the error of the approximation is much smaller. The main reason is that the error-term D_p may have a positive or a negative sign. Some cancellations happen during summation. The assumption of a random distribution would result in a higher order of convergence. In fact, we observed this effect numerically.

Approximation of the Euler Product. Lemma 13 allows us to determine each factor of the Euler product exactly. As we need to know the numerical value of $\tau_{a,b,\mathrm{fin}}^e$, we approximate it by a finite product.

Observe that the factor at a good prime p is simply $(1 - 1/p) \cdot \#V_{a,b}^e(\mathbb{F}_p)/p^3$. In particular, for this factor there are only e^2 values possible. Even more, these numbers had been precomputed using FFT point counting (Algorithm 10). The algorithm below is based on the fact that the vast majority of the factors actually do not need to be computed. They are available from a list.

Algorithm 17 (Compute an approximate value for $\tau_{a,b,\mathrm{fin}}^3$ ($\tau_{a,b,\mathrm{fin}}^4$))
i) *Let p run over all prime numbers such that $p \equiv 2 \pmod 3$ ($p \equiv 3 \pmod 4$) and $p \leq N$ and calculate the product of all values of $(1 - 1/p^4)$.*

ii) *Compute the factor corresponding to $p = 3$ ($p = 2$) by Lemma 13.b).*

iii) *Let p run over all prime numbers such that $p \equiv 1 \pmod 3$ ($p \equiv 1 \pmod 4$) and $p \leq N$. Calculate the product of the factors described below.*

If $p|ab$ then the corresponding factor is given by Lemma 13.b). Otherwise, compute the e-th power residue-symbols of a and b and look up the precomputed factor for this $\mathbb{F}_p$-isomorphism class of varieties in the list.

iv) *Multiply the two products from steps i) and iii) and the factor from step ii) with each other. Correct the product by taking the bad primes $p \equiv 2 \pmod 3$ ($p \equiv 3 \pmod 4$) into consideration.*

Remark 18. When we meet a bad prime p, we have to count $\mathbb{Z}/p^n\mathbb{Z}$-valued points on $V_{a,b}^e$. This is done by an algorithm which is very similar to Algorithm 10.

We used Algorithm 17 to compute the Euler products $\tau_{a,b,\mathrm{fin}}^3$ and $\tau_{a,b,\mathrm{fin}}^4$ for $a, b = 1, \ldots, 100$. We did all calculations for $N = 10^6$. Note that step i) had to be done only once for $e = 3$ and once for $e = 4$. The running-time was a quarter of an hour for either exponent.

The Factor at the Infinite Place. For the quartic $V_{a,b}^4$, we have the integral

$$\omega_{\mathbf{H},\infty}(V_{a,b}^4(\mathbb{R})) = \frac{1}{4\sqrt[4]{a}} \iiiint\limits_R \frac{1}{(by^4 + z^4 + v^4 + w^4)^{3/4}} \, dy \, dz \, dv \, dw$$

over $R := \{(y, z, v, w) \in \mathbb{R}^4 \mid |y|, |z|, |v|, |w| \leq 1 \text{ and } |by^4 + z^4 + v^4 + w^4| \leq a\}$. The integrand is singular in one point. We used a simple substitution to make it sufficiently smooth for numerical integration.

On the other hand, for the cubic $V_{a,b}^3$, we have to consider

$$\omega_{\mathbf{H},\infty}(V_{a,b}^3(\mathbb{R})) = \frac{1}{6\sqrt[3]{a}} \iiiint\limits_R \frac{1}{(by^3 + z^3 + v^3 + w^3)^{2/3}} \, dy \, dz \, dv \, dw$$

for $R := \{(y, z, v, w) \in \mathbb{R}^4 \mid |y|, |z|, |v|, |w| \leq 1 \text{ and } |by^3 + z^3 + v^3 + w^3| \leq a\}$. The difficulty here is the handling of the singularity of the integrand. It is located in the zero set of $by^3 + z^3 + v^3 + w^3$ in R.

Since $(by^3 + z^3 + v^3 + w^3)^{-2/3}$ is a homogeneous function, it is enough to integrate over the boundary of R. This reduces the problem to several three-dimensional integrals of functions having a two-dimensional singular locus. If $a \geq b + 3$ then R is a cube and the boundary of R is easy to describe. We restricted our attention to this case. We smoothed the singularities by separation of Puiseux expansions and substitutions. The resulting integrals were treated by the Gauß-Legendre formula [Kr].

3   On the Geometry of Diagonal Cubic Threefolds

Lemma 19. *Let $V \subset \mathbf{P}^4$ be any smooth hypersurface. Then, every (reduced but possibly singular) surface $S \subset V$ is a complete intersection $V \cap H_d$ with a hypersurface $H_d \subset \mathbf{P}^4$.*

Proof. By the Noether-Lefschetz Theorem, we have $\mathrm{Pic}(V) \cong \mathbb{Z}$. The surface S is a Weil divisor on V. Hence, $\mathscr{O}(S) = \mathscr{O}(d) \in \mathrm{Pic}(V)$ for a certain $d > 0$. The restriction $\Gamma(\mathbf{P}^4, \mathscr{O}(d)) \to \Gamma(V, \mathscr{O}(d))$ is surjective as $H^1(\mathbf{P}^4, \mathscr{O}_V(d - \deg V)) = 0$ [Ha2, Theorem III.5.1.b)].   □

Elliptic Cones. Let $V \subset \mathbf{P}^4(\mathbb{C})$ be the diagonal cubic threefold given by the equation $x^3 + y^3 + z^3 + v^3 + w^3 = 0$. Fix $\zeta \in \mathbb{C}$ such that $\zeta^3 = 1$. Then, for every point $(x_0 : y_0 : z_0)$ on the elliptic curve $F \colon x^3 + y^3 + z^3 = 0$, the line given by $(x : y : z) = (x_0 : y_0 : z_0)$ and $v = -\zeta w$ is contained in V. All these lines together form a cone C_F over F the cusp of which is $(0 : 0 : 0 : -\zeta : 1)$. C_F is a singular model of a ruled surface over an elliptic curve. This shows, there are no other rational curves contained in C_F.

By permuting coordinates, one finds a total of thirty elliptic cones of that type within V. The cusps of these cones are usually named *Eckardt points* [Mu, CG].

We call the lines contained in one of these cones the *obvious lines* lying on V. It is clear that there are an infinite number of lines on V running through each of the thirty Eckardt points.

Proposition 20 (cf. [Mu, Lemma 1.18]). *Let $V \subset \mathbf{P}^4$ be the diagonal cubic threefold given by the equation $x^3 + y^3 + z^3 + v^3 + w^3 = 0$. Then, through each point $p \in V$ different from the thirty Eckardt points there are precisely six lines on V.*

Proof. Let $P = (x_0 : y_0 : z_0 : v_0 : w_0)$. A line l through P and another point $Q = (x : y : z : v : w)$ is parametrized by $(s : t) \mapsto ((sx_0 + tx) : \ldots : (sw_0 + tw))$. Comparing coefficients at $s^2 t$, st^2, and t^3, we see that the condition that l lies on V may be expressed by the three equations below.

$$x_0^2 x + y_0^2 y + z_0^2 z + v_0^2 v + w_0^2 w = 0 \tag{2}$$
$$x_0 x^2 + y_0 y^2 + z_0 z^2 + v_0 v^2 + w_0 w^2 = 0 \tag{3}$$
$$x^3 + y^3 + z^3 + v^3 + w^3 = 0 \tag{4}$$

The first equation means that Q lies on the tangent hyperplane H_P at P while equation (4) just encodes that $Q \in V$. By [Za, Corollary 1.15.b)], $H_P \cap V$ is an irreducible cubic surface.

On the other hand, the quadratic form q on the left hand side of equation (3) is of rank at least 3 as P is not an Eckardt point. Therefore, q is not just the product of two linear forms. In particular, $q|_{H_P} \not\equiv 0$.

As $H_P \cap V$ is irreducible, $Z(q|_{H_P})$ and $H_P \cap V$ do not have a component in common. By Bezout's theorem, their intersection in H_P is a curve of degree 6. $\square$

Remark 21. It may happen that some of the six lines coincide. Actually, it turns out that a line appears with multiplicity > 1 if and only if it is obvious [Mu, Lemma 1.19]. In particular, for a general point P the six lines through it are different from each other.

Under certain exceptional circumstances it is possible to write down all six lines explicitly. For example, if $P = (\sqrt[3]{-4} : 1 : 1 : 1 : 1)$ then the line $(\sqrt[3]{-4t} : (t + s) : (t + is) : (t - s) : (t - is))$ through P lies on V. Permuting the three rightmost coordinates yields all six lines.

4   Detection of Accumulating Subvarieties

The Detection of Q-rational Lines on the Cubics. On a cubic threefold $V_{a,b}^3$, quadratic growth is predicted for the number of Q-rational points of bounded height. Lines are the only curves with such a growth rate.

The moduli space of the lines on a cubic threefold is well-understood. It is a surface of general type [CG, Lemma 10.13]. Nevertheless, we do not know of a method to find all Q-rational lines on a given cubic threefold, explicitly. For that reason, we use the algorithm below which is an irrationality test for the six lines through a given point $(x_0 : y_0 : z_0 : v_0 : w_0) \in V_{a,b}^3(\mathbb{Q})$.

Algorithm 22 (Test the six lines through a given point for irrationality)

i) *Let p run through the primes from 3 to N.*
For each p, solve the system of equations (2), (3), (4) (adapted to $V^3_{a,b}$) in $\mathbb{F}^5_p$.
If the multiples of $(x_0, y_0, z_0, v_0, w_0)$ are the only solutions then output that there
is no $\mathbb{Q}$-rational line through $(x_0 : y_0 : z_0 : v_0 : w_0)$ and terminate prematurely.

ii) *If the loop comes to its regular end then output that the point is suspicious.*
It could possibly lie on a $\mathbb{Q}$-rational line.

Remark 23. We use a very naive $O(p)$-algorithm to solve the system of equations
over $\mathbb{F}_p$. If, say, $x_0 \neq 0$ then it is sufficient to consider quintuples such that $x = 0$.
We parametrize the projective plane given by (2). Then, we compute all points
on the conic given by (2) and (3). For each such point, we compute the cubic
form on the left hand side of (4).

We carried out the irrationality test on every $\mathbb{Q}$-rational point found on any
of the cubics except for the points lying on an obvious line. We worked
with $N = 600$. It turned out that suspicious points are rare and that, at least in
our sample, each of them actually lies on a $\mathbb{Q}$-rational line.

We found only 42 non-obvious $\mathbb{Q}$-rational lines on all of the cubics $V^3_{a,b}$ for
$100 \geq a \geq b \geq 1$ together. Among them, there are only five essentially different ones. We present them in the table below. The list might be enlarged by
two, as $V^3_{21,6}$ and $V^3_{22,5}$ may be transformed into $V^3_{48,21}$ and $V^3_{40,22}$, respectively,
by an automorphism of $\mathbf{P}^4$. Further, each line has six pairwise different images
under the obvious operation of the group S_3.

Table 1. Sporadic lines on the cubic threefolds

a	b	Smallest point	Point s.t. $x = 0$
19	18	(1 : 2 : 3 : -3 : -5)	(0 : 7 : 1 : -7 : -18)
21	6	(1 : 2 : 3 : -3 : -3)	(0 : 9 : 1 : -10 : -15)
22	5	(1 : -1 : 3 : 3 : -3)	(0 : 27 : -4 : -60 : 49)
45	18	(1 : 1 : 3 : 3 : -3)	(0 : 3 : -1 : 3 : -8)
73	17	(1 : 5 : -2 : 11 : -15)	(0 : 27 : -40 : 85 : -96)

Remark 24. It is a priori unnecessary to search for accumulating surfaces, at
least if we assume some conjectures.

First of all, only rational surfaces are supposed to accumulate that many rational
points that it could be seen through our asymptotics. Indeed, a surface which
is abelian or bielliptic may not have more than $O(\log^t B)$ points of height $< B$.
Non-rational ruled surfaces accumulate points in curves, anyway. Further, it is
expected [Pe2, Conjecture 3.6] that $K3$ surfaces, Enriques surfaces, and surfaces
of Kodaira dimension one may have no more than $O(B^\varepsilon)$ points of height $< B$
outside a finite union of rational curves. For surfaces of general type, finally,
expectations are even stronger (Lang's conjecture).

A rational surface S is, up to exceptional curves, the image of a rational map $\varphi\colon \mathbf{P}^2 \dashrightarrow V \subset \mathbf{P}^4$. There is a birational morphism $\varepsilon\colon P \to \mathbf{P}^2$ such that $\overline{\varphi} := \varphi \circ \varepsilon$ is a morphism of schemes. ε is given by a sequence of blowing-ups [Bv, Theorem II.11]. $\overline{\varphi}$ is defined by the linear system $|dH - E|$ where $d := \deg \varphi$, H is a hyperplane section, and E is the exceptional divisor. On the other hand, $K := K_P = -3H + E$. Therefore, if $d \geq 3$ then

$$\mathrm{H}_{\mathrm{naive}}(\overline{\varphi}(p)) = \mathrm{H}_{dH-E}(p) = \mathrm{H}_{3H-\frac{3}{d}E}(p)^{d/3} \geq c \cdot \mathrm{H}_{-K}(p)^{d/3}$$

for $p \notin \mathrm{supp}(E)$. Manin's conjecture implies there are $O((B \log^t B)^{\frac{3}{d}}) = o(B^2)$ points of height $< B$ on the Zariski-dense subset $\overline{\varphi}(P \setminus \mathrm{supp}(E)) \subseteq S$.

It remains to show that there are no rational maps $\varphi\colon \mathbf{P}^2 \dashrightarrow V$ of degree $d \leq 2$. Indeed, under this assumption, $\deg \varphi(\mathbf{P}^2) \leq 4$. This implies, by virtue of Lemma 19, that $\varphi(\mathbf{P}^2)$ is necessarily a hyperplane section $V \cap H$. Zak's theorem [Za, Corollary 1.8] shows that $V \cap H$ contains only finitely many singular points. It is, however, well known that cubic surfaces in 3-space which are the image of $\mathbf{P}^2$ under a quadratic map have a singular line [Bv, Corollary IV.8].

The Detection of Q-rational Conics on the Quartics. On a quartic threefold, linear growth is predicted for the number of Q-rational points of bounded height. The assumption $b > 0$ ensures that there are no $\mathbb{R}$-rational lines contained in $V_{a,b}^4$. The only other curves with at least linear growth one could think about are conics.

We were not able to create an efficient routine to test whether there is a Q-rational conic through a given point. The resulting system of equations seems to be too complicated to handle.

Conics Through Two Points. A conic Q through $(x_0 : y_0 : z_0 : v_0 : w_0)$ and $(x_1 : y_1 : z_1 : v_1 : w_1)$ may be parametrized in the form

$$(s : t) \mapsto ((\lambda x_0 s^2 + \mu x_1 t^2 + xst) : \ldots : (\lambda w_0 s^2 + \mu w_1 t^2 + wst))$$

for some $x, y, z, v, w, \lambda, \mu \in \mathbb{Z}$. The condition that Q is contained in $V_{a,b}^4$ leads to a system G of seven equations in x, y, z, v, w, and $\lambda\mu$. The phenomenon that λ and μ do not occur individually is explained by the fact that they are not invariant under the automorphisms of $\mathbf{P}^1$ which fix 0 and ∞.

Algorithm 25 (Test for conic through two points)

i) *Let p run through the primes from 3 to N.*

In the exceptional case that G could allow a solution such that $p|x, y, z, v, w$ but $p^2 \nmid \lambda\mu$, do nothing. Otherwise, solve G in $\mathbb{F}_p^6$. If $(0,0,0,0,0,0)$ is the only solution then output that there is no Q-rational conic through $(x_0 : y_0 : z_0 : v_0 : w_0)$ and $(x_1 : y_1 : z_1 : v_1 : w_1)$ and terminate prematurely.

ii) *If the loop comes to its regular end then output that the pair is suspicious. It could possibly lie on a Q-rational conic.*

To solve the system G in $\mathbb{F}_p^6$, we use an $O(p)$-algorithm. Actually, comparison of coefficients at s^7t and st^7 yields two linear equations in x, y, z, v, and w.

We parametrize the projective plane I given by them. Comparison of coefficients at s^6t^2 and s^2t^6 leads to a quadric O and an equation $\lambda\mu = q(x,y,z,v,w)/M$ with a quadratic form q over $\mathbb{Z}$ and an integer $M \neq 0$. The case $p|M$ sends us to the next prime immediately. Otherwise, we compute all points on the conic $I \cap O$. For each of them, we test the three remaining equations.

Conics Through Three Points. Three points P_1, P_2, and P_3 on $V_{a,b}^4$ define a projective plane **P**. The points together with the two tangent lines $\mathbf{P} \cap T_{P_1}$ and $\mathbf{P} \cap T_{P_2}$ determine a conic Q, uniquely. It is easy to transform this geometric insight into a formula for a parametrization of Q. We then need a test whether a conic given in parametrized form is contained in $V_{a,b}^4$. This part is algorithmically simple but requires the use of multiprecision integers.

Detecting Conics. For each quartic $V_{a,b}^4$, we tested every pair of $\mathbb{Q}$-rational points of height $< 10\,000$ for a conic through them. The existence of a conic through (P,Q) is equivalent to the existence of a conic through (gP, gQ) for $g \in (\mathbb{Z}/2\mathbb{Z})^4 \rtimes S_3 \subseteq \mathrm{Aut}(V_{a,b}^4)$. This reduces the running time by a factor of about 96. Further, pairs already known to lie on the same conic were excluded from the test.

For each pair (P,Q) found suspicious, we tested the triples (P,Q,R) for R running through the $\mathbb{Q}$-rational points of height $< 10\,000$, until a conic was found. Due to the symmetries, one finds several conics at once. For each conic detected, all points on it were marked as lying on this conic.

Actually, there were a few pairs found suspicious through which no conic could be found. In any of these cases, it was easy to prove by hand that there is actually no $\mathbb{R}$-rational conic passing through the two points. This means, we detected every conic which meets at least two of the rational points of height $< 10\,000$.

The Conics Found. Up to symmetry, we found a total of $1\,533$ $\mathbb{Q}$-rational conics on all of the quartics $V_{a,b}^4$ for $1 \leq a, b \leq 100$ together.

Among them, $1\,410$ are contained in a plane of type $z = v+w$ and $Yx-Xy = 0$ for (X,Y,t) a rational point on the genus one curve $aX^4 - bY^4 = 2t^2$. Further, there are 90 conics which are slight modifications of the above with y interchanged with z, v, or w. This is possible if b is a fourth power.

There is a geometric explanation for the occurrence of these conics. The hyperplane given by $z = v + w$ intersects $V_{a,b}^4$ in a surface S with the two singular points $(0:0:-1:e^{\pm 2\pi i/3}:e^{\mp 2\pi i/3})$. The linear projection $\pi\colon S \dashrightarrow \mathbf{P}^1$ to the first two coordinates is undefined only in these two points. Its fibers are plane quartics which split into two conics as $(v+w)^4 + v^4 + w^4 = 2(v^2 + vw + w^2)^2$. After resolution of singularities, the two conics become disjoint. $\widetilde{S}$ is a ruled surface over a twofold cover of $\mathbf{P}^1$ ramified in the four points such that $ax^4 - by^4 = 0$, i.e. over a curve of genus one.

In the case a is twice a square, a different sort of conics comes from the equations $v = z + Dy$ and $w = Ly$ when (L,D) is a point on the affine genus three curve $C_b\colon L^4 + b = D^4$. We found 28 conics of this type. C_b has a $\mathbb{Q}$-rational point for $b = 5, 15, 34, 39, 65, 80$, and 84. The conics actually admit a $\mathbb{Q}$-rational point for $a = 2, 18, 32$, and 98.

The remaining five conics are given as follows. For $a = 3$, 12, 27, or 48 and $b = 10$, intersect with the plane given by $v = y + z$ and $w = 2y + z$. For $a = 17$ and $b = 30$, put $v = 2x + y$ and $w = x + 3y + z$.

Remark 26. Again, it is not necessary to search for accumulating surfaces. Here, rational maps $\varphi \colon \mathbf{P}^2 \dashrightarrow V \subset \mathbf{P}^4$ such that $\deg \varphi \leq 3$ need to be taken into consideration. We claim, such a map is impossible.

If $\deg \varphi = 3$ then we had $\varphi \colon (\lambda : \mu : \nu) \mapsto (K_0(\lambda, \mu, \nu) : \ \ldots \ : K_4(\lambda, \mu, \nu))$ where $K_0, \ldots, K_4$ are cubic forms defined over $\mathbb{Q}$. $K_0 = 0$ defines a plane cubic which has infinitely many real points, automatically. As the image of φ is assumed to be contained in $V_{a,b}^4$, we have that $K_0(\lambda, \mu, \nu) = 0$ implies $K_1(\lambda, \mu, \nu) = \ldots = K_4(\lambda, \mu, \nu) = 0$ for $\lambda, \mu, \nu \in \mathbb{R}$. By consequence, $K_1, \ldots, K_4$ are divisible by K_0 (or by a linear factor of K_0 in the case it is reducible) and φ is not of degree three.

For $\deg \varphi \leq 2$, we had $\deg \overline{\varphi(\mathbf{P}^2)} \leq 4$ such that $\overline{\varphi(\mathbf{P}^2)} = V \cap H$ is a hyperplane section. Zak's theorem [Za, Corollary 1.8] shows it has at most finitely many singular points. On the other hand, a quartic in $\mathbf{P}^3$ which is the image of a quadratic map from $\mathbf{P}^2$ is a Steiner surface. It is known [Ap, p. 40] to have one, two, or (in generic case) three singular lines.

5   The Final Results

A Technology to Find Solutions of Diophantine Equations. In [EJ1] and [EJ2], we described a modification of D. Bernstein's [Be] method to search efficiently for all solutions of naive height $< B$ of a Diophantine equation of the particular form $f(x_1, \ldots, x_n) = g(y_1, \ldots, y_m)$. The expected running-time of our algorithm is $O(B^{\max\{n,m\}})$. Its basic idea is as follows.

Algorithm 27 (Search for solutions of a Diophantine equation)

i) (*Writing*)
Evaluate f on all points of the cube $\{(x_1, \ldots, x_n) \in \mathbb{Z}^n \mid |x_i| < B\}$ of dimension n. Store the values within a hash table H.

ii) (*Reading*)
Evaluate g on all points of the cube $\{(y_1, \ldots, y_m) \in \mathbb{Z}^m \mid |y_i| < B\}$. For each value, start a search in order to find out whether it occurs in H. When a coincidence is detected, reconstruct the corresponding values of $x_1, \ldots, x_n$ and output the solution.

Remark 28. In the case of a variety $V_{a,b}^e$, the running-time is obviously $O(B^3)$. We decided to store the values of $z^e + v^e + w^e$ into the hash table. Afterwards, we have to look up the values of $ax^e - by^e$.

In this form, the algorithm would lead to a program in which almost the entire running-time is consumed by the writing part. Observe, however, the following particularity of our method. When we search on up to $O(B)$ threefolds, differing only by the values of a and b, simultaneously, then the running-time is still $O(B^3)$.

We worked with $B = 5\,000$ for the cubics and $B = 10\,000$ for the quartics. In either case, we dealt with all threefolds arising for $a, b = 1, \ldots, 100$, simultaneously. For the quartics, the running-time was around four days of CPU time. This is approximately only three times longer than searching on a single three-fold had lasted. For the cubics, a program with integrated line detection took us approximately ten days.

The Result for the Cubics. We counted all $\mathbb{Q}$-rational points of height less than $5\,000$ on the threefolds $V_{a,b}^3$ where $a, b = 1, \ldots, 100$ and $b \leq a$. Note that $V_{a,b}^3 \cong V_{b,a}^3$. Points lying on one of the elliptic cones or on a sporadic $\mathbb{Q}$-rational line in $V_{a,b}$ were excluded from the count. The smallest number of points found is $3\,930\,278$ for $(a, b) = (98, 95)$. The largest numbers of points are $332\,137\,752$ for $(a, b) = (7, 1)$ and $355\,689\,300$ in the case that $a = 1$ and $b = 1$.

On the other hand, for each threefold $V_{a,b}^3$ whereas $a, b = 1, \ldots, 100$ and $b + 3 \leq a$, we calculated the expected number of points and the quotients

$$\# \{ \text{points of height} < B \text{ found} \} \, / \, \# \{ \text{points of height} < B \text{ expected} \}.$$

Let us visualize the quotients by two histograms.

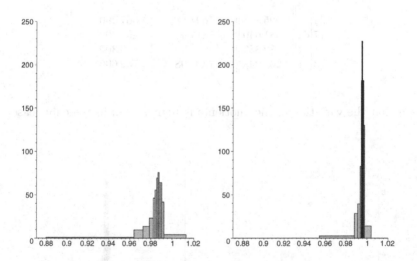

Fig. 1. Distribution of the quotients for $B = 1\,000$ and $B = 5\,000$

The statistical parameters are listed in the table below.

Table 2. Parameters of the distribution in the cubic case

	$B = 1\,000$	$B = 2\,000$	$B = 5\,000$
mean value	0.981 79	0.988 54	0.993 83
standard deviation	0.012 74	0.008 23	0.004 55

The Results for the Quartics. We counted all $\mathbb{Q}$-rational points of height less than $10\,000$ on the threefolds $V_{a,b}^4$ where $a, b = 1, \ldots, 100$. It turns out that on $5\,015$ of these varieties, there are no $\mathbb{Q}$-rational points occurring at all as the equation is unsolvable in $\mathbb{Q}_p$ for some small p. In this situation, Manin's conjecture is true, trivially.

Further, there is the case $(a, b) = (58, 87)$ in which the smallest 96 solutions are the images of $(6\,465 : 637 : 4\,321 : 6\,989 : 17\,719)$ under the obvious operation of the group $(\mathbb{Z}/2\mathbb{Z})^4 \rtimes S_3$. Here, $\tau_{\mathbf{H}}(V_{58,87}^4) \approx 0.002\,722$.

For the remaining varieties, the points lying on a known $\mathbb{Q}$-rational conic in $V_{a,b}$ were excluded from the count. Table 3 shows the quartics sorted by the numbers of points remaining.

Table 3. Numbers of points of height $< 10\,000$ on the quartics

a	b	#points	# not on conic	# expected
29	29	2	2	13.5
58	58	2	2	38.8
51	71	96	96	319.8
87	87	98	98	35.7
⋮	⋮	⋮	⋮	⋮
34	1	995 808	569 088	567 300
17	64	581 640	581 640	564 300
1	14	682 830	598 038	648 300
3	1	1 262 048	739 008	752 600

We see that the variation of the quotients is higher than in the cubic case.

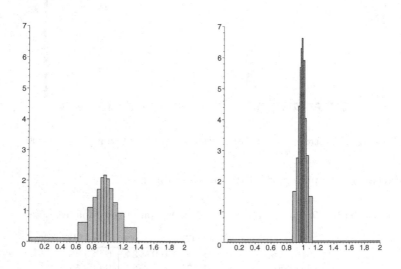

Fig. 2. Distribution of the quotients for $B = 1\,000$ and $B = 10\,000$

The statistical parameters are listed in the table below.

Table 4. Parameters of the distribution in the quartic case

	$B = 1\,000$	$B = 10\,000$
mean value	0.9853	0.9957
standard deviation	0.3159	0.1130

Interpretation of the Result. The results suggest that Manin's conjecture should be true for the two families of threefolds considered. In the cubic case, the standard deviation is by far smaller than in the case of the quartics. This, however, is not very surprising as on a cubic there tend to be much more rational points than on a quartic. This makes the sample more reliable.

Remark 29. The data we collected might be used to test the sharpening of the asymptotic formula (1) suggested by Sir P. Swinnerton-Dyer [S-D].

Question 30. Our calculations seem to indicate that the number of rational points often approaches its expected value from below. Is that more than an accidental effect?

References

[Ap]   Apéry, F.: *Models of the real projective plane*, Vieweg, Braunschweig 1987

[BT]   Batyrev, V. V. and Tschinkel, Y.: *Rational points on some Fano cubic bundles*, C. R. Acad. Sci. Paris **323** (1996), 41–46

[Bv]   Beauville, A.: *Complex algebraic surfaces*, LMS Lecture Note Series 68, Cambridge University Press, Cambridge 1983

[Be]   Bernstein, D. J.: *Enumerating solutions to $p(a) + q(b) = r(c) + s(d)$*, Math. Comp. **70** (2001), 389–394

[Bi]   Birch, B. J.: *Forms in many variables*, Proc. Roy. Soc. Ser. A **265** (1961/1962), 245–263

[CG]   Clemens, C. H. and Griffiths, P. A.: *The intermediate Jacobian of the cubic threefold*, Ann. of Math. **95** (1972), 281–356

[CLR]   Cormen, T., Leiserson, C., and Rivest, R.: *Introduction to algorithms*, MIT Press and McGraw-Hill, Cambridge and New York 1990

[De]   Deligne, P.: *La conjecture de Weil I*, Publ. Math. IHES **43** (1974), 273–307

[EJ1]   Elsenhans, A.-S. and Jahnel, J.: *The Diophantine equation $x^4 + 2y^4 = z^4 + 4w^4$*, Math. Comp. **75** (2006), 935–940

[EJ2]   Elsenhans, A.-S. and Jahnel, J.: *The Diophantine equation $x^4 + 2y^4 = z^4 + 4w^4$ — a number of improvements*, Preprint

[Fo]   Forster, O.: *Algorithmische Zahlentheorie*, Vieweg, Braunschweig 1996

[FMT]   Franke, J., Manin, Y. I., and Tschinkel, Y.: *Rational points of bounded height on Fano varieties*, Invent. Math. **95** (1989), 421–435

[Ha1]   Hartshorne, R.: *Ample subvarieties of algebraic varieties*, Springer, Lecture Notes Math. 156, Berlin-New York 1970

[Ha2]   Hartshorne, R.: *Algebraic geometry*, Springer, Graduate Texts in Mathematics 52, New York-Heidelberg 1977

[H-B]   Heath-Brown, D. R.: *The density of zeros of forms for which weak approxi-mation fails,* Math. Comp. **59** (1992), 613–623

[Hi]   Hirzebruch, F.: *Der Satz von Riemann-Roch in Faisceau-theoretischer For-mulierung: Einige Anwendungen und offene Fragen,* Proc. Int. Cong., Amsterdam 1954, vol. III, 457–473, Erven P. Noordhoff N.V. and North-Holland Publishing Co., Groningen and Amsterdam 1956

[Kr]   Kress, R.: *Numerical analysis,* Springer, Graduate Texts in Mathematics 181, New York 1998

[Mu]   Murre, J. P.: *Algebraic equivalence modulo rational equivalence on a cubic threefold,* Compositio Math. **25** (1972), 161–206

[Pe1]   Peyre, E.: *Hauteurs et mesures de Tamagawa sur les variétés de Fano,* Duke Math. J. **79** (1995), 101–218

[Pe2]   Peyre, E.: *Points de hauteur bornée et géométrie des variétés (d'après Y. Ma-nin et al.),* Séminaire Bourbaki 2000/2001, Astérisque **282** (2002), 323–344

[PT]   Peyre, E. and Tschinkel, Y.: *Tamagawa numbers of diagonal cubic surfaces, numerical evidence,* Math. Comp. **70** (2001), 367–387

[SchSt]   Schönhage, A. and Strassen, V.: *Schnelle Multiplikation großer Zahlen,* Com-puting **7** (1971), 281–292

[Se]   Sedgewick, R.: *Algorithms,* Addison-Wesley, Reading 1983

[S-D]   Swinnerton-Dyer, Sir P.: *Counting points on cubic surfaces II,* in: Geomet-ric methods in algebra and number theory, Progr. Math. 235, Birkhäuser, Boston 2005, 303–309

[Za]   Zak, F. L.: *Tangents and secants of algebraic varieties,* AMS Translations of Mathematical Monographs 127, Providence 1993

Testing Equivalence of Ternary Cubics

Tom Fisher

University of Cambridge, DPMMS, Centre for Mathematical Sciences
Wilberforce Road, Cambridge CB3 0WB, UK
T.A.Fisher@dpmms.cam.ac.uk
http://www.dpmms.cam.ac.uk/~taf1000

Abstract. Let C be a smooth plane cubic curve with Jacobian E. We give a formula for the action of the 3-torsion of E on C, and explain how it is useful in studying the 3-Selmer group of an elliptic curve defined over a number field.

We work over a field K of characteristic zero, with algebraic closure $\overline{K}$.

1   The Invariants of a Ternary Cubic

Let $X_3 = \mathbf{A}^{10}$ be the space of all ternary cubics

$$U(X, Y, Z) = aX^3 + bY^3 + cZ^3 + a_2X^2Y + a_3X^2Z$$
$$+ b_1XY^2 + b_3Y^2Z + c_1XZ^2 + c_2YZ^2 + mXYZ \ .$$

The co-ordinate ring of X_3 is the polynomial ring

$$K[X_3] = K[a, b, c, a_2, a_3, b_1, b_3, c_1, c_2, m] \ .$$

There is a natural action of GL_3 on X_3 given by

$$(gU)(X, Y, Z) = U(g_{11}X + g_{21}Y + g_{31}Z, \ldots, g_{13}X + g_{23}Y + g_{33}Z) \ .$$

The ring of invariants is

$$K[X_3]^{\mathrm{SL}_3} = \{F \in K[X_3] : F \circ g = F \text{ for all } g \in \mathrm{SL}_3(\overline{K})\} \ .$$

A homogeneous invariant F satisfies

$$F \circ g = \chi(g)F \tag{1}$$

for all $g \in \mathrm{GL}_3(\overline{K})$, for some rational character $\chi : \mathrm{GL}_3 \to \mathbf{G}_\mathrm{m}$. But the only rational characters of GL_3 are of the form $\chi(g) = (\det g)^k$ for k an integer. We say that F is an invariant of weight k. Taking g a scalar matrix in (1) shows that F has weight equal to its degree. The following facts are well known: see [1], [10], [15].

F. Hess, S. Pauli, and M. Pohst (Eds.): ANTS 2006, LNCS 4076, pp. 333–345, 2006.
© Springer-Verlag Berlin Heidelberg 2006

Theorem 1.1. *There are invariants c_4, c_6 and Δ of weights 4, 6 and 12, related by $c_4^3 - c_6^2 = 1728\Delta$, with the following properties:*
(i) The invariants of

$$U_E(X, Y, Z) = Y^2 Z + a_1 XYZ + a_3 YZ^2 - X^3 - a_2 X^2 Z - a_4 XZ^2 - a_6 Z^3$$

are given by the standard formulae: see [14, Chap. III].
(ii) The ring of invariants is a polynomial ring in two variables, generated by c_4 and c_6.
(iii) A ternary cubic U is non-singular if and only if $\Delta(U) \neq 0$.
(iv) If the plane cubic $\{U = 0\} \subset \mathbf{P}^2$ is non-singular then it has Jacobian

$$y^2 = x^3 - 27c_4(U)x - 54c_6(U) \ .$$

The Hessian of $U(X, Y, Z)$ is

$$H(X, Y, Z) = (-1/2) \times \begin{vmatrix} \frac{\partial^2 U}{\partial X^2} & \frac{\partial^2 U}{\partial X \partial Y} & \frac{\partial^2 U}{\partial X \partial Z} \\ \frac{\partial^2 U}{\partial X \partial Y} & \frac{\partial^2 U}{\partial Y^2} & \frac{\partial^2 U}{\partial Y \partial Z} \\ \frac{\partial^2 U}{\partial X \partial Z} & \frac{\partial^2 U}{\partial Y \partial Z} & \frac{\partial^2 U}{\partial Z^2} \end{vmatrix} \ .$$

The factor $-1/2$, although not standard, is a choice we find convenient. The Hessian is a polynomial map $H : X_3 \to X_3$ satisfying

$$H \circ g = (\det g)^2 g \circ H$$

for all $g \in \mathrm{GL}_3(\overline{K})$. We say it is a covariant of weight 2. Putting $c_4 = c_4(U)$, $c_6 = c_6(U)$ and $H = H(U)$ we find

$$H(\lambda U + \mu H) = 3(c_4 \lambda^2 \mu + 2c_6 \lambda \mu^2 + c_4^2 \mu^3)U + (\lambda^3 - 3c_4 \lambda \mu^2 - 2c_6 \mu^3)H \ .$$

This formula is classical: see [7], [11]. It is easily verified by restricting to any family of plane cubics covering the j-line. It also gives a convenient way of computing the invariants c_4 and c_6.

2   The 3-Selmer Group

Definition 2.1. *Let U_1 and U_2 be ternary cubics over K.*
(i) U_1 and U_2 are equivalent if $U_2 = \lambda(gU_1)$ for some $\lambda \in K^\times$ and $g \in \mathrm{GL}_3(K)$.
(ii) U_1 and U_2 are properly equivalent if $U_2 = (\det g)^{-1}(gU_1)$ for some $g \in \mathrm{GL}_3(K)$.

Lemma 2.2. *Let U_1 and U_2 be non-singular ternary cubics over K. If U_1 and U_2 are properly equivalent then they have the same invariants. If $K = \overline{K}$ then the converse is also true.*

Proof. The first statement follows from the fact that a homogeneous invariant has weight equal to its degree. For the second statement we may assume

$$U_1(X, Y, Z) = Y^2 Z - (X^3 + a_1 X Z^2 + b_1 Z^3)$$
$$U_2(X, Y, Z) = Y^2 Z - (X^3 + a_2 X Z^2 + b_2 Z^3)$$

for some $a_1, b_1, a_2, b_2 \in \overline{K}$. Since U_1 and U_2 have the same invariants, it follows by Theorem 1.1(i) that $U_1 = U_2$. $\qquad\square$

We consider pairs $(C \to S, \omega)$ where $C \to S$ is a morphism from a smooth curve of genus one C to a Brauer-Severi variety S, and ω is a regular 1-form on C. An isomorphism between $(C_1 \to S_1, \omega_1)$ and $(C_2 \to S_2, \omega_2)$ is a pair of isomorphisms $\phi : C_1 \cong C_2$ and $\psi : S_1 \cong S_2$ such that $\phi^* \omega_2 = \omega_1$ and the diagram

$$
\begin{array}{ccc}
C_1 & \longrightarrow & S_1 \\
\phi \downarrow & & \downarrow \psi \\
C_2 & \longrightarrow & S_2
\end{array}
$$

commutes.

Let $n \geq 2$ be an integer. Let E/K be an elliptic curve with invariant differential ω_E. We map $E \to \mathbf{P}^{n-1}$ via the complete linear system $|n.0_E|$. We recall that objects defined over K are called twists if they are isomorphic over $\overline{K}$.

Lemma 2.3. *The twists of* $(E \to \mathbf{P}^{n-1}, \omega_E)$, *up to K-isomorphism, are parametrised by* $H^1(K, E[n])$.

Proof. The automorphisms α of E with $\alpha^* \omega_E = \omega_E$ are the translation maps. If $\tau_P : E \to E$ is translation by $P \in E(\overline{K})$, we know that $\tau_P^*(n.0_E) \sim n.0_E$ if and only if $nP = 0_E$. So $\mathrm{Aut}(E \to \mathbf{P}^{n-1}, \omega_E) \cong E[n]$. An injective map from the isomorphism classes of twists to $H^1(K, E[n])$ is given by comparing an isomorphism defined over $\overline{K}$ with its Galois conjugates. It remains to prove surjectivity. This follows from the well known facts that the twists of E are parametrised by $H^1(K, \mathrm{Isom}(E))$ and the twists of $\mathbf{P}^{n-1}$ are parametrised by $H^1(K, \mathrm{PGL}_n)$. $\qquad\square$

Remark 2.4. This interpretation of $H^1(K, E[n])$ is a variant of one given in [4], [9]. If $\phi : C \to E$ is an isomorphism of curves defined over $\overline{K}$ with $\phi^* \omega_E = \omega$ then we make C a torsor under E via $(P, Q) \mapsto \phi^{-1}(P + \phi(Q))$. This action depends on ω but not on ϕ.

The obstruction map, defined in [9], is

$$
\mathrm{Ob} : H^1(K, E[n]) \to \mathrm{Br}(K)
$$
$$
(C \to S, \omega) \mapsto [S] \ .
$$

In general this map is not a group homomorphism. Nevertheless we write $\mathrm{ker}(\mathrm{Ob})$ for the inverse image of the identity. We specialise to the case $n = 3$.

Theorem 2.5. *Let $U_E = 0$ be a Weierstrass equation for E. Then the ternary cubics with the same invariants as U_E, up to proper K-equivalence, are parametrised by $\ker(\mathrm{Ob}) \subset H^1(K, E[3])$.*

Proof. A ternary cubic U determines a plane cubic $C = \{U = 0\} \subset \mathbf{P}^2$ and a regular 1-form on C

$$\omega = \frac{Z^2 d(Y/Z)}{\frac{\partial U}{\partial X}(X, Y, Z)}.$$

Conversely, every twist $(C \to S, \omega)$ of $(E \to \mathbf{P}^2, \omega_E)$ with $S \cong \mathbf{P}^2$ arises in this way. In view of Lemmas 2.2 and 2.3 it only remains to show that ternary cubics U_1 and U_2 are properly equivalent if and only if they determine isomorphic pairs $(C_1 \to \mathbf{P}^2, \omega_1)$ and $(C_2 \to \mathbf{P}^2, \omega_2)$. This is immediate from the next lemma, or more precisely the special case of it where $g \in \mathrm{GL}_3(K)$. □

Lemma 2.6. *Let U_1 and U_2 be non-singular ternary cubics, determining pairs $(C_1 \to \mathbf{P}^2, \omega_1)$ and $(C_2 \to \mathbf{P}^2, \omega_2)$. If $gU_1 = U_2$ for some $g \in \mathrm{GL}_3(\overline{K})$ then the isomorphism induced by g, namely*

$$\gamma : C_2 \to C_1 \, ; \quad (X : Y : Z) \mapsto (g_{11}X + g_{21}Y + g_{31}Z : \ldots),$$

satisfies $\gamma^* \omega_1 = (\det g)\omega_2$.

Proof. If the lemma is true for $g_1, g_2 \in \mathrm{GL}_3(\overline{K})$ then it is true for $g_1 g_2$. So it suffices to let g run over a set of generators for $\mathrm{GL}_3(\overline{K})$. The result is already clear for matrices of the form

$$g = \begin{pmatrix} \lambda_1 & 0 & 0 \\ 0 & \lambda_2 & 0 \\ 0 & \mu & \lambda_3 \end{pmatrix}.$$

Then for

$$g = \begin{pmatrix} 1 & 0 & 0 \\ 0 & 0 & 1 \\ 0 & 1 & 0 \end{pmatrix} \quad \text{and} \quad g = \begin{pmatrix} 0 & 1 & 0 \\ 1 & 0 & 0 \\ 0 & 0 & 1 \end{pmatrix}$$

we use the identities

$$\frac{Z}{Y} d\left(\frac{Y}{Z}\right) + \frac{Y}{Z} d\left(\frac{Z}{Y}\right) = 0$$

and

$$\frac{1}{Z^2} \frac{\partial U}{\partial X}(X, Y, Z) d\left(\frac{X}{Z}\right) + \frac{1}{Z^2} \frac{\partial U}{\partial Y}(X, Y, Z) d\left(\frac{Y}{Z}\right) = 0 \, .$$

□

Remark 2.7. The subset $\ker(\mathrm{Ob}) \subset H^1(K, E[3])$ contains the identity and is closed under taking inverses. A ternary cubic U represents the identity if and only if it has a K-rational point of inflection. The inverse of U is $-U$.

Remark 2.8. We claim that if K is a number field then the everywhere locally soluble ternary cubics with the same invariants as U_E, up to proper K-equivalence, are parametrised by the 3-Selmer group $S^{(3)}(E/K)$. It is shown in [9] that $S^{(3)}(E/K) \subset \ker(\text{Ob})$, so this claim is a special case of Theorem 2.5.

This interpretation of $S^{(3)}(E/K)$ becomes more useful if we can find algorithms for performing the following tasks. We write $[U]$ for the proper equivalence class of U.

1. Given U test whether $[U] = 0$.
2. Given U_1, U_2 test whether $[U_1] = [U_2]$. If so find the change of co-ordinates that relates them.
3. Given U_1, U_2, U_3 test whether $[U_1] + [U_2] = [U_3]$.
4. Given U_1, U_2 determine whether there exists U_3 with $[U_1] + [U_2] = [U_3]$. If so compute U_3.

The analogues of these problems for the 2-Selmer group have been solved in [3].

3   The Etale Algebra

Let R be the étale algebra of $E[3]$. It is a product of field extensions of K, one for each orbit for the action of $\text{Gal}(\overline{K}/K)$ on $E[3]$. It is shown in [6], [12] that there is an injective group homomorphism

$$w_1 : H^1(K, E[3]) \to R^\times/(R^\times)^3 \ .$$

According to [4, Paper I, Corollary 3.12] the restriction to $\ker(\text{Ob})$ is given by

$$(C \to \mathbf{P}^2, \omega) \mapsto \alpha = \det M$$

where $M \in \text{GL}_3(R) = \text{Map}_K(E[3], \text{GL}_3(\overline{K}))$ describes the action of $E[3]$ on $C \to \mathbf{P}^2$. (Recall that C is a torsor under E.)

In joint work [4] we describe a method for converting elements of $\ker(\text{Ob})$ represented by $\alpha \in R^\times$ to elements of $\ker(\text{Ob})$ represented by a ternary cubic $U(X, Y, Z)$. In this article we work in the opposite direction. We start with a ternary cubic $U(X, Y, Z)$ and convert it to $\alpha \in R^\times$. We also give a formula for the matrix $M \in \text{GL}_3(R)$. This enables us to solve the problems listed at the end of Sect. 2.

4   The Hesse Family

Let C be a smooth plane cubic with Jacobian E. Let $\zeta \in \overline{K}$ be a primitive cube root of unity. Let S, T be a basis for $E[3]$ with $e_3(S, T) = \zeta$, where e_3 is the Weil pairing. Making a suitable choice of co-ordinates over $\overline{K}$ we may assume

$$M_S = \begin{pmatrix} 1 & 0 & 0 \\ 0 & \zeta & 0 \\ 0 & 0 & \zeta^2 \end{pmatrix}, \quad M_T = \begin{pmatrix} 0 & 0 & 1 \\ 1 & 0 & 0 \\ 0 & 1 & 0 \end{pmatrix} .$$

Then C has equation

$$U(X, Y, Z) = a(X^3 + Y^3 + Z^3) - 3bXYZ .$$

The invariants of this ternary cubic are

$$c_4(a, b) = 3^4(8a^3 + b^3)b$$
$$c_6(a, b) = 3^6(8a^6 + 20a^3b^3 - b^6)$$
$$\Delta(a, b) = -3^9a^3(a^3 - b^3)^3 .$$

The Hessian is

$$H(X, Y, Z) = 27ab^2(X^3 + Y^3 + Z^3) - 27(4a^3 - b^3)XYZ .$$

Taking $0_C = (0 : 1 : -1)$ the elliptic curve $(C, 0_C)$ has Weierstrass equation

$$y^2z = x^3 - 27c_4(a, b)xz^2 - 54c_6(a, b)z^3 .$$

An explicit isomorphism is given by

$$x = -27(4a^3 - b^3)X - 81ab^2(Y + Z)$$
$$y = 972a(a^3 - b^3)(Y - Z) \tag{2}$$
$$z = bX + a(Y + Z) .$$

5   The Syzygetic Triangles

Let $U(X, Y, Z)$ be a non-singular ternary cubic with Jacobian E. The pencil of cubics spanned by U and its Hessian is a twist of the Hesse family. So there are exactly 4 singular fibres, and each singular fibre is a triangle. The sides of each triangle are the fixed lines for the action of M_T on $\mathbf{P}^2$ for some $0 \neq T \in E[3]$. So there is a Galois equivariant bijection between the syzygetic triangles and

$$\mathbf{P}(E[3]) = \frac{E[3] \setminus \{0\}}{\{\pm 1\}} .$$

Lemma 5.1. *Let U be a non-singular ternary cubic with invariants c_4, c_6 and Hessian H. Let $T = (x_T, y_T)$ be a non-zero 3-torsion point on the Jacobian*

$$E : \quad y^2 = x^3 - 27c_4x - 54c_6 .$$

Then the syzygetic triangle corresponding to $\pm T$ has equation

$$\mathcal{T} = \tfrac{1}{3}x_T U + H$$

and this equation satisfies $H(\mathcal{T}) = \tfrac{1}{27}y_T^2 \mathcal{T}$.

Proof. We may assume that U belongs to the Hesse family with T the image of $(0 : \zeta : -\zeta^2)$ under (2). The lemma follows by direct calculation. $\square$

Remark 5.2. The Hessian of a triangle is a non-zero multiple of the triangle. So in Lemma 5.1 we have $y_T \neq 0$. This is no surprise, since a non-zero 3-torsion point on E cannot also be a 2-torsion point.

6   The Invariants of a Triangle

Let S_3 act on $\mathbf{Q}[\alpha_1, \alpha_2, \alpha_3, \beta_1, \beta_2, \beta_3]$ by simultaneously permuting the α_i and the β_i. The ring of invariants has Hilbert series

$$h(t) = \frac{1 + t^2 + 2t^3 + t^4 + t^6}{(1-t)^2(1-t^2)^2(1-t^3)^2} .$$

Let s_1, s_2, s_3 (respectively t_1, t_2, t_3) be the elementary symmetric polynomials in the α_i (respectively β_i). According to MAGMA the primary invariants are $s_1, s_2, s_3, t_1, t_2, t_3$. The remaining coefficients of

$$T_1(X, Y, Z) = \prod_{i=1}^{3}(X + \alpha_i Y + \beta_i Z)$$

are

$$u = \alpha_1(\beta_2 + \beta_3) + \alpha_2(\beta_3 + \beta_1) + \alpha_3(\beta_1 + \beta_2)$$
$$v = \alpha_1\alpha_2\beta_3 + \alpha_1\beta_2\alpha_3 + \beta_1\alpha_2\alpha_3$$
$$w = \alpha_1\beta_2\beta_3 + \beta_1\alpha_2\beta_3 + \beta_1\beta_2\alpha_3 .$$

The secondary invariants are $1, u, v, w, u^2, vw$. So as a $\mathbf{Q}$-algebra, the ring of invariants is generated by the coefficients of T_1. There are 5 relations. These are obtained by writing uv, uw, v^2, w^2, u^3 as linear combinations of the secondary invariants. In fact MAGMA can rewrite any invariant as a $\mathbf{Q}[s_1, s_2, s_3, t_1, t_2, t_3]$-linear combination of the secondary invariants. For example

$$(\alpha_1 - \alpha_2)(\alpha_2 - \alpha_3)(\alpha_3 - \alpha_1) \begin{vmatrix} 1 & \alpha_1 & \beta_1 \\ 1 & \alpha_2 & \beta_2 \\ 1 & \alpha_3 & \beta_3 \end{vmatrix}$$
$$= 2s_1^2 v - s_1 s_2 u - 6 s_1 s_3 t_1 + 2s_2^2 t_1 - 6s_2 v + 9s_3 u .$$

7   Formulae

Let C be a smooth plane cubic defined over K, with Jacobian E. Let L/K be any field extension. Given $T \in E[3](L)$ we aim to compute $M_T \in \mathrm{GL}_3(L)$ describing the action of T on C. We start with an equation $U = 0$ for C. Then we construct the syzygetic triangle $T = \frac{1}{3}x_T U + H$ as described in Lemma 5.1. Making a change of co-ordinates if necessary, we may assume $T(1, 0, 0) \neq 0$. Then factoring over the algebraic closure gives

$$T(X, Y, Z) = r \prod_{i=1}^{3}(X + \alpha_i Y + \beta_i Z) . \tag{3}$$

We put

$$P = \begin{pmatrix} 1 & \alpha_1 & \beta_1 \\ 1 & \alpha_2 & \beta_2 \\ 1 & \alpha_3 & \beta_3 \end{pmatrix}$$

and $\xi = \alpha_1 + \zeta\alpha_2 + \zeta^2\alpha_3$.

Theorem 7.1. *If $\xi \neq 0$ then the matrix*

$$M_T = r\xi P^{-1} \begin{pmatrix} 1 & 0 & 0 \\ 0 & \zeta^2 & 0 \\ 0 & 0 & \zeta \end{pmatrix} P$$

belongs to $\mathrm{GL}_3(L)$ *and describes the action of* T *(or* $-T$*) on* C.

Proof. The required matrix has image in PGL_3 of order 3, and acts on $\mathbf{P}^2$ with fixed lines the sides of $T = 0$. So the second statement is clear. We must check that M_T has coefficients in L.

We write $r^{-1}(\det P)M_T = A + (\zeta - \zeta^2)B$ where A and B are matrices with entries in $\mathbf{Q}[\alpha_1, \alpha_2, \alpha_3, \beta_1, \beta_2, \beta_3]$. We find $\sigma(A) = \mathrm{sign}(\sigma)A$ and $\sigma(B) = B$ for all $\sigma \in S_3$. So the entries of $(\det P)A$ and B are polynomials in the coefficients of $T_1 = r^{-1}T$. As discussed in Sect. 6 we can compute these polynomials using MAGMA.

By (3) we have $H(T) = -r^2(\det P)^2 T$. Comparing with Lemma 5.1 it follows that $y_T = \pm 3(\zeta - \zeta^2)r \det P$. Therefore

$$(\det P)^2 M_T = r(\det P)A \pm \tfrac{1}{3} y_T B$$

and M_T has entries in L as required. □

We write

$$\begin{aligned} T(X, Y, Z) = {}& rX^3 + s_1 X^2 Y + s_2 XY^2 + s_3 Y^3 \\ & + t_1 X^2 Z + t_2 XZ^2 + t_3 Z^3 \\ & + YZ(uX + vY + wZ) \ . \end{aligned} \tag{4}$$

Theorem 7.2. $\det(M_T) = \tfrac{1}{2}(R \pm \tfrac{27r}{y_T} S)$ *where*

$$\begin{aligned} R &= 2s_1^3 - 9rs_1 s_2 + 27r^2 s_3 \\ S &= 2s_1^2 v - s_1 s_2 u - 6s_1 s_3 t_1 + 2s_2^2 t_1 - 6rs_2 v + 9rs_3 u \ . \end{aligned}$$

Proof. Comparing coefficients in (3) and (4) we find

$$\begin{aligned} \det(M_T) &= r^3(\alpha_1 + \zeta\alpha_2 + \zeta^2\alpha_3)^3 \\ &= \tfrac{1}{2}(R - 3(\zeta - \zeta^2)r^3\delta) \end{aligned}$$

where $\delta = (\alpha_1 - \alpha_2)(\alpha_2 - \alpha_3)(\alpha_3 - \alpha_1)$. By the example in Sect. 6 we have $r^3\delta \det P = S$. Finally we recall from the proof of Theorem 7.1 that $y_T = \pm 3(\zeta - \zeta^2)r \det P$. □

Remark 7.3. The formulae of Theorems 7.1 and 7.2 sometimes fail and give zero. (The situation is analogous to the proof of Hilbert's theorem 90 using Lagrange resolvents.) However if they fail for both T and $-T$ then

$$\alpha_1 + \zeta\alpha_2 + \zeta^2\alpha_3 = 0$$

and

$$\alpha_1 + \zeta^2\alpha_2 + \zeta\alpha_3 = 0 \ .$$

From these we deduce $\det P = 0$, contradicting that (3) is the equation of a syzygetic triangle. So if our formula for M_T fails then we can use $(M_{-T})^{-1}$ instead.

8 Galois Actions

The formulae of Sect. 7 are slightly easier to use in the case E does not admit a rational 3-isogeny.

Lemma 8.1. *If E does not admit a rational 3-isogeny and $[U] \neq 0$, then we are guaranteed that $\mathcal{T}(1,0,0) \neq 0$.*

Proof. We recall that $\mathcal{T} = \frac{1}{3}x_T U + H$. By hypothesis $x_T \notin K$. So if $\mathcal{T}(1,0,0) = 0$ then $U(1,0,0) = H(1,0,0) = 0$. But then $(1 : 0 : 0)$ is a K-rational point of inflection and $[U] = 0$. □

Let $G \subset \mathrm{GL}_2(\mathbf{F}_3) \cong \mathrm{Aut}(E[3])$ be the image of Galois.

Lemma 8.2. *If E does not admit a rational 3-isogeny then $-I_2 \in G$.*

Proof. By hypothesis the image of G in $\mathrm{PGL}_2(\mathbf{F}_3) \cong S_4$ acts on $\mathbf{P}^1(\mathbf{F}_3)$ without fixed points. If this image is A_4 or S_4 then G contains $\mathrm{SL}_2(\mathbf{F}_3)$ by [13, IV, §3.4, Lemma 2]. Otherwise G is a 2-group, and so conjugate to a subgroup of the Sylow 2-subgroup generated by

$$ a = \begin{pmatrix} 1 & 1 \\ -1 & 1 \end{pmatrix} \quad \text{and} \quad b = \begin{pmatrix} 1 & 0 \\ 0 & -1 \end{pmatrix}. $$

The only non-trivial subgroups of $\langle a, b \rangle$, not containing $-I_2$, are the conjugates of $\langle b \rangle$. These possibilities for G are again ruled out by the assumption that E does not admit a rational 3-isogeny. □

If $-I_2 \in G$ then our formula for M_T works if and only if our formula for M_{-T} works. According to Remark 7.3 they cannot both fail, so they must both work.

9 Applications

In our examples we take $K = \mathbf{Q}$. Elliptic curves over $\mathbf{Q}$ are referenced by their labellings in [2].

9.1 Testing Proper Equivalence

We are given non-singular ternary cubics U_1 and U_2, and must decide whether they are properly equivalent. First we check that they have the same invariants c_4 and c_6. Then the plane cubics $U_1 = 0$ and $U_2 = 0$ each have Jacobian

$$ E: \quad y^2 = x^3 - 27c_4 x - 54c_6 \ . $$

We compute $\alpha_1, \alpha_2 \in R^\times$ by using Theorem 7.2 once for each orbit for the action of $\mathrm{Gal}(\overline{K}/K)$ on $E[3]$. Then U_1 and U_2 are properly equivalent if and only if $\alpha_1/\alpha_2 \in (R^\times)^3$.

For example the ternary cubics

$$U_1(X, Y, Z) = X^3 - 180Y^3 + 24Z^3 + 8X^2Y - 3X^2Z \\ + 3XY^2 - 148Y^2Z + 76XZ^2 - 280YZ^2 + 59XYZ$$

and

$$U_2(X, Y, Z) = 32X^3 + 48Y^3 + 32Z^3 - 14X^2Y - 17X^2Z \\ + 14XY^2 + 68Y^2Z - 34XZ^2 + 34YZ^2 - 91XYZ$$

each have invariants $c_4 = 1073512497$ and $c_6 = 35173095391575$. The Jacobian is

$$2534e2: \quad y^2 + xy + y = x^3 - x^2 - 22364844x - 40704009937 .$$

A non-trivial 3-torsion point is $T = (x_T, y_T)$ where

$$x_T = \tfrac{1}{12}(289u^6 + 765u^4 + 24567u^2 + 22035)$$
$$y_T = -\tfrac{1}{312}(239307u^7 + 3757u^6 + 638911u^5 + 9945u^4 \\ + 20357181u^3 + 319371u^2 + 45909405u + 286611)$$

and u is a root of $X^8 + 78X^4 - 36X^2 - 507 = 0$. We have $R = \mathbf{Q} \times L$ where $L = \mathbf{Q}(u)$ is a number field of degree 8. The first factor of $\mathbf{Q}$ may be ignored. Using Theorem 7.2 we compute

$$\alpha_1 = \tfrac{144}{13}(548276415600669u^7 - 912344032067546u^6 \\ + 14593793190526 81u^5 - 2428439574347826u^4 \\ + 466500756222102 03u^3 - 77626752951639190u^2 \\ + 104433275464300347u - 173779291524426198)$$

$$\alpha_2 = \tfrac{1152}{13}(23737183831720776u^7 + 38664498064205221u^6 \\ + 63182645951465768u^5 + 102915548856548337u^4 \\ + 2019677284143385464u^3 + 3289767129786200531u^2 \\ + 4521354220053126264u + 73646431321685297 79) .$$

We find $\alpha_1/\alpha_2 = b^3$ where

$$b = \tfrac{1}{31499104}(-35980u^7 + 9880u^6 - 90181u^5 + 294515u^4 \\ - 2820090u^3 + 1603888u^2 - 6288205u + 17147429) .$$

It follows that U_1 and U_2 are properly equivalent.

Remark 9.1. Suppose we are given non-singular ternary cubics U and U' with invariants c_4, c_6 and c'_4, c'_6. To test for equivalence we first find all $\lambda \in K^\times$ satisfying $c'_4 = \lambda^4 c_4$ and $c'_6 = \lambda^6 c_6$. Then for each such λ we test whether λU and U' are properly equivalent.

9.2   Finding Equivalences

We continue with the example of the last subsection and find the change of coordinates relating U_1 and U_2. Following the proof of Theorem 7.1 we compute matrices $M_1, M_2 \in \mathrm{GL}_3(L)$ describing the action of T on $U_1 = 0$ and $U_2 = 0$. Since $\alpha_1/\alpha_2 \in (L^\times)^3$ we may arrange that $\det M_1 = \det M_2$. We are looking for $g \in \mathrm{GL}_3(K)$ with $U_2 = (\det g)^{-1}(gU_1)$. We must have

$$M_1 g^T = c g^T M_2$$

for some $c \in L^\times$. Taking determinants gives $c^3 = 1$. Since L contains no non-trivial cube roots of unity it follows that $c = 1$. Solving for g by linear algebra we find

$$g = \begin{pmatrix} 19 & -1 & 6 \\ -8 & -8 & 0 \\ 22 & -2 & -4 \end{pmatrix} .$$

Remark 9.2. If $E[3](K) \neq 0$ then there will be more than one change of coordinates relating U_1 and U_2. These will correspond to different choices for the constant c. Indeed by the Weil pairing there is an inclusion $E[3](K) \subset \mu_3(R)$.

9.3   Addition of Selmer Group Elements

The rank 0 elliptic curve

$$E = 4343b1 : \quad y^2 + y = x^3 - 325259x - 71398995$$

has Tate-Shafarevich group of analytic order 9. The following two elements of $S^{(3)}(E/\mathbf{Q})$ are visible in the rank 1 elliptic curves $21715a1$ and $117261k1$. (We will explain these calculations more fully in subsequent work. The concept of visibility was introduced in [5].)

$$\begin{aligned} U_1(X,Y,Z) = \; & X^3 + 15Y^3 - 17Z^3 - 8X^2Y + 4X^2Z \\ & + 15XY^2 - 13Y^2Z + 32XZ^2 + 26YZ^2 + 4XYZ \\ U_2(X,Y,Z) = \; & 7X^3 - 13Y^3 - 17Z^3 + 7X^2Y + 3X^2Z \\ & - 4XY^2 - 2Y^2Z + 12XZ^2 - 15YZ^2 - 30XYZ \end{aligned}$$

We use Theorem 7.2 to compute $\alpha_1, \alpha_2 \in R^\times$. We find that α_1, α_2 are independent in $R^\times/(R^\times)^3$. Applying the work of [4] to $\alpha_1\alpha_2$ and α_1/α_2 we obtain

$$\begin{aligned} U_3(X,Y,Z) = \; & -5X^3 + 12Y^3 + 31Z^3 + 3X^2Y - 5X^2Z \\ & + 5XY^2 + 2Y^2Z + 4XZ^2 + 26YZ^2 + 40XYZ \\ U_4(X,Y,Z) = \; & -11X^3 + 8Y^3 - 13Z^3 - 9X^2Y + 11X^2Z \\ & - 15XY^2 - Y^2Z - 16XZ^2 - 3YZ^2 - 38XYZ \; . \end{aligned}$$

Assuming the Birch Swinnerton-Dyer conjecture, we have

$$\mathrm{III}(E/\mathbf{Q}) = \{0, \pm[U_1], \pm[U_2], \pm[U_3], \pm[U_4]\} \; .$$

We have found these equations without the need to compute the class group or unit group of any number field.

9.4   Testing Global Solubility

We show that the ternary cubic

$$U_1(X, Y, Z) = 7X^3 + 9Y^3 + 16Z^3 + 2X^2Y$$
$$+ 5Y^2Z + 5XZ^2 - 7YZ^2 - 31XYZ$$

is insoluble over $\mathbf{Q}$. The Jacobian

$$E = 35882a1: \quad y^2 + xy = x^3 - x^2 - 156926x - 24991340$$

has Mordell-Weil group $E(\mathbf{Q}) \cong \mathbf{Z}/2\mathbf{Z} \times \mathbf{Z}$. A point of infinite order is $P = (693, 13750)$. Embedding $E \subset \mathbf{P}^2$ via $|2.0_E + P|$ we obtain

$$U_2(X, Y, Z) = 15Y^3 + 1254Z^3 + X^2Z - XY^2$$
$$+ 674Y^2Z + 10XZ^2 - 291YZ^2 + XYZ \ .$$

We use Theorem 7.2 to compute $\alpha_1, \alpha_2 \in R^\times$. We then check that α_1, α_2 are independent in $R^\times/(R^\times)^3$. Since U_2 is soluble over $\mathbf{Q}$ and $E(\mathbf{Q})/3E(\mathbf{Q}) \cong \mathbf{Z}/3\mathbf{Z}$, it follows that U_1 is insoluble over $\mathbf{Q}$.

Alternatively this could be checked using the explicit formulae for the covering map given in [1].

9.5   Reduction of Ternary Cubics

It is desirable to be able to replace an integer coefficient ternary cubic by an equivalent one with smaller coefficients. One method, explained to me by Michael Stoll, first computes a certain inner product, and then uses standard lattice reduction techniques. By an inner product on a complex vector space we mean a positive definite Hermitian form. We recall the Weyl unitary trick.

Lemma 9.3. *Let V be an irreducible complex representation of a finite group G. Then (up to scalars) there is a unique G-invariant inner product $\langle \, , \, \rangle : V \times V \to \mathbf{C}$.*

Proof. Let $\langle \, , \, \rangle_0$ be any inner product on V. Then

$$\langle u, v \rangle = \sum_{g \in G} \langle gu, gv \rangle_0$$

is a G-invariant inner product. By Schur's lemma the complex vector space of G-invariant sesquilinear forms on V is 1-dimensional.   □

We now take $C \subset \mathbf{P}^2$ a smooth plane cubic defined over $\mathbf{Q}$ with Jacobian E. The action of $E[3]$ on C extends to $\mathbf{P}^2$ to give $\chi : E[3] \to \mathrm{PGL}_3$. Lifting to SL_3 we obtain a diagram

$$
\begin{array}{ccccccccc}
0 & \longrightarrow & \mu_3 & \longrightarrow & H_3 & \longrightarrow & E[3] & \longrightarrow & 0 \\
& & \| & & \downarrow & & \downarrow \chi & & \\
0 & \longrightarrow & \mu_3 & \longrightarrow & \mathrm{SL}_3 & \longrightarrow & \mathrm{PGL}_3 & \longrightarrow & 0 \ .
\end{array}
$$

The Heisenberg group H_3 is a non-abelian group of order 27. For the reduction of ternary cubics, we use the unique Heisenberg-invariant inner product. Theorem 7.1 gives a convenient way of computing this inner product. Indeed if $M_1 \in \mathrm{SL}_3(\mathbf{R})$ and $M_2 \in \mathrm{SL}_3(\mathbf{C})$ generate the action of $E[3]$ on C then the required inner product on $\mathbf{C}^3$ has Gram matrix

$$\sum_{r=0}^{2} (\overline{M_2}^r)^T \left(\sum_{s=0}^{2} (M_1^s)^T M_1^s \right) M_2^r \ .$$

Acknowledgements

I would like to thank John Cremona, Cathy O'Neil and Michael Stoll for many useful discussions. All computer calculations in support of this work were performed using MAGMA [8]. The examples in Sect. 9 were prepared with the assistance of some programs written by Michael Stoll in connection with the joint work [4].

References

1. S.Y. An, S.Y. Kim, D.C. Marshall, S.H. Marshall, W.G. McCallum and A.R. Perlis. Jacobians of genus one curves. *J. Number Theory* **90** (2001), no. 2, 304–315.
2. J.E. Cremona. *Algorithms for modular elliptic curves*. Cambridge University Press, 1997. See also http://www.maths.nott.ac.uk/personal/jec/ftp/data
3. J.E. Cremona. Classical invariants and 2-descent on elliptic curves. *J. Symbolic Comput.* **31** (2001), no. 1-2, 71–87.
4. J.E. Cremona, T.A. Fisher, C. O'Neil, D. Simon and M. Stoll. *Explicit n-descent on elliptic curves*, I Algebra, II Geometry, III Algorithms. In preparation.
5. J.E. Cremona, B. Mazur. Visualizing elements in the Shafarevich-Tate group. *Experiment. Math.* **9** (2000), no. 1, 13–28.
6. Z. Djabri, E.F. Schaefer and N.P. Smart, Computing the p-Selmer group of an elliptic curve, *Trans. Amer. Math. Soc.* **352** (2000), no. 12, 5583–5597.
7. D. Hilbert. *Theory of algebraic invariants*. Cambridge University Press, 1993.
8. MAGMA is described in W. Bosma, J. Cannon and C. Playoust. The Magma algebra system I: The user language. *J. Symbolic Comput.* **24**, 235–265 (1997). See also http://magma.maths.usyd.edu.au/magma/
9. C. O'Neil. The period-index obstruction for elliptic curves, *J. Number Theory* **95** (2002), no. 2, 329–339.
10. B. Poonen. An explicit algebraic family of genus-one curves violating the Hasse principle. *J. Théor. Nombres Bordeaux* **13** (2001), no. 1, 263–274.
11. G. Salmon. *A treatise on the higher plane curves*. Third edition, Hodges, Foster and Figgis, Dublin, 1879.
12. E.F. Schaefer and M. Stoll. How to do a p-descent on an elliptic curve. *Trans. Amer. Math. Soc.* **356** (2004), no. 3, 1209–1231.
13. J.-P. Serre. *Abelian ℓ-adic representations and elliptic curves*. McGill University lecture notes, W. A. Benjamin, Inc., New York-Amsterdam, 1968.
14. J.H. Silverman. *The arithmetic of elliptic curves*. Springer GTM **106**, 1986.
15. B. Sturmfels. *Algorithms in invariant theory*. Texts and Monographs in Symbolic Computation, Springer-Verlag, Vienna, 1993.

Classification of Genus 3 Curves
in Special Strata of the Moduli Space

Martine Girard and David R. Kohel

School of Mathematics and Statistics, The University of Sydney
{girard, kohel}@maths.usyd.edu.au

Abstract. We describe the invariants of plane quartic curves — nonhyperelliptic genus 3 curves in their canonical model — as determined by Dixmier and Ohno, with application to the classification of curves with given structure. In particular, we determine modular equations for the strata in the moduli space $\mathcal{M}_3$ of plane quartics which have at least seven hyperflexes, and obtain an computational characterization of curves in these strata.

1 Introduction

The classification of curves of genus 0, 1, and 2 is aided by use of various geometric and arithmetic invariants. In this work we consider nonhyperelliptic genus 3 curves, for which the canonical model is an embedding as a projective plane quartic. The work of Hess [5,7] gives generic algorithms for determining the locus of Weierstrass points and for finding whether two curves are isomorphic. Such a generic approach to isomorphism testing works well for curves over finite fields, where a small degree splitting field for the Weierstrass places exists, and when one wants to test only two curves. In this work, we investigate the geometric invariants of nonhyperelliptic genus 3 curves, which are much more suited to classifying curves which are already given in terms of their canonical embeddings.

In particular, plane quartic curves admit explicit formulas for the Weierstrass locus, invariants of Dixmier and Ohno by which the curves may be classified up to isomorphism over an algebraically closed field, and moreover can be classified into strata following Vermeulen's characterization in terms of the number and configuration of Weierstrass points of weight two.

In the generic case, Harris [4] proved that a generic curve of any genus over a field of characteristic zero, is expected to have generic Galois action on the Weierstrass points. Thus in order to establish an isomorphism between the sets of Weierstrass points one needs in general an excessively large degree extension to apply the algorithm of Hess. Thus it becomes essential to exploit any special structure of the Weierstrass points to facilitate this algorithm. In this article we focus on curves whose moduli lie in special strata of the moduli space of genus three curves. We use a classification by invariants to reduce to a trivial calculation of invariants on certain strata of Vermeulen of dimensions 0 and 1.

F. Hess, S. Pauli, and M. Pohst (Eds.): ANTS 2006, LNCS 4076, pp. 346–360, 2006.
© Springer-Verlag Berlin Heidelberg 2006

2 The Weierstrass Locus of Quartic Curves

A nonhyperelliptic curve C of genus 3 can be define via the canonical embedding by a quartic equation $F(X, Y, Z) = 0$ in the projective plane. The Hessian $H(X, Y, Z)$ of the form $F(X, Y, Z)$ is defined by

$$
H = \begin{vmatrix}
\dfrac{\partial^2 F}{\partial X^2} & \dfrac{\partial^2 F}{\partial X \partial Y} & \dfrac{\partial^2 F}{\partial X \partial Z} \\[2ex]
\dfrac{\partial^2 F}{\partial X \partial Y} & \dfrac{\partial^2 F}{\partial Y^2} & \dfrac{\partial^2 F}{\partial Y \partial Z} \\[2ex]
\dfrac{\partial^2 F}{\partial X \partial Z} & \dfrac{\partial^2 F}{\partial Y \partial Z} & \dfrac{\partial^2 F}{\partial Z^2}
\end{vmatrix}.
$$

This form is a sextic, which meets the curve C in the 24 inflection points of the curve (counting multiplicities). These inflection are also the Weierstrass points, hence they may be determined in an elementary way. The inflection points which meet the Hessian with multiplicity 2 are those inflection points which meet their tangent line to multiplicity 4, and are called *hyperflexes*.

The hyperflexes are intrinsic points of the genus 3 curve, since they are Weierstrass of weight 2 – those which have a deficit of 2 in their gap sequences. Thus they are preserved by any isomorphism of curves, and reflect the underlying geometry of the curves (rather than solely of a choice of projective embedding).

We focus in this article on the classification of those curves which have an exceptional number of hyperflexes. The partioning of the Weierstrass points into those of weight 1 and weight 2 can also be efficiently determined since it is the singular subscheme of the intersection $F = H = 0$, defined by the vanishing of the Jacobian minors:

$$
\frac{\partial F}{\partial X} \frac{\partial H}{\partial Y} - \frac{\partial F}{\partial Y} \frac{\partial H}{\partial X} = \frac{\partial F}{\partial X} \frac{\partial H}{\partial Z} - \frac{\partial F}{\partial Z} \frac{\partial H}{\partial X} = \frac{\partial F}{\partial Y} \frac{\partial H}{\partial Z} - \frac{\partial F}{\partial Z} \frac{\partial H}{\partial Y} = 0.
$$

The calculation of the hyperflex locus can be reduced to polynomial factorization, without the need for Gröbner basis calculations. Let R be the resultant $\mathrm{Res}(H, F, Z)$ of degree 24 and let set $G(X, Y) = \mathrm{GCD}(R, R_X, R_Y)$. Then G determines the (X, Y)-coordinates of the hyperflex locus for which $XYZ \neq 0$.

Since plane quartics are canonical embeddings of a genus 3 curve, any isomorphism of such curves is induced by a linear isomorphism of their ambient projective planes. As a result, the problem of determining isomorphisms is reduced to the intersection of a linear algebra problem of finding such an isomorphism and a combinatorial one, of mapping a finite set of Weierstrass points to Weierstrass points. By combining classification of quartics by their moduli invariants into strata determined by the numbers and configurations of hyperflexes, we facilitate the problem of establishing isomorphisms between curves.

3 Quartic Invariants

The j-invariant of an elliptic curve or the Igusa invariants of a genus 2 curve provide invariants by which a curve of genus 1 or 2 can be classified up to isomorphism. Recall that the j-invariant of an elliptic curve is defined in terms of weighted projective invariants E_4, E_6, and Δ such that

$$j = \frac{E_4^3}{\Delta} \text{ and } E_4^3 - E_6^2 = 12^3 \Delta.$$

Similarly Igusa [8] defined weighted invariants J_2, J_4, J_6, J_8, and J_{10} with one relation $J_2 J_6 - J_4^2 = 4 J_8$, from which one can determine a set of absolute invariants (see also Mestre [11]).

For genus $g \geq 2$, the moduli space of curves of genus g is a space of dimension $3g - 3$, thus the determination of generators for the ring of projective invariants becomes increasingly difficult. However, for genus 3, Dixmier [2] provided an explicit set of 7 weighted invariants and proves their algebraic independence over $\mathbb{C}$. The determination of these invariants builds on on the explicit 19th century methods of Salmon [15]). By comparison of the Poincaré series of this ring with that computed by Shioda [16], Dixmier finds that his invariants I_3, I_6, I_9, I_{12}, I_{15}, I_{18}, and I_{27}, determine a subring over which the ring $\mathcal{A}$ of all invariants is free of rank 50.

Recently, Ohno [12] determined a complete set of generators and relations for the full ring of invariants of plane quartics.[1] In particular he shows that there exist six additional invariants J_9, J_{12}, J_{15}, J_{18}, I_{21}, J_{21}, which generate $\mathcal{A}$ as a finite algebraic extension of $\mathbb{C}[I_3, I_6, I_9, I_{12}, I_{15}, I_{18}, I_{27}]$.

Definition of Covariants and Contravariants
In this section we recall the definitions of covariants and contravariants, and basic constructions appearing in Dixmier [2], Ohno [12], and Salmon [15]. We first introduce the definitions of covariant and contravariant, following the modern terminology used in Poonen, Schaefer, and Stoll [14, §7.1]. For other modern treatments of the subject, see Sturmfels [17] and Olver [13].

Let $V = \mathbb{C}^n$ be be equipped with the standard left action of $\mathrm{GL}_n(\mathbb{C})$, which induces a right action on the algebra $\mathbb{C}[x_1, \ldots, x_n] = \mathrm{Sym}(V^*)$. For γ in $\mathrm{GL}_n(\mathbb{C})$ and $F \in \mathbb{C}[x_1, \ldots, x_n]$ we define this action by $F^\gamma(x) = F(\gamma(x))$ for all x in V. We denote $\mathbb{C}[x_1, \ldots, x_n]_d$ the d-th graded components of polynomials homogeneous of degree d.

Definition 1. *A covariant of degree r and order m is a $\mathbb{C}$-linear function*

$$\psi : \mathbb{C}[x_1, \ldots, x_n]_d \longrightarrow \mathbb{C}[x_1, \ldots, x_n]_m,$$

which satisfies

---

[1] Brumer [1] has independently identified a similar set of invariants which conjecturally generate the ring of quartic invariants.

1. $SL_n(\mathbb{C})$-*module homomorphism, i.e.* $\psi(F^\gamma) = \psi(F)^\gamma$ *for all* $\gamma \in SL_n(\mathbb{C})$,
2. *the coefficients of* $\psi(F)$ *depend polynomially in the coefficients of* $x_1^{i_1} \cdots x_{i_n}^n$,
3. *and* $\psi(\lambda F) = \lambda^r \psi(F)$ *for all* $\lambda \in \mathbb{C}$.

We note in particular that the last two conditions imply that ψ is homogeneous of degree r in the coefficients of the degree d form F. An *invariant* is a covariant of order 0.

N.B. One "usually" defines a covariant to satisfy

$$\psi(F^\gamma(\bar{x})) = \det(\gamma)^k \psi(F(x)) \text{ where } \gamma(\bar{x}) = x,$$

or equivalently, $\psi(F^\gamma) = \det(\gamma)^k \psi(F)^\gamma$, for all γ in $GL_n(V)$ and x in V. One defines k to be the *weight* (or *index*) of ψ. Clearly we then have the relation $2k = dr - m$. Applying a scalar matrix γ to an invariant implies that $k \equiv 0 \bmod n$. The definition of Poonen, Schaefer, and Stoll admits the possibility of covariants with a multiplicative character. Following the classical definitions we include the stronger condition above in our definition of covariants.

In order to define a contravariant, we set $\mathbb{C}[u_1, \ldots, u_n] = \text{Sym}(V)$, where $\{u_1, \ldots, u_n\}$ is a basis for V dual to the basis $\{x_1, \ldots, x_n\}$ of V^*. Then $GL_n(\mathbb{C})$ has the right contragradient action on polynomials in $\mathbb{C}[u_1, \ldots, u_n]$, which we denote $G^{\gamma*}$, where γ_* is the inverse transpose of γ.

Definition 2. *A contravariant of degree* r *and order* m *is a* $\mathbb{C}$-*linear function*

$$\psi : \mathbb{C}[x_1, \ldots, x_n]_d \longrightarrow \mathbb{C}[u_1, \ldots, u_n]_m,$$

which satisfies

1. $SL_n(\mathbb{C})$-*module homomorphism, i.e.* $\psi(F^\gamma) = \psi(F)^{\gamma*}$ *for all* $\gamma \in SL_n(\mathbb{C})$,
2. *the coefficients of* $\psi(F)$ *depend polynomially in the coefficients of* $x_1^{i_1} \cdots x_{i_n}^n$,
3. *and* $\psi(\lambda F) = \lambda^{-r} \psi(F)$ *for all* $\lambda \in \mathbb{C}$.

N.B. As noted in [14], we may formally identify $u_1, \ldots, u_n$ with $x_1, \ldots, x_n$, via the isomorphism $V \to V^*$ implied by the choice of basis for V. We nevertheless distinguish the $GL_n(\mathbb{C})$-modules structures by denoting the action by $G \mapsto G^{\gamma*}$ for γ in $GL_n(\mathbb{C})$. In our mathematical exposition we preserve the notational distinction between x_i and u_i.

Covariant and Contravariant Operations
We extend the linear pairing $V \times V^* \to \mathbb{C}$ given by $(u_i, x_j) \mapsto \delta_{ij}$ to a differential operation

$$D : \mathbb{C}[u_1, \ldots, u_n] \times \mathbb{C}[x_1, \ldots, x_n] \to \mathbb{C}[x_1, \ldots, x_n],$$

by identifying a monomial $u_1^{i_1} \cdots u_n^{i_n}$ of total degree m with the operator

$$\frac{\partial^m}{\partial^{i_1} x_i \cdots \partial^{i_n} x_n}.$$

We denote the $D(\psi, \varphi)$ by $D_\psi(\varphi)$. By symmetry, we define a differential operator

$$D : \mathbb{C}[x_1, \ldots, x_n] \times \mathbb{C}[u_1, \ldots, u_n] \to \mathbb{C}[u_1, \ldots, u_n],$$

and denote $D(\varphi, \psi)$ by $D_\varphi(\psi)$. (We resolve the notational ambiguity from the arguments.) We recall as a lemma a classical result.

Lemma 3. *Let φ be a covariant and ψ be a contravariant on $\mathbb{C}[x_1, \ldots, x_n]_d$. Then $D_\varphi(\psi)$ is a contravariant of order $\mathrm{ord}(\psi) - \mathrm{ord}(\varphi)$ and and $D_\psi(\varphi)$ a covariant of order $\mathrm{ord}(\psi) - \mathrm{ord}(\varphi)$, both of degree $\deg(\varphi) + \deg(\psi)$.*

When specializing to ternary forms, we denote (x_1, x_2, x_3) by (x, y, z). For ternary quadratic forms, Dixmier [2] used the additional operations. Let φ be a ternary quadratic form in x, y, z and

$$D(\varphi) = \frac{1}{2} \begin{vmatrix} \dfrac{\partial^2 \varphi}{\partial x^2} & \dfrac{\partial^2 \varphi}{\partial x \partial y} & \dfrac{\partial^2 \varphi}{\partial x \partial z} \\[2mm] \dfrac{\partial^2 \varphi}{\partial x \partial y} & \dfrac{\partial^2 \varphi}{\partial y^2} & \dfrac{\partial^2 \varphi}{\partial y \partial z} \\[2mm] \dfrac{\partial^2 \varphi}{\partial x \partial z} & \dfrac{\partial^2 \varphi}{\partial y \partial z} & \dfrac{\partial^2 \varphi}{\partial z^2} \end{vmatrix}$$

and let $D(\varphi)^*$ be its classical adjoint. Then for φ and ψ covariant and contravariant forms, respectively, we define

$$J_{11}(\varphi, \psi) = \langle D(\varphi), D(\psi) \rangle \text{ and } J_{22}(\varphi, \psi) = \langle D(\varphi)^*, D(\psi)^* \rangle,$$

where $\langle A, B \rangle$ is a matrix dot product, and

$$J_{30}(\varphi, \psi) = J_{30}(\varphi) = \det(D(\varphi)) \text{ and } J_{03}(\varphi, \psi) = J_{03}(\psi) = \det(D(\psi)).$$

The expressions J_{ij} play a role in invariant theory of ternary quadratic forms, but more generally we have the following classical lemma.

Lemma 4. *Let φ be a covariant and ψ be a contravariant on $\mathbb{C}[x, y, z]_d$, each of order 2. Then $J_{ij}(\varphi, \psi)$ is an invariant on $\mathbb{C}[x, y, z]_d$ of degree $i \deg(\varphi) + j \deg(\psi)$.*

In particular we will apply this to describe the construction of the complete invariants of ternary quartics by Dixmier [2] and Ohno [12].

Finally, for two binary forms $F(x, y)$ and $G(x, y)$ of degrees r and s, we define the k-th *transvectant* of is defined to be

$$(F, G)^k = \frac{(r-k)!(s-k)!}{r! s!} \left(\frac{\partial^2}{\partial x_1 \partial y_2} - \frac{\partial^2}{\partial y_1 \partial x_2} \right)^k F(x_1, y_1) G(x_2, y_2) \Big|_{(x_i, y_i) = (x, y)}$$

Lemma 5. *Let $F(x, y) = a_{40}x^4 + 4a_{31}x^3 y + 6a_{22}x^2 y^2 + 4a_{13}xy^3 + a_{04}y^4$ be a binary quartic form, and set $G = (F, F)^2$. Then F has invariants σ and ψ defined by*

$$\sigma = \frac{1}{2}(F, F)^4 = a_{40}a_{04} - 4a_{31}a_{13} + 3a_{22}^2, \text{ and}$$

$$\psi = \frac{1}{6}(F, G)^4 = a_{40}a_{22}a_{04} - a_{40}a_{13}^2 - a_{31}^2 a_{04} + 2a_{31}a_{22}a_{13} - a_{22}^3.$$

The invariant $\sigma^3 - 27\psi^2$ is the discriminant of the form $F(x, y)$ (up to a scalar).

Covariants and Contavariants of Quartics

In this section we use the above construction to define the invariants of Dixmier and Ohno classifying ternary quartics, i.e. genus 3 curves of general type. As noted above, we denote (x_1, x_2, x_3) by (x, y, z), and dual (u_1, u_2, u_3) by (u, v, w). The polynomial rings $k[x, y, z]$ and $k[u, v, w]$ represent coordinate rings of the ambient projective space $\mathbb{P}^2$ and the dual projective space $(\mathbb{P}^2)^*$, respectively, of the quartic $F(x, y, z) = 0$. Where necessary, $k[x_1, x_2, x_3, u_1, u_2, u_3]$ will be the bi-graded coordinate ring of $\mathbb{P}^2 \times (\mathbb{P}^2)^*$.

N.B. In the spirit of the classical literature, we speak of covariants and contravariants of a quartic form $F(x, y, z)$, though formally the covariant or contravariant is a function from $\mathbb{C}[x, y, z]_4$ to $\mathbb{C}[x, y, z]_m$ or $\mathbb{C}[u, v, w]_m$. Similarly, we may express a homogeneous form as

$$F(x_1, \ldots, x_n) = \sum_{i_1, \ldots, i_n} \frac{d!}{i_1! \ldots i_n!} a_{(i_1, \ldots, i_n)} x_1^{i_1} \cdots x_n^{i_n},$$

where the sum is over all indices with $i_1 + \cdots + i_n = d$. The calculation of invariants is thus normalized to be primitive with respect to such *classically integral* forms (as in Lemma 5). In the case of quartics, the constructions and expressions often require the primes 2 and 3 to be invertible, even if the final invariants can be made well-defined in these characteristics. In what follows we follow Dixmier and Ohno in normalizing the expressions to be primitive with respect to the coefficients $\{a_{ijk}\}$, and only at the end provide the scalars by which the invariants must be normalized to be integral with respect to the coefficients a_{ijk} of a integral form

$$F(x, y, z) = \sum_{i, j, k} a_{i,j,k} x^i y^j z^k. \tag{1}$$

In the spirit of Igusa's article on genus 2 curves [8], the determination of a complete set of integral invariants over any ring, and their algorithmic construction, remains open.

The first covariants at our disposal for a form F is the form itself (i.e. the identity covariant) and the Hessian H. We additionally require two contravariants σ and ψ, from which the Dixmier and Ohno invariants are derived by the operations of the previous section. The contravariant σ appears in Salmon [15] (§92 and §292), has degree 2 and order 4, and the construction of the degree 3 and order 6 contravariant ψ appears (in Salmon §92, p.78). Formally intersect $ux + vy + wz = 0$ with the form F, and setting $w = 1$, eliminate z a binary quartic $R(x, y) = F(x, y, -ux - vy)$. Then the invariants σ and ψ of Lemma 5, rehomogenized with respect to w, provide us with the covariants $\sigma(u, v, w)$ and $\psi(u, v, w)$.

We can now define a system of covariants and contravariants for ternary quartics, from the covariants F and H and contravariants σ and ψ.

Covariants	Contravariants
$\tau = 12^{-1}D_\rho(F)$	$\rho = 144^{-1}D_F(\psi)$
$\xi = 72^{-1}D_\sigma(H)$	$\eta = 12^{-1}D_\xi(\sigma)$
$\nu = 8^{-1}D_\eta D_\rho(H)$	$\chi = 8^{-1}D_\tau^2(\psi)$

Subsequently we can define the invariants of Dixmier:

$$I_3 = 144^{-1}D_\sigma(F), \qquad I_9 = J_{11}(\tau,\rho), \qquad I_{15} = J_{30}(\tau),$$
$$I_6 = 4608^{-1}(D_\psi(H) - 8I_3^2), \quad I_{12} = J_{03}(\rho), \qquad I_{18} = J_{22}(\tau,\rho).$$

together with the discriminant I_{27}. Dixmier [2] proved that these invariants are algebraically independent over $\mathbb{C}$ and generate a subring of the ring $\mathcal{A}$ of ternary quartic invariants of index 50. Ohno [12] proved computationally that the additional six invariants

$$J_9 = J_{11}(\xi,\rho), \qquad J_{15} = J_{30}(\xi), \qquad I_{21} = J_{03}(\eta),$$
$$J_{12} = J_{11}(\tau,\eta), \qquad J_{18} = J_{22}(\xi,\rho), \qquad J_{21} = J_{11}(\nu,\eta),$$

generate $\mathcal{A}$; he moreover determined a complete set of algebraic relations for the ring $\mathcal{A} = \mathbb{C}[I_k, J_l]$.

In the following table we summarise the covariant and contravariant degrees and orders, as can be determined from Lemma 3, beginning with the forms F, H, σ and ψ.

	Covariants ord								Contravariants ord					
deg	0	1	2	3	4	5	6	deg	1	2	3	4	5	6
1					F			1						
2								2				σ		
3	I_3						H	3						ψ
4								4		ρ				
5			τ,ξ					5						
6	I_6							6						
7								7		η				
8								8						
9	I_9, J_9							9						
10								10						
11								11						
12	I_{12}, J_{12}							12						
13								13		χ				
14			ν					14						

As noted above, the natural normalization for the invariant to be integral depends whether one considers classically integral forms or integral forms. On integral forms one normalizes the Dixmier–Ohno invariants as follows:

$$2^4 3^2 I_3, \quad 2^{12} 3^6 I_6, \quad 2^{12} 3^8 I_9, \quad 2^{16} 3^{12} I_{12}, \quad 2^{23} 3^{15} I_{15}, \quad 2^{27} 3^{17} I_{18}, \quad 2^{40} I_{27},$$
$$2^{12} 3^7 J_9, \quad 2^{17} 3^{10} J_{12}, \quad 2^{23} 3^{12} J_{15}, \quad 2^{27} 3^{15} J_{18}, \quad 2^{31} 3^{18} I_{21}, \quad 2^{33} 3^{16} J_{21}.$$

We refer to these normalizations of the Dixmier–Ohno invariants as the *integral* Dixmier–Ohno invariants (as opposed to the *classically integral* invariants). *Hereafter we will make these normalizations and write I_{3k} or J_{3l} to denote the above integral invariants.*

In what follows we invert I_3 in order to define six algebraically independent functions on the moduli space of quartic plane curves

$$(i_1, i_2, i_3, i_4, i_5, i_6) = \left(\frac{I_6}{I_3^2}, \frac{I_9}{I_3^3}, \frac{I_{12}}{I_3^4}, \frac{I_{15}}{I_3^5}, \frac{I_{18}}{I_3^6}, \frac{I_{27}}{I_3^9} \right),$$

and those defining an algebraic extension of $\mathbb{C}(i_1, \ldots, i_6)$:

$$(j_1, j_2, j_3, j_4, j_5, j_6) = \left(\frac{J_9}{I_3^3}, \frac{J_{12}}{I_3^4}, \frac{J_{15}}{I_3^5}, \frac{J_{18}}{I_3^6}, \frac{I_{21}}{I_3^7}, \frac{J_{21}}{I_3^7} \right).$$

A complete Magma implementation of the invariants of Dixmier and Ohno (and classical invariant theory constructions) is available from the second author's web page [3].

4   Vermeulen Stratification

In 1983, Vermeulen [18] constructed a stratification of the moduli space $\mathcal{M}_3$ of curves of genus 3 in terms of the number of hyperflexes and their geometric configuration. Similar results were obtained independently around the same time by Lugert [9]. The classification of curves by the structure of their Weierstrass points identifies more subtle structure of the curves than that provided by the automorphism group.

Let $\mathcal{M}_3^\circ$ be $\{[C] \in \mathcal{M}_3, C \text{ non-hyperelliptic}\}$, $M_s = \{[C] \in \mathcal{M}_3^\circ, C \text{ has at least } s \text{ hyperflexes}\}$, and $M_s^\circ = \{[C] \in \mathcal{M}_3^\circ, C \text{ has exactly } s \text{ hyperflexes}\}$. All strata but $\mathcal{M}_3^\circ$ are closed irreducible subvarieties of $\mathcal{M}_3$. Each M_s° is the union of the strata with s hyperflexes. For instance, M_{12}° consists of the two moduli points corresponding to the Fermat curve and the curve $X^4 + Y^4 + Z^4 + 3(X^2Z^2 + X^2Y^2 + Y^2Z^2) = 0$ and M_{11}° and M_{10}° are both empty. In Table 1 below, we summarize the relevant data from Vermeulen's stratification. We denote by s the number of hyperflexes.

By convention the X_i have dimension 3, the Y_i have dimension 2, the Z_i have dimension 1, and the strata represented by Greek letters have dimension zero.

5   Special Strata of Plane Quartics

For the special strata S of $\mathcal{M}_3$ with more than six hyperflexes we determine a parametrization of the stratum and a model for a generic curve $C/\tilde{S}$ for some finite cover $\tilde{S} \to S$. In each case the structure of $\tilde{S} \to S$ is a Galois cover over which the hyperflexes locus splits completely. The Dixmier–Ohno invariants are computed over $\tilde{S}$ by their sequences of covariants and contravariants, rather than evaluation of symbolic expressions.

Table 1. Vermeulen's stratification of $\mathcal{M}_3$

$\mathcal{X}$	s	dim	Substrata		$\mathcal{X}$	s	dim	Substrata
$\mathcal{M}_3^\circ$	0	6	M_1°		Z_8	5	1	$\Theta, \Pi_i, \Sigma, \Psi$
M_1°	1	5	M_2°		Z_2	6	1	$\Pi_i, \Omega_i, \Phi, \Psi$
M_2°	2	4	X_1, X_2, X_3		Z_3	6	1	$\Theta, \Pi_i, \Omega_i, \Psi$
X_2	3	3	Y_1, Y_2, Y_3		Z_5	6	1	Σ, Φ, Ψ
X_3	3	3	Y_1, Y_3, Y_4, Y_5		Z_9	6	1	Ω_i, Φ, Ψ
X_1	4	3	Y_1		Z_4	7	1	Ω_i, Ψ
Y_2	4	2	Z_1, Z_5		Θ	7	0	
Y_3	4	2	$Z_i, 1 \leq i \leq 8$		Π_i	7	0	
Y_4	4	2	$Z_i, i \neq 3, 6, 8$		Z_1	8	1	Φ, Ψ
Y_5	4	2	Z_2, Z_3, Z_6, Z_9		Σ	8	0	
Y_1	5	2	Z_1, Z_2, Z_3, Z_4		Ω	9	0	
Z_6	5	1	$\Theta, \Pi_i, \Omega_i, \Phi$		Φ	12	0	
Z_7	5	1	$\Pi_i, \Sigma, \Omega_i, \Psi$		Ψ	12	0	

5.1 One Dimensional Strata

For each of the one dimensional families Z_j we find an explicit model of the generic curve $\tilde{C}$ over a Galois cover $\tilde{Z}_j$ of Z_j. The models are derived from Vermeulen's geometric characterizations of the hyperflex locus (sending three hyperflexes to $(0:0:1)$, $(0:1:0)$, and $(1:1:1)$, respectively), with the condition that the s hyperflexes split over the function field of $\tilde{Z}_j$. Thus we may view the base space $\tilde{Z}_j$ as a moduli space for tuples $(C, P_1, \ldots, P_s)$ over the given stratum Z_j. Since the (differences of) hyperflexes generate a 4-torsion subgroup of the Jacobian of C, the space $\tilde{Z}_j$ specifies a level 4 structure. For $j \neq 8$, each $\tilde{Z}_j$ is isomorphic to $\mathbb{P}^1/\mathbb{Q}(i)$, and the moduli space Z_j descends to a rational curve $\mathbb{P}^1/\mathbb{Q}$. The space $\tilde{Z}_8 = E_2/\mathbb{Q}(i)$ is the base extension to $\mathbb{Q}(i)$ of the curve

$$y^2 + xy + y = x^3 + x^2 + 1,$$

of conductor 38, and $Z_8 = E_1/\mathbb{Q}$ is the quotient by the $\mathbb{Q}$-rational subgroup of order 5, corresponding to the Galois action on the 5 hyperflexes. We give an explicit description of the models $\tilde{C}$ and $\tilde{Z}_j$ for $j = 4$ and $j = 1$, the two strata with 7 and 8 hyperflexes respectively.

From a particular model $\tilde{C}$ we determine the Dixmier–Ohno invariants following the above construction of its invariants and covariants. Given two invariants, such as $x = I_6/I_3^2$ and $y = I_9/I_3^3$, we solve for a relation $f(x, y) = 0$, and solve for a rational or elliptic parametrization using explicit Riemann–Roch theory in Magma [6,10], and exploit symmetries to descend from $\mathbb{Q}(i)$ to $\mathbb{Q}$.

Stratum Z_4

The moduli space Z_4 is a one-dimensional subspace of $\mathcal{M}_3$, for which we can find a generic curve defined over $\mathbb{Q}(i, t)$ of the form:

$$C : t(t+i)(X^2 - YZ)^2 - YZ(2X - Y - Z)(((i-1)t^2 + 2t + (i+1))X - (it+1)(Y+Z))$$

where $i^2 = -1$ and t is a parameter. Let $\tilde{Z}_4$ be the rational curve with function field $\mathbb{Q}(i,t)$. This model parametrizes the curve plus the triple of Weierstrass points

$$\{(0:0:1),(0:1:0),(1:1:1)\},$$

with tangent lines defined by $Y = 0$, $Z = 0$, and $2X - Y - Z = 0$, respectively, and moreover, the remaining hyperflexes of C split over $\mathbb{Q}(i,t)$. Thus $\tilde{Z}_4$ defines a pointed moduli space parametrizing three hyperflexes, and should determine a Galois cover of the moduli space Z_4 on which the hyperflex locus splits.

A computation of the Dixmier–Ohno invariants reveils an automorphism $t \mapsto i(t-1)/(t+1)$ of order 3 under which the invariants are stable. The degree 3 quotient given by

$$t \mapsto z = \frac{(-(i+1)t^3 + 3t^2 - i)}{(t^2 + (-i+1)t - i)},$$

provides the Galois cover $\tilde{Z}_4 \to Z_4$.

The first of the Dixmier–Ohno invariants (i_1, i_2) can be expressed in terms of the invariant z as:

$$\left(\frac{(z-3)(5z-3)}{4z^2}, \frac{(79z^5 - 1059z^4 - 2670z^3 + 12366z^2 - 13203z + 4455)}{16(z^2 - 12z + 9)z^3} \right).$$

Reciprocally, the expression

$$z = \frac{-165(880i_1^3 + 1336i_1^2 - 558i_1 - 1047)}{(53680i_1^3 - 10560i_1^2 i_2 + 10120i_1^2 - 57750i_1 + 7920i_2 - 6105)} \tag{2}$$

gives z as a rational function in (i_1, i_2) so z generates the function field of Z_4. We note in particular that Z_4 is defined over $\mathbb{Q}$.

Solving for the algebraic relations in (i_1, i_2) and renormalizing, we find a weighted projective equation for Z_4 in terms of the first Dixmier–Ohno invariants:

$$193600I_6^5 - 35776I_3^2 I_6^4 + 86784I_3 I_6^3 I_9 - 96104 0I_3^4 I_6^3$$
$$- 2304I_6^2 I_9^2 + 100608I_3^3 I_6^2 I_9 - 526608I_3^6 I_6^2 - 65376I_3^5 I_6 I_9$$
$$+ 721521I_3^8 I_6 + 1728I_3^4 I_9^2 - 78048I_3^7 I_9 + 515889I_3^{10} = 0. \tag{3}$$

Thus from the invariants I_3, I_6, and I_9 we determine a necessary condition for a given quartic curve to be in the stratum Z_4. The remaining invariants have rational expressions in z, so can be readily computed from the rational expression (2). A comparison with the remaining Dixmier–Ohno invariants verifies or contradicts the hypothesis that a curve with invariants satisfying (3) lies in Z_4.

Given the invariant z for a point in Z_4, we can determine a field of definition for a representative curve with model C above, by solving for a root t of the degree six polynomial

$$2T^6 + 2(z-3)T^5 + (z-3)^2 T^4 + 2(z^2 - 4z + 1)T^3 + 2z^2 T^2 + 2z(z-1)T + (z-1)^2,$$

which is reducible over any field containing a square root of -1.

Stratum Z_1

The moduli space Z_1 is a one-dimensional subspace of $\mathcal{M}_3$ which consists of moduli of curves with 8 hyperflexes and automorphism group D_4. There exists a generic curve over some cover $\tilde{Z}_1$ on which the of Z_1, defined by

$$C : (t^2 + 1)(X^2 - YZ)^2 = YZ(2X - Y - Z)(2tX - Y - t^2Z).$$

By computing the Dixmier–Ohno invariants, we find that there exists a cyclic degree 4 cover $\tilde{Z}_1 \to Z_1$. In particular the transformation $t \mapsto (1 + it)/(t + i)$ of order four maps C to an isomorphic curve. As above, we find an invariant function $z = t + 1/t + u + 1/u - 1/2$, where $u = (1 + it)/(t + i)$, such that the absolute Dixmier–Ohno invariants can be expressed in z. Specifically, the first two are:

$$(i_1, i_2) = \left(\frac{(2z - 9)(2z + 9)}{4z^2}, \frac{(2z + 9)(8z^2 - 24z + 459)}{2^4 z^3} \right),$$

Reciprocally, we find an expression for z in terms of the first invariants

$$z = (-153 i_1 - 171)/(26 i_1 - 12 i_2 + 38),$$

so that z generates the function field of Z_1. The remaining invariants can be expressed in terms of the invariant z:

$$(i_3, i_4, i_5, i_6) = \left(\frac{(2z + 9)(4z - 27)(8z + 9)^2}{2^7 z^4}, \frac{(z + 9)^2(2z - 45)(2z + 9)^2}{4z^5}, \right.$$
$$\left. \frac{(z + 9)(2z + 9)^2(8z + 9)(8z^2 - 129z + 837)}{2^5 z^6}, -\frac{(2z - 7)^3}{2^{17} z^9} \right)$$

and

$$(j_1, \ldots, j_6) = \left(\frac{(2z + 9)(2z^2 - 11z - 9)}{4z^3}, \right.$$
$$\frac{(2z + 9)(4z^3 - 16z^2 - 99z - 1215)}{2^3 z^4}, \frac{(2z + 9)^2(z - 6)^2}{2z^4},$$
$$\frac{(2z + 9)^2(z - 6)(8z + 9)(8z^2 - 49z - 18)}{2^5 z^6}, \frac{(2z + 9)^3(2z - 15)^2(2z^2 - 3z + 27)}{2^4 z^7},$$
$$\left. \frac{(2z + 9)^2(56z^5 - 748z^4 + 1122z^3 + 20907z^2 - 38880z - 374706)}{16z^7} \right).$$

As in the case of Z_4, the moduli space Z_1 is defined over $\mathbb{Q}$. Given the invariant z for a point in Z_1, we can find a curve C which is defined in terms of a root t of the degree four polynomial:

$$(2T^4 - T^3 + 12T^2 - T + 2) - 2z(T^3 + T),$$

defining a cyclic cover of degree 4 over Z_1.

5.2   Zero Dimensional Strata

It remains to classify the strata of dimension zero in terms of moduli, which we denote Θ, Π_1, Π_2, Σ, Ω_1, Ω_2, Φ, and Ψ. In the case of Z_1 and Z_4, the known models for curves in these families were not obviously definable over their field of moduli. With the exception of Θ, Π_i, and Σ below, we will see that these exceptional strata have a curve which may be defined over its field of moduli.

Stratum Θ
The moduli space Θ is a zero-dimensional subspace of $\mathcal{M}_3$ which is represented by the curve C_Θ of the form

$$637(X^2 - YZ)^2 = (2X - Y - Z)(aX + bY + cZ)YZ$$

where

$$a = (-132i - 240)t^2 + (702i + 1068)t - 219i + 59$$
$$b = (-88i - 160)t^2 + (468i + 712)t - 146i - 810$$
$$c = (108i + 428)t^2 + (-806i - 2032)t + 295i + 415$$

with $i^2 = -1$ and $t^3 = ((10 + i)t/2 - 1)(t - 1)$, represented by the Dixmier invariants $(i_1, i_2, i_3, i_4, i_5, i_6)$:

$$\left(-\frac{3^3}{2^27^2}, \frac{3^3 173}{2^47^2}, -\frac{3^5 149}{2^77^2}, \frac{3^5 10223}{2^27^4}, \frac{3^5 5^2 3527}{2^57^4}, \frac{11^4 13^2}{2^{11} 3^{18} 7^5}\right).$$

and the point in $\mathcal{M}_3$ is completely determined by the i_k and the additional invariants $(j_1, j_2, j_3, j_4, j_5, j_6)$ of Ohno:

$$\left(-\frac{3^3 11}{2^3 7^2}, -\frac{3^2 4817}{2^4 7^3}, \frac{3^3}{2 \cdot 7^2}, \frac{3^4 535}{2^5 7^3}, \frac{3^4 55291}{2^7 7^4}, -\frac{3^2 5486023}{2^5 7^5}\right).$$

Strata Π_1, Π_2, Π_3
The strata Π_j are zero-dimensional subspaces of $\mathcal{M}_3$ which are represented by the curves C_{Π_i} of the form

$$49(X^2 - YZ)^2 = YZ(2X - Y - Z)(aX + b(Y + Z))$$

where

$$a = (-52i + 46)t^2 + (49i + 25)t - 82i - 114$$
$$b = (5i - 37)t^2 + (-28i + 19)t + 83i - 6,$$

with $i^2 = -1$ and t satisfying $2t^3 = (-i+1)t^2 + 4it + (i+1)$. The ideal of relations for the absolute Dixmier invariants $(i_1, i_2, i_3, i_4, i_5, i_6)$ for $\Pi = \Pi_1 \cup \Pi_2 \cup \Pi_3$, is the degree three ideal

$$(614656\, i_1^3 + 21952\, i_1^2 - 231516\, i_1 - 62613,$$
$$- 2^5 3^2 2297\, i_2 + 15135904\, i_1^2 - 681236\, i_1 + 418761,$$
$$- 2^7 2297\, i_3 + 7519344\, i_1^2 - 7084828\, i_1 - 3230271,$$
$$2^2 2297\, i_4 + 14936160\, i_1^2 + 20448508\, i_1 + 5686083,$$
$$- 2^7 7^2 2297\, i_5 + 12234260416\, i_1^2 + 10161115868\, i_1 + 1386276669,$$
$$- 2^{21} 3^{18} 7^3 2297\, i_6 + 87127555902240\, i_1^2 - 19953560617372\, i_1 - 29171717887351)$$

We note, in addition, that the curves C_{Π_i} admit an automorphism $\sigma(X,Y,Z) = (X,Z,Y)$, which induces a quotient to a non-CM elliptic curve E_{Π_i}, whose j-invariant satisfies

$$j^3 - \frac{3907322953}{3^4 7^4} j^2 + \frac{429408710168}{3\, 7^4} j - \frac{126488474356752}{7^4} = 0.$$

Stratum Σ

The stratum Σ is represented by the curve

$$X^4 + Y^4 + 6\sqrt{-7} X^2 Y^2 - 3(-1 + \sqrt{-7}) XYZ^2 = (7 + 3\sqrt{-7})/8 Z^4.$$

with absolute Dixmier invariants

$$(i_1, i_2, i_3, i_4, i_5, i_6) = \left(\frac{3^3}{2^2 7}, \frac{1557}{2^4 7}, -\frac{18225}{2^7 7}, -\frac{28403}{2^2 7}, \frac{2419065}{2^5 7^2}, \frac{1}{2^{13} 3^{18} 7^4} \right),$$

and absolute Ohno invariants

$$(j_1, j_2, j_3, j_4, j_5, j_6) = \left(-\frac{159}{56}, -\frac{3249}{112}, 9, \frac{14445}{224}, \frac{166617}{896}, -\frac{2076561}{1568} \right).$$

Strata Ω_1, Ω_2

These strata consists of two curves

$$(X^2 - YZ)^2 = (3 \pm \sqrt{7})(2X - Y - Z)(X - Y - Z)YZ.$$

In terms of the absolute Dixmier invariants $(i_1, i_2, i_3, i_4, i_5, i_6)$, the ideal of relations for $\Omega_1 \cup \Omega_2$ is the degree two ideal

$$(64\, i_1^2 + 64\, i_1 + 9, 1864\, i_1 + 64\, i_2 - 153, 512\, i_3 + 66416\, i_1 + 15435,$$
$$32\, i_4 + 28504\, i_1 + 16695, 64\, i_5 + 383138\, i_1 + 37737,$$
$$1624959306694656\, i_6 + 34973684392\, i_1 + 5920507885).$$

Stratum Φ

The stratum Φ is represented by the Fermat quartic

$$X^4 + Y^4 + Z^4 = 0,$$

with absolute Dixmier invariants

$$(i_1, i_2, i_3, i_4, i_5, i_6) = (0, 0, 0, 0, 0, -2^4 3^{18}),$$

and absolute Ohno invariants

$$(j_1, j_2, j_3, j_4, j_5, j_6) = (0, 0, 0, 0, 0, 0).$$

Stratum Ψ

The stratum Ψ is represented by the quartic

$$X^4 + Y^4 + Z^4 + 3(X^2Z^2 + X^2Y^2 + Y^2Z^2)$$

with absolute Dixmier invariants

$$(i_1, i_2, i_3, i_4, i_5, i_6) = \left(\frac{9}{16}, \frac{3^37^2}{2^7}\frac{3^47^3}{2^{11}}, \frac{3^57^3}{2^7}, \frac{3^67^4}{2^{11}}, -\frac{5^6}{2^{26}3^{18}}\right),$$

and absolute Ohno invariants

$$(j_1, j_2, j_3, j_4, j_5, j_6) = \left(\frac{63}{2^5}, \frac{2457}{2^7}, \frac{9}{2}, \frac{3969}{2^7}, \frac{177957}{2^{12}}, \frac{606879}{2^{10}}\right).$$

6 Galois Structure

Let $K = \mathbb{Q}(\tilde{\mathcal{X}})$ and $k = \mathbb{Q}(\mathcal{X})$ be the function fields of a given cover of strata $\tilde{\mathcal{X}} \to \mathcal{X}$ as schemes over $\mathbb{Q}$. By construction, K is a Galois extension of the field of moduli k which splits the hyperflex locus on a generic curve $\tilde{C}$ over K. We have an explicit action on the fibers of $\tilde{C}/K$ over k, from which we obtain an explicit map $\mathrm{Gal}(K/k) \to \mathrm{PGL}_3(K)$, coming from the embedding in $\mathbb{P}^2_K$. For the strata Ω_j, Φ, and Ψ we have determined a model C over the field of moduli k. In future work we expect to determine whether there exists an obstruction to the existence of a generic curve C/k or determine such a model over its field of moduli.

References

1. A. Brumer, personal communication, 2006.
2. J. Dixmier, On the projective invariants of quartic plane curves, *Adv. in Math.*, **64** (1987), no. 3, 279–304.
3. M. Girard, D. R. Kohel, C. Ritzenthaler, Invariants of plane quartics, magma code available at http://www.maths.usyd.edu.au/u/kohel/alg
4. J. Harris, Galois groups of enumerative problems, *Duke Math. J.*, **46** (1979), 685–724.
5. F. Hess, An algorithm for computing Weierstrass points, *Algorithmic number theory (Sydney, 2002)*, 357–371, *Lecture Notes in Comput. Sci.*, **2369**, Springer, Berlin, 2002.
6. F. Hess, Computing Riemann-Roch spaces in algebraic function fields and related topics, *J. Symbolic Comput.* **33** (2002), no. 4, 425–445.
7. F. Hess, An algorithm for computing isomorphisms of algebraic function fields, *Algorithmic number theory*, 263–271, *Lecture Notes in Comput. Sci.*, **3076**, Springer, Berlin, 2004.
8. J. Igusa, Arithmetic variety of moduli for genus two, *Ann. of Math. (2)*, **72** (1960), 612–649.
9. E. Lugert, Weierstrapunkte kompakter Riemannscher Flächen vom Geschlecht 3, Ph.D. thesis, Friedrich-Alexander-Universität Erlangen-Nürnberg, 1981.

10. Magma   Handbook,   J.   Cannon   and   W.   Bosma,   eds., http://magma.maths.usyd.edu.au/magma/htmlhelp/MAGMA.htm.

11. J.-F. Mestre, Construction de courbes de genre 2 à partir de leurs modules, *Effective methods in algebraic geometry (Castiglioncello, 1990)*, 313–334, *Progr. Math.*, **94**, Birkhäuser Boston, 1991.

12. T. Ohno, Invariant subring of ternary quartics I – generators and relations, preprint.

13. P. J. Olver, *Classical invariant theory*, London Mathematical Society Student Texts, **44**, Cambridge University Press, Cambridge, 1999.

14. B. Poonen, E. Schaefer, and M. Stoll, Twists of $X(7)$ and primitive solutions to $x^2 + y^3 = z^7$, preprint, 2005.

15. G. Salmon, *A treatise on the higher plane curves*, 3rd ed., 1879; reprinted by Chelsea, New York, 1960.

16. T. Shioda, On the graded ring of invariants of binary octavics, *Amer. J. Math.*, **89** (1967), 1022–1046.

17. B. Sturmfels, *Algorithms in invariant theory*, Texts and Monographs in Symbolic Computation, Springer-Verlag, Vienna, 1993.

18. A. M. Vermeulen. *Weierstrass points of weight two on curves of genus three*. PhD thesis, Universiteit van Amsterdam, 1983.

Heegner Point Computations Via Numerical p-Adic Integration

Matthew Greenberg

McGill University, Department of Mathematics and Statistics
805 Sherbrooke St. West, Montreal, Quebec, H3A 2K6, Canada
greenberg@math.mcgill.ca

Abstract. Building on ideas of Pollack and Stevens, we present an efficient algorithm for integrating rigid analytic functions against measures obtained from automorphic forms on definite quaternion algebras. We then apply these methods, in conjunction with the Jacquet-Langlands correspondence and the Cerednik-Drinfeld theorem, to the computation of p-adic periods and Heegner points on elliptic curves defined over $\mathbb{Q}$ and $\mathbb{Q}(\sqrt{5})$ which are uniformized by Shimura curves.

1 Heegner Points on Elliptic Curves

Let $E/\mathbb{Q}$ be an elliptic curve of conductor N. Then by the work of Wiles and his school, there exists a dominant morphism defined over $\mathbb{Q}$,

$$\Phi_N : X_0(N) \to E,$$

witnessing the modularity of E. We may assume that Φ_N sends the cusp at infinity to the origin of E. Let $A \to A'$ be a cyclic N-isogeny of elliptic curves with complex multiplication (henceforth, CM) by the same imaginary quadratic order $\mathfrak{o} \subset K$. Assume that the discriminant of $\mathfrak{o}$ is prime to N. Then by the classical theory of complex multiplication, the point $P = (A \to A')$ represents an element of $X_0(N)(H_{\mathfrak{o}})$, where $H_{\mathfrak{o}}$ is the ring class field attached to the order $\mathfrak{o}$. As Φ_N is defined over $\mathbb{Q}$, the point $\Phi_N(P)$ belongs to $E(H_{\mathfrak{o}})$. Such a point on E is called a *(classical) Heegner point*. These points are of significant interest. In particular, the proof of the conjecture of Birch and Swinnerton-Dyer for elliptic curves over $\mathbb{Q}$ of analytic rank at most one (Gross-Zagier, Kolyvagin) depends essentially on their properties.

These classical Heegner points may be efficiently computed in practice: Let $f_E \in S_2(N)$ be the normalized newform attached to E and let $\tau \in \mathfrak{H}$ represent the point P, where $\mathfrak{H}$ is the complex upper half plane. Then

$$\Phi_N(P) = W\left(\int_\infty^\tau f_E(z)dz\right) = W\left(\sum_{n\geq 1} \frac{a_n(f_E)}{n} q^n\right) \tag{1}$$

where $W : \mathbb{C} \to \mathbb{C}/\Lambda \cong E(\mathbb{C})$ is the Weierstrass uniformization of $E(\mathbb{C})$, the quantity $a_n(f_E)$ is the n-th Fourier coefficient of f_E and $q = e^{2\pi i\tau}$.

F. Hess, S. Pauli, and M. Pohst (Eds.): ANTS 2006, LNCS 4076, pp. 361–376, 2006.

The existence of a point $P = (A \to A')$ on $X_0(N)$ where both A and A' have CM by an order in K implies the validity of the *classical Heegner hypothesis*: that all primes ℓ dividing N are split in K. Due to the theoretical importance of the classical Heegner points, it is natural to desire an analogous systematic construction of algebraic points defined over class fields of imaginary quadratic fields which do not necessarily satisfy this stringent hypothesis, as well as methods to effectively compute these points in practice. Such a generalization requires admitting uniformizations of E by certain *Shimura curves*.

Assume that N is squarefree and $N = N^+N^-$ is factorization of N such that N^- has an even number of prime factors. Let C be the indefinite quaternion $\mathbb{Q}$-algebra ramified precisely at the primes dividing N^- and let S be an Eichler $\mathbb{Z}$-order in C of level N^+. (For basic definitions and terminology concerning quaternion algebras, see [13].) Fix an identification ι_∞ of $C \otimes \mathbb{R}$ with $M_2(\mathbb{R})$ and let Γ_{N^+,N^-} denote the image under ι_∞ of the group of units in S of reduced norm 1. Then Γ_{N^+,N^-} acts discontinuously on $\mathfrak{H}$ with compact quotient $X_{N^+,N^-}(\mathbb{C})$.

By Shimura's theory, the Riemann surface $X_{N^+,N^-}(\mathbb{C})$ has a canonical model X_{N^+,N^-} over $\mathbb{Q}$. This is proved by interpreting X_{N^+,N^-} as a moduli space for abelian surfaces over $\mathbb{Q}$ whose endomorphism rings contain S, together with some auxilliary level N^+-structure. (For a precise description of the moduli problem and the CM theory indicated below, see [14].) An abelian surface A with $S \subset \operatorname{End} A$ is said to have complex multiplication by the order $\mathfrak{o} \subset K$ if $\operatorname{End} A$ contains a subalgebra isomorphic to $\mathfrak{o}$ which commutes with S inside $\operatorname{End} A$. Consequently, there is a natural notion of a "CM-point" on X_{N^+,N^-}. Let $\mathfrak{H}(\mathfrak{o}) \subset \mathfrak{H}$ consist of those points whose images in X_{N^+,N^-} have CM by $\mathfrak{o}$. Then $\mathfrak{H}(\mathfrak{o})$ is Γ_{N^+,N^-}-stable and the quotient $\operatorname{CM}(\mathfrak{o}) := \Gamma_{N^+,N^-}\backslash\mathfrak{H}(\mathfrak{o})$ is a finite subset of $X_{N^+,N^-}(H_\mathfrak{o})$. Since we assume that the discriminant of $\mathfrak{o}$ is prime to N, the set $\operatorname{CM}(\mathfrak{o})$ is nonempty if and only if all rational primes ℓ dividing N^+ (resp. N^-) are split (resp. inert) in the fraction field K of $\mathfrak{o}$. We dub this condition the *Shimura-Heegner hypothesis*.

Let J_{N^+,N^-} denote the Jacobian variety of X_{N^+,N^-}. By the modularity theorem for elliptic curves over $\mathbb{Q}$ together with the Jacquet-Langlands correspondence, there exists a dominant morphism

$$\Phi_{N^+,N^-} : J_{N^+,N^-} \to E \qquad (2)$$

defined over $\mathbb{Q}$. (See [3, Ch. 4] for a discussion of this point.) The uniformization, Φ_{N^+,N^-} maps the set $\operatorname{CM}(\mathfrak{o})$ into $E(H_\mathfrak{o})$. To emphasize their origin, we shall refer to such points on E as *Shimura-Heegner points*.

Shimura formulated a reciprocity law which gives an alternate description of the Galois action on Shimura-Heegner points. Suppose that K satisfies the Shimura-Heegner hypothesis. He showed that there is a natural free action of $\operatorname{Pic}\mathfrak{o}$ on $\operatorname{CM}(\mathfrak{o})$ with $2^{\omega(N)}$ orbits ($\omega(N)$ = number of prime factors of N) such that for

$$P' - P \in \operatorname{Div}^0 \operatorname{CM}(\mathfrak{o}) \subset \operatorname{Div}^0 X_{N^+,N^-}(H_\mathfrak{o}),$$

we have

$$\Phi_{N^+,N^-}((P' - P)^{[\mathfrak{a}]}) = \Phi_{N^+,N^-}(P' - P)^{(\mathfrak{a}, H_\mathfrak{o}/K)}, \qquad (3)$$

where $(-, H_o/K) : \operatorname{Pic} \mathfrak{o} \to \operatorname{Gal} H_o/K$ is the reciprocity map of class field theory.

The phenomenon of elliptic curves being uniformized by Shimura curves generalizes to certain elliptic curves defined over totally real fields. For simplicity, let $F/\mathbb{Q}$ be a real quadratic field with infinite places σ_1 and σ_2 and let $\mathfrak{p}$ be a finite prime of F. Let C be the quaternion F-algebra ramified at $\mathfrak{p}$ and σ_1 and let S be a maximal order of C. Fix an isomorphism $\iota_{\sigma_2} : C \otimes_{\sigma_2} \mathbb{R} \to M_2(\mathbb{R})$ and let Γ be the image under ι_{σ_2} of the group of units in S with reduced norm 1. Then as above, the quotient $\Gamma\backslash\mathfrak{H}$ is a compact Riemann surface which admits a description as the complex points of a Shimura curve X, as well as a corresponding CM theory.

Let $f \in S_2(\mathfrak{p})$ be a Hilbert modular form. Then the Jacquet-Langlands correspondence together with the appropriate analog of the Eichler-Shimura construction asserts the existence of an elliptic curve E/F of conductor $\mathfrak{p}$ and a uniformization $J \to E$, where J is the Jacobian of X, such that the L-functions of E and f match. The images of CM divisors on X in E, also called Shimura-Heegner points, satisfy a reciprocity law analogous to (3). Zhang, generalizing the work of Gross-Zagier, has derived formulas relating heights of these Shimura-Heegner points to special values of derivatives of L-functions.

Unfortunately, since modular forms on non-split quaternion algebras do not admit q-expansions, there is no known explicit formula for the map (2) analogous to (1) which may be exploited to compute these important Shimura-Heegner points in practice. Our goal in this work is to describe and implement a p-adic analytic algorithm for performing such computations. The existence of a general algorithm for performing such Heegner point computations using only classical (i.e. archimedean) analysis remains an open problem, although some progress has been made by N. Elkies [7].

This paper is organized as follows: In §2 we introduce the formalism of p-adic integration. In §§3, 4 we set some basic notation and define our main technical tools – rigid analytic automorphic forms on definite quaternion algebras. In §§5, 6, we discuss (following [1]) how the Jacquet-Langlands correspondence allows one to associate a $\mathbb{Z}$-valued automorphic form to an elliptic curve $E/\mathbb{Q}$ and how one may use the Cerednik-Drinfeld theorem on p-adic uniformization of Shimura curves to give a p-adic integral formula for the Shimura-Heegner points on E introduced above. In §7 we indicate how rigid the rigid analytic automorphic forms introduced in §4 may be expoloited to evaluate this integral formula to high precision. The technical core of the paper is §8 where we adapt ideas of Pollack and Stevens to develop an efficient algorithm for lifting $\mathbb{Z}_p$-valued automorphic forms to rigid analytic automorphic ones, thereby facilitating the efficient calculation of Shimura-Heegner points.

For simplicity, we will develop the above mentioned theory in the situation where the base field is $\mathbb{Q}$, although an analogous theory exists for totally real base fields. We have implemented these methods in Magma to compute Shimura-Heegner points on

1. elliptic curves defined over $\mathbb{Q}$ with conductor $2p$, where p is an odd prime,
2. elliptic curves defined over $\mathbb{Q}(\sqrt{5})$ with degree one prime conductor.

Sample computations are given in §9. The author is more than happy to make this Magma code available to all; those interested should simply contact him.

This work owes much to the ideas of Pollack and Stevens, and the author wishes to thank them for providing him with a draft of [11]. This paper is part of the author's PhD thesis [8], written at McGill University under the supervision of Prof. Henri Darmon, whom the author would like to gratefully acknowledge for his expert guidance, advice and encouragement.

2 p-Adic Integration

Let p be a prime, let T be a complete subring of $\mathbb{C}_p$ and let X be a compact, totally disconnected topological space.

Definition 1. *A T-valued distribution on X is defined to be finitely additive T-valued function on the set of compact-open subsets of X. If the values of of a distribution are p-adically bounded, then we call it a* measure.

Let $\mathbf{D}(X,T)$ denote the set of T-valued measures on X and let $\mathbf{D}_0(X,T)$ denote the subspace of measures μ of total measure zero. If μ is in $\mathbf{D}(X,T)$ and $f : X \to T$ is locally constant, the symbol $\int_X f(x)d\mu(x)$ can be defined in the obvious way. To ease notation, we will sometimes write $\mu(f)$ instead. If μ is a measure, then we may extend μ to a linear functional on the space $\mathcal{C}(X,T)$ of continuous T-valued functions on X.

Suppose now that the distribution μ on X actually takes values in $\mathbb{Z}$ (implying, in particular, that μ is a measure). If $f = \sum_i a_i \mathbf{1}_{E_i}$ is a locally constant function on X, we may define the *multiplicative integral* of f against μ by the formula

$$\fint_X f(x)d\mu(x) = \prod_i a_i^{\mu(E_i)}.$$

By the boundedness of μ, the multiplicative integral extends to a group homomorphism from $\mathcal{C}(X,T^*)$ into T^*.

Let $\mathfrak{H}_p = \mathbb{P}^1(\mathbb{C}_p) - \mathbb{P}^1(\mathbb{Q}_p)$ be the *p-adic upper half-plane*, let μ be a $\mathbb{C}_p$-valued measure on $\mathbb{P}^1(\mathbb{Q}_p)$ and choose points $\tau, \tau' \in \mathfrak{H}_p$. We define a *p-adic line integral* by the formula

$$\int_\tau^{\tau'} \omega_\mu = \int_{\mathbb{P}^1(\mathbb{Q}_p)} \log \frac{x - \tau'}{x - \tau} d\mu(x), \tag{4}$$

where "log" denote the standard branch of the p-adic logarithm. If μ takes values in $\mathbb{Z}$, we may define a multiplicative analog of (4) above by posing

$$\fint_\tau^{\tau'} \omega_\mu = \fint_{\mathbb{P}^1(\mathbb{Q}_p)} \frac{x - \tau'}{x - \tau} d\mu(x). \tag{5}$$

Noting the relation

$$\int_\tau^{\tau'} \omega_\mu = \log \fint_\tau^{\tau'} \omega_\mu,$$

we see that (5) is actually a refinement of (4) as we avoid the choice of a branch of the p-adic logarithm. For motivation behind the formalism of p-adic line integration, see [3, Ch. 6].

3   Rigid Analytic Distributions

In this section, we consider p-adic integration over $\mathbb{Z}_p$. The naive method for computing integrals of the form

$$\int_{\mathbb{Z}_p} v(x)d\mu(x) \tag{6}$$

is of exponential complexity in the sense of [4]. Fortunately, many of the functions $v(x)$ which arise in practice are of a special type. Let

$$\mathbf{A}_{\mathrm{rig}} = \left\{ v(x) = \sum_{n \geq 0} a_n x^n : a_n \in \mathbb{Q}_p, \quad a_n \to 0 \text{ as } n \to \infty \right\}. \tag{7}$$

Elements of $\mathbf{A}_{\mathrm{rig}}$ are rigid analytic functions on the closed unit disk in $\mathbb{C}_p$ which are defined over $\mathbb{Q}_p$.

Definition 2. *Let $\mathbf{D}_{rig}$ be the continuous dual of $\mathbf{A}_{rig}$. Elements of $\mathbf{D}_{rig}$ are called rigid analytic distributions.*

Let $\mu \in \mathbf{D}_{\mathrm{rig}}$. Then by the continuity of μ, the problem of computing (6) is reduced to the calculation of the moments

$$\mu(x^n) = \int_{\mathbb{Z}_p} x^n d\mu(x), \quad n \geq 0.$$

A polynomial time algorithm for calculating such moments was recently discovered by R. Pollack and G. Stevens [11] in the situation where the measure μ is that attached to a cuspidal eigenform form on $\Gamma_0(N)$ as in [10]. Although the main goal of their theory was the study of normalized eigenforms g of weight $k+2$ with $\mathrm{ord}_p\, a_p(g) = k+1$ (a so-called critical slope eigenform) and their p-adic L-functions, we are interested in the so-called ordinary case: $\mathrm{ord}_p\, a_p(g) = 0$. The main objects of study in the [11] are modular symbols. We will develop analogues of their results where the role of the modular symbols is played by automorphic forms on definite quaternion algebras (see §4).

Let $\mathbf{D}^\circ_{\mathrm{rig}}$ be the subset of $\mathbf{D}_{\mathrm{rig}}$ consisting of those distributions with moments in $\mathbb{Z}_p$. The space $\mathbf{D}^\circ_{\mathrm{rig}}$ admits a useful filtration, first introduced by Pollack and Stevens in [11]. Define

$$F^0 \mathbf{D}^\circ_{\mathrm{rig}} = \mathbf{D}^\circ_{\mathrm{rig}},$$
$$F^N \mathbf{D}^\circ_{\mathrm{rig}} = \{\mu \in \mathbf{D}^\circ_{\mathrm{rig}} : \mu(x^j) \in p^{N-j}\mathbb{Z}_p, \quad j = 0, \ldots, N-1\}, \quad N \geq 1.$$

Now let

$$A^N \mathbf{D}^\circ_{\mathrm{rig}} = \mathbf{D}^\circ_{\mathrm{rig}}/F^N \mathbf{D}^\circ_{\mathrm{rig}}, \quad N \geq 1.$$

We call $A^N \mathbf{D}^\circ_{\mathrm{rig}}$ the N-th approximation to the module $\mathbf{D}^\circ_{\mathrm{rig}}$, following the terminology of [11].

4 Automorphic forms on Definite Quaternion Algebras

Let N, N^+, and N^- be as in § 1 and assume the existence of a prime p dividing N^-. Let B be the definite quaternion algebra ramified precisely at the infinite place of $\mathbb{Q}$ together with the primes dividing N^-/p, and let R be an Eichler $\mathbb{Z}$-order in B of level pN^+. Fix an identification ι_p of B_p with $M_2(\mathbb{Q}_p)$ under which R_p corresponds to

$$M_0(p\mathbb{Z}_p) = \left\{ \begin{pmatrix} a & b \\ c & d \end{pmatrix} \in M_2(\mathbb{Z}_p) : c \equiv 0 \pmod{p} \right\}. \tag{8}$$

Let $\hat{\mathbb{Q}}$ be the finite adèles of $\mathbb{Q}$ and let $\hat{\mathbb{Z}} = \prod_\ell \mathbb{Z}_\ell$ be the profinite completion of $\mathbb{Z}$. Let $\hat{B} = B \otimes_\mathbb{Q} \hat{\mathbb{Q}}$ and $\hat{R} = R \otimes_\mathbb{Z} \hat{\mathbb{Z}}$ be the adelizations of B and R, respectively.

Define the semigroup

$$\Sigma_0(p) = \left\{ \begin{pmatrix} a & b \\ c & d \end{pmatrix} \in M_2(\mathbb{Z}_p) : p \mid c, \ d \in \mathbb{Z}_p^*, \text{ and } ad - bc \neq 0 \right\}.$$

Let A be a left $\Sigma_0(p)$-module.

Definition 3. *An automorphic form on B of level R taking values in A is a map $f : \hat{\mathbb{Q}}^* B^* \backslash \hat{B}^* \to A$ such that $u_p f(bu) = f(b)$ for all $u \in \hat{R}^*$.*

We denote the set of such automorphic forms by $\mathcal{S}(B, R; A)$.

The double coset space $B^* \backslash \hat{B}^* / \hat{R}^*$ is in bijection with the set of right ideal classes of the order R, which is finite of cardinality h, say. Writing

$$\hat{B}^* = \coprod_{k=1}^{h} B^* b_i \hat{R}^*, \tag{9}$$

we see that an automorphic form $f \in \mathcal{S}(B, R; A)$ is completely determined by the finite sequence $(f(b_1), \ldots, f(b_h))$.

View B_p as a subring of $\hat{B}$ via the natural inclusion j_p. By the *strong approximation theorem*, $\hat{B}^* = B^* B_p^* \hat{R}^*$, so j_p induces a bijection

$$R[1/p]^* \backslash B_p / R_p^* \to B^* \backslash \hat{B}^* / \hat{R}^*.$$

Letting $\mathcal{S}(B_p, R_p; A)$ be the collection of functions $\varphi : R[1/p]^* \backslash B_p \to A$ such that $u\varphi(zbu) = f(b)$ for all $u \in R_p^*$ and $z \in \mathbb{Q}_p^*$, it is easy to see that j_p induces an isomorphism of $\mathcal{S}(B, R; A)$ with $\mathcal{S}(B_p, R_p; A)$. Since it shall be easier for us to work locally at p rather that adelically, we will work mostly with $\mathcal{S}(B_p, R_p; A)$.

The group $\mathcal{S}(B_p, R_p; A)$ is endowed with the action of a Hecke operator U_p given by

$$(U_p \varphi)(b) = \sum_{a=0}^{p-1} \left(\begin{pmatrix} p & a \\ 0 & 1 \end{pmatrix} \varphi \right)(b) = \sum_{a=0}^{p-1} \begin{pmatrix} p & a \\ 0 & 1 \end{pmatrix} \varphi \left(b \begin{pmatrix} p & a \\ 0 & 1 \end{pmatrix} \right). \tag{10}$$

When the action of $\Sigma_0(p)$ is trivial, an Atkin-Lehner involution W_p also acts on $\mathcal{S}(B_p, R_p; A)$ by the rule

$$W_p\varphi(b) = \varphi\left(b \begin{pmatrix} 0 & 1 \\ p & 0 \end{pmatrix}\right).$$

Other Hecke operators T_ℓ for $\ell \nmid N$ may also be defined using standard adelic formluas.

Automorphic forms whose coefficient module has the trivial $\Sigma_0(p)$-action "are" measures on $\mathbb{P}^1(\mathbb{Q}_p)$: Let $\mathcal{B}$ be the set of balls in $\mathbb{P}^1(\mathbb{Q}_p)$, on which $\mathrm{GL}_2(\mathbb{Q}_p)$ acts transitively from the left inducing an identification of $\mathrm{GL}_2(\mathbb{Q}_p)/\Gamma_0(p\mathbb{Z}_p)\mathbb{Q}_p^*$ with $\mathcal{B}$. Therefore, a form $\varphi \in \mathcal{S}(B_p, R_p; \mathbb{C}_p)$ may be viewed as a $R[1/p]^*$-invariant function on the balls in $\mathbb{P}^1(\mathbb{Q}_p)$. With this interpretation, the value of $U_p\varphi$ on a ball $\mathbf{b}$ is the sum of the values $\varphi(\mathbf{b}^{(i)})$ where the balls $\mathbf{b}^{(i)}$, $i = 1, \ldots, p$, form the standard subdivision of the ball $\mathbf{b}$. The value of $W_p\varphi$ on $\mathbf{b}$ is simply the value of φ on its complement $\mathbb{P}^1(\mathbb{Q}_p) - \mathbf{b}$. Suppose that $U_p\varphi = -W_p\varphi = a_p = \pm 1$. Define a function μ_φ on $\mathcal{B}$ by

$$\mu_\varphi(\gamma\mathbb{Z}_p) = \mathrm{sign}_p\gamma \cdot \varphi(\gamma), \quad \gamma \in B_p^*$$

where

$$\mathrm{sign}_p\gamma = a_p^{\mathrm{ord}_p \det \gamma}. \tag{11}$$

Then μ_φ is a $\mathcal{G}$-invariant measure on $\mathbb{P}^1(\mathbb{Q}_p)$ of total measure zero, where $\mathcal{G} := \ker(\mathrm{sign}_p : R[1/p]^* \to \{\pm 1\})$. Note that if $a_p = 1$, then $\mu_\varphi = \varphi$ and $\mathcal{G} = R[1/p]^*$.

The left action of $\Sigma_0(p)$ on $\mathbb{P}^1(\mathbb{Q}_p)$ induces a right action of $\Sigma_0(p)$ on $\mathbf{A}_{\mathrm{rig}}$. The space $\mathbf{D}_{\mathrm{rig}}$ inherits a left action of $\Sigma_0(p)$ by duality. The spaces $\mathbf{D}_{\mathrm{rig}}^\circ$ and $F^N\mathbf{D}_{\mathrm{rig}}^\circ$, $N \geq 1$ are all easily seen to be $\Sigma_0(p)$-stable. Therefore, the approximation modules $A^N\mathbf{D}_{\mathrm{rig}}$ inherit a $\Sigma_0(p)$-action. Consequently, these modules are all valid coefficient groups for p-adic automorphic forms. We shall refer to elements of $\mathcal{S}(B_p, R_p; \mathbf{D}_{\mathrm{rig}})$ as *rigid analytic automorphic forms*.

5 p-Adic Uniformization

Let $\Gamma_{N+,N-}^{(p)}$ denote the image under ι_p of the elements of $R[1/p]$ of reduced norm 1. The group $\Gamma_{N+,N-}^{(p)}$ acts discontinuously on $\mathfrak{H}_p$ and the quotient $\Gamma_{N+,N-}^{(p)} \backslash \mathfrak{H}_p$, has the structure of a rigid analytic curve $X_{N+,N-}^{(p)}$. The following result, due to Cerednik and Drinfeld, connects this rigid variety with the Shimura curves introduced in § 1.

Theorem 1 ([2,6]). *There is a canonical rigid analytic isomorphism*

$$\mathrm{CD} : X_{N+,N-}^{(p)}(\mathbb{C}_p) \to X_{N+,N-}(\mathbb{C}_p).$$

Let Ω denote the global sections of the sheaf of rigid analytic differential 1-forms on $X_{N+,N-}^{(p)}$.

Proposition 1. *The spaces Ω and $\mathcal{S}(B_p, R_p; \mathbb{C}_p)$ are naturally isomorphic as Hecke-modules.*

(A p-adic residue map and Teitelbaum's p-adic Poisson integral give the mutually inverse isomorphisms proving the theorem. For details, see [3, Ch. 5].) This proposition, together with the Jacquet-Langlands correspondence as invoked in §1, give the following corollary:

Corollary 1. *Choosing an isomorphism of $\mathbb{C}$ with $\mathbb{C}_p$, there is an isomorphism of Hecke-modules*

$$S_2(\Gamma_0(N))^{new\text{-}N^-} \cong \mathcal{S}(B_p, R_p; \mathbb{C}_p).$$

Let $E/\mathbb{Q}$ be an elliptic curve of conductor N and f_E the associated newform. Then by Corollary 1, there is a corresponding form $\varphi_E \in \mathcal{S}(B_p, R_p; \mathbb{C}_p)$ with the same Hecke-eigenvalues as f_E. In fact, we may (and do) assume that φ takes values in $\mathbb{Z}$. Let $\mu_E = \mu_{\varphi_E}$ be the associated measure on $\mathbb{P}^1(\mathbb{Q}_p)$ as constructed in §4.

Consider the map $\Psi : \mathrm{Div}^0\, \mathfrak{H}_p \to \mathbb{C}_p$ given by

$$\Psi(\tau' - \tau) = \oint_\tau^{\tau'} \omega_{\mu_E}.$$

Let Tate : $\mathbb{C}_p^* \to E(\mathbb{C}_p)$ be the Tate parametrization of E and recall the map Φ_{N^+, N^-} of (2). Assume that E is the strong Weil curve for (2) at the cost of replacing it by an isogenous curve.

Proposition 2. *The following diagram is commutative:*

$$
\begin{array}{ccc}
\mathrm{Div}^0\, \mathfrak{H}_p & \xrightarrow{\ \Psi\ } & \mathbb{C}_p^* \\
{\scriptstyle CD}\Big\downarrow & & \Big\downarrow{\scriptstyle Tate} \\
J_{N^+, N^-}(\mathbb{C}_p) & \xrightarrow[\Phi_{N^+, N^-}]{} & E(\mathbb{C}_p)
\end{array}
$$

For a discussion of this result, see [1].

6   A p-Adic Integral Formula for Heegner Points

Let K be an imaginary quadratic field satisfying the Shimura-Heegner hypothesis and $\mathfrak{o}$ be an order in K of conductor prime to N. Let us call an embedding ψ of $\mathfrak{o}[1/p]$ into $R[1/p]$ *optimal* if it does not extend to an embedding of a larger $\mathbb{Z}[1/p]$-order of K. Denote by $\mathcal{E}_p(\mathfrak{o})$ the set of all such embeddings. The Shimura-Heegner hypothesis guarantees that $\mathcal{E}_p(\mathfrak{o})$ is nonempty. For each $\psi \in \mathcal{E}_p(\mathfrak{o})$, the order $\mathfrak{o}[1/p]$ acts on $\mathfrak{H}_p$ via the composite $\iota_p \circ \psi$ with a unique fixed point $\tau_\psi \in \mathfrak{H}_p$ satisfying

$$\alpha \begin{pmatrix} \tau_\psi \\ 1 \end{pmatrix} = \psi(\alpha) \begin{pmatrix} \tau_\psi \\ 1 \end{pmatrix}$$

for all $\alpha \in \mathfrak{o}[1/p]$. Let $\mathfrak{H}_p(\mathfrak{o})$ be the set of all such τ_ψ, and let $\mathrm{CM}_p(\mathfrak{o})$ be its image in $X_{N^+,N^-}^{(p)}$. The set $\mathrm{CM}_p(\mathfrak{o})$ is endowed with a natural action of $\mathrm{Pic}\,\mathfrak{o} = \mathrm{Pic}\,\mathfrak{o}[1/p]$ (see [9]). The sets $\mathrm{CM}(\mathfrak{o})$ and $\mathrm{CM}_p(\mathfrak{o})$ are related through Theorem 1 as follows:

Theorem 2 ([1, Proposition 4.15]). *The map* CD *of Theorem 1 restricts to a* $\mathrm{Pic}\,\mathfrak{o}$*-equivariant bijection from* $\mathrm{CM}_p(\mathfrak{o})$ *onto* $\mathrm{CM}(\mathfrak{o})$.

Combining this theorem with Proposition 2, we see that module of Shimura-Heegner points on E defined over the ring class field attached to $\mathfrak{o}$ is generated by points of the form $\mathrm{Tate}(J(\tau, \tau'))$, $\tau, \tau' \in \mathfrak{H}_p(\mathfrak{o})$, where

$$J(\tau, \tau') = \fint_{\mathbb{P}^1(\mathbb{Q}_p)} \frac{x - \tau'}{x - \tau} d\mu_E(x) \tag{12}$$

7 Computing $J(\tau, \tau')$

Fix a positive integer M and suppose that we wish to compute the quantity $J(\tau, \tau')$ to a precision of p^{-M}. Let

$$M' = \max\{n : \mathrm{ord}_p(p^n/n) < M\}, \quad M'' = M + \left\lfloor \frac{\log M'}{\log p} \right\rfloor. \tag{13}$$

Assume that we may lift the automorphic form φ_E attached to E to a U_p-eigenform $\Phi_E \in \mathcal{S}(B_p, R_p; \mathbf{D}_{\mathrm{rig}}^\circ)$. That such a lift exists and is unique will be proved in §8.1. Moreover, we will see that the disribution $\Phi(1)$ is just $\mu_E|_{\mathbb{Z}_p}$, the restriction of the measure μ_E to the set of compact open subsets of $\mathbb{Z}_p$. Let $\Phi_E^{M''}$ denote the natural image of Φ_E in $\mathcal{S}(B_p, R_p; A^N \mathbf{D}_{\mathrm{rig}}^\circ)$. The goal of this section is to show that knowledge of $\Phi_E^{M''}$ is sufficient to allow for the calculation of $J(\tau, \tau')$ to a precision of p^{-M}. In §8.2, we shall give an algorithm for computing $\Phi_E^{M''}$.

It is easy to see that the points of $\mathfrak{H}_p(\mathfrak{o})$ actually lie in the subset $\mathbb{P}^1(\mathbb{Q}_{p^2}) - \mathbb{P}^1(\mathbb{Q}_p)$ of $\mathfrak{H}_p$, where $\mathbb{Q}_{p^2}$ is the quadratic unramified extension of $\mathbb{Q}_p$. Let $\mathfrak{H}_p^0$ be the set of elements τ in $\mathfrak{H}_p$ whose image under the natural reduction map $\mathbb{P}^1(\mathbb{C}_p) \to \mathbb{P}^1(\bar{\mathbb{F}}_p)$ does not belong to $\mathbb{P}^1(\mathbb{F}_p)$. We assume, without loss of generality, that:

1. τ and τ' reduce to elements of $\mathfrak{H}_p^0$.
2. there exists an element $i \in R[1/p]$ such that $i^2 = -1$.

By assumption 2., we may choose the isomorphism ι_p such that $\iota_p(i) = \begin{pmatrix} 0 & -1 \\ 1 & 0 \end{pmatrix}$. (Instead of assuming the existence of such an i, one could work with the two measures μ_E and $\begin{pmatrix} 0 & -1 \\ 1 & 0 \end{pmatrix}\mu_E$, and thus no generality is lost.)

By the first assumption above, $J(\tau, \tau')$ lies in $\mathbb{Z}_{p^2}^*$ and its Teichmüller representative is the same as that of

$$\prod_{a=0}^{p-1} \left(\frac{a - \tau'}{a - \tau} \right)^{\mu_E(a + p\mathbb{Z}_p)},$$

an easily computed quantity (actually, we need only compute it modulo p). Therefore, it suffices to compute $\log J(\tau, \tau')$.

Write

$$\log J(\tau, \tau') = \sum_{a \in \mathbb{P}^1(\mathbb{F}_p)} \log J_a(\tau, \tau'), \quad \text{where}$$

$$J_a(\tau, \tau') = \oint_{\mathbf{b}_a} \frac{x - \tau}{x - \tau'} d\mu_E(x),$$

and $\mathbf{b}_a$ is the standard residue disk around a. Let

$$J_\infty(\tau) = \oint_{\mathbf{b}_0} (1 + \tau x) d\mu_E(x),$$

$$J_a(\tau) = \oint_{\mathbf{b}_a} (x - \tau) d\mu_E(x), \quad 0 \le a \le p - 1.$$

Then for each $a \in \mathbb{P}^1(\mathbb{F}_p)$, we have

$$J_a(\tau, \tau') = J_a(\tau') / J(\tau).$$

(To prove the above for $a = \infty$, we use assumption 2.)

Straightforward manipulations (see [5, §1.3]) show that the expansions

$$\log J_\infty(\tau) = \sum_{n \ge 1} \frac{(-1)^n}{n} \tau^n \omega(0, n), \tag{14}$$

$$\log J_a(\tau) = \sum_{n \ge 1} \frac{1}{n(a - \tau)^n} \omega(a, n), \quad 0 \le a \le p - 1. \tag{15}$$

are valid, where (following the notation of [5]),

$$\omega(a, n) = \int_{\mathbf{b}_a} (x - a)^n d\mu_E(x), \quad 0 \le a \le p - 1.$$

An examination of formulas (14) and (15) shows that they may be computed to a precision of p^{-M} given the data

$$\omega(a, n) \pmod{p^{M''}}, \quad 0 \le a \le p - 1, \quad 0 \le n \le M'. \tag{16}$$

Proposition 3. *Let* $\Psi \in \mathcal{S}(B_p, R_p; \mathbf{D}_{rig})$ *be a* U_p-*eigenform with eigenvalue* $a_p = \pm 1$. *Then the formula*

$$\int_{a + p\mathbb{Z}_p} (x - a)^n d\Psi(b)(x) = a_p p^n \int_{\mathbb{Z}_p} x^n d\Psi \left(b \begin{pmatrix} p & a \\ 0 & 1 \end{pmatrix} \right)(x).$$

holds for $1 \le a \le p - 1$. *Consequently, the data (16) may be extracted from* $\Phi_E^{M''}$.

Proof.

$$\int_{a+p\mathbb{Z}_p} (x-a)^n d\Psi(b)(x)$$

$$= (a_p U_p \Psi)(b)((x-a)^n \mathbf{1}_{a+p\mathbb{Z}_p}(x))$$

$$= a_p \sum_{d=0}^{p-1} \Psi\left(b\begin{pmatrix} p & d \\ 0 & 1 \end{pmatrix}\right)((d+px-a)^n \mathbf{1}_{a+p\mathbb{Z}_p}(d+px))$$

$$= a_p p^n \Psi\left(b\begin{pmatrix} p & a \\ 0 & 1 \end{pmatrix}\right)(x^n)$$

$$= a_p p^n \int_{\mathbb{Z}_p} x^n d\Psi\left(b\begin{pmatrix} p & a \\ 0 & 1 \end{pmatrix}\right),$$

as desired.

To prove the second statement, take $\Psi = \Phi_E$, $b = 1$ and use the fact that $\Phi_E(1) = \mu_E|_{\mathbb{Z}_p}$. By the definition of $\Phi_E^{M''}$,

$$\Phi_E\left(\begin{pmatrix} p & a \\ 0 & 1 \end{pmatrix}\right)(x^n) \pmod{p^{M''-n}} = \Phi_E^{M''}\left(\begin{pmatrix} p & a \\ 0 & 1 \end{pmatrix}\right)(x^n)$$

for $0 \le a \le p-1$ and $0 \le n \le M''$. Now multiply the above by p^n and apply the first statement of the proposition, noting that $M'' \ge M'$.

8 Lifting U_p-Eigenforms

8.1 Existence and Uniqueness of Lifts

Define the *specialization map*

$$\rho : \mathcal{S}(B_p, R_p; \mathbf{D}_{\mathrm{rig}}) \to \mathcal{S}(B_p, R_p; \mathbb{Q}_p)$$

by the rule $\rho(\Phi)(b) = \Phi(b)(\mathbf{1}_{\mathbb{Z}_p})$. It is easily verified that ρ is $U_{\mathfrak{p}}$-equivariant. Let $\varphi \in \mathcal{S}(B_p, R_p; \mathbb{Q}_p)$ be a U_p-eigenform with eigenvalue $a_p = \pm 1$ and let μ_φ be the associated measure on $\mathbb{P}^1(\mathbb{Q}_p)$ as constructed in §4. The following proposition should be viewed as an analogue of the containment of classical modular forms in the space of p-adic modular forms.

Proposition 4. *The form φ lifts canonically with respect to ρ to a U_p-eigenform Φ satisfying $\Phi(1) = \mu_\varphi$.*

Proof. Define $\Psi : B_{\mathfrak{p}} \to \mathbf{D}(\mathbb{P}^1(\mathbb{Q}_p), \mathbb{Q}_p)$ by $\Psi(b) = (\mathrm{sign}_p b)b^{-1}\varphi$ where $\mathrm{sign}_p b$ is as defined in (11), and let $\Phi : B_p^* \to \mathbf{D}_{\mathrm{rig}}$ be given by $\Phi(b) = \Psi(b)|_{\mathbb{Z}_p}$. The conclusions of the proposition are now easily verified.

The next proposition forms the basis of our algorithm.

Proposition 5. *Let Ψ belong to* $\ker \rho \cap \mathcal{S}(B_p, R_p; F^N \mathbf{D}^\circ_{rig})$. *Then*

$$U_p \Psi \in \ker \rho \cap \mathcal{S}(B_p, R_p; F^{N+1}\mathbf{D}^\circ_{rig}).$$

Proof. By the U_p-equivariance of ρ, its kernel is certainly U_p-stable. For $1 \leq n \leq N$, we have

$$U_p\Psi(b)(x^n) = \sum_{a=0}^{p-1} \begin{pmatrix} p & a \\ 0 & 1 \end{pmatrix} \Psi\left(b \begin{pmatrix} p & a \\ 0 & 1 \end{pmatrix}\right)(x^n)$$

$$= \sum_{a=0}^{p-1} \Psi\left(b \begin{pmatrix} p & a \\ 0 & 1 \end{pmatrix}\right)((px+a)^n)$$

$$= \sum_{k=0}^{n} \sum_{a=0}^{p-1} \binom{n}{k} p^k a^{n-k} \Psi\left(b \begin{pmatrix} p & a \\ 0 & 1 \end{pmatrix}\right)(x^k).$$

Note that the $k = 0$ term in the above sum vanishes as $\Psi \in \ker \rho$. If $1 \leq k \leq n$ and $0 \leq a \leq p-1$, then

$$\binom{n}{k} p^k a^{n-k} \Psi\left(b \begin{pmatrix} p & a \\ 0 & 1 \end{pmatrix}\right)(x^k) \in p^k p^{N-k} \mathbb{Z}_p = p^N \mathbb{Z}_p \subset p^{N+1-n}\mathbb{Z}_p.$$

The result follows.

Let $\Phi \in \mathcal{S}(B_p, R_p; \mathbf{D}_{rig})$ be the lift of φ constructed in Proposition 4. Since the double-coset space $R[1/p]^* \backslash B_p / R_p$ is finite, we may assume without loss of generality that all moments involved are actually in $\mathbb{Z}_p$ (just multiply φ, Φ, and Φ_0 by a suitably chosen scalar $c \in \mathbb{Q}_p$). Let Φ^N be the natural image of Φ in $\mathcal{S}(B_p, R_p; A^N \mathbf{D}^\circ_{rig})$.

Corollary 2

1. (a) Φ^N *is the unique U_p-eigenform in* $\mathcal{S}(B_p, R_p; A^N \mathbf{D}^\circ_{rig})$ *lifting φ.*
   (b) *If Φ^N_0 is any element of* $\mathcal{S}(B_p, R_p; A^N \mathbf{D}^\circ_{rig})$ *lifting φ, then*

$$(a_p U_p)^N \Phi^N_0 = \Phi^N.$$

2. (a) Φ *is the unique U_p-eigenform in* $\mathcal{S}(B_p, R_p; \mathbf{D}_{rig})$ *satisfying $\rho(\Phi) = \varphi$.*
   (b) *If Φ_0 is any element of* $\mathcal{S}(B_p, R_p; \mathbf{D}_{rig})$ *satisfying $\rho(\Phi_0) = \varphi$, then the sequence $\{(a_p U_p)^n \Phi_0\}$ converges to Φ.*

Proof. Statement (2) follows from statement (1) and the relation

$$\mathcal{S}(B_p, R_p; \mathbf{D}_{rig}) = \left(\varprojlim_N \mathcal{S}(B_p, R_p; A^N \mathbf{D}^\circ_{rig})\right) \otimes_{\mathbb{Z}_p} \mathbb{Q}_p.$$

By the above Proposition 5, we have

$$(a_p U_p)^N (\Phi - \Phi_0) \in \mathcal{S}(B_p, R_p; F^N \mathbf{D}^\circ_{rig}).$$

Statement (1) now follows easily.

The unicity result 2(a) of the corollary may viewed as an analogue of assertion that ordinary p-adic modular eigenforms are classical.

By Corollary 2, in order to approximate the moments of $\Phi(b)$ for $b \in B_p^*$, it suffices to produce an initial approximation Φ_0 to Φ and then to apply the U_p-operator repeatedly until the desired accuracy is achieved. Such an initial approximation may be constructed explicitly as follows: Using the decomposition (9), let $S_k = R_p^* \cap b_k^{-1} R[1/p]^* b_k$, which is finite as it is contained in $b_k^{-1} R^* b_k$. (The group of units of a $\mathbb{Z}$-order in a definite quaternion algebra over $\mathbb{Q}$ is finite.) For $z \in \mathbb{Z}_p$, let $\delta_z \in \mathbf{D}_{\mathrm{rig}}$ be the Dirac distribution centred at z, i.e. $\delta_z(f) = f(z)$.

Proposition 6. *There is a unique element Φ_0 of $\mathcal{S}(B_p, R_p; \mathbf{D}_{rig})$ satisfying*

$$\Phi_0(b_k) = \frac{\varphi(b_k)}{\#S_k} \sum_{v \in S_k} v^{-1}\delta_0, \quad 1 \le k \le h. \tag{17}$$

Its moments are given by

$$\Phi(b_k)(x^n) = \frac{\varphi(b_k)}{\#S_k} \sum_{v \in S_k} z_v^n, \quad \text{where} \quad z_v = v \cdot 0.$$

(By $v \cdot 0$ we mean the image of v in $\mathrm{GL}_2(\mathbb{Q}_p)$ acting as a fractional linear transformation on $0 \in \mathbb{P}^1(\mathbb{Q}_p)$.)

Proof. To see that (17) gives a well defined element of $\mathcal{S}(B_p, R_p; \mathbf{D}_{\mathrm{rig}})$, notice that if $\gamma b_k u = b_k$, then v varies over S_k if and only if vu does. The uniqueness is clear.

8.2 Computing the Lifts in Practice

We now turn to the problem of computing Φ^N in practice. Representing an element of $\mathcal{S}(B_p, R_p; A^N \mathbf{D}_{\mathrm{rig}}^\circ)$ is straight-forward. First observe that the correspondence

$$\mu \mapsto (\mu(x^0) \ (\mathrm{mod}\, p^N), \mu(x^1) \ (\mathrm{mod}\, p^{N-1}), \dots, \mu(x^{N-1}) \ (\mathrm{mod}\, p)),$$

for μ in $\mathbf{D}_{\mathrm{rig}}^\circ$, descends to an isomorphism

$$A^N \mathbf{D}_{\mathrm{rig}} \cong \mathbb{Z}/p^N \mathbb{Z} \times \mathbb{Z}/p^{N-1}\mathbb{Z} \times \cdots \times \mathbb{Z}/p\mathbb{Z}.$$

Therefore, an element of $A^N \mathbf{D}_{\mathrm{rig}}^\circ$ may be stored simply as an N-tuple of integers.

The $\Sigma_0(p)$-action on $A^N \mathbf{D}_{\mathrm{rig}}^\circ$ may be computed as follows: Let μ be in $\mathbf{D}_{\mathrm{rig}}$ and let ν be any lift of μ to $\mathbf{D}_{\mathrm{rig}}^\circ$. For any $u = \left(\begin{smallmatrix} a & b \\ c & d \end{smallmatrix}\right) \in \Sigma_0$ and $n \ge 0$, the rational function $\left(\frac{ax+b}{cx+d}\right)^n$ may be expanded in a Taylor series $\sum \alpha_m x^m$, and the moments of $u\nu$ may be computed by "integrating term by term":

$$(u\nu)(x^n) = \nu\left(\left(\frac{ax+b}{cx+d}\right)^n\right) = \sum_{m \ge 0} \alpha_m \nu(x^m).$$

Moreover, by the stability of $F^N \mathbf{D}^\circ_{\mathrm{rig}}$ under $\Sigma_0(p)$, the image of $u\nu$ is $u\mu$. There-fore, the N-tuple representing $u\mu$ may be computed from that representing μ.

Recall the double-coset decomposition (9). Then since an automorphic form $\Psi \in \mathcal{S}(B_p, R_p; A^N \mathbf{D}^\circ_{\mathrm{rig}})$ is completely determined by $\Psi(b_1), \ldots, \Psi(b_h)$, it may be stored simply as a sequence of h N-tuples of integers. Assuming knowledge of the values of φ, the moments of the initial lift Φ_0^N of φ constructed explicitly in Proposition 6 may be computed and thus Φ_0^N may be stored as a sequence of N-tuples as described above.

It remains to describe how, for a form Ψ as above, the data

$$((U_p\Psi)(b_k)(x^0) \ (\mathrm{mod}\, p^N), \ldots, (U_p\Psi)(b_k)(x^{N-1}) \ (\mathrm{mod}\, p)), \quad 1 \le k \le h$$

may be computed from the corresponding data for Ψ. For $1 \le k \le h$ and $0 \le a \le p-1$, find elements $\gamma(k,a) \in R[1/p]^*$, $u(k,a) \in R_p^*$, and $j(k,a) \in \{0,\ldots,p-1\}$ such that

$$b_k \begin{pmatrix} p & a \\ 0 & 1 \end{pmatrix} = \gamma(k,a) b_{j(k,a)} u(k,a), \tag{18}$$

and let $\xi(k,a) = \begin{pmatrix} p & a \\ 0 & 1 \end{pmatrix} u(k,a)^{-1}$. Then $U_p\Psi$ is given by the formula

$$(U_p\Psi)(b_k) = \sum_{a=0}^{p-1} \xi(k,a)\Psi(b_{j(k,a)}). \tag{19}$$

The measures $\Psi(b_{j(k,a)})$ are assumed to be known and the action of the $\xi(k,a)$ on them may be computed as described above. Thus, an algorithm for computing Φ^N from φ may proceed as follows:

1. Compute the elements $\gamma(k,a)$, $j(k,a)$, and $u(k,a)$ as in (18).
2. Compute an initial lift Φ_0^N of φ to $\mathcal{S}(B_p, R_p; A^N \mathbf{D}^\circ_{\mathrm{rig}})$ as in Proposition 6.
3. Compute $(a_p U_p)^N \Phi_0^N$. By Corollary 2, the result is Φ^N.

An operations count similar to that performed in [5, Proposition 2.14] yields the following:

Proposition 7. *The above procedure computes the symbol Φ^N in $O(N^3 p^3 \log N)$ arithmetic operations on integers with size on the order of p^N.*

Remark 1. In [5], the analogous computation with modular symbols requires $O(N^3 p^3 \log N \log p)$ operations. The factor of $\log p$ is absent in our version be-cause we do not need to apply Manin's continued fraction algorithm.

9   Examples

Example 1. Consider the elliptic curve

$$E: \ y^2 + xy + y = x^3 + x^2 - 70x - 279 \qquad (38B2).$$

For this curve, $N^+ = 1$, $N^- = 2 \cdot 19$, $p = 19$, and B is the algebra of rational Hamilton quaternions. The field $K = \mathbb{Q}(\xi)$, where $\xi = (1 + \sqrt{-195})/2$, satisfies the Shimura-Heegner hypothesis. Let $\mathfrak{o} = \mathbb{Z}[\xi]$ be its maximal order. The class number of K is 4 and $\mathrm{Pic}\,\mathfrak{o} \cong (\mathbb{Z}/2\mathbb{Z})^2$. Therefore, $\mathrm{Pic}\,\mathfrak{o}$ has three characters χ_1, χ_2, χ_3 of exact order 2. Let $\tau \in \mathfrak{H}_p(\mathfrak{o})$ be a base point and define divisors

$$\mathfrak{d}_i = \sum_{\alpha \in \mathrm{Pic}\,\mathfrak{o}} \chi_i(\alpha)\tau^\alpha \in \mathrm{Div}^0\,\mathfrak{H}_p(\mathfrak{o}), \quad i = 1, 2, 3.$$

Define a divisor $\mathfrak{d}_0$ (corresponding to the trivial character) by

$$\mathfrak{d}_0 = \sum_{\alpha \in \mathrm{Pic}\,\mathfrak{o}} ((3 + 1 - T_3)\tau)^\alpha$$

where T_3 is the standard Hecke operator. Let

$$P_i = \mathrm{Tate}\left(\oint_{\mathfrak{d}_i} \omega_{\mu_E}\right), \quad , i = 0, 1, 2, 3.$$

be the corresponding Heegner points. We computed the points P_i as described above and these points were recognized (using the LLL-based algorithm described in [5, §1.6]) as

$$P_0 = (-4610/39, 1/1521(-277799\xi + 228034)),$$
$$P_1 = (25/12, -94/9u + 265/72),$$
$$P_2 = (10, -11v),$$
$$P_3 = (1928695/2548, 1/463736(-2397574904w + 1023044339)),$$

where $\quad u = \dfrac{1 + \sqrt{-15}}{2}, \quad v = \dfrac{1 + \sqrt{5}}{2} \quad$ and $\quad w = \dfrac{1 + \sqrt{65}}{2}.$

We remark that $K(u, v, w)$ is in fact the Hilbert class field of K and that $K(u)$, $K(v)$ and $K(w)$ are the fields corresponding to the characters χ_1, χ_2 and χ_3, respectively.

Example 2. Let $\omega = (1 + \sqrt{5})/2$ and let $F = \mathbb{Q}(\omega)$. Consider the elliptic curve

$$E: \ y^2 + xy + \omega y = x^3 - (\omega + 1)x^2 - (30\omega + 45)x - (11\omega + 117)$$

defined over F. The conductor of E is $\mathfrak{p} = (3 - 5\omega)$, a degree one prime of F dividing 31. Here, $N = N^- = \mathfrak{p}$, $N^+ = 1$, and B is the base change to F of the $\mathbb{Q}$-algebra of Hamilton's quaternions. Let K be the CM field $F(\sqrt{2\omega - 15})$. The class group of K is cyclic of order 8 and thus has a unique character χ of exact order 2 and corresponding field $K(\sqrt{-13\omega + 2})$. Let $\tau \in \mathfrak{H}_p(\mathfrak{o})$ be a base point, define a divisor $\mathfrak{d}_\chi$ attached to χ as in Example 1, and let P_χ be the corresponding Heegner point. Our computations yielded a point recognizable (using a higher dimensional variant of the LLL-based algorithm of [5, §1.6]; see also [12]) as the point $(x, y) \in E(F(\sqrt{-13\omega + 2}))$, where

$$x = 1/501689727224078580 \times (-20489329712955302181\omega +$$
$$1590697243182535465)$$
$$y = 1/794580338951539798133856600 \times$$
$$(-2430756213639475197971343 8023\omega -$$
$$5224406254275398040668003 6861)\sqrt{-13\omega + 2} +$$
$$1/1003379454448157160 \times (19987639985731223601\omega$$
$$- 1590697243182535465).$$

References

1. M. Bertolini, H. Darmon, *Heegner points, p-adic L-functions and the Cerednik-Drinfeld uniformization*, Invent. Math. 126 (1996) 413-456.
2. I. Cerednik, *Uniformization of algebraic curves by discrete arithmetic subgroups of* $PGL_2(k_w)$, Math. Sbornik 100 (1976) 59-88.
3. H. Darmon, *Rational points on modular elliptic curves*, CBMS Regional Conference Series in Mathematics, 101. Published for the Conference Board of the Mathematical Sciences, Washington, DC; by the American Mathematical Society, Providence, RI, 2004.
4. H. Darmon, P. Green, *Elliptic curves and class fields of real quadratic fields: algorithms and evidence*, Experimental Mathematics, 11:1 (2002) 37-55.
5. H. Darmon, R. Pollack, *The efficient calculation of Stark-Heegner points via overconvergent modular symbols*, to appear in Israel Journal of Mathematics.
6. V. Drinfeld, *Coverings of p-adic symmetric regions*, Funct. Anal. Appl. 10 (1976) 29-40.
7. N. Elkies, *Shimura curve computations*, In Algorithmic number theory (Ithaca, NY, 1994) Lecture Notes in Comput. Sci., 877, Springer, Berlin, 1994, 122133.
8. M. Greenberg, *Heegner points and rigid analytic modular forms*, PhD thesis, McGill University, In progress.
9. B. H. Gross, *Heights and special values of L-series*, CMS Conference Proc., H. Kisilevsky and J. Labute, eds., Vol. 7 (1987).
10. B. Mazur, P. Swinnerton-Dyer, *Arithmetic of Weil curves*, Invent. Math. 25 (1974) no. 1, 1-61.
11. R. Pollack, G. Stevens, *Computations with overconvergent modular symbols*, in preparation.
12. M. Trifković, *Stark-Heegner points on elliptic curves defined over imaginary quadratic fields*, Submitted.
13. M.F. Vignéras, *Arithmétique des algèbres de quaternions*, Lecture Notes in Mathematics 800, Springer, Berlin, 1980.
14. S. Zhang, *Heights of Heegner points on Shimura curves*, Ann. of Math. (2) 153 (2001), no. 1, 27-147.

Symmetric Powers of Elliptic Curve L-Functions

Phil Martin and Mark Watkins*

University of Bristol
phil_martin_uk@hotmail.com
watkins@maths.usyd.edu.au

Abstract. The conjectures of Deligne, Beĭlinson, and Bloch-Kato assert that there should be relations between the arithmetic of algebro-geometric objects and the special values of their L-functions. We make a numerical study for symmetric power L-functions of elliptic curves, obtaining data about the validity of their functional equations, frequency of vanishing of central values, and divisibility of Bloch-Kato quotients.

1 Introduction and Motivation

There are many conjectures that relate special values of L-functions to the arithmetic of algebro-geometric objects. The celebrated result $\zeta(2) = \pi^2/6$ of Euler [20, §XV] can be reinterpreted as such, but Dirichlet's class number formula [15, §5] is better seen to be the primordial example. Modern examples run the gamut, from conjectures of Stark [43] on Artin L-functions and class field theory, to that of Birch and Swinnerton-Dyer [2] for elliptic curves, to those of Beĭlinson [1,35] related to K-theory, with a passel of others we do not mention. For maximal generality the language of motives is usually used (see [21, §1-4]).

One key consideration is where the special value is taken. The L-function can only vanish inside the critical strip or at trivial zeros; indeed, central values (at the center of symmetry of the functional equation) are the most interesting ones that can vanish, and the order of vanishing is likely related to the rank of a geometric object (note that orders of trivial zeros can be similarly interpreted).

We have chosen to explore a specific family of examples, namely symmetric power L-functions for elliptic curves over $\mathbf{Q}$. The impetus for our work was largely a result [19] of the first author, whose computation of Euler factors in the difficult case of additive primes greatly reduced the amount of hassle needed to do large-scale computations. Previous work includes that of Coates and Schmidt [7] on the symmetric square (with computational data in [47]). and Buhler, Schoen, and Top on the symmetric cube [6]; this latter paper builds on the work of Garrett [23], Harris-Kudla [26], and Gross-Kudla [24] concerning triple products, and contains some numerical data. The computations we describe below are only valid if we assume various conjectures, such as the existence of a functional equation; even without such an assumption, we can take the "numerical coincidences" in our data to be evidence for these conjectures.

---

* Supported during parts of this research by EPSRC grants GR/N09176/01 and GR/T00658/01, the Isaac Newton Institute, the CNRS and the Institut Henri Poincaré, and the MAGMA Computer Algebra Group at the University of Sydney.

F. Hess, S. Pauli, and M. Pohst (Eds.): ANTS 2006, LNCS 4076, pp. 377–392, 2006.

2 L-Functions

We define the symmetric power L-functions of an elliptic curve $E/\mathbf{Q}$ via computing an Euler factor at every prime p. This Euler factor is computed by a process that essentially just takes the symmetric power representation of the standard 2-dimensional Galois representation associated to E, and thus our method is a generalisation of that used by Coates and Schmidt [7] for the symmetric square,[1] following the original description of Serre [39]. We briefly review the theoretical framework, and give explicit formulae for the Euler factors in a later section.

For every prime p choose an auxiliary prime $l \neq 2, p$ and fix an embedding of $\mathbf{Q}_l$ into $\mathbf{C}$. Let E_t denote the t-torsion of E, and $T_l(E) = \varprojlim E_{l^n}$ be the l-adic Tate module of E (we fix a basis). The module $V_l(E) = T_l(E) \otimes_{\mathbf{Z}_l} \mathbf{Q}_l$ has dimension 2 over $\mathbf{Q}_l$ and has a natural action of $\mathrm{Gal}(\overline{\mathbf{Q}}_p/\mathbf{Q}_p)$ [indeed one of of $\mathrm{Gal}(\overline{\mathbf{Q}}/\mathbf{Q})$], and from this we get a representation $\rho_l : \mathrm{Gal}(\overline{\mathbf{Q}}_p/\mathbf{Q}_p) \to \mathrm{Aut}(V_l)$. We write $H_l^1(E) = \mathrm{Hom}_{\mathbf{Q}_l}(V_l(E), \mathbf{Q}_l)$, and take the mth symmetric power of the contragredient of ρ_l, getting

$$\rho_l^m : \mathrm{Gal}(\overline{\mathbf{Q}}_p/\mathbf{Q}_p) \to \mathrm{Aut}\big(\mathrm{Sym}^m\big(H_l^1(E)\big)\big) \subset GL_{m+1}(\mathbf{C}).$$

We write $D_p = \mathrm{Gal}(\overline{\mathbf{Q}_p}/\mathbf{Q}_p)$, let I_p be the inertia group of this extension, and let Frob_p be the element of $D_p/I_p \cong \mathrm{Gal}(\overline{\mathbf{F}}_p/\mathbf{F}_p)$ given by $x \to x^p$. With all of this, we have

$$L(\mathrm{Sym}^m E, s) = \prod_p \det\Big[\mathrm{Id}_{m+1} - \rho_l^m(\mathrm{Frob}_p^{-1})p^{-s} \,\Big|\, \big(\mathrm{Sym}^m\big(H_l^1(E)\big)\big)^{I_p}\Big]^{-1}.$$

For brevity, we write $L_m(E, s) = L(\mathrm{Sym}^m E, s)$, and denote the factors on the right side by $U_m(p; s)$. As mentioned by Coates and Schmidt [7, p. 106], it can be shown that $U_m(p; s)$ is independent of our choices. The analytic theory and conjectures concerning these symmetric power L-functions are described in [42]. In particular, the above Euler product converges in a half-plane, and is conjectured to have a meromorphic continuation to the whole complex plane.

We also need the conductor N_m of this symmetric power representation. We have $N_m = \prod_p p^{f_m(p)}$ where $f_m(p) = \epsilon_m(I_p) + \delta_m(p)$. Here $\epsilon_m(I_p)$ is the codimension of $\big(\mathrm{Sym}^m\big(H_l^1(E)\big)\big)^{I_p}$ in $\mathrm{Sym}^m\big(H_l^1(E)\big)$; we shall see that it can be computed via a character-theoretic argument. The wild conductor $\delta_m(p)$ is 0 unless $p = 2, 3$, when it can be computed as in [39, §2.1] or the appendix of [7].

2.1   Critical Values

The work of Deligne [14, Prop. 7.7ff] tells us when and where to expect critical values; these are a subset of the more-general special values, and are the easiest

---

[1] Note that Buhler, Schoen, and Top [6] phrase their definition of Euler factors differently, as they emphasise that conjecturally the L-function is related to a motive or higher-dimensional variety; however, their definition is really the same as ours.

to consider.[2] When $m = 2v$ with v odd there is a critical value $L_m(E, v+1)$ at the edge of the critical strip, and when $m = 2u - 1$ is odd there is a critical central value $L_m(E, u)$. We let Ω_+, Ω_- be the real/imaginary periods of E for $m \equiv 1, 2 \pmod 4$, and vice-versa for $m \equiv 3 \pmod 4$. In the respective cases of m even/odd we expect rationality (likely with small denominator) of either

$$\frac{L_m(E, v+1)}{(2\pi)^{v+1}} \left(\frac{2\pi N}{\Omega_+ \Omega_-} \right)^{v(v+1)/2} \quad \text{or} \quad \frac{L_m(E, u)(2\pi N)^{u(u-1)/2}}{\Omega_+^{u(u+1)/2} \Omega_-^{u(u-1)/2}}. \tag{1}$$

When m is odd, the order of $L_m(E, s)$ at $s = u$ should equal the rank of an associated geometric object. The Bloch-Kato conjecture [4] relates the quotients in (1) to H^0-groups, Tamagawa numbers, and generalised Shafarevich-Tate groups.[3]

3   Computation of Euler Factors and Local Conductors

We first consider multiplicative and potentially multiplicative reduction for a given prime p; these cases can easily be detected since $v_p(j_E)$, the valuation of the j-invariant, is negative, with the reduction being potentially multiplicative when $p|c_4$. When E has multiplicative reduction, the filtration of [6, §8] implies the local tame conductor ϵ_m is m and $\delta_m(2) = \delta_m(3) = 0$ for all m. The Euler factor is $U_m(p; s) = (1 - a_p^m/p^s)^{-1}$, where $a_p = \pm 1$ is the trace of Frobenius. In the case of potentially multiplicative reduction, for m odd we have $\epsilon_m = m + 1$, and so $U_m(p; s) \equiv 1$, while with m even, we have that $\epsilon_m = m$ and compute that $U_m(p; s) = (1 - 1/p^s)^{-1}$. The wild conductor at $p = 2$ is $\delta_m(2) = \frac{m+1}{2}\delta_1(2)$ for odd m and is zero for even m, while $\delta_m(3) = 0$ for all m.

3.1   Good and Additive Reduction — Tame Conductors

Let E have good or potentially good reduction at a prime p, and choose an auxiliary prime $l \neq 2, p$. The inertia group I_p acts on $V_l(E)$ by a finite quotient in this case. Let $\mathbf{G}_p = \mathrm{Gal}(\mathbf{Q}_p(E_l)/\mathbf{Q}_p)$ and Φ_p be the inertia group of this extension.[4] The work of Serre [40] lists the possibilities for Φ_p. It can be a cyclic group C_d with $d = 1, 2, 3, 4, 6$; additionally, when $p = 2$ it can be Q_8 or $SL_2(\mathbf{F}_3)$, and when $p = 3$ it can be $C_3 \rtimes C_4$. For each group there is a unique faithful 2-dimensional representation Ψ_Φ of determinant 1 over $\mathbf{C}$, which determines $\bar{\rho}_l$.

Our result now only depends on Φ; for a representation Ψ we have the trace relation (which is related to Chebyshev polynomials of the second kind)

---

[2] Critical values conjecturally only depend on periods (which are local objects), while the more-general special values can also depend on (global) regulators from K-theory.

[3] See [18, §7] for an explicit example; note his imaginary period is twice that of our normalisation (and the formula is out by a power-of-2 in any case), and the conductor enters the formula in a different place (this doesn't matter for semistable curves).

[4] The group Φ_p is independent of the choice of l (see [40, p. 312]), while only whether $\mathbf{G}_p$ is abelian matters, and this independence follows as in [7, Lemmata 1.4 & 1.5].

$$\text{tr}(\text{Sym}^m \Psi) = \sum_{k=0}^{m/2} \binom{m-k}{k} \text{tr}(\Psi)^{m-2k} (-\det \Psi)^k, \tag{2}$$

and from taking the inner product of $\text{tr}(\text{Sym}^m \Psi_\Phi)$ with the trivial character we find the dimension of the Φ-fixed subspace of $\text{Sym}^m \big(H_l^1(E) \big)$, which we denote by $\beta_m(\Phi)$. Upon carrying out this calculation, we obtain Table 1, which lists values for $\beta_m(\Phi)$, from which we get the tame conductor $\epsilon_m(\Phi) = m+1-\beta_m(\Phi)$. The wild conductors $\delta_m(p)$ are 0 for $p \geq 5$, and for $p = 2, 3$ are described below.

Table 1. Values of $\beta_m(\Phi)$ for various inertia groups; here $\tilde{m}$ is m modulo 12

$\tilde{m}$	C_2	C_3	C_4	C_6	Q_8	$C_3 \rtimes C_4$	$SL_2(\mathbf{F}_3)$
0	$m+1$	$(m+3)/3$	$(m+2)/2$	$(m+3)/3$	$(m+4)/4$	$(m+6)/6$	$(m+12)/12$
1	0	$(m-1)/3$	0	0	0	0	0
2	$m+1$	$(m+1)/3$	$m/2$	$(m+1)/3$	$(m-2)/4$	$(m-2)/6$	$(m-2)/12$
3	0	$(m+3)/3$	0	0	0	0	0
4	$m+1$	$(m-1)/3$	$(m+2)/2$	$(m-1)/3$	$(m+4)/4$	$(m+2)/6$	$(m-4)/12$
5	0	$(m+1)/3$	0	0	0	0	0
6	$m+1$	$(m+3)/3$	$m/2$	$(m+3)/3$	$(m-2)/4$	$m/6$	$(m+6)/12$
7	0	$(m-1)/3$	0	0	0	0	0
8	$m+1$	$(m+1)/3$	$(m+2)/2$	$(m+1)/3$	$(m+4)/4$	$(m+4)/6$	$(m+4)/12$
9	0	$(m+3)/3$	0	0	0	0	0
10	$m+1$	$(m-1)/3$	$m/2$	$(m-1)/3$	$(m-2)/4$	$(m-4)/6$	$(m-10)/12$
11	0	$(m+1)/3$	0	0	0	0	0

3.2   Good and Additive Reduction — Euler Factors for $p \geq 5$

When $p \geq 5$, a result of Serre [40] tells us that the inertia group is $\Phi = C_d$ where $d = 12/\gcd\big(12, v_p(\Delta_E)\big)$. Note that this gives $d = 1$ when p is a prime of good reduction, which we naturally include in the results of this part. We summarise the results of Martin's work [19] concerning the Euler factors. Note that the result of Dąbrowski [10, Lemma 1.2.3] appears to be erroneous.

There are two different cases for the behaviour of the Euler factor, depending on whether the decomposition group $\mathbf{G}_p = \text{Gal}\big(\mathbf{Q}_p(E_l)/\mathbf{Q}_p\big)$ is abelian. From [36, Prop. 2.2] or [47, Th. 2.1], we get that this decomposition group is abelian precisely when $p \equiv 1 \pmod{d}$. When $\mathbf{G}_p$ is nonabelian we have

$$U_m(p; s) = (1 - (-p)^{m/2}/p^s)^{-A_m} (1 + (-p)^{m/2}/p^s)^{-B_m}, \tag{3}$$

where $A_m + B_m = \beta_m$ and A_m is the dimension of $\big(\text{Sym}^m \big(H_l^1(E) \big)\big)^{\mathbf{G}_p}$. Using $\mathbf{G}_p/\Phi_p \cong C_2$ and $\det\big(\Psi_\Phi(x)\big) = -1$ for $x \in \mathbf{G}_p \backslash \Phi_p$, more character calculations tell us this dimension is $(\beta_m+1)/2$ when β_m is odd and is $\beta_m/2$ when β_m is even. This also holds for the non-cyclic Φ when $p = 2, 3$, for which $\mathbf{G}_p$ is automatically nonabelian. When $\Phi = C_3$ and m is odd, we have $U_m(p; s) = (1 + p^m/p^{2s})^{-\beta_m/2}$.

When $\mathbf{G}_p$ is abelian, we need to compute a Frobenius eigenvalue α_p (whose existence follows from [41, p. 499]). In the case of good reduction, this comes from counting points mod p on the elliptic curve; we have $\alpha_p = (a_p/2) \pm i\sqrt{p - a_p^2/4}$ where $p + 1 - a_p$ is the number of (projective) points on E modulo p. And when $\Phi = C_2$ we count points on the pth quadratic twist of E. In general, we need to re-scale the coefficients of our curve by some power of p that depends on the valuations v_p of the coefficients. Since $p \geq 5$, we can write our curve as $y^2 = x^3 + Ax + B$, and then re-scale by a factor $t = p^{\min(v_p(A)/2, v_p(B)/3)}$ to get a new curve $E^t : y^2 = x^3 + Ax/t^2 + B/t^3$, possibly defined over some larger field. Because of our choice of t, at least one of A/t^2 and B/t^3 will have v_p equal to 0. The reduction $\tilde{E}^t$ modulo some (fractional) power of p is then well-defined and non-singular, and we get α_p from counting points on $\tilde{E}^t$; it turns out that choices of roots of unity will not matter when we take various symmetric powers. Returning back to $U_m(p; s)$, we get that when $\mathbf{Q}_p(E_l)/\mathbf{Q}_p$ is abelian this Euler factor is

$$U_m(p; s) = \prod_{\substack{0 \leq i \leq m \\ d | (2i - m)}} (1 - \alpha_p^{m-i} \bar{\alpha}_p^i / p^s)^{-1}. \tag{4}$$

3.3 Considerations When $p = 3$

Next we consider good and additive reduction for $p = 3$. We first determine the inertia group, using the 3-valuation of the conductor as our main guide. In the case that $v_3(N) = 0$ we have good reduction, while when $v_3(N) = 2$ and $v_3(\Delta)$ is even we have $\Phi = C_2$. Since $\mathbf{G}_3$ is abelian here, the Euler factor is given by (4), while the wild conductor is 0 and tame conductor is obtained from Table 1. When $v_3(N) = 2$ and $v_3(\Delta)$ is odd we have that $\Phi = C_4$ and $\mathbf{G}_3$ is nonabelian. The wild conductor $\delta_m(3)$ is 0, and the Euler factor is given by (3).

When $v_3(N) = 4$ we get $\Phi = C_3$ or C_6, the former case when $4|v_3(\Delta)$. For these inertia groups, the question of whether $\mathbf{G}_3$ is abelian can be resolved as follows (see [47, Th. 2.4]). Let $\hat{c}_4$ and $\hat{c}_6$ be the invariants of the minimal twist of E at 3. In the case that $\hat{c}_4 \equiv 9 \pmod{27}$, we have that $\mathbf{G}_3$ is abelian when $\hat{c}_6 \equiv \pm 108 \pmod{243}$ while if $3^3|\hat{c}_4$ then $\mathbf{G}_3$ is abelian when $\hat{c}_4 \equiv 27 \pmod{81}$. In the abelian case we have $\alpha_3 = \zeta_{12}\sqrt{3}$ up to sixth roots, which is sufficient. The Euler factor is then given by either (3) or (4), the tame conductor can be obtained from Table 1, and the wild conductor (computed as in the appendix of [7]) from Table 3. When $v_3(N) = 3, 5$ we have that $\Phi = C_3 \rtimes C_4$. The Euler factor is given by (3) and the wild conductor can be obtained from Table 3, with the first $C_3 \rtimes C_4$ corresponding to $v_3(N) = 3$, and the second to $v_3(N) = 5$.

3.4 Considerations When $p = 2$

Finally we consider $p = 2$, where first we determine the inertia group. Let M be the conductor of the minimal twist F of E at 2, recalling [47, § 2.1] that in general we need to check four curves to determine this twist. Table 4 then gives the inertia group. The appendix of [7] omits a few of these cases; see [47]. When $\Phi = C_1, C_2$ we can always determine α_p via counting points modulo p on E or a quadratic twist, and $\mathbf{G}_2$ is always abelian. The Euler factor is then as in (4). For

$\Phi = C_3, C_6$ the group $\mathbf{G}_2$ is always nonabelian, and the Euler factor is as in (3). For the case of $\Phi = C_4$ and $p = 2$, the question of whether $\mathbf{G}_2$ is abelian comes down [47, Th. 2.3] to whether the c_4 invariant of F is 32 or 96 modulo 128, it being abelian in the latter case, where we have $\alpha_2 = \zeta_8\sqrt{2}$ up to fourth roots. The Euler factors for this and the two cases of noncyclic Φ are obtained from (3) or (4), while the wild conductors $\delta_m(2)$ are given in Table 2, with the appropriate line being determinable from the conductor of the first symmetric power.

Table 2. Values for $\delta_m(2)$

Φ_2	$m=1$	formula
C_2, C_6	2	$\epsilon_m(C_2)$
C_2, C_6	4	$2\epsilon_m(C_2)$
C_4	6	$2\epsilon_m(C_4) + \epsilon_m(C_2)$
Q_8	3	$\epsilon_m(Q_8) + \frac{1}{2}\epsilon_m(C_2)$
Q_8	4	$\epsilon_m(Q_8) + \epsilon_m(C_2)$
Q_8	6	$\epsilon_m(Q_8)+\epsilon_m(C_4)+\epsilon_m(C_2)$
$SL_2(\mathbf{F}_3)$	1	$\frac{1}{3}\epsilon_m(Q_8) + \frac{1}{6}\epsilon_m(C_2)$
$SL_2(\mathbf{F}_3)$	2	$\frac{1}{3}\epsilon_m(Q_8) + \frac{2}{3}\epsilon_m(C_2)$
$SL_2(\mathbf{F}_3)$	4	$\frac{1}{3}\epsilon_m(Q_8) + \frac{5}{3}\epsilon_m(C_2)$
$SL_2(\mathbf{F}_3)$	5	$\frac{5}{3}\epsilon_m(Q_8) + \frac{5}{6}\epsilon_m(C_2)$

Table 3. Values for $\delta_m(3)$

Φ_3	$m=1$	formula
C_3, C_6	2	$\epsilon_m(C_3)$
$C_3 \rtimes C_4$	1	$\frac{1}{2}\epsilon_m(C_3)$
$C_3 \rtimes C_4$	3	$\frac{3}{2}\epsilon_m(C_3)$

Table 4. Values of Φ_2

$v_2(M)$	Φ_2
0	C_1 if $v_2(N) = 0$ else C_2
2	C_3 if $v_2(N) = 2$ else C_6
3,7	$SL_2(\mathbf{F}_3)$
5	Q_8
8	Q_8 if $2^9|c_6(F)$ else C_4

3.5   The Case of Complex Multiplication

When E has complex multiplication by an order of some imaginary quadratic field K, the situation simplifies since we have $L(E,s) = L(\psi, s-1/2)$ for some[5] Hecke Grössencharacter ψ. For the symmetric powers we have the factorisation

$$L(\mathrm{Sym}^m E, s) = \prod_{i=0}^{m/2} L(\psi^{m-2i}, s - m/2), \qquad (5)$$

where ψ^0 is the ζ-function when $4|m$, and when $2\|m$ it is $L(\theta_K, s)$ for the quadratic character θ_K of the field K. Note that the local conductors and Euler factors for each $L(\psi^j, s)$ can be computed iteratively from (5) since this information is known for the left side from the previous subsections. This factorisation reduces the computational complexity significantly, as the individual conductors will be smaller than their product; however, since there are more theoretical results in this case, the data obtained will often lack novelty. The factorisation (5) also implies that $L_{2u-1}(E, s)$ should vanish to high degree at $s = u$, since each

---

[5] This is defined on ideals coprime to the conductor by $\psi(z) = \chi(|z|)(z/|z|)$ where z is the primary generator of the ideal and χ is generally a quadratic Dirichlet character, but possibly cubic or sextic if $K = \mathbf{Q}(\sqrt{-3})$, or quartic if $K = \mathbf{Q}(\sqrt{-1})$. When taking powers, we take χ^j to be the primitive Dirichlet character which induces χ^j.

term has about a 50% chance of having odd functional equation. We found some examples where $L(\psi^3, s)$, $L(\psi^5, s)$, or $L(\psi^7, s)$ has a double zero at the central point, but we know of no such triple zeros.

4   Global Considerations and Computational Techniques

We now give our method for computing special values of the symmetric power L-functions defined above. To do this, we complete the L-function with a Γ-factor corresponding to the prime at infinity, and then use the (conjectural) functional equation in conjunction with the method of Lavrik [32] to write the special value as a "rapidly-converging" series whose summands involve inverse Mellin transforms related to the Γ-factor. First we digress on poles of our L-functions.

4.1   Poles of L-Functions

It is conjectured that $L_m(E, s)$ has an entire continuation, except when $4|m$ and E has complex multiplication (CM) there is a pole at $s = 1 + m/2$, which is the edge of the critical strip.[6] We give an explanation of this expectation from the standpoint of analytic number theory; it is likely that a different argument could be given via representation theory. We write each Euler factor as $U_m(p; s) = \left(1 - b_m(p)/p^s + \cdots\right)^{-1}$ and as $s \to 1 + m/2$ we have $\log L_m(s) \sim \sum_p b_m(p)/p^s$. We will now compute that the conjectural Sato-Tate distribution [44] implies that the average value of $b_m(p)$ is 0, while for CM curves the Hecke distribution [27] will yield an average value for $b_m(p)$ of $p^{m/2}$ when $4|m$.

Similar to (2), for a good prime p we have $b_m(p) = \sum_{i=0}^{m/2} \binom{m-i}{i} a_p^{m-2i}(-p)^i$. The Sato-Tate and Hecke distributions imply that the average values of the kth power of a_p are given by

$$\langle a_p^k \rangle = (2\sqrt{p})^k \frac{\int_0^\pi (\cos\theta)^k (\sin\theta)^2 \, d\theta}{\int_0^\pi (\sin\theta)^2 \, d\theta} \quad \text{and} \quad \langle a_p^k \rangle_{CM} = (2\sqrt{p})^k \frac{\int_0^\pi (\cos\theta)^k \, d\theta}{2\int_0^\pi d\theta}.$$

We have $\langle a_p^k \rangle = 0$ for k odd; for even k the Wallis formula [46] implies

$$\int_0^\pi (\cos\theta)^k (\sin\theta)^2 \, d\theta = \frac{\pi(k-1)!!}{k!!} - \frac{\pi(k+1)!!}{(k+2)!!} = \frac{\pi(k-1)!!}{(k+2)!!},$$

so that $\langle a_p^k \rangle$ is $(2\sqrt{p})^k \frac{2(k-1)!!}{(k+2)!!}$. An induction exercise shows that this implies $\langle b_m(p) \rangle = 0$ when E does not have CM. We also have $\langle a_p^k \rangle_{CM} = (2\sqrt{p})^k \frac{(k-1)!!}{2 \cdot k!!}$ for even k, and again an inductive calculation shows that $\langle b_m(p) \rangle_{CM} = p^{m/2}$ when $4|m$ and is zero otherwise. This behaviour immediately implies the aforementioned conjecture about the poles of $L_m(E, s)$ at $s = 1 + m/2$.

---

[6] The case of $m = 4$ follows as a corollary of work of Kim [29, Corollary 7.3.4].

4.2 Global Considerations

Let $\Lambda_m(E,s) = C_m^s \gamma_m(s) L_m(E,s)$, where $C_m^2 = N_m/(2\pi)^{m+1}$ for m odd and is twice this for m even. For m odd we write $m = 2u - 1$ and for m even we write $m = 2v$; then from [14, §5.3] we have respectively either

$$\gamma_m(s) = \prod_{i=0}^{u-1} \Gamma(s-i) \quad \text{or} \quad \gamma_m(s) = \Gamma\big(s/2 - \lfloor v/2 \rfloor\big) \prod_{i=0}^{v-1} \Gamma(s-i).$$

When $4|m$ and E has CM, we multiply $\gamma_m(s)$ by $(s-v)(s-v-1)$. We expect $\Lambda_m(E,s)$ to have an entire continuation and satisfy a functional equation $\Lambda_m(E,s) = w_m \Lambda_m(E, m+1-s)$ for some $w_m = \pm 1$. The works of Kim and Shahidi [30] establish parts of this conjecture.[7] We can find w_m via experiment as described in Section 4.4, but we can also try to determine w_m theoretically.

4.3 Digression on Local Root Numbers

The sign w_m can theoretically be determined via local computations as in [12], but this is non-trivial to implement algorithmically, especially when $p = 2, 3$. We expect to have a factorisation $w_m = \prod_p w_m(p)$ where the product is over bad primes p including infinity. For m even, the very general work of Saito [38] can then be used to show[9] that $w_m = +1$, so we assume that m is odd. From [14, §5.3] we have $w_m(\infty) = -\left(\frac{-2}{m}\right)$; combined with the relation $w_m(p) = w_1(p)^m$ for primes p of multiplicative reduction, this gives the right sign for semistable curves. The potentially multiplicative case has $w_m(p) = w_1(p)^{(m+1)/2}$.

In the additive cases, the first author [19] has used the work of Rohrlich [36] to compute the sign for $p \geq 5$. We get that[10] $w_m(p) = w_1(p)^{\epsilon_m(\Phi_p)/2}$, and $w_1(p)$ is listed in [36]. For $p = 2, 3$ the value of $w_1(p)$ is given[11] by Halberstadt [25], and our experiments for higher (odd) powers indicate that $w_m(2) = \eta_2 w_1(2)^{\epsilon_m(\Phi_p)/2}$ where $\eta_2 = -1$ if $v_2(N)$ is odd and $m \equiv 3 \pmod 8$ and else $\eta_2 = +1$, while the expected values of $w_m(3)$ are given in Table 5.

Table 5. Experimental values for $w_m(3)$ (periodic mod 12 in m.)

Φ_3	1	3	5	7	9	11	Φ_3	1	3	5	7	9	11	Φ_3	1	3	5	7	9	11
C_3, C_4	+	+	+	+	+	+	C_6	+	−	−	−	+	+	$C_3 \rtimes C_4$	+	+	−	+	+	+
C_2	−	+	−	+	−	+	C_6	−	+	−	+	−	+	$C_3 \rtimes C_4$	−	−	−	−	−	+

---

[7] The full conjecture[8] follows from Langlands functoriality [31]. In the CM case, the functional equation follows from the factorisation (5) and the work of Hecke [27].

[8] Added in proof: A recent preprint [45] on Taylor's webpage shows the meromorphic continuation and functional equation for all symmetric powers for curves with $j \notin \mathbf{Z}$.

[9] The work of Fröhlich and Queyrut [22] and Deligne [13] might give a direct argument.

[10] Since we are assuming that m is odd, the exponent is just $(m+1)/2$ unless $\Phi_p = C_3$.

[11] Note the third case in Table 1 of [25] needs a *Condition spéciale* of $c_4' \equiv 3 \pmod 4$.

4.4   Computations

From [32], [8, Appendix B], or [16], the assumption of the functional equation $\Lambda_m(E, s) = w_m \Lambda_m(E, m+1-s)$ allows us to compute (to a given precision) any value/derivative $\Lambda_m^{(d)}(E, s)$ in time proportional to $C_m \approx \sqrt{N_m}$. Additionally, numerical tests on the functional equation arise naturally from the method.

We follow [6, § 7, p. 119ff]. Suppose we have $\Lambda_m(s) = w_m \Lambda(m+1-s)$, and the dth derivative is the first one that is nonzero at $s = \kappa$. Our main interest is in $\kappa = u$ for $m = 2u-1$ and $\kappa = v+1$ for $m = 2v$, and we note that $d = 0$ for even m. Via Cauchy's residue theorem, for every real $A > 0$ we have

$$\frac{\Lambda_m^{(d)}(\kappa)}{d!} = \frac{1}{2\pi i}\left(\int_{(\delta)} - \int_{(-\delta)}\right)\frac{\Lambda_m(z+\kappa)}{z^{d+1}}\frac{dz}{A^z},$$

where δ is small and positive and $\int_{(\sigma)}$ is the integral along $\Re z = \sigma$. In the second integral we change variables $z \to -z$ and apply the functional equation. Then we write $\kappa + \lambda = m+1$, move both contours sufficiently far to right (say $\Re z = 2m$) and expand Λ_m in terms of the L-function to get

$$\frac{\Lambda_m^{(d)}(\kappa)}{d!} = \int_{(2m)} C_m^{z+\kappa}\gamma_m(z+\kappa)\sum_{n=1}^{\infty}\frac{b_m(n)}{n^{z+\kappa}}\frac{1}{z^{d+1}}\frac{dz}{2\pi i\,A^z}$$
$$+ (-1)^d w_m \int_{(2m)} C_m^{z+\lambda}\gamma_m(z+\lambda)\sum_{n-1}^{\infty}\frac{b_m(n)}{n^{z+\lambda}}\frac{1}{z^{d+1}}\frac{A^z\,dz}{2\pi i}.$$

Thus we get that

$$\frac{\Lambda_m^{(d)}(\kappa)}{d!} = C_m^{\kappa}\sum_{n=1}^{\infty}\frac{b_m(n)}{n^{\kappa}}F_m^d\left(\kappa; \tfrac{n}{AC_m}\right) + (-1)^d w_m C_m^{\lambda}\sum_{n=1}^{\infty}\frac{b_m(n)}{n^{\lambda}}F_m^d\left(\lambda; \tfrac{nA}{C_m}\right),$$

where

$$F_m^d(\mu; x) = \int_{(2m)}\frac{\gamma_m(z+\mu)}{z^{d+1}x^z}\frac{dz}{2\pi i}.$$

The $F_m^d(\mu; x)$-functions are "rapidly decreasing" inverse Mellin transforms. Note that we have $L_m^{(d)}(\kappa) = \Lambda_m^{(d)}(\kappa)/\gamma_m(\kappa)C_m^{\kappa}$, and so can recover the L-value as desired. The parameter A allows us to test the functional equation; if we compute $\Lambda_m^{(d)}(\kappa)$ to a given precision for $A = 1$ and $A = 9/8$, we expect disparate answers if we have the wrong Euler factors or sign w_m.

We compute $F_m^d(\mu; x)$ as a sum of residues at poles in the left half-plane, the first pole being at $z = 0$, following [11]. We need to calculate Laurent series expansions of the Γ-factors about the poles.[12] We let $\zeta(1)$ denote Euler's constant $\gamma \approx 0.577$, and define $H_1(n) = 1$ for all n, and $H_k(1) = \sum_{i=1}^{k} 1/i$ for all k, and

---

[12] When $\Lambda_m(E, s)$ has a pole the factor $\gamma_m(s)$ has two additional linear factors (which are easily handled). But in this case it is better to use the factorisation (5).

recursively define $H_k(n) = H_{k-1}(n) + H_k(n-1)/k$ for $n, k \geq 2$. At a pole $z = -k$ for k a nonnegative integer, we have the Laurent expansion

$$\Gamma(z) = \frac{(-1)^k}{k!(z+k)}\left(1 + \sum_{n=1}^{\infty} H_k(n)(z+k)^n\right) \exp\left(\sum_{n=1}^{\infty} \frac{(-1)^n \zeta(n)}{n}(z+k)^n\right),$$

and for k a negative integer (these only occur for a few cases) we can use the relation $z\Gamma(z) = \Gamma(z+1)$ to shift. To expand $\Gamma(z/2)$ around an odd integer $z = -k$, we use the duplication formula $\Gamma(z) = \Gamma(z/2)\Gamma\left(\frac{z+1}{2}\right)\frac{\sqrt{\pi}}{2^{z-1}}$ to replace $\Gamma(z/2)$ by a quotient of Γ-factors that can each be expanded as above. The trick works in reverse to expand $\Gamma\left(\frac{z+1}{2}\right)$ about an even integer $z = -k$. We also have the series expansions for 2^z and $1/z$ about $z = -k$ given by

$$2^z = 2^{-k} \sum_{n=0}^{\infty} \frac{(\log 2)^n}{n!}(z+k)^n \quad \text{and} \quad \frac{1}{z} = -\frac{1}{k} - \sum_{n=1}^{\infty} \frac{(z+k)^n}{k^{n+1}} \quad \text{(for } k \neq 0\text{)}.$$

Since these $F_m^d(\mu; x)$ functions are (except for CM) independent of the curve, we pre-computed a large mesh of values and derivatives of these functions, and then in our programme we compute via local power series. Thus, unlike the setting of Dokchitser [16], we are not worried too much about the cost of computing $F_m^d(\mu; x)$ for large x via a massively-cancelling series expansion, since we only do this in our pre-computations. For each implemented function we have its value and first 35 derivatives for all $x = i2^k/32$ for $32 \leq i \leq 63$ for k in some range, such as $-3 \leq k \leq 19$. For sufficiently small x we just use the log-power-series expansion. The choice of 35 derivatives combined with the maximal radius of $x/64$ for expansions about x implies that our maximal precision is around $35 \times 6 = 210$ bits. When working to a lower precision, we need not sum so many terms in the local power series. Note that $F_m^d(\mu; x)$ dies off roughly like $\exp(-x^{2/(m+1)})$, and thus it is difficult to do high precision calculations for m large.

To compute the meshes of inverse Mellin transforms described above, we used PARI/GP [34], which can compute to arbitrary precision. However, PARI/GP was too slow to use when actually computing the L-values; instead we used a C-based adaption of Bailey's quad-double package [28], which provides up to 212 bits of precision while remaining fairly fast.[13]

5   Results

We tested the functional equation (via the above method of comparing the computed values for $A = 1$ and $A = 9/8$) for odd symmetric powers $m = 2u - 1$ at the central point $\kappa = u$, and for even symmetric powers $m = 2v$ at the edge of the critical strip $\kappa = v + 1$. We did this for all non-CM isogeny classes in Cremona's database [9] with conductor less than 130000; this took about 3 months on a cluster of 48 computers (each running at about 1 Ghz).

---

[13] The SYMPOW package can be obtained from www.maths.bris.ac.uk/~mamjw.

Table 6. Test-counts (right) and data for order of vanishing (non-CM isogeny classes)

m	Tested	Order 0	Order 1	2	3	4		m	# tests		m	# tests
1	567735	216912	288128	61787	908	0		6	4953		14	26
3	567735	262751	287281	16782	905	16		8	1259		15	1 even
5	46105	22448	23076	569	12	0		10	190		15	16 odd
7	3573	1931	1616	25	1	0		12	142		16	8
9	947	542	400	5	0	0		13	5 even		17	3 odd
11	134	51	82	1	0	0		13	30 odd		18	2

We computed as many as 10^8 terms of the various L-series for each curve, which was always sufficient to check the functional equation of the third symmetric power to about six decimal digits.[14] In all cases, we found the expected functional equation to hold to the precision of our calculation. The results for the order of vanishing (at the central point) for odd powers appear in the left half of Table 6. The right half lists how many tests[15] we did for other[16] symmetric powers (again to six digits of precision). There are less data for higher symmetric powers due to our imposed limit of 10^8 terms in the L-series computations, but since the symmetric power conductors for curves with exotic inertia groups often do not grow so rapidly, we can still test quite high powers in some cases.

Buhler, Schoen, and Top [6] already listed 2379b1 and 31605ba1 as 2 examples of (suspected) 4th order zeros for the symmetric cube. We found 14 more, but no examples of 5th order zeros. For higher powers, we found examples of 3rd order zeros for the 5th and 7th powers, and 2nd order zeros for the 9th, 11th, and 13th powers, though as noted above, we cannot obtain as much data for higher powers.[17] We list the Cremona labels for the isogeny classes in Table 7.

We also looked at extra vanishings of the 3rd symmetric power in a quadratic twist family. We took E as 11a3:$[0, -1, 1, 0, 0]$ and computed the twisted central value $L_3(E_d, 2)$ or central derivative $L'_3(E_d, 2)$ for fundamental discriminants $|d| < 5000$. We found 58 double zeros (to 9 digits) and one triple zero ($d = 3720$). A larger experiment (for $|d| < 10^5$) for 10 different CM curves found (proportionately) fewer double zeros and no triple zeros.

Finally, we used higher-precision calculations to obtain the Bloch-Kato numbers of equation (1) for various symmetric powers of some non-CM curves of

---

[14] In about 0.3% of the cases, the computations for both the zeroth and first derivatives showed no discrepancy with $A = 1$ and $A = 9/8$; this coincidence is to be expected on probabilistic grounds, and for these cases we computed to higher precision to get an experimental confirmation of the sign of the functional equation.

[15] We need not compute even powers when there is a lack of quadratic-twist-minimality.

[16] We did not test the fourth symmetric power, as the work of Kim [29] proves the validity of the functional equation in this case. Since there is no critical value, a calculation would do little more than verify that our claimed Euler factors are correct.

[17] Given that we only computed the L-value of the 13th symmetric power for five curves of even sign, to find one that has a double-order zero is rather surprising. Higher-order zeros were checked to 12 digits; the smallest "nonzero" value was $\approx 2.9 \cdot 10^{-8}$.

Table 7. Experimentally observed high order vanishings (non-CM isogeny classes)

ord	format is **power:label(s)**
4th	**3**:2379b 5423a 10336d 29862s 31605ba 37352d 46035a 48807b 55053a
	3:59885g 64728a 82215d 91827a 97448a 104160bm 115830a
3rd	**5**:816b 2340i 2432d 3776h 5248a 6480t 7950w 8640bl 16698s 16848r
	5:18816n 57024du **7**:176a
2nd	**7**:128b 160a 192a 198b 200e 320b 360b 425a 576b 726g 756b 1440a
	7:1568i 1600b 2304g 3267f 3600h 3600j 3600n 3888e 4225m 6272d
	7:11552r 15876f 21168g **9**:40a 96a 162b 324d 338b **11**:162b **13**:324c

Table 8. Selected Bloch-Kato numbers for various powers and curves

5th powers

20a2 2^9
37a1 2^9
43a1 $2^7 5$
44a1 2^{17}

6th powers

11a3 $2^4 5$
14a4 $2^9 3$
15a8 2^{10}
17a4 2^{12}
19a3 $2^4 3^3 5^2$
20a2 $2^{17}/3$
24a4 $2^{17}/3$
26a3 $2^7 3 \cdot 5 \cdot 23$
26b1 $2^7 3 \cdot 7^3 \cdot 23$
30a1 $2^{15} 3^3 7$
33a2 $2^{17} 3 \cdot 5 \cdot 7$
34a1 $2^{13} 3^3 59$
35a3 $2^8 3 \cdot 7^2 31$
37a1 $2^9 3^4 7$
37b3 $2^7 3^4 467$
38a3 $2^7 3^4 5 \cdot 11 \cdot 137$
38b1 $2^7 3^2 5^2 13 \cdot 31$
39a1 $2^{20} 3^2 7$
40a3 $2^{20} 7/3$
42a1 $2^{19} 3^2 7 \cdot 19$
43a1 $2^6 3^2 1697$
44a1 $2^{21} 5 \cdot 31/3$
46a1 $2^9 5 \cdot 23 \cdot 30661$
50a1 $2^3 5^{11}/3$
51a1 $2^9 3^3 4517$

7th powers

24a4 $2^{23} 7/3$
37a1 $2^{13} 3 \cdot 5$
43a1 $2^{17} 3 \cdot 5$

9th powers

11a3 2^{12}
14a4 $2^{14} 3^4 5$
15a8 2^{16}
17a4 $2^{16} 3^6 5$
19a3 $2^{19} 3^2 5$
21a4 $2^{20} 5 \cdot 59^2$
24a4 $2^{38}/9$
26a3 $2^{11} 3^4 5 \cdot 7^4$
26b1 $2^{11} 3^2 5 \cdot 7^3 1933^2$
30a1 $2^{16} 3^5 5^5 37^2$
33a2 $2^{24} 5 \cdot 107^2 167^2$
34a1 $2^{23} 3^5 5 \cdot 7^2 53^2$
35a3 $2^{25} 3^4 5$
37b3 $2^{20} 3^2 5 \cdot 7^2 53^2$
38a3 $2^{11} 3^{14} 5 \cdot 19^2$
38b1 $2^{11} 5^8 109^2$
39a1 $2^{40} 3^2 5 7^4$
40a3 0
42a1 $2^{25} 5 \cdot 223^2 241^2$
44a1 $2^{47} 3$
45a1 $2^{16} 3^{19} 5^2 7 \cdot 13^2$
46a1 $2^{14} 3^{10} 5^3 14071^2$
48a4 $2^{43} 5$
50a1 $2^5 3 \cdot 5^{22}$
54a3 $2^9 3^{24}$
54b1 $2^7 3^{25} 5$

10th powers

11a3 $2^{14} 5 \cdot 22453/3$
14a4 $2^{16} 3^3 5 \cdot 6691$
15a8 $2^{26} 5 \cdot 541$
17a4 $2^{23} 3^2 7 \cdot 11 \cdot 227$
19a3 $2^{14} 3^2 47 \cdot 179 \cdot 5023$
20a2 $2^{44} 53/3$
21a4 $2^{28} 3^7 5^2 29$
24a4 $2^{49} 13/9$
26a3 $2^{19} 3^5 7 \cdot 47 \cdot 1787$
26b1 $2^{19} 3^3 5^2 7^3 127 \cdot 2102831$
40a3 $2^{54} 5 \cdot 683$
44a1 $2^{56} 5 \cdot 11 \cdot 215447/3$
50a1 $2^{11} 5^{28} 7/3$
52a2 $2^{44} 3^5 5 \cdot 7 \cdot 19 \cdot 279751$
54b1 $2^{14} 3^5 7$
56a1 $2^{66} 3^2 5 \cdot 11 \cdot 71$
75c1 $2^{14} 5^{28} \cdot 31 \cdot 41 \cdot 61/9$
96b1 $2^{84} 197/3$
99a1 $2^{18} 3^{31} 5^3 7 \cdot 1367$

11th powers

11a3 $2^{26} 5^4/3$
14a4 $2^{23} 3^5 5^2 7^2 11^2$
15a8 $2^{29} 11^2 23^2/3$
17a4 $2^{26} 3^{11}$
21a4 $2^{36} 11^2 211^2/3$
24a4 $2^{57} 13^2/45$
48a4 $2^{70} 11^2/3$
54b1 $2^{20} 3^{41}$
56a1 $2^{74} 11^2/5$
72a1 $2^{58} 3^{28} 59^2/5$

small conductor (see Table 8). More on the arithmetic significance of these quotients will appear elsewhere. In some cases, we were able to lessen the precision because it was known that a large power of a small prime divided the numerator.

5.1 Other Directions

In this work, we looked at symmetric powers for weight 2 modular forms. Delaunay has done some computations [11] for modular forms of higher weight; in that case, the work of Deligne again tells us where to expect critical values, and the experiments confirm that we do indeed get small-denominator rationals after proper normalisation. We looked at critical values at the edge and center of the critical strip, whereas we expect L-functions evaluated at other integers to take special values related to K-theory; see [3,33,17,48] for examples. The programmes written for this paper are readily modifiable to compute other special values. The main advantage that our methods have over those of Dokchitser [16] is that we fixed the Γ-factors and the L-values of interest, which then allowed a large pre-computation for the inverse Mellin transforms; if we wanted (say) to compute zeros of L-functions (as with [37]), our method would not be as useful.

Finally, the thesis of Booker [5] takes another approach to some of the questions we considered. The scope is much more broad, as it considers not only numerical tests of modularity, but also tests of GRH (§3.4), recovery of unknown Euler factors possibly using twists (§5.1), and also high symmetric powers (§7.2).

Acknowledgements

We thank Neil Dummigan for useful comments and his interest in these computations; indeed much of this paper came out of a desire to generalise the numerical experiment in [18, §7]. We thank Frank Calegari for pointing out that we have a factorisation in the CM case, and Erez Lapid for help with local root numbers.

References

1. A. A. Beĭlinson, *Higher regulators and values of L-functions*. J. Soviet Math. **30** (1985), 2036–2070. English translation from Russian original in Современные проблемы математики, Итоги Науки и Техники, **24** (1984), 181–238.

2. B. J. Birch, H. P. F. Swinnerton-Dyer, *Notes on elliptic curves. I. II.* J. reine angew. Math. **212** (1963), 7–25, **218** (1965), 79–108.

3. S. Bloch, D. Grayson, K_2 *and L-functions of elliptic curves: computer calculations*. In *Applications of algebraic K-theory to algebraic geometry and number theory. Part I*. Proceedings of the AMS-IMS-SIAM joint summer research conference held at the University of Colorado, Boulder, Colo., June 12–18, 1983. Edited by S. J. Bloch, R. K. Dennis, E. M. Friedlander and M. R. Stein. Contemporary Mathematics, **55**. American Mathematical Society, Providence, RI (1986), 79–88.

4. S. Bloch, K. Kato, *L-functions and Tamagawa numbers of motives*. In *The Grothendieck Festschrift. Vol. I*. A collection of articles written in honor of the 60th birthday of Alexander Grothendieck. Edited by P. Cartier, L. Illusie, N. M. Katz, G. Laumon, Yu. Manin and K. A. Ribet. Progress in Mathematics, **86**. Birkhäuser Boston (1990), 333–400.

5. A. R. Booker, *Numerical tests for modularity*. Ph. D. dissertation, Princeton (2003). Shortened version appeared in J. Ramunajan Math. Soc., **20** (2005), 283–339.

6. J. Buhler, C. Schoen, J. Top, *Cycles, L-functions and triple products of elliptic curves.* J. Reine Angew. Math. **492** (1997), 93–133.

7. J. Coates, C.-G. Schmidt, *Iwasawa theory for the symmetric square of an elliptic curve.* J. Reine Angew. Math. **375/376** (1987), 104–156.

8. H. Cohen, *Advanced topics in computational number theory.* Graduate Texts in Mathematics, **193**. Springer-Verlag, New York (2000), 578pp.

9. J. E. Cremona, *Algorithms for modular elliptic curves.* Second edition. Cambridge University Press, Cambridge, (1997), 376pp. See Cremona's webpage www.maths.nottingham.ac.uk/personal/jec for latest data and online book.

10. A. Dąbrowski, *On the symmetric power of an elliptic curve.* In *Algebraic K-theory.* Proceedings of the Research Conference held at Adam Mickiewicz University, Poznań, September 4–8, 1995. Edited by G. Banaszak, W. Gajda and P. Krasoń. Contemporary Math., **199**, Amer. Math. Soc., Providence, RI (1996), 59–82.

11. C. Delaunay, *Vérification numérique des conjectures de Deligne* (French). [Numerical verification of conjectures of Deligne.] Preprint (2004), part of his doctoral thesis, available from math.univ-lyon1.fr/~delaunay/these.pdf

12. P. Deligne, *Les constantes des équations fonctionnelles* (French). [The constants of functional equations.] Séminaire Delange-Pisot-Poitou: 1969/70, Théorie des Nombres, Fasc. 2, Exp. **19 bis**, Secrétariat mathématique, Paris (1970), 13pp.
P. Deligne, *Les constantes des équations fonctionnelles des fonctions L* (French). [The constants of functional equations of *L*-functions.] In *Modular functions of one variable, II.* Proceedings of the International Summer School at the University of Antwerp, 1972. Edited by P. Deligne and W. Kuyk. Lecture Notes in Math., **349**, Springer-Verlag (1973), 501–597. Correction in *Modular functions of one variable, IV.* Edited by B. J. Birch and W. Kuyk. Lecture Notes in Math., **476**, Springer-Verlag (1975), 149.

13. P. Deligne, *Les constantes locales de l'équation fonctionnelle de la fonction L d'Artin d'une représentation orthogonale* (French). [The local constants of the functional equation of the Artin *L*-function of an orthogonal representation.] Invent. Math. **35** (1976), 299–316.

14. P. Deligne, *Valeurs de fonctions L et périodes d'intégrales* (French). [Values of *L*-functions and periods of integrals.] In *Automorphic forms, representations, and L-functions,* part 2, Proceedings of the Symposium in Pure Mathematics of the American Mathematical Society (Twenty-fifth Summer Research Institute, Corvallis, OR). Edited by A. Borel and W. Casselman. AMS Proc. Symp. Pure Math. **XXXIII**, no. 2. Amer. Math. Soc., Providence, RI (1979), 313–346. Available online from www.ams.org/online_bks/pspum332

15. P. G. L. Dirichlet, *Vorlesungen über Zahlentheorie* (German). [Lectures on Number Theory.] With supplements by R. Dedekind. First published in 1863. English translation: *Lectures on number theory,* by J. Stillwell, History of Mathematics, **16**. Amer. Math. Soc., Providence, RI; London Math. Soc., London (1999), 275pp.

16. T. Dokchitser, *Computing Special Values of Motivic L-functions.* Experiment. Math. **13** (2004), no. 2, 137–149. Online at arxiv.org/math.NT/0207280

17. T. Dokchister, R. de Jeu, D. Zagier, *Numerical verification of Beilinson's conjecture for K_2 of hyperelliptic curves.* Compositio Math. **142** (2006), no. 2, 339–373. Online at arxiv.org/math.AG/0405040

18. N. Dummigan, *Tamagawa factors for certain semi-stable representations.* Bull. London Math. Soc. **37** (2005), 835–845.

19. N. Dummigan, P. Martin, *Euler factors and root numbers for symmetric powers of elliptic curves,* in preparation.

20. L. Euler. *Introductio in analysin infinitorum. I.* (Latin). [Introduction to the Analysis of the Infinite. Part I.] First published in 1748 (Opera Omnia: Series 1, Volume 1). English translation: *Introduction to the Analysis of the Infinite: Book I,* by J. D. Blanton, Springer-Verlag, New York (1988), 348pp.

21. M. Flach, *The equivariant Tamagawa number conjecture: a survey.* With an appendix by C. Greither. In *Stark's conjectures: recent work and new directions.* Papers from the International Conference on Stark's Conjectures and Related Topics held at Johns Hopkins University, Baltimore, MD, August 5–9, 2002. Edited by D. Burns, C. Popescu, J. Sands, and D. Solomon. Contemp. Math., **358**, Amer. Math. Soc., Providence, RI (2004), 79–125.

22. A. Fröhlich, J. Queyrut, *On the functional equation of the Artin L-function for characters of real representations.* Invent. Math. **20** (1973), 125–138.

23. P. B. Garrett, *Decomposition of Eisenstein series: Rankin triple products.* Ann. of Math. (2) **125** (1987), no. 2, 209–235.

24. B. Gross, S. S. Kudla. *Heights and the central critical values of triple product L-functions.* Compositio Math. **81** (1992), no. 2, 143–209.

25. E. Halberstadt, *Signes locaux des courbes elliptiques en 2 et 3* (French). [Local root numbers of elliptic curves for $p = 2$ or 3: an abridged English version is included.] C. R. Acad. Sci. Paris Sér. I Math. **326** (1998), no. 9, 1047–1052.

26. M. Harris, S. S. Kudla, *The central critical value of a triple product L-function.* Ann. of Math. (2) **133** (1991), no. 3, 605–672.

27. E. Hecke, *Analysis und Zahlentheorie: Vorlesung Hamburg 1920* (German). [Analysis and Number Theory: Hamburg Lectures 1920.] Edited and with a foreword by P. Roquette. Dokumente zur Geschichte der Mathematik [Documents on the History of Mathematics], **3**. Friedr. Vieweg & Sohn, Braunschweig, (1987), 234pp.

28. Y. Hida, X. S. Li, D. H. Bailey, *Quad-Double Arithmetic: Algorithms, Implementation, and Application.* Tech. Report LBNL-46996, Lawrence Berkeley National Laboratory, 2000. See www.eecs.berkeley.edu/~yozo/papers/LBNL-46996.ps.gz Bailey's high-precision libraries are online at crd.lbl.gov/~dhbailey/mpdist

29. H. H. Kim, *Functoriality for the exterior square of* GL_4 *and the symmetric fourth of* GL_2. J. Amer. Math. Soc. **16** (2003), no. 1, 139–183.

30. H. H. Kim, F. Shahidi, *Cuspidality of symmetric powers with applications.* Duke Math. J. **112** (2002), no. 1, 177–197.

31. R. P. Langlands, *Problems in the theory of automorphic forms.* In *Lectures in modern analysis and applications, III.* Edited by C. T. Taam. Lecture Notes in Math., **170**. Springer, Berlin (1970), 18–61. See also his *Letter to André Weil,* January 1967. Online from www.sunsite.ubc.ca/DigitalMathArchive/Langlands

32. A. F. Lavrik, *On functional equations of Dirichlet functions.* Math. USSR, Izv. **1** (1968), 421–432. See also Изв. Акад. Наук СССР, Сер. Мат. **31** (1967), 421–432.

33. J.-F. Mestre, N. Schappacher, *Séries de Kronecker et fonctions L des puissances symétriques de courbes elliptiques sur Q* (French). [Kronecker series and L-functions of symmetric powers of elliptic curves over Q]. In *Arithmetic algebraic geometry.* Papers from the conference held in Texel, April 1989. Edited by G. van der Geer, F. Oort and J. Steenbrink. Progress in Mathematics, **89**. Birkhäuser Boston, Inc., Boston, MA, (1991), 209–245.

34. PARI/GP, version 2.2.11, Bordeaux (France), 2005, pari.math.u-bordeaux.fr

35. M. Rapoport, N. Schappacher, P. Schneider (ed.), *Beilinson's conjectures on special values of L-functions.* Perspectives in Mathematics, **4**. Academic Press, Inc., Boston, MA (1988), 373pp.

36. D. E. Rohrlich, *Variation of the root number in families of elliptic curves.* Compos. Math. **87** (1993), 119–151.

37. M. O. Rubinstein, *L*. C++ class library and command line program for computing zeros and values of L-functions. Includes data. www.math.uwaterloo.ca/~mrubinst

38. T. Saito, *The sign of the functional equation of the L-function of an orthogonal motive.* Invent. Math. **120** (1995), no. 1, 119–142.

39. J.-P. Serre, *Facteurs locaux des fonctions zêta des variétés algébriques (définitions et conjectures)* (French). [Local factors of zeta functions of algebraic varieties (definitions and conjectures).] Séminaire Delange-Pisot-Poitou, 1969/70, Théorie des Nombres, Fasc. 2, Exp. **19**, Secrétariat mathématique, Paris (1970).

40. J.-P. Serre, *Propriétés galoisiennes des points d'ordre fini des courbes elliptiques* (French). [Galois properties of points of finite order of elliptic curves.] Invent. Math. **15** (1972), no. 4, 259–331.

41. J.-P. Serre, J. Tate, *Good reduction of abelian varieties.* Ann. of Math. (2) **88** (1968), 492–517.

42. F. Shahidi, *Symmetric power L-functions for* GL(2). In *Elliptic curves and related topics.* Edited by H. Kisilevsky and M. R. Murty. CRM Proceedings & Lecture Notes, **4**. American Mathematical Society, Providence, RI (1994), 159–182.

43. H. M. Stark, *L-functions at s* = 1. *I. L-functions for quadratic forms. II. Artin L-functions with rational characters. III. Totally real fields and Hilbert's twelfth problem. IV. First derivatives at s* = 0, Adv. in Math. **7** (1971), 301–343, **17** (1975), no. 1, 60–92, **22** (1976), no. 1, 64–84, **35** (1980), no. 3, 197–235.

44. J. T. Tate, *Algebraic cycles and poles of zeta functions.* In *Arithmetical Algebraic Geometry.* Proceedings of a Conference at Purdue Univ. 1963. Edited by O. F. G. Schilling. Harper & Row, New York (1965), 93–110.

45. R. Taylor, *Automorphy for some l-adic lifts of automorphic mod l representations. II.* Preprint available from www.math.harvard.edu/~rtaylor

46. J. Wallis, *Arithmetica infinitorum* (Latin). Originally published in 1656. English translation: *The Arithmetic of Infinitesimals* by J. A. Stedall, Sources and Studies in the History of Mathematics and Physical Sciences Series, Springer (2004), 192pp.

47. M. Watkins, *Computing the modular degree of an elliptic curve,* Experiment. Math. **11** (2002), no. 4, 487–502. Online from www.expmath.org/expmath/contents.html

48. D. Zagier, H. Gangl, *Classical and elliptic polylogarithms and special values of L-series.* In *The arithmetic and geometry of algebraic cycles.* Proceedings of the NATO Advanced Study Institute held as part of the 1998 CRM Summer School at Banff, AB, June 7–19, 1998. Edited by B. B. Gordon, J. D. Lewis, S. Müller-Stach, S. Saito and N. Yui. NATO Science Series C: Mathematical and Physical Sciences, **548**. Kluwer Academic Publishers, Dordrecht (2000), 561–615.

Determined Sequences, Continued Fractions, and Hyperelliptic Curves

Alfred J. van der Poorten

Centre for Number Theory Research, 1 Bimbil Place, Killara, NSW 2071, Australia
alf@math.mq.edu.au

Abstract. In this report I sanitise (in the sense of 'bring some sanity to') the arguments of earlier reports detailing the correspondence between sequences $(M + hS)_{-\infty < h < \infty}$ of divisors on elliptic and genus two hyperelliptic curves, the continued fraction expansion of quadratic irrational functions in the relevant elliptic and hyperelliptic function fields, and certain integer sequences satisfying relations of Somos type. I note that one may often readily determine the coefficients in those relations by elementary linear algebra.

I begin with some musings on here called 'determined sequences', and continue with detail on continued fraction expansion of square roots of polynomials and associated Somos type sequences particularly in the genus 1 and 2 cases.

1   Remarks on Determined Sequences

1.1   Michael Somos' Sequences

The canonical details are given in [8], but for story telling purposes[1] let me introduce the matter as follows. Some fifteen years ago, Michael Somos noticed [8] that the two-sided sequence

$$C_{h-2}C_{h+2} = C_{h-1}C_{h+1} + C_h^2,$$

---

[1] Referee 1 warns me that grossly simplified (in plain language: falsified) stories will not do. More precisely, David Gale [8] reports that Michael Somos discovered the apparent integrality of 1, 1, 1, 1, 1, 1, 3, 5, 9, 23, 75, 421, 1103, 4057, 41783, ..., namely 6-Somos, leading others to investigate 4-Somos and 5-Somos; specifically, Janice Malouf was the first to prove the integrality of 4-Somos. The trouble was that the early integrality proofs (of 4, 5, and 6-Somos) seemingly relied on algebraic accident and could not properly be said to give any explanation. To me, it seemed sufficient to mention a first-hand source allowing readers to replace legend by history. However, all this did provoke me to reread [8] alerting me to a number of interesting facts I had quite forgotten. I use this aside also to report a recent note of Chris Swart and Andy Hone [19] giving an alternative proof that the T_h satisfying (1) are Laurent polynomials in the initial data, and sharper integrality conditions than immediately derivable from [7]. Referee 2 adds that "[this alternative] proof is based on (6) with $t = 1$ and the analogous formula with asymmetric shifts, and there is also a (much more verbose) discussion similar in spirit to §1.2."

F. Hess, S. Pauli, and M. Pohst (Eds.): ANTS 2006, LNCS 4076, pp. 393–405, 2006.

which I refer to as 4-Somos in his honour, apparently takes only integer values if we start from C_{-1}, C_0, C_1, $C_2 = 1$.

Indeed, Somos goes on to investigate also the width 5 sequence, $B_{h-2}B_{h+3} = B_{h-1}B_{h+2} + B_h B_{h+1}$, now with five initial 1s, the width 6 sequence $D_{h-3}D_{h+3} = D_{h-2}D_{h+2} + D_{h-1}D_{h+1} + D_h^2$, and so on, testing whether each, when initiated by an appropriate number of 1s, yields only integers. Naturally, he asks: "What is going on here?"

While 4-Somos (A006720), 5-Somos (A006721), 6-Somos (A006722), and 7-Somos (A006723), do yield only integers; 8-Somos does *not*. The codes in parentheses refer to Neil Sloane's *On-line encyclopedia of integer sequences*.

Fomin and Zelevinsky [7] give an algebraic explanation. For example, their theory of *cluster algebras* entails that a sequence (T_h) satisfying

$$\alpha T_{h-2}T_{h+2} + \beta T_{h-1}T_{h+1} + \gamma T_h^2 = 0, \quad \text{for all } h \in \mathbb{Z}, \tag{1}$$

has the T_h Laurent polynomials in the four initial values over $\mathbb{Z}[\beta/\alpha, \gamma/\alpha]$.

1.2   Self-determining Relations

Suppose we are given a family $\mathcal{T}$ of sequences (T_h) all satisfying a relation (1) with constant coefficients α, β, γ not all zero depending only on the family $\mathcal{T}$. Further, there is no loss of generality in supposing that our family $\mathcal{T}$ contains a *singular* sequence, (W_h) say, here specified by $W_0 = 0$.

Then we readily determine the nontrivial coefficients by noting that

$$\Delta_{0,1,h} = \begin{vmatrix} W_{-2}W_2 & W_{-1}W_1 & 0 \\ W_{-1}W_3 & 0 & W_1^2 \\ W_{h-2}W_{h+2} & W_{h-1}W_{h+1} & W_h^2 \end{vmatrix}$$
$$= W_{-1}W_1^3 W_{h-2}W_{h+2} - W_{-2}W_2 W_1^2 W_{h-1}W_{h+1} - W_{-1}^2 W_1 W_3 W_h^2 = 0. \tag{2}$$

In the context I have in mind, (W_h) is in fact anti-symmetric: $W_{-h} = -W_h$, and clearly $W_1 \neq 0$ must be supposed, so our determination yields

$$W_1^2 T_{h-2}T_{h+2} = W_2^2 T_{h-1}T_{h+1} - W_1 W_3 T_h^2. \tag{3}$$

Whatever, given that there is a relation as described, one readily identifies its coefficients in terms of several initial elements of a, or the, singular sequence in the family.

1.3   Elliptic Sequences

In the sequel I discuss curves $\mathcal{C} : Z^2 - AZ - R = 0$ with polynomial coefficients A and R satisfying $\deg A = g + 1$, $0 < \deg R \leq g$.

Disclaimer. Here and throughout below I disregard the possibility that the given curve is of genus lower than g. In particular, if $\mathcal{C}$ is of genus zero then the continued fraction expansions are different from what we assert generic below.

In the case $g = 1$, set $R = v(X - w)$. The transformation

$$U = Z \qquad V - v = XU \qquad (4)$$

transforms $\mathcal{C}$ into a Weierstrass model $\mathcal{E}$ for an elliptic curve, in effect by taking one of the points, say S, at infinity on $\mathcal{C}$ to $S_\mathcal{E} = (0,0)$ on $\mathcal{E}$.

It is an interesting but non-trivial exercise to confirm that there are well-defined integers U_h, V_h, W_h so that the rational points $hS_\mathcal{E}$ for $h \in \mathbb{Z}$ have co-ordinates of the shape $(U_h/W_h^2, V_h/W_h^3)$ satisfying

$$\gcd(U_h, V_h) = W_{h-1} \quad \text{and} \quad U_h = -W_{h-1}W_{h+1}; \qquad (5)$$

all this up to at most finitely many primes. For details see Rachel Shipsey's thesis [16]. It follows, as long ago observed by Morgan Ward [20], that there is indeed a sequence W_h as above satisfying

$$W_{h-m}W_{h+m}W_t^2 = W_m^2 W_{h-t}W_{h+t} - W_{m-t}W_{m+t}W_h^2 \qquad (6)$$

for all integers h, m, and t.

It turns out [14], given an arbitrary point $M_\mathcal{E}$ on $\mathcal{E}$, that just so the 'denominators' T_h of the 'translated' points $M_\mathcal{E} + hS_\mathcal{E}$ satisfy

$$W_t^2 T_{h-m}T_{h+m} = W_m^2 T_{h-t}T_{h+t} - W_{m-t}W_{m+t}T_h^2. \qquad (7)$$

It is easy to check that (6) is self-determining as h or t varies but to prove the identity requires showing, say by induction on m as in [14], that it is implied by the particular case (3) where $m = 2$. Such an argument does not require an understanding of the genesis of the family of sequences exemplified by (T_h).

Alternatively, one might recognise (T_h) as an 'elliptic sequence', identify the corresponding elliptic curve by analytic means, and prove (7) as coming from an identity satisfied by the relevant $\wp$-function. That's done by Ward [20], and rather more directly by Andy Hone [9].

Below I explain the identification by algebraic methods of M and the curve $\mathcal{C}$, or $\mathcal{E}$. It is intriguing that the recursion (3) depends only on the curve, but that four nonzero initial values of T_h are required both to fix the curve among a class of admissible curves and to find the translation M.

1.4   Division Polynomials

One might remark that there is gain in generality in having changed the transformation, by $U \leftarrow (U - x)$, $V \leftarrow (V - y)$, whereby in effect the co-efficients of $\mathcal{E}$ become polynomials in x and y and S is sent to $S_\mathcal{E} = (x, y)$. The result is that the integers W_h become polynomials $W_h(x, y)$ with the evident property that $W_m(a, b) = 0$ if and only if (a, b) is a point of torsion order dividing m. In other words, $W_h(x, y)$ is the h-th division polynomial. That inter alia entails $\gcd\big(W_r(x, y), W_s(x, y)\big) = W_{\gcd(r,s)}(x, y)$, explaining the division properties of the $W_h(0,0)$ and — conversely — the rapid growth of the coefficients of the division polynomials.

1.5   Hyperelliptic Sequences

The formulas (4) so relate the cubic and quartic models that the recursion relations for division polynomials produced from studying the quartic model coincide with those produced by the more familiar cubic model. One cannot expect that to be so if $g > 1$. It should therefore be no special surprise that sequences obtained by David Cantor [5] by studying Padé approximants to square roots of polynomials of odd degree $2g + 1$ and with constant coefficient say 1, viewed as power series about zero, are not the same as sequences I obtain below from the continued fraction expansion of square roots of monic polynomials of even degree viewed as Laurent series about infinity. Just so, the results obtained in [3] by studying Kleinian σ-functions in genus 2 are not immediately applicable to my discussion below.

2   Continued Fraction of the Square Root of a Polynomial

Suppose $A(X)$ denotes a polynomial of degree $g + 1$ and $R(X)$ a polynomial of positive degree at most g. Then, the equation

$$Z^2 - AZ - R = 0 \tag{8}$$

defines a quadratic irrational integer function Z of degree $g+1$ and with conjugate $\overline{Z}$ of negative degree. Note that this definition makes sense over base fields of arbitrary characteristic.

2.1   Laurent Series

Explicitly, albeit not in characteristic two, set $Y^2 = D(X)$ where D, not a square, is a monic polynomial over some field K and is of degree $2g + 2$ in X. Then we may write

$$D(X) = \bigl(A(X)\bigr)^2 + 4R(X),$$

where A is the polynomial part of the square root Y of D; here $4R$, with $\deg R$ at most g, may be referred to as the *remainder*. We then take

$$Y = A\bigl(1 + 4R/A^2\bigr)^{1/2} = A(X) + c_1 X^{-1} + c_2 X^{-2} + \cdots \tag{9}$$

thereby viewing Y as an element of $K((X^{-1}))$, Laurent series in the variable $1/X$. Note that the degree of such a Laurent series is the degree in X of its leading term. Of course $Z = \frac{1}{2}(Y + A)$ and does make sense in characteristic 2.

2.2   Continued Fraction Expansions

Now, for $h \in \mathbb{Z}$ set

$$Z_h = (Z + P_h)/Q_h \,,$$

where P_h and Q_h are polynomials such that $\deg Z_h > 0$ and $\deg \overline{Z}_h < 0$ — one says that Z_h is *reduced*. It follows that both $\deg P_h \le g - 1$ and $\deg Q_h \le g$.

Further we require that Q_h divides the norm $(Z + P_h)(\overline{Z} + P_h)$; this divisibility condition is equivalent to the requirement that the $K[X]$-module $\langle Q_h, Z + P_h \rangle$ be an ideal of the domain $K[X, Z]$.

Finally, denote by a_h the polynomial part of Z_h. Then the continued fraction expansion of, say, Z_0 is a sequence of lines (or steps)

$$(Z + P_h)/Q_h = a_h - (\overline{Z} + P_{h+1})/Q_h \quad \text{in brief:} \quad Z_h = a_h - \overline{R}_h,$$

where, $-Q_h/(\overline{Z} + P_{h+1}) = (Z + P_{h+1})/Q_{h+1}$. Necessarily

$$P_h + P_{h+1} + A = a_h Q_h \quad \text{and} \quad (Z + P_{h+1})(\overline{Z} + P_{h+1}) = -Q_h Q_{h+1},$$

and one readily verifies that the conditions on the P_h and Q_h are in fact sustained for all h.

This does require a minor miracle, but happily one that is well understood. Because the *complete quotients* Z_h all are reduced it follows that also all the R_h are reduced. It follows that the *partial quotients* a_h, which begin life as the polynomial parts of the Z_h, also are the polynomial parts of the R_h.

Hence also the 'conjugate line'

$$R_h = (Z + P_{h+1})/Q_h = a_h - (\overline{Z} + P_h)/Q_h = a_h - \overline{Z}_h$$

is a line in an admissible continued fraction expansion. Thus we may view the continued fraction expansion as being *bi-directional* infinite.

2.3 Normal Expansion

In the immediate sequel I suppose that the base field K is infinite. Given that, I assert that a generic choice of P_0 and Q_0 is so that *all* the a_h are linear — equivalently, so that all the Q_h are of degree g — indeed, a teeny bit less obviously, so that all the P_h are of their maximal degree $g-1$. That's so because the probability of an element of K being 0 *is* zero. Equivalently, a *generic* divisor of the curve (8) is defined by a g-tuple of elements of an algebraic extension of K.

2.4 The Cases $g = 1$ and $g = 2$ are Atypical

All the conditions just mentioned are equivalent to the nonvanishing of the sequence (d_h) of coefficients of the leading term (of degree $g - 1$) of the polynomials P_h. Accordingly, it is an appropriate goal to attempt to obtain relations involving only the parameter d_h.

Denote a typical zero of Q_h by ω_h and recall the recursion relations

$$P_h + P_{h+1} + A = a_h Q_h \quad \text{and}$$
$$- Q_h Q_{h+1} = (Z + P_{h+1})(\overline{Z} + P_{h+1}) = -R + P_{h+1}(A + P_{h+1}). \quad (10)$$

Thus $P_h(\omega_h) + P_{h+1}(\omega_h) + A(\omega_h) = 0$ and so $R(\omega_h) = -P_{h+1}(\omega_h)P_h(\omega_h)$.

Hence $Q_h(X)$ divides $R(X) + P_{h+1}(X)P_h(X)$, and so

$$C_h(X)/u_h = \big(R(X) + P_{h+1}(X)P_h(X)\big)/Q_h(X) \tag{11}$$

defines a polynomial C_h; here u_h denotes the leading coefficient of Q_h.

One notices that $\deg C_h = \max(g, 2(g-1)) - g$; so C_h is a constant if and only if $g = 1$ or $g = 2$. In the sequel I deal primarily just with these simpler cases.

2.5 More General Formulæ

If $P_h(\varepsilon_h) = 0$, then by (11) we have both

$$C_h(\varepsilon_h)Q_h(\varepsilon_h) = u_h R(\varepsilon_h) \quad \text{and} \quad C_{h+1}(\varepsilon_{h+1})Q_h(\varepsilon_{h+1}) = u_h R(\varepsilon_{h+1});$$

$$\text{and thus} \quad C_{h-1}(\varepsilon_h)C_h(\varepsilon_h)Q_{h-1}(\varepsilon_h)Q_h(\varepsilon_h) = u_{h-1}u_h R(\varepsilon_h)^2. \tag{12}$$

From the recursion formulæ (10),

$$u_{h-1}u_h = -d_h, \text{ and } Q_{h-1}(\varepsilon_h)Q_h(\varepsilon_h) = R(\varepsilon_h). \tag{13}$$

Hence

$$C_{h-1}(\varepsilon_h)C_h(\varepsilon_h) = -d_h R(\varepsilon_h), \tag{14}$$

a formula that seemed inexplicably miraculous when I first stumbled upon it [13] in the case $g = 2$.

If ω is a zero of R we have

$$C_h(\omega)Q_h(\omega) = u_h P_h(\omega)P_{h+1}(\omega) \tag{15}$$

and therefore

$$C_{h-1}(\omega)C_h(\omega)Q_{h-1}(\omega)Q_h(\omega) = u_{h-1}u_h P_{h-1}(\omega)P_h(\omega)^2 P_{h+1}(\omega).$$

By (10) and $u_{h-1}u_h = -d_h$ this is

$$C_{h-1}(\omega)C_h(\omega)P_h(\omega)\big(A(\omega) + P_h(\omega)\big) = d_h P_{h-1}(\omega)P_h(\omega)^2 P_{h+1}(\omega) \tag{16}$$

2.6 What the Continued Fraction Does

It also follows from $Q_h(\omega_h) = 0$ that, for $h \in \mathbb{Z}$, the points $\big(\omega_h, -P_h(\omega_h)\big)$ specify a sequence (M_h) of divisor classes on the Jacobian of the curve $\mathcal{C}$: $Z^2 - AZ - R = 0$.

We may set $M_h = M + S_h$ (so $M = M_0$). It then turns out that $S_h = hS$ — with S the class of the divisor at infinity. In other words, *each step of the continued fraction expansion corresponds to addition of the divisor at infinity.* Comments by David Cantor [4] and Kristin Lauter [10] assist one in accepting this notion. Adams and Razar [1] give a very explicit proof in the elliptic case and Tom Berry [2] provides analogous arguments for general g.

We note that $\deg Q_h \le g$ and $\deg P_h \le g - 1$, generically with equality if the base field is of characteristic zero. I note that because the complete quotients all are reduced, always $\deg P_h < \deg Q_h$.

The pairs $(Q_h, -P_h)$ of polynomials are the respective Mumford representations of divisors $M + hS$ on the hyperelliptic curve $\mathcal{C} : Z^2 - AZ - R = 0$.

3   Continued Fraction Relations

3.1   The Elliptic Case $g = 1$

Here $R = v(X - w)$ and the $P_h(X) = d_h$ are polynomials of degree $g - 1 = 0$. Plainly $C_h = v$. The relation (16) becomes just

$$v^2(A(w) + d_h) = d_{h-1}d_h^2 d_{h+1}. \tag{17}$$

This identity depends only on the given curve, *not* on the 'translation' M.

It follows from (17) that $d_{h+1}d_h + v^2/d_h + d_h d_{h-1}$ is independent of h. A little work then yields

$$d_{h-1}d_h^2 d_{h+1}^2 d_{h+2} = v^2 A(w)d_h d_{h+1} + v^3(v + 2wA(w)). \tag{18}$$

For more detail see [12] or [14]; for the complex function view note Hone [9].

3.2   The Hyperelliptic Case $g = 2$; First Steps

In general we have $R = u(X - \omega)(X - \overline{\omega}) = u(X^2 - vX + w)$, say, and the $P_h(X) = d_h(X + e_h)$ are polynomials of degree one; thus the ε_h above are given by $\varepsilon_h = -e_h$. Evidently, $C_h = d_h d_{h+1} + u$. Here the relations (16)

$$C_{h-1}C_h(A(\omega) + P_h(\omega)) = d_h P_{h-1}(\omega)P_h(\omega)P_{h+1}(\omega)$$

still require an elimination of the (e_h), to be assisted by the 'miraculous' identity

$$C_{h-1}C_h = (d_{h-1}d_h + u)(d_h d_{h+1} + u) = -d_h R(-e_h). \tag{19}$$

However,

$$-d_h R(-e_h) = -u d_h(\omega + e_h)(\overline{\omega} + e_h) = -u P_h(\omega)(\overline{\omega} + e_h).$$

Thus we are to deal with

$$-u(\overline{\omega} + e_h)(A(\omega) + P_h(\omega)) = d_h P_{h-1}(\omega)P_{h+1}(\omega). \tag{20}$$

It now seems natural to multiply by the conjugate equation and to use

$$u P_h(\omega)P_h(\overline{\omega}) = u d_h^2(e_h + \omega)(e_h + \overline{\omega}) = -d_h(d_{h-1}d_h + u)(d_h d_{h+1} + u).$$

But that leaves an e_h on the left. Specifically, one obtains

$$- u C_{h-1}C_h(A(\omega)A(\overline{\omega}) + d_h((\overline{\omega} + e_h)A(\omega) + (\omega + e_h)A(\overline{\omega})))$$
$$- d_h C_{h-1}C_h/u) = d_{h-1}d_h^3 d_{h+1}C_{h-2}C_{h-1}C_h C_{h+1}/u^2. \tag{21}$$

While we do have the identity (19), it is quadratic in e_h and seems unhelpful in eliminating e_h. Of course there is no e_h in the happenstance $A(\omega) + A(\overline{\omega}) = 0$.

3.3   The Special Case $g = 2$, $\deg R = 1$

It's all much easier if $R = v(X - w)$. Then $u = 0$, $\omega = w$ and is rational, and (19) becomes

$$C_{h-1}C_h = d_{h-1}d_h^2 d_{h+1} = d_h v(e_h + w) = vP_h(w). \tag{22}$$

Hence (16) is just

$$v^2 d_{h-1}d_h^2 d_{h+1}\left(vA(w) + d_{h-1}d_h^2 d_{h+1}\right) = d_{h-2}d_{h-1}^3 d_h^5 d_{h+1}^3 d_{h+2}.$$

Recasting this, we obtain Theorem 1 of [13]

$$d_{h-2}d_{h-1}^2 d_h^3 d_{h+1}^2 d_{h+2} = v^2 d_{h-1}d_h^2 d_{h+1} + v^3 A(w). \tag{23}$$

4   Somos Sequences

4.1   Suitable Identities

The identities (17) and (23) are *suitable* in the following sense. It turns out that generically the d_h are rationals increasing in complexity with h at frantic pace: the logarithmic height of d_h is $O(h^2)$. One tames the d_h somewhat by introducing a sequence (T_h) given by the recursive definition

$$T_{h-1}T_{h+1} = d_h T_h^2. \tag{24}$$

That this yields elements integral at all but at most a few exceptional primes is not too difficult to show by elementary means in the elliptic case (see my introductory remarks) and is experientially the case for $g = 2$, no doubt *inter alia* for algebraic reasons of the kind described by Fomin and Zelevinsky [7].

Happily, (24) easily yields $T_{h-1}T_{h+2} = d_h d_{h+1}T_h T_{h+1}$ and then

$$d_{h-1}d_h^2 d_{h+1}T_h^2 = T_{h-2}T_{h+2}, \quad d_{h-1}d_h^2 d_{h+1}^2 d_{h+2}T_h T_{h+1} = T_{h-2}T_{h+3},$$
$$\text{and} \quad d_{h-2}d_{h-1}^2 d_h^3 d_{h+1}^2 d_{h+2}T_h^2 = T_{h-3}T_{h+3}.$$

So the identities (17) and (18) become

$$T_{h-2}T_{h+2} = v^2 T_{h-1}T_{h+1} + v^2 A(w)T_h^2$$
$$\text{and} \quad T_{h-2}T_{h+3} = v^2 A(w)T_{h-1}T_{h+2} + v^3\left(u + 2wA(w)\right)T_h T_{h+1}; \tag{25}$$

and (23) yields

$$T_{h-3}T_{h+3} = v^2 T_{h-2}T_{h+2} + v^3 A(w)T_h^2. \tag{26}$$

4.2   Canonical Examples

4-Somos: Suppose $(C_h) = (\ldots, 2, 1, 1, 1, 1, 2, 3, 7, \ldots)$ with $C_{h-2}C_{h+2} = C_{h-1}C_{h+1} + C_h^2$. One sees that $v = \pm 1$, $w = \mp 2$, $A(w) = 1$, and thus that (C_h) arises from

$$Z^2 - (X^2 - 3)Z - (X - 2) = 0 \quad \text{with } M = (1, -1);$$

equivalently from $\mathcal{E} : V^2 - V = U^3 + 3U^2 + 2U$ with $M_\mathcal{E} = (-1,1)$.

5-Somos: The case $(B_h) = (\dots, 2, 1, 1, 1, 1, 1, 1, 2, 3, 5, 11, \dots)$ with $B_{h-2}B_{h+3} = B_{h-1}B_{h+2} + B_h B_{h+1}$ is trickier. One needs to define $c_h B_{h-1}B_{h+1} = d_h B_h^2$ with $c_h c_{h+1}$ independent of h.

One finds that (B_h) arises from

$$Z^2 - (X^2 - 29)Z + \cdot 48(X + 5) = 0 \quad \text{with } M = (-3, -8);$$

equivalently from $\mathcal{E} : V^2 + UV + 6V = U^3 + 7U^2 + 12U$ with $M_\mathcal{E} = (-2, -2)$. The fact $\gcd(a_3, a_4) = \gcd(6, 12) \neq 1$ may here be thought of as 'necessitating' the width 5 recursion.

By symmetry each respective M is a point of order 2 on its curve.

A Width 6 Example à la Somos. The sequence $(T_h) = (\dots, 2, 1, 1, 1, 1, 1, 1, 2, 3, 4, 8, 17, 50, \dots)$, with

$$T_{h-3}T_{h+3} = T_{h-2}T_{h+2} + T_h^2,$$

may be thought of as arising from the points (thus, divisor classes) $\dots$, $M - S$, M, $M + S$, $M + 2S$, $\dots$ on the Jacobian of the genus 2 hyperelliptic curve

$$\mathcal{C} : Z^2 - (X^3 - 4X + 1)Z - (X - 2) = 0.$$

Here S is the class of the divisor at infinity and M is instanced by the divisor defined by the pair of points $(\varphi, \overline{\varphi})$ and $(\overline{\varphi}, \varphi)$: where φ is the golden ratio. The symmetry dictates that $M - S = -M$ so $2M = S$ on $\mathrm{Jac}(\mathcal{C})$.

4.3   An Identity for $g = 2$

In just the above spirit, multiplying (21) by T_h^3 yields

$$
\begin{aligned}
u^3 A(\omega)A(\overline{\omega})T_h^3 &+ u^2(\mathcal{D}T_{h-1}T_h T_{h+1} + \mathcal{F}d_h e_h T_h^3) \\
&- u^3(T_{h-2}T_{h+1}^2 + T_{h-1}^2 T_{h+2}) - u^4 T_{h-1}T_h T_{h+1} + T_{h-3}T_h T_{h+3} \\
&+ u(T_{h-3}T_{h+1}T_{h+2} + T_{h-2}T_{h-1}T_{h+3}) = 0, \quad (27)
\end{aligned}
$$

a *suitable* expression for $d_h e_h T_h^3$. Mind you, this suitability — in other words: that here multiplication by T_h^3 tames the equation — requires a fortunate coincidence of the first term in the expansion of $u^2 d_h C_{h-1}C_h$ and the last term in the expansion of $d_{h-1}d_h^3 d_{h+1}C_{h-2}C_{h+1}$.

Moreover, in the happenstance $\mathcal{F} = u(A(\omega) + A(\overline{\omega})) = 0$, (27) *is* the sought for relation. Apropós of comments at §1.5 on page 396 above, I note that terms of the shapes $T_{h-3}T_{h+1}T_{h+2}$ and $T_{h-2}T_{h-1}T_{h+3}$, or $T_{h-2}T_{h+1}^2$ and $T_{h-1}^2 T_{h+2}$, do *not* occur in Cantor's recurrence formulas.

4.4   Example

Set $A(X) = X^3 - 7X^2 + 8X + 7$ and $R(X) = u(X - 2)(X - 5)$, noting that $A(2) = 3$ and $A(5) = -3$ so $\mathcal{F} = 0$, $\mathcal{D} = 9u$, $A(2)A(5) = -9$. A computation, done for me by David Gruenewald, confirms that the singular sequence $\ldots$, 2, 1, 1, $W_{-1} = 0$, $W_0 = 0$, $W_1 = 1$, 1, 2, 7, -112, -103, 1803, 132603, -1042153, -31597909, -1759068155, $\ldots$, indeed satisfies

$$-9u^3 W_h^3 + (9u^3 - u^4) W_{h-1} W_h W_{h+1}$$
$$- u^3 (W_{h-2} W_{h+1}^2 + W_{h-1}^2 W_{h+2}) + W_{h-3} W_h W_{h+3}$$
$$+ u(W_{h-3} W_{h+1} W_{h+2} + W_{h-2} W_{h-1} W_{h+3}) = 0, \quad (28)$$

of course with $u = 1$. As always set $Z^2 - AZ - R = 0$. Here, to avoid singular steps in the continued fraction, one expands Z/R, deeming that to provide line 1 of the expansion of Z. The recursion relation allows one to fill gaps in and generally to extend the two-sided sequence.

5   Not Enough Determination

5.1   A Determined Sequence for Every g

During ANTS V, Sydney, Noam Elkies was provoked by remarks of mine to notice that Fay's trisecant formula suggests that hyperelliptic curves

$$Z^2 - AZ - v(X - w) = 0$$

of arbitrary genus g, thus $\deg A = g + 1$ but $\deg R = 1$, yield a Somos relation just on the three terms $T_{h-g-1} T_{h+g+1}$, $T_{h-g} T_{h+g}$, and T_h^2 — incidentally explaining the elliptic case and my $g = 2$ result [13] at (25).

In this case, the singular expansion — that of Z itself — yields a sequence (W_h) with g central zeros occasioned by the partial quotient A of degree $g+1$, followed by a 1, and then $g - 1$ zeros occasioned by the next partial quotient, $(A(X) - A(w))/v(X - w)$, of degree g. For $g = 2k + 1$ odd I set notation so that $W_{-h} = W_h$ and if $g = 2s$ even then $W_{-h} = W_{h+1}$. By the way, after bypassing the singular part of the continued fraction expansion, one computes the singular sequence; then backtracking, using the experimentally discovered recurrence relation, one locates the zero entries.

The determined form of Elkies' remark is

$$W_s^2 W_{5s} T_{h-(2s+1)} T_{h+(2s+1)} = -W_{3s}^2 W_{3s+1} T_{h-2s} T_{h+2s} + W_s W_{3s+1} W_{5s} T_s^2 \quad (29)$$

and respectively

$$W_{k+1}^2 W_{5k+3} T_{h-(2k+2)} T_{h+(2k+2)}$$
$$= -W_{3k+2}^2 W_{3k+3} T_{h-(2k+1)} T_{h+(2k+1)} + W_{k+1} W_{3k+3} W_{5k+3} T_k^2. \quad (30)$$

Incidentally, numbering the lines in singular continued fraction expansions, and thence indexing the W_h, is rather problematic. That would have been so even in the elliptic case were it not that Morgan Ward [20] had already set a notation. Determinations such as the present example assist in leading to a coherent notation.

5.2   Determined Sequences for $g = 2$

My remarks above suggest that the general relation linearly relates the terms $T_{h-3}T_hT_{h+3}$, $T_{h-2}T_hT_{h+2}$, $T_{h-1}T_hT_{h+1}$, T_h^3, $T_{h-3}T_{h+1}T_{h+2} + T_{h-2}T_{h-1}T_{h+3}$, and $T_{h-2}T_{h+1}^2 + T_{h-1}^2T_{h+2}$. That this is so is clear from experiment. Here a determination of the coefficients is not immediately successful because it yields the six coefficients as polynomials in $W_1 = 1$, W_2, ..., W_7. Recall that $W_{-h-1} = W_h$ and $W_{-1} = W_0 = 0$.

However, W_7 plainly is supernumerary since it is given — by the as yet unknown relation — in terms of $W_1 = 1$, W_2, ..., W_6. Indeed, if the last two pairs of terms above have a nonzero coefficient, the unknown relation already gives W_6 in terms of W_2, W_3, W_4, W_5.

If not, we have the simpler case of only four terms all divisible by W_h and then yet more plainly W_6 is given in terms of W_2, W_3, W_4 and W_5. In that special case the determined relation must be

$$W_1W_2W_3W_4T_h^3 - W_2^3W_4T_{h-1}T_hT_{h+1} + W_1W_2W_3^2T_{h-2}T_hT_{h+2}$$
$$= W_1^2W_2W_3T_{h-3}T_hT_{h+3}, \quad (31)$$

a matter of interest if one hopes to detail the source of the 6-Somos sequence.

In the case $g = 1$, I had the foresight already to know the relation and, indeed, to have explicit expressions for $W_1 = 1$, W_2, W_3, W_4, W_5 in terms of the parameters defining the elliptic curve. Here, explicit computation of W_2, W_3, ... in the general case quickly seems to become too messy to be informative. Whatever, I have not as yet disentangled the determination just now sketched, probably out of laziness but principally, I claim, because I am looking for methods and ideas that may generalise to arbitrary genus; hard yakka [2] mucking about with absurd identities is unlikely to do that. I should here also admit that extensive computations of quite general examples by David Gruenewald have helped to 'verify' various guesses of mine, and of his, but have not as yet proved useful in readily identifying the coefficients of the general relation as polynomials in W_2, W_3, W_4, and W_5.

6   Comments

I find the Somos sequences interesting as an infinite base field phenomenon which continues to give meaningful information after reduction or specialisation — after all, the elliptic case will do for this remark, the sequences begin life as 'denominators' yet persist under transformation of the base field to a finite field. That all said, the sequences may well be a distraction, hence my feeling that not enough determination is quite enough.

Indeed, a principal charm of the continued fraction expansions is their encapsulating a sequence of divisors $M + hS$ allowing one ready entrance to open questions dealing with torsion possibilities in higher genus, note for easy example

---

[2] yakka: work [Australian Aboriginal].

404   A.J. van der Poorten

the modular curves mentioned in [11]. In this context I also note that many of
the phenomena touched on above will reappear in studying multi-sequences of
Padé approximants to higher degree algebraic functions.

I also note, as hinted at in my opening comment just above, that studying
curves over an infinite field, say $\mathbb{Q}$ — though computationally hopeless — does
nonetheless give insight into the corresponding curves over finite fields, moreover
all at once for almost all p. I hope to give more emphasis to that thought in
future work.

Thoughtful remarks from the two referees helped me to omit some of the
errors in my remarks.

References

1. William W. Adams and Michael J. Razar, 'Multiples of points on elliptic curves
   and continued fractions', *Proc. London Math. Soc.* **41** (1980), 481–498.
2. T. G. Berry, 'A type of hyperelliptic continued fraction', *Monatshefte Math.*, **145**.4
   (2005), 269–283.
3. Harry W. Braden, Victor Z. Enolskii, and Andrew N. W. Hone, 'Bilinear recurrences and addition formulæ for hyperelliptic sigma functions', *J. Nonlin. Math.
   Phys.* **12**, Supplement 2 (2005), 46–62; also at
   http://www.arxiv.org/math.NT/0501162.
4. D.G. Cantor, 'Computing in the Jacobian of a hyperelliptic curve', *Math. Comp.*
   48.177 (1987), 95–101.
5. David G. Cantor, 'On the analogue of the division polynomials for hyperelliptic
   curves', *J. für Math.* (Crelle), **447**, (1994), 91–145.
6. Graham Everest, Alf van der Poorten, Igor Shparlinski, and Thomas Ward, *Recurrence Sequences*. Mathematical Surveys and Monographs 104, American Mathematical Society, (2003), xiv+318pp.
7. Sergey Fomin and Andrei Zelevinsky 'The Laurent phenomenon', *Adv. in Appl.
   Math.*, **28**, (2002), 119–144. Also 21pp: at
   http://www.arxiv.org/math.CO/0104241.
8. David Gale, 'The strange and surprising saga of the Somos sequences', *The Mathematical Intelligencer* **13**.1 (1991), 40–42; Somos sequence update. *Ibid.* **13**.4, 49–50.
9. A. N. W. Hone, 'Elliptic curves and quadratic recurrence sequences', *Bull. London
   Math. Soc.* **37**, (2005), 161–171.
10. Kristin E. Lauter, 'The equivalence of the geometric and algebraic group laws
    for Jacobians of genus 2 curves', *Topics in algebraic and noncommutative geome-
    try* (Luminy/Annapolis, MD, 2001), 165–171, *Contemp. Math.*, **324**, Amer. Math.
    Soc., Providence, RI.
11. Alfred J. van der Poorten, 'Periodic continued fractions and elliptic curves', in
    High Primes and Misdemeanours: lectures in honour of the 60th birthday of Hugh
    Cowie Williams, Alf van der Poorten and Andreas Stein eds., Fields Institute Com-
    munications **42**, American Mathematical Society, 2004, 353–365.
12. Alfred J. van der Poorten, 'Elliptic sequences and continued fractions', *Journal of
    Integer Sequences*, **8** 05.2.5 (2005), 1–19.
13. Alfred J. van der Poorten, 'Curves of genus 2, continued fractions, and Somos
    sequences', *Journal of Integer Sequences*, **8** 05.3.4, (2005), 1–9.

14. Alfred J. van der Poorten and Christine S. Swart, 'Recurrence relations for elliptic sequences: every Somos 4 is a Somos k', to appear in *Bull. London Math. Soc.*; at http://arxiv.org/math.NT/0412293.
15. Jim Propp, *The Somos Sequence Site.* http://www.math.wisc.edu/~propp/somos.html.
16. Rachel Shipsey, *Elliptic divisibility sequences*, Phd Thesis, Goldsmiths College, University of London, 2000; at http://homepages.gold.ac.uk/rachel/.
17. SLOANE, NEIL. *On-Line Encyclopedia of Integer Sequences.* http://www.research.att.com/~njas/sequences/.
18. Christine Swart, *Elliptic curves and related sequences.* PhD Thesis, Royal Holloway, University of London, 2003; http://www.isg.rhul.ac.uk/alumni/thesis/swart_c.pdf.
19. Christine Swart and Andrew Hone, 'Integrality and the Laurent phenomenon for Somos 4 sequences', 18pp at http://www.arxiv.org/math.NT/0508094.
20. WARD, MORGAN (1948). Memoir on elliptic divisibility sequences *Amer. J. Math.* **70**, 31–74.

Computing CM Points on Shimura Curves
Arising from Cocompact Arithmetic Triangle
Groups

John Voight

Magma Group, School of Mathematics and Statistics
University of Sydney, NSW 2006, Australia
jvoight@gmail.com

Abstract. Let $\Gamma \subset PSL_2(\mathbb{R})$ be a cocompact arithmetic triangle group, i.e. a Fuchsian triangle group that arises from the unit group of a quaternion algebra over a totally real number field. The group Γ acts on the upper half-plane $\mathfrak{H}$; the quotient $X_{\mathbb{C}} = \Gamma \backslash \mathfrak{H}$ is a Shimura curve, and there is a map $j : X_{\mathbb{C}} \to \mathbb{P}^1_{\mathbb{C}}$. We algorithmically apply the Shimura reciprocity law to compute CM points $j(z_D) \in \mathbb{P}^1_{\mathbb{C}}$ and their Galois conjugates so as to recognize them as purported algebraic numbers. We conclude by giving some examples of how this method works in practice.

To motivate what follows, we begin with a description of the classical situation. The subgroup $\Gamma_0(N) \subset \mathrm{SL}_2(\mathbb{Z})$ of matrices which are upper triangular modulo $N \in \mathbb{Z}_{>0}$ acts on the completed upper half-plane $\mathfrak{H}^*$ by linear fractional transformations; the quotient $X_0(N)_{\mathbb{C}} = \Gamma_0(N) \backslash \mathfrak{H}^*$ can be given the structure of a compact Riemann surface. The complex curve $X_0(N)_{\mathbb{C}}$ itself is a moduli space for (generalized) elliptic curves equipped with a cyclic subgroup of order N, and consequently it has a model $X_0(N)_{\mathbb{Q}}$ defined over $\mathbb{Q}$. There exist "special" points on $X_0(N)_{\mathbb{Q}}$, known as *CM points*, where the corresponding elliptic curves have complex multiplication by quadratic imaginary fields K. CM points are defined over abelian extensions H of K, and the *Shimura reciprocity law* explicitly describes the action of the Galois group $\mathrm{Gal}(H/K)$ on them. The image of a CM point under the elliptic modular j-function is known as a *singular modulus*. Gross and Zagier give a formula for the norm of the difference of two singular moduli [6]; the traces of singular moduli arise as the coefficients of modular forms (see e.g. [18]).

In this article, we generalize this situation by replacing the modular curve $X_0(N)$ by a *Shimura curve* $X_0(\mathfrak{N})$, associated to a quaternion algebra defined over a totally real number field F. The curves $X_0(\mathfrak{N})$ we will consider similarly come equipped with a map $j : X_0(\mathfrak{N}) \to \mathbb{P}^1$ as well as CM points defined over abelian extensions H of totally imaginary extensions K of F. Developing ideas of Elkies [5], we can compute these points to high precision as complex numbers, and we generalize his methods by using the Shimura reciprocity law to recognize them as putative algebraic numbers by also computing their conjugates under $\mathrm{Gal}(H/K)$. We may then compute the norms, traces, and other information

F. Hess, S. Pauli, and M. Pohst (Eds.): ANTS 2006, LNCS 4076, pp. 406–420, 2006.

about these CM points, with a view towards a generalized Gross-Zagier formula in this setting.

In §§1–2, we introduce the basic facts about quaternion algebras, Fuchsian groups and Shimura curves that we will use in the sequel. In §3, we outline numerical methods for computing the value of the map j to high precision—this can safely be skipped for the reader willing to accept Proposition 3.2. In §4, we treat the problem of principalization of ideals in maximal orders of quaternion algebras, and in Algorithm 4.4 we solve this problem under hypotheses that hold in our situation. In §5, we define CM points and show in Algorithm 5.2 how to compute these points as putative algebraic numbers using the Shimura reciprocity law. In §6, we briefly discuss relevant Galois descent. Finally, in §7, we give examples of how these algorithms work in practice, and in §8 we tabulate some of our results.

1   Quaternion Algebras

In this section, we introduce quaternion algebras and describe some of their basic properties. A reference for the material in this section is [15]. Throughout, let F be a field with char $F \neq 2$.

A *quaternion algebra* A over F is a central simple algebra of dimension 4 over F, or equivalently, an F-algebra with generators $\alpha, \beta \in A$ such that

$$\alpha^2 = a, \quad \beta^2 = b, \quad \alpha\beta = -\beta\alpha \tag{1}$$

with $a, b \in F^*$.

Example 1.1. The matrix ring $M_2(F)$ is a quaternion algebra over any field F, as is the division ring $\mathbb{H}$ of Hamiltonians over $\mathbb{R}$.

Let A be a quaternion algebra over F. Then A has a unique involution $^-: A \to A$ called *conjugation* such that $\theta + \overline{\theta}, \theta\overline{\theta} \in F$ for all $\theta \in A$, and we define the *reduced trace* and *reduced norm* of θ to be respectively $\mathrm{trd}(\theta) = \theta + \overline{\theta}$ and $\mathrm{nrd}(\theta) = \theta\overline{\theta}$. For A as in (1) and $\theta = x + y\alpha + z\beta + w\alpha\beta \in A$, we have

$$\overline{\theta} = x - (y\alpha + z\beta + w\alpha\beta), \quad \mathrm{trd}(\theta) = 2x, \quad \mathrm{nrd}(\theta) = x^2 - ay^2 - bz^2 + abw^2.$$

Let $K \supset F$ be a field containing F. Then $A_K = A \otimes_F K$ is a quaternion algebra over K, and we say K *splits* A if $A_K \cong M_2(K)$. If $[K : F] = 2$, then K splits A if and only if there exists an F-embedding $K \hookrightarrow A$.

Now let F denote a number field with ring of integers $\mathbb{Z}_F$. Let v be a noncomplex place of F, and let F_v denote the completion of F at v. Then there is a unique quaternion algebra over F_v which is a division ring, up to isomorphism. We say A is *unramified* at v if F_v splits A otherwise say A is *ramified* at v. The algebra A is ramified at only finitely many places v, and we define the *discriminant* of A to be the ideal of $\mathbb{Z}_F$ given by the product of all finite ramified places of A.

A $\mathbb{Z}_F$-*lattice* of A is a finitely generated $\mathbb{Z}_F$-submodule I of A such that $FI = A$. An *order* of A is a $\mathbb{Z}_F$-lattice which is also a subring of A. A *maximal order* of A is an order which is not properly contained in any other order.

2 Shimura Curves Arising from Triangle Groups

In this section we introduce Shimura curves and triangle groups; basic references are [7] and [5].

Let $\mathfrak{H}$ be the complex upper-half plane, equipped with the hyperbolic metric d. The group $PSL_2(\mathbb{R})$ isometrically acts on $\mathfrak{H}$ by linear fractional transformation. Let Γ be a *Fuchsian group*, a discrete subgroup of $PSL_2(\mathbb{R})$ such that the orbit space $X_{\mathbb{C}} = \Gamma \backslash \mathfrak{H}$ has finite hyperbolic area. The quotient space $X_{\mathbb{C}}$ can be given the structure of a Riemann surface of genus g.

The stabilizer $\Gamma_z = \{\gamma \in \Gamma : \gamma(z) = z\}$ of a point $z \in \mathfrak{H}$ is finite and cyclic; a point $z \in \mathfrak{H}$ is an *elliptic point* of order $k \geq 2$ if $\#\Gamma_z = k$. A maximal finite subgroup of Γ is known as an *elliptic cycle*. The set of Γ-orbits with nontrivial stabilizer is finite and in bijective correspondence with the set of elliptic cycles up to conjugation. Choosing a point $z_0 \in \mathfrak{H}$ not fixed by any element of $\Gamma \backslash \{1\}$, we obtain a fundamental domain for Γ given by

$$D = \{z \in \mathfrak{H} : d(z, z_0) \leq d(z, \gamma(z_0)) \text{ for all } \gamma \in \Gamma\}. \tag{2}$$

The domain D is a hyperbolic polygon, a connected, closed hyperbolically convex region bounded by a union of geodesics.

Now let F be a totally real number field with $[F : \mathbb{Q}] = n$ and let A be a quaternion algebra over F such that $A \otimes_{\mathbb{Q}} \mathbb{R} \cong M_2(\mathbb{R}) \times \mathbb{H}^{n-1}$. We fix the unique real place of F at which A is unramified and identify F as a subfield of $\mathbb{R}$ by this embedding; we also fix an isomorphism $\iota_{\infty} : A \otimes_F \mathbb{R} \xrightarrow{\sim} M_2(\mathbb{R})$. Let $\mathcal{O}$ be a maximal order in A (unique up to conjugation in A) and define the subgroup

$$\Gamma^*(1) = \{\iota_{\infty}(\gamma) : \gamma \in A, \ \gamma\mathcal{O} = \mathcal{O}\gamma, \ \mathrm{nrd}(\gamma) \text{ totally positive}\}/\{\pm 1\} \subset PSL_2(\mathbb{R}).$$

The group $\Gamma^*(1)$ is an *arithmetic* Fuchsian group, and as above it gives rise to a Riemann surface $X^*(1)_{\mathbb{C}} = \Gamma^*(1)\backslash\mathfrak{H}$.

An example of this situation is the modular group $\Gamma^*(1) = PSL_2(\mathbb{Z})$ with the usual fundamental domain, which corresponds to $F = \mathbb{Q}$ and $A = M_2(\mathbb{Q})$. We will exclude this well-studied case and assume from now on that A is a division ring, and thus the fundamental domain D and $X^*(1)_{\mathbb{C}}$ are compact.

Suppose that Γ has t elliptic cycles of order $m_1, \ldots, m_t$. Then the group Γ is freely generated by elements $a_1, b_1, \ldots, a_g, b_g, s_1, \ldots, s_t$ subject to the relations

$$s_1^{m_1} = \ldots = s_t^{m_t} = s_1 \cdots s_r [a_1, b_1] \cdots [a_g, b_g] = 1$$

where $[a, b] = aba^{-1}b^{-1}$; the group Γ is said to have *signature* $(g; m_1, \ldots, m_t)$. We further make the assumption that $\Gamma^*(1)$ is a *triangle group*, a Fuchsian group of signature $(0; p, q, r)$ with $p, q, r \in \mathbb{Z}_{\geq 2}$. Therefore we have a presentation

$$\Gamma^*(1) = \langle s_p, s_q, s_r | s_p^p = s_q^q = s_r^r = s_p s_q s_r = 1 \rangle. \tag{3}$$

The fundamental domain D is the union of a *fundamental triangle*, a hyperbolic triangle with angles $\pi/p, \pi/q, \pi/r$ and vertices z_p, z_q, z_r at the fixed points of the generators s_p, s_q, s_r, respectively, together with its image in the reflection in the geodesic connecting any two of the vertices.

By assumption we have $g = 0$ and hence we have a map $j : X^*(1)_{\mathbb{C}} \to \mathbb{P}^1_{\mathbb{C}}$, which is uniquely defined once we assert that the images of the elliptic points z_p, z_q, z_r be $0, 1, \infty$, respectively.

By [13], there are exactly 18 quaternion algebras A (up to isomorphism), defined over one of 13 totally real fields F, that give rise to such a cocompact arithmetic triangle group $\Gamma^*(1)$. (As pointed out in [5, p. 3], already these contain a number of highly interesting curves.) We note that each such F is Galois over $\mathbb{Q}$ and has strict class number 1.

3 Computing Hypergeometric Series

We continue notation from §§1–2. In this section, we address the following problem.

Problem 3.1. Given $z \in \mathfrak{H}$, compute the value $j(z) \in \mathbb{P}^1(\mathbb{C})$.

In other words, in Problem 3.1 we wish to compute the parametrization $j : \Gamma^*(1)\backslash\mathfrak{H} \to X^*(1)$ to large precision over $\mathbb{C}$. The reader who is uninterested in these numerical concerns may safely accept the following proposition and proceed to the next section.

Proposition 3.2. *There exists an explicit algorithm to solve Problem 3.1.*

For the details, we refer the reader to [5] and [17, §5.2].

We provide an outline of the proof of Proposition 3.2. In the first step, we reduce the problem to one in a neighborhood of an elliptic point. Let D be the fundamental domain obtained from the union of the fundamental triangle and its image in the reflection in the geodesic connecting z_p and z_r. We then find $z' \in D$ in the Γ-orbit of z as follows.

Algorithm 3.3. For $z \in \mathfrak{H}$, this algorithm returns an element $z' \in D$ in the Γ-orbit of z.

1. Let z_q' be the image of z_q in the reflection of the geodesic $z_p z_r$.
2. Apply s_r to z until z is in the region R bounded by the geodesics $z_r z_q$ and $z_r z_q'$.
3. If $z \in D$, stop. Otherwise, apply s_p until z is in the region bounded by the geodesics $z_p z_q$ and $z_p z_q'$. Return to Step 2.

Proof. For the proof, we map $\mathfrak{H}$ conformally to the unit disc $\mathfrak{D}$ by the map $z \mapsto (z - z_r)/(z - \overline{z_r})$, which maps $z_r \mapsto 0$. The element s_r now acts by rotation on $\mathfrak{D}$ by $2\pi/r$ about the origin, and the image of R is a sector $S \subset \mathfrak{D}$ with central angle $2\pi/r$.

Since the center of rotation of s_p lies away from the origin, for $z \in S$ as in Step 3 we have $z \in D$ if and only if $|s_p^i z| \geq |z|$ for all $0 < i \leq p$. Thus we see that the algorithm terminates correctly, because we obtain in this way a Γ-orbit with strictly decreasing absolute value, and yet the group Γ acts discontinuously so this orbit is finite.

Since $j(z') = j(z)$, we replace z by z'. Now the point z is near to at least one elliptic point τ of Γ. We apply the linear fractional transformation

$$z \mapsto w = \frac{z - \tau}{z - \overline{\tau}}$$

which maps the upper half-plane $\mathfrak{H}$ to the open unit disc $\mathfrak{D}$ and maps $\tau \mapsto 0$. One easily recovers z from w as

$$z = \frac{\overline{\tau}w - (\overline{\tau} + \tau)}{w - 1}.$$

Next, rather than computing the value $j(z)$ directly, we use the fact that $t = j(z)$ arises as an automorphic function for the group Γ. For the elliptic point τ of order s, there exists a Puiseux series $\phi_\tau(t) \in t^{1/s}\mathbb{C}[[t]]$ given as an explicit quotient of two hypergeometric series, such that

$$w = \phi_\tau(j(z)).$$

To conclude, we use a combination of series reversion and Newton's method which, given z (and therefore w), finds the value $t = j(z)$ such that $w = \phi_\tau(t)$.

4   Principalization of Ideals

In this section, we exhibit in Algorithm 4.4 a way to compute a generator for a principal (right) ideal of a maximal order $\mathcal{O}$ for a certain class of quaternion algebras A. Already, computing the class group and unit group of a number field appears to be a difficult task; consequently, we will be content to provide an effective algorithm that seems to work well in practice, as we are unable to prove any rigorous time bounds. We refer the reader to [15, §III.5] for the background relevant to this section.

Let A be a quaternion algebra defined over a number field F, and let $\mathcal{O} \subset A$ be a maximal order. Let I, J be right ideals of $\mathcal{O}$. We say that I and J are in the same *ideal class*, and write $I \sim J$, if there exists an $\alpha \in A^*$ such that $I = \alpha J$, or equivalently, if I and J are isomorphic as right $\mathcal{O}$-modules. It is clear that $\sim$ defines an equivalence relation on the set of right ideals of $\mathcal{O}$. Since A is non-commutative, the set of ideal classes may not form a group; however, the number h of ideal classes is finite and is independent of $\mathcal{O}$.

We are led to the following problem.

Problem 4.1. Given a right ideal $I \subset \mathcal{O}$, determine if I is a principal ideal and, if so, compute an an element α such that $I = \alpha\mathcal{O}$.

For applications to the situation of Shimura curves, we may assume that A has at least one unramified real place; we say then that A *satisfies the Eichler condition*. For $I \subset \mathcal{O}$ a right ideal, we define $\mathrm{nrd}(I)$ to be the ideal of $\mathbb{Z}_F$ generated by the set $\{\mathrm{nrd}(x) : x \in I\}$.

Proposition 4.2 ([9, Corollary 34.21], [15, Théorème III.5.7]). *Suppose that A satisfies the Eichler condition. Then the map nrd gives a bijection between the set of ideal classes and the class group $\mathrm{Cl}\,\mathbb{Z}_F$.*

In view of Proposition 4.2, the task identifying principal ideals in $\mathcal{O}$ is computationally equivalent to the analogous problem for F. From now on, suppose that F is a totally real field with $[F : \mathbb{Q}] = n$ and that A satisfies the Eichler condition.

Lemma 4.3. *Let $I \subset \mathcal{O}$ be a right ideal, and let $\xi \in I$. Then ξ generates I if and only if $\mathrm{nrd}(\xi)\mathbb{Z}_F = \mathrm{nrd}(I)$, which holds if and only if $|N_{F/\mathbb{Q}}(\mathrm{nrd}(\xi))| = N_{F/\mathbb{Q}}(\mathrm{nrd}(I))$.*

Proof. If one first defines the norm N of a right ideal I of $\mathcal{O}$ as the product of the primes of $\mathcal{O}$ occuring in a composition series for $\mathcal{O}/I$ as a $\mathcal{O}$-module (see [9, 24.1]), then the statement $\xi\mathcal{O} = I$ if and only if $N(\xi\mathcal{O}) = N(I)$ is obvious. Since $[A : F] = 4$, then the norm N is the square of the reduced norm by [9, Theorem 24.11]. The second statement follows in the same way, now in the much easier context of Dedekind domains.

The following algorithm then gives a solution to Problem 4.1 under these hypotheses.

Algorithm 4.4 (Principal ideal testing). Let $I \subset \mathcal{O}$ be a right ideal. This algorithm determines if I is principal and outputs a generator for I if one exists.

1. Compute $\mathrm{nrd}(I) \subset \mathbb{Z}_F$. Test if $\mathrm{nrd}(I) \subset \mathbb{Z}_F$ is principal by [3, §6.5.10]. If not, output a message indicating that I is not principal and terminate the algorithm. Otherwise, let $q = N_{F/\mathbb{Q}}(\mathrm{nrd}(I))$.

2. Find a $\mathbb{Z}$-generating set for I and write these elements in a $\mathbb{Z}$-basis for $\mathcal{O}$. Using the MLLL algorithm [3, 2.6.8], find a $\mathbb{Z}$-basis $B = (\gamma_1, \ldots, \gamma_{4n})$ for I.

3. Let $\sigma_1, \ldots, \sigma_n$ be the n distinct real embeddings $F \hookrightarrow \mathbb{R}$. Embed $I \hookrightarrow \mathbb{R}^{4n}$ as a lattice L via the embedding

$$\mu \mapsto (\sigma_i(\mu_j))_{\substack{i=1,\ldots,n \\ j=1,\ldots,4}}$$

where we write $\mu = \mu_1 + \mu_2\alpha + \mu_3\beta + \mu_4\alpha\beta$ for α, β as in (1). Compute an LLL-reduced basis L' of this lattice with respect to the ordinary inner product on $\mathbb{R}^{4n}$, and let T be the unimodular transformation such that $TL = L'$. Let $B' = TB$ be the basis for I obtained by applying T to the basis B.

4. For each μ in the $\mathbb{Z}$-linear span of B', compute $\mathrm{nrd}(\mu)$. If $|N_{F/\mathbb{Q}}(\mathrm{nrd}(\mu))| = q$, output μ and terminate the algorithm.

The algorithm terminates correctly by Proposition 4.2 if I is not principal and by Lemma 4.3 (and sheer enumeration) if I is principal.

Remark 4.5. The LLL-step proves experimentally to be crucial. We can see this more precisely by the following statement: There exists a $C \in \mathbb{R}_{>0}$ such that for every ideal I of $\mathcal{O}$, the first basis element γ in the *LLL*-reduced basis B' in step 3 of Algorithm 4.4 satisfies

$$|N_{F/\mathbb{Q}}(\mathrm{nrd}(\gamma))| \leq C|N_{F/\mathbb{Q}}(\mathrm{nrd}(I))|.$$

Since any generator $\xi \in I$ has $N_{F/\mathbb{Q}}(\mathrm{nrd}(I)) = |N_{F/\mathbb{Q}}(\mathrm{nrd}(\xi))|$, we conclude that the algorithm produces elements which are very close to being generators. We refer the reader to [17, Proposition 4.4.9] for the proof and a discussion.

5   CM Points and Shimura Reciprocity

In this section, we define CM points and give methods for explicitly computing them. We continue notation from §§1–2.

We first classify quadratic orders over $\mathbb{Z}_F$. The quadratic extensions K of F are classified by Kummer theory as the fields $K = F(\sqrt{D})$ for $D \in F^*/F^{*2}$. A *quadratic order* over $\mathbb{Z}_F$ is a $\mathbb{Z}_F$-algebra which is a domain and a projective $\mathbb{Z}_F$-module of rank 2. In our situation, F has class number 1, hence each such quadratic order is equal as a $\mathbb{Z}_F$-module to $\mathbb{Z}_F \oplus \mathbb{Z}_F\delta$ for some $\delta \in \mathbb{Z}_K$; the discriminant $D \in \mathbb{Z}_F$ of a minimal polynomial for δ is independent of the choice of δ up to an element of $\mathbb{Z}_F^{*2}$. Therefore the set of quadratic orders over $\mathbb{Z}_F$ is in bijection with the set of orbits of

$$\{D \in \mathbb{Z}_F : D \text{ is not a square, } D \text{ is a square modulo } 4\mathbb{Z}_F\}$$

under the action of multiplication by $\mathbb{Z}_F^{*2}$. We denote the order of discriminant $D \in \mathbb{Z}_F$ by O_D. Each such order is contained in a unique maximal order of discriminant d, known as the *fundamental discriminant*, with $D = df^2$ and $f \in \mathbb{Z}_F$ (unique up to $\mathbb{Z}_F^*$). We say that a quadratic order O_D is *totally imaginary* if D is totally negative.

Let O_D be a totally imaginary quadratic order of discriminant $D = df^2$ with field of fractions $K = F(\sqrt{d})$. Suppose that K is a splitting field for A. Then there exists an embedding $\iota_K : K \hookrightarrow A$; more concretely, the map ι_K is given by an element $\mu \in O$ whose minimal polynomial over F has discriminant D. We further assume that the embedding is *optimal*, so that $\iota_K(K) \cap O = O_D$ (see [4]). Let $z = z_D$ be the fixed point of $\iota_K(\mu)$ in $\mathfrak{H}$; we then say z is a *CM point* on $\mathfrak{H}$, and $j(z)$ is a *CM point* on $\mathbb{P}^1(\mathbb{C})$.

Let H_D be the ring class field of K of conductor f. By class field theory, we have the Artin isomorphism

$$\mathrm{Cl}(O_D) \xrightarrow{\sim} \mathrm{Gal}(H_D/K)$$
$$[\mathfrak{p}] \mapsto \mathrm{Frob}_\mathfrak{p}$$

for all primes $\mathfrak{p}$ of K unramified in H_D, where $\mathrm{Cl}(O_D)$ is the group of invertible fractional ideals of O_D modulo principal fractional ideals. For any fractional ideal $\mathfrak{c}$ of K with $\mathfrak{c} \leftrightarrow \sigma$ under the Artin map, by Proposition 4.2 there exists $\xi \in A$ such that

$$\iota_K(\mathfrak{c})O = \xi O,$$

which describes the action of $\mathrm{Gal}(H_D/K)$ on $j(z)$ as indicated in the following theorem known as the *Shimura reciprocity law*.

Theorem 5.1 ([11, p. 59]). *We have $j(z) \in \mathbb{P}^1(H_D)$ and*

$$j(z)^\sigma = j(\iota_\infty(\xi^{-1})(z)).$$

We may now compute the conjugates of $j(z)$ under $\mathrm{Gal}(H_D/K)$.

Algorithm 5.2. This algorithm computes the set

$$\{j(z)^\sigma : \sigma \in \mathrm{Gal}(H_D/K)\} \subset \mathbb{C}. \tag{4}$$

1. Compute a set G of ideals in bijection with the ring class group $\mathrm{Cl}\,\mathcal{O}_D$.
2. Using Algorithm 4.4, for each ideal $\mathfrak{c} \in G$, compute an element $\xi \in \mathcal{O}$ such that $\mathfrak{c}\mathcal{O} = \xi\mathcal{O}$.
3. For each ξ from Step 2, compute $j(\iota_\infty(\xi^{-1})(z))$ according to Proposition 3.2, and output this set.

Remark 5.3. One can compute the set G in step 1 by the natural exact sequence

$$1 \to \frac{(\mathbb{Z}_K/f\mathbb{Z}_K)^*}{\mathbb{Z}_K^*(\mathbb{Z}_F/f\mathbb{Z}_F)^*} \to \mathrm{Cl}\,\mathcal{O}_D \to \mathrm{Cl}\,\mathbb{Z}_K \to 1;$$

a representative set of elements of $\mathrm{Cl}\,\mathcal{O}_D$ can be obtained as cosets of $\mathrm{Cl}\,\mathbb{Z}_K$.

Given a complete set of conjugates t^σ of a purported algebraic number t, we then compute the polynomial

$$f(x) = \prod_{\sigma \in G} (x - t^\sigma)$$

and attempt to recognize the coefficients of this polynomial as elements of F using LLL (see [3, §2.7.2]).

6   Galois Descent

In this section, we discuss the Galois descent properties of CM points z. The computationally-minded reader may proceed to the next section, since these results will not affect the output. We continue notation from §2 and §5.

According to Theorem 5.1, a CM point $j(z)$ of discriminant D is defined over the ring class field H_D of $K = F(\sqrt{D})$. However, the set of conjugates of $j(z)$ may descend to a smaller field.

Proposition 6.1. *Let S be a full set of $\mathrm{Gal}(H_D/K)$-conjugates of $j(z)$ as in (4). Then S is in fact a full set of $\mathrm{Gal}(H_D/F)$-conjugates.*

Suppose that $\sigma(D)/D \in \mathbb{Z}_F^{*2}$ for all $\sigma \in \mathrm{Gal}(F/\mathbb{Q})$. Then H_D is Galois over $\mathbb{Q}$, and S is a full set of $\mathrm{Gal}(H_D/\mathbb{Q})$-conjugates.

The first statement is due to Shimura [12, §9.2]. Unfortunately, the proof of the second statement is too detailed to appear in these pages. For some discussion, see [17, Propositions 5.1.2, 5.4.1], though the proof there is incomplete. We now give a sketch of the proof of Proposition 6.1.

Let $\mathfrak{N}$ be an ideal of $\mathbb{Z}_F$, and define

$$\Gamma(\mathfrak{N}) = \{\iota_\infty(\gamma) : \gamma \in \mathcal{O}^*, \ \mathrm{nrd}(\gamma) = 1, \ \gamma \equiv 1 \ (\mathrm{mod} \ \mathfrak{N})\}. \tag{5}$$

We define $X(\mathfrak{N})_{\mathbb{C}} = \Gamma(\mathfrak{N})\backslash\mathfrak{H}$. Denote by $H(\mathfrak{N})$ the ray class field of F of conductor $\mathfrak{N}$.

The curve $X(\mathfrak{N})$ has an interpretation as a moduli space for a certain class of abelian varieties equipped with level structure, and as a result it has a canonical model defined over a number field. The following is due to Shimura.

Theorem 6.2 ([11, Main Theorem I (3.2)]). *There exists a projective, non-singular curve $X(\mathfrak{N})_{H(\mathfrak{N})}$ defined over $H(\mathfrak{N})$ and a holomorphic map $j_\mathfrak{N} : \mathfrak{H} \to X(\mathfrak{N})_{\mathbb{C}}$, such that the map $j_\mathfrak{N}$ yields an analytic isomorphism*

$$j_\mathfrak{N} : \Gamma(\mathfrak{N})\backslash\mathfrak{H} \xrightarrow{\sim} X(\mathfrak{N})_{\mathbb{C}}.$$

As with the case of modular curves, with additional restrictions on the moduli interpretation, one obtains a curve $X(\mathfrak{N})_F$ defined over F.

Claim. If $\sigma(\mathfrak{N}) = \mathfrak{N}$ for all $\sigma \in \mathrm{Gal}(F/\mathbb{Q})$, then $X(\mathfrak{N})_F$ has a model $X(\mathfrak{N})_{\mathbb{Q}}$ defined over $\mathbb{Q}$.

Let S be the set of ramified places of A. For $\sigma \in \mathrm{Gal}(F/\mathbb{Q})$, let A^σ be the quaternion algebra which is ramified at the set $\sigma(S)$, let $\Gamma^*(1)^\sigma$ be the group associated to this data as in §2, and let $\Gamma(\mathfrak{N})^\sigma$ be defined as in (5) for the ideal $\sigma(\mathfrak{N})$. By functoriality, we see that the Galois-conjugate curve $X(\mathfrak{N})^\sigma$ corresponds exactly to the Shimura curve associated to the quaternion algebra A^σ and ideal $\sigma(\mathfrak{N})$.

It is well-known that any two triangle groups of the same type (i.e. having the same signature) are conjugate under $PSL_2(\mathbb{R})$. From the basic theory of Shimura curves, we see that the groups $\Gamma^*(1)$ and $\Gamma^*(1)^\sigma$ have the same type. So let $\delta \in PSL_2(\mathbb{R})$ be such that $\delta\Gamma^*(1)\delta^{-1} = \Gamma^*(1)^\sigma \subset PSL_2(\mathbb{R})$. Now using that $\sigma(\mathfrak{N}) = \mathfrak{N}$, we show that $\delta\Gamma(\mathfrak{N})\delta^{-1} = \Gamma(\mathfrak{N})^\sigma$. It follows that δ gives an isomorphism of Riemann surfaces $X(\mathfrak{N})_{\mathbb{C}} \xrightarrow{\sim} X(\mathfrak{N})_{\mathbb{C}}^\sigma$, which in fact yields an isomorphism $\phi_\sigma : X(\mathfrak{N})_F \xrightarrow{\sim} X(\mathfrak{N})_F^\sigma$ defined over F. The map ϕ_σ lies over $\mathbb{P}_F^1$ since it must pair up the elliptic points which by the classification we note have distinct orders, and hence must act by the identity. The maps ϕ_σ then give the data necessary for Galois descent to $\mathbb{Q}$ (see [16]).

Now suppose that $\mathfrak{N}$ is prime to the discriminant of A. Then we have an isomorphism $\iota_\mathfrak{N} : \mathcal{O} \otimes_{\mathbb{Z}_F} \mathbb{Z}_{F,\mathfrak{N}} \xrightarrow{\sim} M_2(\mathbb{Z}_{F,\mathfrak{N}})$, unique up to conjugation by an element of $GL_2(\mathbb{Z}_{F,\mathfrak{N}})$, where $\mathbb{Z}_{F,\mathfrak{N}}$ denotes the completion of $\mathbb{Z}_F$ at $\mathfrak{N}$. We then define the subgroup

$$\Gamma_0(\mathfrak{N}) = \{\iota_\infty(\gamma) : \gamma \in \mathcal{O}^*, \ \mathrm{nrd}(\gamma) = 1, \ \iota_\mathfrak{N}(\gamma) \text{ upper triangular modulo } \mathfrak{N}\}.$$

We let $\Gamma_0(\mathfrak{N})\backslash\mathfrak{H} = X_0(\mathfrak{N})_{\mathbb{C}}$. The quotient $X(\mathfrak{N})_{\mathbb{Q}}/H \xrightarrow{\sim} X_0(\mathfrak{N})$ by the (Borel) subgroup H is stabilized by the action of the Galois group on the automorphism group of $X(\mathfrak{N})_{\mathbb{Q}}/X^*(1)_{\mathbb{Q}}$, and hence the quotient morphism is defined over $\mathbb{Q}$ and we have a model $X_0(\mathfrak{N})_{\mathbb{Q}}$ for $X_0(\mathfrak{N})_{\mathbb{C}}$.

For each $\mathfrak{N}$, there exists an element $w_{\mathfrak{N}} \in \mathrm{Aut}(\Gamma_0(\mathfrak{N}))$, known by analogy as an *Atkin-Lehner involution*, defined to be a normalizing element $w_{\mathfrak{N}} \in \mathcal{O}$ with $\mathrm{trd}(w_{\mathfrak{N}}) = 0$ and $\mathrm{nrd}(w_{\mathfrak{N}})\mathbb{Z}_{F,\mathfrak{N}} = \mathfrak{N}$. Putting together the functions $j(z), j(w_{\mathfrak{N}}(z))$, we obtain a birational map of $X_0(\mathfrak{N})$ to an irreducible closed subvariety of $\mathbb{P}_{\mathbb{C}}^1 \times \mathbb{P}_{\mathbb{C}}^1$ of dimension 1, described by a polynomial $\Phi_{\mathfrak{N}}(x,y)$ in the affine open $(\mathbb{P}_{\mathbb{C}}^1 \setminus \{\infty\})^2 = \mathbb{A}_{\mathbb{C}}^2$. By the claim above, the polynomial $\Phi_{\mathfrak{N}}(x,y)$ has coefficients in $\mathbb{Q}$.

To conclude the proof of the proposition, let D be as in Proposition 6.1, and let N be an odd rational prime which splits in F and such that a prime above N is principal in O_D; infinitely many such integers exist by the Chebotarev density theorem. Let $O_D = \mathbb{Z}_F[\mu]$. Then there exists an element $\omega_N \in O_D$ of trace zero and norm $4N$; its image $\omega_N = \iota_K(\mu) \in \mathcal{O}$ is an Atkin-Lehner involution on $X_0(N)$. Obviously μ commutes with ω_N, so $z = \omega_N(z)$, and hence $j(z) = j(\omega_N z)$. Therefore $j(z)$ is a root of the polynomial $\Phi_N(x,x)$, and since this is true of each of the conjugates of $j(z)$ as well, we obtain Proposition 6.1.

7 Examples and Applications

We now give examples of the above algorithms for the class XI of Takeuchi [14]. Let F be the totally real subfield of $\mathbb{Q}(\zeta_9)$, where ζ_9 is a primitive ninth root of unity. Then $[F : \mathbb{Q}] = 3$, and $\mathbb{Z}_F = \mathbb{Z}[b]$, where $b = -(\zeta_9 + 1/\zeta_9)$ satisfies $b^3 - 3b - 1 = 0$. We have $\mathrm{disc}(F/\mathbb{Q}) = 3^4$ and F has strict class number 1.

We choose the unique real place σ for which $\sigma(b) > 0$, and we take A to be the quaternion algebra which is ramified at the other two real places and is unramified at all other places. By Takeuchi [14, Proposition 2], we easily compute that A is isomorphic to the algebra as in (1) with $\alpha^2 = -3$, $\beta^2 = b$.

We fix the isomorphism $\iota_\infty : A \otimes_F \mathbb{R} \xrightarrow{\sim} M_2(\mathbb{R})$, given explicitly as

$$\alpha \mapsto \begin{pmatrix} 0 & 3 \\ -1 & 0 \end{pmatrix} \qquad \beta \mapsto \begin{pmatrix} \sqrt{b} & 0 \\ 0 & -\sqrt{b} \end{pmatrix}.$$

We next compute a maximal order $\mathcal{O}$ of A. Since F has class number 1, we may represent $\mathcal{O}$ as a free $\mathbb{Z}_F$-module. We note that $K = F(\alpha) = F(\sqrt{-3}) = \mathbb{Q}(\zeta_9)$ has ring of integers $\mathbb{Z}_K = \mathbb{Z}[\zeta_9]$, and hence we have an integral element $\zeta \in A$ satisfying $\zeta^9 = 1$. Extending this to a maximal order (a naïve approach suffices here, or see [17, §4.3]), we have $\mathcal{O} = \mathbb{Z}_F \oplus \mathbb{Z}_F \zeta \oplus \mathbb{Z}_F \eta \oplus \mathbb{Z}_F \omega$, where

$$\zeta = -\frac{1}{2}b + \frac{1}{6}(2b^2 - b - 4)\alpha$$

$$\eta = -\frac{1}{2}b\beta + \frac{1}{6}(2b^2 - b - 4)\alpha\beta$$

$$\omega = -b + \frac{1}{3}(b^2 - 1)\alpha - b\beta + \frac{1}{3}(b^2 - 1)\alpha\beta.$$

These elements have minimal polynomials

$$\zeta^2 + b\zeta + 1 = 0, \quad \eta^2 - b = 0, \quad \omega^2 + 2b\omega + b^2 - 4b - 1 = 0.$$

From Takeuchi [14, Table (3)], we know that $\Gamma^*(1)$ is a triangle group with signature $(2, 3, 9)$. Explicitly, we find the elements

$$s_p = b + \omega - 2\eta, \quad s_q = -1 + (b^2 - 3)\zeta + (-2b^2 + 6)\omega + (b^2 + b - 3)\eta, \quad s_r = -\zeta$$

with $s_p, s_q, s_r \in \mathcal{O}_1^*$, satisfying the relations

$$s_p^p = s_q^q = s_r^r = s_p s_q s_r = 1,$$

hence the elliptic elements s_p, s_q, s_r generate $\mathcal{O}_1^*$. The fixed points of these elements are $z_p = 0.395526\ldots i$, $z_q = -0.153515\ldots + 0.364518\ldots i$, and $z_r = i$, and they form the vertices of a fundamental triangle. This is shown in Fig. 1: any shaded (or unshaded) triangle is a fundamental triangle for $\Gamma^*(1)$, and the union of any shaded and unshaded triangle forms a fundamental domain for $\Gamma^*(1)$.

By exhaustively listing elements of $\mathcal{O}$, we enumerate (optimal) embeddings $\iota_D : \mathcal{O}_D \hookrightarrow \mathcal{O}$ for orders with discriminant D of small norm. Using Algorithm 5.2, we compute the CM points for these orders, and the results are listed in Tables 1–4 in §8. This follows in the spirit of the extended history of computing such tables for values of the elliptic j-function (see e.g. [6, pp. 193–194]).

Example 7.1. The field $K = F(\sqrt{-7})$ has class number 1. The element

$$\mu = (-b^2 - b + 2) + (-b^2 + 2b + 5)\zeta + (2b^2 - 2b - 8)\omega + (3b + 6)\eta \in \mathcal{O}$$

has minimal polynomial $x^2 - x + 2$ hence $\mathbb{Z}_F[\mu] = \mathbb{Z}_K = O_{-7}$. The fixed point of $\iota_\infty(\mu)$ in $\mathfrak{H}$ is $-0.32\ldots + 0.14\ldots i$, which is Γ-equivalent to $z = 0.758\ldots i$; and we compute that $j(z) = -9594.703125000\ldots$, which agrees with

$$\frac{-614061}{64} = \frac{-3^5 7^1 19^2}{2^6}$$

to the precision computed (100 digits).

Example 7.2. Now take $K = F(\sqrt{-2})$, with class number 3. We find $\mu \in \mathcal{O}$ satisfying $\mu^2 + 2 = 0$, so $\mathbb{Z}_f[\mu] = \mathbb{Z}_K = O_{-8}$; explicitly,

$$\mu = (-b^2 - b + 1) + (-2b^2 + 2)\zeta + (2b^2 - b - 5)\omega + (-b^2 + b + 1)\eta.$$

We obtain the CM point $j(z) = 17137.9737\ldots$ as well as its Galois conjugates $0.5834\ldots \pm 0.4516\ldots i$. We now identify the minimal polynomial and simplify the resulting number field. Let c be the real root of $x^3 - 3x + 10$; the number field $\mathbb{Q}(c)$ has discriminant $2^3 3^4$. Then $H = K(c)$, and in fact $j(z)$ agrees with

$$\frac{4015647c^2 - 10491165c + 15369346}{4096}$$

to the precision computed (200 digits); we recognize the conjugates as

$$-\frac{4015647c^2 - 10491165c - 54832574}{8192} \pm \frac{-3821175c^2 - 7058934c + 7642350}{4096}\sqrt{-2}.$$

The product of these three conjugates is the rational number

$$\frac{7^2 71^2 199^2}{2^{20}}.$$

Once one has a CM point as a purported algebraic number, it is not clear how to prove directly that such an identification is correct! What one really needs in this situation is a Gross-Zagier formula as in [6], which would identify the set of primes dividing the norm of $j(z) - j(z')$ for CM points z, z'. This is already listed as an open problem in [5, p. 42]. The work in this direction concerning the Arakelov geometry of Shimura curves has dealt with either quaternion algebras over $\mathbb{Q}$ (such as [10], [19], [8]) or $M_2(F)$ with F real quadratic, the case of Hilbert modular forms (see [2]). A nice formulation for the case of cocompact arithmetic triangle groups seems to be in order. It is hoped that the data computed here will be useful in proving such a formula.

8   Tables and Figures

In the following tables, we list the results from the extended example in §7 concerning the $(2, 3, 9)$ triangle group.

Let $D \in \mathbb{Z}_F$ be a totally imaginary discriminant such that $\sigma(D)/D \in \mathbb{Z}_F^{*2}$ for all $\sigma \in \text{Gal}(F/\mathbb{Q})$. Then $K = F(\sqrt{D})$ is Galois over $\mathbb{Q}$ and contains an order O_D of discriminant D. We list in Table 1 for each such small D a polynomial g which is a minimal polynomial for the ring class field H_D of K of conductor $f\mathbb{Z}_F$, where $D = df^2$. In Table 2, we list factorizations of the norms of the CM point $j(z_D) \in \mathbb{P}^1(H_D)$.

In Tables 3–4, we repeat the above without assuming that D is Galois-stable.

Table 1. $\text{Gal}(F/\mathbb{Q})$-stable CM Points: Ring class fields

$-D$	$|N(D)|$	g
$b+2$	3	$\mathbb{Q}$
3	27	$\mathbb{Q}$
4	64	$\mathbb{Q}$
$4(b+2)$	192	$\mathbb{Q}$
$3(b-1)^2$	243	$\mathbb{Q}$
7	343	$\mathbb{Q}$
$5(b+2)$	375	$x^2 + x - 1$
8	512	$x^3 - 3x + 10$
$4(b-1)^2$	576	$x^2 - 3$
11	1331	$x^3 + 6x + 1$
$8(b+2)$	1536	$x^2 - 2$
12	1728	$x^3 - 2$
$9(b+2)$	2187	$x^3 + 3$
$7(b-1)^2$	3087	$x^4 - 2x^3 + 6x^2 - 5x + 1$
15	3375	$x^6 + x^3 - 1$
16	4096	$x^4 - 2x^3 + 6x^2 - 4x + 2$
$8(b-1)^2$	4608	$x^6 - 10x^3 + 1$
$12(b+2)$	5184	$x^6 - 4x^3 + 1$
$13(b+2)$	6591	$x^4 + x^3 - x^2 + x + 1$
19	6859	$x^4 + x^3 + 9x^2 + 2x + 23$
20	8000	$x^6 + 9x^4 + 14x^3 + 9x^2 + 48x + 44$
$11(b-1)^2$	11979	$x^6 - x^3 - 8$

Table 2. Gal($F/\mathbb{Q}$)-stable CM Points: Norms

$-D$	Numerator	Denominator
$b+2$	1	0
3	1	1
4	0	1
$4(b+2)$	71^2	2^7
$3(b-1)^2$	-107^2	2^{15}
7	$-3^5 7^1 19^2$	2^6
$5(b+2)$	-179^2	2^{12}
8	$7^2 71^2 199^2$	2^{20}
$4(b-1)^2$	$-19^4 71^2$	2^{21}
11	$7^2 11^{11} 19^6 307^2 431^2$	$2^{30} 17^9$
$8(b+2)$	$-19^4 71^4 503^2$	$2^{21} 17^9$
12	$-11^2 71^2 503^2 971^2$	$2^{14} 17^9$
$9(b+2)$	$71^2 179^2 863^2 1511^2$	$2^{15} 17^9$
$7(b-1)^2$	$19^8 503^2 1259^2 2267^2$	$2^{24} 5^9 17^9$
15	$-7^4 11^2 127^4 359^2 431^2 1439^2$	$2^{36} 71^7$
16	$3^{14} 7^2 19^8 199^2$	$2^{21} 71^7$
$8(b-1)^2$	$-71^4 127^4 503^2 1871^2 3527^2$	$2^{21} 5^9 53^9$
$12(b+2)$	$19^{12} 71^2 163^4$	$2^{21} 107^5$
$13(b+2)$	$-19^8 179^2 307^4 467^2 647^2 1511^2 5147^2$	$2^{24} 3^9 53^9 107^9$
19	$3^{14} 19^4 71^4 107^4 3943^2$	$2^{45} 179^7$
20	$-11^2 19^{12} 71^2 199^2 379^2 739^2 2179^2 2339^2 4519^2 4751^2 5779^2$	$2^{38} 5^9 17^{18} 179^9$
$11(b-1)^2$	$-19^{12} 127^4 827^2 1223^2 1583^2 4787^2 7127^2$	$2^{39} 17^9 197^9$

Table 3. CM Points: Ring class fields

D	$|N(D)|$	g
$-5b^2 + 9b$	71	F
$-5b^2 + b$	199	F
$8b^2 - 4b - 27$	323	$x^2 + (b^2 - 3)x - b^2 + 3$
$-3b^2 + 5b - 3$	379	F
$7b^2 + b - 28$	503	$x^3 + (-b^2 + b + 2)x + 1$
$5b^2 + 2b - 23$	523	F
$3b^2 + b - 16$	591	$x^2 - bx - 1$
$-8b^2 + 4b + 1$	639	$x^2 + (-b^2 + b + 3)x - b^2 + 1$
$-12b^2 + 16b + 5$	699	$x^2 + (-b^2 + b + 1)x - 1$
$9b^2 - 3b - 31$	739	F
$-4b^2 + 4b - 3$	867	$x^2 + (b^2 - 1)x + 1$
$b^2 - 12$	971	$x^3 + (b^2 - 1)x^2 + (b^2 - 2)x - b^2 + 2$
$8b^2 - 31$	1007	$x^4 + (-b^2 + b + 1)x^3 + bx^2 + (2b^2 - 4b - 2)x - b^2 + 2b + 1$
$-8b^2 + 12b$	1088	$x^4 + (-b - 1)x^3 + (b^2 + b - 1)x^2 + (-b^2 - b + 2)x + 1$
$-4b - 12$	1216	$x^2 - b$
$-7b^2 - 3b - 3$	1387	$x^2 + (b^2 - b - 2)x - b + 1$
$-4b^2 + 11b - 10$	1791	$x^4 + (b^2 - b - 1)x^3 + (b^2 - 2b - 2)x^2 + (b^2 - b - 1)x + 1$
$-11b^2 + 6b + 1$	2179	$x^3 + (b^2 - b - 2)x^2 + (-b^2 + 2)x + b$
$-3b^2 + 4b - 8$	2287	$x^3 - x^2 + (b^2 - b - 3)x - b^2 + 3$
$4b^2 - 23$	2719	$x^3 + (-b^2 + b + 2)x^2 + x - b$
$25b^2 - 12b - 80$	3043	$x^2 - x - b$
$-16b^2 + 24b + 4$	3264	$x^4 + (b - 1)x^2 + 1$

Table 4. CM Points: Norms

D	Numerator	Denominator
$-5b^2+9b$	$19^4 71^1$	2^{18}
$-5b^2+b$	$3^9 19^2 199^1$	2^{18}
$8b^2-4b-27$	$-19^6 107^2 163^4$	2^{45}
$-3b^2+5b-3$	$-3^9 19^4 127^2 379^1$	2^{45}
$7b^2+b-28$	$-19^6 107^2 127^6 271^2 307^2 503^1$	$2^{54} 17^9$
$5b^2+2b-23$	$-3^9 19^4 127^2 523^1$	17^9
$3b^2+b-16$	$19^8 107^2 251^2 359^2$	$2^{36} 17^9$
$-8b^2+4b+1$	$-19^8 71^2 107^2 179^2 251^2 431^2$	$2^{36} 17^9$
$-12b^2+16b+5$	$19^8 71^2 179^2 467^2$	2^{45}
$9b^2-3b-31$	$3^{15} 19^6 163^2 307^2 739^1$	$2^{45} 17^9$
$-4b^2+4b-3$	$-71^2 107^2 179^2 359^2 431^2 467^2$	$2^{45} 17^{10}$
b^2-12	$-19^{12} 127^2 179^2 199^2 251^2 271^4 487^4 971^1$	$2^{90} 17^9$
$8b^2-31$	$19^{12} 71^4 127^2 179^2 251^2 271^2 307^2 359^2 631^4$	$2^{72} 17^{18}$
$-8b^2+12b$	$-19^8 71^4 199^4 379^4 503^2 523^2 739^2$	$2^{63} 17^{18}$
$-4b-12$	$-3^{26} 19^2 71^2 199^4 379^2 523^2$	$2^{63} 17^9$
$-7b^2-3b-3$	$3^{26} 19^6 127^2 271^2 307^2$	$2^{45} 53^9$
$-4b^2+11b-10$	$19^8 71^2 107^2 163^4 431^2 467^2 683^2 719^2 1151^2 1187^2$	$2^{72} 17^9 53^9$
$-11b^2+6b+1$	$3^{33} 107^2 271^2 487^2 991^2 1063^2 2179^1$	$2^{45} 71^9$
$-3b^2+4b-8$	$-3^{33} 19^6 71^4 127^2 487^4 631^2 811^2 2287^1$	$2^{54} 17^{18} 53^9$
$4b^2-23$	$-3^{39} 19^{12} 163^2 179^2 631^2 1459^2 2719^1$	$2^{54} 17^9 53^9 71^5$
$25b^2-12b-80$	$3^{18} 19^8 71^2 127^2 163^2 179^2 251^2 271^2 631^2 811^2 1423^2 1783^2$	$2^{90} 53^9 89^9$
$-16b^2+24b+4$	$19^{16} 503^2 971^2 1619^2 1871^2 1907^2 2339^2 2591^2$	$2^{57} 17^9 53^9 71^1$

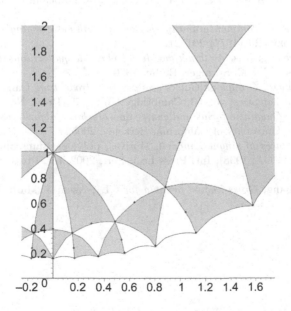

Fig. 1. The translates of a fundamental triangle for $\Gamma^*(1)$

References

1. Montserrat Alsina and Pilar Bayer, *Quaternion orders, quadratic forms, and Shimura curves*, CRM monograph series, vol. 22, American Mathematical Society, Providence, 2004.
2. Jan Hendrik Bruinier, *Infinite products in number theory and geometry*, Jahresber. Deutsch. Math.-Verein. **106** (2004), no. 4, 151–184.
3. Henri Cohen, *A course in computational algebraic number theory*, Graduate texts in mathematics, vol. 138, Springer-Verlag, Berlin, 1993.
4. M. Eichler, *Über die Idealklassenzahl hypercomplexer Systeme*, Math. Z. **43** (1938), 481–494.
5. Noam D. Elkies, *Shimura curve computations*, Algorithmic number theory (Portland, OR, 1998), Lecture notes in Comput. Sci., vol. 1423, Springer, Berlin, 1998, 1–47.
6. Benedict H. Gross and Don B. Zagier, *On singular moduli*, J. Reine Angew. Math. **355** (1985), 191–220.
7. Svetlana Katok, *Fuchsian groups*, University of Chicago Press, Chicago, 1992.
8. Stephen S. Kudla, Michael Rapoport, and Tonghai Yang, *Derivatives of Eisenstein series and Faltings heights*, Compos. Math. **140** (2004), no. 4, 887–951.
9. I. Reiner, *Maximal orders*, Clarendon Press, Oxford, 2003.
10. David Peter Roberts, *Shimura curves analogous to $X_0(N)$*, Harvard Ph.D. thesis, 1989.
11. Goro Shimura, *Construction of class fields and zeta functions of algebraic curves*, Ann. of Math. (2) **85** (1967), 58–159.
12. Goro Shimura, *Introduction to the arithmetic theory of automorphic functions*, Kanô memorial lectures, Princeton University Press, Princeton, 1994.
13. Kisao Takeuchi, *Arithmetic triangle groups*, J. Math. Soc. Japan **29** (1977), no. 1, 91–106.
14. Kisao Takeuchi, *Commensurability classes of arithmetic triangle groups*, J. Fac. Sci. Univ. Tokyo **24** (1977) 201–212.
15. Marie-France Vignéras, *Arithmétique des algèbres de quaternions*, Lecture notes in mathematics, vol. 800, Springer, Berlin, 1980.
16. Helmut Völklein, *Groups as Galois groups: an introduction*, Cambridge studies in advanced mathematics, vol. 53, Cambridge University Press, New York, 1996.
17. John Voight, *Quadratic forms and quaternion algebras: Algorithms and arithmetic*, Ph.D. thesis, University of California, Berkeley, 2005.
18. D. Zagier, *Traces of singular moduli*, Motives, polylogarithms and Hodge theory, Part I (Irvine, CA, 1998), Int. Press Lect. Ser., 2002, Int. Press, Somerville, MA, 211–244.
19. Shou-Wu Zhang, *Gross-Zagier formula for* GL_2, Asian J. Math. **5** (2001), no. 2, 183–290.

Arithmetic of Generalized Jacobians

Isabelle Déchène*

University of Waterloo, Department of Combinatorics and Optimization
Waterloo, Ontario N2L 3G1, Canada
idechene@uwaterloo.ca

Abstract. This paper aims at introducing generalized Jacobians as a new candidate for discrete logarithm (DL) based cryptography. The motivation for this work came from the observation that several practical DL-based cryptosystems, such as ElGamal, the Elliptic and Hyperelliptic Curve Cryptosystems, XTR, LUC as well as CEILIDH can all naturally be reinterpreted in terms of generalized Jacobians. However, usual Jacobians and algebraic tori are thus far the only generalized Jacobians implicitly utilized in cryptography. In order to go one step further, we here study the simplest nontrivial generalized Jacobians of an elliptic curve. In this first of a series of articles, we obtain explicit formulæ allowing to efficiently perform arithmetic operations in these groups. This work is part of our doctoral dissertation, where security aspects are considered in depth. As a result, these groups thus provide the first concrete example of semi-abelian varieties suitable for DL-based cryptography.

Keywords: Public-key cryptography, discrete logarithm problem, generalized Jacobians, semi-abelian varieties, elliptic curves.

1 Introduction and Motivation

Groups where the discrete logarithm problem (DLP) is believed to be intractable are inestimable building blocks for cryptographic applications. They are at the heart of numerous protocols such as key agreements, public-key cryptosystems, digital signatures, identification schemes, publicly verifiable secret sharings, hash functions and bit commitments. The search for new groups with intractable DLP is therefore of great importance.

In 1985, the landmark idea of Koblitz [3] and Miller [5] of using elliptic curves in public-key cryptography would, to say the least, change the perception of many on the tools of number theory that can be of practical use to cryptographers. In 1988, Koblitz [4] generalized this idea by considering Jacobians of hyperelliptic curves, which then led to the broader study of abelian varieties in cryptography. Nearly fifteen years later, Rubin and Silverberg [8] used another family of algebraic groups, namely the algebraic tori, and highlighted their great cryptographic potential.

---

* The research for this paper was done while the author was a Ph.D. student at McGill University under the supervision of Henri Darmon and Claude Crépeau and was supported by the Bell University Laboratories (BUL).

F. Hess, S. Pauli, and M. Pohst (Eds.): ANTS 2006, LNCS 4076, pp. 421–435, 2006.

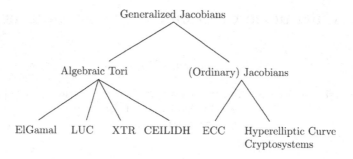

Fig. 1. Relation betweeen DL-based cryptosystems and generalized Jacobians

Now on one hand, Jacobians of curves (of small genus) gained the favor of many over the years, mostly because of the smaller key size needed. This attractive characteristic is in fact possible since we can easily generate curves for which there are no known subexponential time algorithms for solving the corresponding discrete logarithm problem. On the other hand, rational algebraic tori over a finite field offer the convenient advantage of possessing a compact representation of their elements, which then decreases the amount of information needed to be exchanged.

In a nutshell, cryptographers like Jacobians of curves for their security and care about algebraic tori for their efficiency. Thus as far as we can tell, it appears that these two sub-families of algebraic groups somehow possess complementary cryptographic advantages. From a mathematical point of view, however, the overall picture looks quite different. Indeed, with a minimal background in algebraic geometry, they can both be seen as two realizations of a single concept: *generalized Jacobians.*

As a result, several existing DL-based cryptosystems, such as the ElGamal, the Elliptic and Hyperelliptic Curve Cryptosystems, XTR, the Lucas-based cryptosystem LUC as well as the torus-based cryptosystem CEILIDH all possess an underlying structure that can be naturally reinterpreted in terms of generalized Jacobians[1]. Figure 1 provides a simplified view of the interrelation between the cryptosystems and their underlying structures.

This observation then raised the following question at the heart of our research:

Is it possible to use a generalized Jacobian that is neither a usual Jacobian nor an algebraic torus for DL-based cryptography?

An affirmative answer would then widen the class of algebraic groups that are of interest in public-key cryptography. This existence result was established in our doctoral thesis [1] by considering the simplest nontrivial generalized Jacobians of elliptic curves. These test groups are in fact semi-abelian varieties which are extensions (of algebraic groups) of an elliptic curve by the multiplicative group $\mathbb{G}_m$.

---

[1] The interpretation of XTR and LUC in terms of tori can be found in [8, Section 7].

Now recall that there are four main requirements for a group G to be suitable for DL-based cryptography. Namely,

- The elements of G can be easily represented in a compact form,
- The group operation can be performed efficiently,
- The DLP in G is believed to be intractable, and
- The group order can be efficiently computed.

We here address the first, second and fourth of these requirements. For completeness, an outline of the security results, based on [1, Section 5.5], is presented in Section 6.

This paper is organized as follows. In the next section, we give a condensed introduction to generalized Jacobians. In Section 3, we derive a natural representation of the group elements. Using this compact representation, the group law algorithm is obtained in Section 4 and basic properties are presented in Section 5. An outlook is presented in Section 6.

2   Generalized Jacobians: The Essentials

We here present an extremely concise overview[2] of generalized Jacobian varieties [6,7,9]. The underlying idea behind the construction of generalized Jacobians is essentially the same as with the usual Jacobians. That is, starting with your favorite smooth algebraic curve C defined over an algebraically closed field K, one first considers the free abelian group whose elements are (a subgroup of) its divisors of degree zero. A *clever* equivalence relation on these divisors is then defined. The quotient group obtained is then naturally isomorphic to an algebraic group, which we hope to use for cryptographic applications.

Thus the key ingredient in these constructions is the equivalence relation one considers. Loosely speaking, the whole idea behind these equivalence relations is to somehow "measure" how much a divisor $D = \sum_{P \in C} n_P(P)$ differs from a divisor $D' = \sum_{P \in C} n'_P(P)$. Linear equivalence give rise to usual Jacobians. In this case, recall that two divisors D and D' are said to be *linearly equivalent* if $D - D'$ is a principal divisor, say $D - D' = \mathrm{div}(f)$ for some f in the function field $K(C)$ of C. In this case, we write $D \sim D'$. For generalized Jacobians, the equivalence relation will now depend on the choice of an effective divisor[3] $\mathfrak{m} = \sum_{P \in C} m_P(P)$, thereafter called a *modulus*. For a given $f \in K(C)$, it is also a standard notation to write $f \equiv 1 \bmod \mathfrak{m}$ as a shorthand for the requirement $\mathrm{ord}_P(1 - f) \geq m_P$ for each P in the support of $\mathfrak{m}$.

Definition 1. *Let $\mathfrak{m}$ be an effective divisor and let D and D' be two divisors of disjoint support with $\mathfrak{m}$. We say that D and D' are $\mathfrak{m}$-equivalent, and write $D \sim_{\mathfrak{m}} D'$, if there is a function $f \in K(C)^*$ such that $\mathrm{div}(f) = D - D'$ and $f \equiv 1 \bmod \mathfrak{m}$.*

---

[2] For a more detailed exposition in the context of cryptography, see [1, Chapter 4].

[3] That is, each m_P is a nonnegative integer and only finitely many of them are nonzero.

It is a small exercise to verify that this indeed defines an equivalence relation [1, Section 4.2]. Now notice that if two divisors are $\mathfrak{m}$-equivalent, then they must be linearly equivalent as well. Therefore, if we denote by $[D]$ (respectively $[D]_{\mathfrak{m}}$) the class of D under linear equivalence (respectively $\mathfrak{m}$-equivalence), then $[D]_{\mathfrak{m}} \subseteq [D]$. This basic (but nevertheless fundamental) observation will play a key role in Sections 3 and 4, as our prior knowledge about the usual Jacobian will be our main tool for obtaining explicit formulæ for generalized Jacobians.

Next we wish to define the equivalent of the divisor class group for this new equivalence relation. Thus let $\mathrm{Div}_{\mathfrak{m}}(C)$ be the subgroup of $\mathrm{Div}(C)$ formed by all divisors of C of disjoint support with $\mathfrak{m}$. Let also $\mathrm{Div}_{\mathfrak{m}}^0(C)$ be the subgroup of $\mathrm{Div}_{\mathfrak{m}}(C)$ composed of divisors of degree zero. Moreover, let $\mathrm{Princ}_{\mathfrak{m}}(C)$ be the subset of principal divisors which are $\mathfrak{m}$-equivalent to the zero divisor[4]. It is a routine exercise to show that $\mathrm{Princ}_{\mathfrak{m}}(C)$ is a subgroup of $\mathrm{Div}_{\mathfrak{m}}^0(C)$. As a result, the set of $\mathfrak{m}$-equivalence classes is indeed a group. We will therefore consider the quotient group $\mathrm{Div}_{\mathfrak{m}}^0(C)/\mathrm{Princ}_{\mathfrak{m}}(C)$, which will be denoted by $\mathrm{Pic}_{\mathfrak{m}}^0(C)$. At last, we can state the existence theorem of Maxwell Rosenlicht whose complete proof can be found in his original article [7] as well as in [9, Chapter V, in particular Prop. 2 and Thm 1(b)].

Theorem 1 (Rosenlicht). *Let K be an algebraically closed field and C be a smooth algebraic curve of genus g defined over K. Then for every modulus $\mathfrak{m}$, there exists a commutative algebraic group $J_{\mathfrak{m}}$ isomorphic to the group $\mathrm{Pic}_{\mathfrak{m}}^0(C)$. The dimension π of $J_{\mathfrak{m}}$ is given by*

$$\pi = \begin{cases} g & \text{if } \mathfrak{m} = \mathbf{0}, \\ g + \deg(\mathfrak{m}) - 1 & \text{otherwise.} \end{cases} \tag{1}$$

Definition 2. *The algebraic group $J_{\mathfrak{m}}$ is called the* generalized Jacobian *of the curve C with respect to the modulus $\mathfrak{m}$.*

Let's now take a closer look at the relationship between J and $J_{\mathfrak{m}}$. By construction, there are isomorphisms of groups $\varphi : \mathrm{Pic}^0(C) \to J$ and $\psi : \mathrm{Pic}_{\mathfrak{m}}^0(C) \to J_{\mathfrak{m}}$. Furthermore, there is a natural surjective homomorphism $\sigma : \mathrm{Pic}_{\mathfrak{m}}^0(C) \to \mathrm{Pic}^0(C)$ defined by $\sigma([D]_{\mathfrak{m}}) = [D]$. As a result, there is a surjective homomorphism $\tau := \varphi \circ \sigma \circ \psi^{-1}$ from $J_{\mathfrak{m}}$ to J.

An interesting object of study certainly is the kernel $L_{\mathfrak{m}}$ of the map τ since it might give us information about the structure of $J_{\mathfrak{m}}$. First notice that since τ is a homomorphism, then $L_{\mathfrak{m}}$ is a subgroup of $J_{\mathfrak{m}}$. We can then consider the following short exact sequence (of abelian groups[5]):

$$0 \longrightarrow L_{\mathfrak{m}} \overset{\text{inclusion}}{\longrightarrow} J_{\mathfrak{m}} \overset{\tau}{\longrightarrow} J \longrightarrow 0$$

---

[4] The *zero divisor* $\mathbf{0} = \sum_{P \in C} 0(P)$ is the identity element of $\mathrm{Div}(C)$. Thus, $\mathrm{Princ}_{\mathfrak{m}}(C) = [\mathbf{0}]_{\mathfrak{m}} = \{\mathrm{div}(f) \,|\, f \in K(C)^* \text{ and } f \equiv 1 \bmod \mathfrak{m}\}$.

[5] One can also see generalized Jacobians as extensions of *algebraic groups*, which are discussed in [9, Chapter VII]. For the sequel, however, we shall only need to use properties of group extensions.

As a result, *the generalized Jacobian $J_{\mathfrak{m}}$ is an extension of the usual Jacobian J by $L_{\mathfrak{m}}$*. The following theorem of Rosenlicht [7] gives more information about $L_{\mathfrak{m}}$. Complete details can also be found in [9, Sections V.13-V.17].

Theorem 2 (Rosenlicht). *Let C be a smooth algebraic curve defined over an algebraically closed field, J be the Jacobian of C and $J_{\mathfrak{m}}$ be the generalized Jacobian of C with respect to a modulus $\mathfrak{m} = \sum_{P \in C} m_P(P)$ of support $S_{\mathfrak{m}}$. Let also $L_{\mathfrak{m}}$ be the kernel of the natural homomorphism τ from $J_{\mathfrak{m}}$ onto J. Then, $L_{\mathfrak{m}}$ is an algebraic group isomorphic to the product of a torus $T = (\mathbb{G}_m)^{\#S_{\mathfrak{m}}-1}$ by a unipotent group V of the form*

$$V = \prod_{P \in S_{\mathfrak{m}}} V_{(m_P)},$$

where each $V_{(m_P)}$ is isomorphic to the group of matrices of the form:

$$\begin{pmatrix} 1 & a_1 & a_2 & a_3 & \cdots & a_{m_P-1} \\ 0 & 1 & a_1 & a_2 & \cdots & a_{m_P-2} \\ 0 & 0 & 1 & a_1 & \cdots & a_{m_P-3} \\ 0 & 0 & 0 & 1 & \cdots & a_{m_P-4} \\ \vdots & \vdots & \vdots & \vdots & \ddots & \vdots \\ 0 & 0 & 0 & 0 & \cdots & 1 \end{pmatrix}$$

This result allows us (among other things) to easily see why usual Jacobians and algebraic tori are two sub-families of generalized Jacobians.

Usual Jacobians are the generalized Jacobians corresponding to the case where the linear group $L_{\mathfrak{m}}$ is trivial. That is, if the modulus $\mathfrak{m} = \sum_{P \in C} m_P(P)$ with support $S_{\mathfrak{m}}$ was chosen to have degree zero or one. Indeed, if $\mathfrak{m} = 0$, then the condition $f \equiv 1 \bmod \mathfrak{m}$, i.e. $\operatorname{ord}_{P_i}(1 - f) \geq m_i$ for each $P_i \in S_{\mathfrak{m}}$ is vacuously true and therefore, $\mathfrak{m}$-equivalence coincides with linear equivalence. As well, if $\mathfrak{m} = (M)$, then the requirement $f \equiv 1 \bmod \mathfrak{m}$ reduces to $\operatorname{ord}_M(1-f) \geq 1$, which is equivalent to $f(M) = 1$. Hence, $\mathfrak{m}$-equivalence in this case reads $D \sim_{\mathfrak{m}} D'$ iff $\exists f \in K(C)^*$ such that $\operatorname{div}(f) = D - D'$ and $f(M) = 1$. But since $\operatorname{div}(c \cdot f) = \operatorname{div}(f)$ for any nonzero constant c, the condition $f(M) = 1$ is superfluous. It then follows that when $\mathfrak{m} = (M)$, linear and $\mathfrak{m}$-equivalence also define the same divisor classes.

If we are in the situation where $\mathfrak{m} = (P_0) + (P_1) + ... + (P_r)$ with the P_i's distinct, then $L_{\mathfrak{m}}$ is isomorphic to a torus T of dimension r. Moreover, since the usual Jacobian of $\mathbb{P}^1$ is trivial [10, Example II.3.2], it then follows that the generalized Jacobian of $\mathbb{P}^1$ with respect to $\mathfrak{m}$ will be isomorphic to T. As a result, algebraic tori of any dimension can be seen as generalized Jacobians.

With these results at hand, we are now ready to explore the cryptographic potential of these algebraic groups.

3   Compact Representation of the Elements

The explicit family of generalized Jacobians that we consider can now be simply described as follows. Let E be a smooth elliptic curve defined over the finite field $K = \mathbb{F}_q$ with q elements[6] and let $B \in E(\mathbb{F}_q)$ be a point of prime order l. Let also $\mathfrak{m} = (M) + (N)$, where M and N are distinct nonzero points of $E(\mathbb{F}_{q^r})$, where $r \geq 1$ is a chosen integer. Hence, we can let $M = (X_M : Y_M : 1)$ and $N = (X_N : Y_N : 1)$. These are so far the only conditions we impose on $\mathfrak{m}$. Finally, let $J_{\mathfrak{m}}$ be the generalized Jacobian of E with respect to $\mathfrak{m}$. In the light of Theorem 2, this choice of parameters implies that this generalized Jacobian will be an extension of the elliptic curve E by the multiplicative group $\mathbb{G}_{\mathrm{m}}$, which is a nice simple case study since elliptic curves and finite fields already are cherished by cryptographers.

Now, the goal of this section is to obtain a compact representation of the elements of $J_{\mathfrak{m}}$. By a classical result on group extensions, we already know that there is a *bijection of sets* between $J_{\mathfrak{m}}$ and $\mathbb{G}_{\mathrm{m}} \times E$. Hence, each element of $J_{\mathfrak{m}}$ can be conveniently represented as a pair (k, P), where $k \in \mathbb{G}_{\mathrm{m}}$ and $P \in E$. Although the mere existence of this bijection suffices to compactly represent the elements of $J_{\mathfrak{m}}$, understanding this correspondence in depth will prove to be useful in the next section when comes the time to work out explicit formulæ for the group operation on $\mathbb{G}_{\mathrm{m}} \times E$. Indeed, we have by construction that $J_{\mathfrak{m}}$ is isomorphic to $\mathrm{Pic}^0_{\mathfrak{m}}(E)$, and so an explicit bijection of sets $\psi : \mathrm{Pic}^0_{\mathfrak{m}}(E) \to \mathbb{G}_{\mathrm{m}} \times E$ could be used to *"transport"* the known group law on $\mathrm{Pic}^0_{\mathfrak{m}}(E)$ to $\mathbb{G}_{\mathrm{m}} \times E$. Hence, exploring ψ can be seen as the first step towards the obtention of the group law algorithm on $\mathbb{G}_{\mathrm{m}} \times E$.

The official starting point of this exploration will of course be to take advantage of the fact that elliptic curves coincide with their Jacobians. Indeed, we have at our disposal the following well-known isomorphism between E and $\mathrm{Pic}^0(E)$, whose proof can be found for instance in [10, Proposition III.3.4].

Theorem 3. *Let E be a smooth elliptic curve over a perfect field K. Then the map*

$$E \to \mathrm{Pic}^0(E)$$
$$P \mapsto [(P) - (\mathcal{O})]$$

is a group isomorphism with well-defined inverse

$$\mathrm{Pic}^0(E) \to E$$
$$\left[\sum_{P \in E} n_P(P) \right] \mapsto \sum_{P \in E} n_P P.$$

Now let $D = \sum_{P \in E} n_P(P) \in \mathrm{Div}^0_{\mathfrak{m}}(E)$ be given. Under the above isomorphism, the class $[D]$ is mapped to $S = \sum_{P \in E} n_P P \in E$. As a result, $[D] = [(S) - (\mathcal{O})]$, which implies that $D - (S) + (\mathcal{O})$ is a principal divisor, say $D = (S) - (\mathcal{O}) + \mathrm{div}(f)$ for some $f \in \overline{K}(E)^*$. This suggests that $\psi([D]_{\mathfrak{m}}) = (k, S)$, for some $k \in \mathbb{G}_{\mathrm{m}}$. As we will shortly see, the determination of k will involve the computation of

---

[6] For the purpose of constructing the generalized Jacobian, we will view E as being defined over $\mathbb{F}_q$, so that the results of the previous section directly apply here.

$f(M)$ and $f(N)$. If $S \neq M, N$, then $\text{ord}_M(f) = \text{ord}_N(f) = 0$ since D has disjoint support with $\mathfrak{m}$. So in this case, $f(M)$ and $f(N)$ are both defined and nonzero. However, if $S \in \{M, N\}$, then $\text{ord}_S(f) = -1$, which means that f has a pole at S. In this case, the strategy is to use, in place of $(S) - (\mathcal{O})$, another simple divisor linearly equivalent to D which will now have disjoint support with $\mathfrak{m}$. Such a divisor is easily found by appealing to the Abel-Jacobi theorem for elliptic curves, which is easily derived from Theorem 3 [10, Corollary III.3.5].

Theorem 4 (Abel-Jacobi). *Let E be a smooth elliptic curve defined over a perfect field K and $D = \sum_{P \in E} n_P(P) \in \text{Div}(E)$ be given. Then,*

$$D \text{ is principal if and only if } \deg(D) = 0 \text{ and } \sum_{P \in E} n_P P = \mathcal{O}.$$

We therefore have an easy criterion to decide if two divisors are linearly equivalent:

Corollary 1. *Let E be a smooth elliptic curve defined over a perfect field K and let $D_1 = \sum_{P \in E} n_P(P)$, $D_2 = \sum_{P \in E} m_P(P) \in \text{Div}(E)$ be given. Then,*

$$D_1 \sim D_2 \text{ if and only if } \deg(D_1) = \deg(D_2) \text{ and } \sum_{P \in E} n_P P = \sum_{P \in E} m_P P.$$

Now observe that if we *translate* S by a point $T \in E$, we obtain by the above corollary that

$$D \sim (S) - (\mathcal{O}) \sim (S + T) - (T),$$

and thus if $T \notin \{\mathcal{O}, M, N, M - N, N - M\}$, then both $(M + T) - (T)$ and $(N + T) - (T)$ have disjoint support with $\mathfrak{m}$. So from now on, we will assume that such a *'translation point'* T is fixed and publicly known. We can now let

$$R = \begin{cases} \mathcal{O} \text{ if } S \notin \{M, N\}, \\ T \text{ otherwise,} \end{cases}$$

and so there is an $f \in \overline{K}(E)^*$ satisfying

$$D = (S + R) - (R) + \text{div}(f), \tag{2}$$

where the property $\text{ord}_M(f) = \text{ord}_N(f) = 0$ is fulfilled since D has disjoint support with $\mathfrak{m}$. Since this way of writing out a divisor already highlights the point S of E corresponding to D, it thus remains to determine how to *'read'* the corresponding element of $\mathbb{G}_\mathrm{m}$ from (2).

Since any two divisors in an $\mathfrak{m}$-equivalence class are mapped to the same element of $\mathbb{G}_\mathrm{m} \times E$, our approach will be to unravel the definition of $\mathfrak{m}$-equivalence until we can clearly see how to associate an element of $\mathbb{G}_\mathrm{m} \times E$ to each class. So let $D_1 = (S_1 + R_1) - (R_1) + \text{div}(f_1)$, $D_2 = (S_2 + R_2) - (R_2) + \text{div}(f_2) \in \text{Div}_\mathfrak{m}^0(E)$ be given such that

$$R_i = \begin{cases} \mathcal{O} \text{ if } S_i \notin \{M, N\}, \\ T \text{ otherwise,} \end{cases}$$

for $i = 1, 2$. We then have

$D_1 \sim_{\mathfrak{m}} D_2$ iff $\exists f \in \overline{K}(E)^*$ such that $\mathrm{div}(f) = D_1 - D_2$ and $f \equiv 1 \bmod \mathfrak{m}$,

$\qquad$ iff $\exists f \in \overline{K}(E)^*$ such that $\mathrm{div}(f) = (S_1 + R_1) - (S_2 + R_2) + (R_2)$

$\qquad\qquad -(R_1) + \mathrm{div}\left(\dfrac{f_1}{f_2}\right)$ and $\mathrm{ord}_M(1 - f) \geq 1$, $\mathrm{ord}_N(1 - f) \geq 1$,

$\qquad$ iff $S_1 + R_1 - (S_2 + R_2) + R_2 - R_1 = \mathcal{O}$ and $\exists f \in \overline{K}(E)^*$ such that

$\qquad\qquad \mathrm{div}(f) = \mathrm{div}\left(\dfrac{f_1}{f_2}\right)$ and $f(M) = f(N) = 1$,

$\qquad$ iff $S_1 = S_2$, $R_1 = R_2$ and $\exists c \in \overline{K}^*$ such that $\dfrac{f_1(M)}{f_2(M)} = \dfrac{f_1(N)}{f_2(N)} = \dfrac{1}{c}$,

$\qquad$ iff $S_1 = S_2$ and $\dfrac{f_1(M)}{f_2(M)} = \dfrac{f_1(N)}{f_2(N)}$,

$\qquad$ iff $S_1 = S_2$ and $\dfrac{f_1(M)}{f_1(N)} = \dfrac{f_2(M)}{f_2(N)}$.

That means that in order to check whether two given divisors are $\mathfrak{m}$-equivalent, we simply have to test two equalities, one in E and one in $\mathbb{G}_{\mathfrak{m}}$. The obvious candidate for ψ is thus the map

$$\psi : \mathrm{Pic}^0_{\mathfrak{m}}(E) \longrightarrow \mathbb{G}_{\mathfrak{m}} \times E$$
$$[D]_{\mathfrak{m}} \longmapsto (k, S),$$

such that the $\mathfrak{m}$-equivalence class of $D = \sum_{P \in E} n_P(P) \in \mathrm{Div}^0_{\mathfrak{m}}(E)$ corresponds to $S = \sum_{P \in E} n_P P$ and $k = f(M)/f(N)$, where $f \in \overline{K}(E)^*$ is any function satisfying

$$\mathrm{div}(f) = \begin{cases} D - (S) + (\mathcal{O}) & \text{if } S \notin \{M, N\}, \\ D - (S + T) + (T) & \text{otherwise.} \end{cases}$$

Notice that the existence of f is guaranteed by the Abel-Jacobi theorem and that ψ is well-defined since we have just shown that for $D_1 = (S_1 + R_1) - (R_1) + \mathrm{div}(f_1)$, $D_2 = (S_2 + R_2) - (R_2) + \mathrm{div}(f_2)$, $k_1 = f_1(M)/f_1(N)$ and $k_2 = f_2(M)/f_2(N)$, we have:

$\qquad [D_1]_{\mathfrak{m}} = [D_2]_{\mathfrak{m}}$ *implies that* $k_1 = k_2$ *and* $S_1 = S_2$.

Moreover, ψ is injective since we also already know that

$\qquad (k_1, S_1) = (k_2, S_2)$ *implies that* $[D_1]_{\mathfrak{m}} = [D_2]_{\mathfrak{m}}$.

It therefore remains to show that ψ is surjective as well. So given $(k, S) \in \mathbb{G}_{\mathfrak{m}} \times E$, we must find an $f \in \overline{K}(E)^*$ such that $f(M)/f(N) = k$. Using the idea behind the interpolation polynomial of Lagrange, or simply by inspection, we easily see that

$$f(X, Y, Z) = \begin{cases} \dfrac{k(X - X_N Z) + (X_M Z - X)}{(X_M - X_N)Z} & \text{if } X_M \neq X_N, \\ \dfrac{k(Y - Y_N Z) + (Y_M Z - Y)}{(Y_M - Y_N)Z} & \text{otherwise,} \end{cases}$$

fulfills the required conditions (notice that $X_M = X_N$ implies that $Y_M \neq Y_N$ since we assumed that $M \neq N$ and $Z_M = Z_N = 1$). Hence, the divisor

$$D = \begin{cases} (S) - (\mathcal{O}) + \operatorname{div}(f) & \text{if } S \notin \{M, N\}, \\ (S + T) - (T) + \operatorname{div}(f) & \text{otherwise}, \end{cases}$$

is mapped to (k, S), as wanted. We have therefore shown that ψ is the bijection we were looking for.

Proposition 1. *Let E be a smooth elliptic curve defined over $\mathbb{F}_q$, $T \in E \backslash \{\mathcal{O}, M, N, M - N, N - M\}$ and $\mathfrak{m} = (M) + (N)$ with M, $N \in E \backslash \{\mathcal{O}\}$, $M \neq N$ be given. Let also*

$$\psi : \operatorname{Pic}^0_{\mathfrak{m}}(E) \longrightarrow \mathbb{G}_{\mathrm{m}} \times E$$

$$[D]_{\mathfrak{m}} \longmapsto (k, S),$$

be such that the $\mathfrak{m}$-equivalence class of $D = \sum_{P \in E} n_P(P)$ corresponds to $S = \sum_{P \in E} n_P P \in E$ and $k = f(M)/f(N)$, where $f \in \overline{K}(E)^*$ is any function satisfying

$$\operatorname{div}(f) = \begin{cases} D - (S) + (\mathcal{O}) & \text{if } S \notin \{M, N\}, \\ D - (S + T) + (T) & \text{otherwise}. \end{cases}$$

Then, ψ is a well-defined bijection of sets.

Remark 1. Notice that since the zero divisor can be written as

$$\mathbf{0} = (\mathcal{O}) - (\mathcal{O}) + \operatorname{div}(c),$$

where c is any nonzero constant, then $\mathbf{0}$ corresponds to the pair $(1, \mathcal{O})$. That is, $(1, \mathcal{O})$ is the identity element of $J_{\mathfrak{m}}$.

4 Group Law Algorithm

Using the explicit bijection between $\operatorname{Pic}^0_{\mathfrak{m}}(E)$ and $\mathbb{G}_{\mathrm{m}} \times E$ that we just obtained, our next goal is to derive explicit formulæ for the group operation on $\mathbb{G}_{\mathrm{m}} \times E$ induced from $\operatorname{Pic}^0_{\mathfrak{m}}(E)$. First notice that by the theory of group extensions, we already know the basic structure of the addition on $J_{\mathfrak{m}}$. Indeed, we have for any k_1, $k_2 \in \mathbb{G}_{\mathrm{m}}$ and P_1, $P_2 \in E$,

$$(k_1, P_1) + (k_2, P_2) = (k_1 k_2 \cdot \mathbf{c}_{\mathfrak{m}}(P_1, P_2), P_1 + P_2), \tag{3}$$

where $\mathbf{c}_{\mathfrak{m}} : E \times E \to \mathbb{G}_{\mathrm{m}}$ is a 2-cocycle depending on the modulus $\mathfrak{m}$. It thus suffices to make $\mathbf{c}_{\mathfrak{m}}$ explicit.

So given (k_1, P_1) and (k_2, P_2) in $J_{\mathfrak{m}}$, our task is then to compute their sum (k_3, P_3). Notice that there are two distinct cases to study, depending if the use of a 'translation point' T is at all needed. Fortunately, there is an easy criterion to decide when it occurs. Indeed, suppose that the group we consider for cryptographic applications is the subgroup of $J_{\mathfrak{m}}$ generated by the element (k, P). By the addition rule (3), it immediately follows that

$$\text{If } (j, Q) \in \langle (k, P) \rangle, \text{ then } Q \in \langle P \rangle.$$

As a result, if neither M nor N is a multiple of P, then the group operation on $\langle (k, P) \rangle$ will *never* involve points of the form $(*, M)$ or $(*, N)$. Thus, there is no need to employ a translation point in this case. Of course, when either M or N lies in $\langle P \rangle$, then the corresponding addition formulæ will use translation points when appropriate in order to cover all possible cases. This motivates the following definition.

Definition 3. *Let E be an elliptic curve defined over $\mathbb{F}_q$ and $B \in E(\mathbb{F}_q)$ be a given basepoint. Let also M, $N \in E(\overline{\mathbb{F}_q})$ be given. Then the modulus $\mathfrak{m} = (M) + (N)$ is said to be B-unrelated if $M, N \notin \langle B \rangle$. Otherwise, it will be called B-related.*

The aim of this section is to *transport* the addition on $\operatorname{Pic}^0_{\mathfrak{m}}(E)$ to $\mathbb{G}_{\mathrm{m}} \times E$ in order to get explicit equations involving the group laws on $\mathbb{G}_{\mathrm{m}}$ and E in the case of a B-unrelated modulus $\mathfrak{m}$. So given (k_1, P_1), (k_2, P_2) and (k_3, P_3) in $J_{\mathfrak{m}}$ such that

$$(k_1, P_1) + (k_2, P_2) = (k_3, P_3) \text{ and } P_1, P_2, \pm P_3 \notin \{M, N\},$$

our task is to express (k_3, P_3) in terms of (k_1, P_1) and (k_2, P_2). By the explicit bijection between $\operatorname{Pic}^0_{\mathfrak{m}}(E)$ and $\mathbb{G}_{\mathrm{m}} \times E$, the elements (k_1, P_1) and (k_2, P_2) are respectively the image of the $\mathfrak{m}$-equivalence class of $D_1 = (P_1) - (\mathcal{O}) + \operatorname{div}(f_1)$ and $D_2 = (P_2) - (\mathcal{O}) + \operatorname{div}(f_2)$, for some $f_1, f_2 \in \overline{K}(E)^*$ such that $\operatorname{ord}_M(f_1) = \operatorname{ord}_N(f_1) = \operatorname{ord}_M(f_2) = \operatorname{ord}_N(f_2) = 0$, $f_1(M)/f_1(N) = k_1$ and $f_2(M)/f_2(N) = k_2$ (see proof of Proposition 1).

That being said, we can now endow $\mathbb{G}_{\mathrm{m}} \times E$ with the group operation inherited from $\operatorname{Pic}^0_{\mathfrak{m}}(E)$. So basically, all we need to know is to which element of $\mathbb{G}_{\mathrm{m}} \times E$ does $D_3 = D_1 + D_2$ correspond. First, we have by definition that

$$D_3 = (P_1) + (P_2) - 2(\mathcal{O}) + \operatorname{div}(f_1 \cdot f_2), \tag{4}$$

so in order to get the element of $\mathbb{G}_{\mathrm{m}} \times E$ we are looking for, the obvious strategy is to express the right hand side of (4) as $(P_3) - (\mathcal{O}) + \operatorname{div}(f_3)$. By the Abel-Jacobi theorem, we know that

$$(P_1) + (P_2) - 2(\mathcal{O}) \sim (P_1 + P_2) - (\mathcal{O}),$$

and so there is a function $L_{P_1, P_2} \in \overline{K}(E)^*$ satisfying

$$(P_1) + (P_2) - 2(\mathcal{O}) = (P_1 + P_2) - (\mathcal{O}) + \operatorname{div}(L_{P_1, P_2}). \tag{5}$$

Combining (4) and (5) yields

$$D_3 = (P_1 + P_2) - (\mathcal{O}) + \operatorname{div}(f_1 \cdot f_2 \cdot L_{P_1, P_2}).$$

We can thus set $P_3 = P_1 + P_2$ and $f_3 = f_1 \cdot f_2 \cdot L_{P_1,P_2}$. Hence, D_3 corresponds to (k_3, P_3), where

$$k_3 = \frac{f_3(M)}{f_3(N)} = \frac{f_1(M) \cdot f_2(M) \cdot L_{P_1,P_2}(M)}{f_1(N) \cdot f_2(N) \cdot L_{P_1,P_2}(N)} = k_1 \cdot k_2 \cdot \frac{L_{P_1,P_2}(M)}{L_{P_1,P_2}(N)}.$$

That is,

$$(k_1, P_1) + (k_2, P_2) = \left(k_1 \cdot k_2 \cdot \frac{L_{P_1,P_2}(M)}{L_{P_1,P_2}(N)}, P_1 + P_2 \right).$$

Moreover, notice that this addition rule so far agrees with the prediction (3) obtained from group extensions. Hence the 2-cocycle $\mathbf{c_m} : E \times E \to \mathbb{G}_m$ we were seeking is now unveiled:

$$\mathbf{c_m}(P_1, P_2) = \frac{L_{P_1,P_2}(M)}{L_{P_1,P_2}(N)}. \tag{6}$$

The very last step is to make L_{P_1,P_2} explicit. We are thus looking for a function L_{P_1,P_2} satisfying (5), or equivalently,

$$\operatorname{div}(L_{P_1,P_2}) = (P_1) + (P_2) - (P_1 + P_2) - (\mathcal{O}). \tag{7}$$

The natural approach is to consider the line ℓ_{P_1,P_2}, passing through P_1 and P_2, that will inevitably hit $-P_3 = -(P_1 + P_2)$ as well. Then,

$$\operatorname{div}\left(\frac{\ell_{P_1,P_2}}{Z} \right) = (P_1) + (P_2) + (-P_3) - 3(\mathcal{O}). \tag{8}$$

Now in order to introduce the term $-(P_1 + P_2)$ and cancel out $(-P_3)$ at once, we may consider the line $\ell_{P_1+P_2,\mathcal{O}}$ passing through $P_1 + P_2$, $\mathcal{O}$, and a fortiori through $-P_3$. That is,

$$\operatorname{div}\left(\frac{\ell_{P_1+P_2,\mathcal{O}}}{Z} \right) = (P_1 + P_2) + (-P_3) - 2(\mathcal{O}). \tag{9}$$

Subtracting (9) from (8), we get

$$\operatorname{div}\left(\frac{\ell_{P_1,P_2}}{\ell_{P_1+P_2,\mathcal{O}}} \right) = (P_1) + (P_2) - (P_1 + P_2) - (\mathcal{O}). \tag{10}$$

Finally, (7) and (10) imply that L_{P_1,P_2} and $\ell_{P_1,P_2}/\ell_{P_1+P_2,\mathcal{O}}$ differ by a nonzero multiplicative constant:

$$\exists c \in \overline{K}^* \text{ satisfying } L_{P_1,P_2} = c \cdot \frac{\ell_{P_1,P_2}}{\ell_{P_1+P_2,\mathcal{O}}}. \tag{11}$$

Let's point out that our initial conditions $M, N \neq \mathcal{O}$ and $P_1, P_2, P_3 = P_1 + P_2 \notin \{M, N\}$ are sufficient to ensure that $L_{P_1,P_2}(M)$ and $L_{P_1,P_2}(N)$ will both be defined and nonzero, since (7) tells us that the only zeros and poles of L_{P_1,P_2} occur at P_1, P_2, $P_1 + P_2$ and $\mathcal{O}$. Furthermore, we can compute $L_{P_1,P_2}(M)$ and

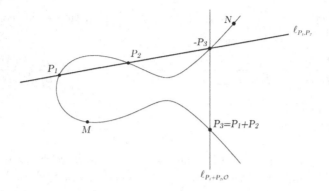

Fig. 2. Unveiling the 2-cocycle $\mathbf{c}_\mathrm{m}$

$L_{P_1,P_2}(N)$ by evaluating $\ell_{P_1,P_2}(M), \ell_{P_1+P_2,\mathcal{O}}(M), \ell_{P_1,P_2}(N)$ and $\ell_{P_1+P_2,\mathcal{O}}(N)$ separately since we also assumed that $-P_3 \neq M, N$.

Therefore, by (6) and (11), it is now legitimate to write

$$\mathbf{c}_\mathrm{m}(P_1, P_2) = \frac{L_{P_1,P_2}(M)}{L_{P_1,P_2}(N)} = \frac{\ell_{P_1,P_2}(M)}{\ell_{P_1+P_2,\mathcal{O}}(M)} \cdot \frac{\ell_{P_1+P_2,\mathcal{O}}(N)}{\ell_{P_1,P_2}(N)}, \tag{12}$$

and our goal is achieved since the 2-cocycle $\mathbf{c}_\mathrm{m}$ is now completely determined. Lastly, since we have some freedom on both the equations of the lines (they are determined up to a constant factor) and on the representatives for the homogeneous coordinates of M and N, we should verify that (12) is well-defined. That is, for $M = (X_M : Y_M : 1)$, $N = (X_N : Y_N : 1)$ and $\lambda_1, \lambda_2, c_1, c_2$ any nonzero constants, we have $M \sim (\lambda_1 X_M : \lambda_1 Y_M : \lambda_1)$, $N \sim (\lambda_2 X_N : \lambda_2 Y_N : \lambda_2)$ and $c_1 \cdot \ell_{P_1,P_2}, c_2 \cdot \ell_{P_1+P_2,\mathcal{O}}$ respectively defining the same line as ℓ_{P_1,P_2} and $\ell_{P_1+P_2,\mathcal{O}}$. Since ℓ_{P_1,P_2} and $\ell_{P_1+P_2,\mathcal{O}}$ are both homogeneous polynomials of degree one, it follows that

$$\frac{c_1 \cdot \ell_{P_1,P_2}(\lambda_1 X_M, \lambda_1 Y_M, \lambda_1)}{c_2 \cdot \ell_{P_1+P_2,\mathcal{O}}(\lambda_1 X_M, \lambda_1 Y_M, \lambda_1)} \cdot \frac{c_2 \cdot \ell_{P_1+P_2,\mathcal{O}}(\lambda_2 X_N, \lambda_2 Y_N, \lambda_2)}{c_1 \cdot \ell_{P_1,P_2}(\lambda_2 X_N, \lambda_2 Y_N, \lambda_2)} =$$

$$\frac{\lambda_1 \cdot \ell_{P_1,P_2}(X_M, Y_M, 1)}{\lambda_1 \cdot \ell_{P_1+P_2,\mathcal{O}}(X_M, Y_M, 1)} \cdot \frac{\lambda_2 \cdot \ell_{P_1+P_2,\mathcal{O}}(X_N, Y_N, 1)}{\lambda_2 \cdot \ell_{P_1,P_2}(X_N, Y_N, 1)} =$$

$$\frac{\ell_{P_1,P_2}(M)}{\ell_{P_1+P_2,\mathcal{O}}(M)} \cdot \frac{\ell_{P_1+P_2,\mathcal{O}}(N)}{\ell_{P_1,P_2}(N)},$$

which confirms that (12) was well-defined. Finally, we are ready to properly write down the group law we just obtained.

Theorem 5. *Let E be a smooth elliptic curve defined over $\mathbb{F}_q$ and let $\mathrm{m} = (M) + (N)$ be given such that M and N are distinct nonzero points of E. If (k_1, P_1) and (k_2, P_2) are elements of J_m fulfilling $P_1, P_2, \pm(P_1 + P_2) \notin \{M, N\}$, then*

$$(k_1, P_1) + (k_2, P_2) = (k_1 k_2 \cdot \mathbf{c}_\mathrm{m}(P_1, P_2), P_1 + P_2), \tag{13}$$

where $\mathbf{c_m} : E \times E \to \mathbb{G}_m$ *is the 2-cocycle given by*

$$\mathbf{c_m}(P_1, P_2) = \frac{\ell_{P_1,P_2}(M)}{\ell_{P_1+P_2,\mathcal{O}}(M)} \cdot \frac{\ell_{P_1+P_2,\mathcal{O}}(N)}{\ell_{P_1,P_2}(N)},$$

and $\ell_{P,Q}$ *denotes the equation of the straight line passing through* P *and* Q *(tangent at the curve if* $P = Q$*).*

The group law for B-related moduli can also be obtained using a similar procedure. This case is fully treated in Section 5.3.2 of [1], where the following result is presented.

Theorem 6. *Let E be a smooth elliptic curve defined over $\mathbb{F}_q$, $\mathbf{m} = (M) + (N)$ be given such that M and N are distinct nonzero points of E and let $T \in E$ be any point such that $T \notin \{\mathcal{O}, M, N, M - N, N - M\}$. Given (k_1, P_1) and (k_2, P_2) in $J_\mathbf{m}$, set $P_3 = P_1 + P_2$ and let, for $i = 1, 2, 3$,*

$$R_i = \begin{cases} T & \text{if } P_i \in \{M, N\}, \\ \mathcal{O} & \text{otherwise.} \end{cases}$$

Then,

$$(k_1, P_1) + (k_2, P_2) = \left(k_1 k_2 \cdot \frac{L(M)}{L(N)}, P_3 \right),$$

where

$$L = \frac{\ell_{P_1,P_2}}{\ell_{P_3,\mathcal{O}}} \cdot \frac{\ell_{P_1+R_1,\mathcal{O}}}{\ell_{P_1,R_1}} \cdot \frac{\ell_{P_2+R_2,\mathcal{O}}}{\ell_{P_2,R_2}} \cdot \frac{\ell_{P_3,R_3}}{\ell_{P_3+R_3,\mathcal{O}}}.$$

As usual, $\ell_{P,Q}$ *denotes the equation of the straight line passing through* P *and* Q *(tangent at the curve if* $P = Q$*).*

5   Basic Properties

We here present a small collection of the basic properties of the group law in these generalized Jacobians, which are easily derived from Theorem 5.

Corollary 2. *Let E be a smooth elliptic curve defined over $\mathbb{F}_q$ and let $\mathbf{m} = (M) + (N)$ be given such that M and N are distinct nonzero points of E. Let also (k, P), (k_1, P_1), $(k_2, P_2) \in J_\mathbf{m}$ be given such that $P_1, P_2, \pm(P_1 + P_2) \notin \{M, N\}$. Then,*

1. $(1, \mathcal{O})$ *is the identity element of* $J_\mathbf{m}$.
2. $\mathbf{c_m}(P_1, P_2) = \mathbf{c_m}(P_2, P_1)$ *(This reflects the fact that* $J_\mathbf{m}$ *is abelian).*
3. *If* $M = (X_M : Y_M : 1)$ *and* $N = (X_N : Y_N : 1)$, *then* $\mathbf{c_m}(P, -P) = \ell_{P,\mathcal{O}}(M)/\ell_{P,\mathcal{O}}(N)$, *and so the inverse of* (k, P) *is given by*

$$-(k, P) = \left(\frac{1}{k} \cdot \frac{\ell_{P,\mathcal{O}}(N)}{\ell_{P,\mathcal{O}}(M)}, -P \right).$$

4. $\mathbf{c_m}(\mathcal{O}, P) = 1$ *for all* $P \in E\backslash\{M, N\}$. *Hence,*

$$(k_1, \mathcal{O}) + (k_2, P) = (k_1 k_2, P).$$

5. *Furthermore,* J_m *contains a subgroup isomorphic to* $\mathbb{G}_m$, *as*

$$(k_1, \mathcal{O}) + (k_2, \mathcal{O}) = (k_1 k_2, \mathcal{O}) \text{ for all } k_1, k_2 \in \mathbb{G}_m.$$

6. *If* $B \in E(\mathbb{F}_q)$ *and* M, $N \in E(\mathbb{F}_{q^r})$ *are such that* $\mathfrak{m}$ *is* B-*unrelated, then* $\mathbb{F}_{q^r}^* \times \langle B \rangle$ *is a subgroup of* J_m.

The only statement that might require a further justification is property 6. Notice that it simply follows from properties 1 and 3, together with the observation that $\ell_{P_1, P_2}(M)$, $\ell_{P_1, P_2}(N) \in \mathbb{F}_{q^r}^*$ whenever P_1, $P_2 \in \langle B \rangle$. We have thus made completely explicit the finite group $\mathbb{F}_{q^r}^* \times \langle B \rangle$ of order $(q^r - 1) \cdot l$ that we wish to use for cryptographic applications.

6   Outlook

Given a smooth elliptic curve E defined over $\mathbb{F}_q$, a point $B \in E(\mathbb{F}_q)$ of prime order l and a B-unrelated modulus $\mathfrak{m} = (M) + (N)$ such that M and N are distinct points of $E(\mathbb{F}_{q^r})$ and $r \geq 1$ is a chosen integer, we now know that $\mathbb{F}_{q^r}^* \times \langle B \rangle$, together with the group law of Theorem 5, is a finite subgroup of J_m for which the elements are compactly represented, the group law efficiently computable and the group order readily determined.

Several other efficiency and security aspects were included in our doctoral dissertation[7] [1]. On one hand, we considered various implementation issues, such as choosing a suitable modulus, speeding up scalar multiplications and selecting parameters such that $\mathbb{F}_{q^r}^* \times \langle B \rangle$ is a cyclic group.

As for security, as soon as $\mathbb{F}_{q^r}^* \times \langle B \rangle$ is a cyclic subgroup of J_m, we obtained the following reductions among discrete logarithm problems:

The DLP in $\mathbb{F}_{q^r}^* \times \langle B \rangle$ *is at least as hard as the DLP in* $\langle B \rangle \subseteq E(\mathbb{F}_q)$
and at least as hard as the DLP in $\mathbb{F}_{q^r}^*$.

Thus from a practical point of view, this result implies that even though generalized Jacobians are newcomers in cryptography, we already know that solving their DLP cannot be easier than solving discrete logarithms in two of the most studied groups used in DL-based cryptography today.

Furthermore, we showed that extracting a discrete logarithm in $\mathbb{F}_{q^r}^* \times \langle B \rangle$ can always be performed by *sequentially* computing a discrete logarithm in E followed by one in $\mathbb{F}_{q^r}^*$. Moreover, it is possible to proceed in parallel when $l \nmid (q^r - 1)$ while this is still an open question in the case of curves suitable for pairing-based cryptography.

Finally, we have also investigated several scenarios involving precomputations in order to further study the DLP in $\mathbb{F}_{q^r}^* \times \langle B \rangle$. To this end, we empirically

---

[7] For which the corresponding articles are currently in preparation.

compared generalized Jacobians with the Classical Occupancy Problem. This preliminary study suggests that none of the proposed scenarios is faster than the known methods described above.

As a result, the generalized Jacobians we considered fulfill the basic requirements for a group to be suitable for DL-based cryptography. It thus provides the first concrete example of semi-abelian varieties that could be used in public-key cryptography.

Note 1. During the preparation of the final version of this article, the author became aware of the doctoral dissertation of Jean-Yves Enjalbert [2]. While this thesis contains an unsuccessful attempt to compactly represent the elements of a generalized Jacobian J_m directly in terms of those of J and L_m, it nevertheless exploits, with cryptographic applications in mind, the link between generalized Jacobians and ray class groups.

Acknowledgments. I would like to thank my thesis co-supervisors Henri Darmon and Claude Crépeau for their guidance and advices. I would also like to thank the Centre for Applied Cryptographic Research (CACR) of the University of Waterloo for providing such a stimulating research environment.

References

1. Isabelle Déchène. *Generalized Jacobians in Cryptography.* PhD thesis, McGill University, 2005.
2. Jean-Yves Enjalbert. *Jacobiennes et cryptographie.* PhD thesis, Université de Limoges, 2003.
3. Neal Koblitz. Elliptic curve cryptosystems. *Mathematics of Computation,* 48(177):203–209, January 1987.
4. Neal Koblitz. Hyperelliptic cryptosystems. *Journal of Cryptology,* 1(3):139–150, 1989.
5. Victor S. Miller. Uses of elliptic curves in cryptography. In Hugh C. Williams, editor, *Advances in Cryptology—CRYPTO '85,* volume 218 of *Lecture Notes in Computer Science,* pages 417–426, Berlin, 1986. Springer-Verlag.
6. Maxwell Rosenlicht. Equivalence relations on algebraic curves. *Annals of Mathematics,* 56:169–191, July 1952.
7. Maxwell Rosenlicht. Generalized Jacobian varieties. *Annals of Mathematics,* 59:505–530, May 1954.
8. Karl Rubin and Alice Silverberg. Torus-based cryptography. In Dan Boneh, editor, *Advances in Cryptology—CRYPTO '03,* volume 2729 of *Lecture Notes in Computer Science,* pages 349–365. Springer-Verlag, 2003.
9. Jean-Pierre Serre. *Algebraic groups and class fields,* volume 117 of *Graduate texts in mathematics.* Springer-Verlag, New-York, 1988.
10. Joseph H. Silverman. *The arithmetic of elliptic curves,* volume 106 of *Graduate Texts in Mathematics.* Springer-Verlag, New York, 1986.

Hidden Pairings and Trapdoor DDH Groups

Alexander W. Dent and Steven D. Galbraith

Information Security Group, Royal Holloway
Egham, Surrey, TW20 0EX, UK
a.dent@rhul.ac.uk, steven.galbraith@rhul.ac.uk

Abstract. This paper suggests a new building block for cryptographic protocols and gives two instantiations of it. The concept is to generate two descriptions of the same group: a public description that allows a user to perform group operations, and a private description that allows a user to also compute a bilinear pairing on the group. A user who has the private information can therefore solve decisional Diffie-Hellman (DDH) problems, and potentially also discrete logarithm problems. Some cryptographic applications of this idea are given.

Both instantiations are based on elliptic curves. The first relies on the factoring assumption for hiding the pairing. The second relies on the difficulty of solving a system of multivariate equations. The second method also potentially gives rise to a practical trapdoor discrete logarithm system.

1 Introduction

It is well-known that the computational operations which may be efficiently performed on a mathematical object depend closely on how the object is described. For example, a cyclic group of order p could be given as the additive group of integers modulo p, as a subgroup of the multiplicative group of some finite field, or as a subgroup of points on an elliptic curve over a finite field $\mathbb{F}_q$ (where $q \neq p$). The discrete logarithm problem can be solved in polynomial time in the first case and is not believed to be solvable in polynomial time in the second two cases.

Public key cryptography takes this idea further, making use of the fact that we can release a public "partial description" of a group that will allow users to compute some operations, while retaining private information that will allow us to compute a greater set of operations.

The paper provides some new examples and applications of partial descriptions of groups. In particular, we consider groups with a "hidden pairing" in the sense that only the holder of some private information can compute a bilinear pairing. We give two instantiations of this idea: one based on elliptic curves modulo an RSA modulus N and another based on Frey's idea of "disguising an elliptic curve". Our attempt to understand the security of these systems has led to the formulation of a number of interesting mathematical questions.

The aims of the paper are to raise awareness of the idea of partial group descriptions, to give some new building blocks for cryptography, and to state a

F. Hess, S. Pauli, and M. Pohst (Eds.): ANTS 2006, LNCS 4076, pp. 436–451, 2006.

number of computational questions which deserve further study. We hope that the ANTS community will find the paper a fruitful source of problems for future study and that further research follows from this work.

1.1 Pairings in Cryptography

The use of pairings has been something of a minor revolution in public key cryptography. First, they were used to attack the discrete logarithm problem in certain elliptic curve groups [8,17]. More recently they have been used as a device with which to build cryptographic primitives (see [1] for a survey). Their usefulness in this latter context is derived from their ability to provide "gap groups": groups in which the decisional Diffie-Hellman (DDH) problem is known to be easy, but in which the computational Diffie-Hellman (CDH) problem is assumed to be difficult to solve (these problems are defined in Section 2).

In this paper, we develop the idea of a "trapdoor DDH group". This is a group whose (public) description allows anyone to compute the group operation and for which there is a private trapdoor which allows a user to solve the DDH problem. Our solutions are based on "hidden pairings", which are pairings on an elliptic curve that can only be computed by an entity in possession of the trapdoor information. It is assumed that the pairing is difficult to compute for anybody not in possession of the trapdoor information. We remark that other researchers have considered the possibility of a hidden pairing (e.g., Rivest [21] and Boneh [3]) but our paper seems to be the first to propose workable solutions to the problem.

The idea of a hidden pairing suggests three applications. First, it implies the existence of trapdoor DDH groups, which could be of direct use in the construction of cryptographic algorithms and protocols.

Second, it could be used to give trapdoor DL groups: groups in which the discrete logarithm (DL) problem is easy to solve for anyone in possession of the trapdoor information, but the DL problem is difficult for anybody not in possession of the trapdoor information. The development of good trapdoor DL groups is a major problem in cryptography. A partial solution to this problem (due to Paillier [20]) provides a trapdoor DL subgroup of $(\mathbb{Z}/N^2\mathbb{Z})^*$. One problem with Paillier's solution is that the trapdoor DL group is only a subgroup of the whole group and it is required to "blind" the trapdoor DL group by elements of $(\mathbb{Z}/N\mathbb{Z})^*$. Some related approaches are Naccache-Stern [18] and Okamoto-Uchiyama [19]. There are several other trapdoor DL proposals in the literature [11,22,25,12] but none of these seem to be practical.

A third application is to cryptographic protocols whose security is proved relative to a "gap assumption", namely that a certain computational problem should be hard even when an oracle for the corresponding decisional problem is provided. This situation arises when one needs the decision oracle as part of the simulation for the security reduction. We stress that the decision oracle is needed only for the simulation, and not for the protocol itself. With current instantiations of gap-DH groups the algorithm to solve the DDH problem is available to all users. Suppose there exists a trapdoor DDH group for which the

CDH problem remains hard even when the adversary has access to the DDH trapdoor. Then a cryptosystem with a security property proven under a gap-DH assumption can be securely implemented with the trapdoor DDH group. In other words, it is possible to prove a scheme secure relative to a gap assumption on a group for which the DDH problem is hard. This may be advantageous if one wishes to prove that a scheme possesses two properties, where one property can only be proven given a gap-DH assumption while the other can only be proven given a DDH assumption. We note that trapdoor discrete log groups would not be useful in this setting, since the gap assumption does not hold.

1.2   Outline

In Section 2 we give precise definitions of trapdoor DDH and DL groups.

After giving the formal definitions, we then propose two methods of instantiating a trapdoor DDH group through the use of hidden pairings. The first method, presented in Section 3, uses elliptic curves over RSA moduli and its security depends on the difficulty of factoring. The advantages of this approach are that it is relatively practical and efficient, and that the security is well understood.

The second proposal, in Section 4, is motivated by Frey's idea of "disguising" an elliptic curve [7]. The advantage of this approach is that it may lead to a relatively efficient trapdoor discrete logarithm system. The disadvantage is that the public key is very large and that the security is less easy to assess.

In Section 5 we describe some possible further applications of the disguised elliptic curve groups proposed in Section 4. In Section 6 we present some cryptographic applications of trapdoor DDH groups.

2   Problem Definitions

In this section we define the relevant computational problems and we formally define trapdoor groups. We use multiplicative notation for groups in this section.

We informally define the DL, CDH and DDH problems. The discrete logarithm problem (DL) in a group G is, given two elements $g, h \in G$, to find the integer a, if it exists, such that $h = g^a$. The computational Diffie-Hellman problem (CDH) in a group G is, given a triple of elements (g, g^a, g^b) in G, to compute the element g^{ab}. The decisional Diffie-Hellman problem (DDH) is, given a quadruple of elements (g, g^a, g^b, g^c), to determine whether $g^c = g^{ab}$.

A trapdoor DDH group is defined as follows.

Definition 1 (Trapdoor DDH Group). *A trapdoor DDH group is defined by*

- *a polynomial-time group generator Gen which takes a security parameter 1^k as input, and outputs a triple (G, g, τ) where G is a group description (including a description, or partial description, of the group operation), g is the generator of a cyclic subgroup of G and τ is some trapdoor information;*
- *a polynomial-time algorithm $\mathcal{DDH}$ which takes as input the group description G, the generator g, the trapdoor information τ and a triple (g^a, g^b, g^c), and outputs 1 if $g^c = g^{ab}$ and 0 otherwise.*

We require that the DDH problem is hard on the group G for any polynomial-time attacker who does not know the trapdoor information τ. Formally, we define the group generator Gen' as the algorithm that computes $(G, g, \tau) = Gen(1^k)$ and outputs (G, g), and insist that the DDH problem is hard for Gen'.

We shall instantiate a trapdoor DDH group using hidden pairings on an elliptic curve in Sections 3 and 4. Here the trapdoor information τ allows the computation of a pairing, but it is difficult to compute the pairing without knowing τ.

Note that the above definition does not precisely state what is meant by the term "group description". In our examples the operations provided by the group description will vary. For example, the ability to compare group elements or test whether a group element is the identity may or may not be provided. It may also be difficult to sample from the group (in a non-trivial way) or hash into the group.

We may go further with one of our constructions and conjecture the existence of trapdoor DL groups: groups in which the discrete logarithm problem is easy to solve for anyone who knows the trapdoor information τ, but difficult for anyone who does not know the trapdoor information.

Definition 2 (Trapdoor DL Group). *A trapdoor DL group is defined by*

- *a polynomial-time group generator Gen which takes a security parameter 1^k as input, and outputs a triple (G, g, τ) where G is a group description (including a description, or partial description, of the group operation), g is the generator of a cyclic subgroup of G and τ is some trapdoor information;*
- *a polynomial-time algorithm $\mathcal{DL}$ which takes as input the group description G, the generator g, the trapdoor information τ and a group element g^a, and outputs a.*

We require that the DL problem is hard on the group G for any polynomial-time attacker who does not know the trapdoor information τ. Formally, we define the group generator Gen' as the algorithm that computes $(G, g, \tau) = Gen(1^k)$ and outputs (G, g), and insist that the DL problem is hard for Gen'.

Applications of such groups are given in Section 6.

3   Hidden Pairings Based on Factoring

Let p_1 and p_2 be randomly generated primes of at least 512 bits in length which are congruent to 3 modulo 4 and for which there exist large primes $r_j \mid (p_j + 1)$ for $j = 1, 2$. The primes r_j should be at least 160-bit integers.

Let $N = p_1 p_2$ and let $E : y^2 = x^3 + x$ be an elliptic curve over $\mathbb{Z}/N\mathbb{Z}$. It is known that E is a supersingular curve (with embedding degree 2) over $\mathbb{F}_{p_j}$, and so $\#E(\mathbb{Z}/N\mathbb{Z}) = (p_1 + 1)(p_2 + 1)$. Let $P = (x_P, y_P) \in E(\mathbb{Z}/N\mathbb{Z})$ be a point of order $r_1 r_2$. For information about elliptic curves over rings see [15,16,9].

The public description of the group is the triple (N, E, P). From the public key one can compute $[a]P \in E(\mathbb{Z}/N\mathbb{Z})$ efficiently. The trapdoor is $\tau = (p_1, p_2, r_1, r_2)$.

Using the trapdoor one can solve the DDH problem by reducing the problem from $E(\mathbb{Z}/N\mathbb{Z})$ to $E(\mathbb{F}_{p_j})$, and solving the DDH problem using the modified Weil or Tate pairing in the usual way [1,8,13,17]. Note that, by the Chinese remainder theorem, a quadruple (P, P_1, P_2, P_3) in $E(\mathbb{Z}/N\mathbb{Z})$ is a valid DDH tuple if and only if the elements reduce modulo p_1 and p_2 to valid DDH tuples in $E(\mathbb{F}_{p_1})$ and $E(\mathbb{F}_{p_2})$. Some other applications of pairings apart from solving DDH (for example, testing subgroup membership) may also be possible in this setting.

One can obviously use other supersingular curves (and therefore have different congruence restrictions on the primes p_j) but there seems to be no reason to use an embedding degree larger than 2 in this situation. One could also use ordinary curves with a low embedding degree (even embedding degree 1), although not all DDH problems are necessarily easy in this setting.

We remark that Boneh, Goh and Nissim [4] have recently proposed a system using pairings which relies on a similar factoring assumption.

3.1 Security Analysis

The claim is that if the trapdoor τ is not known then one cannot solve the DDH problem. Obviously we must ensure that the discrete logarithm and computational Diffie-Hellman problems in $\langle P \rangle$ are hard if the trapdoor is not known. Hence, we insist that the base point P has order of at least 160-bits.

Another obvious attack is to try to factor N using the auxiliary curve data E. Since the order of $E(\mathbb{Z}/N\mathbb{Z})$ is neither smooth nor known to the attacker, it seems these attacks are resisted.

A more subtle attack might be to try and compute the pairing without knowing τ. As mentioned in [9], there is no known way to compute pairings without knowing the order $r_1 r_2$ (or a multiple of the order) of the point P. If the primes r_j are large enough (specifically, $r_j > \sqrt{p_j}$) then knowledge of $r_1 r_2$ is sufficient to factor N. For a discussion of security see [9].

3.2 Additional Features

This system has two potentially useful features. First, one can delegate the ability to compute Weil pairings to a third party, without necessarily revealing the factorisation of the modulus. This can be done by revealing the order $r_1 r_2$ of the point P. It is necessary that $r_1 r_2$ be smaller than $\sqrt{N}$. We refer to [9] for a full security analysis of this situation.

Second, there are variants of the system which allow hashing to the group. To explain this, note that every value $0 \leq x < p_j$ is the x-coordinate of a point $P \in E(\mathbb{F}_{p_j^2})$. Furthermore, the group structure of $E(\mathbb{F}_{p_j^2})$ is $(\mathbb{Z}/(p_j+1)\mathbb{Z})^2$ and so a random point in $E(\mathbb{F}_{p_j^2})$ is very likely to have order divisible by r_j.

The system is then developed using the techniques of Demytko [6] (i.e., working with x-coordinates only). To be precise, we hash messages m, using a cryptographic hash function $Hash$, onto the set of integers $\{x \in \mathbb{Z} : 0 \leq x < N\}$. Write $x_H = Hash(m)$. There is at least one point $H = (x_H, y_H) \in E(R)$, where

R is a ring containing $\mathbb{Z}/N\mathbb{Z}$, with this x-coordinate. It is not necessary to compute the y-coordinate y_H or to know in which extension ring it lies. As is well known, given $a \in \mathbb{Z}$ one can compute the x-coordinate $x([a]H)$ from x_H without ever needing to compute any y-coordinates (see [6]). Most cryptosystems can be implemented in a way which uses x-coordinates only.

The natural generalisation of the DDH problem in this setting is the following. The input is a quadruple $(x(P), x([a]P), x(H), x([b]H))$. This is a generalisation of a Diffie-Hellman-tuple in two ways: H is not necessarily in the subgroup generated by P (in other words, it is what some authors would call a co-DDH problem); H does not necessarily have the same order as P. The DDH problem is then to determine whether $a \equiv b$ modulo the greatest common divisor of the orders of P and H; note that this is the best that can be determined if P and H do not have the same order.

This generalised DDH problem can be solved with the trapdoor information. On receipt of a quadruple of x-coordinates (x_1, x_2, x_3, x_4) modulo N, one performs the following operations for each prime p_j ($j = 1, 2$): First, solve for the y-coordinates (choosing the square root arbitrarily) to obtain points $P_i = (x_i, y_i) \in E(\mathbb{F}_{p_j^2})$. Next, compute the Weil pairings $z_1 = e_n(P_1, \psi(P_4))$ and $z_2 = e_n(P_2, \psi(P_3))$, where $n = p_j + 1$,

$$\psi(x, y) = \begin{cases} (-x, \sigma y) & \text{if } e_n(P_1, P_4) = 1 \\ (x, y) & \text{otherwise} \end{cases}$$

and $\sigma \in \mathbb{F}_{p_j^2}$ is such that $\sigma^2 = -1$. If $z_2 \neq z_1^{\pm 1}$ then declare the input DDH tuple to be invalid for p_j. Otherwise, declare the tuple to be valid for p_j. If the tuple is valid for both p_1 and p_2 then declare the DDH tuple to be valid.

To summarise, by using x-coordinates only, we obtain a trapdoor DDH structure which is not strictly a trapdoor DDH group but which is sufficient for many cryptographic applications. We remark that to delegate pairing computation to a third party in this case requires giving $\#E(\mathbb{Z}/N\mathbb{Z})$, which is equivalent to giving p_1 and p_2, so the third party has the same powers as the owner of the key.

Trapdoor Discrete Logarithms: Given the trapdoor one can reduce the discrete logarithm problem from the elliptic curve to discrete logarithm problems in $\mathbb{F}_{p_j^2}^*$ and then attempt to solve these using an index calculus algorithm (this is just the MOV attack). Since p_j^2 is of the same size as N this will be no easier than factoring N. Hence, it seems that the trapdoor discrete logarithm application is not possible with this system.

4 Hidden Pairings Using a Disguised Elliptic Curve

The proposal in this section is inspired by Frey's idea of disguising an elliptic curve [7]. Essentially, we take the Weil restriction of a supersingular elliptic curve E with respect to $\mathbb{F}_{q^m}/\mathbb{F}_q$ and blind the equations by applying an invertible change of variable. One can then publish a list of multivariate polynomials which perform the group operation on "blinded" points. This gives a "black box group" representation of the elliptic curve. The hope is that a user who is only given the blinded group law can perform point multiplication, but cannot compute pairings on the curve.

In this section we first describe how to obtain systems of multivariate polynomials which represent the group law. We then explain how to "blind" these polynomials using a change of variable. In the later subsections we explain a partial linearisation technique to lower the degree of these polynomials, and discuss several strategies to attack this proposal.

For simplicity, we describe the idea using elliptic curves over finite fields of characteristic 2. Similar techniques can be applied to elliptic curves over finite fields of arbitrary characteristic. Let $E : y^2 z + yz^2 = x^3 + z^3$ over $\mathbb{F}_{q^m}$ where $q = 2^s$ and where ms is odd. Then E is supersingular with embedding degree 2. Suppose there is a large prime $r \mid (q^m + 1)$, some examples are given in the following table (note that the roles of s and m may be interchanged).

s	1	19	23	23	17	31	31	41
m	167	19	17	13	11	11	7	5
$\log_2(r)$	166	200	171	204	161	215	157	160

We now explain how the group operations can be performed. We are required to work with projective representations since we do not want to have divisions in our formulae. Let $(x : y : z)$ be a projective point on the elliptic curve. Then, for example, $[2](x : y : z)$ is given by the point $(u : v : w)$ where

$$u = x^4 z^2$$
$$v = x^6 + x^3 z^3 + yz^5 + z^6$$
$$w = z^6.$$

Choose a vector space basis for $\mathbb{F}_{q^m}$ over $\mathbb{F}_q$ (for example, this could arise from a polynomial representation of $\mathbb{F}_{q^m}$). Every element $x \in \mathbb{F}_{q^m}$ can be represented as an m-tuple $\underline{x} = (x_0, \ldots, x_{m-1})$ of elements in $\mathbb{F}_q$ which correspond to the decomposition of x over the basis. Hence a projective point $P = (x_P : y_P : z_P)$ can be represented non-uniquely as a $3m$-tuple of elements of $\mathbb{F}_q$. The group doubling operation may now be re-written as a list of $3m$ homogeneous polynomial functions of degree 6 in $3m$ variables

$$u_i = f_i(x_0, \ldots, x_{m-1}, y_0, \ldots, y_{m-1}, z_0, \ldots, z_{m-1})$$
$$v_i = f_{m+i}(x_0, \ldots, x_{m-1}, y_0, \ldots, y_{m-1}, z_0, \ldots, z_{m-1})$$
$$w_i = f_{2m+i}(x_0, \ldots, x_{m-1}, y_0, \ldots, y_{m-1}, z_0, \ldots, z_{m-1})$$

where $0 \leq i \leq m - 1$. We denote this system as $(f_i(\underline{x}, \underline{y}, \underline{z}))$.

Similarly, the sum $(x : y : z) + (x' : y' : z')$ on the elliptic curve is given by $(u : v : w)$ where the terms may be computed as $3m$ polynomial functions g_i of degree 10 in $6m$ variables. This gives a system $(\underline{u}, \underline{v}, \underline{w}) = (g_i(\underline{x}, \underline{y}, \underline{z}, \underline{x}', \underline{y}', \underline{z}'))$.

Next an invertible transformation U on the $3m$ variables in $\mathbb{F}_q$ that define a projective point is chosen. In general, this could be non-linear, but we suggest choosing a linear transformation $U \in \mathrm{GL}_{3m}(\mathbb{F}_q)$ since linear maps do not increase the degree of the defining polynomials. As will be explained in Section 4.1, to achieve workable parameter sizes, we must impose that U is defined over $\mathbb{F}_2$ and that U maps the $2m$-dimensional subspace corresponding to the x and z variables onto itself (the proposal works in the general case, but the size of the group description increases). We then "blind" the doubling/addition formulae for the curve by applying the change of variable U to all variables. More precisely, we obtain the "blinded" doubling formulae

$$(\tilde{f}_i(\underline{x}, \underline{y}, \underline{z})) = U(f_i(U^{-1}(\underline{x}, \underline{y}, \underline{z})))$$

and the blinded addition formulae

$$(\tilde{g}_i(\underline{x}, \underline{y}, \underline{z})) = U(g_i(U^{-1}(\underline{x}, \underline{y}, \underline{z}), U^{-1}(\underline{x}', \underline{y}', \underline{z}'))).$$

The public representation of the group consists of the doubling and addition polynomials $\tilde{f}_i, \tilde{g}_i$, a blinded point P and the order r of P (we get P as $U(R)$ where R is a point of order r on the original curve). The values of m and s are implicit in the public key. A user can efficiently compute $[a]P$ using the addition formula and the double and add algorithm in the usual way.

The trapdoor is the original equation of the curve and the inverse transformation to U. A user with the trapdoor can translate blinded $3m$-tuples back to the standard representation of points in $E(\mathbb{F}_{q^m})$. Pairings can then be computed easily. Hence the DDH problem can be solved in the group.

If U maps the $2m$-dimensional subspace corresponding to the x and z variables onto itself, then we can easily recognise representations of the identity element as they will have the form

$$(0, \ldots, 0, y_0, \ldots, y_{m-1}, 0, \ldots, 0).$$

This means that we can check for equality between two points Q_1 and Q_2 in projective coordinates by checking whether $Q_1 - Q_2$ is the identity element. Note that $-Q_2$ can be computed as $[r-1]Q_2$ where r is the order of the base point P. However, it does not seem possible to represent elements in a canonical form, and this may be inconvenient for some cryptographic applications.

We note that it might be interesting to use the above techniques to give a blinded description of a pairing computation algorithm. However, we suspect that the memory requirements will be huge, so we do not pursue this idea further.

4.1 Key Sizes and Optimisations

The public representation of a blinded group is very large. The doubling formulae consist of $3m$ polynomials of degree 6 in $3m$ variables and the addition formulae consist of $3m$ polynomials of degree 10 in $6m$ variables. Since the total number of possible monomials in a homogeneous polynomial of degree d in n variables is

$$\binom{n+d-1}{d} \leq \frac{(n+d-1)^d}{d!},$$

the storage cost for the addition formulae if represented naively would be approximately $3m(6m+9)^{10}/10! \approx 50m^{11}$ elements of $\mathbb{F}_q$. This will quickly become infeasible.

One can slightly improve the storage requirements by applying a partial linearisation (i.e., introducing new variables to represent terms such as x_0^2) and using the fact that squaring is essentially a linear operation, in the sense that there is some matrix S such that if x is represented as $(x_0, \ldots, x_{m-1})$ then x^2 is represented as $(x_0^2, \ldots, x_{m-1}^2)S$. Storage can be further reduced if m and s are coprime and if the basis for $\mathbb{F}_{q^m}/\mathbb{F}_q$ is chosen to be a basis for $\mathbb{F}_{2^m}/\mathbb{F}_2$, since the polynomials will have coefficients in $\mathbb{F}_2$ rather than $\mathbb{F}_q = \mathbb{F}_{2^s}$. We do not give all the details of these optimisations.

Using the above linearisation and the modified projective doubling formulae $u = x^4z^3, v = x^6z + x^3z^4 + yz^6 + z^7, w = z^7$ one needs $7m$ variables (corresponding to $x, y, z, x^2, z^2, x^4, z^4$). The total storage requirement is $3m\binom{7m+2}{3} \approx 172m^4$ bits.

The addition formulae involve many more variables and hence require larger storage. A significant reduction follows from noting that for many (but not all) applications a general addition rule is not required. Instead it is sufficient to publish formulae for addition of the fixed base point P.

The formulae for addition of the fixed base point P to an arbitrary point $(x : y : z)$ can be written projectively as $(x : y : z) + (x_P : y_P : 1) = (x' : y' : z')$ where

$$x' = z(x - x_Pz)(y - y_Pz)^2 - (x - x_Pz)^4$$
$$y' = z(y - y_Pz)^3 + (y - y_Pz)x(x - x_Pz)^2 + (y_P + 1)(x - x_Pz)^3z$$
$$z' = z(x - x_Pz)^3.$$

Note that this formula gives a correct result only if $(x : y : z)$ does not represent P or the identity. We can also apply the methods mentioned earlier about partial linearisation, although since P is defined over $\mathbb{F}_{q^m}$ these polynomials do not have coefficients in $\mathbb{F}_2$.

One can show that the resulting formulae for addition by a fixed point have degree 3 in $8m$ variables and so require storage bounded by $3m(8m+2)^3/6 \approx 256m^4$ elements of $\mathbb{F}_q$ (in this case we cannot reduce to polynomials over $\mathbb{F}_2$). Hence, the total storage for the group description is roughly $(172 + 256s)m^4$ bits.

It should be noted that using this "partial group description" leads to smaller public keys, but also removes some of the functionality associated with the group. In particular, it means that we cannot compute $[a]Q$ for an arbitrary point $Q \neq P$ and we cannot test equality of group elements (see Section 5 for details).

While the above optimisations are significant, compared with the system proposed in Section 3, the public key for this scheme is very large. For transformations defined over $\mathbb{F}_2$ with $m = 5$ the public key is already 814 kilobytes. For $m = 7$ and 11 the values are 2.3 and 8 megabytes respectively. The value $m = 167$ gives a totally infeasible public group description of 41 gigabytes!

4.2 Security Analysis

There are a number of ways to attack this system. As before, one can just try to solve the discrete logarithm problem in the group using the baby-step-giant-step method. We normally impose the restriction that the group order be at least 160 bits to thwart such an attack. It follows that we should have $ms > 160$.

Another attack would be to try to compute a pairing using the blinded description of the group operation. To run Miller's algorithm one needs to obtain functions, defined over $\mathbb{F}_{q^m}$, corresponding to the straight lines in the elliptic curve addition rule. It seems hard to achieve this without being able to invert the blinding.

A natural attack is to try to find the invertible transformation U. The number of elements of $\mathrm{GL}_n(\mathbb{F}_q)$ is

$$(q^n - 1)(q^n - q)(q^n - q^2) \cdots (q^n - q^{n-1}) = \prod_{i=1}^{n} q^{n-i}(q^i - 1) > q^{n(n-1)}.$$

We are essentially choosing U from $\mathrm{GL}_{2m}(\mathbb{F}_2) \times \mathrm{GL}_m(\mathbb{F}_2)$ (indeed, since the equations are projective we could replace GL with PGL, but over $\mathbb{F}_2$ they are the same). The number of choices for U is at least $2^{2m(2m-1)+m(m-1)} = 2^{5m^2-3m}$. The smallest case we consider is when $m = 5$, for which there are already more than 2^{110} choices for U. Hence the system is resistant against brute-force search for U.

The most plausible way to find U is to reduce the problem to solving a system of multivariate polynomial equations. We discuss such an attack here. We assume that an adversary knows not just the public key but also the original system of polynomials defining the group operation. This strong attack scenario is plausible because there are good implementation reasons for certain choices of polynomial basis etc., so one may as well assume that an adversary can simulate the key generation process up to the choice of U. Hence, the security relies on the difficulty of computing the transformation U given the systems of polynomials defining the doubling and addition rules.

The obvious attack is to represent the coefficients of the transformation U as unknowns and to obtain a system of equations among these variables. One natural way to obtain equations is by matching known points in the domain and image. We are given an explicit point P in the image, but we do not have the representation of P in the domain. Hence, as long as the original point P remains private, this attack seems to be hard.

Instead, we may obtain equations in the variables of U by relating the doubling/addition rules on the original curve with the published doubling/addition rules. More precisely, one writes down the doubling/addition systems on the original curve as $3m$-tuples of multivariate polynomials, and then transforms using the "generic" matrix U to get enormous $3m$-tuples of multivariate polynomials. By equating coefficients of monomials with the published doubling/addition rules one obtains a system of multivariate polynomial equations in the unknown entries of U. The degree of the equations depends on the degree of the original system of polynomials; when attacking our proposed partially linearised group description one gets degree 3 equations.

One could apply Gröbner basis or linearisation techniques to find a solution of this system and hence deduce the matrix U. Linearisation is the most natural approach, since the system is already partially linearised. However, the required number of equations to solve the degree 3 system in $(3m)^2$ variables will be roughly $9^6 m^6/6 \approx 88000 m^6$, whereas we only start with around $428 m^4$ equations.

We suggest that parameters satisfying $ms > 160$ with sufficiently large m are secure against multivariate attacks on this system. We recognise that further research into the problem of recovering U is needed before we can have confidence in the the security of this system. In particular, it is important to determine which values for m are secure: we expect that $m = 3$ is too small to achieve the desired level of security and that $m \geq 11$ is sufficient. We hope that our work motivates others to consider this problem.

It is interesting to note that one cannot blind finite fields securely using the above method. A full description of an attack and its generalisation to the torus $T_2 \subset \mathbb{F}_{p^2}^*$ is given in [10], where there is also a discussion of why the methods do not seem to apply to the elliptic curve case.

One might also hope to achieve greater security levels by using a more general initial curve equation, for example $y^2 + Ay = x^3 + Bx + C$ for some $A, B, C \in \mathbb{F}_{q^m}$. All such non-singular equations are isomorphic over $\overline{\mathbb{F}}_2$ to $y^2 + y = x^3 + 1$. An isomorphism of a Weierstrass equation of this form is a linear map. But since the coefficients of the isomorphism may lie in an extension field, this linear change of variable is not necessarily already included in the above analysis. Instead, one would have to perform the linearisation or Gröbner basis methods over small degree extensions of $\mathbb{F}_q$. We do not discuss this idea further as we do not expect it to significantly add to the security.

4.3   Additional Features

The scheme as presented does not seem to have the additional feature of the scheme of Section 3, that one can delegate pairing computations to a third party without revealing all the private information.

Hashing to the group cannot be done when using the system presented above. Instead, one could follow Section 3 and restrict to using x-coordinates only. In this case we could publish a blinded representation for the x-coordinate only addition rules on projective points of the form $(x : z)$. The security and efficiency of this variant deserve further analysis.

Trapdoor Discrete Logarithms: If we apply the Weil pairing to the (unblinded) elliptic curve group, then we map the elliptic curve into $\mathbb{F}_{q^{2m}}$, where $q^m \approx 2^{160}$. A 320-bit discrete logarithm computation in a characteristic 2 finite field is quite feasible. The current world record for the solution of a characteristic 2 discrete logarithm problem is over 600 bits [23,14]. We are told [24] that the relation finding and linear algebra computations for a 320-bit discrete logarithm would take less than a week. Once the linear algebra stage of the index calculus algorithm has been completed we may store the reduced matrix and solve individual discrete logarithm problems in a matter of seconds. Hence, this method does give a completely practical trapdoor discrete logarithm system.

Note that, despite being quite practical, this trapdoor DL system is not polynomial time. Hence it does not satisfy the definition in Section 2.

5 Partial Group Descriptions

In the previous section, as a way of minimising the group description, we suggested publishing just the operations of doubling and addition by a fixed point P. Note that both these operations are unary, whereas a general group description requires a binary operation. We now make some comments about this idea.

Let G be a group written additively and suppose the published description of G comprises just an element P and unary operations $f(Q) = Q + Q$, $g(Q) = Q + P$. In general, there seems to be no way to obtain the sum $Q_1 + Q_2$ of two general points from the operations f and g.

Similarly, using the double-and-add algorithm, one can compute $[a]P$ for any positive integer a. But it does not seem to be possible to compute $[a]Q$ for a randomly chosen element $Q \in G$, unless we know a positive integer b such that $Q = [b]P$.

To summarise, the description of the group G is sufficient for some computations, but it does not satisfy the usual computational definition of a group law.

A natural question is whether the discrete logarithm problem is harder in such a group than in a generic group. We first consider the case where a binary predicate is available which determines whether two given inputs represent the same group element.

The baby-step-giant-step algorithm can be implemented in such a group. This algorithm attempts to solve the discrete logarithm of a point Q to the base P. Let r be the order of P and define $M = \lfloor \sqrt{r} \rfloor$. The standard description of the algorithm is to compute and store a list of "baby steps" $P, [2]P, [3]P, \ldots, [M]P$ and find a collision with the list of "giant steps" $Q, Q - P', Q - [2]P', \ldots$ where $P' = [M]P$. The obstacles are that, a priori, one cannot compute $-P'$ and one cannot compute $Q + (-P')$.

Instead, one can formulate the baby-step-giant-step algorithm as follows. If $Q = [\lambda]P$ then we can write $\lambda = jM - i$ for some $1 \leq j \leq M + 2$ and some $0 \leq i \leq M$. Hence, we compute and store the baby steps $Q, Q + P, Q + [2]P, \ldots, Q + [M]P$ by successively applying the operation g. Then we compute the giant steps $[jM]P$ using the double-and-add algorithm for each value of j and check if there is a match using the predicate. Note that each giant step is now a full point multiplication, rather than a single addition. Nevertheless, the final complexity is still $\tilde{O}(\sqrt{r})$.

To implement a random walk method, such as Pollard rho, we require canonical representatives of group elements. Hence it seems that such methods cannot be implemented on disguised projective elliptic curves.

With the partial group description coming from a disguised elliptic curve we do not have an equality predicate or canonical representatives of group elements. Hence the best algorithm for solving the discrete logarithm problem seems to be brute-force search! This suggests that there could be applications where we can safely reduce the group size to 80 bits.

6 Simple Applications of Trapdoor Groups

In this section we present a few simple cryptographic applications for trapdoor DDH and trapdoor DL groups. We will use multiplicative notation for groups. We will assume that the order r of g in G is known (or that it is possible to compute it efficiently). If this is not the case, then we may still use all of the following applications by taking r to be much larger than the order of g.

6.1 A Simple Identification Scheme Based on Trapdoor DDH Groups

The simplest application of trapdoor DDH groups is to create an identification scheme. Here a central authority Charlie wishes to identify a user Alice.

- At the time of registration, Alice generates a trapdoor DDH group $(G, g, \tau) = Gen(1^k)$ and gives Charlie (G, g, r), where r is the order of the element g.
- When Charlie wishes to identify Alice, he randomly selects a bit $\sigma \in \{0, 1\}$ and integers a and b from $\{1, 2, \ldots, r\}$, and computes $A = g^a$ and $B = g^b$. If $\sigma = 0$ then he computes $C = g^{ab}$, otherwise he randomly chooses a value $c \in \{1, 2, \ldots, r\}$ such that $c \neq ab \bmod r$ and computes $C = g^c$. Charlie then sends the triple (A, B, C) to Alice.
- Alice receives the challenge triple (A, B, C), and checks whether it is a valid DDH triple using the trapdoor information τ. If (A, B, C) is a DDH triple, then Alice sends $\sigma' = 0$ to Charlie; otherwise Alice sends $\sigma' = 1$.
- Charlie receives a bit σ' from Alice, and accepts Alice's identity if $\sigma = \sigma'$.

It is obvious that any attacker that fools the identification scheme with probability $1/2 + \epsilon$ has advantage at least $\epsilon + 1/r$ in breaking the DDH problem in the trapdoor group, therefore ϵ is negligible. Clearly, if this scheme is meant to be practical, then the identification scheme would have to be run multiple times before Alice's identity is actually accepted.

6.2 A Two-User Designated Verifier Signatures

The Boneh-Lynn-Shacham (BLS) signature scheme [2] becomes a designated verifier signature scheme if implemented in a trapdoor DDH group. Note that, for this application, we require the ability to hash into the group and compute full group operations, so this scheme is most easily implemented using the approach of Section 3.

We briefly present the scheme. Alice wishes to sign a message in such a way that no party (except maybe Bob) can fake her signature and only Bob can verify her signatures.

- Bob generates a trapdoor DDH group $(G, g, \tau) = Gen(1^k)$ and publishes his public parameters (G, g, r), where r is theorder of the element g. Bob also publishes a hash function *Hash* which maps bit strings of arbitrary length into the group G.

- Alice randomly chooses a private key $x \in \{1, 2, \ldots, r\}$ and publishes her public key g^x.
- Alice may now sign a message m by computing $Hash(m)^x$.
- Bob verifies a signature σ on a message m by checking that $(g, g^x, Hash(m), \sigma)$ is a valid DDH tuple using the trapdoor information τ.

One can easily prove, in the random oracle model, that no attacker may verify Alice's signature without knowing the trapdoor τ. Furthermore, in the random oracle model, it is easy to show that an attacker that manages to fake a signature from Alice with probability ϵ, and using at most q_H random oracle queries and q_S signing oracle queries, can solve the CDH problem with probability at least $\epsilon/(q_H + q_S + 1)$.

One problem with this scheme, however, is that Alice's public key is an element of G. Hence, she can only choose a private key *after* Bob has published his public key and she must choose a new public key for each entity with which she wishes to communicate.

6.3   An Encryption Scheme Based on the Discrete Logarithm Problem

We present an encryption scheme whose security depends upon the difficulty of solving the discrete logarithm problem in an arbitrary trapdoor DL group. For simplicity we present this encryption scheme as a KEM [5].

- Bob, who wishes to be able to receive encrypted messages, generates a trapdoor DL group $(G, g, \tau) = Gen(1^k)$ and publishes (G, g, r) as his public key, where r is the order of the element g. Bob also publishes a key derivation function KDF which maps elements of the set $\{1, 2, \ldots, r\}$ onto bit-strings of the appropriate key length.
- Alice, who wishes to compute a symmetric key for use in sending an encrypted message to Bob, randomly generates an integer $x \in \{1, 2, \ldots, r\}$. She computes $C = g^x$ and $K = KDF(x)$, and sends C to Bob (along with the encryption of a message computed using the DEM and the key K).
- Bob recovers first x from C using the trapdoor information τ. Bob then computes key $K = KDF(x)$.

It is not difficult to see that, in the random oracle model, an attacker who has an advantage ϵ in breaking the IND-CCA2 security of this scheme, can be used to construct an algorithm that solves the discrete logarithm problem in G with probability ϵ.

7   Conclusions

We have suggested the concept of a hidden pairing, which gives rise to a trapdoor DDH group, and potentially a trapdoor DL group. We have suggested two possible ways to implement such an idea. Our work suggests several problems for further study, which we list below. We hope that the ANTS community will be motivated to study some of these problems further.

- Can the storage requirement for the public group description of a disguised elliptic curve be reduced?
- For which values of m is a disguised elliptic curve secure against Gröbner basis or linearisation attacks?
- Is there a way to perform Miller's algorithm to compute pairings on a disguised elliptic curve?
- Are there cryptosystems which can be securely implemented using an 80-bit partial group law?
- Do there exist partial group descriptions for other groups which may allow interesting cryptographic functionalities?
- Are there further cryptographic applications of hidden pairings?

Acknowledgements

We thank Dan Boneh, David Mireles, Kenny Paterson and the anonymous referees for their comments.

References

1. I. Blake, G. Seroussi and N. P. Smart, Advances in elliptic curve cryptography, Cambridge (2005).
2. D. Boneh, B. Lynn, and H Shacham, Short signatures from the Weil pairing, in C. Boyd (ed.), ASIACRYPT 2001, Springer-Verlag LNCS 2248 (2001) 514–532.
3. D. Boneh, Personal communication, July 1, 2005.
4. D. Boneh, E.-J. Goh, and K. Nissim, Evaluating 2-DNF formulas on ciphertexts, in J. Kilian (ed.), TCC 2005, Springer LNCS 3378 (2005) 325–341.
5. R. Cramer and V. Shoup, Design and analysis of practical public-key encryption schemes secure against adaptive chosen ciphertext attack, *SIAM Journal on Computing*, **33**, No. 1 (2004) 167–226.
6. N. Demytko, A new elliptic curve based analogue of RSA, in T. Helleseth (ed.), EUROCRYPT 1993, Springer-Verlag LNCS 765 (1994) 40–49.
7. G. Frey, How to disguise an elliptic curve (Weil descent), Talk at ECC 1998. Slides available from:
   http://www.cacr.math.uwaterloo.ca/conferences/1998/ecc98/frey.ps
8. G. Frey and H.-G. Rück, A remark concerning m-divisibility and the discrete logarithm in the divisor class group of curves, *Math. Comp.*, **62**, No. 206 (1994) 865–874.
9. S. D. Galbraith and J. F. McKee, Pairings on elliptic curves over finite commutative rings, in N. P. Smart (ed.), Cryptography and Coding: 10th IMA International Conference, Springer-Verlag LNCS 3796 (2005) 392–409.
10. S. D. Galbraith, Disguising tori and elliptic curves, preprint 2006.
11. D. M. Gordon, Designing and detecting trapdoors for discrete log cryptosystems, in E. F. Brickell (ed.), CRYPTO 92, Springer-Verlag LNCS 740 (1993) 66–75.

12. D. Hühnlein, M. J. Jacobson, D. Weber, Towards practical non-interactive public key cryptosystems using non-maximal imaginary quadratic orders, in D. R. Stinson and S. Tavares (eds.), SAC 2000, Springer-Verlag LNCS 2012 (2001) 275–297.

13. A. Joux and K. Nguyen, Separating Decision Diffie–Hellman from Diffie–Hellman in cryptographic groups, *J. Crypt.* **16** (2003) 239–248.

14. A. Joux and R. Lercier, Discrete logarithms in $GF(2^{607})$ and $GF(2^{613})$, posting to the Number Theory Mailing List, 23 Sep 2005.

15. H. W. Lenstra Jr., Factoring integers with elliptic curves, *Annals of Mathematics*, **126** (1987) 649–673.

16. H. W. Lenstra Jr., Elliptic curves and number theoretic algorithms, *Proc. International Congr. Math.*, Berkeley 1986, AMS (1988) 99–120.

17. A. J. Menezes, T. Okamoto and S. A. Vanstone, Reducing elliptic curve logarithms to logarithms in a finite field, *IEEE Trans. Inf. Theory*, **39**, No. 5 (1993) 1639–1646.

18. D. Naccache and J. Stern, A new public-key cryptosystem based on higher residues, ACM Conference on Computer and Communications Security (1998) 59–66.

19. T. Okamoto and S. Uchiyama, A new public key cryptosystem as secure as factoring, in K. Nyberg (ed.), EUROCRYPT '98, Springer-Verlag LNCS 1403 (1998) 308–318.

20. P. Paillier, Public-key cryptosystems based on composite degree residuosity classes, in J. Stern (ed.), EUROCRYPT 1999, Springer-Verlag LNCS 1592 (1999) 223-238.

21. R. L. Rivest, Homework 4 of course "6.897 Selected Topics in Cryptography", May 2004.
    http://theory.lcs.mit.edu/classes/6.897/spring04/hw4.txt

22. E. Teske, An elliptic curve trapdoor scheme, *J. Crypt.*, 19 (2006) 115–133.

23. E. Thomé, Computation of discrete logarithms in $GF(2^{607})$, in C. Boyd (ed.), ASIACRYPT 2001, Springer-Verlag LNCS 2248 (2001) 107–124.

24. E. Thomé, Personal communication, January 9, 2006.

25. S. Vanstone and R. J. Zuccherato, Elliptic curve cryptosystem using curves of smooth order over the ring $\mathbb{Z}_n$, *IEEE Trans. Inf. Theory*, Vol. 43, No. 4 (1997) 1231–1237.

Constructing Pairing-Friendly Elliptic Curves with Embedding Degree 10

David Freeman

University of California, Berkeley
dfreeman@math.berkeley.edu

Abstract. We present a general framework for constructing families of elliptic curves of prime order with prescribed embedding degree. We demonstrate this method by constructing curves with embedding degree $k = 10$, which solves an open problem posed by Boneh, Lynn, and Shacham [6]. We show that our framework incorporates existing constructions for $k = 3$, 4, 6, and 12, and we give evidence that the method is unlikely to produce infinite families of curves with embedding degree $k > 12$.

1 Introduction

A cryptographic pairing is a bilinear map between two groups in which the discrete logarithm problem is hard. In recent years, such pairings have been applied to a host of previously unsolved problems in cryptography, the most important of which are one-round three-way key exchange [13], identity-based encryption [5], and short digital signatures [6].

The cryptographic pairings used to construct these systems in practice are based on the Weil and Tate pairings on elliptic curves over finite fields. These pairings are bilinear maps from an elliptic curve group $E(\mathbb{F}_q)$ to the multiplicative group of some extension field $\mathbb{F}_{q^k}$. The parameter k is called the *embedding degree* of the elliptic curve. The pairing is considered to be secure if taking discrete logarithms in the groups $E(\mathbb{F}_q)$ and $\mathbb{F}_{q^k}^*$ are both computationally infeasible.

For optimal performance, the parameters q and k should be chosen so that the two discrete logarithm problems are of approximately equal difficulty when using the best known algorithms, and the order of the group $\#E(\mathbb{F}_q)$ should have a large prime factor r. For example, a pairing is considered secure against today's best attacks when $r \sim 2^{160}$ and $k \sim$ 6-10, depending on the application. In order to vary the security level or adapt to future improvements in discrete log technology, we would like to have a supply of elliptic curves at our disposal for arbitrary q and k.

Many researchers have examined the problem of constructing elliptic curves with prescribed embedding degree. Menezes, Okamoto, and Vanstone [16] showed that a supersingular elliptic curve must have embedding degree $k \leq 6$, and furthermore $k \leq 3$ in characteristic not equal to 2 or 3. Miyaji, Nakabayashi, and

F. Hess, S. Pauli, and M. Pohst (Eds.): ANTS 2006, LNCS 4076, pp. 452–465, 2006.

Takano [17] have given a complete characterization of ordinary elliptic curves of prime order with embedding degree $k = 3$, 4, or 6, while Barreto and Naehrig [2] give a construction for curves of prime order with $k = 12$. There is a general construction, originally due to Cocks and Pinch [8], for curves of arbitrary embedding degree k, but in this construction the sizes of the field $\mathbb{F}_q$ and the subgroup of prime order r are related by $q \approx r^2$, which leads to inefficient implementation. Recent efforts (cf. [7], [10]) have focused on reducing the ratio $\rho = \log q / \log r$ for arbitrary k, but no additional examples have been found with ρ small enough to allow for curves of prime order.

The focus of this paper is the construction of ordinary elliptic curves of prime order with prescribed embedding degree. In Section 2 we present a general framework for constructing such curves and give conditions under which this method will give us infinite families of elliptic curves. The method is based on the Complex Multiplication method of curve construction [19] and is implicit in the constructions of several other researchers. Our contribution is to gather all of the relevant results in one place and to define terminology that makes it apparent that these various constructions are all instances of the same general method.

Our main contribution appears in Section 3, where we show how the method of Section 2 can be used to construct curves with embedding degree $k = 10$. We give examples of such curves over fields of cryptographic size, solving an open problem posed by Boneh, Lynn, and Shacham [6].

In Section 4 we show how the existing constructions of elliptic curves of prime order with embedding degree $k = 3$, 4, 6, or 12 can be explained via the framework of Section 2. In Section 5, we show that for $k > 6$, our method is not likely to give additional infinite families of elliptic curves with the specified embedding degree. We note, however, that examples of such families exist for $k = 10$ and $k = 12$, and we ask in Section 6 if such examples can be constructed in a systematic fashion.

2   A Framework for Constructing Pairing-Friendly Elliptic Curves

In this section we describe a general framework for constructing elliptic curves of a given embedding degree k. This framework is implicit in the constructions of Miyaji, Nakabayashi, and Takano [17]; Barreto, Lynn, and Scott [1]; Cocks and Pinch [8] (as explained in [4]); and Brezing and Weng [7]. After stating the relevant results, we define terminology that will allow us to show that these constructions are all specific cases of the same general method.

To construct our elliptic curves, we parameterize the number of points on the curve and the size of the field of definition by polynomials $n(x)$ and $q(x)$, respectively. For each x_0 that gives prime values for $n(x_0)$ and $q(x_0)$, we can use the Complex Multiplication method to construct an elliptic curve with the desired properties. The main result of this section is Theorem 2.7, which gives a criterion for the existence of infinite families of such good parameters.

We begin by giving a formal definition of embedding degree.

Definition 2.1. *Let E be an elliptic curve defined over a finite field $\mathbb{F}_q$, and let n be a prime dividing $\#E(\mathbb{F}_q)$. The embedding degree of E with respect to n is the smallest integer k such that n divides $q^k - 1$.*

Equivalently, k is the smallest integer such that $\mathbb{F}_{q^k}$ contains μ_n, the group of nth roots of unity in $\overline{\mathbb{F}}_q$. We often ignore n when stating the embedding degree, as it is usually clear from the context.

If we fix a target embedding degree k, we wish to solve the following problem: find a prime (power) q and an elliptic curve E defined over $\mathbb{F}_q$ such that $n = \#E(\mathbb{F}_q)$ is prime and E has embedding degree k. Furthermore, since we may wish to construct curves over fields of different sizes, we would like to be able to specify (approximately) the number of bits of q in advance.

We follow the strategy of Barreto and Naehrig [2] in parameterizing the trace of the curves to be constructed. Namely, we choose some polynomial $t(x)$, which will be the trace of Frobenius for our hypothetical curve, and construct polynomials $q(x)$ and $n(x)$ that are possible orders of the prime field and the elliptic curve group, respectively. More precisely, if $q(x_0)$ is prime for some x_0, we can use the Complex Multiplication method [3], [19] to construct an elliptic curve over $\mathbb{F}_{q(x_0)}$ with $n(x_0)$ points and embedding degree k.

Theorem 2.2. *Fix a positive integer k, and let $\Phi_k(x)$ be the kth cyclotomic polynomial. Let $t(x)$ be a polynomial with integer coefficients, let $n(x)$ be an irreducible factor of $\Phi_k(t(x) - 1)$, and let $q(x) = n(x) + t(x) - 1$. Let $f(x) = 4q(x) - t(x)^2$. Fix a positive square-free integer D, and suppose (x_0, y_0) is an integer solution to the equation $Dy^2 = f(x)$ for which*

1. *$q(x_0)$ is prime, and*
2. *$n(x_0)$ is prime.*

If D is sufficiently small, then there is an efficient algorithm to construct an elliptic curve E defined over $\mathbb{F}_{q(x_0)}$ such that $E(\mathbb{F}_{q(x_0)})$ has prime order $n(x_0)$ and E has embedding degree at most k.

Proof. By hypothesis, we have a solution (x_0, y_0) to the equation $Dy^2 = f(x)$ for which $q(x_0)$ is prime. If D is sufficiently small then the construction of an elliptic curve E over $\mathbb{F}_{q(x_0)}$ with $\#E(\mathbb{F}_{q(x_0)}) = n(x_0)$ is standard via the Complex Multiplication method; see [3] or [19] for details. Since $n(x_0)$ is prime, $E(\mathbb{F}_{q(x_0)})$ has prime order, and it remains only to show that E has embedding degree at most k. Barreto, Lynn, and Scott [1, Lemma 1] show that E having embedding degree k is equivalent to $n(x_0)$ dividing $\Phi_k(t(x_0) - 1)$ and $n(x_0)$ not dividing $\Phi_i(t(x_i) - 1)$ for $i < k$. Since we have chosen the polynomial $n(x)$ to divide $\Phi_k(t(x) - 1)$, $n(x_0)$ is guaranteed to divide $q(x_0)^k - 1$, and the embedding degree of E is thus at most k. ☐

Remark 2.3. The fact that $n(x)$ does not divide $\Phi_i(t(x) - 1)$ as polynomials for $i < k$ does not guarantee that $n(x_0)$ does not divide $\Phi_i(t(x_0) - 1)$ as integers for some $i < k$. However, this latter case will be rare in practice, and thus the embedding degree of a curve constructed via the method of Theorem 2.2 will usually be k.

Remark 2.4. If we wish to construct curves whose orders are not necessarily prime but merely have a large prime factor, we may relax condition (2) of the theorem accordingly, and the same analysis holds.

In practice, to construct an elliptic curve with embedding degree k one chooses polynomials $t(x)$, $n(x)$, and $q(x)$ satisfying the conditions of Theorem 2.2 and tests various values of x until $n(x)$ and $q(x)$ are prime. If the distributions of the values of the polynomials $n(x)$ and $q(x)$ are sufficiently random, the Prime Number Theorem tells us that we should have to test roughly $\log n(x_1) \log q(x_1)$ values of x near x_1 until we find an x_0 that gives a prime value for both polynomials. Since the distribution of prime values of polynomials is not well understood in general, it will be hard to prove theorems that explicitly construct infinite families of elliptic curves of prime order. Rather, we will be slightly less ambitious and search for polynomials as in Theorem 2.2 that will give us the desired elliptic curves whenever the polynomials take on prime values. We incorporate this approach into the following definition.

Definition 2.5. *Let $t(x)$, $n(x)$, and $q(x)$ be polynomials with integer coefficients. For a given positive integer k and positive square-free integer D, the triple (t, n, q) represents a family of curves with embedding degree k if the following conditions are satisfied:*

1. $n(x) = q(x) + 1 - t(x)$.
2. $n(x)$ *and* $q(x)$ *are irreducible.*
3. $n(x)$ *divides* $\Phi_k(t(x) - 1)$, *where* Φ_k *is the kth cyclotomic polynomial.*
4. *The equation* $Dy^2 = 4q(x) - t(x)^2$ *has infinitely many integer solutions* (x, y).

Defining a family of curves in this way gives us a simple criterion for constructing elliptic curves with embedding degree k. This criterion is implicit in the Barreto-Naehrig construction of curves with $k = 12$ and $D = 3$ [2].

Corollary 2.6. *Suppose (t, n, q) represents a family of curves with embedding degree k for some D. Then for each x_0 such that $n(x_0)$ and $q(x_0)$ are both prime, there is an elliptic curve E defined over $\mathbb{F}_{q(x_0)}$ such that $\#E(\mathbb{F}_{q(x_0)})$ is prime, and E has embedding degree at most k.*

In practice, for any $t(x)$ we can easily find $n(x)$ and $q(x)$ satisfying conditions (1), (2), and (3) of Definition 2.5; the difficulty arises in choosing the polynomials so that $Dy^2 = 4q(x) - t(x)^2$ has infinitely many integer solutions. In general, if $f(x)$ is a square-free polynomial of degree at least 3, then there will be only a finite number of integer solutions to the equation $Dy^2 = f(x)$ (cf. Proposition 2.10). Thus we conclude that (t, n, q) can represent a family of curves only if $f(x)$ has some kind of special form.

We now show that if $f(x)$ is quadratic, then one integral solution to the equation $Dy^2 = f(x)$ will give us infinitely many solutions. This is the technique that Miyaji, et al. [17] use to produce curves with embedding degree 3, 4, or 6, and we will use the same technique in Section 3 to construct curves with embedding degree 10. The idea is as follows: we complete the square to write the equation $Dy^2 = f(x)$ as $u^2 - D'v^2 = T$ for some constant T, and observe that (u, v) is a solution to this equation if and only if $u + v\sqrt{D'}$ has norm T in the real quadratic field $\mathbb{Q}(\sqrt{D'})$. By Dirichlet's unit theorem, there is a one-dimensional set of norm-one integral elements of this field; multiplying each of these units by our element of norm T gives an infinite family of elements of norm T. We then show that a certain fraction of these elements can be converted back to solutions of the original equation.

Theorem 2.7. *Fix an integer $k > 0$, and choose polynomials $t(x)$, $n(x)$, $q(x) \in \mathbb{Z}[x]$ satisfying conditions (1), (2), and (3) of Definition 2.5. Let $f(x) = 4q(x) - t(x)^2$. Suppose $f(x) = ax^2 + bx + c$, with $a, b, c \in \mathbb{Z}$, $a > 0$, and $b^2 - 4ac \neq 0$. Let D be a square-free integer such that aD is not a square. If the equation $Dy^2 = f(x)$ has a solution (x_0, y_0) in the integers, then (t, n, q) represents a family of curves with embedding degree k.*

Proof. Completing the square in the equation $Dy^2 = f(x)$ and multiplying by $4a$ gives

$$aD(2y)^2 = (2ax + b)^2 - (b^2 - 4ac). \tag{2.1}$$

If we write $aD = D'r^2$ with D' square-free and make the substitutions $u = 2ax + b$, $v = 2ry$, $T = b^2 - 4ac$, the equation becomes

$$u^2 - D'v^2 = T. \tag{2.2}$$

Note that since aD is not a square, we have $D' > 1$.

Under the above substitution, a solution (x_0, y_0) to the original equation $Dy^2 = f(x)$ gives an element $u_0 + v_0\sqrt{D'}$ of the real quadratic field $\mathbb{Q}(\sqrt{D'})$ with norm T. Furthermore, this solution satisfies the congruence conditions

$$\begin{aligned} u_0 &\equiv b \pmod{2a} \\ v_0 &\equiv 0 \pmod{2r}. \end{aligned} \tag{2.3}$$

We wish to find an infinite set of solutions (u, v) satisfying the same congruence conditions, for we can transform such a solution into an integer solution to the original equation. To find such solutions we employ Dirichlet's unit theorem [20, §1.7], which tells us that the integer solutions to the equation $\alpha^2 - D'\beta^2 = 1$ are in one-to-one correspondence with the real numbers $\alpha + \beta\sqrt{D'} = \pm(\alpha_0 + \beta_0\sqrt{D'})^n$ for some fixed (α_0, β_0) and any integer n. The real number $\alpha_0 + \beta_0\sqrt{D'}$ is either a fundamental unit of the real quadratic field $\mathbb{Q}(\sqrt{D'})$ or (if the norm of the fundamental unit is -1) the square of a fundamental unit.

Reducing the coefficients of $\alpha_0 + \beta_0\sqrt{D'}$ modulo $2a$ gives an element $z = \bar{\alpha}_0 + \bar{\beta}_0\bar{x}$ of the ring

$$R = \mathbb{Z}[x] / (2a, x^2 - D'). \tag{2.4}$$

Furthermore, since $(\alpha_0 + \beta_0\sqrt{D'})(\alpha_0 - \beta_0\sqrt{D'}) = 1$, z is invertible in R, i.e. $z \in R^*$. Since R^* is a finite group of size less than $4a^2$, there is an integer

$m < 4a^2$ such that $z^m = 1$ in R^*.[1] Lifting back up to the full ring $\mathbb{Z}[\sqrt{D'}]$, we see that $(\alpha_0 + \beta_0\sqrt{D'})^m = \alpha_1 + \beta_1\sqrt{D'}$ for integers α_1, β_1 satisfying

$$\begin{aligned} \alpha_1 &\equiv 1 \pmod{2a}, \\ \beta_1 &\equiv 0 \pmod{2a}. \end{aligned} \tag{2.5}$$

Now for any integer n we can compute integers (u, v) such that

$$u + v\sqrt{D'} = (u_0 + v_0\sqrt{D'})(\alpha_1 + \beta_1\sqrt{D'})^n. \tag{2.6}$$

We claim that (u, v) satisfy the congruence conditions (2.3). To see this, let $\alpha_n + \beta_n\sqrt{D'} = (\alpha_1 + \beta_1\sqrt{D'})^n$. The conditions (2.5) imply that $\alpha_n \equiv 1 \pmod{2a}$ and $\beta_n \equiv 0 \pmod{2a}$. Combining this observation with the formulas

$$\begin{aligned} u &= \alpha_n u_0 + \beta_n v_0 D' \\ v &= \alpha_n v_0 + \beta_n u_0, \end{aligned} \tag{2.7}$$

we see that $u \equiv u_0 \equiv b \pmod{2a}$ and $v \equiv v_0 \pmod{2a}$. Furthermore, $v_0 \equiv 0 \pmod{2r}$ and $2r$ divides $2a$ (since $aD = D'r^2$ and D is square-free, so we conclude that $v \equiv 0 \pmod{2r}$.

The new solution (u, v) thus satisfies the congruence conditions (2.3). Any integer n gives such a solution, so by setting $x = (u - b)/2a$ and $y = v/2r$ for each such (u, v), we have generated an infinite number of integer solutions to the equation $Dy^2 = f(x)$. This is condition (4) of Definition 2.5; by hypothesis (t, n, q) satisfy conditions (1), (2), and (3), so we conclude that (t, n, q) represents a family of curves with embedding degree k.   □

Remark 2.8. More generally, we may find an infinite family of curves in the case where $f(x) = g(x)^2 h(x)$, with $h(x)$ quadratic. Specifically, if we let $y = y'g(x)$, then given one integral solution (x, y') to the equation $Dy'^2 = h(x)$ we may use the method of Theorem 2.7 to find an infinite number of solutions. However, we currently know of no examples for which $f(x)$ is of this form.

Theorem 2.7 tells us that if $f(x)$ is quadratic and square-free, we may get a family of curves of the prescribed embedding degree for *each* D. If $f(x)$ is instead a linear function times a square, then we may still get a family of curves, but for only a single D. This is the method that Barreto and Naehrig [2] use to construct curves with $k = 12$ (see Section 4.2).

Proposition 2.9. Fix an integer $k > 0$, and let $n(x)$, $t(x)$, and $q(x)$ be polynomials in $\mathbb{Z}[x]$ satisfying conditions (1), (2), and (3) of Definition 2.5. Let $f(x) = 4q(x) - t(x)^2$, and suppose $f(x) = (Ax + D)g(x)^2$ for some positive integer D and some polynomial $g(x)$. Then (t, n, q) represents a family of curves with embedding degree k.

---

[1] In fact, since z is an element of the norm-one subgroup of R^*, m is bounded above by $2^s a$, where s is the number of distinct primes dividing $2a$. A more detailed study of the group R^* appears in an earlier draft of this paper [11].

Proof. For any integer v, we set $x = ADv^2 + 2Dv$ and let $y = (Av + 1)g(x)$. An easy computation shows that (x, y) is a solution to the equation $Dy^2 = f(x)$, so if D is square-free then condition (4) is satisfied for the integer D. If D is not square-free then we may absorb its square factors into y, and condition (4) is satisfied for the largest square-free factor D' of D.   □

We conclude this section with a partial converse to Theorem 2.7; namely, if the degree of $f(x)$ is at least 3, then we are unlikely to find an infinite family of curves.

Proposition 2.10. *Let (t, n, q) be polynomials with integer coefficients satisfying conditions (1), (2), and (3) of Definition 2.5, and let $f(x) = 4q(x) - t(x)^2$. Suppose $f(x)$ is square-free and $\deg f(x) \geq 3$. Then (t, n, q) does not represent a family of elliptic curves with embedding degree k.*

Proof. Since $f(x)$ is square-free (i.e. has no double roots) and has degree at least 3, the equation $Dy^2 = f(x)$ defines a smooth affine plane curve of genus $g \geq 1$. By Siegel's Theorem (cf. [23, Theorem IX.4.3] and [9, §I.2]) such curves have a finite number of integral points, so condition (4) is not satisfied.   □

3 Elliptic Curves with Embedding Degree 10

In this section, we use the method of Section 2, and Theorem 2.7 in particular, to construct elliptic curves of prime order with embedding degree 10. Our key observation is that since the hypotheses of Theorem 2.7 require $f(x) = 4n(x) - (t(x) - 2)^2$ to be quadratic, we should choose $n(x)$ and $t(x)$ in such a way that the high-degree terms of $t(x)^2$ cancel out those of $4n(x)$; in particular, the degree of $t(x)$ must be half the degree of $n(x)$. We have discovered that for $k = 10$ there is a choice of $n(x)$ and $t(x)$ such that this is possible. The resulting construction of elliptic curves with embedding degree 10 solves an open problem posed by Boneh, Lynn, and Shacham [6, §4.5].

We begin by recalling that to construct a curve with embedding degree k, we must choose the number of points $n(x)$ and the trace $t(x)$ such that $n(x)$ is an irreducible factor of $\Phi_k(t(x) - 1)$, where Φ_k is the kth cyclotomic polynomial. If $k = 10$ and $t(x)$ is linear then $\Phi_k(t(x) - 1)$ is an irreducible quartic polynomial, so there is no hope of $f(x) = 4n(x) - (t(x) - 2)^2$ being quadratic. If $k = 10$ and $t(x)$ is quadratic, Galbraith, McKee, and Valença [12] show that in this case $\Phi_k(t(x) - 1)$ either is irreducible of degree 8 or factors into two irreducible quartic polynomials. They then show that there is an infinite set of $t(x)$ such that the latter occurs, and that these $t(x)$ are parameterized by the rational points of a certain elliptic curve. By experimenting with some of the examples given by Galbraith, et al., we discovered that $t(x) = 10x^2 + 5x + 3$ leads to a quadratic $f(x)$.

Theorem 3.1. *Fix a positive square-free integer D relatively prime to 15. Define $t(x)$, $n(x)$, and $q(x)$ by*

$$t(x) = 10x^2 + 5x + 3$$
$$n(x) = 25x^4 + 25x^3 + 15x^2 + 5x + 1$$
$$q(x) = 25x^4 + 25x^3 + 25x^2 + 10x + 3.$$

If the equation $u^2 - 15Dv^2 = -20$ has a solution with $u \equiv 5 \pmod{15}$, then (t, n, q) represents a family of curves with embedding degree 10.

Proof. It is easy to verify that conditions (1)-(3) of Definition 2.5 hold. Condition (4) requires an infinite number of integer solutions to $Dy^2 = f(x)$, where $f(x) = 4q(x) - t(x)^2$. The key observation is that for this choice of t and n,

$$f(x) = 4q(x) - t(x)^2 = 15x^2 + 10x + 3. \tag{3.1}$$

Multiplying by 15 and completing the square transforms the equation we wish to solve into

$$D'y^2 = (15x + 5)^2 + 20, \tag{3.2}$$

where $D' = 15D$. Integer solutions to this equation correspond to integer solutions to $u^2 - D'v^2 = -20$ with $u \equiv 5 \pmod{15}$. By Theorem 2.7, if one such solution exists then an infinite number exist, so (t, n, q) represents a family of curves with embedding degree 10. □

To use the above result to construct curves with embedding degree 10, we choose a D and search for solutions to the equation $u^2 - 15Dv^2 = -20$ that give prime values for q and n. The following lemma, proposed by Mike Scott, speeds up this process by restricting the values of D that we can use.

Lemma 3.2. *Let $q(x)$ be as in Theorem 3.1. If (x, y) is an integer solution to $Dy^2 = 15x^2 + 10x + 3$ such that $q(x)$ is prime, then $D \equiv 43$ or $67 \pmod{120}$.*

Proof. If $x \equiv 0$ or $2 \pmod 3$ then $q(x)$ is divisible by 3, while if x is odd then $q(x)$ is even. Thus if $q(x)$ is prime, then $x \equiv 4 \pmod 6$.

To deduce the stated congruence for D, we consider the equation $Dy^2 = 15x^2 + 10x + 3$ modulo 3, 5, and 8. To begin, we have $Dy^2 \equiv x \equiv 1 \pmod 3$, so $D \equiv 1 \pmod 3$. Next, we have $Dy^2 \equiv 3 \pmod 5$, so $y^2 \equiv 1$ or $4 \pmod 5$ and $D \equiv 2$ or $3 \pmod 5$. Finally, since x is even we see that $Dy^2 = 3 \pmod 8$, and thus $y^2 \equiv 1 \pmod 8$ and $D \equiv 3 \pmod 8$. Combining these results via the Chinese remainder theorem, we conclude that $D \equiv 43$ or $67 \pmod{120}$. □

After reading an earlier draft of this paper [11], Mike Scott used Theorem 3.1 and Lemma 3.2 to find examples of elliptic curves with embedding degree 10 via the following algorithm.

Algorithm 3.3. *Let (t, n, q) be as in Theorem 3.1. The following algorithm takes inputs* MaxD, MinBits, *and* MaxBits, *and outputs pairs (D, x) such that $D <$ * MaxD, *the number of bits in $q(x)$ is between* MinBits *and* MaxBits, *and (D, x) satisfy the conditions of Corollary 2.6 with $k = 10$.*

1. *Set D to be a positive integer such that $D \equiv 43$ or $67 \pmod{120}$ and $15D$ is square-free.*
2. *Use the Continued Fraction algorithm [21] to compute a fundamental unit γ of the ring of integers in $\mathbb{Q}(\sqrt{15D})$. Let $\delta = \gamma^2$ if γ has norm -1, $\delta = \gamma$ otherwise.*
3. *Use the algorithm of Lagrange, Matthews [15], and Mollin [18] to find fundamental solutions (u, v) to the equation $u^2 - 15Dv^2 = -20$. (See also [21].)*
4. *For each fundamental solution (u, v) found in (3):*
   (a) *If $\log_2 u > (\mathtt{MaxBits} + 11)/4$, go to the next fundamental solution.*
   (b) *If $u \equiv \pm 5 \pmod{15}$ and $\log_2 u > (\mathtt{MinBits} + 11)/4$, then:*
      i. *Let $x = (-5 \pm u)/15$.*
      ii. *If $q(x)$ and $n(x)$ are prime, output (D, x).*
   (c) *Multiply $u + v\sqrt{15D}$ by δ to get a new u, and return to step (a).*
5. *Increase D. If $D < \mathtt{MaxD}$, return to step (1); otherwise terminate.*

Remark 3.4. The bounds on $\log_2 u$ in Step 4 can be explained as follows: since $q(x) = 25x^4 + O(x^3)$ and $x = (-5 \pm u)/15$, $q(x)$ grows roughly like $u^4/2025$. We conclude that $\log_2 q(x) \approx 4\log_2 u - 11$, so we require u in the algorithm to satisfy

$$\frac{\mathtt{MinBits} + 11}{4} < \log_2 u < \frac{\mathtt{MaxBits} + 11}{4}.$$

In our construction of Algorithm 3.3, the specific parameters of Theorem 3.1 have allowed us to simplify the procedure described in the proof of Theorem 2.7. The requirement that $15D$ be square-free implies that $r = 1$, and the fact that $b = 10$ is even allows us to remove the factors of 2 in the congruence moduli of equations (2.3). Thus in Step 4 we need only to find (u, v) with $u^2 - 15Dv^2 = -20$ and $u \equiv \pm 5 \pmod{15}$. Given this requirement, we see that the only restriction on the unit $\delta = \alpha + \beta\sqrt{15D}$ in Step 4c is that $\alpha \not\equiv 0 \pmod 3$, which must be true since $\alpha^2 - 15D\beta^2 = 1$. Thus our choice of $\delta = \gamma$ or γ^2 will always give new solutions (u, v) with $u \equiv \pm 5 \pmod{15}$; i.e. the parameter m of Theorem 2.7 is equal to 1.

In practice the fundamental unit γ computed in Step 2 will usually be very large, in which case we may skip Step 4c altogether. For example, computations with PARI indicate that when $D \approx 10^9$, γ has at least 100 bits 99.5% of the time and at least 200 bits 98.9% of the time.

Scott ran Algorithm 3.3 with inputs $\mathtt{MaxD} = 2 \cdot 10^9$, $\mathtt{MinBits} = 148$, and $\mathtt{MaxBits} = 512$. For each (D, x) output by the algorithm, one may then use the Complex Multiplication method (cf. [3], [19]) to construct an elliptic curve over $\mathbb{F}_{q(x)}$ whose number of points is $n(x)$. By Theorem 2.2 this curve has embedding degree at most 10, and in practice we find that the embedding degree is exactly 10. Below are two examples of elliptic curves that Scott constructed in this manner.

Example 3.5. (A 234-bit curve.) Running Algorithm 3.3 with $D = 1227652867$ produces the following example. Let q, n, A, B be as follows:

$q = 1821165080396947206449326434737595004593425469665709042072623004320380 3$

$n = 1821165080396947206449326434737594977603315574395203075045003378230665 1$

$A = -3$

$B = 1574866809491340118477796447352285908690083127492294897332068499590327 5.$

Then q and n are 234-bit prime numbers such that the curve $y^2 = x^3 + Ax + B$ defined over $\mathbb{F}_q$ has n points. Since $n \mid q^{10} - 1$ and $n \nmid q^i - 1$ for $i < 10$, this curve has embedding degree 10.

Example 3.6. (A 252-bit curve.) Running Algorithm 3.3 with $D = 1039452307$ produces the following example. Let q, n, A, B be as follows:

$q = 646231099734881696220312491050525208267333884696643120163526269440282546164 3$

$n = 646231099734881696220312491050525208251256184615662859556277645930629210126 1$

$A = -3$

$B = 494653816664025137427462882026969414424918177601315486328808621207680852814 1.$

Then q and n are 252-bit prime numbers such that the curve $y^2 = x^3 + Ax + B$ defined over $\mathbb{F}_q$ has n points. Since $n \mid q^{10} - 1$ and $n \nmid q^i - 1$ for $i < 10$, this curve has embedding degree 10.

Ideally, the bit size of curves with embedding degree 10 should be chosen so that the discrete logarithm in the finite field $\mathbb{F}_{q^{10}}$ is approximately of the same difficulty as the discrete logarithm problem on an elliptic curve of prime order over $\mathbb{F}_q$. Using the best known discrete logarithm algorithms, this happens when q has between 220 and 250 bits [3]. The curves in Examples 3.5 and 3.6 have been selected so that their bit sizes are close to this range and their complex multiplication discriminants D are not much larger than 10^9. The equation for a curve with this size D can be computed in about a week on today's fastest PCs.

In practice, it appears that curves with small embedding degree, prime order, and small complex multiplication discriminant D are quite rare. Luca and Shparlinski [14] come to this conclusion for curves with embedding degree 3, 4, or 6 (the so-called MNT curves) through a heuristic analysis of the MNT construction. Since our construction of curves with embedding degree 10 is similar to the MNT construction (cf. Section 4.1), a similar analysis should hold for our $k = 10$ curves. The experimental evidence supports this reasoning: Scott's execution of Algorithm 3.3 with MaxD $= 2 \cdot 10^9$ found only 23 curves with prime orders between 148 and 512 bits [22].

If we relax the condition on $n(x)$ in step 4(b)ii of Algorithm 3.3 and write $n = kr$ with r a large prime and k a small cofactor, then we may find a larger number of suitable curves. Scott also ran this version of the algorithm and found 101 curves with r between 148 and 512 bits, k at most 16 bits, and $D < 2 \cdot 10^9$ [22].

4   Elliptic Curve Families with Small Embedding Degree

In this section we show how the existing constructions of ordinary elliptic curves of prime order with embedding degree 3, 4, or 6 [17] or embedding degree 12 [2] can be explained via the framework of Section 2. The former uses Theorem 2.7, while the latter employs Proposition 2.9.

4.1   MNT Elliptic Curves

Miyaji, Nakabayashi, and Takano [17] have classified all ordinary elliptic curves of prime order with embedding degree 3, 4, and 6. Their theorem is as follows:

Theorem 4.1 ([17]). *Let E be an ordinary elliptic curve over $\mathbb{F}_q$ such that $\#E(\mathbb{F}_q) = n = q + 1 - t$ is prime and E has embedding degree $k = 3, 4$, or 6. Then there exists an integer x such that t, n, and q are of the form specified in the following table:*

k	t	n	q
3	$-1 \pm 6x$	$12x^2 \mp 6x + 1$	$12x^2 - 1$
4	$-x$ or $x + 1$	$x^2 + 2x + 2$ or $x^2 + 1$	$x^2 + x + 1$
6	$1 \pm 2x$	$4x^2 \mp 2x + 1$	$4x^2 + 1$

This theorem fits into the framework of Section 2 as follows. To find an infinite family of curves via Theorem 2.7, we require $f(x)$ to be quadratic. Since $\deg \Phi_k(x) = 2$ for $k = 3$, 4, or 6, if we let $t(x)$ be any linear polynomial and $n(x)$ be the (irreducible) quadratic $\Phi_k(t(x) - 1)$ (with any constant factor divided out), then $f(x) = 4n(x) - (t(x) - 2)^2$ is quadratic. If $q(x) = n(x) + t(x) - 1$ is also irreducible and the equation $Dy^2 = f(x)$ has one solution, then (t, n, q) satisfy the hypotheses of Theorem 2.7 and thus represent a family of curves with embedding degree k. Miyaji, et al. arrive at their stronger result by using the fact that $\#E(\mathbb{F}_q)$ is prime to show that any values of t, n, and q that give rise to such a curve must be of the specified form.

4.2   Elliptic Curves with Embedding Degree 12

Finally, we note that the Barreto-Naehrig construction [2] of curves with embedding degree 12 falls under the case of Proposition 2.9. Specifically, if $t(x) = 6x^2 + 1$, then $\Phi_{12}(t(x) - 1) = n(x)n(-x)$, where $n(x) = 36x^4 + 36x^3 + 18x^2 + 6x + 1$, and

$$f(x) = 4n(x) - (t(x) - 2)^2 = 3(6x^2 + 4x + 1)^2. \tag{4.1}$$

Since $q(x) = 36x^4 + 36x^3 + 12x^2 + 6x + 1$ is also irreducible, Proposition 2.9 tells us that if we set $D = 3$, then (t, n, q) represents a family of curves with embedding degree 12.

5   Higher Embedding Degrees

To construct families of elliptic curves with prescribed embedding degree, the method of Section 2 requires us to find an infinite number of integer solutions to an equation of the form $Dy^2 = f(x)$. In this section, we give evidence that in general the degree of $f(x)$ is large, and thus by Proposition 2.10 we are unlikely to find an infinite family of curves. We begin with a lemma that restricts the possible degrees of the polynomial $n(x)$; the lemma generalizes a result of Galbraith, et al. [12, Lemma 1].

Lemma 5.1. *Fix k, let $t(x)$ be a polynomial, and let $n(x)$ be an irreducible factor of $\Phi_k(t(x) - 1)$. Then the degree of n is a multiple of $\varphi(k)$, where φ is the Euler phi function.*

Proof. Suppose $t(x)$ has degree d, so $\deg \Phi_k(t(x) - 1) = d\varphi(k)$. Let θ be a root of $n(x)$, and let $\omega = t(\theta) - 1$. Then $\Phi_k(\omega) = 0$, so ω is a primitive kth root of unity. We thus have the inclusion of fields $\mathbb{Q}(\theta) \supset \mathbb{Q}(\omega) \supset \mathbb{Q}$. Since $[\mathbb{Q}(\theta) : \mathbb{Q}] = \deg n(x)$ and $[\mathbb{Q}(\omega) : \mathbb{Q}] = \varphi(k)$, we conclude that $\varphi(k)$ divides $\deg n(x)$. □

The key observation that allowed us to construct families of elliptic curves with embedding degree 10 was that if $f(x)$ is quadratic and $n(x)$ has degree greater than 2, then the polynomial $t(x)$ must be chosen so that the high degree terms of $t(x)^2$ cancel out those of $4n(x)$. The following proposition shows that this is in fact the only way to construct such families.

Proposition 5.2. *Suppose (t, n, q) represents a family of curves with embedding degree k, and suppose further that $f(x) = 4n(x) - (t(x) - 2)^2$ is square-free. If $\varphi(k) \geq 4$, then*

$$\deg t(x) = \frac{1}{2} \deg n(x) = \frac{1}{2} \deg q(x). \tag{5.1}$$

Furthermore, if a is the leading coefficient of $t(x)$, then $a^2/4$ is the leading coefficient of $n(x)$ and $q(x)$.

Proof. Since $\varphi(k) \geq 4$, by Lemma 5.1 $\deg n(x) \geq 4$, and since $f(x)$ is square-free, by Proposition 2.10 $\deg f(x) \leq 2$. Since $f(x) = 4n(x) - (t(x) - 2)^2$, we conclude that $\deg t(x) = \frac{1}{2} \deg n(x)$, and since $n(x) = q(x) + 1 - t(x)$, we see that $\deg n(x) = \deg q(x)$. The observation about the leading coefficients follows immediately. □

As an immediate corollary, we see that if $k > 6$ (so $\varphi(k) \geq 4$) then choosing a linear $t(x)$ will not in general give us an infinite family of curves, whereas if $k > 12$ (so $\varphi(k) \geq 6$) then choosing a quadratic $t(x)$ will not in general give us an infinite family of curves.

Proposition 5.2 tells us that for embedding degrees k with $\varphi(k) \geq 4$, to find an infinite family of curves we will have to choose $t(x_0)$ of degree at least 2 such that $\phi_k(t(x) - 1)$ is not irreducible. Galbraith, McKee, and Valença [12] observe that this is hard even for quadratic $t(x)$, and as the degree increases the problem will only become more difficult. An alternative would be to choose t and n such that $f(x)$ has a square factor; this appears to be just as difficult, but has not been studied in depth.

6   Conclusion

We have seen in Section 2 that the current methods for constructing families of elliptic curves of prime order with prescribed embedding degree can all be subsumed under a general framework. In Section 3 we showed how this framework can be used to construct curves with embedding degree 10 and we gave examples of such curves, which have not previously appeared in the literature. In Section 4 we showed how this framework incorporates the existing constructions for embedding degrees $3, 4, 6$, and 12.

In Section 5 we showed that our method can only produce an infinite family of curves if a certain polynomial $f(x)$ either is quadratic or has a square factor. These two conditions have been achieved for $k = 10$ and $k = 12$, respectively, but these two examples appear to be special cases, and in general we have not found a way to achieve either of these two conditions. The success of our method in producing curves with embedding degree greater than 12 depends on our ability to control the behavior of $f(x)$, which leads to the following important open problem.

Problem 6.1. Given an integer k such that $\varphi(k) \geq 4$, find polynomials $t(x)$ and $n(x)$ such that

1. $n(x)$ is an irreducible factor of $\Phi_k(t(x) - 1)$, where Φ_k is the kth cyclotomic polynomial, and
2. $f(x) = 4n(x) - (t(x) - 2)^2$ is either quadratic or of the form $g(x)^2 h(x)$, with $\deg h(x) \leq 2$.

Acknowledgments

Research for this paper was conducted during a summer internship at Hewlett-Packard Laboratories, Palo Alto. I thank Vinay Deolalikar for suggesting this topic and for providing advice and support along the way. I also thank Gadiel Seroussi for bringing me to HP and for supporting my research.

I thank Paulo Barreto, Steven Galbraith, Ed Schaefer, and Mike Scott for their valuable feedback on earlier versions of this paper. I am especially indebted to Mike Scott, who used the method presented in Section 3 to compute examples of elliptic curves of cryptographic size with embedding degree 10. Two of these curves now appear in this paper as Examples 3.5 and 3.6.

References

1. P.S.L.M. Barreto, B. Lynn, M. Scott, "Constructing elliptic curves with prescribed embedding degrees," in *SCN 2002*, ed. S. Cimato, C. Galdi, G. Persiano, Springer LNCS **2576** (2003) 257-267.
2. P.S.L.M. Barreto, M. Naehrig, "Pairing-friendly elliptic curves of prime order," in *SAC 2005*, ed. B. Preneel, S. Tavares, Springer LNCS **3897** (2006) 319-331.
3. I. Blake, G. Seroussi, N. Smart, *Elliptic Curves in Cryptography*, LMS Lecture Note Series **265**, Cambridge University Press, 1999.
4. I. Blake, G. Seroussi, N. Smart, eds., *Advances in Elliptic Curve Cryptography*, LMS Lecture Note Series **317**, Cambridge Unviersity Press, 2005.
5. D. Boneh, M. Franklin, "Identity based encryption from the Weil pairing," in *CRYPTO '01*, ed. J. Kilian, Springer LNCS **2139** (2001), 213-229.
6. D. Boneh, B. Lynn, H. Shacham, "Short signatures from the Weil pairing," in *ASIACRYPT '01*, ed. C. Boyd, Springer LNCS **2248** (2001), 514-532.
7. F. Brezing, A. Weng, "Elliptic curves suitable for pairing based cryptography," *Designs, Codes, and Cryptography* **37** (2005) 133-141.
8. C. Cocks, R.G.E. Pinch, "Identity-based cryptosystems based on the Weil pairing," unpublished manuscript, 2001.
9. G. Cornell, J. Silverman, eds., *Arithmetic Geometry,* Springer, New York 1986.
10. S. Cui, P. Duan, C.W. Chan, "A new method of building more non-supersingular elliptic curves," in *ISH 2005*, ed. O. Gervasi et al., Springer LNCS **3481** (2005), 657-664.
11. D. Freeman, "Constructing families of pairing-friendly elliptic curves," Hewlett-Packard Laboratories technical report HPL-2005-155 (2005), available at http://www.hpl.hp.com/techreports/2005/HPL-2005-155.html.
12. S. Galbraith, J. McKee, P. Valença, "Ordinary abelian varieties having small embedding degree," in *Proceedings of a Workshop on Mathematical Problems and Techniques in Cryptology*, ed. R. Cramer, T. Okamoto, CRM Barcelona (2005) 29-45.
13. A. Joux, "A one round protocol for tripartite Diffie-Hellman," in *ANTS-IV*, ed. W. Bosma, Springer LNCS **1838** (2000), 385-394.
14. F. Luca, I. Shparlinski, "Elliptic curves with low embedding degree," preprint, available at http://eprint.iacr.org/2005/363.
15. K. Matthews, "The diophantine equation $x^2 - Dy^2 = N$, $D > 0$, in integers," *Expositiones Mathematicae* **18** (2000), 323-331.
16. A. Menezes, T. Okamoto, S. Vanstone, "Reducing elliptic curve logarithms to logarithms in a finite field," *IEEE Transactions on Information Theory* **39** (1993), 1639-1646.
17. A. Miyaji, M. Nakabayashi, S. Takano, "New explicit conditions of elliptic curve traces for FR-reduction," *IEICE Transactions on Fundamentals* **E84-A(5)** (2001), 1234-1243.
18. R. Mollin, *Fundamental Number Theory with Applications*, CRC Press, Boca Raton, 1998.
19. F. Morain, "Building cyclic elliptic curves modulo large primes," in *EUROCRYPT '91*, ed. D. W. Davies, Springer LNCS **547** (1991) 328-336.
20. J. Neukirch, *Algebraic Number Theory*, Springer, Berlin 1999.
21. J. Robertson, "Solving the generalized Pell equation," unpublished manuscript (2004), available at http://hometown.aol.com/jpr2718/pell.pdf.
22. M. Scott, personal communication, 7 November 2005.
23. J. Silverman, *The Arithmetic of Elliptic Curves*, Springer GTM **106**, 1986.

Fast Bilinear Maps from the Tate-Lichtenbaum Pairing on Hyperelliptic Curves

Gerhard Frey[1],* and Tanja Lange[2],*

[1] Institute for Experimental Mathematics (IEM), University of Duisburg-Essen
Ellernstrasse 29, D-45326 Essen, Germany
frey@iem.uni-due.de
[2] Department of Mathematics, Technical University of Denmark
Matematiktorvet 303, 2800 Kgs. Lyngby, Denmark
tanja@hyperelliptic.org

Abstract. Pairings on elliptic curves recently obtained a lot of attention not only as a means to attack curve based cryptography but also as a building block for cryptosystems with special properties like short signatures or identity based encryption.

In this paper we consider the Tate pairing on *hyperelliptic curves* of genus g. We give mathematically sound arguments why it is possible to use particular representatives of the involved residue classes in the second argument that allow to compute the pairing much faster, where the speed-up grows with the size of g. Since the curve arithmetic takes about the same time for small g and constant group size, this implies that $g > 1$ offers advantages for implementations. We give two examples of how to apply the modified setting in pairing based protocols such that all parties profit from the idea.

We stress that our results apply also to non-supersingular curves, e. g. those constructed by complex multiplication, and do not need distortion maps. They are also applicable if the co-factor is nontrivial.

Keywords: Public key cryptography; pairings, hyperelliptic curves, fast computation, Tate pairing.

1 Introduction

Until recently, pairings on elliptic and hyperelliptic curves have been studied for attacks only. Hence, not much effort was put in efficient implementations. With the proposals of tripartite key exchange [26] and identity based encryption [8,9,39] which are both based on bilinear maps they gained a lot of attention as so far the Weil and Tate pairing on elliptic and hyperelliptic curves are the only efficient instantiations of bilinear maps.

In the sequel special cases and improvements of the general implementation [18,19,31] were proposed for elliptic [3,15,16,21,25,6] and hyperelliptic curves

---

* The work described in this paper has been supported in part by the European Commission through the IST Programme under Contract IST-2002-507932 ECRYPT.

F. Hess, S. Pauli, and M. Pohst (Eds.): ANTS 2006, LNCS 4076, pp. 466–479, 2006.

[12,14,16], to mention just a few. The eta-T pairing [2] suggests using supersingular binary curves of genus 2 and gives formulas for efficient implementation for this case. Their proposal makes use of the special choice we are presenting here but they do not give arguments why it is an allowed restriction in the general case.

Apart from that paper mainly direct generalizations of the methods for elliptic curves have been proposed for hyperelliptic curves and the practical interest in hyperelliptic curves is due to the fact that larger embedding degrees k can be obtained for supersingular hyperelliptic curves than for supersingular elliptic curves [20,37]. We like to stress that this is a result for *supersingular curves* and that the important size is the security multiplier k/g which has the same maximal value of 6 for elliptic curves and curves of genus 2. Using complex multiplication one can construct *ordinary* curves with embedding degree in the desired range for elliptic curves [5,4,13,33].

Currently an extension degree of 6 is suitable for the applications but for long-term security or higher security requirements far larger values are needed which cannot be provided by supersingular curves.

Our method speeds up the pairing computation by a factor of about g for the applications of pairings in protocols. Thus, there is no gain for elliptic curves but for hyperelliptic curves which gives an advantage to the application of hyperelliptic curves of genus $g > 1$. In brief we have the following setting:

Let $\mathbb{F}_q, q = p^d, p$ prime, be a finite field and let $C/\mathbb{F}_q$ be a hyperelliptic curves over $\mathbb{F}_q$ of genus g. For every extension degree k, the group of $\mathbb{F}_{q^k}$-rational points $J_C(\mathbb{F}_{q^k})$ on the Jacobian J_C of C is isomorphic to the divisor class group of degree zero $\mathrm{Pic}^0_{C \cdot \mathbb{F}_{q^k}}$ of C over $\mathbb{F}_{q^k}$. We fix a subgroup of $J_C(\mathbb{F}_q)$ of some prime order ℓ. As otherwise the Rück attack [38] is successful in solving the *Discrete Logarithm Problem* (DLP) in this group, we assume $\gcd(q, \ell) = 1$.

We first introduce the mathematical background of the Tate-Lichtenbaum pairing T_ℓ. Generically a divisor class can be represented by g points on the curve but we show that one can restrict the second argument of T_ℓ to the set of divisors $P - P_\infty$ with P an $\mathbb{F}_{q^k}$-rational point of the curve C and still obtain a non-trivial pairing.

As the computation of the pairing involves evaluating a function at the second argument of the pairing, our modification gives a speed-up by a factor of g in this step. Finally we consider applications in protocols and deal with parameter choices.

For a special curve and special extension degrees such that the Jacobian has almost prime order, Duursma and Lee [14] already suggest to use points as input to the pairing instead of divisors of full degree. However, this is not possible in general and they do not show that this leads to a non-trivial pairing. Also later publications use this approach without validating the assumption. It is obviously justified if the group order of $J_C(\mathbb{F}_q)$ is prime and C is supersingular with a distortion map, because then the divisor class of $P - P_\infty$ has prime order and due to the non-degeneracy of the pairing and the distortion map the distorted image is not in the kernel of the pairing.

If the curve is constructed via complex multiplication one usually has to sacrifice some bits in the co-factor of the group order, as at the same time one wants to control the size of the prime p, obtain a fixed optimal embedding degree k and have a small cofactor. Our results imply that in this case a divisor as applied by Duursma and Lee does not work generically as the first argument of the pairing since the order of such a divisor is a multiple of ℓ. Therefore, their fast way of using pairings is not applicable and a distortion map does not exist on ordinary curves.

On a supersingular curve with a non-trivial co-factor one cannot find a divisor class of full order such that the representing divisor has only one affine point in the support. This means that in general applications [14] would need to use a divisor of full degree as first input and as they apply distortion maps also the second argument has full degree which means that for the case considered in our contribution their scheme is less efficient.

Our method works in general, is proved to work, and has the advantage that one does not need distortion maps. This last item means that also non-supersingular curves can be applied efficiently in pairings – and that is a very interesting case for applications since in software implementations, prime field arithmetic is faster than binary and larger embedding degrees could be obtained by generalizing MNT curves. So far no non-supersingular hyperelliptic curves with larger security parameters were found but there is ongoing research in this direction. At the same time this is also the case which needs a proof that the special choices are possible. Our paper is a purely theoretical one due to the lack of satisfying non-supersingular curves; for supersingular ones the distortion maps are likely to lead to faster computations.

2   Tate-Lichtenbaum Pairing on Hyperelliptic Curves

Within the scope of this paper we can only give a short introduction to the topic. More details and general background on hyperelliptic curves can be found in [1,17,30,41].

In this paper we concentrate on hyperelliptic curves, including elliptic curves. We consider hyperelliptic curves which have exactly one $\mathbb{F}_q$-rational point P_∞ at infinity. Hence, for C a genus g curve, an affine part is given by

$$C_a : y^2 + h(x)y = f(x), \; h, f \in \mathbb{F}_q[x],$$

where f is monic of degree $\deg(f) = 2g = 1$ and $\deg(h) \leq g$.

For working in the group we use the isomorphism of the $\mathbb{F}_q$-rational points $J_C(\mathbb{F}_q)$ with the *divisor class group of degree zero divisors* Pic_C^0 over $\mathbb{F}_q$. To fix notation, the letter P always means a point on the curve and not on the Jacobian. For explanations we represent points on the Jacobian as reduced divisor classes $\bar{D}$ with at most g affine points in the support. The group arithmetic is carried out using Cantor's algorithm [11,28] or the explicit formulae which exist for genus $g = 1, 2, 3, 4$ [7,23,29,34,36,42,43,44] to double and add in the group.

For $g = 2$, [29] gives a complete study of addition and doubling for all different numbers of affine points in the input divisors and it is easily seen that a lower number leads to less field operations per group operation. On the other hand, a generic group element has g points and hence the special cases will not occur by accident. In [27] the authors observed that one can enforce this situation by choosing the base of the DL system in this form. However, they leave it open to show that there are base points of this form. And in fact one cannot expect to find a point $P \in C(\mathbb{F}_q)$ such that the divisor class of $P - P_\infty$ has order ℓ, but only a multiple of ℓ. But as we shall show in our applications concerning the computation of the Tate pairing this special case will appear naturally and thus we already point out here that each scalar multiplication of such a divisor class is faster using a double-and-add algorithm from left to right.

In [19] the Tate pairing was introduced in the form due to Lichtenbaum. Let $\ell \mid |J_C(\mathbb{F}_q)|$ and let k be the smallest integer such that $\ell \mid (q^k - 1)$ meaning that the ℓ-th roots of unity are contained in $\mathbb{F}_{q^k}$ but in no smaller field over $\mathbb{F}_q$. As usual we refer to k as the *embedding degree*. Naturally, the improvement presented in the following also speeds up attacks using the Tate pairing. Hence, except for the section about applications in protocols we also allow $k = 1$, i.e. the case that the ℓ-th roots of unity are defined over the ground field. The ℓ-torsion points $J_C(\mathbb{F}_{q^k})[\ell]$ are the points defined over $\mathbb{F}_{q^k}$ having order dividing ℓ.

The Tate-Lichtenbaum pairing is induced by a map

$$T_\ell : J_C(\mathbb{F}_{q^k})[\ell] \times J_C(\mathbb{F}_{q^k}) \to \mathbb{F}_{q^k}^* / \mathbb{F}_{q^k}^{*\ell}$$

defined in the following way: Let $\bar{D}_1 \in J_C(\mathbb{F}_{q^k})[\ell]$ and $\bar{D}_2 \in J_C(\mathbb{F}_{q^k})$. To compute $T_\ell(\bar{D}_1, \bar{D}_2)$ one uses that $\bar{D}_1$ has order ℓ, i.e. there is a function F_{D_1} such that $\ell D_1 \sim \mathrm{div}(F_{D_1})$, where D_1 represents the class $\bar{D}_1$. Let $\bar{D}_2$ be represented by a divisor D_2 such that no point in the support of D_2 occurs in the support of D_1. Then

$$T_\ell(\bar{D}_1, \bar{D}_2) = F_{D_1}(D_2).$$

This means that for $D_2 = \sum_{i=1}^m P_i - \sum_{j=1}^m Q_j$ one has

$$F_{D_1}(D_2) = \frac{\prod_{i=1}^m F_{D_1}(P_i)}{\prod_{j=1}^m F_{D_1}(Q_j)}. \tag{1}$$

One can show that the pairing is well defined, i.e. it does not depend on the choices of D_1 and D_2 if the image is taken modulo ℓ-th powers in $\mathbb{F}_{q^k}$. An important property of the Tate pairing is that it is non-degenerate in the first argument, i.e. for a fixed $\bar{D}_1 \in J_C(\mathbb{F}_{q^k})[\ell]$ the pairing is not constant. The kernel of the second argument are the classes in $\ell J_C(\mathbb{F}_{q^k})$. Therefore, many definitions use $J_C(\mathbb{F}_{q^k})/\ell J_C(\mathbb{F}_{q^k})$ as second domain or state the isomorphic group $J_C(\mathbb{F}_{q^k})[\ell]$. Our improvement is due to a clever choice of the representatives.

For applications one often uses the ℓ-torsion points over the field $\mathbb{F}_q$ only. Furthermore, one modifies the pairing to assume a unique value by raising the result to the power of $(q^k - 1)/\ell$. The result is a unique ℓ-th root of unity. It has been observed for elliptic and hyperelliptic curves that for $k > 1$ one can as well use the standard representation of $\bar{D}_2$ as $D_2 = \sum_{i=1}^{m} P_i - mP_\infty$ even though P_∞ occurs in both supports. Namely, $F_{D_1}(D_2) \equiv F_{D_1}(\sum_{i=1}^{m} P_i)$ modulo ℓ-th powers. This means that the denominator in (1) is not computed. Obviously this saves half of the work.

Remark 1. We like to point out that the final powering to map from $\mathbb{F}_{q^k}^*/\mathbb{F}_{q^k}^{*\ell}$ to $\mathbb{F}_{q^k}^*[\ell]$ is only needed if the uniqueness of the result is required. For most applications one can postpone the powering till a unique value is needed.

To compute the pairing one uses the double-and-add method on $\bar{D}_1$ to recursively obtain $F_{D_1}(D_2)$. The basic step is as follows: for two divisor classes $\bar{E}_1, \bar{E}_2$ represented by E_1, E_2 one finds a divisor E_3 and a function G on C defined over $\mathbb{F}_{q^k}$ such that $E_1 + E_2 + E_3 = \operatorname{div}(G)$. Then $\bar{E}_1 \oplus \bar{E}_2 = -\bar{E}_3$ is the usual computation of the addition in the divisor class group. All algorithms to compute the group operation in the divisor or ideal class group implicitly compute the function G.

This leads to the following algorithm for computing the pairing which was proposed by Miller [31,32] for elliptic curves:

Algorithm 1
IN: $\ell = \sum_{i=0}^{l-1} \ell_i 2^i$, $\bar{D}_1 \in J_C(\mathbb{F}_{q^k})[\ell], \bar{D}_2 \in J_C(\mathbb{F}_{q^k})$, represented by D_1 and $D_2 = D_2' - rP_\infty$
OUT: $T_\ell(\bar{D}_1, \bar{D}_2)$

1. $T \leftarrow D_1$ and $F \leftarrow 1$
2. for $i = l-2$ downto 0 do
3. $\bar{T} \leftarrow [2]\bar{T}$ $\operatorname{div}(G) = 2T - ([2]T)$
4. $F \leftarrow F^2 G(D_2')$
5.   if $\ell_i = 1$ then
6. $\bar{T} \leftarrow \bar{T} \oplus \bar{D}_1$ $\operatorname{div}(G) = T + D_1 - (T \oplus D_1)$
7. $F \leftarrow F G(D_2')$
8. return $(F)^{\frac{q^k-1}{\ell}}$

Hence, a pairing is computed by computing $\ell\bar{D}_1$ with the double-and-add method and additionally updating F. It is possible to use windowing methods and precomputations to speed up the computation of $\ell\bar{D}_1$ by reducing the number of additions.

The function G is the quotient of a function of degree g and the linear functions used to mirror on the x-axis. Hence, using Horner's scheme g multiplications in $\mathbb{F}_{q^k}$ are needed to evaluate the numerator of G at one point followed by the evaluations of the linear functions and a division. The second part can be avoided if the extension degree k is even and the divisor $D_2 = [u, v]$ is chosen such that u is defined over $\mathbb{F}_{q^{k/2}}$, where $[u, v]$ is the Mumford representation of $\bar{D}_2$.

Note, that one need not factor u but taking care of all $r \leq g$ points in the support of D_2' needs $O(rg)$ multiplications in the larger field and for a randomly chosen $\bar{D}_2$ one has $r = g$.

3   Random Points Are Not Special

In cryptographic applications one usually encounters the scenario

$$T_\ell : J_C(\mathbb{F}_q)[\ell] \times J_C(\mathbb{F}_{q^k}) \to \mathbb{F}_{q^k}^*[\ell],$$

where the parameters are chosen large enough that the DLP in $J_C(\mathbb{F}_q)[\ell]$ and the DLP in $\mathbb{F}_{q^k}^*$ are hard and ideally no larger than necessary to obtain optimal speed of the implementation. Assuming current suggestions of group sizes of 160 bit and field sizes of 1024 bit this means that $k \sim 6g$. For higher security requirements the ratio between k and g grows as the DLP in finite fields is subexponential while for curves of low genus it is assumed to be exponential. In the following we assume $k = O(g)$.

The aim of this section is to show that the value of the pairing is non-trivial if one restricts the second argument to the embedding of $C(\mathbb{F}_{q^k})$ into $J_C(\mathbb{F}_{q^k})$; for $k = 2m$ we can even restrict to the embedding of $C_1(\mathbb{F}_{q^k})$, the subset of points $(x, y) \in C(\mathbb{F}_{q^k})$ for which $x \in \mathbb{F}_{q^m}$ but $y \notin \mathbb{F}_{q^m}$, into $J_C(\mathbb{F}_{q^k})$. The proofs make use of combinatorial arguments and the Hasse-Weil bounds.

The advantage is that in Algorithm 1 the evaluation of G is faster. As $\bar{D}_1$ is defined over $\mathbb{F}_q$ the doublings and additions are comparably fast and G is defined over $\mathbb{F}_q$. The costs basically equal the scalar multiplication $\ell \bar{D}_1$. The task of evaluating $G(D_2')$ is sped-up by a factor of g as D_2' consists of only 1 point instead of g points. Note that the improvement over the standard Tate-pairing computation involving $D_2 = \sum_{i=1}^m P_i - \sum_{j=1}^m Q_j$ is even larger and that the computation of the pairing is dominated by the evaluations $G(D_2')$ of the intermediate functions G.

3.1   Definitions and First Properties

Let α be a fixed real number with $0 < \alpha < 1$. Let G be a finite abelian group of order n.

Definition 1. *Let S be a subset of G and $\langle S \rangle$ the smallest subgroup of G containing S. Then S is called α-exceptional (in G) if $\frac{|G|}{|\langle S \rangle|} > n^\alpha$.*

A trivial but useful observation is

Lemma 1. *Let G' be a subgroup of G of order $< n^{1-\alpha}$, and assume that S is not α-exceptional. Then $S \not\subseteq G'$.*

Proof. If $S \subset G'$ then $|\langle S \rangle| < n^{1-\alpha}$ and so $\frac{|G|}{|\langle S \rangle|} > n^\alpha$ which is a contradiction. □

Corollary 1. *Assume that ℓ is a prime with ℓ^k dividing n and $\ell > n^\alpha$. Assume that S is not α-exceptional.*
Then S contains an element whose order is a multiple of ℓ^k.

Proof. If $k = 0$ we have nothing to prove. So assume that $k \geq 1$.

Take G' as the group generated by all elements of G whose order divides $\frac{n}{\ell}$. Then $\mid G' \mid \leq n/\ell < n^{1-\alpha}$. □

Corollary 2. *Assume that φ is a group homomorphism of G with image group G' and $\mid G' \mid > n^{\alpha}$. If S is not α-exceptional then there is an element $x \in S$ which is not in the kernel of φ.*

For applications we have in mind we look at the following situation.

Proposition 1. *Let G_0 be a cyclic group of prime order $\ell > n^{\alpha}$ and let*

$$b : G_0 \times G \to G_1$$

be a $\mathbb{Z}$-bilinear map which is non-degenerate in the first variable, i.e. if $b(P, Q) = e_{G_1}$ for all $Q \in G$ then $P = e_{G_0}$.

If $S \subset G$ is not α-exceptional then there is an element $Q \in S$ such that for all $P \in G_0 \setminus \{e_{G_0}\}$ we get

$$b(P, Q) \text{ has order } \ell.$$

Proof. Take $P \in G_0 \setminus \{e_{G_0}\}$ and define the homomorphism $\varphi_P : G \to G_1$ by

$$\varphi_P(Q) := b(P, Q).$$

Since b is non-degenerate in the first variable its image has order $\ell > n^{\alpha}$ and so by Corollary 2 there is an element $Q \in S$ that is not contained in the kernel of φ_P and hence $b(P, Q)$ has order ℓ. □

3.2 Application to J_C

Proposition 2. *There are effectively computable (and reasonable small) numbers c_0, c depending only on g such that for $q^k > c$ we get:*
If $\alpha \geq \frac{\log(g)}{k \log(q)} + \frac{c}{g \cdot k \cdot q^{-k/2} \log(q)}$ then the set $C(\mathbb{F}_{q^k})$ is not α-exceptional.
If $S \subset C(\mathbb{F}_{q^k})$ generates $\langle C(\mathbb{F}_{q^k}) \rangle$ then S is not α-exceptional.

Proof. The basic ingredients for the proof are the Hasse-Weil bounds for points on C and the fact that J_C is rational over $\mathbb{F}_q$. Our philosophy is that g is fixed and q^k becomes asymptotically large.

One knows that $|C(\mathbb{F}_{q^k})| \geq q^k - 2g q^{k/2}$. We use that in every divisor class we have a uniquely determined divisor of the form $D = P_1 + \cdots + P_r - r P_\infty$, with $r \leq g$. Simple combinatorial considerations show that we can produce

$$(1/g!)(q^k - 2g q^{k/2})^g - c'_1(g) q^{k(g-1)} \geq 1/g! q^{gk} - c_1(g) q^{k(g-1/2)}$$

reduced divisor of degree $\leq g$ with points $P_i \in C(\mathbb{F}_{q^k})$ with a constant $c_1(g)$ not depending on q and k.

On the other side the Weil bounds yield that

$$n := |J_C(\mathbb{F}_{q^k})| \leq q^{gk} + c_2(g)q^{k(g-1/2)}.$$

We are looking for reals α such that $C(\mathbb{F}_{q^k})$ is not α-exceptional. By definition this means that $m > n^{1-\alpha}$.

For this it is sufficient that

$$1/g!q^{gk} - c_2(g)q^{k(g-1/2)} > (q^{gk} + c_2(g)q^{k(g-1/2)})^{1-\alpha}$$

or

$$\alpha \log(q^{gk} + c_2(g)q^{k(g-1/2)})$$
$$> \log(q^{gk} + c_2(g)q^{k(g-1/2)}) - \log(1/g!q^{gk} - c_2(g)q^{k(g-1/2)}).$$

We use that $\log(q^{gk} \pm c \cdot q^{k(g-1/2)}) = gk\log(q) \pm c \cdot q^{-k/2} + O(q^{-k})$ and get that it is sufficient to have

$$\alpha gk \log(q) > log(g!) + c(g)q^{-k/2}$$

with a positive number $c(g)$ depending only on g.

Hence, for $\alpha > \frac{\log(g)}{k\log(q)} + \frac{c(g)}{g \cdot q^{k/2}\log(q)}$ and q sufficiently large we get that $C(\mathbb{F}_{q^k})$ is not α-exceptional. □

Together with Proposition 1 this shows that the suggested pairing taking as second input only one point instead of a divisor of full degree is non-degenerate in the first argument. We now estimate the size of the set S generating $\langle C(\mathbb{F}_{q^k}) \rangle$.

Corollary 3. *Let ℓ be a prime, $s \in \mathbb{N}$ and ℓ^s dividing the exponent of $J_C(\mathbb{F}_{q^k})$. Assume that $\log(\ell) > \frac{2\log(g)}{k\log(q)}$ and $q^k > c(g)^2/g^2$. Assume that S is a subset of $C(\mathbb{F}_{q^k})$ which generates $\langle \{P - P_\infty; P \in C(\mathbb{F}_{q^k})\} \rangle$.*

Then there are points points $P \in S$ such that the order of the class of $P - P_\infty$ is divisible by ℓ^s.

Proof. Let G_1 be the subgroup of $J_C(\mathbb{F}_q)$ containing all elements whose order is not divisible by ℓ^s. By assumption its index in $J_C(\mathbb{F}_q)$ is a multiple of ℓ, and so we can apply Corollary 2 and conclude that there is an element $P \in S$ such that the class of $P - P_\infty$ is not contained in G_1. □

We have seen that if q^k is not very small compared with g the order of $\langle C(\mathbb{F}_{q^k}) \rangle$ can be estimated (very roughly) by q^{gk} and hence this group can be generated by $s \leq gk\log_2(q)$ elements.

It is a well known fact from algorithmic group theory that s "randomly chosen" elements in $C(\mathbb{F}_{q^k})$ generate $\langle C(\mathbb{F}_{q^k}) \rangle$.

Corollary 4. *Let the assumptions of Corollary 3 be satisfied.*

Assume that S is a randomly chosen finite subset in $C(\mathbb{F}_{q^k})$ with at least $gk\log_2(q)$ elements. Then S contains an element whose order is divisible by ℓ^s.

So we obtained a lower bound on the *probability* of finding a point $P \in C(\mathbb{F}_{q^k})$ such that $T_\ell(\bar{D}_1, \overline{P - P_\infty})$ is nontrivial, namely $\frac{1}{gk \log_2(q)}$. This bound is likely to be too large but it is the first proof that the pairing is non-degenerate.

We end this section by an important special case. If $k = 2m$, one can restrict the choice of the point even further such that only the y-coordinate is in $\mathbb{F}_{q^k}$ and $x \in \mathbb{F}_{q^m}$ which speeds up the evaluation of G in Algorithm 1 even further. Noticing that the points $(x, y) \in C(\mathbb{F}_{q^k})$ for which $x \in \mathbb{F}_{q^m}$ but $y \notin \mathbb{F}_{q^m}$ are on the quadratic twist of $C/\mathbb{F}_{q^m}$ one can use the Weil bounds for $C(\mathbb{F}_{q^m})$ instead of $C(\mathbb{F}_{q^k})$ in the previous results and obtain:

Proposition 3. *Let the embedding degree k be even $k = 2m$ and let q be sufficiently large. Let $C_1(\mathbb{F}_{q^k})$ be the subset of points $(x, y) \in C(\mathbb{F}_{q^k})$ for which $x \in \mathbb{F}_{q^m}$ but $y \notin \mathbb{F}_{q^m}$. Then for a random element $\bar{D} \in J_C(\mathbb{F}_q)[\ell]$ the value $T_\ell(\bar{D}, \overline{P - P_\infty})$ is nontrivial.*

4   Applications

In this section we show how one can use the proposed choice of $\overline{P - P_\infty}$ as second argument for the applications. We first provide modifications of the usual protocols. As examples we consider ID-based encryption and short signatures. We assume our reader to be familiar with these protocols and only mention the differences.

Throughout this section we assume $k > g$ as this is always the situation in applications. This means that a point from $C(\mathbb{F}_{q^k})$ which is not defined over a subfield cannot occur in the support of D_1 and hence, the pairing of $T_\ell(\bar{D}_1, \overline{P - P_\infty})$ is non-trivial and we can use the point P as evaluation point in Algorithm 1. To have a shorthand we define $T'_\ell(\bar{D}_1, P) = T_\ell(\bar{D}_1, \overline{P - P_\infty})$ and use $\tilde{T}_\ell$ resp. $\tilde{T}'_\ell$ to include the final powering.

We state the general case for arbitrary k here. If k is even we can further improve the speed by choosing $P \in C_1(\mathbb{F}_{q^k})$ instead of in $C(\mathbb{F}_{q^k})$ as shown in Proposition 3. Note that the following protocols do *not* require a distortion map.

ID-Based Encryption. We assume that each participant can be uniquely identified by a bitstring ID. We use a different hash function as one would use in the direct generalization of the protocol proposed in [8]. Namely we assume that there is a hash function $h_1 : \{0, 1\}^* \to C(\mathbb{F}_{q^k})$, such that one can uniquely associate a point to every identity. This can be done by hashing ID to an element of $\mathbb{F}_{q^k}$ and then increasing it until it is the x-coordinate of a point. The result $h_1(\text{ID})$ is used as public key of ID. We point out that the image space is large enough as one assumes that the DLP is hard in $\mathbb{F}^*_{q^k}$.

Let E_K and D_K denote en- and decryption under the key K and let $\mathbf{K}$ be a key derivation function which operates on the output of the pairing. The trusted authority publishes the curve $C/\mathbb{F}_q$, the embedding degree k and a generator $\bar{D}_1 \in J_C(\mathbb{F}_q)[\ell]$, and the public key $\bar{D}_{\text{TA}} = [a_{\text{TA}}]\bar{D}$ keeping secret the private key a_{TA}.

Algorithm 2 (Identity Based Encryption)
IN: message m, $(J_C(\mathbb{F}_q)[\ell], \bar{D}, \bar{D}_{\mathrm{TA}})$, identity of recipient ID
OUT: ciphertext $(\bar{D}_r, c)$.

1. $r \in_R \mathbb{N}$
2. $\bar{D}_r \leftarrow [r]\bar{D}$
3. $S \leftarrow \tilde{T}'_\ell(\bar{D}_{\mathrm{TA}}, h_1(\mathrm{ID}))^r$
4. $\leftarrow E_{\mathbf{K}(S)}(m)$
5. return $(\bar{D}_r, c)$

Hence, the sender profits from the modified pairing as T'_ℓ is faster to compute.

The private key of ID is obtained from TA as $\bar{D}_{\mathrm{ID}} = [a_{\mathrm{TA}}]\overline{h_1(\mathrm{ID}) - P_\infty}$ which is also computed faster as usual.

To decrypt the recipient computes

$$S \leftarrow \tilde{T}_\ell(\bar{D}_r, \bar{D}_{\mathrm{ID}}) \text{ and } m \leftarrow D_{\mathbf{K}(S)}(c).$$

As

$$\tilde{T}_\ell(\bar{D}_r, \bar{D}_{\mathrm{ID}}) = \tilde{T}_\ell(\bar{D}_r, \overline{h_1(\mathrm{ID}) - P_\infty})^{a_{\mathrm{TA}}} = \tilde{T}'_\ell(\bar{D}_{\mathrm{TA}}, h_1(\mathrm{ID}))^r$$

the scheme works as specified.

Short Signatures. Pointcheval and Okamoto [35], show how a Gap-DH group can be used to design a signature scheme. Boneh, Lynn, and Shacham [10] give a realization using supersingular elliptic curves with the Tate pairing as bilinear structure.

The system parameters are the group $J_C(\mathbb{F}_q)[\ell]$, the embedding degree k, and a hash function $h_1 : \{0,1\}^* \to J_C(\mathbb{F}_q)[\ell]$, which would also be needed in the direct generalization of [10] to hyperelliptic curves. As basis of the system a point $P \in C(\mathbb{F}_{q^k})$ is fixed and it is checked that for one (and therefore for all) divisor classes $\bar{D}_1 \in J_C(\mathbb{F}_q)[\ell]$ the pairing $T'_\ell(\bar{D}_1, P)$ is non-trivial. Since this is done at the set-up of the system even a lower probability of success would be acceptable.

The signers public key is given by $\bar{D}_A = [a]\overline{P - P_\infty}$. This operation is sped-up but it is used only once. To sign message m one computes $S = [a]h_1(m)$ which is an element of $J_C(\mathbb{F}_q)$ and compresses the result [24,40]. For fixed size of ℓ one has a smaller q for larger g by Hasse-Weil, thus the space requirement is equal to that of a point on an elliptic curve with the same group size. Hence, the signatures *are short*.

The verifier accepts if $(T_\ell(h_1(m), \bar{D}_A)/T'_\ell(S, P))^{(q^k-1)/\ell}$ equals 1.

This means that the verifier has the advantage that the first pairing is faster to compute and the signer needs to compute in $\mathbb{F}_q$ only which offers huge advantages.

The fact that T'_ℓ is non-degenerate when applied to P implies that $\overline{P - P_\infty}$ has order a multiple of ℓ. So the DLP is at least as hard as in $J_C(\mathbb{F}_q)[\ell]$ and no extra weakness is introduced.

5   Parameter Choices

While for elliptic curves and curves of genus 2 the generic attacks on the DLP like Pollard rho are the fastest, index calculus attacks become more powerful with increasing genus and the double large prime variant [22] is currently the most powerful attack for medium sized genera.

Non-supersingular curves are particularly interesting over prime fields because of the faster field arithmetic and because the Coppersmith method of solving the DLP in $\mathbb{F}_{q^k}^*$ is very efficient for small characteristic.

Assuming a security level of 2^{80} (corresponding to 160-bit ECC) and assuming that this corresponds to 1024-bit DL in finite fields we obtain the following recommendations for the size of q and of k. Both parameters are rounded to the nearest integer and q is determined based on [22] and then k is computed. Rounding errors imply that some adjustment for q is necessary, e. g. for elliptic curves and $k = 6$ we need to choose a slightly larger base field of $\log_2 q = 171$ bits to ensure a large enough resulting field.

genus	$g = 1$	$g = 2$	$g = 3$	$g = 4$
Pollard's rho	$q^{1/2}$	q	$q^{3/2}$	q^2
Double large prime	–	–	$q^{4/3}$	$q^{3/2}$
$\log_2 q$	160	80	60	54
k	6	13	17	20

Even though the timings for arithmetic on genus 4 curves are slower than on smaller genus curves the advantage of gaining a factor of 4 in the pairing computation should outweigh that drawback. On the other hand we do not know how to construct a genus 4 curve with embedding degree 20 and the CM theory is not developed yet.

6   Conclusion

In this paper we have shown how to speed up pairing-based protocols on hyperelliptic curves by applying the Tate-Lichtenbaum pairing to special divisors in $J_C(\mathbb{F}_{q^k})$. We made clear that these choices are possible in the applications and lead to savings in ID-based encryption and short signatures on hyperelliptic curves of genus $g > 1$. The same advantages apply to tripartite key-exchange, hierarchical encryption and encryption with keyword search. We like to stress that this improvement for computing the Tate-Lichtenbaum pairing on hyperelliptic curves cannot be used for elliptic curves.

References

1. R. Avanzi, H. Cohen, C. Doche, G. Frey, T. Lange, K. Nguyen, and F. Vercauteren. *The Handbook of Elliptic and Hyperelliptic Curve Cryptography.* CRC, 2005.
2. P. S. L. M. Barreto, S. D. Galbraith, C. O hEigeartaigh, and M. Scott. Efficient pairing computation on supersingular abelian varieties. preprint, 2004.

3. P. S. L. M. Barreto, H. Y. Kim, B. Lynn, and M. Scott. Efficient algorithms for pairing-based cryptosystems. In *Advances in Cryptology – Crypto 2002*, volume 2442 of *Lecture Notes in Comput. Sci.*, pages 354–368. Springer-Verlag, Berlin, 2002.

4. P. S. L. M. Barreto, B. Lynn, and M. Scott. Constructing elliptic curves with prescribed embedding degrees. In *Security in Communication Networks – SCN 2002*, volume 2576 of *Lecture Notes in Comput. Sci.*, pages 257–267. Springer-Verlag, Berlin, 2003.

5. P. S. L. M. Barreto and M. Naehrig. Pairing-friendly elliptic curves of prime order. preprint, 2005.

6. I. F. Blake, K. Murty, and G. Xu. Refinements of Miller's algorithm for computing Weil/Tate pairing. preprint, 2004.

7. I. F. Blake, G. Seroussi, and N. P. Smart. *Elliptic curves in cryptography*. London Mathematical Society Lecture Note Series. 265. Cambridge University Press, 1999.

8. D. Bleichenbacher and A. Flammenkamp. An efficient algorithm for computing shortest addition chains.

9. D. Boneh and M. Franklin. Identity based encryption from the Weil pairing. *SIAM J. Comput.*, 32(3):586–615, 2003.

10. D. Boneh, B. Lynn, and H. Shacham. Short signatures from the Weil pairing. In *Advances in Cryptology – Asiacrypt 2001*, volume 2248 of *Lecture Notes in Comput. Sci.*, pages 514–532. Springer-Verlag, Berlin, 2002.

11. D. G. Cantor. Computing in the Jacobian of a hyperelliptic curve. *Math. Comp.*, 48:95–101, 1987.

12. Y. Choie and E. Lee. Implementation of Tate Pairing on Hyperelliptic Curves of Genus 2. In *Information Security and Cryptology - ICISC 2003*, volume 2971 of *Lect. Notes Comput. Sci.*, pages 97–111. Springer, 2004.

13. R. Dupont, A. Enge, and F. Morain. Building curves with arbitrary small MOV degree over finite prime fields. *J. Cryptology*, 18(2):79–89, 2005.

14. I. Duursma and H.-S. Lee. Tate-pairing implementations for tripartite key agreement. 2003.

15. K. Eisenträger, K. Lauter, and P. L. Montgomery. Fast elliptic curve arithmetic and improved Weil pairing evaluation. In *Topics in Cryptology – CT-RSA 2003*, volume 2612 of *Lecture Notes in Comput. Sci.*, pages 343–354. Springer-Verlag, Berlin, 2003.

16. K. Eisenträger, K. Lauter, and P. L. Montgomery. Improved Weil and Tate pairings for elliptic and hyperelliptic curves. In *Algorithmic Number Theory Symposium – ANTS VI*, volume 3076 of *Lecture Notes in Comput. Sci.*, pages 169–183. Springer-Verlag, Berlin, 2004.

17. G. Frey and T. Lange. Mathematical background of public key cryptography. Technical Report 10, IEM Essen, 2003. To appear in Séminaires et Congrès.

18. G. Frey, M. Müller, and H. G. Rück. The Tate pairing and the discrete logarithm applied to elliptic curve cryptosystems. *IEEE Trans. Inform. Theory*, 45(5):1717–1719, 1999.

19. G. Frey and H. G. Rück. A remark concerning m-divisibility and the discrete logarithm problem in the divisor class group of curves. *Math. Comp.*, 62:865–874, 1994.

20. S. D. Galbraith. Supersingular curves in cryptography. In *Advances in Cryptology – Asiacrypt 2001*, volume 2248 of *Lecture Notes in Comput. Sci.*, pages 495–513. Springer-Verlag, Berlin, 2001.

21. S.D. Galbraith, K. Harrison, and D. Soldera. Implementing the Tate pairing. In *Algorithmic Number Theory Seminar ANTS-V*, volume 2369 of *Lecture Notes in Comput. Sci.*, pages 324–337. Springer, 2002.

22. P. Gaudry, N. Thériault, E. Thomé, and C. Diem. A double large prime variation for small genus hyperelliptic index calculus. preprint, last update 21 Nov 2005, 2005.

23. C. Guyot, K. Kaveh, and V. M. Patankar. Explicit algorithm for the arithmetic on the hyperelliptic Jacobians of genus 3. *J. Ramanujan Math. Soc.*, 19:119–159, 2004.

24. F. Hess, G. Seroussi, and N. P. Smart. Two topics in hyperelliptic cryptography. In *Selected Areas in Cryptography – SAC 2000*, volume 2259 of *Lecture Notes in Comput. Sci.*, pages 181–189. Springer-Verlag, Berlin, 2001.

25. T. Izu and T. Takagi. Efficient computations of the Tate pairing for the large MOV degrees. In *Information Security and Cryptology – ICISC 2002*, volume 2587 of *Lecture Notes in Comput. Sci.*, pages 283–297. Springer-Verlag, Berlin, 2003.

26. A. Joux. A one round protocol for tripartite Diffie–Hellman. In *Algorithmic Number Theory Symposium – ANTS IV*, volume 1838 of *Lecture Notes in Comput. Sci.*, pages 385–394. Springer-Verlag, 2000.

27. M. Katagi, I. Kitamura, T. Akishita, and T. Takagi. Novel efficient implementations of hyperelliptic curve cryptosystems using degenerate divisors. In *Workshop on Information Security Applications – WISA 2004*, volume 3325 of *Lecture Notes in Comput. Sci.*, pages 347–361. Springer-Verlag, Berlin, 2004.

28. N. Koblitz. Hyperelliptic cryptosystems. *J. Cryptology*, 1:139–150, 1989.

29. T. Lange. Formulae for arithmetic on genus 2 hyperelliptic curves. *Appl. Algebra Engrg. Comm. Comput.*, 15(5):295–328, 2005.

30. D. Lorenzini. *An invitation to arithmetic geometry*, volume 9 of *Graduate studies in mathematics*. AMS, 1996.

31. V. S. Miller. Short programs for functions on curves, 1986. IBM, Thomas J. Watson Research Center.

32. V.C. Miller. The Weil Pairing, and Its Efficient Calculation. *J. Cryptology*, 17:235–261, 2004.

33. A. Miyaji, M. Nakabayashi, and S. Takano. New explicit conditions of elliptic curve traces for FR-reduction. *IEICE Trans. Fundamentals*, E84-A(5):1234–1243, 2001.

34. Y. Miyamoto, H. Doi, K. Matsuo, J. Chao, and S. Tsuji. A fast addition algorithm of genus two hyperelliptic curve. In *Symposium on Cryptography and Information Security – SCIS 2002*, pages 497–502. In Japanese.

35. T. Okamoto and D. Pointcheval. The gap-problems: a new class of problems for the security of cryptographic schemes. In *Public Key Cryptography – PKC 2001*, volume 1992 of *Lect. Notes Comput. Sci.*, pages 104–118. Springer-Verlag, 2001.

36. J. Pelzl, T. Wollinger, J. Guajardo, and C. Paar. Hyperelliptic curve cryptosystems: Closing the performance gap to elliptic curves. In *Cryptographic Hardware and Embedded Systems CHES 2003*, volume 2779 of *Lect. Notes Comput. Sci.*, pages 351–365. Springer, 2003.

37. R. L. Rivest and R. D. Silverman. Are "strong" primes needed for RSA? preprint, 1997.

38. H. G. Rück. On the discrete logarithm problem in the divisor class group of curves. *Math. Comp.*, 68:805–806, 1999.

39. R. Sakai, K. Ohgishi, and M. Kasahara. Cryptosystems based on pairing. In *Symposium on Cryptography and Information Security – SCIS 2000*, 2000.

40. C. Stahlke. Point compression on Jacobians of hyperelliptic curves over $\mathbb{F}_q$. preprint, 2004.

41. H. Stichtenoth. *Algebraic Function Fields and Codes*. Springer, 1993.
42. H. Sugizaki, K. Matsuo, J. Chao, and S. Tsujii. An Extension of Harley algorithm addition algorithm for hyperelliptic curves over finite fields of characteristic two. Technical Report ISEC2002-9(2002-5), IEICE, 2002.
43. M. Takahashi. Improving Harley algorithms for Jacobians of genus 2 hyperelliptic curves. In *Symposium on Cryptography and Information Security – SCIS 2002*. In Japanese.
44. T. Wollinger. *Software and Hardware Implementation of Hyperelliptic Curve Cryptosystems*. PhD thesis, Ruhr-University of Bochum, 2004.

High Security Pairing-Based Cryptography Revisited

R. Granger, D. Page, and N.P. Smart

Dept. Computer Science, University of Bristol
Merchant Venturers Building Woodland Road, Bristol, BS8 1UB, UK
{granger, page, nigel}@cs.bris.ac.uk

Abstract. The security and performance of pairing based cryptography has provoked a large volume of research, in part because of the exciting new cryptographic schemes that it underpins. We re-examine how one should implement pairings over ordinary elliptic curves for various practical levels of security. We conclude, contrary to prior work, that the Tate pairing is more efficient than the Weil pairing for all such security levels. This is achieved by using efficient exponentiation techniques in the cyclotomic subgroup backed by efficient squaring routines within the same subgroup.[1]

1 Introduction

In commercial cryptographic software libraries one typically employs Occam's Razor in limiting the number of implemented primitives and schemes to a minimum. Occam's razor being a philosophical approach to science, often rephrased as "If you have two equally likely solutions to a problem, pick the simplest". The advantages of this approach are threefold: it reduces the programming, maintainence and security validation workload; it enables one to specialise and hence highly optimise the core operations; and it reduces the library footprint and usage of system resources. Around the time it was first proposed, one of the main criticisms levelled at standard elliptic curve cryptography was that there were too many options; it was hard for non-experts to decide on and construct the types of field and curve needed to satisfy performance and security constraints. Two decades later, pairing based cryptography is in a similar state in the sense that there are a huge range of parameterisation options, algorithmic choices and subtle trade-offs between the two. Hence, there is a real need to focus on a family of parameters which are flexible but offer efficient arithmetic and allow one to focus on a limited number of cases.

---

[1] The work described in this paper has been supported in part by the European Commission through the IST Programme under Contract IST-2002-507932 ECRYPT. The information in this document reflects only the author's views, is provided as is and no guarantee or warranty is given that the information is fit for any particular purpose. The user thereof uses the information at its sole risk and liability.

F. Hess, S. Pauli, and M. Pohst (Eds.): ANTS 2006, LNCS 4076, pp. 480–494, 2006.

Generally speaking, a pairing is a non-degenerate bilinear map

$$t : \mathbb{G}_1 \times \mathbb{G}_2 \longrightarrow \mathbb{G}_T.$$

Here we assume this pairing takes the concrete form

$$\hat{t} : E(\mathbb{F}_p) \times \overline{E}(\mathbb{F}_{p^{k/2}}) \longrightarrow \mathbb{F}_{p^k}^{\times}$$

where $\overline{E}$ is the quadratic twist of an elliptic curve E defined over $\mathbb{F}_{p^{k/2}}$. We restrict our attention to the case of ordinary elliptic curves and assume that $\#E(\mathbb{F}_p)$ is divisible by a large prime n which also divides $p^k - 1$ i.e., n is the order of the subgroups on which the pairing based protocols will be based. We let the respective subgroups of order n of the three groups involved be denoted $\mathbb{G}_1 \mathbb{G}_2$ and $\mathbb{G}_T$ as is common in various papers on the subject.

Koblitz and Menezes [10] introduced the concept of pairing friendly fields. These are Kummer extensions of $\mathbb{F}_p$ defined by the polynomial

$$f(X) = X^k + f_0$$

for a values of $p \equiv 1 \pmod{12}$ and $k = 2^i 3^j$. Generally one assumes that k is even, which aids in efficiency due to the well known denominator elimination trick. Following [10] we particularly focus on the cases $k = 6$, 12 and 24. We let $f(\theta) = 0$ and define $\mathbb{F}_{p^k} = \mathbb{F}_p[\theta]$. Many protocols based on pairings perform arithmetic in the cyclotomic subgroup of $\mathbb{F}_{p^k}^{\times}$, which is the subgroup of order $\Phi_k(p)$, where $\Phi_k(p)$ is the kth cyclcotomic polynomial evaluated at p. We denote this subgroup by $G_{\Phi_k(p)}$; the group $\mathbb{G}_T$ in the pairing above is contained in $G_{\Phi_k(p)}$. Hence if one is to implement pairing based protocols efficiently with such fields then one needs to be able to implement arithmetic efficiently.

The conclusion of [10] is that for high security levels the Weil pairing is to be preferred over the Tate pairing. The main result of this paper is that by optimising the exponentiation method used in the Tate pairing calculation one can in fact conclude the opposite: that in all cases the Tate pairing is the more efficient algorithm for all practical security levels.

In addition, we also look at efficient arithmetic in the group $G_{\Phi_k(p)}$ which will speed up both the Tate pairing and various protocols. This is inspired by work of Lenstra and Stam [14,15] who introduce such efficient arithmetic in a specific family of finite fields of degree six, which are different from the pairing friendly fields. In particular, by restricting to $k = 6$ Lenstra and Stam present algorithms for arithmetic in the cyclotomic extension of $\mathbb{F}_p$ defined by the polynomial

$$g(X) = X^6 + X^3 + 1$$

when $p \equiv 2$ or $5 \pmod{9}$. We shall call such constructions cyclotomic fields of degree 6 in this paper. Lenstra and Stam present efficient squaring routines both for the finite field $\mathbb{F}_{p^6}$ and for the cyclotomic subgroup $G_{\Phi_6(p)}$ of order $\Phi_6(p)$, again $\mathbb{G}_T$ is contained in $G_{\Phi_6(p)}$. We let $g(\zeta) = 0$ and define $\mathbb{F}_{p^6}$, in this case, by $\mathbb{F}_p[\zeta]$. We shall describe how the use of cyclotomic fields, as opposed to the pairing friendly fields, can provide more efficient pairing algorithms when $k = 6$. We present an analogue of these results for pairing friendly fields which provides some efficiency improvement, but not as much as that achieved by Lenstra and Stam for cyclotomic fields of degree six. We leave it as an open research problem to generalise the results of Lenstra and Stam to cyclotomic fields of degree different from six. The only generalisation known is for fields of degree $6 \cdot 5^m$ [8], for which Lenstra and Stam's technique trivially applies.

The paper is organised as follows. In Section 2 we recap on the most efficient field arithmetic known for the two cases of finite fields mentioned above. In Section 3 we briefly recap on some standard formulae for the cost of elliptic curve operations. In Section 4 we recap on the model for estimating the cost pairings which was proposed by Koblitz and Menezes. Then in Section 5 we detail the implications of this model for our choice of finite fields.

2   Finite Field Operations

We let $m, \overline{M}, M$ (resp. $s, \overline{S}, S$) denote the time for multiplication (resp. squaring) in the fields $\mathbb{F}_p, \mathbb{F}_{p^{k/2}}$ and $\mathbb{F}_{p^k}$. In our analysis we shall assume that addition operations are cheap, however in a practical implementation for certain bit sizes the operation counts and algorithm choices we give may not be optimal due to this simplifying assumption.

We first note that if one is computing products (resp. squares) of polynomials of degree $2^i 3^j - 1$ over $\mathbb{F}_p$ then using the Karatsuba and Toom-Cook methods for multiplication and squaring this requires $v(k)$ multiplications (resp. squarings) in the field $\mathbb{F}_p$, where $v(k) = 3^i 5^j$.

2.1   Pairing Friendly Fields

As before we let $k = 2^i 3^j \geq 6$, let p denote a prime congruent to 1 modulo 12 and modulo k and define $\mathbb{F}_{p^k}$ via the polynomial $f(X) = X^k + f_0$. We assume throughout that f_0 has been chosen so that multiplication by f_0 can be performed quickly by simple additions rather than a full multiplication. Arithmetic in the subfield $\mathbb{F}_{p^{k/2}}$ is performed using the polynomial $X^{k/2} + f_0$, and mapping between the two representations is relatively straightforward.

The best algorithms for multiplication and squaring in $\mathbb{F}_{p^k}$ and $\mathbb{F}_{p^{k/2}}$ are the standard ones based on Karatsuba and Toom-Cook. Hence, in this case we obtain

$$\overline{M} = M/3 \quad \overline{S} = S/3$$

and

$$M \approx v(k)m \quad S \approx v(k)s$$

where $v(k) = 3^i 5^j$.

Inversion in the field $\mathbb{F}_{p^k}$ is computed by reduction to inversion in the subfield $\mathbb{F}_{p^{k/2}}$. If we let $\alpha = \sum_{i=0}^{k-1} a_i \theta^i \in \mathbb{F}_{p^k}$ then we can write

$$\alpha = \alpha_0 + \alpha_1 \theta$$

where $\alpha_0, \alpha_1 \in \mathbb{F}_{p^{k/2}}$ and are given by

$$\alpha_0 = \sum_{i=0}^{k/2-1} a_{2i}\theta^{2i} \text{ and } \alpha_1 = \sum_{i=0}^{k/2-1} a_{2i+1}\theta^{2i}.$$

We can thus compute

$$\Delta = \alpha_0^2 - \theta^2 \alpha_1^2,$$

and

$$\alpha^{-1} = \frac{\alpha_0 - \alpha_1 \theta}{\Delta}.$$

Inversion in $\mathbb{F}_{p^k}$ is therefore accomplished using two squarings, one inversion, and two multiplications in $\mathbb{F}_{p^{k/2}}$. Similarly, using the same idea one can reduce inversion in a cubic extension to three squarings, eleven multiplications and one inversion in the base field [9]. Iterating down through the subfields, for pairing-friendly fields inversion can thus be performed with just one inversion in $\mathbb{F}_p$, and a handful of multiplications. We summarize these costs, for the extensions which will interest us,

$$\begin{aligned}
I_2 &= 2s + 2m + \iota, & I_3 &= 3s + 11m + \iota, \\
I_4 &= 8s + 8m + \iota, & I_6 &= 13s + 21m + \iota, \\
I_8 &= 26s + 26m + \iota, & I_{12} &= 43s + 51m + \iota, \\
I_{24} &= 133s + 141m + \iota.
\end{aligned}$$

where I_ι denotes the cost of inversion in $\mathbb{F}_{p^\iota}$ and ι denotes the cost of inversion in $\mathbb{F}_p$.

The Frobenius operation in pairing friendly fields is also efficiently computed as follows. If we define $\mathbb{F}_{p^k} = \mathbb{F}_p[\theta]/(f(\theta))$ then the Frobenius operation on the polynomial generator θ can be easily determined via

$$\theta^p = \theta^{k(p-1)/k+1} = (-f_0)^{(p-1)/k}\theta.$$

For later use we let $g = (-f_0)^{(p-1)/k} \in \mathbb{F}_p$ hence $\theta^p = g \cdot \theta$. Also now note that powers of the Frobenius operation are also easy to compute via

$$\theta^{p^i} = g^i \cdot \theta.$$

We also note that since k is even and $-f_0$ is a quadratic non-residue that we have

$$g^{k/2} = (-f_0)^{(p-1)/2} = -1.$$

To summarize we give the operation counts for the various cases are described in Table 1.

We now turn to the case of arithmetic in the subgroup $G_{\Phi_k(p)}$. For this subgroup we have that inversion comes for free. Let $\alpha \in G_{\Phi_k(p)}$, then since $\Phi_k(p)$ divides $p^{k/2} + 1$ we have that

$$\alpha^{-1} = \alpha^{p^{k/2}}.$$

Table 1. Cost of operations in $\mathbb{F}_{p^{k/2}}$ and $\mathbb{F}_{p^k}$

	$\mathbb{F}_{p^{k/2}}$		$\mathbb{F}_{p^k}$	
k	Mul	Sqr	Mul	Sqr
6	$5m$	$5s$	$15m$	$15s$
12	$15m$	$15s$	$45m$	$45s$
24	$45m$	$45s$	$135m$	$135s$

This leads to an inversion operation which can be performed using only $k/2$ negations in $\mathbb{F}_{p^k}$.

We can also improve the performance of squaring in this subgroup using a trick originally proposed by Lenstra and Stam [14,15] in the context of finite extension fields defined by cyclotomic polynomials of degree 6. We first define

$$\alpha = \sum_{i=0}^{k-1} a_i \theta^i$$

where we now think of the coefficients a_i as variables. We then compute symbolically $\alpha^{p^{k/3}}$ and $\alpha^{p^{k/6}}$. One can then derive a set of equations defining the elements of the group $G_{\Phi_k(p)}$ via

$$\alpha^{p^{k/3}} \cdot \alpha - \alpha^{p^{k/6}} = \sum_{i=0}^{k-1} v_i \theta^i.$$

The variety defined by $v_0 = v_1 = \cdots = v_{k-1} = 0$ defines the set of elements of $G_{\Phi_k(p)}$. This follows since

$$\Phi_k(X) = X^{k/3} - X^{k/6} + 1$$

for all values of k arising in pairing friendly fields. As an example for the case $k = 6$ we obtain the set of equations

$$v_0 = -a_0 + a_0{}^2 + f_0 a_5 a_1 - f_0 a_3{}^2 + f_0 a_2 a_4,$$
$$v_1 = g \cdot (-a_1 + 2 f_0 a_5 a_2 - f_0 a_3 a_4 + a_0 a_1),$$
$$v_2 = (1 - g) \cdot (a_2 - a_1{}^2 + a_0 a_2 - f_0 a_5 a_3 + f_0 a_4{}^2),$$
$$v_3 = a_3 + 2 a_0 a_3 - a_2 a_1 + f_0 a_5 a_4,$$
$$v_4 = g \cdot (a_0 a_4 + f_0 a_5{}^2 + a_3 a_1 - a_2{}^2 + a_4),$$
$$v_5 = (1 - g) \cdot (-a_5 + a_0 a_5 - 2 a_4 a_1 + a_3 a_2).$$

Note that for any $k \times k$ matrix Γ that

$$\alpha^2 = a^2 + b \cdot (\Gamma \cdot v^t),$$

where $b = (1, \theta, \theta^2, \ldots, \theta^{k-1})$ and $v = (v_0, v_1, \ldots, v_{k-1})$. Hence, to find different forms of the squaring operation we simply need to select a matrix Γ which produces equations for squaring which are efficient.

A choice for Γ which seems to work well for $k = 6, 12$ and 24 is to set $\Gamma = \mathrm{diag}(d_1, d_2, d_3, d_1, d_2, d_3, \ldots, d_1, d_2, d_3)$ where

$$d_1 = 2, \quad d_2 = 2g^{k/6} - 2, \quad d_3 = -2g^{k/6}.$$

In this case for $k = 6$ we obtain the following formulae for squaring, if $\beta = \sum_{i=0}^{5} b_i \theta^i = \alpha^2$,

$$b_0 = -3 f_0 a_3{}^2 + 3 a_0{}^2 - 2 a_0,$$
$$b_1 = -6 f_0 a_5 a_2 + 2 a_1,$$
$$b_2 = -3 f_0 a_4{}^2 + 3 a_1{}^2 - 2 a_2,$$
$$b_3 = 6 a_0 a_3 + 2 a_3,$$
$$b_4 = 3 a_2{}^2 - 3 f_0 a_5{}^2 - 2 a_4,$$
$$b_5 = 6 a_4 a_1 + 2 a_5.$$

The formulae for $k = 12$ and $k = 24$ can be found in the Appendix.

Ignoring multiplication by f_0 and by small constants we then derive the Table 2 detailing the comparative cost of squaring in both $\mathbb{F}_{p^k}$ and the subgroup $G_{\Phi_k(p)}$. Hence, we see that we have a significant improvement in the squaring operation for the subgroup $G_{\Phi_k(p)}$ although this improvement decreases as k increases.

Table 2. Cost of squaring in $\mathbb{F}_{p^k}$ and $G_{\Phi_k(p)}$ for various values of k

k	$\mathbb{F}_{p^k}$	$G_{\Phi_k(p)}$
6	$15s$	$6s + 3m$
12	$45s$	$12s + 18m$
24	$135s$	$24s + 84m$

2.2 Cyclotomic Fields of Degree 6

We recap on the techniques of [14,15] for the finite fields $\mathbb{F}_p[\zeta]$, with $p \equiv 2$ (mod 9). Elements in $\mathbb{F}_{p^6}$ are represented in the basis $\{\zeta, \zeta^2, \zeta^3, \zeta^4, \zeta^5, \zeta^6\}$. Using this representation multiplication in $\mathbb{F}_{p^6}$ can be performed using 15 multiplications in $\mathbb{F}_p$ (note that [14] gives the figure as 18 multiplications as the paper only considers Karatsuba and not Toom-Cook multiplication).

Squaring can be performed more efficiently using the fact that if we write $\alpha = \alpha_0 \zeta + \alpha_1 \zeta^4$, where α_i are polynomials in ζ of degree at most two, then one has

$$\alpha^2 = (\alpha_0 - \alpha_1)(\alpha_0 + \alpha_1)\zeta^2 + (2\alpha_0 - \alpha_1)\alpha_1 \zeta^5.$$

Since, the α_i are of degree at most two this above formulae requires 10 multiplication in $\mathbb{F}_p$ to perform a squaring operation in $\mathbb{F}_{p^6}$.

Arithmetic in the subfield $\mathbb{F}_{p^3}$ is performed as in [9]. We set $\psi = \zeta + \zeta^{-1}$ and define $\mathbb{F}_{p^3} = \mathbb{F}_p[\psi]$. As a basis for $\mathbb{F}_{p^3}$ we take $\{1, \psi, \psi^2 - 2\}$. Via Toom-Cook multiplication (resp. squaring) requires 5 multiplications (resp. squares) in $\mathbb{F}_p$.

As noted in Section 2.1 inversion in $\mathbb{F}_{p^3}$ can be performed in 11 multiplications in $\mathbb{F}_p$ and one inversion in $\mathbb{F}_p$.

Using this subfield inversion an inversion operation can be defined for $\mathbb{F}_{p^6}$. This inversion is carried out, in the language of [9], by mapping our $\mathbb{F}_{p^6}$ element to the representation F_2 and then performing the inversion in that representation and then mapping back to our representation. The conversion between representations requires four $\mathbb{F}_p$ multiplications, whilst the inversion in the F_2 representation requires $4\overline{S}$ plus application of the inversion in $\mathbb{F}_{p^3}$. Hence, requiring a total of 26 multiplications in $\mathbb{F}_p$ and one inversion in $\mathbb{F}_p$.

We now turn to the subgroup $G_{\Phi_6(p)}$. As before, inversion comes for free via the operation of the Frobenius map. Multiplication is performed just as for the full finite field, however squaring can be performed significantly faster using the equations contained in [14,15]. If we let $\alpha = \sum_{i=0}^{5} a_i \zeta^{i+1} \in G_{\Phi_6(p)}$ and set $\beta = \sum_{i=0}^{5} b_i \zeta^{i+1} = \alpha^2$ then we have

$$b_0 = 2a_1 + 3a_4(a_4 - 2a_1),$$
$$b_1 = 2a_0 + 3(a_0 + a_3)(a_0 - a_3),$$
$$b_2 = -2a_5 + 3a_5(a_5 - 2a_2),$$
$$b_3 = 2(a_1 - a_4) + 3a_1(a_1 - 2a_4),$$
$$b_4 = 2(a_0 - a_3) + 3a_3(2a_0 - a_3),$$
$$b_5 = -2a_2 + 3a_2(a_2 - 2a_5).$$

Hence, squaring requires six $\mathbb{F}_p$ multiplications. The operation counts for the various cases are as summarised by Table 3.

Table 3. Operation counts when $k = 6$

$\mathbb{F}_{p^3}$		$\mathbb{F}_{p^6}$		$G_{\Phi_6(p)}$	
Mul	Sqr	Mul	Sqr	Mul	Sqr
$5m$	$5m$	$15m$	$10m$	$15m$	$6m$

2.3   Exponentiation in $G_{\Phi_k(p)}$

Finally, we address the issue of exponentiation, by an exponent e, of elements in the cyclotomic subgroup $G_{\Phi_k(p)}$ of $\mathbb{F}_{p^k}^{\times}$ which has order divisible by n. Using Lucas sequences [13] this can be accomplished in time

$$C_{\text{Luc}}(e) = (\overline{M} + \overline{S}) \log_2 e.$$

However, one could also use exponentiation via standard signed sliding window methods [4] since inversion is cheap in $G_{\Phi_k(p)}$. If $e \leq p$ then the best way to perform the exponentiation, using windows of width at least r, will take time

$$C_{\text{SSW}}(e) = \overline{\overline{S}}(1 + \log_2 e) + M \left(\frac{\log_2 e}{r + 2} + (2^{r-2} - 1) \right)$$

where $\overline{\overline{S}}$ denotes the time needed to perform a squaring operation in $G_{\Phi_k(p)}$. We also need to store 2^{r-2} elements during the exponentiation algorithm.

When $e \geq p$, as is the case in the final powering of the algorithm to compute the Tate pairing, one uses the fact that we can perform the Frobenius operation on $G_{\Phi_k(p)}$ for free. Thus we write e in base p, and perform a simultaneous exponentiation. Using the techniques of Avanzi [1], we can estimate the time needed to perform such a multi-exponentiation by

$$C_{\mathrm{bigSSW}}(e) = (d + \log_2 p)\overline{\overline{S}} + \left(d(2^{r-1} - 1) + \frac{\log_2 e}{r + 2} - 1 \right) M$$

using windows of width r, where $d = \lceil \log_2 e / \log_2 p \rceil$. The precomputation storage can be reduced using techniques described in [2].

Note that for $k = 6$ one can also use XTR [11,16] to gain a slight efficiency advantage over these methods if this is desirable [9], at a cost of altering particular protocols accordingly since multiplication is not straightforward in this case. For $k = 12$ and 24, one can also employ XTR defined over $\mathbb{F}_{p^2}$ and $\mathbb{F}_{p^4}$ respectively [12], however further work is required to determine if arithmetic can be made as efficient as in the original scheme for cases of interest in pairing-based cryptography.

3   Elliptic Curve Operations

In pairing based protocols we also need to conduct elliptic curve group operations. These are either on the main base curve $E(\mathbb{F}_p)$, or on the twisted curve $\overline{E}(\mathbb{F}_{p^{k/2}})$. We assume these curves take the form

$$E(\mathbb{F}_p) : Y^2 = X^3 - 3X + B$$

and

$$\overline{E}(\mathbb{F}_{p^{k/2}}) : \chi Y^2 = X^3 - 3X + B$$

where χ is a quadratic non-residue in $\mathbb{F}_{p^{k/2}}$ for which multiplication by χ is for free. Whether one should use affine or standard Jacobian projective coordinates are used, depends on the ratio ι/m and on the size of the $k/2$. It turns out that in some instances arithmetic in $\overline{E}(\mathbb{F}_{p^{k/2}})$ is better performed in affine coordinates. The various point addition and doubling times are summarized in the Table 4.

Table 4. Elliptic curve operation counts over $\mathbb{F}_p$ and $\mathbb{F}_{p^{k/2}}$

	$E(\mathbb{F}_p)$	$\overline{E}(\mathbb{F}_{p^{k/2}})$	
		Projective	Affine
Addition (A)	$12m + 4s$	$12\overline{M} + 4\overline{S}$	$2\overline{M} + 1\overline{S} + I_{k/2}$
Mixed Addition (A_M)	$8m + 3s$	$8\overline{M} + 3\overline{S}$	-
Doubling (D)	$4m + 4s$	$4\overline{M} + 4\overline{S}$	$2\overline{M} + 2\overline{S} + I_{k/2}$

We assume that exponentiation is performed via a signed sliding window method and mixed/affine addition

$$\text{EC}_{\text{SSW}}(e) = D(1 + \log_2 e) + A_M \left(\frac{\log_2 e}{r+2} + 2^{r-2} - 1 \right).$$

where the exact optimal choice for r depends on the size of e.

In some instances we wish to multiply by a random element in $\mathbb{Z}_n$, however in other instances (for example in the MapToPoint operation within the Boneh–Franklin encryption scheme [5]) we need to multiply by the cofactor. If we let $\log_2 p = \rho \cdot \log_2 n$ then the quantity 2^ρ measures how big the elliptic curve cofactor is for the curve $E(\mathbb{F}_p)$; a similar measure for the twisted curve is $(k\rho/2 - 1)\log_2 n$.

4   Application to Pairing Based Cryptography

In this section we wish to investigate the application of our techniques to pairing based cryptography in particular we focus on the case of non-supersingular curves of embedding degree $k \geq 6$. We follow the methodology of Koblitz and Menezes [10] which we recap on here, however we express our formulae in terms of total number of $\mathbb{F}_p$ operations as opposed to operations per bit. This is because this enables us to compare our sliding windows method in a more accurate manner and to also compare how other components of the protocols are affected by the choice of field.

Following Koblitz and Menezes we look at the cost of computing a Full-Miller operation or a Miller-Lite operation. The cost of these two operations, assuming projective coordinates are used, is

$$C_{\text{Full}} = (km + 4\overline{S} + 6\overline{M} + S + M)\log_2 n$$
$$C_{\text{Lite}} = (4s + (k+7)m + S + M)\log_2 n.$$

In some instances one can more easily compute the Full-Miller algorithm by using affine coordinates in the main loop. In this case the cost is given by

$$C_{\text{Full}} = (2\overline{S} + 2\overline{M} + I_{k/2} + km + S + M)\log_2 n.$$

In computing the Tate pairing one executes one Miller-Lite operation and then an exponentiation for an exponent given by $\Phi_t(p)/n$ in the subgroup $G_{\Phi_k(p)}$. The bit length of $\Phi_t(p)/n$ is estimated by

$$\phi(k)\log_2 p - \log_2 n,$$

which can be expressed as

$$(\phi(k)\rho - 1)\log_2 n.$$

Thus a Tate pairing computation requires time

$$C_{\text{Tate}} = C_{\text{Luc}}(\Phi_t(p)/n) + C_{\text{Lite}}$$

or

$$C_{\text{Tate}} = C_{\text{bigSSW}}(\Phi_t(p)/n) + C_{\text{Lite}}.$$

In both of the above formulae for the Tate pairing we have ignored the inversion needed to take the input of Miller-Lite into the subgroup $G_{\Phi_k(p)}$, this is consistent with the analysis of Koblitz and Menezes but does slightly underestimate the cost in both cases.

The Weil pairing as pointed out by Koblitz and Menezes, could be more efficient, as it does not require an exponentiation by a large number. It requires time

$$C_{\text{Weil}} = C_{\text{Lite}} + C_{\text{Full}} + S.$$

We shall show in all cases of cryptographic relevance that the Weil pairing is always slower than the Tate pairing.

5   Results

In what follows we make the simplifying assumption that $m \approx s$. We wish to investigate what happens to pairing based protocols as the security level increases. The parameter sizes we fix on to illustrate our discussion we give in Table 5. We do not discuss how such curves are generated, nor do we make use of special properties of the curves. For example when $k = 12$ with current technology one can only achieve $n \approx p$ by using the method of Barreto and Naehrig [3]. This results in curves with complex multiplication by $D = -3$, our analysis takes no account of the special optimizations which can be applied to such curves.

Table 5. Parameter sizes for various security levels

Case	Security	k	$\log_2 n$	$\log_2 p$
A	80	6	160	160
B	128	6	256	512
C	128	12	256	256
D	192	6	384	1365
E	192	12	384	683
F	256	6	512	2560
G	256	12	512	1280
H	256	24	512	640

For each case we first present, in Table 6, the operation counts, in terms of multiplications in $\mathbb{F}_p$, for the operations which do not appear to depend on the exact finite field we choose to use, namely the elliptic curve operations. We denote by (r) the size of the windows which produces the smallest operation count, the column n corresponds to exponentiation by a random integer of size n, whilst c corresponds to multiplication by the relevant cofactor. We limit window sizes to at most 9 bits, as otherwise the required look up table is likely to be prohibitively expensive. So as to get some idea about the relative merits of projective vs affine

Table 6. Operation count for elliptic curve calculations

Case	$E(\mathbb{F}_p)$		$\overline{E}(\mathbb{F}_{p^{k/2}})$	
	n	c	n	c
A	1614 (4)	-	8071 (4)	15739 (5)
B	2535 (5)	2535 (5)	12676 (5)	60767 (8)
C	2535 (5)	-	34813 (5)(A)	169000 (7)(A)
D	3760 (6)	9369 (6)	18802 (5)	172356 (8)
E	3760 (5)	2946 (5)	51801 (5)(A)	483113 (8)(A)
F	4973 (6)	19236 (7)	24865 (6)	329585 (9)
G	4973 (6)	7373 (6)	68671 (6)(A)	926142 (9)(A)
H	4973 (6)	1229 (4)	164573 (6)(A)	$2.2 \cdot 10^6$ (9)(A)

coordinates we made the assumption that $\iota/m \approx 10$ and in the table if the best performance for a given parameter set was using affine coordinates with give the multiplication count for this curve and denote this by an (A). We see that when $k \geq 12$ that it may make sense to use affine coordinates for the arithmetic in $\mathbb{G}_2$.

We now turn to the operations which depend on the field representation, i.e. whether we use a pairing friendly or a cyclotomic field extension. There are two operations which are important, the pairing computation itself and exponentiation in $G_{\Phi_k(p)}$ by an element of $\mathbb{Z}_n$. The pairing computation can itself either be computed by the Weil or Tate pairings. The results, in terms of estimated multiplications in $\mathbb{F}_p$, are presented in Table 7. The (r) in the Tate column denotes the window size in the final exponentiation step, if Lucas sequences are faster we denote this by (L) and the operation count is for the application of Lucas sequences. In all cases using the Weil pairing method which used affine coordinates in the Full-Miller operation loop was the most efficient.

We see that for all fields the Tate pairing is always more efficient than the Weil pairing, at least for the security sizes that are likely to be used in practice. This is more due to the use of the efficient exponentiation algorithm as compared to the

Table 7. Operation count for pairing and $G_{\Phi_k(p)}$ calculations

Case	Pairing Friendly			Cyclotomic Field		
	Pairing		Exp in	Pairing		Exp in
	Weil	Tate	$G_{\Phi_k(p)}$	Weil	Tate	$G_{\Phi_k(p)}$
A	19855	9120 (L)	1411 (4)	18250	8247 (3)	1411 (4)
B	31759	18738 (5)	2195 (5)	29194	15916 (5)	2195 (5)
C	83757	43703 (4)	3502 (5)	-	-	-
D	47631	34664 (6)	3237 (5)	43786	29643 (6)	3237 (5)
E	125613	81751 (5)	5093 (5)	-	-	-
F	63503	56677 (6)	4263 (6)	58378	46431 (6)	4263 (6)
G	167469	127831 (6)	6633 (6)	-	-	-
H	446087	331078 (5)	13743 (6)	-	-	-

efficient squaring algorithm for $G_{\Phi_k(p)}$. In addition Lucas sequences are only more efficient than the signed sliding window method for very small security parameters. We also did some calculations with fixed k and increasing p to very large levels and always found that the Weil pairing was slower than the Tate pairing.

To compare the different values of k we need to estimate the relative difference in time needed to compute a multiplication in $\mathbb{F}_p$, for the different sizes of p. If we assume that each finite field multiplication is performed using a standard interleaved Montgomery multiplication then the total number of 32-bit by 32-bit multiplication instructions which are needed to be performed per $\mathbb{F}_p$ multiplication is given by

$$2 \cdot t \cdot (t+1),$$

where $t = \log_2 p / 32$. This leads us to Table 8, where we present the number of 32-bit by 32-bit multiplication instructions needed for the various operations.

Table 8. 32×32 bit multiplications required for various operations

	Curve Operations				Pairing Friendly		Cyclotomic Field	
	$E(\mathbb{F}_p)$		$\overline{E}(\mathbb{F}_{p^{k/2}})$			Exp in		Exp in
Case	n	c	n	c	Pairing	$G_{\Phi_k(p)}$	Pairing	$G_{\Phi_k(p)}$
A	$9.7 \cdot 10^4$	-	$4.8 \cdot 10^5$	$9.4 \cdot 10^4$	$5.4 \cdot 10^5$	$8.4 \cdot 10^4$	$4.9 \cdot 10^5$	$8.4 \cdot 10^5$
B	$1.3 \cdot 10^6$	$1.3 \cdot 10^6$	$6.8 \cdot 10^6$	$3.3 \cdot 10^7$	$1.0 \cdot 10^7$	$1.1 \cdot 10^6$	$8.6 \cdot 10^6$	$1.1 \cdot 10^6$
C	$3.6 \cdot 10^5$	-	$5.0 \cdot 10^6$	$2.4 \cdot 10^7$	$6.2 \cdot 10^6$	$1.1 \cdot 10^6$	-	-
D	$1.4 \cdot 10^7$	$3.4 \cdot 10^7$	$7.0 \cdot 10^7$	$6.4 \cdot 10^8$	$1.3 \cdot 10^8$	$1.2 \cdot 10^7$	$1.1 \cdot 10^8$	$1.2 \cdot 10^7$
E	$3.6 \cdot 10^6$	$2.8 \cdot 10^6$	$4.9 \cdot 10^7$	$4.6 \cdot 10^8$	$7.8 \cdot 10^7$	$4.8 \cdot 10^6$	-	-
F	$6.4 \cdot 10^7$	$2.5 \cdot 10^8$	$3.2 \cdot 10^8$	$4.3 \cdot 10^9$	$7.3 \cdot 10^8$	$5.5 \cdot 10^7$	$6.0 \cdot 10^8$	$5.5 \cdot 10^7$
G	$1.6 \cdot 10^7$	$2.4 \cdot 10^7$	$2.2 \cdot 10^8$	$3.0 \cdot 10^9$	$4.2 \cdot 10^8$	$2.2 \cdot 10^7$	-	-
H	$4.0 \cdot 10^6$	$1.0 \cdot 10^6$	$1.4 \cdot 10^8$	$1.8 \cdot 10^9$	$2.8 \cdot 10^8$	$1.1 \cdot 10^7$	-	-

From Table 8 one can see that the main advantage in using values of k which are larger than 6 is in the basic elliptic curve operations over $\mathbb{F}_p$, rather than in the pairing computation. For the pairing computation one gains some advantage for using large values of k, but this is not as pronounced as for the elliptic curve operations.

However, elliptic curve operations are relatively cheap in comparison to pairing calculation and so the performance improvement will not be so pronounced. Except, for protocols in which one party only needs to perform elliptic curve operations in $\mathbb{F}_p$, such as the encryptor in the Sakai–Kasahara KEM [7]. It does however imply that for pairing based protocols one should not neglect selecting parameter values which speed up the elliptic curve operations and not just the pairing calculation.

However, our estimates are on the conservative side for arithmetic in cyclotomic fields at high security levels. This is for a number of reasons. The overhead in not having to deal with different values of k and f_0 means that the library overhead in using cyclotomic fields of degree six will be less than for pairing friendly fields. Recall, we have not given accurate cycle counts, but simply estimated the number of multiplication instructions needed.

One should also bear in mind that larger values of k mean that one can shrink the bandwidth required in communication if one is communicating elements in $E(\mathbb{F}_p)$, since a larger value of k corresponds to a smaller value of p.

Acknowledgements

The authors would like to thank Mike Scott, for a some insightful comments which improved this paper.

References

1. R.M. Avanzi. *On Multi-exponentiation in Cryptography.* In Cryptology ePrint Archive, Report 2002/154, 2002.
2. R. M. Avanzi and P. Mihailescu. *Generic efficient arithmetic algorithms for PAFFs (Processor Adequate Finite Fields) and related algebraic structures.* In Selected Areas in Cryptology – SAC 2003, Springer-Verlag LNCS 3006, 320–334, 2004.
3. P.S.L.M. Barreto and M. Naehrig. Pairing-friendly elliptic curves of prime order. In *Selected Areas in Cryptography – SAC 2005*, Springer-Verlag LNCS 3897, 319–331, 2006.
4. I.F. Blake, G. Seroussi and N.P. Smart. *Elliptic Curves in Cryptography.* Cambridge University Press, 1999.
5. D. Boneh and M. Franklin. *Identity-based encryption from the Weil pairing.* SIAM Journal of Computing, **32**, 586–615, 2003.
6. F. Brezing and A. Weng. *Elliptic Curves Suitable for Pairing Based Cryptography.* Designs, Codes and Cryptography, **37**, 133–141, 2005.
7. M. Cheng, L. Chen, J. Malone-Lee and N.P. Smart. *An Efficient ID-KEM Based On The Sakai-Kasahara Key Construction.* To appear, 2006.
8. M. van Dijk, R. Granger, D. Page, K. Rubin, A. Silverberg, M. Stam and D. Woodruff. *Practical cryptography in high dimensional tori.* In Advances in Cryptology – EUROCRYPT 2005, Springer-Verlag LNCS 3494, 234–250, 2005.
9. R. Granger, D. Page and M. Stam. *A Comparison of CEILIDH and XTR.* In Algorithmic Number Theory Symposium – ANTS VI, Springer-Verlag LNCS 3076, 235–249, 2004.
10. N. Koblitz and A. Menezes. *Pairing-based Cryptography at High Security Levels.* In Cryptography and Coding, Springer-Verlag LNCS 3796, 13–36, 2005.
11. A. K. Lenstra and E. Verheul. *The XTR Public Key System.* In Advances in Cryptology – CRYPTO 2000, Springer-Verlag LNCS 1880, 1–19, 2000.
12. S. Lim, S. Kim, I. Yie, J. Kim and H. Lee. *XTR extended to $GF(p^{6m})$.* In Selected Areas in Cryptography – SAC 2001, Springer-Verlag LNCS 2259, 301–312, 2001.
13. M. Scott and P.S.L.M. Barreto. *Compressed Pairings.* In Advances in Cryptology – CRYPTO 2004, Springer-Verlag LNCS 3152, 140–156, 2004.
14. M. Stam. *Speeding up Subgroup Cryptosystems.* PhD Thesis, T.U. Eindhoven, 2003.
15. M. Stam and A. Lenstra. *Efficient Subgroup Exponentiation in Quadratic and Sixth Degree Extensions.* In Cryptographic Hardware and Embedded Systems – CHES 2002, Springer-Verlag LNCS 2523, 318–332, 2002.
16. M. Stam and A. K. Lenstra. *Speeding Up XTR.* In Advances in Cryptology – ASIACRYPT 2001, Springer-Verlag LNCS 2248, 125–143, 2001.

A   Squaring Formulae for $G_{\Phi_k(p)}$ for Pairing Friendly Fields and $k = 12$ and $k = 24$

A.1 $k = 12$

$$b_0 = 3\,a_0{}^2 - 6\,f_0 a_3 a_9 - 3\,f_0 a_6{}^2 - 2\,a_0,$$
$$b_1 = -6\,f_0 a_{11} a_2 - 6\,f_0 a_5 a_8 + 2\,a_1,$$
$$b_2 = 3\,a_1{}^2 - 6\,f_0 a_4 a_{10} - 3\,f_0 a_7{}^2 - 2\,a_2,$$
$$b_3 = -6\,f_0 a_6 a_9 + 6\,a_0 a_3 + 2\,a_3,$$
$$b_4 = -3\,f_0 a_8{}^2 - 6\,f_0 a_{11} a_5 + 3\,a_2{}^2 - 2\,a_4,$$
$$b_5 = -6\,f_0 a_7 a_{10} + 6\,a_4 a_1 + 2\,a_5,$$
$$b_6 = -3\,f_0 a_9{}^2 + 6\,a_0 a_6 + 3\,a_3{}^2 - 2\,a_6,$$
$$b_7 = 6\,a_5 a_2 - 6\,f_0 a_{11} a_8 + 2\,a_7,$$
$$b_8 = -3\,f_0 a_{10}{}^2 + 3\,a_4{}^2 + 6\,a_7 a_1 - 2\,a_8,$$
$$b_9 = 6\,a_6 a_3 + 6\,a_0 a_9 + 2\,a_9,$$
$$b_{10} = -3\,f_0 a_{11}{}^2 + 3\,a_5{}^2 + 6\,a_8 a_2 - 2\,a_{10},$$
$$b_{11} = 6\,a_{10} a_1 + 6\,a_7 a_4 + 2\,a_{11}.$$

A.2 $k = 24$

$$b_0 = -2\,a_0 + 3\,a_0{}^2 - 3\,f_0 a_{12}{}^2 - 6\,f_0 a_3 a_{21} - 6\,f_0 a_6 a_{18} - 6\,f_0 a_9 a_{15},$$
$$b_1 = 2\,a_1 - 6\,f_0 a_2 a_{23} - 6\,f_0 a_5 a_{20} - 6\,f_0 a_8 a_{17} - 6\,f_0 a_{11} a_{14},$$
$$b_2 = -2\,a_2 - 6\,f_0 a_4 a_{22} - 6\,f_0 a_{10} a_{16} - 3\,f_0 a_{13}{}^2 - 6\,f_0 a_7 a_{19} + 3\,a_1{}^2,$$
$$b_3 = 2\,a_3 - 6\,f_0 a_9 a_{18} + 6\,a_0 a_3 - 6\,f_0 a_6 a_{21} - 6\,f_0 a_{12} a_{15},$$
$$b_4 = -2\,a_4 - 3\,f_0 a_{14}{}^2 - 6\,f_0 a_5 a_{23} - 6\,f_0 a_8 a_{20} - 6\,f_0 a_{11} a_{17} + 3\,a_2{}^2,$$
$$b_5 = 2\,a_5 - 6\,f_0 a_{13} a_{16} - 6\,f_0 a_{10} a_{19} + 6\,a_4 a_1 - 6\,f_0 a_7 a_{22},$$
$$b_6 = -2\,a_6 + 6\,a_0 a_6 - 3\,f_0 a_{15}{}^2 - 6\,f_0 a_{12} a_{18} - 6\,f_0 a_9 a_{21} + 3\,a_3{}^2,$$
$$b_7 = 2\,a_7 - 6\,f_0 a_{14} a_{17} - 6\,f_0 a_{11} a_{20} - 6\,f_0 a_8 a_{23} + 6\,a_2 a_5,$$
$$b_8 = -2\,a_8 - 6\,f_0 a_{10} a_{22} - 6\,f_0 a_{13} a_{19} + 3\,a_4{}^2 + 6\,a_7 a_1 - 3\,f_0 a_{16}{}^2,$$
$$b_9 = 2\,a_9 + 6\,a_0 a_9 - 6\,f_0 a_{12} a_{21} - 6\,f_0 a_{15} a_{18} + 6\,a_6 a_3,$$
$$b_{10} = -2\,a_{10} - 3\,f_0 a_{17}{}^2 - 6\,f_0 a_{11} a_{23} - 6\,f_0 a_{14} a_{20} + 6\,a_2 a_8 + 3\,a_5{}^2,$$
$$b_{11} = 2\,a_{11} - 6\,f_0 a_{13} a_{22} - 6\,f_0 a_{16} a_{19} + 6\,a_{10} a_1 + 6\,a_7 a_4,$$
$$b_{12} = -2\,a_{12} + 6\,a_0 a_{12} + 6\,a_9 a_3 - 3\,f_0 a_{18}{}^2 - 6\,f_0 a_{15} a_{21} + 3\,a_6{}^2,$$
$$b_{13} = 2\,a_{13} - 6\,f_0 a_{17} a_{20} - 6\,f_0 a_{14} a_{23} + 6\,a_5 a_8 + 6\,a_2 a_{11},$$
$$b_{14} = -2\,a_{14} + 6\,a_{13} a_1 + 6\,a_{10} a_4 - 3\,f_0 a_{19}{}^2 - 6\,f_0 a_{16} a_{22} + 3\,a_7{}^2,$$
$$b_{15} = 2\,a_{15} + 6\,a_0 a_{15} + 6\,a_{12} a_3 + 6\,a_9 a_6 - 6\,f_0 a_{18} a_{21},$$
$$b_{16} = -2\,a_{16} - 3\,f_0 a_{20}{}^2 - 6\,f_0 a_{17} a_{23} + 6\,a_5 a_{11} + 6\,a_2 a_{14} + 3\,a_8{}^2,$$

$$b_{17} = 2\,a_{17} - 6\,f_0 a_{19} a_{22} + 6\,a_{13} a_4 + 6\,a_{10} a_7 + 6\,a_{16} a_1,$$

$$b_{18} = -2\,a_{18} + 6\,a_0 a_{18} + 6\,a_{15} a_3 + 6\,a_{12} a_6 - 3\,f_0 a_{21}{}^2 + 3\,a_9{}^2,$$

$$b_{19} = 2\,a_{19} - 6\,f_0 a_{20} a_{23} + 6\,a_5 a_{14} + 6\,a_8 a_{11} + 6\,a_2 a_{17},$$

$$b_{20} = -2\,a_{20} - 3\,f_0 a_{22}{}^2 + 6\,a_{19} a_1 + 6\,a_{16} a_4 + 6\,a_{13} a_7 + 3\,a_{10}{}^2,$$

$$b_{21} = 2\,a_{21} + 6\,a_0 a_{21} + 6\,a_{18} a_3 + 6\,a_{15} a_6 + 6\,a_{12} a_9,$$

$$b_{22} = -2\,a_{22} - 3\,f_0 a_{23}{}^2 + 6\,a_5 a_{17} + 6\,a_8 a_{14} + 6\,a_2 a_{20} + 3\,a_{11}{}^2,$$

$$b_{23} = 6\,a_{16} a_7 + 6\,a_{13} a_{10} + 6\,a_{19} a_4 + 6\,a_{22} a_1 + 2\,a_{23}.$$

Efficiently Computable Endomorphisms for Hyperelliptic Curves

David R. Kohel and Benjamin A. Smith

School of Mathematics and Statistics, The University of Sydney
{kohel, bens}@maths.usyd.edu.au

Abstract. Elliptic curves have a well-known and explicit theory for the construction and application of endomorphisms, which can be applied to improve performance in scalar multiplication. Recent work has extended these techniques to hyperelliptic Jacobians, but one obstruction is the lack of explicit models of curves together with an efficiently computable endomorphism. In the case of hyperelliptic curves there are limited examples, most methods focusing on special CM curves or curves defined over a small field. In this article we describe three infinite families of curves which admit an efficiently computable endomorphism, and give algorithms for their efficient application.

Keywords: Hyperelliptic curve cryptography, efficiently computable endomorphisms.

1 Introduction

The use of efficiently computable endomorphisms for speeding up point multiplication on elliptic curves is well-established for elliptic curves and more recently has been used for hyperelliptic curves. Koblitz [10] proposed τ-adic expansions of the Frobenius endomorphism on curves over a small finite fields. Gallant, Lambert, and Vanstone [6] later proposed using an expression

$$[k]P = [k_0]P + [k_1]\phi(P)$$

on more general curves to evaluate multiplication by k on a point P, using an efficiently computable endomorphism ϕ. Various improvements and combinations of these methods have been proposed for both elliptic and hyperelliptic curves [11,17,2].

One feature of elliptic curves, not available for multiplicative groups of finite fields, is the freedom to choose a parameter: geometrically they form a one-dimensional family, parametrized by the j-invariant. Restriction to curves of a special form destroys this degree of freedom. While no proof exists that special curves, CM curves or Koblitz curves are less insecure, these nonrandom curves can be qualitatively distinguished from their nonrandom cousins in terms of their endomorphism rings. Thus preference is often given to curves randomly selected over a large finite field when performance is not the determining issue.

F. Hess, S. Pauli, and M. Pohst (Eds.): ANTS 2006, LNCS 4076, pp. 495–509, 2006.

In contrast, hyperelliptic curves of genus g admit a much larger degree of freedom. In genus 2, they form a three dimensional family: curves with different classifying triple of invariants (j_1, j_2, j_3) can not be isomorphic over any extension field. Until the recent work of Takashima [19], the only curves proposed for cryptographic use with efficiently computable endomorphisms are either the CM curves with exceptional automorphisms — the analogues of elliptic curves $y^2 = x^3 + a$ or $y^2 = x^3 + ax$ — or Koblitz curves — curves defined over a small field with point on the Jacobian taken over a large prime degree extension (see Park et al. [15] for the former and Lange [11] for the latter). Besides the notable exceptions of CM curves with exceptional automorphisms, curves with CM have been exploited for point counting but not for their endomorphism ring structure, for lack of a constructive theory of efficiently computable endomorphisms.

In this work, we address the problem of effective algorithms for endomorphisms available on special families of curves. We describe three families, of dimensions 1, 1, and 2 respectively, of curves whose Jacobians admit certain *real* endomorphisms. First, we introduce the general framework for constructing endomorphisms via correspondences derived from covering curves. Subsequently, we provide a one-dimensional family derived from Artin–Schreier covers, then describe a construction of Tautz, Top, and Verberkmoes [20] for a one-dimensional family of curves with explicit endomorphisms deriving from cyclotomic covers. Finally, we describe an elegant construction of Mestre [14] from which we obtain a two-dimensional family of curves whose Jacobians admit explicit endomorphisms, derived from covers of elliptic curves. In each case we develop explicit algorithms for efficient application of the endomorphism, suitable for use in a GLV decomposition. Independently, Takashima [19] provided an efficient algorithm for endomorphisms in the latter family (in terms of variants of Brumer and Hashimoto) with real multiplication by $(1 + \sqrt{5})/2$. These families provide a means of generating curves randomly selected within a large family, yet which admit efficiently computable endomorphisms.

2   Arithmetic on Hyperelliptic Jacobians

In the sequel we denote by X/k a hyperelliptic curve of genus g_X in the form

$$v^2 = f(u) = u^{2g_X+1} + c_{2g_X} u^{2g_X} + \cdots + c_0,$$

with each c_i in k, which we require to be a field of characteristic not 2. The Jacobian of X, denoted $\mathrm{Jac}(X)$, is a g_X-dimensional variety whose points form an abelian group. Let $\mathcal{O}$ denote the point at infinity of X. Each point P on $\mathrm{Jac}(X)$ may be represented by a divisor on X, that is, as a formal sum of points

$$P = \sum_{i=1}^{m} [P_i] - m[\mathcal{O}] = \sum_{i=1}^{m} [(u_i, v_i)] - m[\mathcal{O}],$$

where $m \leq g_X$. We say such a divisor is *semi-reduced* if $(u_i, v_i) \neq (u_j, -v_j)$ for all $i \neq j$. For a point to be defined in $\mathrm{Jac}(X)(k)$, its divisor must be Galois-stable;

the representation as a divisor has the disadvantage that the individual points (u_i, v_i) may be defined only over some finite extension K/k. Thus, for computations, we use instead the Mumford representation for divisors, identifying P with the ideal class

$$P = [(a(u), v - b(u))],$$

where a and b are polynomials in $k[u]$ such that $a(u) = \prod_i (u - u_i)$ and $v_i = b(u_i)$ for all i. In this guise, addition of points P and Q is an ideal product, followed by a reduction algorithm to produce a unique "reduced" ideal representing $P + Q$. Cantor [1] provides algorithms to carry out these operations.

Algorithm 1. Given a semi-reduced representative $(a(u), v - b(u))$ for a point P on the Jacobian of a hyperelliptic curve $X : v^2 = f(u)$, returns the reduced representative of P.

> **function** CANTORREDUCTION$((a(u), v - b(u)))$
>   **while** $\deg(a) > g_X$ **do**
> $a := (f - b^2)/a;$
> $b := -b \bmod a;$
>   **end while;**
> $a := a/\text{LEADINGCOEFFICIENT}(a);$
>   **return** $(a, v - b(u));$
> **end function;**

Each iteration of Algorithm 1 replaces a with a polynomial of degree $\max(2g_X + 1 - \deg(a), \deg(a) - 2)$. It follows that Algorithm 1 will produce a reduced representative for the ideal class $[(a(u), v - b(u))]$ after $\lceil (\deg(a) - g_X)/2 \rceil$ iterations.

3   Explicit Endomorphisms

Let C be a curve with an automorphism ζ, and let $\pi : C \to X$ be a covering of X. We have two coverings, π and $\pi \circ \zeta$, from C to X; together, they induce a map η of divisors

$$\eta := (\pi \circ \zeta)_* \pi^* : \text{Div}(X) \to \text{Div}(X),$$

where

$$\pi^*([P]) = \sum_{Q \in \pi^{-1}(P)} e_\pi(Q)[Q] \quad \text{and} \quad (\pi \circ \zeta)_*([Q]) = [\pi(\zeta(Q))].$$

This map on divisors induces an endomorphism of the Jacobian $\text{Jac}(X)$, which we also denote η.

In our constructions, we take π to be the quotient by an involution σ of C, so that π is a degree-2 covering, and $\pi = \pi \circ \sigma$. Thus

$$\pi^*([P]) = [Q] + [\sigma(Q)]$$

for any point Q in $\pi^{-1}(P)$. We will take ζ to be an automorphism of C of prime order p, such that $\langle \zeta, \sigma \rangle$ is a dihedral subgroup of the automorphism group of C: that is, $\sigma\zeta = \zeta^{-1}\sigma$. The following proposition describes the resulting endomorphism $\eta = (\pi \circ \zeta)_* \circ \pi^*$.

Proposition 2. *Let C be a curve with an involution σ and an automorphism ζ of prime order p such that $\sigma\zeta = \zeta^{-1}\sigma$. Let $\pi : C \to X := C/\langle\sigma\rangle$ be the quotient of C by the action of σ, and let $\eta := (\pi\circ\zeta)_* \circ \pi^*$ be the endomorphism of $\mathrm{Jac}(X)$ induced by ζ. The subring $\mathbb{Z}[\eta]$ of $\mathrm{End}(\mathrm{Jac}(X))$ is isomorphic to $\mathbb{Z}[\zeta_p + \zeta_p^{-1}]$, where ζ_p is a primitive p^{th} root of unity over $\mathbb{Q}$.*

Proof. The subring $\mathbb{Z}[\zeta_* + \zeta_*^{-1}]$ of $\mathrm{Jac}(C)$ is isomorphic to $\mathbb{Z}[\zeta_p + \zeta_p^{-1}]$, since p is prime. The statement follows upon noting that the following diagram commutes.

$$
\begin{array}{ccc}
\mathrm{Jac}(C) & \xrightarrow{\ \zeta_* + \zeta_*^{-1}\ } & \mathrm{Jac}(C) \\
\Big\downarrow{\scriptstyle \pi_*} & & \Big\downarrow{\scriptstyle \pi_*} \\
\mathrm{Jac}(X) & \xrightarrow{\ \eta = \pi_*\zeta_*\pi^*\ } & \mathrm{Jac}(X)
\end{array}
$$

To see this, observe that for any Q in $\mathrm{Jac}(C)$ we have

$$
\begin{aligned}
\eta(\pi_*(Q)) &= \pi_*\zeta_*\pi^*\pi_*(Q) \\
&= \pi_*\zeta_*(1 + \sigma_*)(Q) \\
&= \pi_*(\zeta_* + \sigma_*\zeta_*^{-1})(Q) \\
&= \pi_*(\zeta_* + \zeta_*^{-1})(Q),
\end{aligned}
$$

since $\pi^*\pi_* = (1 + \sigma_*)$ and $\pi_*\sigma_* = \pi_*$. See also Ellenberg [4, §2]. $\square$

Suppose C, X, π, ζ and η are as in Proposition 2. Our aim is to give an explicit realization of the endomorphism η of $\mathrm{Jac}(X)$, in the form of a map on ideal classes. To do this, we form the algebraic correspondence

$$
Z := (\pi \times (\pi \circ \zeta))(C) \subset X \times X.
$$

Let π_1 and π_2 be the restrictions to Z of the projections from $X \times X$ to its first and second factors, respectively; then $\eta = (\pi_2)_* \circ \pi_1^*$. We will give an affine model for Z as the variety cut out by an ideal in $k[u_1, v_1, u_2, v_2]/(v_1^2 - f(u_1), v_2^2 - f(u_2))$; for this model, the maps π_1 and π_2 are defined by $\pi_i(u_1, v_1, u_2, v_2) = (u_i, v_i)$.

Suppose that Z is defined by an ideal $(v_2 - v_1, E(u_1, u_2))$, where E is quadratic in u_1 and u_2 (this will be the case in each of our constructions). If (u, v) is a generic point on X, then $\pi_1^*([[(u, v)]])$ is the effective divisor on Z cut out by $(v_2 - v, E(u, u_2))$. Therefore, if e_1 and e_2 are the solutions in $\overline{k(u)}$ to the quadratic equation $E(u, x) = 0$ in x, then

$$
\eta([[(u, v)]]) = (\pi_2)_*\pi_1^*([[(u, v)]]) = [(e_1, v)] + [(e_2, v)].
$$

It remains to translate this description of the action of η in terms of points into a map on ideal classes.

Suppose $[(a(u), v - b(u))]$ is a point on $\mathrm{Jac}(X)$. Extending the above, we have

$$
\begin{aligned}
\eta([[(a(u), v - b(u))]]) &= [(a(e_1), v - b(e_1))] + [(a(e_2), v - b(e_2))] \\
&= [(N(a), v - \frac{(f(u) + N(b))}{T(b)} \bmod N(a)],
\end{aligned}
$$

where $N(a) = a(e_1)a(e_2)$, $N(b) = b(e_1)b(e_2)$, and $T(b) = b(e_1) + b(e_2)$.[1] Since functions $T(a)$, $N(b)$ and $T(b)$ are symmetric polynomials in e_1 and e_2, we can write each as a polynomial in the elementary symmetric functions $e_1 + e_2$ and e_1e_2. Moreover, $e_1 + e_2$ and e_1e_2 are elements of $k(u)$: if $E(u, x) = E_2(u)x^2 + E_1(u)x + E_0(u)$, then $e_1 + e_2 = -E_1/E_2$ and $e_1e_2 = E_0/E_2$.

Definition 3. *For any polynomial $a(x)$ over k, we define $T(a) = a(e_1) + a(e_2)$ and $N(a) = a(e_1)a(e_2)$, and for $i, j \geq 0$ we define*

$$t_i := e_1^i + e_2^i, \quad n_i := (e_1e_2)^i \quad and \quad n_{i,j} := e_1^i e_2^j + e_1^j e_2^i.$$

Note that t_i and n_{ij} are elements of $k(u)$ and that

$$T\left(\sum_{i=0}^{g_X} a_i x^i \right) = \sum_{i=0}^{g_X} a_i t_i, \tag{1}$$

and

$$N\left(\sum_{i=0}^{g_X} a_i x^i \right) = \sum_{i=0}^{g_X} \sum_{i=0}^{g_X} a_i a_j n_{i,j}. \tag{2}$$

The following elementary lemma provides simple recurrences for the construction of the sequences $\{t_i\}$ and $\{n_{i,j}\}$.

Lemma 4. *The elements t_i, n_i and $n_{i,j}$ satisfy the following recurrences:*

1. *$n_{i+1} = (e_1e_2)n_i$ for $i \geq 0$, with $n_0 := 1$;*
2. *$t_{i+1} = (e_1 + e_2)t_i - (e_1e_2)t_{i-1}$ for $i \geq 1$, with $t_0 = 2$ and $t_1 = (e_1 + e_2)$;*
3. *$n_{i,i} = n_i$ and $n_{i,j} = n_i t_{j-i}$ for $i \geq 0$ and $j > i$.*

Equations (1) and (2) above express T and N in terms of the functions t_i and $n_{i,j}$, which depend *only* upon t_1 and n_1 by Lemma 4. Thus, given $t_1 = e_1 + e_2$ and $n_1 = n_{1,1} = e_1e_2$, the recurrences of Lemma 4 give a simple and fast algorithm for computing the maps T and N. If we further assume that T and N will only be evaluated at polynomials a and b from reduced ideal class representatives $(a(u), v - b(u))$, then we need only compute the t_i and $n_{i,j}$ for $0 \leq i \leq j \leq g_X$.

Algorithm 5. Given functions t_1 and n_1 in $k(u)$, together with the genus g_X of a curve X, returns the maps T and N of Definition 3.

    function RATIONALMAPS(t_1, n_1, g_X)
        $n_0 := 1$;
        $t_0 := 2$;
        for i in $[1, \ldots, g_X]$ **do**
            $n_{i+1} := n_1 n_i$;
            $t_{i+1} := t_1 t_i - n_1 t_{i-1}$;

---

[1] The modular inversion of $T(b)$ should be carried out after clearing denominators and removing common factors from $N(a)$, $T(b)$, and $f(u) + N(b)$ (generically, $N(a)$ and $T(b)$ are coprime). Proposition 6 below makes this precise.

```
        end for;
        for i in [1,...,gₓ] do
            n_{i,i} := n_i;
            for j in [i+1,...,gₓ] do
                n_{i,j} := n_i t_{j-i};
            end for;
        end for;
        T := (∑_{i=0}^{gₓ} a_i X^i ⟼ ∑_{i=0}^{gₓ} a_i t_i);
        N := (∑_{i=0}^{gₓ} a_i X^i ⟼ ∑_{i=0}^{gₓ} ∑_{j=i}^{gₓ} a_i a_j n_{i,j});
        return T, N;
    end function;
```

The following proposition shows that the maps T and N may be used to compute $\eta([(a(u), v - b(u))])$ for all points $[(a(u), v - b(u))]$ of $\text{Jac}(X)$.

Proposition 6. *Let η be the endomorphism of $\text{Jac}(X)$ induced by a correspondence $V(v_2 - v_1, E_2(u_1)u_2^2 + E_1(u_1)u_2 + E_0(u_1))$ on $X \times X$; set $t_1 = -E_1/E_2$ and $n_1 = E_0/E_2$, and let T and N be the maps of Definition 3. If $(a(u), v - b(u))$ is the reduced representative of a point P of $\text{Jac}(X)$, then $\eta(P)$ is represented by*

$$\left(\frac{E_2^{gₓ} N(a)}{G}, v - \left(\frac{(f + N(b))/G}{T(b)/G} \mod \frac{E_2^{gₓ} N(a)}{G} \right) \right),$$

where $G = \gcd(E_2^{gₓ} N(a), E_2^{gₓ} T(b))$. Algorithm 1 computes the reduced representative of $\eta(P)$ after at most $\lceil gₓ/2 \rceil$ iterations of its main loop.

Proof. We have

$$\begin{aligned}
\eta([(a(u), v - b(u))]) &= [(a(e_1), v - b(e_1))(a(e_2), v - b(e_2))] \\
&= [(N(a), v^2 - T(b)v + N(b))] \\
&= [(E_2^{gₓ})(N(a), T(b)v - (f + N(b)))]. \\
&= [(E_2^{gₓ} N(a), E_2^{gₓ} T(b)v - E_2^{gₓ}(f + N(b)))].
\end{aligned}$$

It is easily verified that $E_2^{gₓ} N(a)$, $E_2^{gₓ} T(b)$ and $E_2^{gₓ}(f + N(b))$ are polynomials, and that if $G = \gcd(E_2^{gₓ} N(a), E_2^{gₓ} T(b))$, then G also divides $E_2^{gₓ}(f + N(b))$. Therefore

$$\begin{aligned}
\eta([(a(u), v - b(u))]) &= [(G)(E_2^{gₓ} N(a)/G, E_2^{gₓ} T(b)v/G - E_2^{gₓ}(f + N(b))/G))] \\
&= [(E_2^{gₓ} N(a)/G, E_2^{gₓ} T(b)v/G - E_2^{gₓ}(f + N(b))/G)] \\
&= [(E_2^{gₓ} N(a)/G, v - I \cdot E_2^{gₓ}(f + N(b))/G)],
\end{aligned}$$

where I denotes the inverse of $E_2^{gₓ}(f + N(b))/G$ modulo $E_2^{gₓ} N(a)/G$, proving the first claim. Now, if $(a(u), v - b(u))$ is the reduced representative of P, then $\deg(a) \le gₓ$, so the degree of $E_2^{gₓ} N(a)$ is at most $2gₓ$. After each iteration of Algorithm 1, the degree of a becomes $\max(2gₓ + 1 - \deg(a), \deg(a) - 2)$, and the algorithm terminates when $\deg(a) \le gₓ$; this occurs after $\lceil gₓ/2 \rceil$ iterations.   □

The following algorithm applies Proposition 6 to compute the image of a point of $\text{Jac}(X)$ under η. This gives an explicit realization of η as a map on ideal classes.

Algorithm 7. Given a point P on the Jacobian of a curve $X : v^2 = f(u)$ and rational maps T and N derived for an endomorphism η of $\text{Jac}(X)$ using Algorithm 5, returns the reduced ideal class representative of $\eta(P)$.

> **function** $\text{EVALUATE}(P = (a(u), v - b(u)), T, N)$
> $a' := N(a)$;
> $d := T(b)$;
> $E := \text{LCM}(\text{DENOMINATOR}(a'), \text{DENOMINATOR}(d))$;
> $G := \text{GCD}(\text{NUMERATOR}(a'), \text{NUMERATOR}(d))$;
> $a' := E \cdot a'/G$;
> $d := E \cdot d/G$;
> $I := d^{-1} \pmod{a'}$;
> $b' := I \cdot E \cdot (f + N(b))/G \pmod{a'}$;
>   **return** $\text{CANTORREDUCTION}((a', v - b'))$;
> **end function;**

Remark 8. In the families of curves described below in Sections 4 and 5 below, T and N are polynomial maps, and we may take $E = 1$ in Algorithm 7.

4   Applications I: Curves with Artin–Schreier Covering

In this section we construct a family of curves X_p in one free parameter t for each prime $p \geq 5$, and determine explicit endomorphisms deriving from a cover by the Artin–Schreier curve defined over $\mathbb{F}_p$ by

$$C_p : y^p - y = x + \frac{t}{x}.$$

The eigenvalues of Frobenius in this family are described by classical Kloosterman sums [21].

An analogous family $y^2 = x^p - x + t$ was described by Duursmaa and Sakurai [3], for which the automorphism $x \mapsto x + 1$ was proposed for efficient scalar multiplication. In constrast to our family, every member of this family is isomorphic over a base extension to the supersingular curve $y^2 = x^p - x$.

4.1   Construction of the Artin–Schreier Covering

The curve C_p has automorphisms ζ (of order p) and σ (of order 2), defined by

$$\zeta(x, y) = (x, y + 1) \quad \text{and} \quad \sigma(x, y) = (-t/x, -y).$$

Let X_p be the quotient of C_p by $\langle \sigma \rangle$, with affine model

$$X_p : v^2 = f(u) = u(u^{(p-1)/2} - 1)^2 - 4t.$$

The quotient map $\pi : C_p \to X_p$ is a covering of degree 2, sending (x, y) to $(u, v) = (y^2, x - t/x)$. Observe that X_p is a family of curves of genus $(p - 1)/2$.

The automorphism ζ of C_p induces an endomorphism $\eta := (\pi \circ \zeta)_* \pi^*$ on $\mathrm{Jac}(X_p)$, whose minimal polynomial equals that of $\eta_p = \zeta_p + \zeta_p^{-1} \in \mathbb{C}$. The endomorphism η is induced by the correspondence $Z := (\pi \circ \zeta \times \pi)(C_p)$ on $X_p \times X_p$, for which we may directly compute an affine model

$$Z = V(v_2 - v_1, u_2^2 + u_1^2 - 2u_1u_2 - 2u_2 - 2u_1 + 1).$$

Setting $t_1 := 2(u+1)$ and $n_1 := (u-1)^2$ and applying Algorithm 5, we obtain polynomial maps T and N such that η is realized by $P \mapsto \mathrm{EVALUATE}(P, T, N)$, using Algorithm 7. The first few t_i and $n_{i,j}$ derived in Algorithm 5 are given in Table 1 below.

Proposition 9. *The Jacobian $\mathrm{Jac}(C_p)$ is isogenous to $\mathrm{Jac}(X_p)^2$, and its endomorphism ring contains an order in $\mathbb{M}_2(\mathbb{Q}(\eta_p))$.*

Proof. The automorphisms ζ and σ determine a homomorphic image of the group algebra $A = \mathbb{Q}[\langle \zeta, \sigma \rangle]$ in $\mathrm{End}^\circ(\mathrm{Jac}(C_p))$. But A is a semisimple algebra of dimension $2p$ over $\mathbb{Q}$, whose simple quotients are of dimensions 1, 1, and $2\varphi(p)$. Moreover, $\zeta + \zeta^{-1}$ is in the centre of A and generates a subring isomorphic to $\mathbb{Q} \times \mathbb{Q}(\eta_p)$. Since ζ and σ do not commute, it follows that the latter algebra is isomorphic to $\mathbb{M}_2(\mathbb{Q}(\eta_p))$.

Let e_1 and e_2 be the central idempotents associated to the quotients of dimensions 1. On each associated abelian variety $e_i \mathrm{Jac}(C_p)$, the automorphism ζ acts trivially, thus maps through the Jacobian of the genus 0 quotient $C_p/\langle \zeta \rangle$; it follows that the image of A in $\mathrm{End}^\circ(\mathrm{Jac}(C_p))$ is isomorphic to $\mathbb{M}_2(\mathbb{Q}(\eta_p))$.

Let $\epsilon_1 = 1 + \sigma$ and $\epsilon_2 = 1 - \sigma$. Noting that

$$\epsilon_i^2 = 2\epsilon_i, \quad \epsilon_1\epsilon_2 = 0, \text{ and } \epsilon_1 + \epsilon_2 = 2,$$

we let $A_1 = \epsilon_{1*}\mathrm{Jac}(C_p)$ and $A_2 = \epsilon_{2*}\mathrm{Jac}(C_p)$ be subabelian varieties of $\mathrm{Jac}(C_p)$ such that $\mathrm{Jac}(C_p) = A_1 + A_2$, and $A_1 \cap A_2$ is finite. Since $\zeta - \zeta^{-1}$ determines an isogeny $\psi = \zeta_* - \zeta_*^{-1}$ of $\mathrm{Jac}(C_p)$ to itself, the relation

$$(\zeta - \zeta^{-1})\epsilon_1 = \epsilon_2(\zeta - \zeta^{-1}),$$

implies that $\psi(A_1) = \epsilon_{2*}\psi(\mathrm{Jac}(C_p)) = A_2$, so that A_1 and A_2 are isogenous. But π_* is an isogeny of A_1 to $\mathrm{Jac}(X_p)$, whence $\mathrm{Jac}(C_p) \sim \mathrm{Jac}(X_p)^2$. $\square$

Corollary 10. *The Jacobian $\mathrm{Jac}(X_p)$ has a rational p-torsion point. In particular, $\mathrm{Jac}(X_p)$ is not a supersingular abelian variety.*

Proof. The curve C_p has two rational points fixed by ζ, whose difference determines a point in $\ker(1 - \zeta_*)$. But

$$(1 - \zeta)(1 - \zeta^2) \cdots (1 - \zeta^{p-1}) = p,$$

so $\ker(1 - \zeta_*)$ is contained in $\mathrm{Jac}(X_p)[p]$. If $\chi(T)$ and $\xi(T)$ are the characteristic polynomials of Frobenius on $\mathrm{Jac}(C_p)$ and $\mathrm{Jac}(X_p)$, respectively, then $\chi(T) = \xi(T)^2$. Since $|\mathrm{Jac}(C_p)(k)| = \chi(1)$ is divisible by p, so is $|\mathrm{Jac}(X_p)(k)| = \xi(1)$. $\square$

Remark 11. In fact, it is possible to show that the p-rank of $\mathrm{Jac}(X_p)$ is exactly equal to 1, so the Jacobians are neither ordinary nor supersingular.

Table 1. Artin–Schreier covers: t_i and $n_{i,j}$ for $0 \le i \le j \le 3$

t_0	2		$n_{0,3}$	$2(u^3 + 15u^2 + 15u + 1)$
t_1	$2(u+1)$		$n_{1,1}$	$(u-1)^2$
t_2	$2(u^2 + 6u + 1)$		$n_{1,2}$	$2(u-1)^2(u+1)$
t_3	$2(u^3 + 15u^2 + 15u + 1)$		$n_{1,3}$	$2(u-1)^2(u^2+6u+1)$
$n_{0,0}$	1		$n_{2,2}$	$(u-1)^4$
$n_{0,1}$	$2(u+1)$		$n_{2,3}$	$2(u-1)^4(u+1)$
$n_{0,2}$	$2(u^2+6u+1)$		$n_{3,3}$	$(u-1)^6$

4.2 Hyperelliptic Curves of Genus 2 with Real Multiplication by η_5

For $p = 5$, the construction above yields a one-parameter family of genus 2 hyperelliptic curves defined by

$$X_5 : v^2 = f_5(u) = u(u^2 - 1)^2 + t,$$

whose Jacobian has endomorphism ring containing $\mathbb{Z}[\eta_5] \cong \mathbb{Z}[x]/(x^2 + x - 1)$.

Each point P of $\mathrm{Jac}(X_5)$ may be represented by an ideal $(a(u), v - b(u))$ with a and b of degrees 2 and 1 respectively: hence, suppose $a(u) = a_2u^2 + a_1u + a_0$ and $b(u) = b_1u + b_0$. Applying Algorithm 5, we see that

$$N(a) = a_2^2 n_{2,2} + a_2a_1n_{1,2} + a_1^2n_{1,1} + a_2a_0n_{0,2} + a_1a_0n_{0,1} + a_0^2n_{0,0},$$
$$N(b) = b_1^2n_{1,1} + b_1b_0n_{0,1} + b_0^2n_{0,0}, \text{ and}$$
$$T(b) = 2b_1(u+1) + 2b_0,$$

with the $n_{i,j}$ as in Table 1. The endomorphism η is then explicitly realized by $\eta : P \mapsto \text{EVALUATE}(P, T, N)$, using Algorithm 7.

Remark 12. The Igusa invariants of the curve X_5 determine the weighted projective point $(J_2 : J_4 : J_6 : J_8 : J_{10}) = (3 : 2 : 0 : 4 : 4t^2)$. In particular, the curves determine a one-dimensional subvariety of the moduli space of genus 2 curves.

4.3 Hyperelliptic Curves of Genus 3 with Real Multiplication by η_7

For $p = 7$, we derive a family of genus 3 hyperelliptic curves

$$X_7 : v^2 = u(u^3 - 1)^2 + 3t,$$

and an endomorphism η of $\mathrm{Jac}(X_7)$ with $\mathbb{Z}[\eta] \cong \mathbb{Z}[\zeta_7 + \zeta_7^{-1}]$ by Proposition 2. Applying Algorithm 5, we derive polynomial maps T and N, which we use with Algorithm 7 to realize η as $\eta : P \mapsto \text{EVALUATE}(P, T, N)$.

5 Applications II: Curves with Cyclotomic Covering

In this section we develop explicit endomorphisms for the one dimensional families of hyperelliptic curves with real multiplication based on cyclotomic coverings, as defined in Tautz, Top, and Verberkmoes [20].

5.1   Construction of the Cyclotomic Covering

Let $n \geq 2$, and let ρ_n and ρ_{2n} be primitive n^{th} and $2n^{\text{th}}$ roots of unity over k such that $\rho_{2n}^2 = \rho_n$; also set $\tau_n = \rho_n + \rho_n^{-1}$. Consider the family of hyperelliptic curves of genus n over k in one free parameter t defined by

$$C_n : y^2 = x(x^{2n} + tx^n + 1).$$

The curve C_n has an automorphism ζ of order $2n$ and an involution σ, defined by

$$\zeta : (x,y) \longmapsto (\rho_n x, \rho_{2n} y) \quad \text{and} \quad \sigma : (x,y) \longmapsto \left(x^{-1}, x^{-(n+1)}y\right),$$

respectively; note that ζ^n is the hyperelliptic involution $(x,y) \mapsto (x,-y)$. We define $X_n := C_n/\langle\sigma\rangle$ to be the quotient of C_n by the action of σ. The curve X_n has an an affine model

$$X_n : v^2 = f_n(u) = D_n(u,1) + t,$$

where $D_n(u,1)$ is the nth Dickson polynomial of the first kind with parameter[2] 1, defined recursively by

$$D_n(u,1) = uD_{n-1}(u,1) - D_{n-2}(u,1) \tag{3}$$

for $n \geq 2$, with $D_0(u,1) = 2$ and $D_1(u,1) = u$. Dickson polynomials and their properties are described in [12]; for our purposes, it is enough to know that

$$D_n(u + u^{-1}, 1) = u^n + u^{-n} \tag{4}$$

(this is easily verified by induction), which further implies

$$D_{nm}(u,1) = D_n(D_m(u,1), 1). \tag{5}$$

Remark 13. When n is odd, our curves C_n and X_n coincide with the curves $\mathcal{D}_n$ and $\mathcal{C}_n$ of [20]; for even n, our families instead coincide with the curves described in the remark of [20, page 1058].

The quotient projection $\pi : C_n \to X_n$ is a covering of degree 2. Equation (4) above shows that it is defined by

$$\pi : (x,y) \longmapsto (u,v) = (x + x^{-1}, x^{-(n+1/2)}y).$$

The automorphism ζ of C_n induces an endomorphism $\eta = (\pi\circ\zeta)_* \circ \pi^*$ of $\mathrm{Jac}(X_n)$. If n is prime, then Proposition 2 implies that $\mathbb{Z}[\eta] \cong \mathbb{Z}[\zeta_n + \zeta_n^{-1}]$, where ζ_n is an nth root of unity over $\mathbb{Q}$.

---

[2] Dickson polynomials are generally defined with a parameter a in k, by the recurrence

$$D_n(u,a) = uD_{n-1}(u,a) - aD_{n-2}(u,a).$$

It is easily shown that the curve defined by $v^2 = D_n(u,a) + t$ for any nonzero a is a twist of X_n. When $a = 0$, we obtain a one-dimensional family of curves with complex multiplication by $\mathbb{Z}[\zeta_n]$; these curves are described in [16, §6.4].

The endomorphism η is induced by the correspondence $Z := (\pi \circ \zeta \times \pi)(C_n)$ on $X_n \times X_n$, for which we directly compute an affine model

$$Z = V(v_2 - v_1, u_2^2 + u_1^2 - \tau_n u_1 u_2 + \tau_n^2 - 4).$$

Setting $t_1 := \tau_n u$ and $n_1 := u^2 + \tau_n^2 - 4$, we apply Algorithm 5 to obtain maps $T : k[u] \to k[u]$ and $N : k[u] \to k[u]$ such that the endomorphism η is realized by $P \mapsto \text{EVALUATE}(P, T, N)$, using Algorithm 7. The first few t_i and $n_{i,j}$ derived in Algorithm 5 are given in Table 2 below.

Table 2. Cyclotomic covers: t_i and $n_{i,j}$ for $0 \le i \le j \le 3$

t_0	2		$n_{0,3}$	$\tau_n((\tau_n^2 - 3)u^2 - 3(\tau_n^2 - 4))u$
t_1	$\tau_n u$		$n_{1,1}$	$u^2 + \tau_n^2 - 4$
t_2	$(\tau_n^2 - 2)u^2 - 2(\tau_n^2 - 4)$		$n_{1,2}$	$\tau_n(u^2 + \tau_n^2 - 4)u$
t_3	$\tau_n(\tau_n^2 - 3)u^3 - 3\tau_n(\tau_n^2 - 4)u$		$n_{1,3}$	$(\tau^2 - 2)u^4 + (\tau^2 - 4)^2(u^2 - 2)$
$n_{0,0}$	1		$n_{2,2}$	$(u^2 + \tau_n^2 - 4)^2$
$n_{0,1}$	$\tau_n u$		$n_{2,3}$	$\tau_n(u^2 + \tau_n^2 - 4)^2)u$
$n_{0,2}$	$(\tau_n^2 - 2)u^2 - 2(\tau_n^2 - 4)$		$n_{3,3}$	$(u^2 + \tau_n^2 - 4)^3$

The elliptic curve $C_1 : y^2 = x(x^2 + tx + 1)$ is obviously covered by C_n, and is therefore a factor of $\text{Jac}(C_n)$. The following analogue of Theorem 9 holds for this cyclotomic family, and is proved similarly.

Proposition 14. *The Jacobian $\text{Jac}(C_n)$ is isogenous to $C_1 \times \text{Jac}(X_n)^2$ for n prime, and its endomorphism ring contains an order in $\mathbb{Q} \times \mathbb{M}_2(\mathbb{Q}(\eta_n))$.*

Remark 15. If n is a prime other than 5, then [20, Corollary 6] implies that $\text{Jac}(X_n)$ is absolutely simple for general values of t over a field of characteristic 0. For $n = 5$, we find that the condition of Stoll [18] (see [5, §14.4]) for $\text{Jac}(X_5)$ to be absolutely simple is satisfied by X_5 with $t = 1$ at $p = 11$. Conversely, if $n = pm$, for $p > 2$ and $m > 1$, then identity (5) above gives a covering $X_n \to X_p$ of degree m, defined by $(u, v) \mapsto (D_m(u, 1), v)$. It follows that $\text{Jac}(X_n)$ has a factor isogenous to $\text{Jac}(X_p)$, and so is not simple.

5.2 Hyperelliptic Curves of Genus 2 with Real Multiplication by η_5

Consider the case $n = 5$. Equation (3) shows that $D_5(u, 1) = u^5 - 5u^3 + 5u$, so the curve $X_5 = C_5/\langle \sigma \rangle$ is the curve of genus 2 defined by the affine model

$$X_5 : v^2 = f_5(u) = u^5 - 5u^3 + 5u + t.$$

Each point on $\text{Jac}(X_5)$ has a representative in the form $(a(u), v - b(u))$, with $\deg a = 2$ and $\deg b = 1$; so suppose $a(u) = a_2 u^2 + a_1 u + a_0$ and $b(u) = b_1 u + b_0$. Applying Algorithm 5, we obtain maps T and N such that

$$N(a) = a_2^2 u^4 + a_2 a_1 \tau_5 u^3 + (2a_2^2(\tau_5^2 - 4) + a_1^2 + a_2 a_0(\tau_5^2 - 2))u^2$$
$$\quad + a_1(a_2(\tau_5^2 - 4) + a_0)\tau_5 u + ((\tau_5^2 - 4)(a_2^2(\tau_5^2 - 4) + a_1^2 - 2a_2 a_0) + a_0^2),$$
$$N(b) = b_1^2 u^2 + b_1 b_0 \tau_5 u + b_1^2(\tau_5^2 - 4) + b_0^2, \text{ and}$$
$$T(b) = \tau_5 b_1 u + 2b_0.$$

The endomorphism η is then explicitly realized by $\eta : P \mapsto \text{EVALUATE}(P, T, N)$, using Algorithm 7.

Remark 16. The weighted projective Igusa invariants of the generic curve are:

$$(140 : 550 : 640t^2 - 60 : 22400t^2 - 77725 : 256t^4 - 2048t^2 + 4096).$$

In particular, this family corresponds to a one-dimensional subvariety in the moduli space.

5.3   Hyperelliptic Curves of Genus 3 with Real Multiplication by η_7

In the case $n = 7$, we derive a family of curves

$$X_7 : v^2 = u^7 - 7u^5 + 14u^3 - 7u + t,$$

and an endomorphism η of $\text{Jac}(X_7)$ with $\mathbb{Z}[\eta] \cong \mathbb{Z}[\zeta_7 + \zeta_7^{-1}]$ by Proposition 2. Applying Algorithm 5, we derive polynomial maps T and N, which we may then use with Algorithm 7 to realize η as $\eta : P \mapsto \text{EVALUATE}(P, T, N)$.

6   Applications III: Curves from Elliptic Coverings

In [14], Mestre constructs a series of two dimensional families of hyperelliptic curves with explicit real endomorphisms, which are similarly realized by explicit correspondences. For the case η_5, Takashima [19] independently developed an explicit algorithm and complexity analysis for two and three dimensional families[3] referred to as Mestre–Hashimoto and Brumer–Hashimoto (see [8]).

6.1   Hyperelliptic Curves of Genus 2 with Real Multiplication by η_5

Let s and t be free parameters, and consider the family of curves defined by

$$X_5 : v^2 = f_5(u) = u^4(u - s) - s(u + 1)(u - s)^3 + s^3 u^3 - tu^2(u - s)^2.$$

Mestre shows that $\text{Jac}(X_5)$ has an endomorphism η satisfying $\eta^2 + \eta - 1 = 0$, induced by the correspondence Z with affine model

$$Z = V(v_2 - v_1, u_1^2 u_2^2 + s(s - 1)u_1 u_2 - s^2(u_1 - u_2) + s^3).$$

We will derive an explicit form for η. Since X_5 is a curve of genus 2, each point of $\text{Jac}(X_5)$ may be represented by an ideal $(a(u), v - b(u))$ with $a = a_2 u^2 + a_1 u + a_0$ and $b = b_1 u + b_0$. Setting $t_1 = -s((s - 1)u_2 - s)/u_2^2$ and $n_1 = s^2(u_2 + s)/u_2^2$, we apply Algorithm 5 to derive maps T and N such that

$$N(a) = a_2^2 n_{2,2} + a_2 a_1 n_{1,2} + a_1^2 n_{1,1} + a_2 a_0 a_2 n_{0,2} + a_1 a_0 n_{0,1} + a_0^2 n_{0,0},$$
$$N(b) = b_1^2 n_{1,1} + b_1 b_0 n_{0,1} + b_0^2 n_{0,0}, \text{ and}$$
$$T(b) = -b_1 s((s - 1)u - s)/u^2 + 2b_0,$$

with the $n_{i,j}$ given in the table below.

---

[3] The moduli of genus 2 curves with real multiplication by η_5 form a two dimensional subvariety of the moduli space of genus 2 curves, so this three dimensional family contains one dimensional fibres of geometrically isomorphic curves.

$n_{0,0}$	1	$n_{1,1}$	$s^2(u+s)/u^2$
$n_{0,1}$	$-s((s-1)u-s)/u^2$	$n_{1,2}$	$-s^3(u+s)((s-1)u-s)/u^4$
$n_{0,2}$	$s^2(((s-1)u-s)^2-2u^2(u+s))/u^4$	$n_{2,2}$	$s^4(u+s)^2/u^4$

The endomorphism η is then explicitly realized by $\eta : P \mapsto \text{EVALUATE}(P, T, N)$, using Algorithm 7.

6.2 Hyperelliptic Curves of Genus 3 with Real Multiplication by η_7

Let s and t be free parameters, and consider the family of hyperelliptic genus 3 curves defined by

$$X_7 : v^2 = f_7(u) = \phi_7(u) - t\,\psi_7(u)^2$$

where $\psi_7(u) := u(u - s^3 + s^2)(u - s^2 + s)$ and

$$
\begin{aligned}
\phi_7(u) := {} & u\psi_7(u)^2 + s(s-1)(s^2 - s + 1)(s^3 + 2s^2 - 5s + 1)u^5 \\
& - s^3(s-1)^2(6s^4 - 11s^3 + 12s^2 - 11s - 1)u^4 \\
& + s^4(s-1)^3(s^2 - s - 1)(s^3 + 2s^2 + 6s + 1)u^3 \\
& - s^6(s-1)^4(s+1)(3s^2 - 5s - 3)u^2 \\
& + s^8(s-1)^5(s^2 - 3s - 3)u + s^{10}(s-1)^6.
\end{aligned}
$$

Mestre shows that $\text{Jac}(X_7)$ has an endomorphism η satisfying $\eta^3 + \eta^2 - 2\eta - 1 = 0$, induced by the correspondence $Z = V(v_2 - v_1, E)$ on $X_7 \times X_7$, where

$$E = u_1^2 u_2^2 - s^2(s-1)(s^2 - s - 1)u_1 u_2 - s^4(s-1)^2(u_1 + u_2) + s^6(s-1)^3.$$

Since X_7 is a curve of genus 3, each point on $\text{Jac}(X_7)$ may be represented by an ideal $(a(u), v - b(u))$, where a and b are polynomials of degree 3 and 2, respectively. Setting

$$
\begin{aligned}
t_1 &= s^2(s-1)((s^2 - s - 1)u + s^2(s-1))/u^2 \quad \text{and} \\
n_1 &= -s^4(s-1)^2(u + s^2(s-1))/u^2,
\end{aligned}
$$

we apply Algorithm 5 to derive maps T and N from $k[u]$ into $k(u)$; the elements $n_{i,j}$ computed by Algorithm 5 are given in the table below.

$n_{0,0}$	1
$n_{0,1}$	$s^2(s-1)((s^2 - s - 1)u + s^2(s-1))/u^2$
$n_{0,2}$	$(s-1)^2 s^4(2u^3 + (s^4 - 3s^2 + 2s + 1)u^2$
	$\qquad + 2(s-1)(s^2 - s - 1)s^2 u + (s-1)^2 s^4)/u^4$
$n_{1,1}$	$-s^4(s-1)^2(u + s^2(s-1))/u^2$
$n_{1,2}$	$(s^6(s-1)^3(s^2 - s - 1)u^2 + s^9(s-1)^5 u + s^{10}(s-1)^5)/u^4$
$n_{2,2}$	$s^8(s-1)^4(u + s^2(s-1))^2/u^4$

The endomorphism η is then explicitly realized by $\eta : P \mapsto \text{EVALUATE}(P, T, N)$, using Algorithm 7.

7   Construction of Curves of Cryptographic Proportions

The curves presented here not only admit efficiently computable endomorphisms, they also permit random selection of curve parameters in a large family. For example, let $\mathbb{F}_{5^{37}} = \mathbb{F}_5[\xi]$ be extension of $\mathbb{F}_5$ such that $\xi^{37} + 4\xi^2 + 3\xi + 3 = 0$, and take

$$t = 3\xi^5 + \xi^4 + 3\xi^3 + \xi^2 + 2\xi + 3.$$

This gives a curve $X : v^2 = u(u^2 - 1)^2 + t$ in the Artin–Schreier family whose Jacobian has nearly prime group order

$$|\mathrm{Jac}(X)(\mathbb{F}_5[\xi])| = 5 \cdot n,$$

with prime cofactor

$$n = 105879118406770168967463702534053156545601790341311.$$

Such curves are amenable to efficient point counting techniques using Monsky-Washnitzer cohomology [7,9]. If y is a square root of t, then $(0, y)$ is a point on X; let $P = [(u, v - y)]$ be the corresponding point on J. Then $Q = [5](P)$ generates a cyclic group of order n, on which $[\eta]$ satisfies

$$([\eta_5]^2 + [\eta_5] - 1)(Q) = [(1)]$$

and in particular, $[\eta_5](5P) = [m](5P)$, where

$$m = 33689405394100488551926661702895689897261990766730 1$$

is one of the two roots of $x^2 + x - 1 \bmod n$.

Acknowledgement. The authors thank K. Takashima for providing an advance draft of his article [19], and for references to the work of Hashimoto.

References

1. D. Cantor, Computing in the Jacobian of a hyperelliptic curve, *Math. Comp.*, **48** (1987), 95–101.
2. M. Ciet, T. Lange, F. Sica, and J.-J. Quisquater, Improved algorithms for efficient arithmetic on elliptic curves using fast endomorphisms, *Advances in Cryptology — EUROCRYPT 2003*, 387–400, LNCS **2656**, Springer, Berlin, 2003.
3. I. Duursma and K. Sakurai, Efficient algorithms for the Jacobian variety of hyperelliptic curves $y^2 = x^p - x + 1$ over a finite field of odd characteristic p. *Coding theory, cryptography and related areas (Guanajuato, 1998)*, 73–89, Springer, Berlin, 2000.
4. J. Ellenberg, Endomorphism algebras of Jacobians. *Advances in Mathematics* **162** (2001), 243–271.
5. J. W. S. Cassels and E. V. Flynn, *Prolegomena to a middlebrow arithmetic of curves of genus 2*, London Mathematical Society Lecture Note Series **230**, Cambridge University Press, Cambridge, 1996.

6. R. Gallant, R. Lambert, and S. Vanstone, Faster point multiplication on elliptic curves with efficient endomorphisms, *Advances in Cryptology — CRYPTO 2001*, 190–200, LNCS **2139**, Springer, 2001.
7. P. Gaudry and N. Gürel, Counting points in medium characteristic using Kedlaya's algorithm, *Experiment. Math.* **12** (2003), no. 4, 395–402.
8. K.-I. Hashimoto, On Brumer's family of RM-curves of genus two. *Tohoku Math. J. (2)* **52**, (2000), no. 4, 475–488.
9. K. Kedlaya, Counting points on hyperelliptic curves using Monsky-Washnitzer cohomology, *J. Ramanujan Math. Soc.* **16** (2001), no. 4, 323–338.
10. N. Koblitz, CM-curves with good cryptographic properties. *Advances in Cryptology — CRYPTO '91*, 279–287, LNCS **576**, Springer, 1992.
11. T. Lange, *Efficient arithmetic on hyperelliptic Koblitz curves*, Ph.D. Thesis, 2001.
12. R. Lidl, G. L. Mullen and G. Turnwald, *Dickson polynomials*, Pitman monographs and surveys in pure and applied mathematics **65**, Longman Scientific & Technical, 1993.
13. A. J. Menezes and S. A. Vanstone, The implementation of elliptic curve cryptosystems, *Advances in Cryptology – Auscrypt 1990*, 2–13, *LNCS* **453**, Springer-Verlag, 1990.
14. J.-F. Mestre, Familles de courbes hyperelliptiques à multiplications réelles, *Arithmetic algebraic geometry (Texel, 1989)*, 193–208, *Progress in Math.*, **89**, Birkhäuser Boston, Boston, MA, 1991.
15. Y.-H. Park, S. Jeong, J. Lim, Speeding up point multiplication on hyperelliptic curves with efficiently-computable endomorphisms, *Advances in cryptology — EUROCRYPT 2002 (Amsterdam)*, 197–208, LNCS **2332**, Springer, 2002.
16. B. A. Smith, *Explicit endomorphisms and correspondences*, Ph.D. Thesis, The University of Sydney, 2005.
17. J. A. Solinas, Efficient arithmetic on Koblitz curves, *Des. Codes Cryptogr*, **19** (2000), no. 2-3, 195–249.
18. M. Stoll, Two simple 2-dimensional abelian varieties defined over Q with Mordell-Weil group of rank at least 19, *C. R. Acad. Sci. Paris Sér. I Math.* **321** (1995), no. 10, 1341–1345.
19. K. Takashima, A new type of fast endomorphisms on Jacobians of hyperelliptic curves and their cryptographic application. *IEICE Trans. Fundamentals*, **E89-A** (2006), no. 1, pp. 124-133.
20. W. Tautz, J. Top, A. Verberkmoes, Explicit hyperelliptic curves with real multiplication and permutation polynomials, *Canad. J. Math.*, **43** (1991), no. 5, 1055–1064.
21. A. Weil, On some exponential sums, *Proc. Nat. Acad. Sci.*, **34**, (1948), 204–207.

Construction of Rational Points on Elliptic Curves over Finite Fields

Andrew Shallue[1] and Christiaan E. van de Woestijne[2]

[1] University of Wisconsin-Madison, Mathematics Department
480 Lincoln Dr, Madison, WI 53706-1388 USA
shallue@math.wisc.edu
[2] Universiteit Leiden, Mathematisch Instituut
Postbus 9512, 2300 RA Leiden, The Netherlands
cvdwoest@math.LeidenUniv.nl

Abstract. We give a deterministic polynomial-time algorithm that computes a nontrivial rational point on an elliptic curve over a finite field, given a Weierstrass equation for the curve. For this, we reduce the problem to the task of finding a rational point on a curve of genus zero.

1 Introduction

Elliptic curves over finite fields have been in the centre of attention of cryptographers since the invention of ECC, and in that of number theorists for a much longer time. It is not very hard to show that, unless the base field is extremely small, such curves always have rational points other than O, the point at infinity. However, it is a different question how to construct such rational points efficiently.

Until now, this was possible only using an obvious probabilistic method: given an equation for the curve, substitute random values for all coordinates but one and see if the remaining univariate equation can be solved for the last coordinate. If so, a probabilistic polynomial factorisation algorithm will give the last coordinate and a rational point has been found. The challenge for a deterministic algorithm has been up at least since 1985, when R. Schoof posed it in [8].

In a recent publication [11], however, M. Skałba proved that, given a cubic polynomial $f(x) = x^3 + Ax + B$ over a field F with characteristic unequal to 2 or 3, with $A \neq 0$, we have the identity

$$f(X_1(t^2))f(X_2(t^2))f(X_3(t^2)) = U(t)^2 \tag{1}$$

for some nonconstant univariate rational functions X_1, X_2, X_3, U over F. Such functions are given explicitly in his paper [11, Theorem 1]. We do not reproduce them here, as both their degree and their coefficients are large; if $X_1 X_2 X_3 = N/D$ for coprime polynomials N and D in $F[t]$, then $\deg N \leq 26$ and $\deg D \leq 25$, depending on the characteristic of F.

Now assume that $\mathbb{F}$ is a finite field and that the curve E is defined over $\mathbb{F}$ by the equation $y^2 = f(x)$, with f as above. The multiplicative group $\mathbb{F}^*$ is cyclic,

F. Hess, S. Pauli, and M. Pohst (Eds.): ANTS 2006, LNCS 4076, pp. 510–524, 2006.

and therefore, as Skałba notes, if we specialise t in (1) to some value t_0 in $\mathbb{F}$, we find that at least one of the $f(X_i(t_0^2))$ is a square in $\mathbb{F}^*$. However, no efficient deterministic algorithm is known to date to take the square root.

In this paper, we show how to go on from this point to obtain a complete efficient deterministic algorithm for constructing rational points on curves given by cubic Weierstrass equations over finite fields. We will reprove Skałba's result to obtain, for the case of finite fields of odd characteristic, a parametrisation as in (1) that is *invertible* as a rational map (Lemmas 6 and 7 below).

The construction of this parametrisation in the case of odd characteristic rests on the ability to solve deterministically and efficiently equations of the form

$$ax^2 + by^2 = c \tag{2}$$

over finite fields, for which an algorithm will be given in Section 2 (Theorem 4).

In Section 2, we also give a deterministic algorithm that, given nonzero elements a_0, a_1, a_2 in a finite field such that their product is the square of a given element, computes a square root of one of them, in polynomial time. It is clear that such an algorithm is the missing step to construct a rational point on E, when an equation of the form (1) is given.

An analogon of (1) for finite fields of characteristic 2 will be used to obtain a point finding algorithm for elliptic curves in this case as well.

The main result is as follows:

Theorem 1. *There exists a deterministic algorithm that, given a finite field $\mathbb{F}$ of q elements and a cubic Weierstrass equation $f(x, y)$ over $\mathbb{F}$:*

(i) *detects if $f(x, y)$ is singular, and if so, computes the singular points and gives a rational parametrisation of all rational points on the curve $f(x, y) = 0$;*

(ii) *if $f(x, y)$ is nonsingular and $|\mathbb{F}| > 5$, computes an explicit rational map ρ from the affine line over $\mathbb{F}$ to an affine threefold V that is given explicitly in terms of the coefficients of f;*

(iii) *given a rational point on the threefold V, computes a rational point on the elliptic curve $E : f(x, y) = 0$, in such a way that at least $(q-4)/8$ rational points on E are obtained from the image of the map ρ, and at least $(q-4)/3$ if $\mathbb{F}$ has characteristic 2;*

and performs all these tasks in time polynomial in $\log q$.

From the proofs in this paper, such an algorithm can be explicitly constructed; the running time of this algorithm is not much worse than that of a probabilistic point generation algorithm. We plan to give an explicit algorithm, with detailed running time bounds, in a forthcoming publication.

After Section 2 on how to solve diagonal quadratic equations, we give some generalities on Weierstrass equations in Section 3 and show how to parametrise the solutions of a *singular* Weierstrass equation in Section 4. The nonsingular case, where the given equation indeed defines an elliptic curve, is split into two cases: in Section 5, we prove Theorem 1 for base fields of odd characteristic, whereas base fields of characteristic 2 are considered in Section 6.

2   Quadratic Equations

Before turning to cubic equations, we first give the necessary algorithms for solving quadratic equations. Theorem 3 is concerned with taking square roots, while Theorem 4 is about equations of the form (2). These results, which are taken from the second author's Ph.D. thesis [12], are new and deterministic efficient algorithms have been unknown to date.

We write $v_2(a)$ to denote the number of factors 2 in a nonzero integer a; if a is a nonzero element of a finite field $\mathbb{F}$, we write $\mathrm{ord}(a)$ to denote the order of a in the multiplicative group $\mathbb{F}^*$.

Lemma 2. *There exists a deterministic algorithm that, given a finite field $\mathbb{F}$ of q elements, and nonzero elements a and z of $\mathbb{F}$ such that either*

(i) $v_2(\mathrm{ord}\, a) < v_2(\mathrm{ord}\, z)$, or
(ii) $\mathrm{ord}\, a$ is odd,

computes a square root of a, in time polynomial in $\log q$.

Proof. We construct a deterministic adaptation of the Tonelli-Shanks algorithm; for the latter, see Section 1.5.1 in [5], for example.

It is easy to prove that to compute a square root of a nonzero element $a \in \mathbb{F}$, it is sufficient to have a generator z of the 2-Sylow subgroup of $\mathbb{F}^*$. Usually, such a generator is obtained by guessing a nonsquare element n and computing $z = n^u$, where we write $q - 1 = 2^e \cdot u$ such that u is an odd integer; this is the only probabilistic part of the Tonelli-Shanks algorithm.

The proof is as follows: a^u is in the 2-Sylow subgroup, and hence there exists an integer k such that $z^k = a^u$. The integer k is even if and only if a is a square in $\mathbb{F}$; furthermore, it is clear that $z^{k/2}$ is a square root of a^u, and from a square root of a^u it is easy to compute a square root of a, because u is odd and hence a^{u+1} is an obvious square. Thus, the real task of the Tonelli-Shanks algorithm is the computation of the integer k.

However, the only thing that is used about z is the fact that

$$a^u = z^k$$

for some even integer k; and for such a k to exist, it is only necessary that either $a^u = 1$, or the group generated by a^u inside 2-Syl$\,\mathbb{F}^*$ is strictly contained in the group generated by z. But these conditions correspond to our assumptions $v_2(\mathrm{ord}\, a) = 0$ and $v_2(\mathrm{ord}\, a) < v_2(\mathrm{ord}\, z)$, respectively. Therefore, if instead of a 2-Sylow subgroup generator we use any element whose order contains enough factors 2, the Tonelli-Shanks algorithm as given in [5] works just as well, while the nondeterministic part of guessing a nonsquare element is eliminated.   ◆

Theorem 3. *There exists a deterministic algorithm that, given a finite field $\mathbb{F}$ of q elements, and nonzero elements a_0, a_1, a_2, b of $\mathbb{F}$ such that $a_0 a_1 a_2 = b^2$, returns an i in $\{0, 1, 2\}$ and a square root of a_i, in time polynomial in $\log q$.*

Proof. After changing the order of the a_i, we may assume that $v_2(\operatorname{ord} a_0) \geq v_2(\operatorname{ord} a_1) \geq v_2(\operatorname{ord} a_2)$. If $v_2(\operatorname{ord} a_0) > v_2(\operatorname{ord} a_1)$, then by Lemma 2 we may use a_0 as a substitute for a 2-Sylow subgroup generator, and compute a square root of a_1; and if $v_2(\operatorname{ord} a_1) > v_2(\operatorname{ord} a_2)$, the same holds for a_1 and a_2.

Thus, consider the case where $v_2(\operatorname{ord} a_0) = v_2(\operatorname{ord} a_1) = v_2(\operatorname{ord} a_2)$. Then it follows that, say, $a_0 a_1$ has fewer factors 2 in its order than a_2, and we can compute $\sqrt{a_0 a_1}$; but by the given relation among the a_i, we have $a_0 a_1 = a_2/b^2$, and we compute a square root of a_2.   ◆

Theorem 4. *There exists a deterministic algorithm that, given a finite field $\mathbb{F}$ of q elements, and nonzero elements a, b, c of $\mathbb{F}$, computes $x, y \in \mathbb{F}$ such that*

$$ax^2 + by^2 = c.$$

Proof. We may of course assume that $c = 1$. Now if $v_2(\operatorname{ord}(a)) > v_2(\operatorname{ord}(b))$, we can use the algorithm in Lemma 2 to take a square root of b, and the problem is solved by taking $x = 0$ and $y = 1/\sqrt{b}$; and analogously if b has the larger order. If $v_2(\operatorname{ord}(a)) = v_2(\operatorname{ord}(b)) =_{\text{def}} w$, we distinguish three cases: $w = 0$, $w = 1$, and $w > 1$.

If $w = 0$, then we can still compute square roots of both a and b by means of Lemma 2, and we are done. If $w > 1$, then $v_2(\operatorname{ord}(-ab)) < w$, so that, after computing $\sqrt{-ab}$, we may assume $b = -a$. The equation $ax^2 - ay^2 = 1$ is easily solved by putting $x + y = 1$ and $x - y = 1/a$ and solving the linear system.

The case $w = 1$ is the hardest. Both $-a$ and $-b$ have odd order, so we may take their square roots by Lemma 2 and obtain the equation $-x^2 - y^2 = 1$. One sees that this is equivalent to

$$x^2 + y^2 + z^2 = 0.$$

For this, we developed a fast algorithm in Section 5.5 of [12]. A slower, but also deterministic, algorithm for this problem can be found in [4], and also the algorithm given in the second proof of Corollary 1 in [11] can be adapted to this case, by using Lemma 2.   ◆

Remarks. It is well known that (2) is always solvable; this follows already from the fact that the cardinalities of the sets $\{ax^2 \mid x \in \mathbb{F}\}$ and $\{c - by^2 \mid y \in \mathbb{F}\}$ add up to more than q, and therefore these sets must meet.

The algorithm for solving (2) given above is a special case of the main algorithm from [12]; this algorithm can solve diagonal equations of the form

$$a_1 x_1^n + \ldots + a_n x_n^n = b$$

over finite fields.

In finite fields of characteristic 2, the above results are trivial, since all elements have odd order. However, over such fields many quadratic equations cannot be reduced to the diagonal form (2), and this yields new difficulties. We refer to Section 6 for a discussion of this case.

3   Weierstrass Equations

Let $\mathbb{F}$ be a finite field, let q be its number of elements, and let E be the affine curve given by the *Weierstrass equation*

$$y^2 + a_1 xy + a_3 y = x^3 + a_2 x^2 + a_4 x + a_6, \tag{3}$$

where the a_i are in $\mathbb{F}$. The curve E also has one point at infinity, with homogeneous coordinates $(0 : 1 : 0)$, which is called $\mathcal{O}$.

If E is nonsingular, then the projective closure $\tilde{E}$ of E is a smooth projective curve of genus 1 over $\mathbb{F}$ with a specified rational point, so it is an *elliptic curve* over $\mathbb{F}$, and every elliptic curve over $\mathbb{F}$ may be given in this way [10, Proposition III.3.1]. The set of rational points on $\tilde{E}$ has a natural abelian group structure, with the point $\mathcal{O}$ as identity element.

We will be interested in methods to construct rational points on $\tilde{E}$ other than $\mathcal{O}$, or to show that no other points exist. By Hasse's bound [10, V.1.4], we know that the number N of rational points on $\tilde{E}$ satisfies

$$|q + 1 - N| \le 2\sqrt{q}.$$

From this, it is easily verified that $\tilde{E}$ has at least 2 rational points whenever $q \ge 5$. On the other hand, if $q \le 4$, curves over $\mathbb{F}$ exist with only the trivial rational point $\mathcal{O}$, such as the curve $y^2 = x^3 - x - 1$ over $\mathbb{F}_3$, and the curve $y^2 + y = x^3 + \alpha$ over $\mathbb{F}_4 = \mathbb{F}_2(\alpha)$.

Normal Forms. The equation (3) may be simplified depending on the characteristic of the base field. We give these forms in detail as we will use their properties later on; these formulas are given in Section III.1 and Appendix A of [10].

If the characteristic of $\mathbb{F}$ is not 2 or 3, then a linear change of coordinates transforms (3) into

$$y^2 = x^3 + Bx + C =_{\text{def}} f(x). \tag{4}$$

For this form of the equation, the important associated quantities Δ (the *discriminant*) and j (the *j-invariant*) are easily computed: we have

$$\Delta = -16(4B^3 + 27C^2), \qquad j = -1728(4B)^3/\Delta.$$

Now E is singular if and only if $\Delta = 0$, and thus if and only if the right hand side $f(x)$ of (4) has a repeated zero; it has j-invariant 0 if and only if $\Delta \ne 0$ and $B = 0$.

In characteristic 3, we must admit a third coefficient; we can transform (3) into

$$y^2 = x^3 + Ax^2 + Bx + C =_{\text{def}} f(x), \tag{5}$$

with associated quantities

$$\Delta = A^2 B^2 - A^3 C - B^3, \qquad j = A^2/\Delta.$$

Again, E is singular if and only if f has a double zero. Also, we find that for a nonsingular equation we have $j = 0$ if and only if $A = 0$.

In characteristic 2, no coefficient of (3) can be omitted in all cases. However, we can obtain one of the following two normal forms, depending on whether a_1 is zero:

$$Y^2 + a_3Y = X^3 + a_4X + a_6 \qquad \text{if } a_1 = 0 \text{ initially,} \qquad (6)$$
$$Y^2 + XY = X^3 + a_2X^2 + a_6 \qquad \text{if } a_1 \neq 0 \text{ initially.} \qquad (7)$$

In these normal forms, we have $\Delta = (a_3)^4$ and $\Delta = a_6$, respectively, which gives an easy criterion for singularity of E. Furthermore, for nonsingular equations, the two cases correspond to j being respectively zero or nonzero.

4   Singular Weierstrass Equations

For completeness, we show how to detect deterministically whether E is singular and, if it is, how to find points on it. We continue to assume that $\mathbb{F}$ is a finite field, although the only thing we really use in this section is the assumption that the base field is perfect.

If the singularity test is positive, the projective closure $\tilde{E}$ has genus 0 and a unique singular point, which is rational over $\mathbb{F}$ provided $\mathbb{F}$ is perfect. We can use this point to find a rational parametrisation of all *nonsingular* points on E. It follows that the construction of rational points on a singular E is easy. Furthermore, the constructions given below give rise to efficient deterministic algorithms, whenever the operations of the field $\mathbb{F}$ are deterministically and efficiently computable, including the operation of taking a pth root if char $\mathbb{F} = p$.

We distinguish the cases of characteristic equal to 2 and unequal to 2.

Odd Characteristic. Let char $\mathbb{F}$ be unequal to 2, and let E be given by $y^2 = f(x)$ for some cubic polynomial f over $\mathbb{F}$. If f has a double zero x_2, then $(x_2, 0)$ is the unique singular point on E. Such a double zero must be in $\mathbb{F}$, as f has degree 3, and also the third zero of f must be rational.

Let $d = \gcd(f, f')$, where f' is the derivative of f. If d is constant, then f does not have a double zero and E is nonsingular. If d is linear, then its unique zero gives the double zero x_2. If d is quadratic, then char $\mathbb{F} \neq 3$ and f has a triple zero, which is equal to the unique zero of the linear polynomial d', the derivative of d. If d is cubic, then char $\mathbb{F} = 3$ and f has a triple zero $x_2 = \sqrt[3]{C} = C^{3^{m-1}}$, where m is the order of 3 modulo $|\mathbb{F}| - 1$.

Assume E is singular; by an $\mathbb{F}$-linear change of variables, we may assume that the singularity is at $(0,0)$, and hence E is given by $y^2 = x^3 + Ax^2$ for some $A \in \mathbb{F}$. Now we parametrise E by projecting lines from the singular point: any such line has the form $y = \ell x$ with $\ell \in \mathbb{F}$, and it intersects E twice in $(0,0)$ and once more in $(\ell^2 - A, \ell^3 - A\ell)$. This provides a rational parametrisation of E, which is clearly computable efficiently and deterministically.

Characteristic 2. Now let char $\mathbb{F}$ be 2, and let E be given by the generic cubic Weierstrass equation (3). We have $\frac{\partial}{\partial y} = a_1x + a_3$, and hence E can be singular in two ways.

The first is to have $a_1 = a_3 = 0$; we get $\frac{\partial}{\partial x} = x^2 + a_4$, and the singular point will be $(\sqrt{a_4}, \sqrt{a_6})$, which we move to $(0,0)$ by a translation. We have already seen that the equation becomes $y^2 = x^3 + Ax^2$ for some $A \in \mathbb{F}$. We parametrise E just as in the case of characteristic not 2, and find that $\ell \mapsto (\ell^2 + A, \ell^3 + A\ell)$ is a rational parametrisation, computable efficiently and deterministically.

The second has $a_1 \neq 0$, and there we may assume $a_3 = a_4 = 0$ and $a_1 = 1$ by linear substitutions; by the equation $0 = \frac{\partial}{\partial x} = y + x^2$, we find that E has a singularity at $(0,0)$ if and only if in addition $a_6 = 0$, and that E is nonsingular otherwise. Assume E is singular; we now get the equation $y^2 + xy = x^3 + Ax^2$, for some $A \in \mathbb{F}$. The same way of parametrising shows that $\ell \mapsto (\ell^2 + \ell + A, \ell^3 + \ell^2 + A\ell)$ is a rational parametrisation, computable efficiently and deterministically.

Remark. For a singular Weierstrass equation, there even exists a parametrisation that is also a group homomorphism, but this map uses another affine patch of the equation and need not always be defined over the base field $\mathbb{F}$ (see Proposition 2.5 in [10]).

5   Elliptic Curves in Odd Characteristic

In this section, we prove Theorem 1 under the assumption that the base field $\mathbb{F}$ is a finite field of odd characteristic and that E is the curve given by a nonsingular Weierstrass equation (4) or (5). In particular, we let f be a cubic monic polynomial over $\mathbb{F}$ without double zeros. The considerations up to Lemma 7 in fact work over any field of characteristic not 2.

Let V denote the threefold

$$f(x_1)f(x_2)f(x_3) = y^2, \tag{8}$$

which, geometrically speaking, is the quotient of $E \times E \times E$ by the action of a Klein 4-group of automorphisms, namely those automorphisms that act as -1 on two components and as the identity on the third. We will obtain an explicit birational map from the affine line to a curve on V; see Lemmas 6 and 7 below.

Let $R = \mathbb{F}[x]/(f)$ be the residue class ring of polynomials over $\mathbb{F}$ modulo f; as f has no multiple zeros, the ring R is a finite étale algebra over $\mathbb{F}$ (cf. [3], Section V.6, especially Theorem 4 in V.6.7, and Section V.8). We denote by θ the class of x modulo f; thus θ generates R as an $\mathbb{F}$-algebra. If g is a polynomial in $\mathbb{F}[x]$ of degree d, then the *homogenisation* $g^{\mathrm{hom}} \in \mathbb{F}[x,y]$ of g is defined to be $y^d g(x/y)$.

Lemma 5. *For any $u, v, w \in \mathbb{F}$ satisfying $u + v + w + A = 0$, we have*

$$f(u)f(v)f(w) = (uv + uw + vw - B)^3 f\left(\frac{uvw + C}{uv + uw + vw - B}\right). \tag{9}$$

Proof. Let $\phi : \mathbb{F}^3 \to R$ be the map sending (u, v, w) to $(u - \theta)(v - \theta)(w - \theta)$. For any $u, v, w \in \mathbb{F}$, we have

$$\phi(u, v, w) = uvw - (uv + uw + vw)\theta + (u + v + w)\theta^2 - \theta^3$$

$$= (uvw + C) - (uv + uw + vw - B)\theta + (u + v + w + A)\theta^2, \quad (10)$$

because $f(\theta) = 0$.

Let H_A be the subspace of $\mathbb{F}^3$ of triples (u, v, w) satisfying $u + v + w + A = 0$. Then ϕ maps H_A into the subspace R_{lin} of R of elements that are *linear* in θ. Now if $\alpha - \beta\theta \in R_{\text{lin}}$, with $\alpha, \beta \in \mathbb{F}$, then we have

$$\text{Norm}_{R/\mathbb{F}}(\alpha - \beta\theta) = f^{\text{hom}}(\alpha, \beta) = \beta^3 f\left(\frac{\alpha}{\beta}\right). \quad (11)$$

In particular, $\text{Norm}(\alpha - \theta) = f(\alpha)$. Thus by taking norms, equation (10) is mapped to (9). ◆

Remark. The formula in the Lemma is a bit misleading in the sense that u, v, w will not perform the same functions as x_1, x_2, x_3 in (8).

Lemma 6. *Put* $h(u, v) = u^2 + uv + v^2 + A(u + v) + B$, *and define*

$$S : y^2 h(u, v) = -f(u), \quad (12)$$

$$\psi : (u, v, y) \mapsto \left(v, -A - u - v, \, u + y^2, \, f(u + y^2)h(u, v)\, y^{-1}\right). \quad (13)$$

Then ψ *is a rational map from the surface* S *to* V *that is invertible on its image.*

Proof. We break the symmetry in (9) by putting $w = -A - u - v$. We find $uv + uw + vw - B = -u^2 - uv - v^2 - A(u + v) - B = -h(u, v)$, and $uvw + C = u(uv + uw + vw - B) + u^2(-v - w) + uB + C = -uh(u, v) + f(u)$.

Now let (u, v, y) be a rational point on S such that $f(u) \neq 0$; it follows that $y \neq 0$ and $h(u, v) \neq 0$, as well. Then applying Lemma 5 with u, v, and $-A - u - v$ and using the equation of S twice gives us

$$f(v)f(-A - u - v)y^2 = h(u, v)^2 f\left(\frac{-uh(u, v) + f(u)}{-h(u, v)}\right) = h(u, v)^2 f(u + y^2). \quad (14)$$

We multiply by $f(u + y^2)$ and divide by y^2 to see that we have a rational point on the threefold V.

From the definition of the map, it is clear that u, v, y can be computed from the image of (u, v, y) on V, so that ψ is invertible on its image. ◆

Lemma 7. *There exists a deterministic algorithm that, given a finite field* $\mathbb{F}$ *of* q *elements, where* q *is odd, a nonsingular cubic Weierstrass equation* $y^2 = f(x)$ *over* $\mathbb{F}$, *and an element* $u \in \mathbb{F}$ *such that*

$$f(u) \neq 0 \quad \text{and} \quad \tfrac{3}{4}u^2 + \tfrac{1}{2}Au + B - \tfrac{1}{4}A^2 \neq 0,$$

computes a rational map

$$\phi : \mathbb{A}^1 \to S$$

defined over $\mathbb{F}$ *that is invertible on its image, in time polynomial in* $\log q$. *Here the surface* S *is as defined in* (12).

Proof. Note that we may assume $A = 0$ whenever char $\mathbb{F} \neq 3$; this could facilitate reading the proof.

We fix a $u \in \mathbb{F}$ that satisfies the requirements given above; then the equation (12) of the surface S specialises to a nondegenerate quadratic equation

$$\left[y(v + \tfrac{1}{2}u + \tfrac{1}{2}A)\right]^2 + \left[\tfrac{3}{4}u^2 + \tfrac{1}{2}Au + B - \tfrac{1}{4}A^2\right] y^2 = -f(u), \qquad (15)$$

which is of the form (2) for the variables $z = y(v + \tfrac{1}{2}u + \tfrac{1}{2}A)$ and y.

Now use Theorem 4 to compute a rational point (z_0, y_0) on (15), and let $t \mapsto (\alpha(t), \beta(t))$ be the corresponding rational parametrisation of the conic (15), still for the variables (z, y) (see [9, Sect. 1.2] or [6, Sect. 1.1]). We have $v = z/y - u/2 - A/2$; therefore the map

$$\phi : t \mapsto \left(u, \frac{\alpha(t)}{\beta(t)} - \frac{u}{2} - \frac{A}{2}, \beta(t)\right) \qquad (16)$$

parametrises all rational points on S with the given u-coordinate, except $(u, z_0/y_0 - u/2 - A/2, y_0)$, because this point corresponds to $t = \infty$.   ◆

After having given the ingredients of the construction of rational points on the threefold V, we ask ourselves how many rational points will be found in this way. The bound of $(q-4)/16$ given by Lemma 9 can probably be improved.

Definition 8. *We define two points $P = (x_1, x_2, x_3, y)$ and $P' = (x_1', x_2', x_3', y')$ on V to be* disjoint *if the sets $\{x_1, x_2, x_3\}$ and $\{x_1', x_2', x_3'\}$ are disjoint.*

Lemma 9. *Let $\mathbb{F}$ be a finite field of q elements, let $u_0 \in \mathbb{F}$ satisfy the requirements of Lemma 7, and let $\phi : \mathbb{A}^1 \to S$ be the corresponding map. Let ψ be the map from Lemma 6.*

Then there is a subset $T \subseteq \mathbb{F}$ of cardinality at least $(q-4)/16$, such that for all distinct $t, t' \in T$, the points $\psi \circ \phi(t)$ and $\psi \circ \phi(t')$ are disjoint.

Proof. Let u_0 be as in the Lemma; we fix it for the whole proof. The corresponding map ϕ is well-defined except perhaps in two values of t where $\beta(t) = 0$, and two others where $(\alpha(t), \beta(t))$ lies at infinity. It follows that the image of ϕ contains at least $q-4$ points.

Let $\psi : S \to V$ be the map from Lemma 6; for two points $P = (u_0, v, y)$ and $P' = (u_0, v', y')$ on S, we want to find sufficient conditions for $\psi(P)$ and $\psi(P')$, or, equivalently, the sets $\{v, -A - u_0 - v, u_0 + y^2\}$ and $\{v', -A - u_0 - v', u_0 + y'^2\}$, to be disjoint.

Note that $v \mapsto -A - u - v$ and $y \mapsto -y$ are automorphisms of S; these automorphisms generate a Klein 4-group G. If P and P' share an orbit under G, then $\psi(P)$ and $\psi(P')$ cannot be disjoint. Note there is at most one orbit under G for any given value of y^2, as $u = u_0$ is assumed to be fixed.

Assume now $\psi(P)$ and $\psi(P')$ are not disjoint. A case-by-case analysis shows that y'^2 is equal to one of y^2, $v - u_0$, $-A - 2u_0 - v$, or $-f(u_0)/h(u, u + y^2)$, where (12) is used to derive the last option.

Let us define a graph on the set of G-orbits on S with $u = u_0$ by putting an edge between two distinct orbits X and X' if there are non-disjoint points $P \in \psi(X)$ and $P' \in \psi(X')$. The above reasoning shows that in this graph, every vertex has at most three neighbours. We want to find a maximal set Σ of pairwise nonadjacent vertices, meaning that if $X \neq X' \in \Sigma$, then all points in $\psi(X)$ are disjoint from all points in $\psi(X')$. Such a set Σ can be constructed greedily by selecting any vertex, adding it to Σ, deleting it and its neighbours with all the incident edges from the graph, and repeating this process until no vertices remain. As we include at least every fourth G-orbit, and as the orbits contain at most 4 points, we see that at least a fraction of $1/16$ of the points in the image of ϕ have pairwise disjoint images under ψ. ◆

Proof of Theorem 1 (odd characteristic). Let $\mathbb{F}$ be a finite field of cardinality greater than 5, so that there exists some $u \in \mathbb{F}$ satisfying the conditions of Lemma 7; we fix such a u for the rest of the proof.

We first show how to compute rational points on the elliptic curve E, which we assume to be given by an equation $y^2 = f(x)$, for some cubic polynomial f with no double roots. By composing the maps ψ from Lemma 6 and ϕ from Lemma 7, we can compute rational points on the threefold V. Then, given a rational point $P = (x_1, x_2, x_3, y)$ on V, we apply the algorithm from Theorem 3 to $f(x_i)$ for $i = 1, 2, 3$ to compute a square root c of $f(x_i)$ for, say, $i = i_0$. Having done this, we see that (x_{i_0}, c) is a rational point on the elliptic curve E.

The next question is whether two different points on V can lead to the same point on E. This is rather subtle; it is even the case that one point on V can lead to several points on E, for example when $f(x_i)$ has odd order for $i = 1, 2, 3$. However, it is clear that if two points on V are *disjoint* in the sense defined above, then they can only give rise to different points on E. Indeed, the x-coordinate of the point on E computed from $P = (x_1, x_2, x_3, y)$ is either x_1, x_2, or x_3. We can therefore use Lemma 9 to show that, if we let the argument t of $\psi \circ \phi$ run through all of $\mathbb{F}$, then at least $(q-4)/16$ valid x-coordinates of points on E follow from the obtained rational points on V. This gives $(q-4)/8$ rational points on E, as claimed. ◆

Remark. It is an interesting question whether the surface S given in Lemma 6 is rational over the ground field $\mathbb{F}$. This question is addressed in [7], for any base field of characteristic different from 2. If we homogenise the equation for S given in (15), we obtain a diagonal ternary quadratic form over the function field $\mathbb{F}(u)$, whose coefficients have degrees 0, 2, and 3. Using the notation and definitions given in [7], we see that the equation has minimal *index* 6 if we use the weights $(3, 2, 1)$ for the variables, whereas a rational surface of this form must have index at most 3 for some weight vector. Therefore, unless some factors of the discriminant of the equation are removable, S is not rational over $\mathbb{F}$.

6 Elliptic Curves in Characteristic 2

In this section we complete the proof of Theorem 1 under the assumption that the characteristic of the base field $\mathbb{F}$ is 2 and that E is given by a nonsingular Weierstrass equation.

Recall that by [10, Appendix A] we know that E has a Weierstrass equation of one of two following forms:

$$Y^2 + a_3 Y = X^3 + a_4 X + a_6 \quad \text{if } j(E) = 0,$$
$$Y^2 + XY = X^3 + a_2 X^2 + a_6 \quad \text{if } j(E) \neq 0.$$

In the case when $\mathbb{F}$ is finite of order 2^r, let Tr stand for the trace map from $\mathbb{F}$ to $\mathbb{F}_2$, which is defined by

$$\mathrm{Tr}_{\mathbb{F}/\mathbb{F}_2}(x) := x + x^2 + x^{2^2} + \cdots + x^{2^{r-1}} .$$

For motivation, consider the problem of finding rational points on

$$Y^2 + Y = f(X).$$

Lemma 10. *If f is linear in X, then there exists a deterministic polynomial-time algorithm that returns a point of $Y^2 + Y = f(X)$ over a finite field $\mathbb{F}$.*

Proof. It is well known that the valid X-coordinates are exactly $x \in \mathbb{F}$ satisfying $\mathrm{Tr}(f(x)) = 0$ [2, Sect. 6.6]. First precompute $a \in \mathbb{F}$ such that $\mathrm{Tr}(f(a)) = 1$. Since $x \mapsto \mathrm{Tr}(f(x))$ is a linear map over $\mathbb{F}_2$, we can deterministically compute the required a using linear algebra. Now, one of x or $x + a$ must be a valid X-coordinate.

Given such an x, it remains to solve for Y. Here we have an advantage over the case of odd characteristic in that there exist deterministic polynomial-time algorithms for solving quadratics ([2, Chap. 6], [1, Sect. 7.4]).   ◆

For more general f, the new idea is to look for points on the threefold

$$f(x_1) + f(x_2) + f(x_3) = y^2 + y .$$

Elements of the form $y^2 + y$ are exactly those in $\mathrm{Ker}(\mathrm{Tr})$, and form an index two subgroup of $\mathbb{F}^+$. Thus one of the three terms must itself be of the form $y^2 + y$.

With this in mind, we define

$$g(x) = x^{-2} \cdot (x^3 + a_2 x^2 + a_6), \text{ and}$$
$$h(x) = x^3 + a_4 x + a_6 .$$

Now let V_1 and V_2 be threefolds given by the equations

$$V_1 : g(x) + g(y) + g(z) = w^2 + w$$
$$V_2 : h(x) + h(y) + h(z) = w^2 + a_3 w .$$

These have the same geometric definition as the threefold V given in the previous section.

As in the odd characteristic case, we will construct a computable rational map from a parametrisable surface to the appropriate threefold. Once we have a point on the threefold it will be easy to get rational points on E. The surfaces we need are given by the equations

$$S_1 : x + y + xy(x + y)^{-1} + a_2 = w^2 + w$$
$$S_2 : x^2 y + y^2 x + a_6 = w^2 + a_3 w \ .$$

Lemma 11. *Let $\mathbb{F}$ be a field of characteristic 2. There exist rational maps $\phi_1 :$ $S_1 \to V_1$ and $\phi_2 : S_2 \to V_2$ over $\mathbb{F}$ which are invertible on their images, given by*

$$\phi_1 : (x, y, w) \mapsto (x, \ y, \ xy(x + y)^{-1}, \ w)$$
$$\phi_2 : (x, y, w) \mapsto (x, \ y, \ x + y, \ w) \ .$$

Proof. First consider ϕ_1, the map that will be used in the case when $j(E) \neq 0$. Recall that $g(x) = x + a_2 + a_6 x^{-2}$. We have

$$
\begin{aligned}
g(x) + g(y) + g\left(\frac{xy}{x+y}\right) &= x + y + \frac{xy}{x+y} + 3a_2 + a_6\left(\frac{1}{x} + \frac{1}{y} + \frac{x+y}{xy}\right)^2 \\
&= x + y + \frac{xy}{x+y} + a_2 \\
&= w^2 + w
\end{aligned}
$$

since (x, y, w) is a point on S_1. Hence $(x, \ y, \ xy(x + y)^{-1}, \ w)$ is a point on V_1.

Next consider ϕ_2, the map that will be used when $j(E) = 0$. We have

$$
\begin{aligned}
h(x) + h(y) + h(x + y) &= x^3 + a_4 x + y^3 + a_4 y + (x + y)^3 + a_4(x + y) + 3a_6 \\
&= x^2 y + y^2 x + a_6 \\
&= w^2 + a_3 w
\end{aligned}
$$

since (x, y, w) is a point on S_2.

Note that given a point in the image of one of these maps we can trivially find its preimage on the surface, so that both maps are invertible on their images. ♦

Remark. A useful geometric interpretation of these maps is that the image of ϕ_1 is contained in the intersection of V_1 with $x^{-1} + y^{-1} + z^{-1} = 0$, while the image of ϕ_2 is contained in the intersection of V_2 with $x + y + z = 0$.

These maps now play a critical role in the following main theorem.

Theorem 12. *There exists a deterministic polynomial-time algorithm that, given a finite field $\mathbb{F}$ of characteristic 2 with more than 4 elements and an elliptic curve E over $\mathbb{F}$, computes a nontrivial rational point on E.*

Proof. There are two cases to consider, since E can either have j-invariant zero or nonzero. In both cases our strategy is to deterministically find points on the appropriate surface, map them to the threefold, and from there get a point on E.

First assume that $j(E) \neq 0$. For arbitrary c the equation

$$x + y + \frac{xy}{x+y} = c$$

is equivalent to the genus 0 curve $C: x^2 + y^2 + xy + c(x+y) = 0$ except when $x = y$. However, if (x,y) is a point on C with $x = y$ then it must be the point $(0,0)$, so not much is lost. We have the generic solution $(0,c)$ and from this get all points of C through the rational parametrisation

$$y = tx + c$$
$$x = \frac{tc}{1 + t + t^2} \ .$$

Thus we have a family of rational points on S_1 parametrised by t and w which can be mapped to points on V_1 via ϕ_1. It now remains to compute rational points of E.

For $a \in \mathbb{F}^*$ consider the set

$$\{u^2 + au \mid u \in F\}.$$

This set is an additive subgroup of $\mathbb{F}^+$ of index 2, so if $g(x) + g(y) + g(z) = w^2 + w$ then at least one of $g(x)$, $g(y)$, $g(z)$ is itself of the form $u^2 + u$. Discover which it is, call it x, and deterministically solve the quadratic to find u. From $u^2 + u = x^{-2}(x^3 + a_2 x^2 + a_6)$ we now have

$$(ux)^2 + x(ux) = x^3 + a_2 x^2 + a_6$$

and hence a point on E.

Suppose instead that $j(E) = 0$. We wish to compute points on S_2. Taking $y = u^2$, we transform the equation for S_2 as following:

$$xy(x+y) + a_6 = w^2 + a_3 w$$
$$x^2 u^2 + a u^4 + a_6 = w^2 + a_3 w$$
$$a_3 xu + xu^4 + a_6 = (w + xu)^2 + a_3(w + xu) \ .$$

Now, choose y and compute its square root u (possible deterministically since squaring is an automorphism). There are at most four bad choices of y to avoid, corresponding to the roots of $u^4 + a_3 u$. If $u^4 + a_3 u \neq 0$, the equation $x(a_3 u + u^4) + a_6 = z^2 + a_3 z$ is linear in x and hence for any given z, we easily compute the unique value for x. Now the point $(x, y, z + xu)$ is a point on S_2, which we map to V_2 via ϕ_2.

It remains to find a point on E. Mirroring the argument in the previous case, one of $h(x)$, $h(y)$, and $h(z)$ has the form $u^2 + a_3 u$. Discover which it is, call it x, and solve the quadratic $u^2 + a_3 u = h(x)$ for u. Output (x, u) as a rational point on E.   ♦

Remark. This argument can be generalised to work over any perfect characteristic 2 field, but only gives an algorithm when the maps $u \mapsto u^2$ and $u \mapsto u^2 + au$ are algorithmically invertible.

An important question to analyze is how many of the $\mathbb{F}$-rational points of E are obtained by this algorithm. The next theorem will demonstrate that the number is quite large, in particular at least a constant proportion. We define disjointness for points on V_1 as in Definition 8.

Theorem 13. *Let $\mathbb{F}$ be a finite field of order $q = 2^r$ with $q > 4$. The number of disjoint points of V_1 that arise from Theorem 12 is at least $(q-4)/6$.*

Proof. Throughout, assume that the parameter w from Theorem 12 is fixed. Allowing different values could improve the bound, but that analysis has not yet been done.

It was noted before that S_1 can be transformed into a genus 0 curve C : $x^2 + y^2 + xy + c(x + y) = 0$, with C having only gained the point $(0,0)$. Let $C'(\mathbb{F})$ be the points of C except for $(0,0)$, $(c,0)$, and $(0,c)$.

It can easily be confirmed that if (x,y) is a point on C', then $\sigma_1(x,y) = (x, xy(x+y)^{-1})$ and $\sigma_2(x,y) = (y,x)$ are points on C'. We conclude that the group $G = \langle \sigma_1, \sigma_2 \rangle$ acts on $C'(\mathbb{F})$, is isomorphic to Sym(3), and splits the points of C' into orbits of size 6. For the last statement, note that $x = y$ implies $(x,y) = (0,0)$ and $y = xy(x+y)^{-1}$ implies $y = 0$. Thus the stabiliser in Sym(3) of any point has index 6, giving an orbit of size 6.

Any coordinate only appears in its orbit, and each orbit yields the same set $(x, y, xy(x+y)^{-1})$. Thus each orbit when mapped via ϕ_1 yields a disjoint point on V_1.

It remains to count the number of orbits. If r is odd, $t^2 + t + 1$ is irreducible over $\mathbb{F}$ and hence all $t \in \mathbb{F}$ are valid. Thus C has $q+1$ points, but after discarding $(0,0)$, $(c,0)$, and $(0,c)$ we are left with $(q-2)/6$ orbits. If r is even, $t^2 + t + 1$ splits and hence there are $q - 2$ valid t, leaving us with $(q-4)/6$ orbits.   ◆

Remark. We note that the case with $j(E) = 0$ yields a similar bound, since fixing w in S_2 yields a curve of genus 0 that also breaks up into orbits of size 6, each element of the orbit resulting in the same triples $(x, y, x+y)$.

Proof of Theorem 1 (even characteristic). Let $\mathbb{F}$ be a finite field of order $q = 2^r$ with $q > 4$, and let E be a nonsingular elliptic curve over $\mathbb{F}$. From Theorem 12 we obtain a deterministic polynomial-time algorithm that computes points on E. From Theorem 13 we see that this algorithm results in at least $(q-4)/6$ disjoint points on the threefold. This yields at least $(q-4)/6$ x-coordinates of E, and hence at least $(q-4)/3$ points of E.

This completes the proof of Theorem 1.   ◆

Remark. If $\mathbb{F}$ is too small we simply check all pairs $(x,y) \in \mathbb{F}^2$ and obtain the set $E(\mathbb{F})$. This also holds for $\mathbb{F}$ of odd characteristic.

Acknowledgement. We would like to thank our respective advisors Eric Bach and Hendrik Lenstra for helpful discussions.

References

1. Bach, E. and Shallit, J.: Algorithmic Number Theory. The MIT Press, Cambridge (1996)
2. Berlekamp, E.R.: Algebraic coding theory. McGraw-Hill Book Co., New York (1968)
3. Bourbaki, N.: Algebra II. Chapters 4–7. Elements of Mathematics. Springer-Verlag, Berlin (2003) Translated from the 1981 French edition by P. M. Cohn and J. Howie, Reprint of the 1990 English edition
4. Bumby, R.T.: Sums of four squares. In: Number theory (New York, 1991–1995). Springer, New York (1996) 1–8
5. Cohen, H.: A course in computational algebraic number theory. Volume 138 of Graduate Texts in Mathematics. Springer-Verlag, Berlin (1993)
6. Reid, M.: Undergraduate algebraic geometry. Volume 12 of London Mathematical Society Student Texts. Cambridge University Press, Cambridge (1988)
7. Schicho, J.: Proper parametrization of surfaces with a rational pencil. In: Proceedings of the 2000 International Symposium on Symbolic and Algebraic Computation (St. Andrews), New York, ACM (2000) 292–300 (electronic)
8. Schoof, R.: Elliptic curves over finite fields and the computation of square roots mod p. Math. Comp. **44**(170) (1985) 483–494
9. Shafarevich, I.R.: Basic algebraic geometry. 1. Second edn. Springer-Verlag, Berlin (1994) Varieties in projective space, Translated from the 1988 Russian edition and with notes by Miles Reid.
10. Silverman, J.H.: The arithmetic of elliptic curves. Volume 106 of Graduate Texts in Mathematics. Springer-Verlag, New York (1992). Corrected reprint of the 1986 original
11. Skałba, M.: Points on elliptic curves over finite fields. Acta Arith. **117**(3) (2005) 293–301
12. van de Woestijne, C.: Deterministic equation solving over finite fields. PhD thesis, Universiteit Leiden (2006)

20 Years of ECM

Paul Zimmermann[1] and Bruce Dodson[2]

[1] LORIA/INRIA Lorraine
615 rue du jardin botanique, BP 101, F-54602 Villers-lès-Nancy, France
zimmerma@loria.fr
[2] Dept. of Math., 14 E. Packer Ave., Lehigh University, Bethlehem, PA 18015 USA
bad0@lehigh.edu

Abstract. The Elliptic Curve Method for integer factorization (ECM) was invented by H. W. Lenstra, Jr., in 1985 [14]. In the past 20 years, many improvements of ECM were proposed on the mathematical, algorithmic, and implementation sides. This paper summarizes the current state-of-the-art, as implemented in the GMP-ECM software.

Introduction

Before ECM was invented by H. W. Lenstra, Jr. in 1985 [14], Pollard's ρ algorithm and some variants were used, for example to factor the eighth Fermat number F_8 [8]. As soon as ECM was discovered, many researchers worked hard to improve the original algorithm or efficiently implement it. Most current improvements to ECM were already invented by Brent and Montgomery in the end of 1985 [5,18][1].

In [5], Brent describes the "second phase" in two flavours, the "P−1 two-phase" and the "birthday paradox two-phase". He already mentions Brent-Suyama's extension, and the possible use of fast polynomial evaluation in stage 2, but does not yet see how to use the Fast Fourier Transform (FFT). At that time (1985), ECM could find factors of about 20-30 digits only; however Brent predicted: *"we can forsee that p around 10^{50} may be accessible in a few years time"*. This happened in September 1998, when Conrad Curry found a 53-digit factor of $2^{677} - 1$ with Woltman's MPRIME program. According to Fig. 1, which displays the evolution of the ECM record since 1991, and extrapolates it using Brent's formula $\sqrt{D} = (Y - 1932.3)/9.3$, a 100-digit factor — which corresponds to the current GNFS record (RSA-200) — could be found by ECM around 2025, i.e., in another 20 years.

In [18], Montgomery gives a unified description of P−1, P+1 and ECM. He already mentions the "FFT continuation" suggested by Pollard for P−1. A major improvement was proposed by Montgomery with the "FFT extension" [19], which enables one to significantly speed up stage 2.

---

[1] The first version of Brent's paper is from September 24, 1985 — revised December 10, 1985 — and Montgomery's paper was received on December 16, 1985.

F. Hess, S. Pauli, and M. Pohst (Eds.): ANTS 2006, LNCS 4076, pp. 525–542, 2006.
© Springer-Verlag Berlin Heidelberg 2006

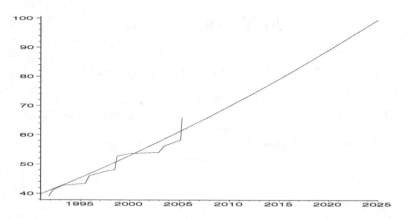

Fig. 1. Graph of ecm records since 1991 (digits vs year), and extrapolation until 2025

Several efficient implementations have been made, in particular by Brent [6], Montgomery (ECMFFT), and Woltman (PRIME95/MPRIME). Already in 1986, Montgomery found a 36-digit factor of the Lucas number L_{464}.

Many large factors have been found by ECM. Among others we can cite the 40-digit prime in the factorization of the tenth Fermat number [7] (the two smaller factors were found by other methods):

$$F_{10} = 45592577{\cdot}6487031809{\cdot}46597757852200185432645607430767781 92897{\cdot}p_{252}.$$

The smallest unfactored Fermat number, F_{12}, is out of reach for NFS-based methods (Number Field Sieve), so the main hope to factor it rests on ECM.

The aim of this paper is to describe the state-of-the-art in the ECM domain, and in particular the algorithms implemented in the GMP-ECM software. §1 recalls the ECM algorithm and defines the notation used in the rest of the paper, while §2 describes the algorithms used in Stage 1 of ECM, and §3 those in Stage 2. Finally, §4 exhibits nice factors found by ECM, and discusses further possible improvements.

1   The ECM Method

Notations. In the whole paper, n denotes the number to be factored, p a (possibly unknown) prime factor of n, and π a prime; the function $\pi(x)$ denotes the number of primes less than or equal to x. All arithmetic operations are implicitly performed modulo n. We assume n has l words in the machine word base β — usually $\beta = 2^{32}$ or 2^{64} —, i.e., $\beta^{l-1} \leq n < \beta^l$. Depending on the context, we write $M(d)$ for the cost of multiplying two d-bit integers, or two degree-d polynomials — where operations on the coefficients count $O(1)$. The notation $\lfloor x \rceil$ stands for $\lfloor x + 1/2 \rfloor$.

This section is largely inspired by [7] and [18]. Consider a field K of characteristic other than 2 or 3. An elliptic curve E is the set of points $(X, Y) \in K$ such that

$$Y^2 = X^3 + AX + B,$$

where $A, B \in K$, and $4A^3 + 27B^2 \neq 0$, plus a "point at infinity" denoted O_E. The curve E admits a group structure, where the addition of two points can be effectively computed, and O_E is the neutral element.

For a computer implementation, it is more efficient to use Montgomery's form $E_{a,b}$ with $a^2 \neq 4$ and $b \neq 0$:

$$by^2 = x^3 + ax^2 + x,$$

which can obtained from Weierstrass form above by the change of variables $X \to (3x + a)/(3b), Y \to y/b, A \to (3 - a^2)/(3b^2), B \to (2a^3/9 - a)/(3b^3)$. Moreover, one usually prefers a homogeneous form:

$$by^2z = x^3 + ax^2z + xz^2, \tag{1}$$

where the triple $(x : y : z)$ represents the point $(x/z : y/z)$ in affine coordinates.

The ECM method starts by choosing a random curve $E_{a,b}$ and a random point $(x : y : z)$ on it. All computations are done modulo the number n to factor, as if $\mathbb{Z}/n\mathbb{Z}$ were a field. The only operation which may fail is when computing the inverse of a nonzero residue x modulo n, if $\gcd(x, n) \neq 1$. But then a factor of n is found, the program outputs it and exits.

Here is a high-level description of the ECM algorithm (recall π denotes a prime):

Algorithm ECM.
Input: an integer n not divisible by 2 nor 3, and integer bounds $B_1 \leq B_2$.
Output: a factor of n, or FAIL.
Choose a random elliptic curve $E_{a,b}$ mod n and a point $P_0 = (x_0 : y_0 : z_0)$ on it.
[Stage 1] Compute $Q := \prod_{\pi \leq B_1} \pi^{\lfloor (\log B_1)/(\log \pi) \rfloor} P_0$ on $E_{a,b}$
[Stage 2] For each π, $B_1 < \pi \leq B_2$,
        compute $(x_\pi : y_\pi : z_\pi) = \pi Q$ on $E_{a,b}$
        $g \leftarrow \gcd(n, z_\pi)$
        if $g \neq 1$, output g and exit
    output FAIL.

Suyama's Parametrization. Suyama's parametrization works as follows. Choose a random integer $\sigma > 5$ (we might also consider a rational value); usually a random 32-bit value is enough, but when running many curves on the same number, one might want to use a larger range. Then compute $u = \sigma^2 - 5$, $v = 4\sigma$, $x_0 = u^3$ mod n, $z_0 = v^3$ mod n, $a = (v - u)^3(3u + v)/(4u^3v) - 2$ mod n. One can check that Eq. (1) holds with for example $b = u/z_0$ and $y_0 = (\sigma^2 - 1)(\sigma^2 - 25)(\sigma^4 - 25)$. This parametrization is widely used, and therefore enables one to reproduce factorizations found by different programs.

In fact, the values of b and y are not needed; all the arithmetic operations involve x and z only. Indeed, for a given pair (x, z), at most two values of y give a valid point $(x : y : z)$ on $E_{a,b}$ according to Eq. (1). When there are two solutions, they are y and $-y$, and ignoring the y-coordinate identifies P and $-P$. As will be seen later, this is precisely what we want. We then write $P = (x : : z)$.

1.1   Why Does ECM Work?

Let p be a prime factor of n, and consider the elliptic curve $E_{a,b}$ mod p. Hasse's theorem says that the order g of $E_{a,b}$ mod p satisfies

$$|g - (p + 1)| < 2\sqrt{p}.$$

When a and b vary, g essentially behaves as a random integer in $[p+1-2\sqrt{p}, p+1+2\sqrt{p}]$, with some additional conditions imposed by the type of curve chosen. For example Suyama's parametrization ensures 12 divides g: Montgomery's form (1) ensures 4 divides g, Suyama gives the additional factor 3.

ECM will find the factor p — which is not necessarily the smallest factor of n — when g is (B_1, B_2)-smooth, i.e., when the largest prime factor of g is less or equal to B_2, and its second largest prime factor less or equal to[2] B_1. The factor p will be found in stage 1 when g is B_1-smooth — i.e., all its prime factors are less or equal to B_1 —, and in stage 2 otherwise.

Remark. If two or more factors of n have a (B_1, B_2)-smooth group order for the chosen curve, they will be found simultaneously, which means that ECM will output their product, which can even be n if all its prime factors have a (B_1, B_2)-smooth group order. This should not be considered a failure: instead check whether the factor is a prime power, and if not restart the same curve with smaller B_1, B_2 to split the different prime factors.

1.2   Complexity of ECM

The expected time used by ECM to find a factor p of a number n is

$$O(L(p)^{\sqrt{2}+o(1)} M(\log n)),$$

where $L(p) = e^{\sqrt{\log p \log \log p}}$, and $M(\log n)$ representes the complexity of multiplication modulo n. The second stage enables one to save a factor of $\log p$ — which is absorbed by the $o(1)$ term above. Mathematical and algorithmic improvements act on the $L(p)^{\sqrt{2}+o(1)}$ factor, while arithmetic improvements act on the $M(\log n)$ factor.

---

[2] The definition of (B_1, B_2)-smoothness used in Algorithm ECM above and by most software is slightly different: all primes $\pi \leq B_1$ should appear to a power $\pi^k \leq B_1$, and similarly for B_2; in practice this makes little difference.

2   Stage One

Stage 1 computes $Q := \prod_{\pi \le B_1} \pi^{\lfloor (\log B_1)/(\log \pi) \rfloor} P_0$ on $E_{a,b}$. That big product is not computed as such. Instead, we use the following loop:

$$Q \leftarrow P_0$$
for each prime $\pi \le B_1$
    compute k such that $\pi^k \le B_1 < \pi^{k+1}$
    for $i := 1$ to k do
        $Q \leftarrow \pi \cdot Q$.

The multiplication $\pi \cdot Q$ on the elliptic curve is done using additions $(P, Q \to P+Q)$ and duplications $(P \to 2P)$.

To add two distinct points $(x_P :: z_P)$ and $(x_Q :: z_Q)$, one uses the following formula, where $(x_{P-Q} :: z_{P-Q})$ corresponds to the difference $P - Q$:

$$x_{P+Q} = 4z_{P-Q} \cdot (x_P x_Q - z_P z_Q)^2, \qquad z_{P+Q} = 4x_{P-Q} \cdot (x_P z_Q - z_P x_Q)^2.$$

This can be computed using 6 multiplications (among which 2 are squares) as follows:

$$\begin{aligned}
u &\leftarrow (x_P + z_P)(x_Q - z_Q) & v &\leftarrow (x_P - z_P)(x_Q + z_Q) \\
w &\leftarrow (u+v)^2 & t &\leftarrow (u-v)^2 \\
x_{P+Q} &\leftarrow z_{P-Q} \cdot w & z_{P+Q} &\leftarrow x_{P-Q} \cdot t.
\end{aligned}$$

To duplicate a point $(x_P :: z_P)$, one uses the following formula:

$$x_{2P} = (x_P^2 - z_P^2)^2, \qquad z_{2P} = (4x_P z_P)[(x_P - z_P)^2 + d(4x_P z_P)], \qquad (2)$$

where $d = (a+2)/4$, with a from Eq. (1). This formula can be implemented using 5 multiplications (including 2 squares) as follows:

$$\begin{aligned}
u &\leftarrow (x_P + z_P)^2 & v &\leftarrow (x_P - z_P)^2 & t &\leftarrow d(u-v) + v \\
x_{2P} &\leftarrow uv & z_{2P} &\leftarrow (u-v)t.
\end{aligned}$$

Since the difference $P - Q$ is needed to compute $P + Q$, this is a special case of addition chains, called "Lucas chains" by Montgomery, who designed an heuristic algorithm "PRAC" to compute them [16] (see §2.2).

2.1   Residue Arithmetic

To obtain an efficient implementation of ECM, an efficient underlying arithmetic is important. The main operations to be performed are additions, subtractions and multiplications modulo the number n to be factored. Other operations (divisions, gcds) are rare, or can be replaced by modular multiplications. Since additions and subtractions have cost $O(\log n)$, the main operation to be optimized is the modular multiplication: given $0 \le a, b < n$, compute $c = ab \bmod n$.

We distinguish two cases: classical $O(\log^2 n)$ arithmetic, and subquadratic arithmetic. On a Pentium 4, GMP-4.2 switches to Karatsuba's algorithm up from 23 words, i.e., about 220 decimal digits. Since ECM is often used to factor numbers smaller than this, it is worth optimizing classical arithmetic.

For special numbers, like factors of $\beta^k \pm 1$, one may use ad-hoc routines. Assume for example $dn = \beta^k - 1$. The product $c = ab$ of two residues can be reduced as follows: write $c = c_0 + c_1 \beta^k$, where $0 \le c_0, c_1 < \beta^k$; then $c = c_0 + c_1 \bmod n$. Instead of reducing a $2l$-word integer c (recall n has l words), we reduce $c_0 + c_1$, which has k words only (plus possibly one carry bit). Alternatively, if the cofactor d is small, one can reduce c modulo $\beta^k - 1$ only, and perform multiplications on k words instead of l words. GMP-ECM implements such a special reduction for large divisors of $2^k \pm 1$, using the latter method. It also uses special code for Fermat numbers $2^{2^k} + 1$: indeed, GMP fast multiplication code precisely uses Schönhage-Strassen algorithm, i.e., multiplication modulo $2^m + 1$ [21].

Efficient Assembly Code. While using clever high-level algorithms may give a speedup of 10% or 20%, at the expense of several months to invent and implement those algorithms, a twofold speedup may be obtained in a few days, just rewriting one of the assembly routines for integer arithmetic[3].

GMP-ECM is based on the GNU MP library (GMP for short) [11], thus benefits from the portability of GMP, and from the efficiency of its assembly routines (found in the mpn layer). A library dedicated to modular arithmetic — or even better to computations on elliptic curves — might yet be faster. Since all operations are done on numbers of the same size, we might use a library with special assembly code for each word size, up to some reasonable small size.

Quadratic Arithmetic. In the quadratic domain, up to 200-300 digits depending on the processor, the best current solution is to use Montgomery representation [17]: The number n to be factored having l words in base β, each residue a is replaced by $a' = \beta^l a \bmod n$. Additions and subtractions are unchanged, multiplications are replaced by the REDC operation: $\text{REDC}(a, b) := ab\beta^{-l} \bmod n$. This operation can be efficiently implemented on modern computers, and unlike classical division does not require any correction.

There are two ways to implement REDC: (i) either interleave the multiplication and the reduction as in algorithm MODMULN from [18], (ii) or perform them separately. The latter way enables one to use the efficient GMP assembly code for base-case multiplication. One first computes $c = ab$, having at most $2l$ words in base β. The reduction $r := c \bmod n$ is performed with the following GMP code, which is exactly that of version 6.0.1 of GMP-ECM, with variable names changed to match the above notations (the mpn_ functions are described in the GMP documentation [11]):

```
static void
ecm_redc_basecase (mpz_ptr r, mpz_ptr c, mpmod_t modulus)
{
  mp_ptr rp = PTR(r), cp = PTR(c);
```

---

[3] The first author indeed noticed a speedup of more than 2 with GMP-ECM, when Torbjörn Granlund rewrote the UltraSparc assembly code for GMP.

```
mp_srcptr np = PTR(modulus->orig_modulus);
mp_limb_t cy;
mp_size_t j, L = modulus->bits / __GMP_BITS_PER_MP_LIMB;

for (j = ABSIZ(c); j < 2 * L; j++)
  cp[j] = 0;
for (j = 0; j < L; j++, cp++)
  cp[0] = mpn_addmul_1 (cp, np, L, cp[0] * modulus->Nprim);
cy = mpn_add_n (rp, cp, cp - L, L);
if (cy != 0)
  mpn_sub_n (rp, rp, np, L);
MPN_NORMALIZE (rp, L);
SIZ(r) = SIZ(c) < 0 ? -L : L;
}
```

The main idea — independently discovered by Kevin Ryde and the first author — is to store the carry words from mpn_addmul_1 in the low l words of c, just after they are set to zero by REDC. In such a way, one replaces l expensive carry propagations by one call to mpn_add_n.

Subquadratic Arithmetic. For large numbers, subquadratic arithmetic is needed. Again, one can use either the classical representation, or Montgomery representation. In both cases, the best known algorithms require $2.5M(l)$ for a l-word modular multiplication: $M(l)$ for the multiplication $c := ab$, and $1.5M(n)$ for the reduction $c \bmod n$ using Barrett's algorithm [1], or its least-significant-bit (LSB) variant for $c\beta^{-l} \bmod n$. LSB-Barrett is exactly REDC, where β is replaced by β^l [20]: after the precomputation of $m = -n^{-1} \bmod \beta^l$, compute $d = cm \bmod \beta^l$, and $(c + dn)\beta^{-l}$. Since all reductions are done modulo the same n, the precomputation of m is amortized and does not impact the average cost. The $1.5M(n)$ reduction cost is obtained using the "wrap-around" trick for the last multiply dn (see §3.2), since the low part is known to be equal to $-c \bmod \beta^l$.

2.2   Evaluation of Lucas Chains

A Lucas chain is an addition chain in which the sum $i + j$ of two terms can appear only if $|i - j|$ also appears. (This condition is needed for the point addition in homogeneous coordinates, see §2.) For example $1 \to 2 \to 3 \to 5 \to 7 \to 9 \to 16 \to 23$ is a Lucas chain for 23.

The basic idea of Montgomery's PRAC algorithm [16] is to find a Lucas chain using some heuristics. Assume for example we want to generate $1009 \cdot P$. To generate a sequence close to optimal, a natural idea is to use as previous term $1009/\phi \approx 624$, where $\phi = (1 + \sqrt{5})/2$ is the golden ratio, but this requires $1009 - 624 = 385$ to be a term in the sequence. We get $1009 \to 624 \to 385 \to 239 \to 146 \to 93 \to 53 \to 40 \to 13$. At this point we cannot continue using the same transform $(d, e) \to (e, d - e)$.

To generate $\pi \cdot P$, Montgomery starts with $(d, e) = (\pi, \lfloor \pi/\alpha \rfloor)$, with $\alpha = \phi$, and iteratively uses 9 different transforms to reduce the pair (d, e), each transform using from 1 to 4 point additions or duplicates, to finally reach $d = 1$. (PRAC actually generates a dual of the chain.)

Montgomery improved PRAC as follows: instead of using $\alpha = \phi$ only, try several values of α, and keep the one giving the smallest cost in terms of modular multiplications. The α's are chosen so that after a few steps, the remaining values (d, e) have a ratio near ϕ, i.e., $\alpha = (a\phi + b)/(c\phi + f)$ with small a, b, c, f. If $r = \lfloor \pi/\alpha \rfloor$, the idea is to share the partial quotients different from 1 among the first and last terms from the continued fraction of π/r, hoping to have small trailing quotients.

Fig. 2 gives 10 such values of α, the first partial quotients of their continued fraction, and the total cost — in terms of curve additions or duplicates — of PRAC for all primes up to B_1, for $B_1 = 10^6$ and 10^8. For a given row, all values of α above and including this row are assumed to be used. The gain using those 10 values instead of only $\alpha = \phi$ is 3.72% for $B_1 = 10^6$, 3.74% for $B_1 = 10^8$, and the excess with respect to the lower bounds given by Theorem 8 of [16] — 2114698 for $B_1 = 10^6$ and 210717774 for $B_1 = 10^8$ — is 3.7% and 5.1% respectively.

α	first partial quotients	$B_1 = 10^6$	$B_1 = 10^8$
$\phi \approx 1.61803398875$	$1, 1, 1, \ldots$	2278430	230143294
$(\phi + 7)/5 \approx 1.72360679775$	$1, 1, 2, 1, \ldots$	2240333	226235929
$(\phi + 2311)/1429 \approx 1.618347119656$	$1, 1, 1, 1, 1, 1, 1, 2, 1, \ldots$	2226042	224761495
$(6051 - \phi)/3739 \approx 1.617914406529$	$1, 1, 1, 1, 1, 1, 1, 1, 2, 1, \ldots$	2217267	223859686
$(129 - \phi)/79 \approx 1.612429949509$	$1, 1, 1, 1, 1, 2, 1, \ldots$	2210706	223226409
$(\phi + 49)/31 \approx 1.632839806089$	$1, 1, 1, 1, 2, 1, \ldots$	2205612	222731604
$(\phi + 337)/209 \approx 1.620181980807$	$1, 1, 1, 1, 1, 1, 2, 1, \ldots$	2201615	222335307
$(19 - \phi)/11 \approx 1.580178728295$	$1, 1, 1, 2, 1, \ldots$	2198400	222013974
$(883 - \phi)/545 \approx 1.617214616534$	$1, 1, 1, 1, 1, 1, 1, 2, 1, \ldots$	2195552	221729046
$3 - \phi \approx 1.38196601125$	$1, 2, 1, \ldots$	2193683	221533297

Fig. 2. Total cost of PRAC with several α's, for all $\pi < B_1$ (using the best double-precision approximation of α)

3   Stage Two

All of P−1, P+1 and ECM work in an Abelian group G. For P−1, G is the multiplicative group of nonzero elements of $GF(p)$ where p is the factor to be found; for P+1, G is a multiplicative subgroup of $GF(p^2)$; for ECM, G is an elliptic curve $E_{a,b} \bmod p$. In all cases, the calculations in G reduce to arithmetic operations — additions, subtractions, multiplications, divisions — in $\mathbb{Z}/n\mathbb{Z}$. The only computation that may fail is the inversion $1/a \bmod n$, but then a non-trivial factor of n is found, unless $a = 0 \bmod n$. A unified description of stage 2 is possible [18]; for sake of clarity, we here prefer to focus on ECM.

3.1   Overall Description

Stage 1 of ECM computes a point Q on an elliptic curve E. In case it fails, i.e., $\gcd(n, z_Q) = 1$, we hope there exists a prime π in the stage 2 range $[B_1, B_2]$

such that $\pi Q = O_E \bmod p$. In such a case, while computing $\pi Q = (x : y)$ in Weierstrass coordinates[4], a non-trivial gcd will yield the prime factor p of n. A *continuation* of ECM — also called stage two, phase two, or step two — tries to find those matches. The first main idea is to avoid computing every πQ, using a "meet-in-the-middle" — or baby-step, giant step — strategy: one computes σQ and τQ such that $\pi = \sigma \pm \tau$. If $\sigma Q = (x_\sigma : y_\sigma)$ and $\tau Q = (x_\tau : y_\tau)$, then $\sigma Q + \tau Q = O_E \bmod p$ implies $x_\sigma = x_\tau \bmod p$. It thus suffices to compute $\gcd(x_\sigma - x_\tau, n)$ to obtain[5] the factor p.

Two classes of continuations differ in the way they choose σ and τ. The *birthday paradox continuation* takes $\sigma \in S$ and $\tau \in T$, with S and T two large sets, which are either random or geometric progressions, hoping that $S+T$ covers most primes in $[B_1, B_2]$, and usually other larger primes. Brent suggests taking $T = S$.

We focus here on the *standard continuation*, which takes S and T in arithmetic progressions, and guarantees that all primes π in $[B_1, B_2]$ are hit. Assume for simplicity that $B_1 = 1$. Choose a composite integer $d < B_2$, then all primes up to B_2 can be written

$$\pi = \sigma + \tau,$$

with $\sigma \in S = \{i \cdot d, 0 \le i \cdot d < B_2\}$, and $\tau \in T = \{j, 0 < j < d, \gcd(j, d) = 1\}$. Computing values of σQ and τQ costs $O(B_2/d + d)$ elliptic curve operations, which is $O(\sqrt{B_2})$ for $d \approx \sqrt{B_2}$. Choosing d with many small factors also reduces the cost. The main problem is how to evaluate all $x_\sigma - x_\tau$ for $\sigma \in S$, $\tau \in T$, and take their gcd with n.

A crucial observation is that for ECM, if $jQ = (x : y)$, then $-jQ = (x : -y)$. Thus jQ and $-jQ$ share the same x-coordinate. In other words, if one computes $x_i - x_j$ corresponding to the prime $\pi = i \cdot d + j$, one will also hit $i \cdot d - j$ — which may be prime or not — for free. This can be exploited in two ways: Either restrict to $j \le d/2$, as proposed by Montgomery [18]; or restrict j to the "positive" residues prime to d, for example if d is divisible by 6, one can restrict to $j = 1 \bmod 6$. This is what is used in GMP-ECM.

3.2 Fast Polynomial Arithmetic

Classical implementations of the standard continuation cover primes in $[B_1, B_2]$, and therefore require $\Theta(\pi(B_2))$ operations, assuming $B_1 \ll B_2$. The main idea of the "FFT continuation" is to use fast polynomial arithmetic to compute all $x_\sigma - x_\tau$ — or their product mod n — in less than $\pi(B_2)$ operations. It would be better to call it "fast polynomial arithmetic continuation", since any sub-quadratic algorithm works, not only the FFT.

Here again, two variants exist. They share the idea that what one really wants is:

$$h = \prod_{\sigma \in S} \prod_{\tau \in T} (x_\sigma - x_\tau) \bmod n, \tag{3}$$

---

[4] It is simpler to describe stage 2 in Weierstrass coordinates.

[5] Unless $x_\sigma = x_\tau \bmod n$ too, but if we assume x_σ and x_τ to be uniformly distributed, this has probability p/n only.

since if any $\gcd(x_\sigma - x_\tau, n)$ is non-trivial, so will be $\gcd(h, n)$. Eq. (3) computes many $x_\sigma - x_\tau$ that do not correspond to prime values of $\sigma \pm \tau$, but the gain of using fast polynomial arithmetic largely compensates for this fact.

Let $F(X)$ (respectively $G(X)$) be the polynomial whose roots are the x_τ (respectively x_σ). Both F and G can be computed in $O(M(d) \log d)$ operations over $\mathbb{Z}/n\mathbb{Z}$ with the "product tree" algorithm and fast polynomial multiplication [3,22], where d is the cardinal of the sets S and T (see §3.1). The "POLYGCD" variant interprets h as the resultant $\mathrm{Res}(F, G)$, which reduces to a polynomial gcd. It is known that the gcd of two degree-d polynomials can be computed in $O(M(d) \log d)$, too. The "POLYEVAL" variant interprets h as

$$h = \pm \prod_{\tau \in T} G(x_\tau) \bmod n,$$

thus it suffices to evaluate G at all roots x_τ of F. This problem is known as "multipoint polynomial evaluation", and can be solved in $O(M(d) \log d)$ with a "remainder tree" algorithm [3,22].

Algorithm POLYEVAL is faster, since it admits a smaller multiplicative constant in front of the $M(d) \log d$ asymptotic complexity. However, it needs — with the current state of art — to store $\Theta(d \log d)$ coefficients in $\mathbb{Z}/n\mathbb{Z}$, instead of only $O(d)$ for POLYGCD.

Fast Polynomial Multiplication. Several algorithms are available to multiply polynomials over $(\mathbb{Z}/n\mathbb{Z})[x]$. Previous versions of GMP-ECM used Karatsuba, Toom 3-way and 4-way for polynomial multiplication, and division was performed using the Borodin-Moenck-Jebelean-Burnikel-Ziegler algorithm [9]. To multiply degree-d polynomials with the FFT, we need to find $\omega \in \mathbb{Z}/n\mathbb{Z}$ such that $\omega^{d/2} = -1 \bmod n$, which is not easy, if possible at all.

Montgomery [19] suggests performing several FFTs modulo small primes — chosen so that finding a primitive d-root of unity is easy — and then recovering the coefficients by the Chinese Remainder Theorem. This approach was recently implemented by Dave Newman in GMP-ECM. On some processors, it is faster than the second approach described below; however, it requires implementing a polynomial arithmetic over $\mathbb{Z}/p\mathbb{Z}$, for p a small prime (typically fitting in a machine word).

The second approach uses the "Kronecker-Schönhage trick"[6]. Assume we want to multiply two polynomials $p(x)$ and $q(x)$ of degree less than d, with coefficients $0 \le p_i, q_i < n$. Choose $\beta^l > dn^2$, and create the integers $P = p(\beta^l)$ and $Q = q(\beta^l)$. Now multiply P and Q using fast integer arithmetic (integer FFT for example). Let $R = PQ$. The coefficients of $r(x) = p(x)q(x)$ are simply obtained by reading R as $r(\beta^l)$. Indeed, the condition $\beta^l > dn^2$ ensures that consecutive coefficients of $r(x)$ do not "overlap" in R. It just remains to reduce the coefficients modulo n.

The advantage of the Kronecker-Schönhage trick is that no algorithm has to be implemented for polynomial multiplication, since one directly relies on

---

[6] The idea of using this trick is due to Dave Newman; a similar algorithm is attributed to Robbins in [19, §3.4].

fast integer multiplication. Division is performed in a similar way, with Barrett's algorithm: first multiply by the pseudo-inverse of the divisor — which is invariant here, namely $F(X)$ when using $k \geq 2$ blocks, see below —, then multiply the resulting quotient by the divisor. A factor of two can be saved in the latter multiplication, by using the "wrap-around" or "$x^d + 1$" trick[7], assuming the integer FFT code works modulo $2^m + 1$ [2].

3.3 Stage 2 Blocks

For a given stage 2 bound B_2, computing the product and remainder trees may be relatively expensive. A workaround is to split stage 2 into $k > 1$ blocks [19]. Let $B_2 = kb_2$, and choose $d \approx \sqrt{b_2}$ as in §3.1. The set $S = \{i \cdot d, 0 \leq i \cdot d < b_2\}$ of §3.1 is replaced by $S_1, \ldots, S_k$ that cover all multiples of d up to B_2, and correspond to polynomials $G_1, \ldots, G_k$. The set T remains unchanged, and still corresponds to the polynomial F. Instead of evaluating G at all roots of F, one evaluates $H = G_1 G_2 \cdots G_k$ at all roots of F. Indeed, if one of the G_l vanishes at a root of F, the same holds for H. Moreover, it suffices to compute $H \mod F$, which can be done by $k - 1$ polynomial multiplications and divisions modulo F.

Assume a product tree costs $qM(d)\log d$, and a remainder tree $rM(d)\log d$. With a single block ($k = 1$), we compute two product trees — for F and G —, and one remainder tree, all of size d, with a total cost of $(2q+r)M(d)\log d$. With k blocks, we compute $k + 1$ product trees for $F, G_1, \ldots, G_k$, and one remainder tree, all of degree about $d/\sqrt{k}$. Assuming $M(d)$ is quasi-linear, and neglecting all other costs in $O(M(d))$, the total cost is $\frac{(k+1)q+r}{\sqrt{k}}M(d)\log d$. The optimal value of k then depends on the ratio r/q. Without caching Fourier transforms, the best known ratio is $r/q = 2$ using Bernstein's "scaled remainder trees" [3]. Each node of the product tree corresponds to one product of degree l polynomials, while the corresponding node of the remainder tree corresponds to two "middle products" [4,12]. For $r/q = 2$, the theoretical optimal value is $k = 3$, with a cost of $3.46qM(d)\log d$, instead of $4qM(d)\log d$ for $k = 1$. In some cases, one may want to use a larger number k of blocks for a given stage 2 range, in order to decrease the memory usage.

3.4 Brent-Suyama's Extension

Brent-Suyama's extension increases the probability of success of stage 2, with a small additional cost. Recall stage 2 succeeds when the largest factor π of the group order can be written as $\pi = \sigma \pm \tau$, where points σQ and τQ have been computed for σ, τ in sets S and T respectively. The idea of Brent and Suyama [5] is to compute $\sigma^e Q$ and $\tau^e Q$ instead, or more generally $f(\sigma)Q$ and $f(\tau)Q$ for some odd or even integer polynomial $f(x)$, as suggested by Montgomery [18]. If $\pi = \sigma \pm \tau$, then π divides one of $f(\sigma) \pm f(\tau)$. Thus all primes π up to B_2 will still be hit, but other larger primes may be hit too, especially if

---

[7] If the upper or lower half of a $2m$-bit product is known, computing it modulo $2^m + 1$ easily yields the other half.

$f(x) \pm f(y)$ has many algebraic factors. This is the case for $f(x) = x^e$, but also for Dickson polynomials as suggested by Montgomery in [19]. GMP-ECM uses Dickson polynomials of parameter $\alpha = -1$ with the notation from [19]: $D_1 = x$, $D_2 = x^2 + 2$, and $D_{e+2} = x D_{e+1} + D_e$ for $e \geq 1$, which gives $D_3(x) = x^3 + 3x$, $D_4(x) = x^4 + 4x^2 + 2$.

To efficiently compute the values of $f(\sigma)Q$, we use the "table of differences" algorithm [18, §5.9]. For example, to evaluate x^3 we form the following table:

1	8	27	64	125	216
	7	19	37	61	91
		12	18	24	30
			6	6	6

Once the entries in boldface have been computed[8], one deduces the corresponding points over the elliptic curve, for example here $1Q$, $7Q$, $12Q$ and $6Q$. Then each new value of $x^e Q$ is obtained with e point additions: $1Q + 7Q = 8Q$, $7Q + 12Q = 19Q$, ... One has to switch to Weierstrass coordinates, since if iQ and jQ are in the difference table, $|i - j|Q$ is not necessarily, for example $5Q = 12Q - 7Q$ is not here. As mentioned in [19], the e point additions in the downward diagonals are performed in parallel, using Montgomery's trick to perform one modular inverse only, at the cost of $O(e)$ extra multiplications. Efficient ways to implement Brent-Suyama's trick for P-1 and P$+1$ are described in [18].

Note that since Brent-Suyama's extension depends on the choice of the stage 2 parameters $(k, d, \dots)$, extra-factors found may not be reproducible with other software, or even different versions of the same software.

3.5   Montgomery's $d_1 d_2$ Improvement

A further improvement is proposed by Montgomery in [18]. Instead of sieving primes of the form $\pi = id + j$ as in §3.1, use a double sieve with d_1 coprime to d_2:

$$\pi = id_1 + jd_2.$$

(The description in §3.1 corresponds to $d_1 = d$ and $d_2 = 1$.) Each $0 < \pi \leq B_2$ can be written uniquely as $\pi = id_1 + jd_2$ with $0 \leq j < d_1$: take $j = -\pi/d_2 \bmod d_1$, then $i = (\pi - jd_2)/d_1$.

To sieve all primes up to B_2, take $S = \{id_1, -d_1 d_2 < id_1 \leq B_2, \gcd(i, d_2) = 1\}$ and $T = \{jd_2, 0 \leq j < d_1, \gcd(j, d_1) = 1\}$. In comparison to §3.1: (i) the lower bound for id_1 is now $-d_1 d_2$ instead of 0, but this has little effect if $d_1 d_2 \ll B_2$; (ii) the additional condition $\gcd(i, d_2) = 1$ reduces the size of S by a factor $1/d_2$.

When using several blocks, the extra values of i mentioned in (i) occur for the first block only, whereas the speedup in (ii) holds for all blocks. In fact, since the size of T yields the degree of the polynomial arithmetic — i.e., $\phi(d_1)/2$ with

---

[8] Over the integers, and not over the elliptic curve as the first author did in a first implementation!

the remark at end of §3.1 — and we want S to have the same size, this means we can enlarge the block size b_2 by a factor $1/d_2$ for free.

This improvement was implemented in GMP-ECM by Alexander Kruppa, up from version 6.0, with d_2 being a small prime. The following table gives for several factor sizes, the recommended stage 1 bound B_1, the corresponding effective stage 2 bound B_2', the ratio B_2'/B_1, the number k of blocks, the parameters d_1 and d_2, the degree $\phi(d_1)/2$ of polynomial arithmetic, the polynomial used for Brent-Suyama's extension, and finally the expected number of curves. All values are the default ones used by GMP-ECM 6.0.1 for the given B_1.

digits	B_1	B_2'	B_2'/B_1	k	d_1	d_2	$\phi(d_1)/2$	poly.	curves
40	$3 \cdot 10^6$	4592487916	1531	2	150150	17	14400	$D_6(x)$	2440
45	$11 \cdot 10^6$	30114149530	2738	2	371280	11	36864	$D_{12}(x)$	4590
50	$43 \cdot 10^6$	198654756318	4620	2	1021020	19	92160	$D_{12}(x)$	7771
55	$110 \cdot 10^6$	729484405666	6632	2	1891890	17	181440	$D_{30}(x)$	17899
60	$260 \cdot 10^6$	2433583302168	9360	2	3573570	19	322560	$D_{30}(x)$	43670
65	$850 \cdot 10^6$	15716618487586	18490	2	8978970	17	823680	$D_{30}(x)$	69351

As an example, with $B_1 = 3 \cdot 10^6$, the default B_2 value used for ECM is[9] $B_2 = 4592487916$ (i.e., about $1531 \cdot B_1$) with $k = 2$ blocks, $d_1 = 150150$, $d_2 = 17$. This corresponds to polynomial arithmetic of degree $\phi(150150)/2 = 14400$. With those parameters and the degree-6 Dickson polynomial, 2440 curves are expected to find a 40-digit prime factor.

4   Results and Open Questions

Largest ECM Factor. Records given in this section are as of January 2006. The largest prime factor found by ECM is a 66-digit factor of $3^{466} + 1$ found by the second author on April 6, 2005:

$p_{66} = 709601635082267320966424084955776789770864725643996885415676682297.$

This factor was found using GMP-ECM, with $B_1 = 110 \cdot 10^6$ and $\sigma = 1875377824$; the corresponding group order, computed with the Magma system [15], is:

$g = 2^2 \cdot 3 \cdot 11243 \cdot 336181 \cdot 844957 \cdot 1866679 \cdot 6062029 \cdot 7600843 \cdot 8046121 \cdot 8154571 \cdot 13153633 \cdot 249436823.$

The largest group order factor is only about $2.3B_1$, and much smaller than the default $B_2' = 729484405666$ (see above table).

We can reproduce this lucky curve with GMP-ECM 6.0.1, here on an Opteron 250 at 2.4Ghz, with improved GMP assembly code from Torbjörn Granlund[10]:

---

[9] The printed value is 4016636513, but the effective value is slightly larger, since "good" values of B_2 are sparse.

[10] Almost the same speed is obtained with Gaudry's assembly code at http://www.loria.fr/~gaudry/mpn_AMD64/.

```
GMP-ECM 6.0.1 [powered by GMP 4.1] [ECM]
Input number is 180241397103940772078159779297801504017708653303813750145082169906990204420366728928
9127\
4814402760531304131590067861951398548382931195190615371324248478807099289879585509160103
8513 (180 digits)
Using MODMULN
Using B1=110000000, B2=680270182898, polynomial Dickson(30), sigma=1875377824
Step 1 took 748990ms
B2'=729484405666 k=2 b2=364718554200 d=1891890 d2=17 dF=181440, i0=42
Expected number of curves to find a factor of n digits:
20      25      30      35      40      45      50      55      60      65
2       4       10      34      135     617     3155    17899   111395  753110
Initializing tables of differences for F took 501ms
Computing roots of F took 29646ms
Building F from its roots took 27847ms
Computing 1/F took 13902ms
Initializing table of differences for G took 656ms
Computing roots of G took 25054ms
Building G from its roots took 27276ms
Computing roots of G took 24723ms
Building G from its roots took 27184ms
Computing  G * H took 8041ms
Reducing   G * H mod F took 12035ms
Computing polyeval(F,G) took 64452ms
Step 2 took 262345ms
Expected time to find a factor of n digits:
20      25      30      35      40      45      50      55      60      65
29.45m  1.06h   2.88h   9.63h   1.58d   7.23d   36.93d  209.51d 3.57y   24.15y
********** Factor found in step 2: 709601635082267320966424084955776789770864725643996885415676682297
Found probable prime factor of 66 digits: 709601635082267320966424084955776789770864725643996885415676682297
Probable prime cofactor 2540036383696390063049462605801550334164274148410764601894236335648589609705
2304\
48527170095214007673747373786652729 has 114 digits
Report your potential champion to Richard Brent <rpb@comlab.ox.ac.uk>
(see ftp://ftp.comlab.ox.ac.uk/pub/Documents/techpapers/Richard.Brent/champs.txt)
```

Several comments can be made about this verbose output. First we see that
the effective stage 2 bound $B_2' = 729484405666$ is indeed larger than the "requested" one $B_2 = 680270182898$. The stage 2 parameters k, $d(= d_1)$, d_2 and
the Dickson polynomial $D_{30}(x)$ are those of the 55-digit row in the above table
(dF is the polynomial degree, and i_0 the starting index in $id_1 + jd_2$). Initializ-
ing the table of differences — i.e., computing the first downward diagonal for
Brent-Suyama's extension — is clearly cheap with respect to "Computing roots
of F/G", which corresponds to the computation of the values x_σ and x_τ, together
with the whole table of differences. "Building F/G from its roots" corresponds
to the product tree algorithm; "Computing 1/F" is the precomputation of the
inverse of F for Barrett's algorithm. "Computing G * H" corresponds to the
multiplication $G_1 G_2$, and "Reducing G * H mod F" to the reduction of $G_1 G_2$
modulo F: we clearly see the 1.5 factor announced in §3.2. "Computing polye-
val(F,G)" stands for the remainder tree algorithm: the ratio with respect to the
product tree is slightly larger than the theoretical value of 2. Finally the total
stage 2 time is only 35% of the stage 1 time, for a stage 2 bound 6632 times
larger!

Largest P−1 and P+1 Factors. The largest prime factor found by P−1 is
a 58-digit factor of $2^{2098} + 1$, found by the first author on September 28, 2005
with $B_1 = 10^{10}$ and $B_2 = 13789712387045$:

$p_{58} = 1372098406910139347411473978297737029649599583843164650153,$
$p_{58} - 1 = 2^3 \cdot 3^2 \cdot 1049 \cdot 1627 \cdot 139999 \cdot 1284223 \cdot 7475317 \cdot 341342347 \cdot 2456044907 \cdot 9909876848747.$

The largest prime factor found by P+1 is a 48-digit factor of the Lucas number $L(1849)$, found by Alexander Kruppa on March 29, 2003 with $B_1 = 10^8$ and $B_2 = 52337612087$:

$$p_{48} = 884764954216571039925598516362554326397028807829,$$
$$p_{48} + 1 = 2 \cdot 5 \cdot 19 \cdot 2141 \cdot 30983 \cdot 32443 \cdot 35963 \cdot 117833 \cdot 3063121 \cdot 80105797 \cdot 2080952771.$$

Other P−1 or P+1 Factors. The authors performed complete runs on the about 1000 composite numbers from the regular Cunningham table with P−1 and P+1 [23]. The largest run used $B_1 = 10^{10}$, $B_2 \approx 1.3 \cdot 10^{13}$, polynomial x^{120} for P−1, and $B_1 = 4 \cdot 10^9$, $B_2 \approx 1.0 \cdot 10^{13}$, polynomial $D_{30}(x)$ for P+1.

A total of 9 factors were found by P−1 during these runs, but strangely no factor was found by P+1. Nevertheless, the authors believe that the P−1 and (especially) P+1 methods are not used enough. Indeed, if one compares the current records for ECM, P−1 and P+1, of respectively 66, 58 and 48 digits (http://www.loria.fr/~zimmerma/records/Pminus1.html), there is no theoretical reason why the P±1 records would be smaller, especially if one takes into account that the P±1 arithmetic is faster.

Largest ECM Group Order Factor. The largest group order factor of a lucky elliptic curve is 81325590104999, for a 47-digit factor of $5^{430} + 1$ found by the second author on December 27, 2005:

$$p_{47} = 29523508733582324644807542345334789774261776361,$$

with $B_1 = 260 \cdot 10^6$ and $\sigma = 610553462$; the corresponding group order is:

$$g = 2^2 \cdot 3 \cdot 13 \cdot 347 \cdot 659 \cdot 163481 \cdot 260753 \cdot 9520793 \cdot 25074457 \cdot 81325590104999.$$

This factor is a success for Brent-Suyama's extension, since the largest factor of g is much larger than B_2 (about $33.4B_2$). The degree-30 Dickson polynomial was used here, with $\sigma = 92002 \cdot 1891890$ and $\tau = 1518259 \cdot 17$, i.e., $d_1 = 1891890$ and $d_2 = 17$.

From January 1st, 2000 to January 19th, 2006, a total of 619 prime factors of regular Cunningham numbers were found by ECM, P+1 or P−1 [10]. Among those 619 factors, 594 were found by ECM with known B_1 and σ values. If we denote by g_1 the largest group order factor of each lucky curve, Fig. 3 shows an histogram of the ratio $\log(g_1/B_1)$. Most ECM programs use $B_2 = 100B_1$. Since $\log 100 \approx 4.6$, we see that they miss about half the factors that could be found using the FFT continuation.

Save and Resume Interface. George Woltman's PRIME95 implementation of ECM uses the same parametrization as GMP-ECM (see §1). PRIME95 runs on x86 architectures, and factors only base-2 Cunningham numbers so far, but Stage 1 of PRIME95 is much faster than GMP-ECM, thanks to some highly-tuned assembly code. Since PRIME95 does not implement the "FFT continuation" yet, a public interface was designed to perform stage 1 with PRIME95, and stage 2 with GMP-ECM. The first factor found by this collaboration between PRIME95

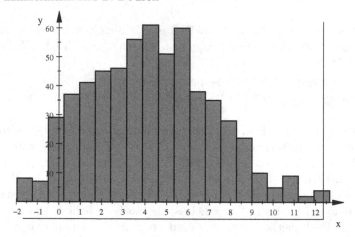

Fig. 3. Histogram of $\log(g_1/B_1)$ for 594 Cunningham factors found by ECM

and GMP-ECM was obtained by Patrik Johansson, who found a 48-digit factor of $2^{731}-1$ on March 30th, 2003, with $B_1 = 11000000$ and $\sigma = 7706350556508580$:

$$p_{48} = 223192283824457474300157944531480362369858813007.$$

This save/resume interface may have other applications:

- after a stage 1 run, we may split a huge stage 2 on several computers. Indeed, GMP-ECM can be given a range $[l, h]$ as stage 2 range, meaning that all primes $l \le \pi \le h$ are covered. The total cpu time will be slightly larger than with a single run, due to the fact that several product/remainder trees will be computed, but the real time may be drastically decreased;
- when using P±1, previous stage 1 runs with smaller B_1 values can be reused. If one increases B_1 by a factor of 2 after each run, a factor of 2 will be saved on each stage 1 run.

Library Interface. Since version 6, GMP-ECM also includes a library, distributed under the GNU Lesser General Public License (LGPL). This library enables other applications to call ECM, P+1 or P−1 directly at the C-language level. For example, the Magma system uses the library since version V2.12, released in July 2005 [15].

Open Questions. The implementation of the "FFT continuation" described here is fine for moderate-size numbers (say up to 1000 digits) but may be too expensive for large inputs, for example Fermat numbers. In that case, one might want to go back to the classical standard continuation. Montgomery proposes in [18] the PAIR algorithm to hit all primes in the stage 2 range with small sets S and T. This algorithm was recently improved by Alexander Kruppa in [13], by choosing nodes in a partial cover of a bipartite graph.

Although many improvements have been made to stage 2 in the last years, the real bottleneck remains stage 1. The main question is whether it is possible to break the sequentiality of stage 1, i.e., to get a $o(B_1)$ cost. Any speedup to stage 1 is welcome: Alexander Kruppa suggested (personal communication) designing a sliding window variant in affine coordinates. Another idea is to save one multiply per duplicate by forcing d to be small in Eq. (2), as pointed out by Montgomery; Bernstein suggests to use $(16d + 18)y^2 = x^3 + (4d + 2)x^2 + x$ with starting point $(2 : 1)$. Computer experiments indicate that these curves have, on average, 3.49 powers of 2 and 0.78 powers of 3, while Suyama's family has 3.46 powers of 2 and 1.45 powers of 3.

Finally, is it possible to design a "stage 3", i.e., hit two large primes in stage 2? How much would it increase the probability of finding a factor?

Acknowledgements. Most of the ideas described here are due to other people: many thanks of course to H. W. Lenstra, Jr., for inventing that wonderful algorithm, to Peter Montgomery and Richard Brent for their great improvements, to George Woltman who helped to design the save/resume interface, and of course to the other developers of GMP-ECM, Alexander Kruppa, Jim Fougeron, Laurent Fousse, and Dave Newman. Part of the success of GMP-ECM is due to the GMP library, developed mainly by Torbjörn Granlund. The second author wishes to acknowledge computational support from Lehigh University, including access to NSF-funded Major Research Instrumentation. James Wanless, Sam Wagstaff, Richard Brent, Alexander Kruppa, Torbjörn Granlund, Peter Montgomery and the anonymous referee pointed out typos in previous versions of the paper. Finally, many thanks to all users of GMP-ECM, those who found large factors as well as the anonymous users who did not (yet) find any!

References

1. BARRETT, P. Implementing the Rivest Shamir and Adleman public key encryption algorithm on a standard digital signal processor. In *Advances in Cryptology, Proceedings of Crypto'86* (1987), A. M. Odlyzko, Ed., vol. 263 of *Lecture Notes in Computer Science*, Springer-Verlag, pp. 311–323.
2. BERNSTEIN, D. J. Removing redundancy in high-precision Newton iteration. http://cr.yp.to/fastnewton.html, 2004. 13 pages.
3. BERNSTEIN, D. J. Scaled remainder trees. http://cr.yp.to/papers.html#scaledmod, 2004. 8 pages.
4. BOSTAN, A., LECERF, G., AND SCHOST, E. Tellegen's principle into practice. In *Proceedings of the 2003 international symposium on Symbolic and algebraic computation* (Philadelphia, PA, USA, 2003), pp. 37–44.
5. BRENT, R. P. Some integer factorization algorithms using elliptic curves. *Australian Computer Science Communications 8* (1986), 149–163. http://web.comlab.ox.ac.uk/oucl/work/richard.brent/pub/pub102.html.
6. BRENT, R. P. Factor: an integer factorization program for the IBM PC. Tech. Rep. TR-CS-89-23, Australian National University, 1989. 7 pages. Available at http://wwwmaths.anu.edu.au/~brent/pub/pub117.html.
7. BRENT, R. P. Factorization of the tenth Fermat number. *Mathematics of Computation 68*, 225 (1999), 429–451.

8. BRENT, R. P., AND POLLARD, J. M. Factorization of the eighth Fermat number. *Mathematics of Computation 36* (1981), 627–630.
9. BURNIKEL, C., AND ZIEGLER, J. Fast recursive division. Research Report MPI-I-98-1-022, MPI Saarbrücken, 1998.
10. CHARRON, T., DAMINELLI, N., GRANLUND, T., LEYLAND, P., AND ZIMMERMANN, P. The ECMNET Project. http://www.loria.fr/~zimmerma/ecmnet/.
11. GRANLUND, T. *GNU MP: The GNU Multiple Precision Arithmetic Library*, 4.2 ed., 2006. http://www.swox.se/gmp/#DOC.
12. HANROT, G., QUERCIA, M., AND ZIMMERMANN, P. The middle product algorithm, I. Speeding up the division and square root of power series. *AAECC 14*, 6 (2004), 415–438.
13. KRUPPA, A. Optimising the enhanced standard continuation of the P−1 factoring algorithm. Diplomarbeit Report, Technische Universität München, 2005. http://home.in.tum.de/~kruppa/DA.pdf, 55 pages.
14. LENSTRA, H. W. Factoring integers with elliptic curves. *Annals of Mathematics 126* (1987), 649–673.
15. The Magma computational algebra system. http://magma.maths.usyd.edu.au/, 2005. Version V2.12.
16. MONTGOMERY, P. L. Evaluating recurrences of form $x_{m+n} = f(x_m, x_n, x_{m-n})$ via Lucas chains, 1983. Available at ftp.cwi.nl:/pub/pmontgom/Lucas.ps.gz.
17. MONTGOMERY, P. L. Modular multiplication without trial division. *Mathematics of Computation 44*, 170 (1985), 519–521.
18. MONTGOMERY, P. L. Speeding the Pollard and elliptic curve methods of factorization. *Mathematics of Computation 48*, 177 (1987), 243–264.
19. MONTGOMERY, P. L. *An FFT Extension of the Elliptic Curve Method of Factorization.* PhD thesis, University of California, Los Angeles, 1992. ftp.cwi.nl:/pub/pmontgom/ucladissertation.psl.gz.
20. PHATAK, D. S., AND GOFF, T. Fast modular reduction for large wordlengths via one linear and one cyclic convolution. In *Proceedings of 17th IEEE Symposium on Computer Arithmetic (ARITH'17), Cape Cod, MA, USA* (2005), IEEE Computer Society, pp. 179–186.
21. SCHÖNHAGE, A., AND STRASSEN, V. Schnelle Multiplikation großer Zahlen. *Computing 7* (1971), 281–292.
22. VON ZUR GATHEN, J., AND GERHARD, J. *Modern Computer Algebra.* Cambridge University Press, 1999.
23. WAGSTAFF, S. S. The Cunningham project. http://www.cerias.purdue.edu/homes/ssw/cun/.
24. WILLIAMS, H. C. A $p+1$ method of factoring. *Mathematics of Computation 39*, 159 (1982), 225–234.

An Index Calculus Algorithm for Plane Curves of Small Degree

Claus Diem

Mathematisches Institut, Universität Leipzig
Augustusplatz 10-11, 04109 Leipzig, Germany
diem@math.uni-leipzig.de

Abstract. We present an index calculus algorithm which is particularly well suited to solve the discrete logarithm problem (DLP) in degree 0 class groups of curves over finite fields which are represented by plane models of small degree. A heuristic analysis of our algorithm indicates that asymptotically for varying q, "almost all" instances of the DLP in degree 0 class groups of curves represented by plane models of a fixed degree $d \geq 4$ over $\mathbb{F}_q$ can be solved in an expected time of $\tilde{O}(q^{2-2/(d-2)})$.

Additionally we provide a method to represent "sufficiently general" (non-hyperelliptic) curves of genus $g \geq 3$ by plane models of degree $g + 1$. We conclude that on heuristic grounds, "almost all" instances of the DLP in degree 0 class groups of (non-hyperelliptic) curves of a fixed genus $g \geq 3$ (represented initially by plane models of bounded degree) can be solved in an expected time of $\tilde{O}(q^{2-2/(g-1)})$.

1 Introduction

In recent works by Gaudry, Thomé, Thériault and the author ([13]) as well as Nagao ([22]), a double large prime variation for index calculus in degree 0 class groups of curves of small genus over finite fields has been introduced.

In this work, we present a different double large prime variation algorithm which is particularly well suited for the computation of the discrete logarithm problem (DLP) in degree 0 class groups of curves which are represented by plane models of *small degree*.

A heuristic analysis of our algorithm indicates (see Section 4):

Heuristic Result 1. *Let $d \geq 4$ be fixed. Let us consider the DLP in degree 0 class groups of curves of a fixed genus $g \leq (d-1)(d-2)/2$ represented by plane models of degree d over finite fields $\mathbb{F}_q$. Then "almost all" instances of the DLP in such groups can be solved in an expected time of $\tilde{O}(q^{2-\frac{2}{d-2}})$.*

Here, the $\tilde{O}$-notation means that we suppress logarithmic factors.

Additionally to the index calculus algorithm, we present a method to find plane models of degree $g + 1$ of "sufficiently general" (non-hyperelliptic) curves of genus $g \geq 3$ (see Section 6).

F. Hess, S. Pauli, and M. Pohst (Eds.): ANTS 2006, LNCS 4076, pp. 543–557, 2006.

By applying our algorithm to such a plane model, we obtain that on heuristic grounds "almost all" instances of the DLP in degree 0 class groups of (non-hyperelliptic) curves of a fixed genus $g \geq 3$ (initially represented by plane models of bounded degree) can be solved in an expected time of

$$\tilde{O}(q^{2-\frac{2}{g-1}}).$$

This result should be compared with the following provable result which can be obtained with a variant of one of the algorithms in [13] (see [7]).

Let $g \geq 2$ be fixed. Then the DLP in cyclic degree 0 class groups of curves of genus g represented by plane models of bounded degree can with a randomized algorithm be solved in an expected time of $\tilde{O}(q^{2-2/g})$.

An important special case for our algorithm is constituted by the DLP in degree 0 class groups of *non-hyperelliptic curves of genus 3* over finite fields $\mathbb{F}_q$: Every such curve can (via the canonical embedding) be represented as a plane quartic. By applying our algorithm to such a model, we obtain a heuristic running time of $\tilde{O}(q)$.

This result is of particular importance because the DLP in degree 0 class groups of non-hyperelliptic genus 3 curves has recently received considerable attention as a potential cryptographic primitive; it is studied in detail in the related article [10] in which also some experimental data is presented.

Even though the DLP in degree 0 class groups of non-hyperelliptic curves of genus larger than 3 has not received much attention as a potential cryptographic primitive, our algorithm has yet another important application in cryptanalysis:

The method of "covering attacks" (a.k.a. Weil descent attacks) (cf. [8, Appendix], [9], [17], [12, Section 4.4]) allows to transfer the DLP in groups of rational points of certain elliptic curves (or in degree 0 class groups of certain curves of small genus) over extension fields into the DLP in degree 0 class groups of curves of rather small genus over smaller fields. The results in the present work suggest that it is advantageous for the attack if the resulting curves are not hyperelliptic.

2   Setting and First Remarks

Preliminaries
In this work, if not stated otherwise, a *curve* is always non-singular, projective and geometrically irreducible.

In the presentation above we implicitly used the following conventions concerning the representation of curves, divisors and divisor classes:

Let q be a prime power. We let $\mathbb{P}^2_{\mathbb{F}_q} := \mathrm{Proj}(\mathbb{F}_q[X,Y,Z])$; we thus have the canonical "homogeneous coordinate system" $X, Y, Z \in \Gamma(\mathbb{P}^2_{\mathbb{F}_q}, \mathcal{O}(1))$.

We think of every curve in question as being the normalization of a possibly singular curve in $\mathbb{P}^2_{\mathbb{F}_q}$. We distinguish the two by calling the latter one a *plane model* of the curve, denoted by $\mathcal{C}_{pm}$. We use a defining homogeneous polynomial to represent the plane model (and thus the curve itself).

By a *divisor* on a curve $\mathcal{C}$ over $\mathbb{F}_q$ we mean a divisor over $\mathbb{F}_q$. We think of divisors as being represented as a formal sum of closed points in $\mathcal{C}$. (This is called the *free representation* in [16].)

For some divisor D on $\mathcal{C}$, we denote the corresponding divisor class by $[D]$. We denote the degree 0 class group of $\mathcal{C}$ over $\mathbb{F}_q$ by $\mathrm{Cl}^0(\mathcal{C})$.

For fixed genus g and $q \gg 0$, $\mathcal{C}(\mathbb{F}_q)$ is non-empty; we assume that this is the case and fix some $P_0 \in \mathcal{C}(\mathbb{F}_q)$. An effective divisor D on $\mathcal{C}$ is called *maximally reduced along* P_0 if the linear system $|D - P_0|$ is empty. By the Riemann-Roch theorem, maximally reduced divisors have degree $\leq g$, and $D \mapsto [D] - \deg(D) \cdot [P_0]$ defines a bijection between the effective maximally reduced divisors and the elements of the degree 0 class group $\mathrm{Cl}^0(\mathcal{C})$ (see [16, Prop. 8.2.]).

It is by now a classical result that with this representation of the elements of the degree 0 class group, the arithmetic in $\mathrm{Cl}^0(\mathcal{C})$ can – for curves represented by plane models of bounded degree – be carried out in randomized polynomial time (cf. e.g. [26], [18], [16], [20], [19]).

Further Notation and Conventions
We use the same notation for functions on $\mathbb{P}^2_{\mathbb{F}_q}$, their restriction to $\mathcal{C}_{pm}$, their pull-back to $\mathcal{C}$ as well as the induced element in the function field $\mathbb{F}_q(\mathcal{C})$. Moreover, if $\varphi : \mathcal{C} \longrightarrow \mathbb{P}^2_{\mathbb{F}_q}$ is the (fixed) morphism from $\mathcal{C}$ to $\mathbb{P}^2_{\mathbb{F}_q}$, we use the same notation for elements of $\Gamma(\mathbb{P}^2_{\mathbb{F}_q}, \mathcal{O}(1))$ and their pull-backs to $\Gamma(\mathcal{C}, \varphi^*(\mathcal{O}(1)))$.

We identify zero-dimensional closed subschemes on $\mathcal{C}$ with effective divisors. To distinguish the divisor of zeros of an element of $W \in \Gamma(\mathbb{P}^2_{\mathbb{F}_q}, \mathcal{O}(1))$ from the divisor of zeros of the induced element in $\Gamma(\mathcal{C}, \varphi^*(\mathcal{O}(1)))$, we write $\mathrm{div}_\mathcal{C}(W)$ for the latter. (See [15, II, §7] for information about the divisor of zeros.)

Calculating the Group Order
We assume that the order of the degree 0 class group is known. From a theoretical point of view this is however not an obstacle because it can be shown that the L-polynomials of curves over $\mathbb{F}_q$ represented by plane models of bounded degree can be calculated in (deterministic) polynomial time in $\log(q)$. (This result follows from [24, Theorem H] which in turn relies on Pila's extension of the point counting algorithm by Schoof ([25]) to abelian varieties ([23]).) Moreover, in cryptographic situations, the order of the cyclic subgroup in question is always known, and this suffices for practical applications of our algorithm.

Overview over the New Algorithm
Our algorithm can be viewed as a variant of the recent double large prime variation algorithms by Gaudry, Thomé, Thériault and the author ([13]) as well as Nagao ([22]) (see also [3]).

The main difference is that we use principal divisors to construct the graph of large prime relations, whereas in [13] and [22] random linear combinations of the two input elements in the degree 0 class group have been used.

More concretely, we find relations by intersecting the plane model with lines running through two elements of the factor base. We advice the reader to have the following *intuitive idea* about the algorithm and its heuristic analysis in mind: Every line which runs through the non-singular part of the plane model defines a divisor of degree d on the curve. If we now intersected the plane model with arbitrary lines, heuristically we would obtain a running time which is analogous to the running time of the previous double-large prime-variation algorithms with g substituted by d. As we however only consider lines which already run through two points of the factor base, we obtain a running time which is analogous to the running time of the previous algorithms with g substituted by $d - 2$.

We recall that there are two algorithms in [13]: the "full algorithm" and the "simplified algorithm". Our algorithm is closer to the "full algorithm" but there is an essential difference: In the full algorithm in [13], recombined relations over the factor base are already obtained during the construction of the graph. In contrast, we first try to construct a sufficiently dense graph, and after that we construct what is known as a *shortest path tree*. Then we use random linear combinations of the two input elements to generate recombined relations over the factor base with the help of the tree.

The Heuristic Nature of Our Results
The analysis of the algorithm presented in this work is heuristic. It is conceivable that there is a sequence of instances which violates the stated running times. This is why we talk about "almost all" instances.

A rigorous interpretation of our claims can be given as follows:

Let us fix the degree d and the genus $g \leq (d-1)(d-2)/2$. Now for a prime power q, let $S(q)$ be the set of all instances of the DLP in curves of genus $g \leq (d-1)(d-2)/2$ over $\mathbb{F}_q$ represented by plane models of degree d. (With the representations described above.)

The (conjectural) claim is now that there exist subsets $S_1(q)$ of $S(q)$ with $\#S_1(q)/\#S(q) \longrightarrow 1$ $(q \longrightarrow \infty)$ such that the instances in $S_1(q)$ can be solved in the stated time.

Above, we also used the term "sufficiently general". This term will be defined in Section 6.

Historical Remarks and Comparison
The idea to use principal divisors to generate relations in class groups is not new. For example, the same approach was taken in the work by Adleman, DeMarrais, Huang ([1]), in which the first algorithm with a heuristic subexponential running time for the computation of the DLP in degree 0 class groups of hyperelliptic curves of large genus was given.

We note that to our knowledge, all known index calculus algorithms which rely on the consideration of principal divisors are analyzed only heuristically. With our two-step procedure to generate relations we have however eliminated a crucial hypothesis which previously occurred in the analyses of such algorithms: the hypothesis that "sufficiently many" of the relations generated are linearly independent.

3   The Algorithm

We consider curves over $\mathbb{F}_q$ represented by plane models of a fixed degree $d \geq 4$. Let $\mathcal{C}$ be such a curve with a fixed plane model $\mathcal{C}_{pm}$ in $\mathbb{P}^2_{\mathbb{F}_q}$, given by

$$F(X, Y, Z) = 0 \,.$$

Let $a, b \in \mathrm{Cl}^0(\mathcal{C})$ such that $b \in \langle a \rangle$. The goal is to compute an $x \in \mathbb{N}$ with $x \cdot a = b$.

Let $D_\infty := \mathrm{div}_{\mathcal{C}}(Z)$. Note that this is a divisor of degree d on $\mathcal{C}$. (This divisor will appear in the description of the algorithm, it is however not necessary to compute it.)

Let $\mathcal{C}_{ns}$ be the non-singular part of $\mathcal{C}_{pm}$.

We now describe how the partial relations used to construct the graph of large prime relations are obtained.

The following classical statement from the theory of linear systems is crucial:

Lemma 1. *Let $W \in \Gamma(\mathbb{P}^2_{\mathbb{F}_q}, \mathcal{O}(1))$ ($W \neq 0$), and let $D := \mathrm{div}_{\mathcal{C}}(W)$. Then D is linearly equivalent to D_∞.*

Sketch of the proof. $D - D_\infty$ is the principal divisor of $\frac{W}{Z} \in \mathbb{F}_q(\mathcal{C})$.   □

As a reformulation of this we obtain: Let $c_X, c_Y, c_Z \in \mathbb{F}_q$, not all 0, and let L be the line defined by $c_X X + c_Y Y + c_Z Z = 0$. Let $D := L \cap \mathcal{C}_{pm}$ be the (scheme-theoretic) intersection. If then D is contained in $\mathcal{C}_{ns}$, we can regard D as a divisor on $\mathcal{C}$, and we have

$$[D] - [D_\infty] = 0 \,. \tag{1}$$

Lemma 2. *Given $c_X, c_Y, c_Z \in \mathbb{F}_q$, not all 0, one can decide in randomized polynomial time in $\log(q)$ if the support of the intersection of $\mathcal{C}_{pm}$ with L consists of $\mathbb{F}_q$-rational points of $\mathcal{C}_{ns}$ and – if this is the case – compute the (completely split) intersection divisor D.*

Proof. Let us (w.l.o.g.) assume that $c_Y = 1$. Then the point $(0 : 1 : 0)$ does not lie on L. The homogeneous polynomial $F(X, -c_X X - c_Z Z) \in \mathbb{F}_q[X, Z]$ now defines the image of the intersection under the projection to the (X, Z)-coordinates (with multiplicities). The support of the intersection of $\mathcal{C}_{pm}$ with L consists of $\mathbb{F}_q$-rational points of $\mathcal{C}_{pm}$ if and only if this polynomial factors completely. This factorization can be computed in randomized polynomial time in $\log(q)$. The Y-coordinates of the intersection points can then easily be obtained by using the equation for the line L. Finally, one can check whether the intersection points lie in $\mathcal{C}_{ns}$ by evaluating the partial derivatives of F.   □

Let us now fix a *factor base* $\mathcal{F} = \{F_1, F_2, \ldots\} \subset \mathcal{C}_{ns}(\mathbb{F}_q)$. Let $\mathcal{L} := \mathcal{C}_{ns}(\mathbb{F}_q) - \mathcal{F}$ be the set of the so-called *large primes*. Analogously to [13] we define:

Definition 1. *A relation (1) (with $D \geq 0$) is called a* Full *relation if D is a sum of elements of the factor base. It is called an* FP *relation if D is a sum of elements of the factor base and the non-trivial multiple of one large prime. It is called a* PP *relation if D is the sum of elements of the factor base and non-trivial multiples of two large primes.*

In the first phase of the algorithm, we construct a *graph of large prime relations* on $\mathcal{L} \cup \{*\}$ using FP and PP relations.

We find such relations by intersecting the curve with lines $L : c_X X + c_Y Y + c_Z Z = 0$ $(c_X, c_Y, c_Z \in \mathbb{F}_q)$ running through two points of the factor base.

For the construction of the graph of large prime relations, we proceed as follows:

If we have a Full relation, we do nothing. If we have an FP relation with a large prime P, we consider the edge between $*$ and P, if we have a PP relation with two large primes P and Q, we consider the edge between P and Q. If the edge does not yet occur in the graph, we insert it, labeled with the data for the relation.

Remark 1. The graph we construct here can have many cycles. In contrast, the graph constructed in the "full algorithm" in [13] is acyclic.

After having constructed a graph with a sufficiently large connected component containing $*$, we construct what is known as a *shortest path tree* with root $*$.

Definition 2. *Let G be an undirected (unweighted) graph, and let $*$ be a vertex in G. Then a* shortest path tree *with root $*$ is a tree on a subset of the set of vertices of G with the following properties:*

- *The vertices in T are the vertices in the connected component of $*$ in G.*
- *For any vertex V in T, the distance between $*$ and V in G is equal to the distance between $*$ and V in T.*

Notation 1. *The set of vertices of a tree T is also denoted by T.*

It is easy to construct a shortest-path tree algorithmically with the so-called breadth-first search (see [6, Section 22.2]).

As written in Section 2, for every element $c \in \mathrm{Cl}^0(\mathcal{C})$ there is a unique along F_1 maximally reduced effective divisor D such that $[D] - \deg(D) \cdot [F_1] = c$ (here as above, F_1 is the first element of the factor base).

We use this representation of the elements of the degree 0 class group and proceed as in Phase 2 of the "simplified algorithm" in [13]. Provided that the degree 0 class group is cyclic and generated by a this means that we consider random linear combinations of the inputs a and b which we try to express as sums of elements of $\mathcal{F} \cup T$. We then use the tree to substitute the vertices of T involved by sums of (possibly negative) multiples of elements in the factor base and D_∞. Finally, we solve the DLP with an algorithm from sparse linear algebra.

We are now ready to give the complete algorithm. For simplicity we thereby assume that the group order ℓ is prime. (If the group is cyclic but not of prime order or the group is arbitrary but its structure is known, Steps 5 and 6 should be modified according to the descriptions in [13] and [11].)

The Algorithm

Input: A curve $\mathcal{C}/\mathbb{F}_q$, given by a plane model of degree d,
the group order $\ell := \#\mathrm{Cl}^0(\mathcal{C})$ and two elements $a, b \in \mathrm{Cl}^0(\mathcal{C})$ with $\langle a \rangle = \mathrm{Cl}^0(\mathcal{C})$.

1. Enumerate $\mathcal{C}_{ns}(\mathbb{F}_q)$ and choose a factor base $\mathcal{F} = \{F_1, F_2, \ldots\}$ uniformly at random from the set of all subsets of $\mathcal{C}_{ns}(\mathbb{F}_q)$ with $\lceil (4 \cdot (d-2)!)^{1/(d-2)} \cdot q^{1-1/(d-2)} \rceil$ elements.
   (If $\mathcal{C}_{ns}(\mathbb{F}_q)$ has fewer elements, terminate.)
2. Construct a graph G on $\mathcal{L} \cup \{*\}$ (where $\mathcal{L} := \mathcal{C}_{ns}(\mathbb{F}_q) - \mathcal{F}$) as follows:
   For all $i < j$ do
      Compute the line L through F_i and F_j.
      If $D := L \cap \mathcal{C}_{pm}$ is contained in $\mathcal{C}_{ns}$
      and splits completely into points of $\mathcal{C}_{ns}(\mathbb{F}_q)$, then
         if it defines an FP or a PP relation, then
            if the corresponding edge does not yet occur in the graph, then
               insert the edge in the graph.
3. Construct a shortest path tree T with root $*$ in G.
4. If T has less than $\frac{1}{\log(q)} \cdot q$ vertices or the depth of T is $> \log^2(q)$, go back to 1.
5. Construct a sparse matrix R over $\mathbb{Z}/\ell\mathbb{Z}$ as follows:
   For $i = 1, \ldots, \#\mathcal{F} + 1$ do
      Repeat
         Choose uniformly and independently randomly α_i and β_i and compute the unique along F_1 maximally reduced effective divisor D with
         $[D] - \deg(D) \cdot [F_1] = \alpha_i a + \beta_i b$.
      Until D splits into elements of $\mathcal{F} \cup T$.
      Use the tree T to substitute these elements
      by sums of multiples of elements of $\mathcal{F} \cup \{D_\infty\}$.
      If this substitution leads to the relation $\sum_j r_{i,j}[F_j] + r_i[D_\infty] = \alpha_i a + \beta_i b$,
      store $(r_{i,j})_j$ as the i-th row of R.
6. Compute a non-zero vector γ over $\mathbb{Z}/\ell\mathbb{Z}$ with $\gamma R = 0$ with an algorithm from sparse linear algebra.
7. If $\sum_i \gamma_i \beta_i \in (\mathbb{Z}/\ell\mathbb{Z})^*$, let

$$ x \longleftarrow -\frac{\sum_i \gamma_i \alpha_i}{\sum_i \gamma_i \beta_i}, $$

otherwise go back to 5.
Output x.

Proposition 1. *If the algorithm outputs x, we have $x \cdot a = b$.*

Proof. With the notation in Steps 5 and 6, we have

$$ \sum_i \gamma_i \alpha_i a + \sum_i \gamma_i \beta_i b = \sum_{i,j} \gamma_i r_{i,j}[F_j] + \sum_i \gamma_i r_i[D_\infty] = \sum_i \gamma_i r_i[D_\infty]. $$

As $\sum_i \gamma_i \alpha_i a + \sum_i \gamma_i \beta_i b$ has degree 0, we have $\sum_i \gamma_i r_i = 0$, i.e. $\sum_i \gamma_i \alpha_i a + \sum_i \gamma_i \beta_i b = 0$. This implies $x \cdot a = b$.   □

550   C. Diem

4 Heuristic Analysis

The following heuristic analysis is for fixed degree d and fixed genus $g \leq (d-1)(d-2)/2$ and $q \longrightarrow \infty$. We note that even though the genus is bounded if we fix the degree (which suffices for our heuristic analysis), we fix the genus additionally to the degree because we want to derive statements on almost all instances for every fixed degree and genus.

A "randomized" factor base as in Step 1 can be found in an expected time of $\tilde{O}(q)$ as follows:

First all points of $\mathcal{C}_{ns}(\mathbb{F}_q)$ are enumerated. By iterating over the (X, Z)-coordinates and considering the possible Y-coordinates, this can be done in a time of $\tilde{O}(q)$. After this, a factor base as in Step 1 of the algorithm can be constructed by uniformly randomly choosing points of $\mathcal{C}(\mathbb{F}_q)$. The expected running time is then again in $\tilde{O}(q)$.

We now come to the task to analyze the size of the tree T as well as its depth. This task seems to be very difficult, and our analysis relies on several heuristic assumptions. A key technique of our approach is to use the randomization of the factor base and to rely on a heuristic comparison of the graph which is constructed in Step 2 with an appropriate "random graph".

We will use these notations:

Definition 3. *Let $(a_n)_{n \in \mathbb{N}}$ and $(b_n)_{n \in \mathbb{N}}$ be two sequences of real numbers. Then we write*
$$a_n \gtrsim b_n$$
if $\liminf \frac{a_n}{b_n} \geq 1$.

Definition 4. *For $P, Q \in \mathcal{C}_{ns}(\mathbb{F}_q)$ with $P \neq Q$, let p_{PQ} be the probability that $P, Q \in \mathcal{L}$ and the unordered pair $\{P, Q\}$ occurs as an edge in the graph (if we choose the factor base uniformly at random from the set of all factor bases with $\lceil (4 \cdot (d-2)!)^{1/(d-2)} \cdot q^{1-1/(d-2)} \rceil$ elements). Let*
$$p_{av} := \frac{1}{\#\mathcal{C}_{ns}(\mathbb{F}_q) \cdot (\#\mathcal{C}_{ns}(\mathbb{F}_q) - 1)} \cdot \sum_{P,Q \in \mathcal{C}_{ns}(\mathbb{F}_q) \text{ with } P \neq Q} p_{PQ} \ .$$

Note that p_{av} can be seen as the *average probability* that an (unordered) pair of distinct points in $\mathcal{C}_{ns}(\mathbb{F}_q)$ occurs as an edge in the graph.

Lemma 3. *For $P, Q \in \mathcal{C}_{ns}(\mathbb{F}_q)$ with $P \neq Q$ such that the line through P and Q intersects $\mathcal{C}_{pm}$ only in $\mathcal{C}_{ns}$ and the intersection divisor splits completely into a sum of distinct points of $\mathcal{C}_{ns}(\mathbb{F}_q)$, we have*
$$p_{PQ} \sim 4 \cdot (d-2)! \cdot \frac{1}{q} \ .$$

Proof. Let $D = P + Q + R$ be the intersection divisor. Then the probability p_{PQ} is equal to the probability that the factor base contains all $d-2$ points from R and does not contain P and Q.

The probability p_{PQ} is thus

$$\frac{\binom{\#\mathcal{C}_{ns}(\mathbb{F}_q)-d}{\lceil(4\cdot(d-2)!)^{1/(d-2)}\cdot q^{1-1/(d-2)}\rceil-(d-2)}}{\binom{\#\mathcal{C}_{ns}(\mathbb{F}_q)}{\lceil(4\cdot(d-2)!)^{1/(d-2)}\cdot q^{1-1/(d-2)}\rceil}} \ .$$

For $q \longrightarrow \infty$ this is asymptotically equivalent to

$$\left(\frac{(4\cdot(d-2)!)^{1/(d-2)}\cdot q^{1-1/(d-2)}}{q}\right)^{d-2} = 4\cdot(d-2)!\cdot\frac{1}{q} \ . \qquad \square$$

By the Hasse-Weil bounds, there are $\sim q^d$ divisors of degree d on $\mathcal{C}$ of whose $\sim \frac{1}{d!}q^d$ split completely. The probability that a uniformly randomly chosen divisor on $\mathcal{C}$ of degree d is completely split is thus asymptotically equal to $\frac{1}{d!}$. This motivates:

Heuristic Assumption 1. *For almost all instances, the probability that a uniformly randomly chosen divisor in the linear system $|D_\infty|$ is completely split is $\geq \frac{1}{2}\cdot\frac{1}{d!}$.*

Remark 2. In the case non-hyperelliptic curves of genus 3 (given as plane quartics), it is possible to prove via an effective Chebotarev theorem that the probability that a uniformly randomly chosen divisor in $|D_\infty|$ is completely split is asymptotically equal to $\frac{1}{4!}$. Thus Heuristic Assumption 1 is satisfied in this case (see [10]).

Proposition 2. *Under Heuristic Assumption 1, for almost all instances, $p_{av}\cdot q \gtrsim 2$, and the expected number of edges in the graph of large prime relations is $\gtrsim q$.*

Proof. We restrict ourselves to instances for which Heuristic Assumption 1 is satisfied.

We first note that the number of divisors in $|D_\infty|$ which split completely into sums of distinct points of $\mathcal{C}_{ns}(\mathbb{F}_q)$ is $\gtrsim \frac{1}{2}\frac{1}{d!}\cdot q^2$.

Indeed, the number of completely split divisors is by assumption $\geq \frac{1}{2}\frac{1}{d!}\cdot\frac{q^3-1}{q-1}$. By the formulae for the arithmetic and the geometric genus, the number of singular points in $(\mathcal{C}_{pm})_{\overline{\mathbb{F}}_q}$ is $\leq (d-1)(d-2)/2$, thus the number of lines in $\mathbb{P}^2_{\mathbb{F}_q}$ through singular points is in $O(q)$. Moreover, every divisor in $|D_\infty|$ which has the form $\sum_P n_P P$ with $n_P \geq 2$ for some $P \in \mathcal{C}_{ns}(\mathbb{F}_q)$ is defined by a line in $\mathbb{P}^2_{\mathbb{F}_q}$ which is tangential to $\mathcal{C}_{pm}$. This means that the total number of such divisors is also in $O(q)$.

For any divisor D in $|D_\infty|$ which splits completely into a sum of distinct points of $\mathcal{C}_{ns}(\mathbb{F}_q)$, there are $d\cdot(d-1)$ ordered pairs of distinct points in $\mathcal{C}_{ns}(\mathbb{F}_q)$ in the support of the divisor. Each of these pairs of points fulfills the assumptions of Lemma 3 (and conversely, any pair of points fulfilling the assumptions of

Lemma 3 determines uniquely such a divisor D). Thus there are $\gtrsim \frac{1}{2 \cdot d!} \cdot d(d - 1) \cdot q^2 = \frac{1}{2 \cdot (d-2)!} \cdot q^2$ ordered pairs of distinct points of $\mathcal{C}_{ns}(\mathbb{F}_q)$ which fulfill the assumption of Lemma 3.

The average probability p_{av} is thus

$$\underset{\sim}{\gtrsim} \frac{1}{q^2} \cdot \left(\frac{1}{2 \cdot (d-2)!} \cdot q^2 \right) \cdot \left(4 \cdot (d-2)! \cdot \frac{1}{q} \right) = \frac{2}{q} \ .$$

If one multiplies the average probability p_{av} by the number of unordered pairs of points of $\mathcal{C}_{ns}(\mathbb{F}_q)$, one obtains the claimed asymptotic lower bound on the expected number of edges.   □

It does not seem to be easy to study the number of vertices in the connected component of $*$ of G (which is equal to the number of vertices in the tree T) as well as the depth of the tree.

We note however the following result from the theory of random graphs: Let $G(n, p)$ denote a random graph on n vertices in which each unordered pair of vertices appears (independently of the other pairs of vertices) as an edge with probability p (this is called a *Bernoulli random graph* in [27]). Then we have (see [4, Theorem 6.11] together with [4, Theorem 2.2 a)] as well as [5]):

Proposition 3. *Let $c > 1$ be a constant. Then for $p \cdot n \geq c$, with probability converging to 1 for $n \longrightarrow \infty$, $G(n, p)$ has a "giant connected component" of size $\Theta(n)$, and the diameter of the graph is in $O(\log(n))$.*

We now have the following situation: As in the conclusion of Proposition 2, let $p_{av} \cdot q \gtrsim 2$. Then with probability converging to 1, a random graph $G(\#\mathcal{L} \cup \{*\}, p_{av})$ has a "giant connected component" of size $\Theta(q)$ and diameter $O(\log(q))$.

Clearly, there are three essential differences between Bernoulli random graphs and the situation we have here:

1. In contrast to Bernoulli random graphs, many of the pairs of vertices are never drawn.
2. In contrast to Bernoulli random graphs, the probabilities of two pairs of vertices appearing as edges in the graph are not independent.
3. In contrast to Bernoulli random graphs, we have the "special vertex" $*$ which heuristically occurs in much more edges than the vertices in $\mathcal{L}$.

The analysis now relies on the heuristic assumption that analogous to a random graph $G(\#\mathcal{L} \cup \{*\}, p_{av})$, for almost all instances, "sufficiently often" our graph has a "giant connected component" of "sufficient size" and "sufficiently small" diameter containing $*$. As an approach to cope with possible distortions, we require only that with a probability of $\tilde{\Omega}(1)$, we have $\geq \frac{1}{\log(q)} \cdot q$ vertices and the maximal distance to $*$ is $\leq \log^2(q)$ (cf. the conditions in Step 4). (The $\tilde{\Omega}$-notation should be understood analogously to the $\tilde{O}$-notation.)

The above considerations motivate:

Heuristic Result 2. *For almost all instances, Step 5 of the algorithm is reached after at most $\tilde{O}(1)$ iterations of 1 – 4.*

As there are $\Theta(q^{2-2/(d-2)})$ iterations within Step 2, this step has a running time of $\tilde{O}(q^{2-2/(d-2)})$.

With the breadth-first algorithm, given a graph on n vertices with m edges represented by numbers whose bit-length is polynomial in $\log(n)$, a shortest-path tree can be computed in a time of $\tilde{O}(n+m)$. As the graph clearly contains $O(q)$ vertices and $O(q^{2-2/(d-2)})$ edges, the running time of Step 3 is in $\tilde{O}(q^{2-2/(d-2)})$.

This means that on the basis of Heuristic Result 2, for almost all instances, Step 5 of the algorithm can be reached in a time of $\tilde{O}(q^{2-2/(d-2)})$.

Under the assumption that the degree 0 class group is cyclic or the group structure is known, the rest of the algorithm can be analyzed rigorously. For simplicity, as in the description of the algorithm, we stick to the case that the degree 0 class group has prime order ℓ. For modifications for the general case, we refer to [11] and [13].

We have the following general lemma.

Lemma 4. *Let us consider curves $\mathcal{C}$ over $\mathbb{F}_q$ of a fixed genus g together with a point $P_0 \in \mathcal{C}(\mathbb{F}_q)$ and a set of rational points $S \subset \mathcal{C}(\mathbb{F}_q)$ such that $\#S = \tilde{\Omega}(q)$. Then there are $\tilde{\Omega}(q^g)$ effective divisors D which split completely into sums of elements of S and are maximally reduced along P_0.*

Proof. If D is a non-special effective divisor of degree g, then the unique effective divisor D' which does not have P_0 in its support and satisfies $D' + (g - \deg(D')) \cdot P_0 = D$ is maximally reduced along P_0. Clearly there are $\tilde{\Omega}(q^g)$ effective divisors of degree g which split completely into sums of elements of S, and by the Hasse-Weil bounds, there are only $O(q^{g-1/2})$ special divisors of degree g. This is asymptotically negligible against $\tilde{\Omega}(q^g)$. □

This lemma implies that with a probability of $\tilde{\Omega}(1)$ one choice of α_i and β_i in Step 5 leads to a divisor D which splits over $\mathcal{F} \cup T$. Step 5 then has an expected running time of $\tilde{O}(q^{1-1/(d-2)})$.

Because of the condition that the depth of T is $\leq \log^2(q)$, the expected average number of elements in each row of the relation matrix is in $O(\log(q)^2)$. This implies that Step 6 has a running time of $\tilde{O}(q^{2-2/(d-2)})$. Finally, as argued in [13], $\sum_i \gamma_i \beta_i$ is uniformly randomly distributed over the group $\mathbb{Z}/\ell\mathbb{Z}$.

All in all, we have the following heuristic result:

Heuristic Result 3. *For almost all instances, the DLP in $\mathrm{Cl}^0(\mathcal{C})$ can be computed in a time of $\tilde{O}(q^{2-2/(d-2)})$.*

This is essentially the heuristic result stated in the introduction.

We note however that in the introduction we did not assume that the group is cyclic or the group structure is known. We have to make an additional heuristic assumption if the relation generation takes place in a proper subgroup of the degree 0 class group.

5   Practical Aspects

In this section, we briefly discuss some practical aspects of our algorithm and possible variants for concrete computations.

1. For practical purposes it might be advisable not to first construct the graph, then the shortest path tree and then to use this tree to derive relations via random linear combinations of the input values. Instead, one can proceed as follows:
   - First, one computes representatives of multiples αa and βb of the input values a and b which split completely into sums of points of $\mathcal{C}_{ns}(\mathbb{F}_q)$.
   - One chooses the factor base, thereby inserting the points in $\mathcal{C}_{ns}(\mathbb{F}_q)$ for the representatives for αa and βb.
   - One generates relations by considering lines through points of the factor base as described in Section 3 but otherwise one proceeds as in the "full algorithm" of [13]. This means that every time one would obtain a cycle, one does not insert the corresponding edge in the graph but instead tries to use this cycle to obtain a relation over the factor base.
   - One stops if one has found enough "sufficiently light" cycles. Then one solves the DLP via linear algebra.

   With this approach only for the initial computation of multiples of a and b one needs an algorithm for arithmetic in the degree 0 class group. If this initial computation is not time-critical, this might simplify the implementation.

   The approach presented above is particularly advantageous if g is much larger than d (for example if the plane model itself is non-singular and therefore $g = (d-1)(d-2)/2$. Note that the initial computation of multiples of a and b might even dominate the running time.

2. The number of points in the factor base ($\lceil (4 \cdot (d-2)!)^{1/(d-2)} \cdot q^{1-1/(d-2)} \rceil$) was chosen such that we expect the graph of large prime relations to be large enough for fixed degree d and $q \gg 0$. It might be necessary to choose the factor base slightly larger for concrete computations. This applies in particular if one follows the variant presented above.

3. If every pair of points in the factor base is considered to generate the graph, a line through the factor base defining a PP relation is usually considered $\binom{d-2}{2}$ times. To decrease the occurrence of such "repeated selections", it might be advisable to choose the factor base larger than necessary.

4. For the graph of large prime relations to be large enough, one needs at least about q divisors in $|D_\infty|$ which split completely. This implies that *one should have $q > d!$ if one applies the algorithm*. The case $d! \approx q$ can be considered as a boundary case. In this case, one could try to apply the variant presented in Point 1 by choosing the factor base equal to $\mathcal{C}_{ns}(\mathbb{F}_q)$ and ignoring the large prime variation.

5. To reduce the storage requirements, it might be advisable to combine our relation generation with the "simplified algorithm" of [13], i.e. when constructing the graph of large prime relations, one discards all edges which are not connected to $*$. The factor base then has to be enlarged by a logarithmic factor.

6 Finding Plane Models of Degree g+1

In this section we start off with some curve C of genus ≥ 3 over an "effective field" k. The goal is to find a plane model of degree $g + 1$ (provided such a model exists). In order to bound the time for computation of this plane model we assume that the curve C is initially given by a plane model of bounded degree.

The idea is to define a morphism $C \longrightarrow \mathbb{P}^2_k$ via a special linear system of degree $g + 1$. The case of non-hyperelliptic genus 3 curves is particularly easy: the canonical system $|K|$ itself defines an embedding into $\mathbb{P}^2_k$. For the general case we have the following proposition (see Point (b) in the introduction of [14]):

Proposition 4. *A general linear system of degree d and (projective) dimension ≥ 2 on a general curve of genus g has dimension 2, no base-points and defines a morphism to $\mathbb{P}^2$ which is birational onto its image.*

Here as usual, by a *general curve* we mean a curve which is obtained by base-change from the curve corresponding to the generic point of the (coarse) moduli space $\mathcal{M}_g$. (This space exists by [21, Corollary 7.14.].) A *general effective divisor* of degree d is the divisor on $C_{k(C_d)}$ corresponding to the generic point of C_d. Here, following [2] and [14], C_d denotes the d-fold symmetric power of C.

Let us say that a property holds for *sufficiently general* curves (of a prescribed genus) and / or for *sufficiently general* linear systems of divisors (of a prescribed degree and dimension) if it holds for curves and divisors in an open part of the corresponding moduli space.

We can then conclude that the linear system of any sufficiently general linear system of degree d and dimension ≥ 2 on any sufficiently general curve defines a morphism to $\mathbb{P}^2$ which is birational onto its image. (As usual, the morphism is unique up to an automorphism of $\mathbb{P}^2$.)

Following [2] and [14], let us denote the locus of complete linear systems of degree d and (projective) dimension $\geq n$ (in a twist of the Jacobian) by $W_d^n(C)$. We have the following proposition.

Proposition 5. *Let C be any curve of genus $g \geq 3$. Then we have birational morphisms*

$$C_{g-3} \longrightarrow W_{g-3}^0(C) \longrightarrow W_{g+1}^2(C)$$
$$D \mapsto |D| \mapsto |K - D|.$$

In particular, for any (sufficiently) general effective divisor D of degree $g-3$ on a (sufficiently) general curve C, $|K - D|$ has no base points, (projective) dimension 2 and defines a morphism to $\mathbb{P}^2$ which is birational onto its image.

Proof. The morphism $C_{g-3} \longrightarrow W_{g-3}^0(C)$, $D \mapsto |D|$ is birational because for any curve, the linear system of any general effective divisor of degree $< g$ has dimension 0.

By the Riemann-Roch theorem ([15, IV, Theorem 1.3]) and the fact that $\deg(K) = 2g - 2$, we have an isomorphism $W_{g-3}^0(C) \longrightarrow W_{g+1}^2(C)$, $|D| \mapsto |K - D|$. $\qquad\square$

Remark 3. Not every curve of genus g has a plane model of degree $g + 1$. For example, no hyperelliptic curve has such a model.

We have the following method to compute plane models of degree $g + 1$:

Computation of a Plane Model of Degree $g + 1$

Input: Any curve $\mathcal{C}/k$.

1. Compute a canonical divisor K on $\mathcal{C}$.
2. Select any effective divisor D on $\mathcal{C}$ of degree $g - 3$.
3. Compute a basis $b_1, \ldots, b_n$ of the Riemann-Roch space $L(K - D)$.
4. If the basis has more than 3 elements, terminate.
5. Compute a homogeneous polynomial $F(X, Y, Z) \in k[X, Y, Z]$ of minimal degree with $F(b_1, b_2, b_3) = 0$.
6. If $\deg(F) < g + 1$, terminate.
7. Output $(F; b_1, b_2, b_3)$.

The necessary computations of divisors and Riemann-Roch spaces can be carried out with the algorithms in [16]. Step 5 can be performed by computing (successively for $i = 1, \ldots, g + 1$) the functions $b_1^{i_1} \cdot b_2^{i_2} \cdot b_3^{i_3}$ with $i = i_1 + i_2 + i_3$ and trying to find a linear relation between them. The latter task is a linear algebra problem. Over finite fields we have:

Proposition 6. *There exists a specification of the above method such that for curves over finite fields $\mathbb{F}_q$ initially represented by plane models of bounded degree the expected running time is polynomial in $\log(q)$.*

Example 1. At the end of [8], an elliptic curve $\mathcal{E}$ over $\mathbb{F}_{p^7}$ with $p = 10000019$ is given such that the DLP in $\mathcal{E}(\mathbb{F}_{p^7})$ can be transferred into a DLP in the degree 0 class group of a certain curve $\mathcal{C}$ of genus 7 over $\mathbb{F}_p$. An explicit equation for $\mathcal{C}$ is also given. Using the method outlined above, we computed various degree 8 models of this curve.

Acknowledgments

With great pleasure, I thank G. Frey, P. Gaudry, F. Hess, K. Khuri-Makdisi, J. Pila, J. Scholten, N. Thériault, E. Thomé and E. Viehweg for discussions and comments. I also thank the anonymous referee for suggestions.

References

[1] L. Adleman, J. DeMarrais, and M.-D. Huang. A Subexponential Algorithm for Discrete Logarithms over the Rational Subgroup of the Jacobians of Large Genus Hyperelliptic Curves over Finte Fields. In L. Adleman and M.-D. Huang, editors, *Algorithmic Number Theory – ANTS I*, LNCS, pages 28–40, Berlin, 1994. Springer-Verlag.

[2] E. Arbarello, M. Cornalba, P. Griffiths, and J. Harris. *Geometry of Algebraic Curves*. Springer-Verlag, 1985.

[3] R. Avanzi and N. Thériault. Index Calculus for Hyperelliptic Curves. In H. Cohen and G. Frey, editors, *Handbook of Elliptic and Hyperelliptic Curve Cryptogrpahy*, chapter 21. Chapman & Hall/CRC, Boca Raton, 2006.

[4] B. Bollobas. *Random Graphs*. Cambridge University Press, Cambridge, 2001.

[5] F. Chung and L. Lu. The diameter of sparse random graphs. *Adv. in Appl. Math.*, 26:257–279, 2001.

[6] T. Cormen, C. Leiserson, R. Rivest, and C. Stein. *Introduction to Algorithms*. McGraw-Hill and The MIT Press, 2001. Second Edition.

[7] C. Diem. Index calculus with double large prime variation for arbitrary curves of small genus. Forthcoming.

[8] C. Diem. The GHS Attack in odd Characteristic. *J. Ramanujan Math. Soc.*, 18:1–32, 2003.

[9] C. Diem and J. Scholten. Cover attacks. A report for the AREHCC project, available under http://www.arehcc.com/documents.htm, 2003.

[10] C. Diem and E. Thomé. Index calculus in class groups of non-hyperelliptic curves of genus 3. Forthcoming.

[11] A. Enge and P. Gaudry. A general framework for subexponential discrete logarithm algorithms. *Acta. Arith.*, 102:83–103, 2002.

[12] S. Galbraith and A. Menezes. Algebraic curves and cryptography. *Finite fields and applications*, 11:544–577, 2005.

[13] P. Gaudry, E. Thomé, N. Thériault, and C. Diem. A double large prime variation for small genus hyperelliptic index calculus. accepted for publication in Math. Comp., 2005.

[14] P. Griffiths and J. Harris. On the variety of special linear systems on a general algebraic curve. *Duke Math. J.*, 47(1):233–272, 1980.

[15] R. Hartshorne. *Algebraic Geometry*. Springer-Verlag, New York, 1977.

[16] F. Heß. Computing Riemann-Roch spaces in algebraic function fields and related topics. *J. Symbolic Computation*, 11, 2001.

[17] F. Heß. Weil descent attacks. In G. Seroussi, I. Blake, and N. Smart, editors, *Advances in Elliptic Curve Cryptography*. Cambridge University Press, 2004.

[18] M.-D. Huang and D. Ierardi. Efficient Algorithms for the Riemann-Roch Problem and for Addition in the Jacobian of a Curve. *J. Symbolic Computation*, 18:519–539, 1994.

[19] K. Khuri-Makdisi. Asymptotically fast group operations on Jacobians of general curves. available on the arXiv under math.NT/0409209, 2004.

[20] K. Khuri-Makdisi. Linear algebra algorithms for divisors on an algebraic curve. *Math. Comp.*, 73:333–357, 2004.

[21] D. Mumford. *Geometric Invariant Theory*. Springer-Verlag, Berlin, 1965.

[22] K. Nagao. Improvement of Thériault Algorithm of Index Calculus of Jacobian of Hyperelliptic Curves of Small Genus. Cryptology ePrint Archive, Report 2004/161, http://eprint.iacr.org/2004/161, 2004.

[23] J. Pila. Frobenius maps of abelian varieties and fining roots of unity in finite fields. *Math. Comp.*, 55:745–763, 1990.

[24] J. Pila. Counting points on curves over families in polynomial time. available on the arXiv under math.NT/0504570, 1991.

[25] R. Schoof. Elliptic curves over finite fields and the compuation of square roots mod *p*. *Math. Comp.*, 44:483–494, 1985.

[26] E. Volcheck. Computing in the Jacobian of a Plane Algebraic Curve. In L. Adleman and M.-D. Huang, editors, *Algorithmic Number Theory – ANTS I*, LNCS, pages 221–233, Berlin, 1994. Springer-Verlag.

[27] N. Wormald. Random Graphs. In I. Gross and J. Yellen, editors, *Handbook of Graph Theory*, chapter 8.2. CRC Press, Boca Raton, 2004.

Signature Calculus and Discrete Logarithm Problems

Ming-Deh Huang[1] and Wayne Raskind[2]

[1] Department of Computer Science, University of Southern California
Los Angeles, CA 90089-0781, USA
huang@pollux.usc.edu
[2] Department of Mathematics, University of Southern California
Los Angeles, CA 90089-2532, USA
raskind@math.usc.edu

Abstract. Index calculus has been successful in many cases for treating the discrete logarithm problem for the multiplicative group of a finite field, but less so for elliptic curves over a finite field. In this paper we seek to explain why this might be the case from the perspective of arithmetic duality and propose a unified method for treating both problems which we call *signature calculus*.

1 Introduction

Let A be a finite abelian group and x an element of A. Let y be in the subgroup generated by x, so that $y = nx$ for some positive integer n. Recall that the *discrete logarithm problem* (DLP) is to determine n in a computationally efficient way. The computational complexity of solving this problem when the bit size of the inputs is large is the basis of many public-key encryption schemes used today. Two of the most important examples of finite abelian groups that are used in public-key cryptography are the multiplicative group of a finite field and the group of points on an elliptic curve over a finite field (see [K] and [Mill] for the original papers and [KMV] for a survey of work as of 2000).

In what follows below, we will assume that ℓ is a large prime number dividing the order of A and that x is an element of order ℓ. For p a prime number and q a power of p, we denote by $\mathbb{F}_q$ the finite field with q elements and by $\mathbb{F}_q^*$ its multiplicative group of nonzero elements.

Index calculus has been successful in many cases for treating the discrete logarithm problem for the multiplicative group of a finite field (see e.g. [Mc], §5 or [SWD]), but less so for elliptic curves over a finite field (see e.g. [HKT] or [JKSST]). In this paper we seek to "explain" why this might be the case from the perspective of arithmetic duality and propose a unified method for treating both problems which we call *signature calculus*.

We will give more details below, but in this introduction we will simply say that to address the DLP in an abelian algebraic group, we use a lifting of the group over a number field and use the reciprocity law of global class field theory.

F. Hess, S. Pauli, and M. Pohst (Eds.): ANTS 2006, LNCS 4076, pp. 558–572, 2006.

We explain in detail how this works when the group is either the multiplicative group of a finite field or the group of points of an elliptic curve over a finite field. The idea is to construct a suitable "test" element, which is a Dirichlet character in the multiplicative group case and a principal homogeneous space in the elliptic curve case. This element pairs with a point of the group to give an equation between the local terms of this pairing. We thus "shift" the computation to other places where it is expected to be easier. In §2, we define the *signature* of these test elements, and in §3, we prove the equivalence of computing the signature with the respective DLP.

The unifying approach based on global duality provides an ideal setting to investigate the feasibility of the index calculus method for both types of discrete logarithm problems. Following the equivalence results we show that in this setting, the index calculus method arises quite naturally for the discrete-log problem in the multiplicative case and the corresponding signature computation problem. In contrast, a similar method cannot be fashioned for the elliptic curve case. The success in one case and the lack thereof in the other is due to the difference in the nature of the pairings involved. In the multiplicative case, a Dirichlet character which is unramified at a place can nevertheless pair nontrivially with local non-units of the place. This makes it possible for small primes to play a role in forming relations among values of local pairings. In the elliptic curve case, a locally unramified principal homogeneous space at a good reduction place is simply trivial. For bad reduction places not dividing ℓ, only the group of components of the special fibre of the Néron model of the elliptic curve over the ring of integers plays a role, and the order of this group is unlikely to be divisible by ℓ. As a result only primes of large norm can play a role in forming relations among values of local pairings.

The computational complexity of signature calculus is an intriguing question, since the objects involved (Dirichlet characters and principal homogeneous spaces) and their associated field extensions are huge, but the signature that is sought is small. Although we show that the testing Dirichlet characters and principal homogeneous spaces exist, it remains an interesting question how they can be explicitly constructed. The question in the multiplicative case is easier to handle. In that case we also derive a concrete number theoretical characterization of the character signature by working out the local pairings using norm residue symbols. For the elliptic curve case we have a partial solution for the construction.

The idea of using global methods in this way was originally proposed by Frey [F], whom we thank for inspiration, helpful discussions and for inviting us to present our work at the Elliptic Curve Cryptography (ECC) conference in Bochum in September 2004. Methods of this type have also been used by Frey and Rück [FR], and by Nguyen [N].

Due to space limitations some of the proofs will be omitted or sketched in this extended abstract. Detailed proofs appear in [HR1-3].

2   Global Framework

2.1   Notation and Review of Algebraic Number Theory

This subsection is meant primarily to fix notation and to recall some basic concepts from algebraic number theory. Let K be an algebraic number field. We fix an algebraic closure $\overline{K}$ of K and let $G = Gal(\overline{K}/K)$. An equivalence class of absolute values on K will be denoted by v and called a *place*. For each v, we denote by K_v the completion of K with respect to the corresponding absolute value. As most of our discussion will pertain to abelian groups that are ℓ-torsion, where ℓ is an odd prime number, we shall ignore the real places for the most part.

Recall that the Brauer group $Br(K)$ is an abelian group that classifies the equivalence classes of central simple algebras over K, where two such algebras A and B are equivalent if there are matrix algebras $M_n(K), M_m(K)$ such that

$$A \otimes_K M_n(K) \cong B \otimes_K M_m(K).$$

We have that $Br(K_v) \cong \mathbb{Q}/\mathbb{Z}$ if v is nonarchimedean, $Br(\mathbb{R}) \cong \mathbb{Z}/2\mathbb{Z}$ and $Br(\mathbb{C}) = 0$. We can describe $Br(K)$ in terms of Galois cohomology by

$$Br(K) \cong H^2(G, \overline{K}^*).$$

One of the most important results in algebraic number theory is the exact sequence:

$$0 \to Br(K) \to \bigoplus_v Br(K_v) \to \mathbb{Q}/\mathbb{Z} \to 0.$$

This is the beginning of the theory of *global duality*, which shows how to relate the arithmetic of K with that of all of the K_v. The following subsections review this theory briefly in the context in which we shall use it (see [HR1] for more details).

2.2   Reciprocity Law for the Multiplicative Group

Let K^* denote the set of nonzero elements of K, which is an abelian group under multiplication. A Dirichlet character χ of K is a homomorphism of $G = Gal(\overline{K}/K)$ into $\mathbb{Q}/\mathbb{Z}$, which we view as an element of the Galois cohomology group $H^1(G, \mathbb{Q}/\mathbb{Z})$. For $a \in K^*$ we denote by $< \chi, a >$ the cup product $\partial(\chi) \cup a \in H^2(G, \overline{K}^*)$ (see [S], Ch. XIV, §1 for notation). For each nonarchimedean place v of K we can consider χ as a character of $G_v = Gal(\overline{K}_v/K_v)$ and denote it by χ_v. Given $a \in K_v^*$, we can do the same cup-product construction as we did above for K to get an element

$$< \chi_v, a >_v \in H^2(G_v, \overline{K}_v^*) \cong \mathbb{Q}/\mathbb{Z}.$$

If v is a place where χ is unramified and a is a unit at v, then $< \chi_v, a_v >_v = 0$. Thus $< \chi_v, a_v >_v = 0$ for all but finitely many v. We then have the *reciprocity law*: for any Dirichlet character χ of K and any $a \in K^*$,

$$\sum_v < \chi_v, a_v >_v = 0 \in \mathbb{Q}/\mathbb{Z}.$$

2.3 Reciprocity Law for Elliptic Curves

Let E be an elliptic curve over K. Recall that a *principal homogeneous space* of E over K is a curve F of genus 1 over K together with a simply transitive group action of E on F. The isomorphism classes of principal homogeneous spaces are classified by the group $H^1(G, E(\overline{K}))$, where $G = Gal(\overline{K}/K)$. A principal homogeneous space is trivial if and only if it has a rational point over K, in which case it is isomorphic to E over K. Thus any principal homogeneous space becomes isomorphic to E over a finite extension of K. Let Q be a point of E. Then for $\alpha \in H^1(G, E)$ and $Q \in E(K)$, we consider the pairing

$$< \alpha, Q > \in Br(K).$$

This are not as easy to describe explicitly as in the case of the multiplicative group and we refer to [HR1] for a general description. We can make a similar definition for

$$< \alpha_v, Q_v >_v \in Br(K_v) \cong \mathbb{Q}/\mathbb{Z}$$

over the nonarchimedean fields K_v. We then have that $< \alpha_v, Q_v >_v = 0$ for almost all v and the *reciprocity law*:

$$\sum_v < \alpha_v, Q_v >_v = 0 \in \mathbb{Q}/\mathbb{Z}.$$

2.4 Duality

Let S be a finite set of places of K including the places dividing ℓ, and G_S be the Galois group of a maximal extension of K that is unramified outside S. Let μ_ℓ be the Galois module of ℓ-th roots of unity. Then we have the *Poitou-Tate exact sequence* (see [HR1]):

$$0 \to H^0(G_S, \mu_\ell) \to \bigoplus_{v \in S} H^0(K_v, \mu_\ell) \to H^2(G_S, \mathbb{Z}/\ell\mathbb{Z})^* \to H^1(G_S, \mu_\ell)$$

$$\to \bigoplus_{v \in S} H^1(K_v, \mu_\ell) \to H^1(G_S, \mathbb{Z}/\ell\mathbb{Z})^* \to H^2(G_S, \mu_\ell) \to \bigoplus_{v \in S} H^2(K_v, \mu_\ell)$$

$$\to H^0(G_S, \mathbb{Z}/\ell\mathbb{Z})^* \to 0.$$

Here, for an abelian group A, A^* denotes $\mathrm{Hom}(A, \mathbb{Q}/\mathbb{Z})$. We are mainly interested in the part

$$(*) \ H^1(G_S, \mu_\ell) \to \bigoplus_{v \in S} H^1(K_v, \mu_\ell) \to H^1(G_S, \mathbb{Z}/\ell\mathbb{Z})^*$$

which will be used in proving the existence of our testing Dirichlet characters. If the order of the class group of K is not divisible by ℓ, then the last map in this sequence is surjective (see [HR2]).

Recall the Shafarevich-Tate group

$$\mathrm{III}(E) = \ker[H^1(K, E) \to \bigoplus_{all \ v} H^1(K_v, E)],$$

where the sum runs over all places of K. It is conjectured that $\text{III}(E)$ is finite for any elliptic curve over a number field, but this is not known, in general. It has been proved in many cases for E of small rank. In what follows we will need to assume this.

Let $\mathcal{E}$ be a smooth proper model of E over an open subset U of the ring of integers of K on which ℓ is invertible and put $S = X - U$. We have the exact sequence (see [HR1]):

$$(**) \; E(K)^{(\ell)} \to \bigoplus_{v \in S} E(K_v)^{(\ell)} \to H^1(U, \mathcal{E})\{\ell\}^* \to \text{III}(E)\{\ell\} \to 0.$$

Here (ℓ) denotes completion with respect to subgroups of ℓ-power index and $\{\ell\}$ denotes the ℓ-primary part. This sequence is usually called the *Cassels-Tate exact sequence*.

3   Existence of Testing Dirichlet Characters and Principal Homogeneous Spaces

3.1   The Multiplicative Case - Dirichlet Characters

Throughout this subsection, let p, ℓ be rational primes with $p \equiv 1 \pmod{\ell}$ and $\ell > 2$. Let $K/\mathbb{Q}$ be a real quadratic extension where p and ℓ split. Let α be a fundamental unit of K. Let Σ be the set of all places over ℓ and p, together with all the archimedean places. For any place u of K let P_u denote the prime ideal corresponding to u.

Proposition 1. *Let S be a subset of Σ that contains both places over ℓ and both archimedean places. Suppose*

1. *$\ell \nmid h_K$ where h_K is the class number of K;*
2. *either $\alpha^{l-1} \not\equiv 1 \pmod{P_w^2}$ for some $w \in S$ over ℓ, or $\alpha^{\frac{p-1}{\ell}} \not\equiv 1 \pmod{P_w}$ for some $w \in S$ over p (that is, locally α is not an ℓ-th power at either a place over ℓ or a place over p).*

Then the $\mathbb{F}_\ell$-dimension of $H^1(G_S, \mathbb{Z}/\ell\mathbb{Z})$ equals $n(S) - 1$ where $n(S)$ is the number of finite places in S.

Corollary 1. *Let S be the set consisting of one place u over ℓ, one place v over p, and both archimedean places. Suppose*

1. *$\ell \nmid h_K$ where h_K is the class number of K;*
2. *$\alpha^{l-1} \not\equiv 1 \pmod{P_u^2}$;*
3. *$\alpha^{\frac{p-1}{\ell}} \not\equiv 1 \pmod{P_v}$.*

Then the $\mathbb{F}_\ell$-dimension of $H^1(G_S, \mathbb{Z}/\ell\mathbb{Z})$ is one.

See [HR2] for a proof of Proposition 1 and Corollary 1 which involves the Poitou-Tate sequence.

Assuming the conditions in Corollary 1, then $H^1(G_S, \mathbb{Z}/\ell\mathbb{Z})$ is isomorphic to $\mathbb{Z}/\ell\mathbb{Z}$. Every nontrivial character in it is ramified at u and v and unramified at all other finite places; moreover, $< \chi, \alpha >_u \neq 0$ and $< \chi, \alpha >_v \neq 0$, and $< \chi, \alpha >_u + < \chi, \alpha >_v = 0$. This group of characters corresponds to a unique cyclic extension K_S of degree ℓ over K which is ramified at u and v and unramified at all other finite places.

For a ring R, we denote by R^* the group of units of R.

At u, we take the class of $1 + \ell$ as the generator of the group $O_u^*/O_u^{*\ell} \cong \mathbb{Z}_\ell^*/\mathbb{Z}_\ell^{*\ell} \cong \mathbb{Z}/\ell\mathbb{Z}$. For $\chi \in H^1(G_S, \mathbb{Z}/\ell\mathbb{Z})$, we call $\sigma_u(\chi) = < \chi, 1 + \ell >_u$ the u-signature of χ.

Let $g \in \mathbb{Z}$ so that $g \bmod p$ generates the multiplicative group of $\mathbb{F}_p$. Then the class of g generates $O_v^*/O_v^{*\ell} \cong \mathbb{Z}_p^*/\mathbb{Z}_p^{*\ell} \cong \mathbb{Z}/\ell\mathbb{Z}$. For $\chi \in H^1(G_S, \mathbb{Z}/\ell\mathbb{Z})$, we call $\sigma_v(\chi) = < \chi, g >_v \neq 0$ the v-signature of χ.

We call the pair $(\sigma_u(\chi), \sigma_v(\chi))$ the *signature* of χ. Since $\sigma_u(\chi)\sigma_v(\chi)^{-1} \in \mathbb{Z}/\ell\mathbb{Z}$ is the same for $\chi \in H^1(G_S, \mathbb{Z}/\ell\mathbb{Z})$, it depends on K_S alone and we call it the *ramification signature* of K_S with respect to $1 + \ell$ and g.

3.2   The Elliptic Curve Case - Principal Homogeneous Spaces

Lemma 1. *Let K_v be a local field with finite residue field k. Let E be an elliptic curve defined over K_v with good reduction.*

1. *Suppose the characteristic of k is ℓ. Then $H^1(K_v, E)[\ell] \cong \mathbb{Z}/\ell\mathbb{Z}$ if $K_v \cong \mathbb{Q}_\ell$ and $\ell \nmid \#\tilde{E}(k)$.*
2. *Suppose the characteristic of k is not ℓ. Then*
   (a) *$H^1(K_v, E)[\ell] = 0$ if $\ell \nmid \#\tilde{E}(k)$;*
   (b) *$H^1(K_v, E)[\ell] \cong \mathbb{Z}/\ell\mathbb{Z}$ if $\ell \mid \#\tilde{E}(k)$ but $\ell^2 \nmid \#\tilde{E}(k)$.*

For the rest of this section, let p, ℓ be odd, rational primes. Let $K/\mathbb{Q}$ be a real quadratic extension. Let $X = \mathrm{Spec}(\mathcal{O}_K)$. Let E be an elliptic curve defined over K. Let Σ be the set of all places at which E has bad reduction, together with all the archimedean places. Let $\mathcal{E}$ be a smooth proper model of E over the open subset $X - \Sigma$.

Proposition 2. *Let S be a finite set of places of K containing all bad reduction places of E and the places above ℓ. Then if $\text{Ш}(E)\{\ell\} = 0$, we have the exact sequence:*

$$E(K)/\ell \to \prod_{v \in S} E(K_v)/\ell \to (H^1(\mathcal{O}_S, \mathcal{E})[\ell])^* \to 0.$$

Proof: Consider the Cassels-Tate exact sequence

$$E(K)^{(\ell)} \to \prod_{v \in S} E(K_v)^{(\ell)} \to H^1(\mathcal{O}_S, \mathcal{E})\{\ell\}^* \to \text{Ш}(E)\{\ell\} \to 0.$$

Lemma 2. *Let B be a torsion abelian group. Then we have*

$$B[\ell]^* \cong B^*/\ell B^*$$

and

$$B\{\ell\}^* \cong B^{*(\ell)}$$

The proposition follows from the lemma, the assumption that $\text{III}(E)\{\ell\} = 0$ and the Cassels-Tate sequence above.

For the remainder of this section we assume that p and ℓ split in K, and E has good reduction at p and ℓ, with $\#\tilde{E}(\mathbb{F}_p) = \ell$ and $\ell \neq \#\tilde{E}(\mathbb{F}_\ell)$. Moreover we assume that ℓ is sufficiently large so that $E(L)[\ell]$ is trivial for all quadratic extensions L of $\mathbb{Q}$. Finally, we assume that the discriminant of E is small compared to ℓ, which implies that for a bad reduction place v not dividing ℓ, we have $E(K_v)/\ell = 0$ (see [HR3] for more details about this last assumption).

Proposition 3. *Let S be a finite set of places of K containing all bad reduction places of E and the places above ℓ, but no other places away from ℓ and p. Suppose*

1. $\text{III}(E)\{\ell\} = 0$;
2. *the map $E(K)/\ell \to E(K_u)/\ell \oplus E(K_{u'})/\ell$ is an isomorphism, where u and u' are the two places of K over ℓ.*

Then the $\mathbb{F}_\ell$-dimension of $H^1(\mathcal{O}_S, \mathcal{E})[\ell]$ equals $n(S) - 2$ where $n(S)$ is the number of finite places in $S - \Sigma$.

Proof: Since $\text{III}(E)\{\ell\} = 0$, we have the exact sequence

$$E(K)/\ell \to \prod_{v \in S} E(K_v)/\ell \to (H^1(\mathcal{O}_S, \mathcal{E})[\ell])^* \to 0$$

by Proposition 2. The middle group in the sequence $\prod_{v \in S} E(K_v)/\ell$ is isomorphic to the direct sum of $n(S)$ copies of $\mathbb{Z}/\ell\mathbb{Z}$ by Lemma 1. Since the map

$$E(K)/\ell \to E(K_u)/\ell \oplus E(K_{u'})/\ell \cong \mathbb{Z}/\ell\mathbb{Z} \oplus \mathbb{Z}/\ell\mathbb{Z}$$

is an isomorphism, it follows that the image of the map

$$E(K)/\ell \to \prod_{v \in S} E(K_v)/\ell$$

is isomorphic to $\mathbb{Z}/\ell\mathbb{Z} \oplus \mathbb{Z}/\ell\mathbb{Z}$. Hence the $\mathbb{F}_\ell$-dimension of $H^1(\mathcal{O}_S, \mathcal{E})[\ell]$ equals $n(S) - 2$.

Corollary 2. *With assumptions as in Proposition 3, let S be the set consisting of all bad reduction places of E, together with the two places u and u' over ℓ, and one place v over p. Suppose*

1. $\text{III}(E)\{\ell\} = 0$;
2. *the map $E(K)/\ell \to E(K_u)/\ell \oplus E(K_{u'})/\ell$ is an isomorphism.*

Then the $\mathbb{F}_\ell$-dimension of $H^1(\mathcal{O}_S, \mathcal{E})[\ell]$ is one. Moreover every nontrivial element of $H^1(\mathcal{O}_S, \mathcal{E})[\ell]$ is ramified at v.

For $w \in \{u, u', v\}$, let $\rho_w \in E(K_w) - \ell E(K_w)$, so that the class of ρ_w generates $E(K_w)/\ell$. Then Corollary 2 implies that $< \chi, \rho_v >_v \neq 0$ for any nontrivial $\chi \in H^1(\mathcal{O}_S, \mathcal{E})[\ell]$. Moreover, $(\frac{<\chi, \rho_u>_u}{<\chi, \rho_v>_v}, \frac{<\chi, \rho_{u'}>_{u'}}{<\chi, \rho_v>_v})$ is the same for all such χ. We call this pair the *signature* of $H^1(\mathcal{O}_S, \mathcal{E})[\ell]$ with respect to ρ_u, $\rho_{u'}$ and ρ_v.

Suppose, in addition to the map $E(K)/\ell \rightarrow E(K_u)/\ell \oplus E(K_{u'})/\ell$ being an isomorphism, that the map $E(K)/\ell \rightarrow E(K_v)/\ell$ is nontrivial. In this case we may form the ρ_w's as follows. Let $Q, R \in E(K)$ so that their classes generate $E(K)/\ell$. Suppose without loss of generality that the class of Q is nontrivial in $E(K_u)/\ell$ and the class of R is nontrivial in $E(K_{u'})/\ell$. As $E(K)/\ell \rightarrow E(K_v)/\ell$ is nontrivial, the class of either Q or R is nontrivial in $E(K_v)/\ell$. Suppose without loss of generality the class of Q is nontrivial in $E(K_v)/\ell$. Then we may take $\rho_v = Q$, $\rho_u = Q$ and $\rho_{u'} = R$.

4   Discrete Logarithms and Signature Computations

4.1   DLP and Signature Computation

DL Problem: Given p, ℓ, g and a, where p and ℓ are prime with $p \equiv 1 \pmod{\ell}$ and $p \not\equiv 1 \pmod{\ell^2}$, g is a generator for the multiplicative group $\mathbb{F}_p^*$, and $a \in \mathbb{F}_p^*$, to compute $m \bmod \ell$ where $a = g^m$ in $\mathbb{F}_p$.

Signature Computation Problem: Suppose we are given K, p, ℓ, u, v, α and g, where $K = \mathbb{Q}(\sqrt{D})$ is a real quadratic field, ℓ, p are primes that split in K and the class number of K is not divisible by ℓ, u is a place of K over ℓ, v is a place of K over p, α is a unit of K such that $\alpha^{l-1} \not\equiv 1 \pmod{P_u^2}$ and $\alpha^{\frac{p-1}{\ell}} \not\equiv 1 \pmod{P_v}$, and g is a generator for $\mathbb{F}_p^*$. Then compute the ramification signature, with respect to $1 + \ell$ and g, of the cyclic extension of degree ℓ over K which is ramified at u, v and unramified elsewhere.

Theorem 1. *The problems DL and Signature Computation are random polynomial time equivalent.*

For the proof of the theorem, we first give a random polynomial time reduction from DL to Signature Computation. This part of the proof depends on some heuristic assumption which will be made clear below.

Let $a = g^m$ in $\mathbb{F}_p$ where m is to be computed. If $a^{\frac{p-1}{\ell}} = 1$, then $m \equiv 0 \pmod{\ell}$. So suppose $a^{\frac{p-1}{\ell}} \neq 1$. We lift a to some unit α of a real quadratic field K such that $\alpha \equiv a \pmod{v}$ for some place v of K over p. This can be done as follows.

1. Compute $b \in \mathbb{F}_p$ such that $ab = 1$ in $\mathbb{F}_p$.
2. $c \leftarrow 2^{-1}(a + b)$; $d \leftarrow 2^{-1}(a - b)$. Note that $c^2 - d^2 = 1$, and $a = c + d$. We may assume $d \neq 0$ otherwise $a^2 = 1$ and $m = (p - 1)/2$ or $p - 1$.
3. Treat d as an integer. Let $\gamma \in \bar{\mathbb{Q}}$ be such that $\gamma^2 = 1 + d^2$.
4. Check if $1 + d^2$ is a quadratic residue modulo ℓ. Otherwise substitute $d + rp$ for d for random r until the condition is met. This is to make sure that ℓ splits in K.

5. $\gamma^2 = 1 + d^2 \equiv c^2 \pmod{p}$ implies $\gamma \equiv c \pmod{v}$, and $\gamma \equiv -c \pmod{v'}$ where v and v' are the two places of K over p.
6. Let $\alpha = \gamma + d$. Then $\alpha \equiv c + d \equiv a \pmod{v}$. Note that the norm of α is $d^2 - \gamma^2 = -1$, so α is a unit of K.

We make the heuristic assumption that it is likely for K to satisfy the conditions in Corollary 1 for v and a place u of K over ℓ. (Note that the second condition is satisfied since $\alpha \equiv a \pmod{v}$ and $a^{\frac{p-1}{\ell}} \neq 1$.) We argue below that computing the discrete logarithm m where $a = g^m$ is reduced to solving the Signature Computation problem on input K, p, ℓ, u, v, α and g, where $K = \mathbb{Q}(\gamma)$ with $\gamma^2 = 1 + d^2$, $\alpha = \gamma + d$, u and v are as constructed above. A simple analysis shows that the expected time complexity in constructing these objects is $O(\log^3 p)$.

For $\chi \in H^1(K, \mathbb{Z}/\ell\mathbb{Z})$ that is ramified at u and v, and unramified elsewhere, we have
$$0 = <\chi, \alpha>_u + <\chi, \alpha>_v .$$
Moreover since $\alpha^{\frac{p-1}{\ell}} \neq 1 \pmod{v}$, α generates $O_v^* / O_v^{*\ell}$, so $<\chi, \alpha>_v \neq 0$, and it follows that $<\chi, \alpha>_u \neq 0$.

In general for a field k and $a, b \in k^*$, we write $a \sim_l b$ if $a/b \in k^{*\ell}$. We have $\alpha \sim_l g^m$ in K_v since $\alpha \equiv a \equiv g^m \pmod{v}$. Hence
$$<\chi, \alpha>_v = <\chi, g^m>_v = m <\chi, g>_v .$$
Write $\alpha = \xi(1 + y\ell) \pmod{\ell^2}$ with $\xi^{\ell-1} = 1$ after identifying α with its isomorphic image in $\mathbb{Q}_\ell$. Then $\alpha \sim_\ell (1 + \ell)^y$, and
$$0 = <\chi, \alpha>_u = <\chi, (1 + \ell)^y>_u = y <\chi, 1 + \ell>_u .$$

Hence we have
$$<\chi, \alpha>_u + <\chi, \alpha>_v = y <\chi, 1 + \ell>_u + m <\chi, g>_v .$$

So $y\sigma_u(\chi) + m\sigma_v(\chi) = 0$. From this we see that if the ramification signature $\sigma_u(\chi)(\sigma_v(\chi))^{-1}$ is determined then m is determined. The expected time in this reduction is $O(\log^3 p)$.

Next we give a random polynomial time reduction from Signature Computation on input K, p, ℓ, u, v, α and g, to DL on input p, ℓ, g and a where $\alpha \equiv a \pmod{v}$.

Call the oracle to DL on input p, ℓ, g and a to compute m such that $g^m = a \pmod{p}$. Then $\alpha \equiv g^m \pmod{v}$.

Write $\alpha = \xi(1 + y\ell) \pmod{\ell^2}$ with $\xi^{\ell-1} = 1$ after identifying α with its isomorphic image in $\mathbb{Q}_\ell$. Then $\alpha \sim_\ell (1 + \ell)^y$. Again, $\xi \bmod \ell^2$ and hence y can be computed efficiently in time $O(||\alpha|| \log \ell + \log^3 \ell) = O(||\alpha|| \log p + \log^3 p)$.

For $\chi \in H^1(K, \mathbb{Z}/\ell\mathbb{Z})$ that is ramified at u and v, and unramified elsewhere, we have as before $<\chi, \alpha>_v = <\chi, g^m>_v = m <\chi, g>_v$, and $<\chi, \alpha>_u = <\chi, (1 + \ell)^y>_u = y <\chi, 1 + \ell>_u$. Hence
$$0 = <\chi, \alpha>_u + <\chi, \alpha>_v = y <\chi, 1 + \ell>_u + m <\chi, g>_v$$
from this we can determine the signature $\sigma_u(\chi)(\sigma_v(\chi))^{-1}$. The expected running time in this reduction is $O(||\alpha|| \log p + \log^3 p)$.

4.2   ECDL and Signature Computation

ECDL: Given p, ℓ, $\tilde{E}$, $\tilde{Q}$ and $\tilde{R}$, where p and ℓ are prime, $\tilde{E}$ is an elliptic curve defined over $\mathbb{F}_p$ with $\#\tilde{E}(\mathbb{F}_p) = \ell$, and non-zero points $\tilde{Q}, \tilde{R} \in \tilde{E}(\mathbb{F}_p)$, to determine m so that $\tilde{R} = m\tilde{Q}$.

Homogeneous Space Signature Computation: Suppose we are given p, ℓ, K, E, v, Q, R, where p and ℓ are prime, K is a quadratic field where p and ℓ both split, E is an elliptic curve defined over K with $\mathrm{III}(E)\{\ell\} = 0$ and the discriminant of E being prime to ℓ, v is a place of K over p, Q and $R \in E(K)$ such that $Q \not\equiv 0 \pmod{\ell E(K_v)}$ and the images of R and Q in $E(K_u)/\ell \oplus E(K_{u'})/\ell$ are independent, where u and u' are the two places of K over ℓ. Then compute the signature of $H^1(\mathcal{O}_S, \mathcal{E})[\ell]$ with respect to $\rho_v = Q$, $\rho_u = Q$ and $\rho_{u'} = R$, where S is the set consisting of u, u', v and all places of bad reduction of E. (Note that ρ_w generates $E(K_w)/\ell E(K_w)$ for $w = u, u', v$.)

Theorem 2. *The problems ECDL and Homogeneous Space Signature Computation are random polynomial time equivalent.*

For the proof of the theorem, we first give a random polynomial time reduction from ECDL to Homogeneous Space Signature Computation. This part of the proof depends on some heuristic assumption which will be made clear below.

Given $\tilde{E}/\mathbb{F}_p$ where $\tilde{E}(\mathbb{F}_p)[\ell] = <\tilde{Q}>$, and $\tilde{R}$, we are to compute m so that $\tilde{R} = m\tilde{Q}$. Steps 1-3 of the reduction construct an instance p, ℓ, K, E, v, Q, R for the Homogeneous Space Signature Computation problem.

1. Construct $E/\mathbb{Q}$ with $Q \in E(\mathbb{Q})$ such that $\tilde{Q} = Q \bmod p$. This can be done as follows. Suppose $\tilde{E}$ is specified by an affine equation $y^2 = x^3 + \bar{a}x + b$ where $\bar{a} = a \bmod p$, $\bar{b} = b \bmod p$ with $0 \le a, b < p$ and $\tilde{Q} = (u \bmod p, v \bmod p)$ with $0 < u, v < p$. Choose a random integer r, $0 \le r < p$, and let $Q = (u, v + rp)$. Let $b_r = (v + rp)^2 - (u^3 + au)$. Then $Q \in E_r(\mathbb{Q})$ where E_r is the elliptic curve with the affine equation $y^2 = x^3 + ax + b_r$. Set $E = E_r$. The point Q cannot be torsion for otherwise it would have to be in $E(\mathbb{Q})[\ell]$, which has no non-zero point since ℓ is big. The height of Q is far smaller than that of a point in $\ell E(\mathbb{Q})$, so Q is not in $\ell E(\mathbb{Q})$. Since $\tilde{E}(\mathbb{F}_p)[\ell] \cong \mathbb{Z}/\ell\mathbb{Z}$, $E(\mathbb{Q}_p)/\ell \cong \tilde{E}(\mathbb{F}_p)/\ell \cong \mathbb{Z}/\ell\mathbb{Z}$ and the class of Q generates $E(\mathbb{Q}_p)/\ell$.

2. Check that E has good reduction at ℓ and that $|\tilde{E}(\mathbb{F}_\ell)|$ is not divisible by ℓ. Otherwise, go back to 1. to find a different E.

3. Lift $\tilde{R}$ to $R \in E(K)$ where $K/\mathbb{Q}$ is a quadratic extension in which p and ℓ both split. This can be done as follows. Suppose E is defined by the affine equation $y^2 = x^3 + ax + c$. Suppose $\tilde{R} = (\mu \bmod p, \nu \bmod p)$ with $0 < \mu, \nu < p$. Choose a random positive integer $r < p$. Set $\mu_r = \mu + rp$. Let β be a root of $y^2 = \mu_r^3 + a\mu_r + c$. Then (μ_r, β) is a lift of $\tilde{R}$ in $E(K)$ where $K = \mathbb{Q}(\beta)$. By construction p splits in K,

$$E(K_v)/\ell \cong E(\mathbb{Q}_p)/\ell \cong \tilde{E}(\mathbb{F}_p)/\ell \cong \mathbb{Z}/\ell\mathbb{Z}$$

and $R - mQ \in \ell E(K_v)$. Check that ℓ splits in K and that the images of R and Q in $E(K_u)/\ell \oplus E(K_{u'})/\ell$ are independent; otherwise repeat the above steps with a different r until a suitable K is found. Say the class of Q is nontrivial in $E(K_u)/\ell$ and the class of R is nontrivial in $E(K_{u'})/\ell$.

4. Call the oracle for the Homogeneous Space Signature Computation on input p, ℓ, E, K, Q, R, v to compute the signature (α, β) of $H^1(\mathcal{O}_S, \mathcal{E})[\ell]$ with respect to $\rho_v = Q$, $\rho_u = Q$ and $\rho_{u'} = R$ (where S is the set consisting of u, u', v and all places of bad reduction of E). Then for all nontrivial $\chi \in H^1(\mathcal{O}_S, \mathcal{E})[\ell]$, $\alpha = \frac{<\chi,Q>_u}{<\chi,Q>_v}$ and $\beta = \frac{<\chi,R>_{u'}}{<\chi,Q>_v}$.

5. Identify K_u with $\mathbb{Q}_\ell$ and compute n so that $R \equiv nQ \pmod{\ell E(K_u)}$ as follows. Compute $d = |\tilde{E}(\mathbb{F}_\ell)|$. Observe that dQ and dR are both in $E_1(\mathbb{Q}_\ell)$. Compute n such that $n(dQ) \equiv (dR) \pmod{\ell}$ in $E_1(\mathbb{Q}_\ell)$. Then $d(nQ - R) = \ell Z$ for some $Z \in E_1(\mathbb{Q}_\ell)$. Since d is not divisible by ℓ, $d^{-1} \in \mathbb{Z}_\ell$, so $nQ - R = d^{-1}\ell Z = \ell(d^{-1}Z) \in \ell E(\mathbb{Q}_\ell)$.

6. Now

$$0 = \sum_{w \in \{v,u,u'\}} <\chi, R>_w$$
$$= m <\chi, Q>_v + n <\chi, Q>_u + <\chi, R>_{u'}.$$

From this we get $m + n\alpha + \beta \equiv 0 \pmod{\ell}$. Hence m can be determined.

We make the heuristic assumption that it is likely for E and K to satisfy the conditions in Proposition 3. Note that by construction $E(\mathbb{Q})$ is of rank at least one. The points Q and R are likely to be integrally independent in $E(K)$ as they both have small height by construction. So $E(K)$ is likely to be of rank at least two and we make the heuristic assumption that with nontrivial probability its rank is exactly two. Moreover, since $Q \in E(\mathbb{Q})$ and $R \in E(K) - E(\mathbb{Q})$, the images of Q and R are likely to be independent in $E(K_u)/\ell \oplus E(K_{u'})/\ell$, heuristically speaking. The expected running time of this reduction is dominated by Step 2 where the number of rational points on the reduction of E mod ℓ is counted. The running time of that step is $O(\log^8 \ell)$ [Sc], hence it is $O(\log^8 p)$.

Next we give a random polynomial time reduction from Homogeneous Space Signature Computation with input p, ℓ, E, K, Q, R, v to ECDL with input p, ℓ, $\tilde{E}$, $\tilde{Q}$, $\tilde{R}$, where $\tilde{E}$ is the reduction of E mod v, $\tilde{Q}$ (resp. $\tilde{R}$) is the reduction of Q (resp. R) mod v.

For any nontrivial $\chi \in H^1(K, E)[\ell]$ that is unramified away from u, u' and v, we have

$$< \chi, Q >_v + < \chi, Q >_u + < \chi, Q >_{u'} = 0,$$
$$< \chi, R >_v + < \chi, R >_u + < \chi, R >_{u'} = 0.$$

Suppose $Q = a_w \rho_w \pmod{\ell E(K_w)}$ and $R = b_w \rho_w \pmod{\ell E(K_w)}$ for $w = u, u', v$. Note that from Lemma 1, a_v and b_v can be computed by solving ECDL on the reduction of E modulo v. On the other hand a_w, b_w for $w = u, u'$, can be computed in a manner as described in Step 5 above.

Then we get

$$a_v < \chi, \rho_v >_v + a_u < \chi, \rho_u >_u + a_{u'} < \chi, \rho_{u'} >_{u'} = 0,$$
$$b_v < \chi, \rho_v >_v + b_u < \chi, \rho_u >_u + b_{u'} < \chi, \rho_{u'} >_{u'} = 0$$

Condition (2) of Proposition 3 implies that the two relations above are linearly independent. From these we can compute the the the signature of χ; that is $\left(\frac{<\chi, \rho_u>_u}{<\chi, \rho_v>_v}, \frac{<\chi, \rho_{u'}>_{u'}}{<\chi, \rho_v>_v} \right)$. The expected running time of this reduction can be shown to be $O(\log^4 p) + O(M \log p)$ where M is the maximum of the length of R, Q and D.

5 Feasibility of Index Calculus

In reducing the discrete-log problems to the signature computations, the basic idea is to lift elements from a finite field $\mathbb{F}_p$ to a global field K where discrete logarithms are preserved at a place over p, then pair the elements with testing Dirichlet characters in the multiplicative case, or principal homogeneous spaces in the elliptic curve case. The reciprocity laws then allow us to distribute information of the discrete logarithms among a set of places. This set of places depends on the choice of a Dirichlet character (resp. homogeneous space) and the manner of lifting. We will demonstrate how the classical index calculus method emerges in this context as the result of one particular choice of Dirichlet character and method of lifting. We will derive a similar index calculus method for the signature computation problem of Dirichlet characters. We will discuss why a similar method cannot work for principal homogeneous spaces.

5.1 Classical Index Calculus from the Perspective of Arithmetic Duality

Let p and ℓ be odd primes such that $p \equiv 1 \pmod{\ell}$ but $p \not\equiv 1 \pmod{\ell^2}$. Suppose $t = s^n$ in $\mathbb{F}_p^*[\ell]$ and n is to be computed, given s and t. Let K be a number field with a place v over p such that the residue field $\mathbb{F}_v$ is isomorphic to $\mathbb{F}_p$. Let $\alpha, \beta \in O_K$ be lifting of s and $r = s^a t$ (with a random) so that $\alpha \equiv s \pmod{v}$ and $\beta \equiv r \pmod{v}$. Then the relation $r = s^{n+a}$ is preserved at v in the sense that $\beta = \alpha^{n+a} \gamma^\ell$ for some $\gamma \in O_v$. Therefore for all $\chi \in H^1(K, \mathbb{Z}/l\mathbb{Z})$, $< \chi, \beta >_v = (a + n) < \chi, \alpha >_v$. It follows that

$$(a + n) < \chi, \alpha >_v = < \chi, \beta >_v = - \sum_{u \neq v} < \chi, \beta >_u .$$

Note that if χ is ramified at v, then $< \chi, \alpha >_v \neq 0$ since the class of α generates $O_v^* / O_v^{*\ell} \cong \mathbb{F}_p^* / \mathbb{F}_p^{*\ell} \cong \mathbb{F}_p^*[\ell]$.

In particular by choosing $K = \mathbb{Q}$, lifting s to s (considered as an integer), targeting the lifting r to some $\beta = \prod_q q^{e_q} s$ which is smooth over a factor base, and choosing χ to be ramified only at p,

$$(a + n) < \chi, s >_p = < \chi, \beta >_p = - \sum_q e_q < \chi, q >_q .$$

If β is B-smooth then we get a linear relation modulo ℓ of n and $< \chi, q >_q$ ($< \chi, s >_p)^{-1}$, $q < B$, and $O(B)$ relations will allow us to solve for the unknown quantities, including n. What we have derived is in essence the classical index calculus method.

5.2   Index Calculus for Signature Computation of Dirichlet Characters

Suppose we are given a real quadratic field K, primes ℓ, p, places u, v satisfying the conditions in Proposition 1. Let $K = \mathbb{Q}(\alpha)$ with $\alpha^2 \in \mathbb{Z}_{>0}$. To compute the signature of $\chi \in H^1(K, \mathbb{Z}/\ell\mathbb{Z})$ that is ramified precisely at u and v, we generate random algebraic integers $\beta = r\alpha + s$ with $r, s \in \mathbb{Z}$ so that $r\alpha + s \equiv g \pmod{v}$ and $\beta \sim (1 + \ell)^a$ at u for some a. Now suppose the norm of β is B-smooth for some integer B. Then

$$0 = \sum_w < \chi, \beta >_w = < \chi, g >_v + a < \chi, 1 + \ell >_u + \sum_w e_w < \chi, \pi_w >_w,$$

where w in the last sum ranges over all places of K of norm less than B, π_w is a local parameter at w, and e_w is the valuation of β under w. Hence we have obtained a $\mathbb{Z}/\ell\mathbb{Z}$-linear relation on $(< \chi, g >_v)^{-1} < \chi, 1 + \ell >_u$, and $(< \chi, g >_v)^{-1} < \chi, \pi_w >_w$. With $O(B)$ relations we can solve for all these unknowns, in particular the signature $(< \chi, g >_v)^{-1} < \chi, 1 + \ell >_u$.

5.3   The Elliptic Curve Case

We see that one important reason why index calculus is viable in the multiplicative case is due to the fact that locally unramified Dirichlet characters can be paired nontrivially with non-units. For the elliptic curve case, pairing a principal homogeneous space χ and a global point α yields similarly a relation:

$$0 = \sum_v < \chi, \alpha >_v.$$

However from Lemma 1 we see that in the sum above we have nontrivial contribution from a place $v \nmid \ell$ (and where E has good reduction) only if ℓ divides $\#\tilde{E}(\mathbb{F}_v)$. Since $\#\tilde{E}(\mathbb{F}_v)$ is of the order $\#\mathbb{F}_v$, which is the norm of v, we see that the finite places of good reduction that are involved in the sum are all of large norm. As for the bad reduction places, the heuristic assumption that we discussed just before Proposition 3 implies that these will not play any role in this sum, since it will be likely that $E(K_v)/\ell = 0$ for such places v, because v is of small norm. This explains why the index calculus method is lacking in the case of elliptic curve discrete logarithm problem.

6   Further Results

In the case of multiplicative groups of finite fields, we derive a concrete number theoretical characterization of the character signature. Let K, ℓ, p, u, v, S be as in Proposition 1. Let $g \in \mathbb{Z}$ so that $g \bmod p$ generates the multiplicative group of $\mathbb{F}_p$. Let w be the place of $K(\mu_\ell)$ over v such that $g^{\frac{p-1}{\ell}} \equiv 1 \pmod{w}$. Let $M = K_S$ be the cyclic extension corresponding to $H^1(G_S, \mathbb{Z}/\ell\mathbb{Z})$. Then $M(\mu_\ell) = K(\mu_\ell)(A^{\frac{1}{\ell}})$

for some $A \in K(\mu_\ell)^*$. A nontrivial $\chi \in H^1(G_S, \mathbb{Z}/\ell\mathbb{Z})$ corresponds to some $A \in K(\mu_\ell)$ through $H^1(K(\mu_\ell), \mathbb{Z}/\ell\mathbb{Z}) \cong H^1(K(\mu_\ell), \mu_\ell) \cong K(\mu_\ell)^*/\ell$, so that for all σ in the absolute Galois group of K, $\chi(\sigma) = i$ iff $\sigma(A^{\frac{1}{\ell}})/A^{\frac{1}{\ell}} = \zeta^i$.

Proposition 4. *If we identify $K(\mu_\ell)_w$ with $\mathbb{Q}_p$ and K_u with $\mathbb{Q}_\ell$, then $A \sim^\ell p^m$ in $\mathbb{Q}_p^{ur}$ where $m = \sigma_v(\chi) = < \chi, g >_v$, and $A \sim^\ell \zeta^n$ in $\mathbb{Q}_\ell(\mu_\ell)^{ur}$ where $n = \sigma_u(\chi) = < \chi, 1 + \ell >_u$.*

See [HR1] for a proof of the proposition that involves computation using norm residue symbols (see [S]).

For the elliptic curve case we have a partial solution for explicit construction of the testing principal homogeneous spaces (see [HR3]).

References

[F]   G. Frey, *Applications of arithmetical geometry to cryptographic constructions*, In Proceedings of the Fifth International Conference on Finite Fields and Applications. Springer Verlag, page 128-161, 1999; Preprint also available at http://www.exp-math.uni-essen.de/zahlentheorie/preprints/Index.html.

[FR]   G. Frey and H.-G. Rück, A remark concerning m-divisibility and the discrete logarithm in the divisor class group of curves, *Mathematics of Computation*, 62(206):865–874, 1994.

[HKT]   M.-D. Huang, K. L. Kueh, and K.-S. Tan *Lifting elliptic curves and solving the elliptic curve discrete logarithm problem* In ANTS, Lecture Notes in Computer Science, Volume 1838 Springer-Verlag, 2000.

[HR1]   M.-D. Huang and W. Raskind, *Global duality and the discrete logarithm problem*, preprint 2006. http://www-rcf.usc.edu/~mdhuang/papers.html

[HR2]   M.-D. Huang and W. Raskind, *Signature calculus and the the discrete logarithm problem for the multiplicative group case*, preprint 2006. http://www-rcf.usc.edu/~mdhuang/papers.html

[HR3]   M.-D. Huang and W. Raskind, *Signature calculus and the discrete logarithm problem for elliptic curves*, preprint 2006

[JKSST] M.J. Jacobson, N. Koblitz, J.H. Silverman, A. Stein, and E. Teske. Analysis of the Xedni calculus attack. Design, Codes and Cryptography, 20 41-64, 2000

[K]   N. Koblitz *Elliptic curve cryptosystems* Mathematics of Computation, 48 203-209, 1987.

[KMV]   N. Koblitz, A. Menezes and S. Vanstone *The state of elliptic curve cryptography*, Design, Codes and Cryptography, 19, 173-193 (2000)

[Ma]   B. Mazur, *Notes on the étale cohomology of number fields*, Ann. Sci. École Normale Supérieure 6 (1973) 521-556

[Mc]   K. McCurley, *The discrete logarithm problem*, in Cryptology and Computational Number Theory, C. Pomerance, editor, Proceedings of Symposia in Applied Mathematics, Volume 42, 49-74, 1990

[Mill] V. Miller *Uses of elliptic curves in cryptography*, In Advances in Cryptology: Proceedings of Crypto 85, Lecture Notes in Computer Science, volume 218, 417-426. Springer-Verlag, 1985.

[MET]   J.S. Milne, *Étale Cohomology*, Princeton Mathematical Series, Volume 33, Princeton University Press 1980

[MAD]   J.S. Milne, *Arithmetic Duality Theorems*, Perspectives in Mathematics, Volume 1., Academic Press 1986

[N]   K. Nguyen, Thesis, Universität Essen, 2001

[S]   J.-P. Serre, *Local Fields*, Graduate Texts in Mathematics, Volume 67, Springer Verlag 1979.

[Sc]   R. Schoof, *Counting points on elliptic curves over finite fields*, Journal de Théorie des Nombres de Bordeaux 7 (1995), 219-254.

[SWD]   O. Schirokauer, D.Weber, and T. Denny *Discrete logarithms: The effectiveness of the index calculus method* In ANTS II, volume 1122 of Lecture Notes in Computer Science. Springer-Verlag, 1996.

Spectral Analysis of Pollard Rho Collisions

Stephen D. Miller[1,*] and Ramarathnam Venkatesan[2]

[1] Einstein Institute of Mathematics, The Hebrew University
Givat Ram, Jerusalem 91904, Israel
and
Department of Mathematics, Rutgers University, Piscataway, NJ 08854, USA
miller@math.huji.ac.il
[2] Microsoft Research, Cryptography and Anti-piracy Group
1 Microsoft Way, Redmond, WA 98052, USA
and
Cryptography Research Group, Microsoft Research India
Scientia - 196/36 2nd Main, Sadashivnagar, Bangalore 560 080, India
venkie@microsoft.com

Abstract. We show that the classical Pollard ρ algorithm for discrete logarithms produces a collision in expected time $O(\sqrt{n}(\log n)^3)$. This is the first nontrivial rigorous estimate for the collision probability for the *unaltered* Pollard ρ graph, and is close to the conjectured optimal bound of $O(\sqrt{n})$. The result is derived by showing that the mixing time for the random walk on this graph is $O((\log n)^3)$; without the squaring step in the Pollard ρ algorithm, the mixing time would be exponential in $\log n$. The technique involves a spectral analysis of *directed* graphs, which captures the effect of the squaring step.

Keywords: Pollard Rho algorithm, discrete logarithm, random walk, expander graph, collision time, mixing time, spectral analysis.

1   Introduction

Given a finite cyclic group G of order n and a generator g, the Discrete Logarithm Problem (DLOG) asks to invert the map $y \mapsto g^y$ from $\mathbb{Z}/n\mathbb{Z}$ to G. Its presumed difficulty serves as the basis for several cryptosystems, most notably the Diffie-Hellman key exchange and some elliptic curve cryptosystems. Up to constant factors, the Pollard ρ algorithm is the most efficient and the only version with small memory known for solving DLOG on a general cyclic group – in particular for the group of points of an elliptic curve over a finite field.

We quickly recall the algorithm now. First one randomly partitions G into three sets S_1, S_2, and S_3. Set $x_0 = h$, or more generally to a random power $g^{r_1}h^{r_2}$. Given x_k, let $x_{k+1} = f(x_k)$, where $f : G \to G$ is defined by

---

* Partially supported by NSF grant DMS-0301172 and an Alfred P. Sloan Foundation Fellowship.

$$f(x) = \begin{cases} gx, & x \in S_1 \,; \\ hx, & x \in S_2 \,; \\ x^2, & x \in S_3 \,. \end{cases} \tag{1.1}$$

Repeat until a collision of values of the $\{x_k\}$ is detected (this is done using Floyd's method of comparing x_k to x_{2k}, which has the advantage of requiring minimal storage). We call the underlying directed graph in the above algorithm (whose vertices are the elements of G, and whose edges connect each vertex x to gx, hx, and x^2) as the *Pollard ρ Graph*. At each stage x_k may be written as $g^{a_k y + b_k}$, where $h = g^y$. The equality of x_k and x_ℓ means $a_k y + b_k = a_\ell y + b_\ell$, and solving for y (if possible) recovers the DLOG of $h = g^y$.

The above algorithm *heuristically* mimics a random walk. Were that indeed the case, a collision would be found in time $O(\sqrt{n})$, where n is the order of the group G. (The actual constant is more subtle; indeed, Teske [13] has given evidence that the walk is somewhat worse than random.)

The main result of this paper is the first rigorous nontrivial upper bound on the collision time. It is slightly worse than the conjectured $O(\sqrt{n})$, in that its runtime is $\widetilde{O}(\sqrt{n})$, i.e. off from $O(\sqrt{n})$ by at most a polynomial factor in $\log n$. As is standard and without any loss of generality, we tacitly make the following

assumption: the order $|G| = n$ is prime.   (1.2)

Theorem 1. *Fix $\varepsilon > 0$. Then the Pollard ρ algorithm for discrete logarithms on G finds a collision in time $O_\varepsilon(\sqrt{n} \, (\log n)^3)$ with probability at least $1 - \varepsilon$, where the probability is taken over all partitions of G into three sets S_1, S_2, and S_3.*

In the black-box group model (i.e. one which does not exploit any special properties of the encoding of group elements), a theorem of Shoup [11] states that any DLOG algorithm needs $\Omega(\sqrt{n})$ steps. Hence, aside from the probabilistic nature of the above algorithm and the extra factor of $(\log n)^3$, the estimate of Theorem 1 is sharp.

It should be noted that finding a collision does not necessarily imply finding a solution to DLOG; one must also show the resulting linear equation is non-degenerate. Since $n = |G|$ is prime this is believed to happen with overwhelming probability, much more so than for the above task of finding a collision in $O(\sqrt{n})$ time. This was shown for a variant of the Pollard ρ algorithm in [6], but the method there does not apply to the original algorithm itself. Using more refined techniques we are able to analyze this question further; the results of these investigations will be reported upon elsewhere.

This paper is the first analysis of the unmodified Pollard ρ Graph, including the fact that it is *directed*. One can obtain the required rapid mixing result for directed graphs by (a) assuming that rapid mixing holds for the undirected version, and (b) adding self-loops to each vertex. However, one still needs to prove (a), which in our situation is no simpler. In addition, the loops and loss of direction cause short cycles, which lead to awkward complications in the context of studying collisions.

Technically, analyzing directed graphs from a spectral point of view has the well known difficulty that a spectral gap is not equivalent to rapid mixing. A natural generalization of the spectral gap is the operator norm gap of the adjacency matrix, which suffices for our purposes (see Section 2). For a recent survey of mixing times on directed graphs, see [9].

The Pollard ρ graph is very similar to the graphs introduced by the authors in [8]. These graphs, which are related to expander graphs, also connect group elements x to $f(x)$ via the operations given in (1.1) – in particular they combine the operations of multiplication and squaring. The key estimate, a spectral bound on the adjacency operator on this graph, is used to show its random walks are rapidly mixing. Though the Pollard ρ walk is only *pseudorandom* (i.e., x_{k+1} is determined completely from x_k by its membership in S_1, S_2, or S_3), we are solely interested here in proving that it has a collision. The notions of random walk and pseudorandom walk (with random assignments of vertices in the sets S_i) coincide until a collision occurs.

1.1   Earlier Works

Previous experimental and theoretical studies of the Pollard ρ algorithm and its generalizations all came to the (unproven) conclusion that it runs in $O(\sqrt{n})$ time; this is in fact the basis for estimating the relative bit-for-bit security of elliptic curve cryptosystems compared to others, e.g. RSA. For an analysis of DLOG algorithms we refer the reader to the survey by Teske [14], and for an analysis of random walks on abelian groups, to the one by Hildebrand [4]. For the related Pollard ρ algorithm for factoring integers, Bach [1] improved the trivial bound of $O(n)$ by logarithmic factors.

An important statistic of the involved graphs is the *mixing time* τ, which loosely speaking is the amount of time needed for the random walk to converge to the uniform distribution, when started at an arbitrary node.[1] The existing approaches to modeling Pollard ρ can be grouped into two categories:

1. *Birthday attack in a totally random model:* each step is viewed as a move to a random group element, i.e. a completely random walk. In particular one assumes that the underlying graph has mixing time $\tau = 1$ and that its degree equals the group size; in reality the actual Pollard ρ graph has degree only 3. The $O(\sqrt{n})$ collision time is immediate for random walks of this sort.
2. *Random walk in an augmented graph:* The Pollard ρ graph is modified by increasing the number of generators k, but removing the squaring step. One then models the above transitions as random walks on directed abelian Cayley graphs. To ensure the mixing time is $\tau = O(\log |G|)$, however, the graph degree must grow at least logarithmically in $|G|$. The importance of τ stems from the fact that, typically, one incurs a overhead of multiplicative factor of τ^{const} in the overall algorithm.

---

[1] There are many inequivalent notions of mixing time (see [7]). Mixing time is only mentioned for purposes of rough comparison between different graphs; whatever we need about it is proved directly. Similarly, the reader need not recall any facts about expander graphs, which are mentioned only for motivation.

Teske [13], based on Hildebrand's results [4] on random walks on the cyclic group $\mathbb{Z}/m\mathbb{Z}$ with respect to steps of the form $x \mapsto x + a_i$, $i \leq k$, shows that the mixing time of an algorithm of the second type is on the order of $n^{\frac{2}{k-1}}$; she gives supporting numerics of random behavior for k large. In particular, without the squaring step the Pollard ρ walk would have mixing time on the order of n^2, well beyond the expected $O(\sqrt{n})$ collision time. This operation is an intriguing and cryptographically[2] important aspect of the Pollard ρ algorithm, and makes it inherently *non-abelian*: the Pollard ρ graphs are not isomorphic to any abelian Cayley graphs. Its effect cannot be accounted for by any analysis which studies only the additive structure of $\mathbb{Z}/m\mathbb{Z}$.

The present paper indeed analyzes the *exact* underlying Pollard ρ graph, without any modifications. We are able to show that the inclusion of the squaring step reduces the mixing time τ from exponential in $\log n$, to $O((\log n)^3)$ — see the remark following Proposition 2.

Our result and technique below easily generalize from the unmodified Pollard ρ algorithm, which has only 2 non-squaring operations, to the generalized algorithms proposed by Teske [13] which involve adding further such operations. Furthermore, it also applies more generally to additional powers other than squares. We omit the details, since the case of interest is in fact the most difficult, but have included a sketch of the argument at the end of the paper.

2   Rapid Mixing on Directed Graphs

In the next two sections we will describe some results in graph theory which are needed for the proof of Theorem 1. Some of this material is analogous to known results for *undirected* graphs (see, for example, [2]); however, since the literature on spectral analytic aspects of directed graphs is relatively scarce, we have decided to give full proofs for completeness.

The three properties of subset expansion, spectral gap, and rapid mixing are all equivalent for families of undirected graphs with fixed degree. This equivalence, however, fails for directed graphs. Although a result of Fill [3] allows one to deduce rapid mixing on directed graphs from undirected analogs, it involves adding self-loops (which the Pollard ρ graph does not have) and some additional overhead. In any event, it requires proving an estimate about the spectrum of the undirected graph. We are able to use the inequality [8, (A.10)], which came up in studying related undirected graphs, in order to give a bound on the operator norm of the directed graphs. This bound, combined with Lemma 1, gives an estimate of $\tau = O((\log n)^3)$ for the mixing time of the Pollard ρ graph.

Let Γ denote a graph with a finite set of vertices V and edges E. Our graphs will be directed graphs, meaning that each edge has an orientation; an edge from v_1 to v_2 will be denoted by $v_1 \to v_2$. Assume that Γ has *degree* k, in other words that each vertex has exactly k edges coming in and k edges coming out of it.

---

[2] In this version one can derive a secure hash function [5] whose security is based on the difficulty of the discrete logarithm problem; here the input describes the path taken in the graph from a fixed node, and the hash value is the end point.

The *adjacency* operator A acts on $L^2(V) = \{f : V \to \mathbb{C}\}$ by summing over these k neighbors:

$$(Af)(v) \;=\; \sum_{v \to w} f(w). \qquad (2.1)$$

Clearly constant functions, such as $\mathbb{1}(v) \equiv 1$, are eigenfunctions of A with eigenvalue k. Accordingly, $\mathbb{1}$ is termed the trivial eigenfunction and k the trivial eigenvalue of A. Representing A as a $|V| \times |V|$ matrix, we see it has exactly k ones in each row and column, with all other entries equal to zero. It follows that $\mathbb{1}$ is also an eigenfunction with eigenvalue k of the adjoint operator A^*

$$(A^*f)(v) \;=\; \sum_{w \to v} f(w), \qquad (2.2)$$

and that all eigenvalues λ of A or A^* satisfy the bound $|\lambda| \leq k$.

The subject of *expander graphs* is concerned with bounding the (undirected) adjacency operator's restriction to the subspace $L_0 = \{f \in L^2(V) \mid f \perp \mathbb{1}\}$, i.e. the orthogonal complement of the constant functions under the L^2-inner product. This is customarily done by bounding the nontrivial eigenvalues away from k. However, since the adjacency operator A of a directed graph might not be self-adjoint, the operator norm can sometimes be a more useful quantity to study. We next state a lemma relating it to the rapid mixing of the random walk. To put the statement into perspective, consider the k^r random walks on Γ of length r starting from any fixed vertex. One expects a uniformly distributed walk to land in any fixed subset S with probability roughly $\frac{|S|}{|V|}$. The lemma gives a condition on the operator norm for this probability to in fact lie between $\frac{1}{2}\frac{|S|}{|V|}$ and $\frac{3}{2}\frac{|S|}{|V|}$ for moderately large values of r. This can alternatively be thought of as giving an upper bound on the mixing time.

Lemma 1. *Let Γ denote a directed graph of degree k on n vertices. Suppose that there exists a constant $\mu < k$ such that $\|Af\| \leq \mu\|f\|$ for all $f \in L^2(V)$ such that $f \perp \mathbb{1}$. Let S be an arbitrary subset of V. Then the number of paths of length $r \geq \frac{\log(2n)}{\log(k/\mu)}$ which start from any given vertex and end in S is between $\frac{1}{2}k^r \frac{|S|}{|V|}$ and $\frac{3}{2}k^r \frac{|S|}{|V|}$.*

Proof. Let y denote an arbitrary vertex in V, and χ_S and $\chi_{\{y\}}$ the characteristic functions of S and $\{y\}$, respectively. The number of paths of length r starting at y and ending in S is exactly the $L^2(V)$-inner product $\langle \chi_S, A^r \chi_{\{y\}} \rangle$. Write

$$\chi_S \;=\; \frac{|S|}{n}\mathbb{1} + w \quad \text{and} \quad \chi_{\{y\}} \;=\; \frac{1}{n}\mathbb{1} + u, \qquad (2.3)$$

where $w, u \perp \mathbb{1}$. Because $\mathbb{1}$ is an eigenfunction of A^*, A preserves the orthogonal complement of $\mathbb{1}$, and thus

$$\|A^r u\| \;\leq\; \mu\|A^{r-1}u\| \;\leq\; \cdots \;\leq\; \mu^r\|u\|. \qquad (2.4)$$

Also, by orthogonality

$$\|w\| \leq \|\chi_S\| = \sqrt{|S|} \quad \text{and} \quad \|u\| \leq \|\chi_{\{y\}}\| = 1. \tag{2.5}$$

We have that $A^r \chi_{\{y\}} = \frac{1}{n} k^r \mathbb{1} + A^r u$, so the inner product may be calculated as

$$\langle \chi_S, A^r \chi_{\{y\}} \rangle = \frac{|S|}{n} k^r + \langle w, A^r u \rangle. \tag{2.6}$$

It now suffices to show that the absolute value of the second term on the right-hand side is bounded by half of the first term. Indeed,

$$|\langle w, A^r u \rangle| \leq \|w\| \|A^r u\| \leq \mu^r \sqrt{|S|}, \tag{2.7}$$

and

$$\mu^r \sqrt{|S|} \leq \frac{1}{2n} k^r \sqrt{|S|} \leq \frac{1}{2} k^r \frac{|S|}{n} \tag{2.8}$$

when $r \geq \frac{\log(2n)}{\log(k/\mu)}$. $\qquad\qquad\qquad\qquad\qquad\qquad\qquad\qquad\qquad\qquad\square$

3   Collisions on the Pollard ρ Graph

In this section, we prove an operator norm bound on the Pollard ρ graph that is later used in conjunction with Lemma 1. These graphs are closely related to an undirected graph studied in [8, Theorem 4.1]. We will start by quoting a special case of the key estimate of that paper, which concerns quadratic forms. At first glance, the analysis is reminiscent of the of the Hilbert inequality from analytic number theory (see [10, 12]), but where the quadratic form coefficients are expressed as $1/\sin(\mu_j - \mu_k)$.

Let n be an odd integer and $\lambda_k = |\cos(\pi k/n)|$ for $k \in \mathbb{Z}/n\mathbb{Z}$. Consider the quadratic form $Q : \mathbb{R}^{n-1} \to \mathbb{R}$ given by

$$Q(x_1, \ldots, x_{n-1}) := \sum_{k=1}^{n-1} x_k \, x_{2k} \, \lambda_k, \tag{3.1}$$

in which the subscripts are interpreted modulo n.

Proposition 1. *There exists an absolute constant $c > 0$ such that*

$$|Q(x_1, \ldots, x_{n-1})| \leq \left(1 - \frac{c}{(\log n)^2}\right) \sum_{k=1}^{n-1} x_k^2. \tag{3.2}$$

Proof. Let γ_k be arbitrary positive quantities (which will be specified later in the proof). Since

$$\gamma_k \, x_k^2 + \gamma_k^{-1} \, x_{2k}^2 \pm 2 \, x_k \, x_{2k} = \left(\gamma_k^{1/2} \, x_k \pm \gamma_k^{-1/2} \, x_{2k}\right)^2 \geq 0, \tag{3.3}$$

one has that

$$|Q(x)| \leq \frac{1}{2} \sum_{k=1}^{n-1} \left(\gamma_k\, x_k^2 + \gamma_k^{-1}\, x_{2k}^2 \right) \lambda_k = \frac{1}{2} \sum_{k=1}^{n-1} x_k^2 \left(\gamma_k\, \lambda_k + \gamma_{\bar{2}k}^{-1} \lambda_{\bar{2}k} \right),$$

(3.4)

where $\bar{2}$ denotes the multiplicative inverse to 2 modulo n. The proposition follows if we can choose γ_k and an absolute constant $c > 0$ such that

$$\gamma_k\, \lambda_k + \gamma_{\bar{2}k}^{-1} \lambda_{\bar{2}k} < 2 - \frac{c}{(\log n)^2} \qquad \text{for all } 1 \leq k < n.$$

(3.5)

Now we come to the definition of the γ_k. We set $\gamma_k = 1$ for $n/4 \leq k \leq 3n/4$; the definition for the set of other nonzero indices S is more involved. For $s \geq 0$, define

$$t_s = 1 - s\, \frac{d}{(\log n)^2},$$

where $d > 0$ is a small constant that is chosen at the end of the proof. Given an integer ℓ in the range $-n/4 < \ell < n/4$, we define $u(\ell)$ to be order to which 2 divides ℓ. For the residues $k \in S$, which are all equivalent modulo n to some integer ℓ in the interval $-n/4 < \ell < n/4$, we define $\gamma_k = t_{u(\ell)}$. Note also that $\lambda_k \leq 1/\sqrt{2}$ for $k \notin S$, and is always ≤ 1. With these choices the lefthand side of (3.5) is bounded by

$$\gamma_k\, \lambda_k + \gamma_{\bar{2}k}^{-1} \lambda_{\bar{2}k} \leq \begin{cases} \frac{1}{\sqrt{2}} + \frac{1}{\sqrt{2}}, & k, \bar{2}k \notin S \\ \frac{1}{\sqrt{2}} + \gamma_{\bar{2}k}^{-1}, & k \notin S,\ \bar{2}k \in S \\ \gamma_k + \frac{1}{\sqrt{2}}, & k \in S,\ \bar{2}k \notin S \\ \gamma_k + \gamma_{\bar{2}k}^{-1}, & k, \bar{2}k \in S. \end{cases}$$

(3.6)

In the last case, the residues k and $\bar{2}k$ both lie in S. The integer $\ell \equiv \bar{2}k \pmod{n}$, $-n/4 < \ell < n/4$, of course satisfies the congruence $2\ell \equiv k \pmod{n}$. Since $k \in S$, 2ℓ is the unique integer in $(-n/4, n/4)$ congruent to k. That means $\gamma_k = t_{s+1}$ and $\gamma_{\bar{2}k} = t_s$ for some positive integer $s = O(\log n)$. A bound for the last case in (3.6) is therefore $t_{s+1} + t_s^{-1} = 2 - d/(\log n)^2 + O(s^2 d^2/(\log n)^4)$. We conclude in each of the four cases that, for d sufficiently small, there exists a positive constant $c > 0$ such that (3.5) holds. $\square$

The Pollard ρ graph, introduced earlier, is the graph on $\mathbb{Z}/n\mathbb{Z}$ whose edges represent the possibilities involved in applying the iterating function (1.1):

$$\Gamma \text{ has vertices } V = \mathbb{Z}/n\mathbb{Z} \text{ and directed edges } x \to x+1,\ x \to x+y,$$
$$\text{and } x \to 2x \text{ for each } x \in V \text{ (where } y \neq 1 \text{)}.$$

(3.7)

Proposition 2. *Let A denote the adjacency operator of the graph (3.7) and assume that n is prime. Then there exists an absolute constant $c > 0$ such that*

$$\|Af\| \leq \left(3 - \frac{c}{(\log n)^2} \right) \|f\|$$

(3.8)

for all $f \in L^2(V)$ such that $f \perp \mathbb{1}$.

S.D. Miller and R. Venkatesan

Proof. Let $\chi_k : \mathbb{Z}/n\mathbb{Z} \to \mathbb{C}$ denote the additive character given by $\chi_k(x) = e^{2\pi i k x/n}$. These characters, for $1 \le k < n$, form a basis of functions $L_0 = \{f \in L^2 \mid f \perp \mathbb{1}\}$. The action of A on this basis is given by

$$A\chi_k \;=\; d_k \, \chi_k + \chi_{2k} \quad , \qquad \text{where} \;\; d_k \;=\; e^{2\pi i k/n} + e^{2\pi i k y/n}. \tag{3.9}$$

One has that $|d_k| = 2|\cos(\frac{\pi k(y-1)}{n})| = 2\lambda_{k(y-1)}$. Using the inner product relation

$$\langle \chi_k, \chi_\ell \rangle \;=\; \begin{cases} n, & k = \ell \\ 0, & \text{otherwise}, \end{cases} \tag{3.10}$$

we compute that $\|f\|^2 = n \sum |c_k|^2$, where $f = \sum_{k \ne 0} c_k \chi_k$. Likewise,

$$\|Af\|^2 \;=\; \langle Af, Af \rangle \;=$$
$$\sum_{k,\ell \ne 0} c_k \, \overline{c_\ell} \, [\langle d_k\chi_k, d_\ell\chi_\ell \rangle + \langle \chi_{2k}, \chi_{2\ell} \rangle + \langle d_k\chi_k, \chi_{2\ell} \rangle + \langle \chi_{2k}, d_\ell\chi_\ell \rangle]$$
$$\le \; n \left(5 \sum |c_k|^2 + 2 \sum |c_k||c_{2k}||d_{2k}| \right). \tag{3.11}$$

Note that $|d_k| = 2\lambda_{k(y-1)}$, and that $y-1$ and 2 are invertible in $\mathbb{Z}/n\mathbb{Z}$, by assumption in (3.7). The result now follows from (3.2) with the choice of $x_{2(y-1)k} = |c_k|$.
$\qquad\square$

Remark: the above Proposition, in combination with Lemma 1, is the source of the $\tau = O((\log n)^3)$ mixing time estimate for the Pollard ρ graph that we mentioned in the introduction.

Proof (of Theorem 1). Consider the set S of the first $t = \lfloor \sqrt{n} \rfloor$ iterates $x_1, x_2, \ldots, x_t$. We may assume that $|S| = t$, for otherwise a collision has already occurred in the first $\sqrt{n}$ steps. Lemma 1 and Proposition 2 show that the probability of a walk of length $r \gg (\log n)^3$ reaching S from any fixed vertex is at least $1/(2\sqrt{n})$. Thus the probabilities that $x_{t+r}, x_{t+2r}, x_{t+3r}, \ldots, x_{t+kr}$ lie in S are all, independently, at least $1/(3t)$. One concludes that for k on the order of $3bt$, b fixed, the probability that none of these points lies in S is at most $(1 - \frac{1}{3t})^{3bt} \approx e^{-b}$, which is less than ε for large values of b.

Generalizations: the analysis presented here extends to generalized Pollard ρ graphs in which each vertex x is connected to others of the form xg_i, for various group elements g_i, along with powers x^{r_j}. This can be done as follows. First of all, if r-th powers are to be used instead of squares, then the subscript $2k$ in (3.1) must be changed to rk. The key bound on (3.2), stated here for $r = 2$, in fact holds for any fixed integer $r > 1$ which is relatively prime to n [8, Appendix]. Thus changing the squaring step to $x \to x^r$ does not change the end results. Secondly, the proof of the bound (3.8) requires only some cancellation in (3.11). If additional operations are added, the cross terms from which the cancellation was derived here are still present. Thus Proposition 2 is remains valid, only with the 3 replaced by the degree of the graph. Provided this degree (= the total number of operations) is fixed, the graph still has rapid mixing.

It is unclear if including extra power operations speeds up the discrete logarithm algorithm. However, the rapid mixing of such random walks may have additional applications, such as to the stream ciphers in [8].

Acknowledgements. The authors wish to thank R. Balasubramanian, Michael Ben-Or, Noam Elkies, David Jao, László Lovász, and Prasad Tetali for helpful discussions and comments.

References

1. Eric Bach, *Toward a theory of Pollard's rho method*, Inform. and Comput. **90** (1991), no. 2, 139–155.
2. Béla Bollobás, *Modern graph theory*, Graduate Texts in Mathematics, vol. 184, Springer-Verlag, New York, 1998.
3. J.A. Fill, *Eigenvalue bounds on convergence to stationarity for nonreversible Markov chains, with an application to the exclusion process*, Ann. Appl. Probab. **1** (1991), 62-87.
4. Martin Hildebrand, *A survey of results on random walks on finite groups*, Probab. Surv. **2** (2005), 33–63 (electronic).
5. Jeremy Horwitz, *Applications of Cayley Graphs, Bilinearity, and Higher-Order Residues to Cryptology*, Ph.D. Thesis, Stanford University, 2004. http://math.scu.edu/~jhorwitz/pubs/.
6. Jeremy Horwitz and Ramarathnam Venkatesan, *Random Caylcy digraphs and the discrete logarithm*, Algorithmic number theory (Sydney, 2002), 2002, pp. 416–430.
7. László Lovász and Peter Winkler, *Mixing times*, Microsurveys in discrete probability (Princeton, NJ, 1997), 1998, pp. 85–133.
8. Stephen D. Miller, Ilya Mironov, and Ramarathnam Venkatesan, *Fast and Secure Stream Cipher Designs Using Rapidly Mixing Random Walks and Revolving Buffers* (2005). Preprint.
9. Ravi Montenegro and Prasad Tetali, *Mathematical Aspects of Mixing Times in Markov Chains*, Foundations and Trends in Theoretical Computer Science, 2006.
10. Hugh L. Montgomery, *Ten lectures on the interface between analytic number theory and harmonic analysis*, CBMS Regional Conference Series in Mathematics, vol. 84, Published for the Conference Board of the Mathematical Sciences, Washington, DC, 1994.
11. Victor Shoup, *Lower bounds for discrete logarithms and related problems*, Advances in cryptology—EUROCRYPT '97 (Konstanz), 1997, pp. 256–266. Updated version at http://www.shoup.net/papers/dlbounds1.pdf.
12. J. Michael Steele, *The Cauchy-Schwarz master class*, MAA Problem Books Series, Mathematical Association of America, Washington, DC, 2004. An introduction to the art of mathematical inequalities.
13. Edlyn Teske, *On random walks for Pollard's rho method*, Math. Comp. **70** (2001), no. 234, 809–825.
14. _____, *Square-root algorithms for the discrete logarithm problem (a survey)*, Public-key cryptography and computational number theory (Warsaw, 2000), 2001, pp. 283–301.

Hard Instances of the Constrained Discrete Logarithm Problem

Ilya Mironov[1], Anton Mityagin[2], and Kobbi Nissim[3,*]

[1] Microsoft Corp, SVC-5, 1065 La Avenida, Mountain View, CA 94041, USA
mironov@microsoft.com
[2] Department of Computer Science and Engineering, University of California
San Diego, 9500 Gilman Drive, La Jolla, CA 92093-0404, USA
mityagin@gmail.com
[3] 28 Haroozim st., Ramat-Gan, 52525, Israel
kobbi@cs.bgu.ac.il

Abstract. The discrete logarithm problem (DLP) generalizes to the constrained DLP, where the secret exponent x belongs to a set known to the attacker. The complexity of generic algorithms for solving the constrained DLP depends on the choice of the set. Motivated by cryptographic applications, we study explicit construction of sets for which the constrained DLP is hard. We draw on earlier results due to Erdös et al. and Schnorr, develop geometric tools such as generalized Menelaus' theorem for proving lower bounds on the complexity of the constrained DLP, and construct explicit sets with provable non-trivial lower bounds.

1 Introduction

One of the most important assumptions in modern cryptography is the hardness of the discrete logarithm problem (DLP). The scope of this paper is restricted to groups of prime order p, where the DLP is the problem of computing x given (g, g^x) for x chosen uniformly at random from $\mathbb{Z}_p$ (see the next section for notation). In some groups the DLP is believed to have average complexity of $\Theta(\sqrt{p})$ group operations. The *constrained DLP* is defined as the problem of computing x given (g, g^x) where x is chosen uniformly at random from a publicly known set $S \subseteq \mathbb{Z}_p$.

For the standard DLP there is a well-understood dichotomy between generic algorithms, which are oblivious to the underlying group, and group-specific algorithms. By analogy, we distinguish between generic and group-specific algorithms for the constrained DLP. In this paper we concentrate on the former kind, i.e., generic algorithms. Our main tool for analysis of generic algorithms is the Shoup-Nechaev generic group model [Sho97, Nec94].

The main motivation of our work is the fundamental nature of the problem and the tantalizing gap that exists between lower and upper bounds on the

---

* The work was done in Microsoft Research (Silicon Valley Campus).

F. Hess, S. Pauli, and M. Pohst (Eds.): ANTS 2006, LNCS 4076, pp. 582–598, 2006.

constrained DLP. A trivial generalization of Shoup's proof shows that the DLP constrained to any set $S \subseteq \mathbb{Z}_p$ has generic complexity $\Omega(\sqrt{|S|})$ group operations. On the other hand, Schnorr demonstrates that the DLP constrained to a *random* S of size $\sqrt{p}$ has complexity $\tilde{\Theta}(\sqrt{p}) = \tilde{\Theta}(|S|)$ with high probability [Sch01]. *Explicit* constructions of small sets with high complexity, or any complexity better than the square root lower bound were conspicuously absent.

The importance of improving the square root lower bound for concrete subsets of $\mathbb{Z}_p$ is implicit in [Yac98, HS03, SJ04], which suggest exponentiation algorithms that are faster than average for exponents sampled from certain subsets. These algorithms either rely on heuristic assumptions of security of the DLP constrained to their respective sets or use the square root lower bound to the detriment of their efficiency. For example, Yacobi proposes to use "compressible" exponents whose binary representation contains repetitive patterns [Yac98], which can be exploited by some algorithms for fast exponentiation. However, without optimistic assumptions about the complexity of the DLP constrained to this set the method offers no advantage over the sliding window exponentiation. Another method of speeding up exponentiation is to generate an exponent together with a short addition chain for it [Knu97, Ch. 4.6.3]. Absent reliable methods of sampling addition chains with uniformly distributed last elements, this approach depends on the hardness of the DLP on a non-uniform distribution.

The main technical contribution of our work is the proof that the DLP constrained to a set S, which is chosen from an easily sampleable family of sets of cardinality $p^{1/12-\varepsilon}$, has complexity $\Omega(|S|^{3/5})$ with probability $1 - 6p^{-12\varepsilon}$. At a higher level of abstraction we develop combinatorial techniques to bound the complexity of the constrained DLP, which is a global property, using the set's local properties. We view our work as a step towards better understanding the constrained DLP and possibly designing fast exponentiation algorithms tuned to work on exponents from "secure" subsets.

The structure of the paper is as follows. In Section 2 we present a number of results which are known but otherwise scattered in the literature. In Sections 3 and 4 we give new constructions of sets with provable lower bounds on various families of algorithms for solving the constrained DLP.

1.1   Notation

We use the standard notation for asymptotic growth of functions, where

$$O(g) = \{f\colon \mathbb{N} \to \mathbb{R}^+ \mid \exists c, n_0 > 0 \text{ s. t. } 0 \le f(n) \le cg(n) \text{ for all } n > n_0\};$$
$$\Omega(g) = \{f\colon \mathbb{N} \to \mathbb{R}^+ \mid \exists c, n_0 > 0 \text{ s. t. } cg(n) \le f(n) \text{ for all } n > n_0\};$$
$$\Theta(g) = \{f\colon f = O(g) \text{ and } g = O(f)\};$$

$\tilde{O}, \tilde{\Omega}, \tilde{\Theta}$ — same as O, Ω, Θ with logarithmic factors ignored;

$\mathbb{Z}_p$ — the field of residues modulo prime p;

$x \in_R S$ — x chosen uniformly at random from S.

1.2 Previous Work

Algorithms for solving number-theoretic problems can be grouped into two main classes: generic attacks, applicable in any group, and specific attacks designed for particular groups. The generic attacks on discrete logarithm include the baby-step giant-step attack [Sha71], Pollard's rho and lambda algorithms [Pol78] as well as their parallelized versions [vOW99, Pol00], surveyed in [Tes01]. The specific attacks have sprouted into a field in their own right, surveyed in [SWD96, Odl00].

A combinatorial view on generic attacks on the DLP was first introduced by Schnorr [Sch01]. He suggested the concept of the generic DL-complexity of a subset $S \subseteq \mathbb{Z}_p$ defined as the minimal number of generic operations required to solve the DLP for any element of $\{g^x \mid x \in S\}$. He showed that the generic DL-complexity of random sets of size $m < \sqrt{p}$ is $m/2 + o(1)$. In part our work is an extension of Schnorr's paper. The combinatorial approach to the DLP was further advanced by [CLS03] which gave a characterization of generic attacks on the entire group of prime order.

Systematically the constrained DLP has been studied for two special cases: Exponents restricted to an interval and exponents with low Hamming weight. Pollard's kangaroo method has complexity proportional to the square root of the size of the interval [Pol00]. The running time of a simple Las Vegas baby-step giant-step attack on low-weight exponents is $O(\sqrt{t}\binom{n/2}{t/2})$, where n is the length and t is the weight of the exponent [Hei93] (for a deterministic version see [Sti02], which credits it to Coppersmith). See [CLP05] for cryptanalysis of a similar scheme in a group of unknown order.

Erdös and D. Newman studied the BSGS-1 complexity (in our notation) and asked for constructions of sets with a high (better than a random subset's) BSGS-1 complexity [EN77].

1.3 Generic Algorithms

The generic group model introduced by Shoup and Nechaev [Sho97, Nec94] provides access to a group G via a random injective mapping $\sigma : G \to \Sigma$, which *encodes* group elements. The group operation is implemented as an oracle that on input $\langle \sigma(g), \sigma(h), \alpha, \beta \rangle$ outputs $\sigma(g^\alpha h^\beta)$ (for the sake of notation brevity we roll three group operations, group multiplication, group inversion, and group exponentiation, in one). Wlog, we restrict the arguments of the queries issued by algorithms operating in this model to encodings previously output by the oracle.

The discrete logarithm problem for groups of prime order has a trivial formalization in the generic group model:

Given $p, \sigma(g), \sigma(g^x)$ where g has order p and $x \in_R \mathbb{Z}_p$, find x.

The proof sketch of the theorem below, which is essentially the original one due to Shoup, is reproduced here because it lays the ground for a systematic study of complexity of algorithms in the generic group model.

Theorem 1 ([Sho97]). *Let $\mathcal{A}$ be a probabilistic algorithm and m be the number of queries made by $\mathcal{A}$. $\mathcal{A}$ solves the discrete logarithm problem in a group of prime order p with probability*

$$\Pr[\mathcal{A}(p, \sigma(g), \sigma(g^x)) = x] < \frac{(m+2)^2}{2p} + \frac{1}{p}.$$

The probability space is x, $\mathcal{A}$'s coin tosses, and the random function σ.

Proof (sketch). Instead of letting $\mathcal{A}$ interact with a real oracle, consider the following game played by a simulator. The simulator keeps track of two lists of equal length L_1 and L_2: the list of encodings $\sigma_1, \ldots, \sigma_{m+2} \in \Sigma$ and the list of linear polynomials $a_1 x + b_1, \ldots, a_{m+2} x + b_{m+2} \in \mathbb{Z}_p[x]$. Initially L_1 consists of two elements σ_1 and σ_2, which are the two inputs of $\mathcal{A}$, and L_2 consists of 1 and x. When $\mathcal{A}$ issues a query $\langle \sigma_i, \sigma_j, \alpha, \beta \rangle$, the simulator fetches the polynomials $a_i x + b_i$ and $a_j x + b_j$ from L_2, computes $a = \alpha a_i + \beta a_j$ and $b = \alpha b_i + \beta b_j$ and looks up $ax + b$ in L_2. If $ax + b = a_k x + b_k$ for some k, the simulator returns σ_k as the answer to the query. Otherwise, the simulator generates a new element $\sigma \in_R \Sigma \setminus L_1$, appends σ to L_1 and $ax + b$ to L_2, and returns σ.

$\mathcal{A}$ terminates by outputting some $y \in \mathbb{Z}_p$. The game completes as follows:

1. The simulator randomly selects $x^* \in_R \mathbb{Z}_p$.
2. Compute $a_i x^* + b_i$ for all $i \le m+2$. If $a_i x^* + b_i = a_j x^* + b_j$ for some $i \ne j$, the simulator fails.
3. $\mathcal{A}$ succeeds if and only if $x^* = y$.

Observe that the game played by the simulator is indistinguishable from the transcript of $\mathcal{A}$'s interaction with the actual oracle unless the simulator fails in step 2. Since for any two distinct polynomials $a_i x + b_i$ and $a_j x + b_j$ the probability that $a_i x^* + b_i = a_j x^* + b_j$ is at most $1/p$, the probability that step 2 fails is at most $(m+2)^2/2p$. Finally, we observe that the probability that $\mathcal{A}$ wins the game in step 3 is exactly $1/p$, which completes the proof. $\square$

It follows from the proof that the probability of success of any probabilistic adaptive algorithm for solving the discrete logarithm in $\mathbb{Z}_p$ in the generic group model can be computed given the list of the linear polynomials induced by its queries. This observation leads us to the concept of generic complexity defined in the next section.

2   Generic Complexity

Definition 1 (Intersection set). *For a set of pairs $L \subseteq \mathbb{Z}_p^2$, we define its intersection set*

$$I(L) = \{x \in \mathbb{Z}_p \mid \exists (a, b), (a', b') \in L \text{ s.t. } ax + b = a'x + b' \text{ and } (a, b) \ne (a', b')\}.$$

The set of pairs from the above definition corresponds to the set of queries asked by the generic algorithm. Its intersection set is the set of inputs on which the simulator from the proof of Theorem 1 fails.

Definition 2 (*L* **recognizes an** α**-fraction of** *S*). *For* $L \subseteq \mathbb{Z}_p^2$, $S \subseteq \mathbb{Z}_p$, *and* $0 < \alpha \leq 1$ *we say that* L *recognizes an* α*-fraction of the set* S *if*

$$|S \cap I(L)| \geq \alpha|S|.$$

Definition 3 (**Generic complexity**). *The set* $S \subseteq \mathbb{Z}_p$ *is said to have* generic α*-complexity* m *denoted as* $\mathcal{C}_\alpha(S)$ *if* m *is the smallest cardinality of a set* L *recognizing an* α*-fraction of* S.

Our definition of generic complexity is slightly different from a similar concept of the generic DL-complexity put forth by Schnorr. We only require that the intersection set $I(L)$ covers a constant fraction of S rather than the entire set [Sch01]. Our definition better matches the standard practice of cryptanalysis, when an attack is considered successful if it succeeds on a nontrivial fraction of the inputs. Moreover, our bounds exhibit different scaling behavior as a function of α, and by parametrizing the definition with α we make the dependency explicit.

Proposition 1 ([Sch01]). *For any* $S \subseteq \mathbb{Z}_p$ *the generic* α*-complexity of the set* S *is bounded as*

$$\sqrt{2\alpha|S|} < \mathcal{C}_\alpha(S) \leq \alpha|S|/2 + 3.$$

Proof. The lower bound follows from the fact that for any $L \subseteq \mathbb{Z}_p^2$ the cardinality of the intersection set is bounded as $|I(L)| < |L|^2/2$. Therefore, in order to cover at least an α-fraction of the set, $|L|^2/2$ must be more than $\alpha|S|$.

The upper bound is attained by the following construction. If $2m \geq \alpha|S|$ and $\{x_1, \ldots, x_{2m}\} \subseteq S$, then an α-fraction of S is recognized by L of size $m + 2$ defined as

$$L = \left\{ (0,0), (0,1), \left(\frac{1}{x_2 - x_1}, \frac{x_1}{x_1 - x_2}\right), \ldots, \left(\frac{1}{x_{2m} - x_{2m-1}}, \frac{x_{2m-1}}{x_{2m-1} - x_{2m}}\right) \right\},$$

since x_i and x_{i-1} are the x-coordinates of the points of intersection of the line $\left(\frac{1}{x_i - x_{i-1}}, \frac{x_{i-1}}{x_{i-1} - x_i}\right)$ with lines $y = 0$ and $y = 1$ respectively. $\square$

Proposition 2. $\sqrt{2\alpha p} < \mathcal{C}_\alpha(\mathbb{Z}_p) \leq 2\lceil\sqrt{\alpha p}\rceil$.

Proof. The lower bound on $\mathcal{C}_\alpha(\mathbb{Z}_p)$ is by Proposition 1. The upper bound is given by the set $L = \{(0, i), (1, -\lambda i) \mid 0 \leq i < \lambda\}$, where $\lambda = \lceil\sqrt{\alpha p}\rceil$. Indeed,

$$I(L) = \bigcup_{0 \leq i,j < \lambda} I(\{(0, i), (1, -\lambda j)\}) = \bigcup_{0 \leq i,j < \lambda} \{\lambda j + i\},$$

which covers $[0, \alpha p)$. $\square$

A tighter (up to a constant factor) bound in the general case and exact values for $\mathcal{C}_1(\mathbb{Z}_p)$ for small primes $p < 100$ appear in [CLS03].

Since the generic complexity is a monotone property, it follows that for any set $S \subseteq \mathbb{Z}_p$

$$\mathcal{C}_\alpha(S) \leq \min(\alpha|S|/2 + 3, 2\lceil\sqrt{\alpha p}\rceil).$$

Now we are ready to establish the connection between the generic complexity of a set and the discrete logarithm problem.

Theorem 2. *Let $S \subseteq \mathbb{Z}_p$, $\mathcal{A}_S$ be a generic algorithm that makes at most $m < \sqrt{p}$ queries and outputs a number from $\mathbb{Z}_p$. Suppose*

$$\Pr[\mathcal{A}_S(\sigma(g), \sigma(g^x)) = x] > \alpha + \frac{1}{|S| - (m+2)^2},$$

where the probability is taken over $\mathcal{A}$'s random tape, the oracle answers, and $x \in_R S$. Then necessarily

$$m \geq \mathcal{C}_\alpha(S).$$

The above bound is nearly tight, i.e., for any set S there is a generic algorithm whose query complexity is $\mathcal{C}_\alpha(S)$ and probability of success is at least $\alpha + 1/(|S| - \alpha|S|)$.

Proof (sketch). The proof essentially follows that of Theorem 1. Let L be a set of pairs (a_i, b_i) constructed by the simulator. Unless x belongs to its intersection set, the output of $\mathcal{A}_S$ is independent of x. In this case its probability of success is at most $1/(|S \setminus I(L)|) \leq 1/(|S| - (m+2)^2)$. The probability that x belongs to the intersection set of a set of size m is less than α as long as $m < \mathcal{C}_\alpha(S)$.

The tightness property follows from the definition of generic complexity. Let L be the set of pairs of size $\mathcal{C}_\alpha(S)$ so that $|S \cap I(L)| \geq \alpha|S|$. Query the oracle $\langle \sigma(g^x), \sigma(g), a, b \rangle$ for all pairs $(a, b) \in L$. With probability $|I(L) \cap S|/|S|$ there is a collision that gives away x, otherwise make a guess that succeeds with probability $1/|S \setminus I(L)|$. □

Notice that the theorem above is unconditional and the adversary is computationally unbounded. In particular, the adversary is given full access to S and can design an S-specific algorithm. As long as the algorithm has only oracle access to the group, $\mathcal{C}_\alpha(S)$ is a lower bound on the number m of oracle queries needed by the algorithm to succeed with probability at least $\alpha + 1/(|S| - (m+2)^2)$.

We know that $\mathcal{C}_\alpha(S)$ can be negligible compared to $|S|$ (for instance, according to Proposition 2, when $S = \mathbb{Z}_p$, $|S| = p$ but its generic complexity is $O(\sqrt{p})$). Since the generic complexity is intimately related to the query complexity of any discrete logarithm-solving algorithm, we would like to build sets with higher generic complexity. The next theorem demonstrates that for a fixed p a random set of size less than $\sqrt{p}$ has a near-linear generic complexity.

Theorem 3. *For a random subset $S \subseteq_R \mathbb{Z}_p$ of size p^ε for some constant $\varepsilon \leq 1/2$ its generic α-complexity is at least*

$$\mathcal{C}_\alpha(S) > \frac{\alpha|S|}{2 \ln p}$$

with probability $1 - 1/p$ for large enough p.

Proof. The proof is by a counting argument. We shall bound the number of the sets S of size $k = p^\varepsilon$ whose α-fraction can be recognized by a set L of size δk, when $\varepsilon \leq 1/2$ and $\delta = \alpha/(2 \ln p)$. Suppose $|L| = \delta k$ and $|I(L) \cap S| \geq \alpha k$, where S

is to be constructed. There are $\binom{p^2}{\delta k}$ subsets $L \subseteq \mathbb{Z}_p^2$ of size δk. The intersection set $I(L)$ has size at most $(\delta k)^2$ and contains at least αk elements which belong to S. There are thus at most $\binom{p^2}{\delta k}\binom{(\delta k)^2}{\alpha k}$ distinct possibilities for these αk elements. The $(1-\alpha)$-fraction of S can be chosen arbitrarily from $\mathbb{Z}_p$, in $\binom{p}{(1-\alpha)k}$ many ways. In total the number of subsets S of generic complexity δk and cardinality k is bounded by $\binom{p^2}{\delta k}\binom{(\delta k)^2}{\alpha k}\binom{p}{(1-\alpha)k}$. Using Stirling's approximation formula one can verify that this number is less than a $1/p$-fraction of the number of subsets of $\mathbb{Z}_p$ of size $k = p^\varepsilon$ for constant ε and large enough p.   □

The bottom line of the theorem we just proved is that hard sets (where the discrete logarithm is hard to compute using a generic algorithm) are easy to come by. In fact, almost any set has high generic complexity (also previously observed in [Sch01]).

In what follows we sharply lower the amount of randomness that is required to provide any non-trivial guarantee of generic complexity.

3   More Complexities and Lower Bounds

Many sets of group elements with special properties may be attacked using a baby-step giant-step method. In this method the attacker first computes $g^{c_1}, \ldots, g^{c_m}$ (giant steps) and then compares them against $g^{a_1 x + b_1}, \ldots, g^{a_m x + b_m}$ (baby steps). Any collision between a baby step and a giant step gives away x. We define the complexity of this method along the lines of the generic complexity from the previous section.

Definition 4 (Intersection set-2). *For a set of pairs $L \subseteq \mathbb{Z}_p^2$ and a set of points $C \subseteq \mathbb{Z}_p$, we define their* intersection set *as*

$$I(L, C) = \{x \in \mathbb{Z}_p \mid \exists (a, b) \in L, c \in C \text{ s.t. } a \neq 0 \text{ and } ax + b = c\}.$$

Definition 5 (Baby-step giant-step complexity.). *The set $S \subseteq \mathbb{Z}_p$ is said to have the* baby-step giant-step α-complexity *(BSGS complexity for short) m denoted as $\mathcal{C}_\alpha^{\mathrm{bsgs}}(S)$ if m is the smallest integer such that there exist $L \subseteq \mathbb{Z}_p^2$ and $C \subseteq \mathbb{Z}_p$, with $|L| = |C| = m$ and $|I(L, C) \cap S| \geq \alpha |S|$.*

An important particular case of the baby-step giant-step method is when all lines defined by L are parallel (i.e., all $a_i = 1$).

Definition 6 (BSGS-1 complexity). *The set $S \subseteq \mathbb{Z}_p$ has BSGS-1 α-complexity m denoted as $\mathcal{C}_\alpha^{\mathrm{bsgs1}}(S)$ if m is the smallest integer such that there exist $L \subseteq \{1\} \times \mathbb{Z}_p$ and $C \subseteq \mathbb{Z}_p$, with $|L| = |C| = m$ and $|I(L, C) \cap S| \geq \alpha |S|$.*

Equivalently, $\mathcal{C}_\alpha^{\mathrm{bsgs1}}(S)$ is the smallest integer n such that there exist $X, Y \subseteq \mathbb{Z}_p$ with $n = |X| = |Y|$ and $|S \cap (X - Y)| \geq \alpha |S|$, where $X - Y$ is the set of pairwise differences between X and Y.

The problem of computing $\mathcal{C}_\alpha^{\mathrm{bsgs1}}(S)$ is superficially similar to a number of problems in additive number theory concerned with studying properties of $X - Y$.

However, our goal is fundamentally different since we require that $X - Y$ cover a large fraction of S rather than be its exact equal. To the best of our knowledge, the only paper in the literature directly applicable to bounding $C_1^{\text{bsgs}1}(S)$ is a 1977 paper by Erdös and Newman [EN77]. They proved analogues of our Theorems 3 and 6 and bounded the BSGS-1 complexity (in our notation) of the set of small squares $\{x^2 \mid x < \sqrt{p}\}$, which has the order of $p^{1/3 - c/\log\log p}$. They leave open the problem of constructing sets with a strictly linear BSGS-1 complexity (without the $1/\log p$ factor).

The BSGS and BSGS-1 complexities provide useful upper bounds for the generic complexity.

Proposition 3. $\frac{1}{2}C_\alpha(S) \leq C_\alpha^{\text{bsgs}}(S) \leq C_\alpha^{\text{bsgs}1}(S)$.

Proof. Let $C' = \{0\} \times C = \{(0, c) \mid c \in C\}$. Then $I(L, C) \subseteq I(L \cup C')$, which implies the first inequality. The second inequality follows from the fact that any BSGS-1 attack is also a BSGS attack. $\qquad\square$

Consider, for example, the baby-step giant-step attacks on exponents with low Hamming weight [Hei93, Sti02]. Define $S_\lambda = \{x \in \mathbb{Z}_p \mid \nu(x) = \lambda|x|\}$, where $\nu(x)$ is the number of ones in the binary representation of x. If $\lambda = 1/4$, an elementary analysis of the attacks shows that

$$C_1^{\text{bsgs}1}(S_{0.25}) = O(p^{0.406}).$$

Following Propositions 2, 3, and Theorem 3 the BSGS-1 α-complexity of a set of cardinality less than $\sqrt{p}$ lies between $\sqrt{\alpha|S|/2}$ and $2\sqrt{\alpha p}$, where the lower bound is trivial and the upper bound is approximated up to a logarithmic factor by almost any subset of size $\sqrt{p}$. In this section we give an explicit construction of a set with a non-trivial BSGS-1 complexity. We start by stating without proof an important combinatorial lemma known as the Zarankiewicz problem [Zar51]:

Theorem 4. *[Bol98, Ch. IV.2] Let $Z(n, s, t)$ be the maximum number of ones that can be arranged in an $n \times n$ matrix such that there is no all-one $t \times s$ (possibly disjoint) submatrix. Then*

$$Z(n, s, t) < s^{1/t}n^{2-1/t}.$$

Notice that the asymptotic of the bound on $Z(n, s, t)$ depends on the smallest of the two dimensions of the prohibited all-one submatrix. It is known that the bound is tight (up to a constant factor) for $t = 2, 3$.

Our second combinatorial tool is the upper bound due to A. Naor and Verstraëte on the number of edges in a bipartite graph without cycles of length $2k$ (C_{2k}-free graph):

Theorem 5 ([NV05]). *The maximum number of edges in a C_{2k}-free (n, n)-bipartite graph is less than $2kn^{1+1/k}$.*

When $k = 2$ the two theorems overlap. Indeed, a 0-1 matrix is also a bipartite graph, where the rows and columns form the vertex set and the non-zero elements indicate adjacency of corresponding vertices. In this case an all-one 2×2 submatrix represents a cycle of length 4 in the graph. Our theorems fully reflect this relationship: Theorem 6 can be proved using either the Zarankiewicz or the Naor-Verstraëte bound; its generalization Theorem 7 makes use of C_{2k}-free graphs, while Theorems 8 and 9 apply the Zarankiewicz bound.

Theorem 6. *Suppose $S \subseteq \mathbb{Z}_p$ has the property that all pairwise sums of different elements of S are distinct. Then*

$$C_\alpha^{\mathrm{bsgs1}}(S) > (\alpha|S|/\sqrt{2})^{2/3}.$$

Proof. Take $X, Y \subseteq \mathbb{Z}_q$, such that $n = |X| = |Y|$ and $|S \cap (X - Y)| > \alpha|S|$. Consider an $n \times n$ matrix M, whose rows and columns are labeled with elements of X and Y respectively. For each element $s \in S \cap (X - Y)$ find one pair $x \in X$ and $y \in Y$ such that $s = x - y$ and set the (x, y) entry of the matrix to one. Since $X - Y$ covers at least an α-fraction of S, the number of ones in the matrix is at least $\alpha|S|$.

We claim that M does not contain an all-one 2×2 submatrix. Assume the opposite: The submatrix given by elements x_1, x_2 and y_1, y_2 has four ones. It follows that all four $s_{11} = x_1 - y_1$, $s_{12} = x_1 - y_2$, $s_{21} = x_2 - y_1$, $s_{22} = x_2 - y_2 \in S$. Then $s_{11} + s_{22} = s_{12} + s_{21}$, which contradicts the assumption that all pairwise sums of elements of S are distinct. Applying the Zarankiewicz bound for the case $s = t = 2$, we prove that

$$\alpha|S| < Z(n, 2, 2) < \sqrt{2}n^{2-1/2} = \sqrt{2}n^{3/2},$$

which implies that $n = C_\alpha^{\mathrm{bsgs1}}(S) > (\alpha|S|/\sqrt{2})^{2/3}$. $\qquad\square$

The sets where sums of pairs of different elements are distinct are known in combinatorics as weak Sidon sets. They are closely related to (strong) Sidon sets, also called B_2 sequences, where all pairwise sums (of not necessarily different elements) are distinct (for a comprehensive survey see [O'B04] that includes more than 120 bibliographic entries). Explicit constructions of Sidon subsets of $\{1, \ldots, n\}$ due to Singer and Ruzsa have cardinality at least $\sqrt{n} - n^{.263}$ [Sin38, BC63, Ruz93, BHP01].

We additionally require that the sums be different modulo p. The size of such sets is bounded from above by $p^{1/2} + 1$ [HHÖ04, Theorem 3]. The easiest shortcut to constructing weak modular Sidon sets is to take a strong Sidon subset of $\{0 \ldots \lfloor p/2 \rfloor\}$ (see also [O'B02, Ch. 3] and [GS80]). Denser Sidon sets may be constructed for primes of the form $p = q^2 + q + 1$, where q is also prime [Sin38]. Existence of infinitely many such primes is implied by Schnitzel's Hypothesis H and their density follows from the even stronger Bateman-Horn conjecture [Guy04, A]. Interestingly, modular Sidon sets are useful not only in constructing sets with high complexity, via Theorem 6, but also for solving the discrete logarithm problem in $\mathbb{Z}_p$ [CLS03].

4   Beyond the Basics

Theorem 6 can be generalized to make use of Sidon sets of higher order. First, we prove that if all k-wise sums of elements of S are distinct (counting permutations of the same k-tuple only once), then there is a bound on the BSGS-1 complexity. Second, we provide a result that there exist such sets of size $\Theta(p^{1/k})$.

Theorem 7. *Suppose $S \subseteq \mathbb{Z}_p$ is such that all k-wise sums of different elements of S are distinct (excluding permutation of the summands). Then*

$$\mathcal{C}_\alpha^{\mathrm{bsgs1}}(S) > (\alpha|S|/(2k))^{k/(k+1)}.$$

Proof. Take $X, Y \subseteq \mathbb{Z}_p$, such that $\mathcal{C}_\alpha^{\mathrm{bsgs1}}(S) = |X| = |Y| = n$ and $|S \cap (X-Y)| > \alpha|S|$. Instead of the matrix as in Theorem 6, consider a bipartite graph $G(X, Y)$, where there is an edge (x, y) if and only if $x - y \in S$ (keep only one edge per element of S).

We claim that there are no $2k$-cycles (without repetitive edges) in the bipartite graph G. Assume the opposite: There is a cycle $(x_1, y_1, \ldots, x_k, y_k, x_1, y_1)$. Consider two sums: $(x_1 - y_1) + (x_2 - y_2) + \cdots + (x_k - y_k)$ and $(x_2 - y_1) + (x_3 - y_2) + \cdots + (x_k - y_{k-1}) + (x_1 - y_k)$. Not only are the two sums equal, they also consist of k elements of S each, and these elements are all distinct (as every element of S appears as an edge of G at most once). A contradiction is found.

The number of edges in an (n, n)-bipartite graph without $2k$-cycles is less than $2kn^{1+1/k}$ (Theorem 5). Therefore $\alpha|S| < 2kn^{1+1/k}$, and $n = \mathcal{C}_\alpha^{\mathrm{bsgs1}}(S) > (\alpha|S|/(2k))^{k/(k+1)}$. $\square$

Bose and Chowla give a construction for subsets of $\{1, \ldots, q^k\}$ of prime size q whose k-wise sums are distinct (in integers, not modulo p) [BC63]. By choosing the largest prime q less than $p^{1/k}$ (which, for large p is more than $p^{1/k} - p^{0.525/k}$ [BHP01]) an interval of length q^k/k with a $1/k$ proportion of the set's elements, we guarantee that all k-sums are distinct in $\mathbb{Z}_p$ as well. Unfortunately, [BC63] does not provide an efficient sampling algorithm.

Along the lines of Theorem 6 we prove that other verifiable criteria imply non-trivial bounds on the BSGS and generic complexity.

Theorem 8. *Suppose $S \subseteq \mathbb{Z}_p$ is such that for any distinct $x_1, x_2, y_1, y_2, z_1, z_2 \in S$:*

$$\det \begin{pmatrix} x_1 - y_1 & x_2 - y_2 \\ y_1 - z_1 & y_2 - z_2 \end{pmatrix} \neq 0. \tag{1}$$

Then

$$\mathcal{C}_\alpha^{\mathrm{bsgs}}(S) > (\alpha|S|/\sqrt{3})^{2/3}.$$

Proof. Take $L \subseteq \mathbb{Z}_p^2$ and $C \subseteq \mathbb{Z}_p$, such that $|L| = |C| = n$ and $|I(L, C) \cap S| > \alpha|S|$. As in Theorem 6 consider the $n \times n$ matrix M whose rows and columns are labeled with elements of L and C respectively. For each element $s \in S \cap I(L, C)$

set one entry in row (a,b) and column c to one, where $s = (c-b)/a$. Thus, the total number of ones in the matrix is exactly $m = |I(L,C) \cap S|$. If there is a 2×3 all-one submatrix in M, then property (1) does not hold (three parallel lines divide two other lines proportionally). The Zarankiewicz bound implies that

$$\alpha|S| < Z(n,3,2) < \sqrt{3}n^{2-1/2} = \sqrt{3}n^{3/2}.$$

Hence $\mathcal{C}_\alpha^{\text{bsgs}}(S) = n \geq (\alpha|S|/\sqrt{3})^{2/3}$. $\qquad\square$

Constructing a large explicit subset of $\mathbb{Z}_p$ satisfying the condition of the previous theorem is a difficult problem. Fortunately, the probability that a random 6-tuple of $\mathbb{Z}_p$ elements fails to satisfy (1) is $2/p$ [Sch80]. This observation motivates the following definition:

Definition 7 ($\mathcal{S}(N,k)$ family). *Let $\mathcal{S}(N,k) = \{x_1, \ldots, x_N\}$ be a family of subsets of $\mathbb{Z}_p$, where $x_1, \ldots, x_N \colon \mathcal{K} \mapsto \mathbb{Z}_p$ are k-wise independent random variables ($\mathcal{K}$ is the probability space).*

Properties of $\mathcal{S}(N,k)$ are established in the following proposition:

Proposition 4. *1. $\mathcal{S}(N,k)$ can be defined over $\mathcal{K} = \mathbb{Z}_p^k$.*
2. For $k > 1$, $\Pr_{S \in_R \mathcal{S}(N,k)}[|S| \neq N] < N^2/p$.
3. If $h \in \mathbb{Z}[y_1, \ldots, y_k]$ and $d = \deg(h) > 0$, then

$$\Pr_{S \in \mathcal{S}(N,k)}[\exists \text{ distinct } z_1, \ldots, z_k \in S \text{ with } h(z_1, \ldots, z_k) = 0] < N^k d/p.$$

Proof. 1. To construct $\mathcal{S}(N,k)$ we use a well-known k-universal family of functions (following [CW77]). Let the probability space be $\mathcal{K} = \mathbb{Z}_p^k$ and $f_a(x) = a_{k-1}x^{k-1} + \cdots + a_0$ for $a = (a_0, \ldots, a_{k-1}) \in \mathcal{K}$. Define the random variables $x_i = f_a(i) \colon \mathcal{K} \to \mathbb{Z}_p$ for $1 \leq i \leq N$. We claim that the variables $x_1, \ldots, x_N$ are k-wise independent. This follows from the system $f_a(i_1) = y_1, \ldots, f_a(i_k) = y_k$ having a unique solution $a \in \mathcal{K}$ for any distinct $i_1, \ldots, i_k \in \{1, \ldots, N\}$ and $y_1, \ldots, y_k \in \mathbb{Z}_p$. Notice that the elements of any $S \in \mathcal{S}(N,k)$ can be easily sampled and enumerated.
2. Let I_{ij} be the indicator variable, which is equal to 1 when $x_i = x_j$ and 0 otherwise. The cardinality of $S = \{x_1, \ldots, x_N\}$ is at least $N - \sum_{i<j} I_{ij}$. Since x_i and x_j are independent for all $i \neq j$, $E[I_{ij}] = 1/p$. By linearity of expectation, the expected value $E[\sum_{i<j} I_{ij}] < N^2/p$. By Markov's inequality $\Pr[|S| \neq N] = \Pr[\sum_{i<j} I_{ij} \geq 1] < N^2/p$.
3. Let $I_{i_1,\ldots,i_k}$ for all distinct $1 \leq i_1, \ldots, i_k \leq N$ be the indicator variable that is 1 if and only if $h(x_{i_1}, \ldots, x_{i_k}) = 0$. By independence of the variables and [Sch80] $E[I_{i_1,\ldots,i_k}] \leq 2/p$, which by linearity of expectation and Markov's inequality implies that $\Pr_S[\exists \text{ distinct } x_1, \ldots, x_k \in S \text{ with } h(x_1, \ldots, x_k) = 0] \leq \Pr_S[\sum_{i_1,\ldots,i_k} I_{i_1,\ldots,i_k} \geq 1] < N^k d/p$. $\qquad\square$

It follows that a randomly chosen set from the family $\mathcal{S}(p^{1/6-\varepsilon}, 6)$ has size $p^{1/6-\varepsilon}$ with probability at least $1 - p^{-2/3}$ and satisfies the condition of Theorem 8 with probability at least $1 - 2p^{-6\varepsilon}$.

To apply a similar argument to the all-powerful generic complexity, we may show that for small constants m_1 and m_2, the projections on the x axis of the intersection points of an irregular m_1 by m_2 grid (in which lines need not be parallel) satisfy a certain relationship. Next, a set S, where any $m_1 m_2$-tuple avoids this relationship, is to be constructed.

Let us see first why this argument works for some values of m_1 and m_2, and then improve the parameters. Let $m_1 = 4$ and $m_2 = 5$. There are 9 lines that can be described using 18 parameters. On the other hand, there are 20 points that form the intersection set of these lines. Each of the 20 intersection points imposes a linear equation on the parameters, and hence the system is overdetermined (even if we exclude linearly dependent equations). In particular, this implies that the probability that a random 20-tuple of elements of $\mathbb{Z}_p$ is coverable by a 4×5 grid is negligible. We refine this argument in the following proposition.

Proposition 5 (Bipartite Menelaus' theorem). *Consider seven lines $l_{x,y,z}$, $l_{1,2,3,4}$ forming an irregular grid, and their twelve intersection points. Let x_i, y_i, z_i be projections on the x axis of the intersection points of l_i with lines l_x, l_y, l_z. Then the following holds:*

$$\det \begin{pmatrix} x_1 - y_1 & x_1 - z_1 & z_1(x_1 - y_1) & y_1(x_1 - z_1) \\ x_2 - y_2 & x_2 - z_2 & z_2(x_2 - y_2) & y_2(x_2 - z_2) \\ x_3 - y_3 & x_3 - z_3 & z_3(x_3 - y_3) & y_3(x_3 - z_3) \\ x_4 - y_4 & x_4 - z_4 & z_4(x_4 - y_4) & y_4(x_4 - z_4) \end{pmatrix} = 0. \tag{2}$$

Proof. Denote the 4×4 matrix in (2) by M. Observe that if any of the seven lines is vertical, (2) follows immediately. Indeed, if $l_y = \{x = \text{const}\}$, then $y_1 = y_2 = y_3 = y_4$ and the second and the fourth columns of M are linearly dependent. Moreover, $\det M$ is invariant under permutations of l_x, l_y, and l_z, which takes care of vertical l_y or l_z. If l_i is vertical for some $1 \le i \le 4$, then the ith row of M is all-zero, and $\det M = 0$.

If none of the lines is vertical, we can write down equations for all of them in the Cartesian coordinates. Let $l_{x,y,z} = \{a_{x,y,z}x + b_{x,y,z}\}$ and $l_i = \{c_i x + d_i\}$ for $1 \le i \le 4$. Each intersection point imposes an equation on the parameters of the two lines incident with it, a total of 12 equations in 14 unknowns. However, the system always has a trivial solution, when all lines are equal. Rewrite the system using new variables: $\tilde{a}_y = a_y - a_x$, $\tilde{b}_y = b_y - b_x$, $\tilde{a}_z = a_z - a_x$, $\tilde{b}_z = b_z - b_x$, $\tilde{c}_1 = c_1 - a_x$, $\tilde{d}_1 = d_1 - b_x$, etc. The result is a homogenous system of 12 linear equations in 12 new variables. It has a non-zero solution if and only if its matrix is singular (only non-zero elements are shown):

$$M' = \begin{pmatrix} \cdot & \cdot & \cdot & \cdot & \cdot & \cdot & -x_1 & -1 & \cdot & \cdot & \cdot & \cdot & \cdot \\ \cdot & \cdot & \cdot & \cdot & \cdot & \cdot & \cdot & \cdot & -y_1 & -1 & \cdot & \cdot \\ \cdot & \cdot & \cdot & \cdot & \cdot & \cdot & \cdot & \cdot & \cdot & \cdot & -z_1 & -1 \\ x_2 & 1 & \cdot & \cdot & \cdot & \cdot & -x_2 & -1 & \cdot & \cdot & \cdot & \cdot \\ y_2 & 1 & \cdot & \cdot & \cdot & \cdot & \cdot & \cdot & -y_2 & -1 & \cdot & \cdot \\ z_2 & 1 & \cdot & \cdot & \cdot & \cdot & \cdot & \cdot & \cdot & \cdot & -z_2 & -1 \\ \cdot & \cdot & x_3 & 1 & \cdot & \cdot & -x_3 & -1 & \cdot & \cdot & \cdot & \cdot \\ \cdot & \cdot & y_3 & 1 & \cdot & \cdot & \cdot & \cdot & -y_3 & -1 & \cdot & \cdot \\ \cdot & \cdot & z_3 & 1 & \cdot & \cdot & \cdot & \cdot & \cdot & \cdot & -z_3 & -1 \\ \cdot & \cdot & \cdot & \cdot & x_4 & 1 & -x_4 & -1 & \cdot & \cdot & \cdot & \cdot \\ \cdot & \cdot & \cdot & \cdot & y_4 & 1 & \cdot & \cdot & -y_4 & -1 & \cdot & \cdot \\ \cdot & \cdot & \cdot & \cdot & z_4 & 1 & \cdot & \cdot & \cdot & \cdot & -z_4 & -1 \end{pmatrix}.$$

One can verify that $\det(M) = \det(M')$.   □

In the full version of the paper we give a geometric proof of the proposition, deriving (2) directly, and explain the connection with classic Menelaus' theorem. We also give an alternative statement of the theorem, which puts it in the realm of projective geometry.

Proposition 5 is the "minimal" condition that holds for the x-coordinates of the intersection points of two sets of lines in general position. Indeed, it follows from the proof that for any assignment of distinct values to the eleven variables $x_{1,2,3,4}$, $y_{1,2,3,4}$, $z_{1,2,3}$ there is a configuration of lines whose intersection points project to those variables. Other configurations with as many or fewer intersection points do not produce any conditions either. For example, six lines intersecting two lines can project to any collection of twelve points.

All geometric arguments (Theorems 6, 7, and 8, Proposition 5) are illustrated in Fig. 1.

Theorem 9. *If S is chosen from $\mathcal{S}(p^{1/12-\varepsilon}, 12)$, then with probability at least $1 - 6p^{-12\varepsilon}$*

$$\mathcal{C}_\alpha(S) > (\alpha|S|/\sqrt[3]{4})^{3/5}.$$

Proof. Consider the set of lines $L \subseteq \mathbb{Z}_p^2$ such that $\mathcal{C}_\alpha(S) = |I(L) \cap S|$ and $n = |L|$. As in Theorem 8, we apply the Zarankiewicz bound to the $n \times n$ matrix, only now both the rows and the columns are labeled with elements of the set L. Similarly, only one occurrence of an element of S as the x-coordinate of the intersection of two distinct lines is recorded in the matrix.

According to Proposition 4, S avoids solutions to the equation (2), whose left-hand side is a multivariate polynomial of total degree 6, with probability greater than $1 - 6p^{-12\varepsilon}$. Therefore the probability that there exist 12 points in S that can be the intersection set of two groups of lines consisting of 3 and 4 lines respectively is less than $6p^{-12\varepsilon}$. Finally, as before, $\alpha|S| < Z(n, 4, 3) < 4^{1/3}n^{2-1/3} = \sqrt[3]{4}n^{5/3} = \sqrt[3]{4}\mathcal{C}_\alpha(S)^{5/3}$.   □

Unlike the proofs of Theorems 6, 7, and 8, where the classes in which the lines are grouped arise naturally, the use of bipartite Menelaus' theorem in the analysis of

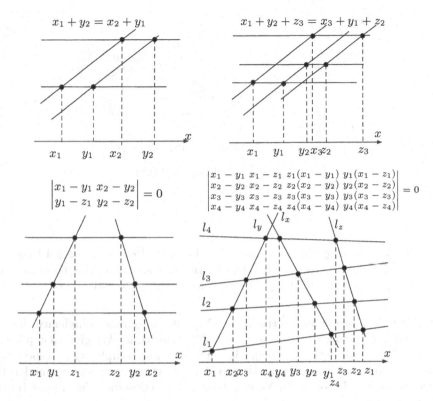

Fig. 1. "Prohibited" configurations (Theorems 6, 7 and 8, Proposition 5)

generic complexity above might appear less motivated. In fact, classic Menelaus' theorem imposes a simple condition (a cubic equation) on the intersection set of four lines. It is the second step of the argument, where we translate absence of a certain submatrix (subgraph) into sparseness of the entire matrix, which becomes problematic: Unless H is bipartite, H-free graphs on n vertices may have as many as $\Theta(n^2)$ edges according to the celebrated Turán theorem [Bol98, Ch. IV.2].

5   Conclusion

In this paper we develop a theory of lower bounds in the generic group model on the discrete logarithm problem constrained to a subset $S \subseteq \mathbb{Z}_p$ known to the attacker (constrained DLP). We give a first concrete construction of a set whose generic complexity is more than the square root of its size (Theorem 9). There exists an apparent gap between our explicit construction ($|S| = p^{1/12}$ and $\mathcal{C}_1(S) = |S|^{3/5}$) and a random set of size $p^{1/2}$ whose complexity is almost linear in its size. Bridging this gap constitutes an interesting open problem whose solution would shed some light on the intrinsic difficulty of the discrete logarithm

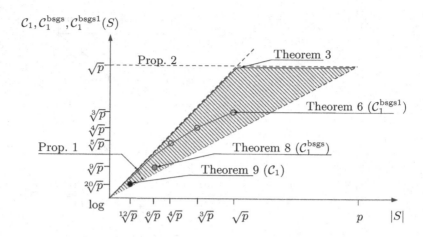

Fig. 2. Generic complexities and bounds (in logscale). Propositions 2 and 1 bound the triangle that contains all possible values for generic complexity. The theorems point to lower bounds provable for complexities of their respective constructions.

problem. We also define restricted versions of the generic complexity that capture the complexity of baby-step-giant-step algorithms. We give an explicit, deterministic construction of a collection of sets, whose complexity in respect to the weakest family of baby-step-giant-step algorithms becomes near-optimal as their size decreases (Theorem 7). Various bounds and constructions are put together in Fig. 2.

Acknowledgments. The authors thank Fan Chung, Kevin O'Bryant, Imre Ruzsa, Constantin Shramov, and the anonymous reviewer of ANTS VII for their advice and helpful comments.

References

[BC63]   Raj C. Bose and Sarvadaman Chowla. Theorems in the additive theory of numbers. *Comment. Math. Helv.*, 37:141–147, 1962–1963.

[BHP01]   R. C. Baker, G. Harman, and J. Pintz. The difference between consecutive primes. II. *Proc. London Math. Soc. (3)*, 83(3):532–562, 2001.

[Bol98]   Béla Bollobás. *Modern graph theory*, volume 184 of *Graduate texts in mathematics*. Springer, 1998.

[CLP05]   Jean-Sébastien Coron, David Lefranc, and Guillaume Poupard. A new baby-step giant-step algorithm and some applications to cryptanalysis. In Josyula R. Rao and Berk Sunar, editors, *CHES 2005*, volume 3659 of *Lecture Notes in Computer Science*, pages 47–60. Springer, 2005.

[CLS03]   M. Chateauneuf, Alan Ling, and Douglas R. Stinson. Slope packings and coverings, and generic algorithms for the discrete logarithm problem. *J. Comb. Designs*, 11(1):36–50, 2003.

[CW77]   Larry Carter and Mark N. Wegman. Universal classes of hash functions. In *STOC 1977*, pages 106–112, 1977.

[EN77]   Paul Erdös and Donald J. Newman. Bases for sets of integers. *J. Number Theory*, 9(4):420–425, 1977.

[GS80]   Ronald L. Graham and Neal J.A. Sloane. On additive bases and harmonious graphs. *SIAM J. Algebraic and Discrete Methods*, 1:382–404, 1980.

[Guy04]   Richard K. Guy. *Unsolved Problems in Number Theory*. Springer-Verlag, third edition, 2004.

[Hei93]   Rafi Heiman. A note on discrete logarithms with special structure. In Rainer A. Rueppel, editor, *Advances in Cryptology—EUROCRYPT '92*, volume 658 of *Lecture Notes in Computer Science*, pages 454–457. Springer, 1993.

[HHÖ04]   Harri Haanpää, Antti Huima, and Patric R. J. Östergård. Sets in $\mathbb{Z}_n$ with distinct sums of pairs. *Discrete Applied Mathematics*, 138(1–2):99–106, 2004.

[HS03]   Jeffrey Hoffstein and Joseph H. Silverman. Random small Hamming weight products with applications to cryptography. *Discrete Applied Mathematics*, 130(1):37–49, 2003.

[Knu97]   Donald E. Knuth. *Seminumerical Algorithms*, volume 2 of *The Art of Computer Programming*. Addison-Wesley, third edition, 1997.

[Nec94]   Vassiliy I. Nechaev. Complexity of a determinate algorithm for the discrete logarithm. *Math. Notes*, 55(2):165–172, 1994.

[NV05]   Assaf Naor and Jacques Verstraëte. A note on bipartite graphs without $2k$-cycles. *Probability, Combinatorics and Computing*, 14(5–6):845–849, 2005.

[O'B02]   Kevin O'Bryant. *Sidon Sets and Beatty Sequences*. PhD thesis, U. of Illinois in Urbana-Champaign, 2002.

[O'B04]   Kevin O'Bryant. A complete annotated bibliography of work related to Sidon sequences. *Electr. J. Combinatorics*, DS11, July 2004.

[Odl00]   Andrew M. Odlyzko. Discrete logarithms: The past and the future. *Des. Codes Cryptography*, 19(2/3):129–145, 2000.

[Pol78]   John M. Pollard. Monte Carlo methods for index computation (mod p). *Mathematics of Computation*, 32:918–924, 1978.

[Pol00]   John M. Pollard. Kangaroos, monopoly and discrete logarithms. *J. Cryptology*, 13(4):437–447, 2000.

[Ruz93]   Imre Z. Ruzsa. Solving a linear equation in a set of integers. Part I. *Acta Arith.*, 65:259–282, 1993.

[Sch80]   Jacob T. Schwartz. Fast probabilistic algorithms for verification of polynomial identities. *J. ACM*, 27(4):701–717, 1980.

[Sch01]   Claus-Peter Schnorr. Small generic hardcore subsets for the discrete logarithm. *Inf. Process. Lett.*, 79(2):93–98, 2001.

[Sha71]   Daniel Shanks. Class number, a theory of factorization, and genera. In Donald J. Lewis, editor, *1969 Number Theory Institute*, volume 20 of *Proceedings of Symposia in Pure Mathematics*, pages 415–440, Providence, Rhode Island, 1971. American Mathematical Society.

[Sho97]   Victor Shoup. Lower bounds for discrete logarithms and related problems. In Walter Fumy, editor, *Advances in Cryptology—EUROCRYPT '97*, volume 1233 of *Lecture Notes in Computer Science*, pages 256–266. Springer, 1997.

[Sin38]   James Singer. A theorem in finite projective geometry and some applications to number theory. *Trans. Amer. Math. Soc.*, 43:377–385, 1938.

[SJ04]   Yaron Sella and Markus Jakobsson. Constrained and constant ratio hash functions. Manuscript, 2004.

598   I. Mironov, A. Mityagin, and K. Nissim

[Sti02]   Douglas R. Stinson. Some baby-step giant-step algorithms for the low Hamming weight discrete logarithm problem. *Math. Comput.*, 71(237):379–391, 2002.
[SWD96]   Oliver Schirokauer, Damian Weber, and Thomas F. Denny. Discrete logarithms: The effectiveness of the index calculus method. In H. Cohen, editor, *ANTS-II*, volume 1122 of *Lecture Notes in Computer Science*, pages 337–361. Springer, 1996.
[Tes01]   Edlyn Teske. Square-root algorithms for the discrete logarithm problem (a survey). In *Public-Key Cryptography and Computational Number Theory*, pages 283–301, 2001.
[vOW99]   Paul C. van Oorschot and Michael J. Wiener. Parallel collision search with cryptanalytic applications. *J. Cryptology*, 12(1):1–28, 1999.
[Yac98]   Yacov Yacobi. Fast exponentiation using data compression. *SIAM J. Comput.*, 28(2):700–703, 1998.
[Zar51]   Kazimierz Zarankiewicz. Problem P 101. *Colloq. Math.*, 2:301, 1951.

Author Index

Lecture Notes in Computer Science

For information about Vols. 1–3981

please contact your bookseller or Springer